Digital Innovations in Agriculture —Volume I

Digital Innovations in Agriculture —Volume I

Editors

Gniewko Niedbała
Sebastian Kujawa

MDPI

Basel • Beijing • Wuhan • Barcelona • Belgrade • Novi Sad • Cluj • Manchester

Editors
Gniewko Niedbała
Poznań University of Life Sciences
Poznań, Poland

Sebastian Kujawa
Poznań University of Life Sciences
Poznań, Poland

Editorial Office
MDPI
St. Alban-Anlage 66
4052 Basel, Switzerland

This is a reprint of articles from the Special Issue published online in the open access journal *Agriculture* (ISSN 2077-0472) (available at: https://www.mdpi.com/journal/agriculture/special_issues/Digital_Innovations_Agriculture).

For citation purposes, cite each article independently as indicated on the article page online and as indicated below:

Lastname, A.A.; Lastname, B.B. Article Title. *Journal Name* **Year**, *Volume Number*, Page Range.

Volume I
ISBN 978-3-0365-8848-3 (Hbk)
ISBN 978-3-0365-8849-0 (PDF)
doi.org/10.3390/books978-3-0365-8849-0

Set
ISBN 978-3-0365-8846-9 (Hbk)
ISBN 978-3-0365-8847-6 (PDF)

Contents

About the Editors

Gniewko Niedbała

Gniewko Niedbała is a Full Professor at the Department of Biosystems Engineering, Poznań University of Life Sciences, Poland. He defended his doctorate in 2006 at the August Cieszkowski Agricultural University of Poznań and habilitated in 2019 at the Poznań University of Life Sciences. Presently, he is working on using artificial neural networks and machine learning in many aspects of agriculture and agronomy. Between 2012 and 2016, he was a Board Member of the National Centre for Research and Development, Poland. He ranked in the World's Top 2% of most cited scientists in 2020. He was a Guest Editor for a Special Issue of Agriculture: "Neural Networks in Agriculture and Digital Innovations in Agriculture". Additionally, he is a Member of the Editorial Board of Agronomy and a Member of the Reviewer Boards of Agriculture, Water, and Land. In addition, he has authored over 150 journal and conference papers and book chapters related to artificial intelligence in agriculture.

Sebastian Kujawa

Sebastian Kujawa is employed as an Assistant Professor in the Department of Biosystems Engineering, Poznań University of Life Sciences, Poland. In 2009, he received his PhD degree from the Poznań University of Life Sciences. His scientific activity concerns the applications of computer image analysis and machine learning in developing methods for condition assessments of dynamic biosystems. Dr. Kujawa is an author of over 50 publications in scientific journals and peer-reviewed conference proceedings. He is a member of the Board of the Polish Society for Information Technology in Agriculture (POLSITA). He has served as a Guest Editor of a Special Issue in Agriculture: "Neural Networks in Agriculture, and Digital Innovations in Agriculture".

agriculture

Editorial
Digital Innovations in Agriculture

Gniewko Niedbała *,† and Sebastian Kujawa *,†

Department of Biosystems Engineering, Faculty of Environmental and Mechanical Engineering, Poznań University of Life Sciences, Wojska Polskiego 50, 60-627 Poznań, Poland
* Correspondence: gniewko.niedbala@up.poznan.pl (G.N.); sebastian.kujawa@up.poznan.pl (S.K.)
† These authors contributed equally to this work.

1. Introduction

Digital agriculture, defined as the analysis and collection of various farm data, is constantly evolving. Data collected on an ongoing basis from fields, machinery, weather stations, sensors, and systems are used to perform a wide range of tasks and to make optimal decisions in the running of agricultural production. Platforms or management software for agriculture have tremendous potential. Digital agriculture relies heavily on detailed image analysis, artificial neural networks, machine learning, the Internet of Things (IoT), and big data [1,2]. The aforementioned digital agriculture techniques can be successfully used for qualitative assessments of agricultural crops, diagnoses of plant diseases, yield predictions, classification issues, and intelligent weed controls [3–7]. With the rapid developments of precision farming and digital agriculture, more and more farms are turning to tools based on artificial intelligence. The purpose of this Special Issue was to publish high-quality research and review papers that cover issues related to digital innovations in agriculture. Below, we have included highlights from all of the articles published in this Special Issue.

2. Papers in This Special Issue

In the research described in the first article [8], the authors attempted to find correlations between several selected neural network models and statistical methods commonly used in agriculture. The comparison has a universal dimension—it applies to crop production, livestock production, and the quality of the natural environment. The authors emphasize that artificial neural networks are a convenient, fast, and accurate tool; therefore, they have become very popular. The authors recommend their use in preparing analyses for agriculture.

In the second article [9] , the authors performed a review of the current state of research on the use of artificial intelligence, the Internet of Things, and hyperspectral imaging for crop disease detection. In addition, they compared several different techniques for diagnosing disease symptoms in plants. Convolutional neural network models proved to be the most efficient and effective methods for locating visual patterns in images.

In the third article [10], the authors systematically analysed climate services for agriculture in Africa. They reviewed 50 literature items from 20 African countries to develop such services. It was found that the development of such services is still in the early stages, as innovations in mobile telephony and Internet services integrated with climate services are presently undergoing a trial stage. The article confirms the need to integrate indigenous and scientific knowledge systems in creating climate information in Africa.

In the fourth review article [11], the authors managed the discussion of Findable, Accessible, Interoperable, and Reusable (FAIR) data in sustainable agriculture. With digital agriculture generating more and more data, there is a need to create systems for collecting reliable information. In conclusion, the authors proposed a method for identifying the main problems in making FAIR data available.

Citation: Niedbała, G.; Kujawa, S. Digital Innovations in Agriculture. *Agriculture* **2023**, *13*, 1686. https://doi.org/10.3390/agriculture13091686

Received: 23 August 2023
Revised: 24 August 2023
Accepted: 25 August 2023
Published: 26 August 2023

The fifth comprehensive review article [12] is an interesting summary of the relationships between sustainable agriculture and the Internet of Things (IoT). The authors hope to involve robots, cloud computing, and artificial intelligence in agricultural production even more than has been done before. In addition, the paper evaluates selected tools and equipment used in wireless sensor applications in IoT agriculture. The challenges of combining the technology with conventional agricultural operations were pointed out.

In the sixth paper [13], the authors attempted to determine whether milk flow characteristics can be considered biomarkers of lameness incidence in cows. A study was conducted on more than 100 head of dairy cows, and both cortisol concentrations in the animals' blood and milking characteristics were evaluated. The conducted experiment confirmed the thesis that milk flow characteristics can act as biomarkers of lameness in dairy cows.

The seventh article [14] deals with the simulation of fuel consumption depending on the load level of the tractor engine. A 95 kW engine with a partial power shift transmission was tested using the PTO dyno test method. It was observed that the engine load and fuel consumption were directly proportional to the engine load levels. Statistical analyses of the results indicated an exponential relationship between fuel consumption and engine load levels. The published research provides an opportunity to design an innovative agricultural tractor with a higher fuel efficiency.

In the eighth article [15], the authors made predictions regarding the reproductive success of multiparous dairy cows among a primiparous population, based on selected parameters generated by sensors. They tested an automatic milking system during the pregnancy of multiparous dairy cows and evaluated the system's accuracy based on blood parameters. In the group of non-pregnant cows, more samples with elevated cortisol were observed. Other interesting correlations indicated the risk of mastitis or oxidative stress in cows, depending on the presence or absence of pregnancy.

The ninth article [16] dealt with image analysis techniques for the optimization of a peanut planting space in Hainan. The study's authors used PlanetScope images with a spatial resolution of 3 m. The research was based on the construction of three models for peanut planting area extraction, based on the support vector machine (SVM), BP neural network (BPNN) and random forest (RF) classification algorithms. The results confirmed the effectiveness of PlanetScope imagery in solving agricultural problems. The RF model successfully optimized the planting of peanuts.

In the tenth article [17], the authors used deep learning neural networks, specifically the LC-DenseFCN point surveillance algorithm, to count chickens in a closed-circuit camera environment. Compared to conventional counting and other popular techniques, the proposed optimized method offers the potential for the fast and accurate counting of chickens residing at high densities.

The eleventh article [18] concerned on the identification of risk factors for detecting lameness in cows using natural biosensors. Before the study, the authors hypothesized that the formation of inline biomarkers would depend on the lameness of dairy cows in early lactations. Among other things, the study indicated that low milk fat content was maintained from before the onset of the disease until the very day of its confirmation.

In the twelfth article [19], the authors presented a study to assess the impact of lameness on the feeding attributes of dairy cows. The study used a sensor-nose band that a dairy cow wore just after calving. They found that lameness affected cows' feeding preferences and changes in biomarkers. For example, the lowest eating time was found on the day of diagnosis, and the highest on the ninth day before lameness was detected.

The thirteenth article [20] discusses developing a lettuce growth model using U-Net. *Arabidopsis* plants were used as the model-training material. The DL model developed for *Arabidopsis* works well for modelling the growth of other target species.

In the fourteenth article [21], the authors developed a predictive yield model for rice. Phenological data and hyperspectral imaging data were used to build the model. The final model was based on the values of vegetation indices: NDVI, EVI, SAVI, and REP.

The generated model was characterized by a high accuracy ($R^2 = 0.84$). In addition, the optimal time intervals for predicting rice yields were identified. The results show the high usefulness of data from late vegetative and flowering stages.

In the fifteenth article [22], the authors dealt with the individual identification of cattle faces using neural networks, a deep learning method. The activity involved an image analysis technique and modern RetinaNet technology. The selected algorithm allowed for a very high identification precision of up to 99.8%, and a very short average processing time of 0.0438 s per image.

In the sixteenth article [23], the authors determined the management zones within a field using remote sensing for variable rates of nitrogen fertilization of wheat. Simplified and hybrid models were used, and machine learning was also involved. The methodology for creating the models was enriched with information on phenological phases and the occurrence of agricultural droughts. The results showed that agronomic and climatic information allow for improving and optimizing the designation of management zones.

The seventeenth article [24] addresses an innovative method for monitoring soil nutrients using hyperspectral remote sensing. Linear and nonlinear algorithms were used to perform the task, including machine learning, LASSO, and GBDT algorithms. The results showed that LASSO and GBDT algorithms can improve the quality of TN, TP, and TK soil content estimation. This has great application significance in agriculture.

In the eighteenth article [25], the authors used the method of multiple regression and artificial neural networks to model the essential oil content and yield of trans-anethole obtained from fennel. The study aimed to identify the most accurate predictive tool. The results showed that artificial neural networks made more accurate predictions than regression methods. The published research may be useful for breeders of plants in the *Apiaceae* family to model the various complex polygenic traits of crop plants, which is important from the points of view of industry, herbal uses, pharmaceuticals, etc.

The nineteenth article [26] uses a four-stage image processing algorithm to identify and count *Metis plana* Walker, an oil palm pest from Malaysia. The solution described by the authors was designed to distinguish between live and dead bagworm larvae using motion detection. Highly accurate results were obtained, i.e., 73–100% accuracy was achieved at a camera distance of 30 cm in close conditions. Using deep learning with Faster R-CNN in the methodology is a feasible, practical, and reliable method for bagworm detection. The above research is of great practical importance.

In the twentieth article [27], the authors detected and identified sows using image analysis. Two accurate models based on deep learning, Mask-RCNN and UNet-Attention, were developed. A very high recognition rate of 96.8% for a specific individual was obtained by using the Mask-RCNN model. The proposed system allowed for recognition of various behaviours, such as eating, drinking, etc.

The twenty-first article [28] deals with modifying and improving classification methods that can be applied to agriculture. The authors applied multi-temporal data fusion to specific types of images (i.e., MS and SAR) using a dynamic time-warping method in paddy rice classification. Three different types of SPOT6 satellite images and nine Sentinel-1A synthetic aperture radar images were used. Using neural networks made it possible to significantly improve the overall accuracy of the collected images. In turn, the best performance of the classification results was obtained from the decision tree.

In the twenty-second article [29], the authors presented the concept of modelling the mechanical properties of fresh and stored large cranberry fruit using multiple linear regression and artificial neural network models. The analyses aimed to determine the apparent elasticity index of large cranberry fruit variables relating to harvest time, water content, storage time and conditions. Using neural modelling techniques allowed for more accurate prediction of the elasticity of the tested material compared to classical regression techniques.

The twenty-third article [30] presents an innovative approach for fast identification and counting of pig herds in complex image segmentation. The proposed DeepLab V3+ network model was created based on the neural networks, a deep learning method. The approach used made it possible to obtain the values of comprehensive evaluation indicators at a very high level of 86%.

In the twenty-fourth article [31], artificial neural networks were used to identify the key meteorological factors affecting the harvest date and yield of soybeans (*Glycine max* L. Merrill) variety Augusta. To perform the task, the most important dates of successive development stages, meteorological data, and yield data were collected. The ranking of factors shaping yield and harvest date was obtained using the sensitivity analysis of a neural network. The study results are highly practical in nature, indicating "difficult for soybeans" periods, during which special care should be taken to ensure they remain in good condition.

The twenty-fifth article [32] evaluates the performance of various communication systems used through the Internet of Things in agricultural production. The authors conducted a detailed analysis of the specific protocols and simulation tools used to improve connectivity and connection quality. Using two gateways with Adaptive Data technology can increase the network delivery without changing energy consumption. The results presented are promising for broader data optimization.

In the twenty-sixth article [33], the authors used an improved YOLOX-S algorithm for robots picking kiwi fruit based on image analyses. The analysis results proved that the authors' improved model helped improve the precision of kiwi fruit detection, reduced the number of model parameters, and improved the system's speed. This approach can also be applied to other fruits or objects in general, so-called small, targets.

The twenty-seventh article [34] proposed a tea recognition method based on a light convolutional neural network and support vector machine (L-CNN-SVM). The purpose of this method was to recognize tea using wavelet numerical data generated by wavelet decomposition and reconstruction of the time–frequency signal. First, a redundant discrete wavelet transform was used to decompose the wavelet components of hyperspectral images of three teas (black, green, and yellow), which were used to construct the datasets. Second, a lightweight CNN model was improved to generate a tea recognition model. The results showed that the tea recognition results based on the L-CNN-SVM method outperformed MobileNet v2+RF, MobileNet v2+KNN, MobileNet v2+AdaBoost, AlexNet, and MobileNet v2. For the recognition results of the three teas using a LL plus HL plus LH wavelet component reconstruction, the overall accuracy rate reached 98.7%.

The twenty-eighth article [35] uses a fractional differential order for the soil hyperspectral inversion of iron oxide. Among other things, the content and movement of iron oxide in soil informs numerous degradation processes. Analysing the content of this component in the soil is difficult, because its analytical spectra overlap with the infrared components of organic matter. In addition to a specific spectral transformation, a soil iron oxide content prediction was made using artificial intelligence. It was shown that using a fractional-order differential transformation can significantly improve the results for this type of analysis.

In the twenty-ninth article [36], the authors used supervised classifiers for feature analyses, in order to evaluate the accuracy of a maturity analysis of fruit palm fruit. For this purpose, unconstrained remote sensing, advanced multivariate techniques, and artificial neural networks were used. Measurements were made in real-time. A very high image processing efficiency of more than 93% was achieved.

The thirtieth article [37] deals with improved optimization systems using the NSGA-III algorithm to improve the precision of two-way fertilizer applications. A test of the rotational speed and opening length of a bidirectional granular fertilizer applicator was conducted. A fertilizer rate prediction model based on machine learning was developed. Applying the new method led to a halving of the absolute error. The presented research significantly improves the quality of fertilizer spreading.

In the thirty-first article [38], the authors proposed PLSR models for predicting of the content of functional components in *Brassica juncea* based on hyperspectral imaging. The contents of chlorophyll, carotenoids, phenols, glucosinolates, and anthocyanins were studied. Incorporating SNV and first-order derivative pre-processing with spectral data into the study methodology yielded models with low prediction errors.

In the thirty-second article [39], the authors collected a dataset of whitefly-infested cotton leaves containing 5135 images divided into two main classes, namely healthy and unhealthy. They then used a compact convolutional transform (CCT) approach to classify the image dataset. Experimental results showed the effectiveness of the proposed CCT-based approach compared to other state-of-the-art approaches. The produced model achieved an accuracy of 97.2%.

The thirty-third article [40] presents a fuzzy certification of wheat quality. This analysis is based on a fuzzy model for wheat analysis. The authors developed a MATLAB application, with which they modelled perceptions concerning wheat's main physical and chemical characteristics, obtaining a wheat batch quality index. The generated algorithm makes it possible to obtain and use a global quality index, which is applicable not only in the commercial sphere as a quality reference and for pricing, but also as a measure for evaluating processing capabilities.

In the thirty-fourth article [41], a multi-objective logistic distribution path optimization model with time constraints was constructed, and a genetic algorithm was used to optimize the commercial distribution path for fresh agricultural products. Combining the genetic algorithm with a real case to be studied, the study aimed to solve enterprises' narrow distribution paths and promote the model's application to similar enterprises with similar characteristics. The results show that (1) the commercial distribution path scheme optimized by the genetic algorithm can reduce distribution centre distribution costs and improve customer satisfaction, and (2) the genetic algorithm can bring economic benefits and reduce transportation losses in trade for trade distribution centres with the same spatial and quality characteristics as distribution centres for fresh agricultural products.

The thirty-fifth article [42] presents a data science approach that agglomerates the soil parameter space into a limited number of soil process functional units (SPUs) that can run agricultural process models. In reality, two unsupervised classification methods were developed to generate a 3D multidimensional data product consisting of SPUs, each defined by a multidimensional parameter distribution along a depth profile from 0 to 100 cm.

In the thirty-sixth article [43], the authors hypothesised that an automated body condition scoring system could indicate health and pregnancy success in cows. Therefore, this study aimed to determine the association of automated recorded body condition score (BCS) with pregnancy and inline biomarkers, such as milk beta-hydroxybutyrate (BHB), milk lactate dehydrogenase (LDH), milk progesterone (mP4), and milk yield (MY) in dairy cows.

In the thirty-seventh article [44], an intelligent control system for temperature and humidity in a piggery was proposed, based on machine learning and a fuzzy control algorithm. Sensors were used to collect data on temperature and humidity values and store these data in chronological order. These data formed a time series to train the GRU model, which was used to predict the curve of temperature and humidity changes in the piggery over the next 24 h.

In the thirty-eighth article [45], the authors proposed a lightweight maize disease identification model called the Double Fusion block with Coordinate Attention Network (DFCANet). DFCANet mainly consists of two components: Double Feature Fusion with Coordinate Attention and Down-Sampling (DS) modules. The results show that DFCANet has an average recognition accuracy of 98.47%.

The thirty-ninth article [46] is about a high-accuracy model for predicting blueberry yields, trained using structurally innovative datasets. Data were collected between 2016 and 2021, and included agronomic, climatic, soil data, and satellite data on vegetation. In addition, vegetation periods by BBCH scale and aggregates were considered. Of the 11 models, the Extreme Gradient Boosting algorithm performed best, with a MAPE prediction error of 12.48%.

The fortieth paper [47] proposed a novel scheme to optimise the decision-making capability of a combination of control and PID controller parameters, in order to improve the feasibility and practicality of variable fertiliser applicators. Firstly, an EDEM was applied to obtain the minimum acceptable bore length and the appropriate gap between the spiral vanes and the discharge cavity wall, followed by calibration experiments to establish a fertiliser rate-fitting model using polynomial fitting. Secondly, a modified sparrow search algorithm (SSA) with a chaotic operator and a mutation section of the DE algorithm was used to optimise the control combination using accuracy, homogeneity, and control time as evaluation criteria.

In the forty-first article [48], the research aimed to develop linear and non-linear models for predicting the protein content percentage of pea seeds, and to conduct a comparative analysis of the performance of these models. The analyses also focused on identifying the variables with the greatest influence on protein content. The research involved machine learning (specifically, artificial neural networks) and multiple linear regression (MLR) methods. The input parameters of the models were weather, agronomic, and phyto-phenological data from 2016 to 2020. The neural model (N1) performed better than the multiple regression (RS) model. The RMS error of the N1 model was 0.838, while the RS model obtained a mean error value of 2.696. The MAPE error for the N1 and RS models was 2.721 and 8.852, respectively. The sensitivity analysis conducted for the best neural network showed that the independent variables most influencing the protein content of pea seeds were soil abundance of magnesium, potassium, and phosphorus.

The forty-second article [49] aimed to investigate the automatic detection of Basal Stem Rot (BSR) at the seedling stage, using a pre-trained deep-learning model and hyperspectral images. The oil palm seedling image was divided into three regions to determine whether there were significant spectral changes at different leaf positions. How background images affect detection performance was also investigated. The segmented images of the plant sapling were automatically generated using a convolutional neural network (RCNN) based on the mask region. Three models were used for BSR detection: a convolutional neural network with a depth of 16 layers (VGG16) trained on the segmented image, and the VGG16 and Mask RCNN models trained on the original images. The results show that the VGG16 model performed best in terms of accuracy (91.93%), precision (94.32%), recall (89.26%), and F1 score (91.72%).

In the forty-third article [50], an automatic classification of the larval stage of the cutworm, starting from the second (S2) to the fifth (S5) stage, was proposed using a structure based on transfer learning. Five different CNN architectures, namely VGG16, ResNet50, ResNet152, DenseNet121, and DenseNet201, were used to categorise the larval stages. Of the five models used, the DenseNet121 model had the best classification accuracy of 96.18%. In addition, all developmental stages from S2 to S5 could be identified with high accuracy (94.52–97.57%), precision (89.71–95.87%), sensitivity (87.67–96.65%), specificity (96.51–98.61%), and F1 score (88.89–96.18%).

In the forty-fourth article [51], a study was carried out that created a dataset containing images of the leaves of cash crops, which were divided into two basic categories, namely healthy and unhealthy. The next step was to train a deep model to identify healthy and unhealthy leaves. The trained YOLOv5 model was used to identify stains in exclusive and public datasets. This study quickly and accurately identified even a small disease patch using YOLOv5. This research aimed to provide the best hyper-parameters for classifying and detecting healthy and unhealthy parts of leaves in exclusive and public datasets. The trained YOLOv5 model achieves a 93% accuracy on the test set.

The forty-fifth article [52] concerned research on the trajectory of agricultural machinery. The paper proposed a multi-node path planning algorithm based on the Improved Whale Optimised ACO (ACO) algorithm, called IWOA-ACO. The algorithm first introduces an inverse learning strategy, a non-linear convergence rate, and an adaptive inertia factor to improve the global and local convergence ability. The simulation results show that, in a flat environment, the length and energy consumption of the planned IWOA-ACO path are the same as those of Particle Swarm Optimisation ACO (PSO-ACO), and are 0.61% less than those of WOA-ACO. Furthermore, in a bumpy environment, the length and energy consumption of the IWOA-ACO planned path is 1.91% and 4.32% less than for PSO-ACO, and 1.95% and 1.25% less than for WOA-ACO.

The forty-sixth article [53] dealt with the development and evaluation of the Maize Yield Prediction System (MYPS), which uses a Short Message Service (SMS) and the Internet to enable rural farmers and government officials to predict end-of-season maize yields in Tanzania. The system uses Long Short-Term Memory (LSTM) deep learning models to predict end-of-season maize yields at the district level, based on remote sensing (NDVI) and climate data. The deep learning models are very effective in yield prediction, achieving a mean absolute percentage error (MAPE) of 3.656% and 6.648%, respectively, on test data.

The forty-seventh article [54] analysed the performance of pea (*Pisum sativum* L.) seed yield prediction using a linear (MLR) and a non-linear (ANN) model. The study used meteorological, agronomic, and phytophysical data from 2016 to 2020. The neural model (N2) generated highly accurate predictions of pea seed yield, with a correlation coefficient of 0.936 and RMS and MAPE errors of 0.443 and 7.976, respectively. The model significantly outperformed the multiple linear regression model (RS2), which had an RMS error of 6.401 and a MAPE error of 148.585. The sensitivity analysis conducted for the neural network showed that the traits with the greatest impact on pea seed yield were the onset of maturity, harvest date, total rainfall, and mean air temperature.

The forty-eighth paper [55] trained and compared the performance of two machine learning methods, a multivariate regression network and a ResNet-50-based neural network, for predicting plant biomass and determining plants' relative growth rates in aeroponic cultivation. The training dataset consisted of images of 57 plants taken at two different angles every hour over five days. The results show that images taken from a top-down perspective give better results for the multivariate regression network. In contrast, images taken from the side are better for the ResNet-50 neural network. The best biomass estimates are obtained from the multivariate regression model trained on the top camera images using a moving average filter, giving a mean square error of 0.0466 g. The best estimates of relative growth rate are obtained from the ResNet-50 trained on images from both cameras, giving a mean square error of 0.1767 g/(g-day).

The forty-ninth article [56] aimed to evaluate different vegetation indices to predict the growth rates and harvest points of lettuce. Twenty-five genotypes of biofortified green lettuce were evaluated. Green leaf index (GLI), normalised green–red difference index (NGRDI), spectral slope saturation index (SI), and total colour index (HUE) were calculated from images taken 1, 8, 18, 24 and 36 days after transplanting (vegetative state). Averages were compared using the Scott-Knott test ($p \leq 0.05$), and simple linear regression models were generated to monitor growth rates, yielding R^2 values ranging from 62% to 99%. Multivariate analysis confirmed genetic dissimilarity, with a correlation coefficient of 88.49%.

Finally, in the fiftieth article [57], a machine vision system was developed using a dataset of 7328 high-density images (1229 pixels per centimetre) of planthoppers collected in the field using sticky light traps. The dataset included four planthopper classes: brown planthopper (BPH), green leafhopper (GLH), white-backed planthopper (WBPH), and zigzag leafhopper (ZIGZAG). Five deep CNN models—ResNet-50, ResNet-101, ResNet-152, VGG-16, and VGG-19—were applied and tuned to classify the planthopper species. The experimental results indicated that the ResNet-50 model performed the best overall, achieving average values of 97.28% for accuracy, 92.05% for precision, 94.47% for recall, and 93.07% for the F1 score.

3. Conclusions

The Special Issue "Digital Innovations in Agriculture" brings fascinating insights into the agricultural sector's future. The use of advanced ICT, data analytics, and artificial intelligence makes it possible to achieve sustainability, increase production efficiency, and improve animal husbandry. Digital innovations have the potential to revolutionise agriculture, and their implementation can contribute to a more sustainable, productive, and knowledge-based agricultural space. We hope that these papers will stimulate further research into this domain.

Funding: This research received no external funding.

Institutional Review Board Statement: Not applicable.

Data Availability Statement: Not applicable.

Conflicts of Interest: The authors declare no conflict of interest.

References

1. Osinga, S.A.; Paudel, D.; Mouzakitis, S.A.; Athanasiadis, I.N. Big Data in Agriculture: Between Opportunity and Solution. *Agric. Syst.* **2022**, *195*, 103298. [CrossRef]
2. Niedbała, G.; Wróbel, B.; Piekutowska, M.; Zielewicz, W.; Paszkiewicz-Jasińska, A.; Wojciechowski, T.; Niazian, M. Application of Artificial Neural Networks Sensitivity Analysis for the Pre-Identification of Highly Significant Factors Influencing the Yield and Digestibility of Grassland Sward in the Climatic Conditions of Central Poland. *Agronomy* **2022**, *12*, 1133. [CrossRef]
3. Xu, K.; Shu, L.; Xie, Q.; Song, M.; Zhu, Y.; Cao, W.; Ni, J. Precision Weed Detection in Wheat Fields for Agriculture 4.0: A Survey of Enabling Technologies, Methods, and Research Challenges. *Comput. Electron. Agric.* **2023**, *212*, 108106. [CrossRef]
4. Taoumi, H.; Lahrech, K. Economic, Environmental and Social Efficiency and Effectiveness Development in the Sustainable Crop Agricultural Sector: A Systematic in-Depth Analysis Review. *Sci. Total Environ.* **2023**, *901*, 165761. [CrossRef]
5. Shoaib, M.; Shah, B.; EI-Sappagh, S.; Ali, A.; Ullah, A.; Alenezi, F.; Gechev, T.; Hussain, T.; Ali, F. An Advanced Deep Learning Models-Based Plant Disease Detection: A Review of Recent Research. *Front. Plant Sci.* **2023**, *14*, 1158933. [CrossRef] [PubMed]
6. van Klompenburg, T.; Kassahun, A.; Catal, C. Crop Yield Prediction Using Machine Learning: A Systematic Literature Review. *Comput. Electron. Agric.* **2020**, *177*, 105709. [CrossRef]
7. Niedbała, G.; Niazian, M.; Sabbatini, P. Modeling *Agrobacterium*-mediated gene transformation of tobacco (*Nicotiana tabacum*)-a model plant for gene transformation studies. *Front. Plant Sci.* **2021**, *12*, 1454. [CrossRef]
8. Boniecki, P.; Sujak, A.; Niedbała, G.; Piekarska-Boniecka, H.; Wawrzyniak, A.; Przybylak, A. Neural Modelling from the Perspective of Selected Statistical Methods on Examples of Agricultural Applications. *Agriculture* **2023**, *13*, 762. [CrossRef]
9. Orchi, H.; Sadik, M.; Khaldoun, M. On Using Artificial Intelligence and the Internet of Things for Crop Disease Detection: A Contemporary Survey. *Agriculture* **2021**, *12*, 9. [CrossRef]
10. Ofoegbu, C.; New, M. Evaluating the Effectiveness and Efficiency of Climate Information Communication in the African Agricultural Sector: A Systematic Analysis of Climate Services. *Agriculture* **2022**, *12*, 160. [CrossRef]
11. Ali, B.; Dahlhaus, P. The Role of FAIR Data towards Sustainable Agricultural Performance: A Systematic Literature Review. *Agriculture* **2022**, *12*, 309. [CrossRef]
12. Dhanaraju, M.; Chenniappan, P.; Ramalingam, K.; Pazhanivelan, S.; Kaliaperumal, R. Smart Farming: Internet of Things (IoT)-Based Sustainable Agriculture. *Agriculture* **2022**, *12*, 1745. [CrossRef]
13. Juozaitienė, V.; Antanaitis, R.; Urbonavičius, G.; Urbutis, M.; Tušas, S.; Baumgartner, W. Can Milk Flow Traits Act as Biomarkers of Lameness in Dairy Cows? *Agriculture* **2021**, *11*, 227. [CrossRef]
14. Siddique, M.A.A.; Baek, S.-M.; Baek, S.-Y.; Kim, W.-S.; Kim, Y.-S.; Kim, Y.-J.; Lee, D.-H.; Lee, K.-H.; Hwang, J.-Y. Simulation of Fuel Consumption Based on Engine Load Level of a 95 KW Partial Power-Shift Transmission Tractor. *Agriculture* **2021**, *11*, 276. [CrossRef]
15. Antanaitis, R.; Juozaitienė, V.; Malašauskienė, D.; Televičius, M.; Urbutis, M.; Zamokas, G.; Baumgartner, W. Prediction of Reproductive Success in Multiparous First Service Dairy Cows by Parameters from In-Line Sensors. *Agriculture* **2021**, *11*, 334. [CrossRef]
16. Jin, Y.; Guo, J.; Ye, H.; Zhao, J.; Huang, W.; Cui, B. Extraction of Arecanut Planting Distribution Based on the Feature Space Optimization of PlanetScope Imagery. *Agriculture* **2021**, *11*, 371. [CrossRef]
17. Cao, L.; Xiao, Z.; Liao, X.; Yao, Y.; Wu, K.; Mu, J.; Li, J.; Pu, H. Automated Chicken Counting in Surveillance Camera Environments Based on the Point Supervision Algorithm: LC-DenseFCN. *Agriculture* **2021**, *11*, 493. [CrossRef]
18. Antanaitis, R.; Juozaitienė, V.; Urbonavičius, G.; Malašauskienė, D.; Televičius, M.; Urbutis, M.; Džermeikaitė, K.; Baumgartner, W. Identification of Risk Factors for Lameness Detection with Help of Biosensors. *Agriculture* **2021**, *11*, 610. [CrossRef]
19. Antanaitis, R.; Juozaitienė, V.; Urbonavičius, G.; Malašauskienė, D.; Televičius, M.; Urbutis, M.; Baumgartner, W. Impact of Lameness on Attributes of Feeding Registered with Noseband Sensor in Fresh Dairy Cows. *Agriculture* **2021**, *11*, 851. [CrossRef]
20. Chang, S.; Lee, U.; Hong, M.J.; Jo, Y.D.; Kim, J.-B. Lettuce Growth Pattern Analysis Using U-Net Pre-Trained with Arabidopsis. *Agriculture* **2021**, *11*, 890. [CrossRef]
21. Nazir, A.; Ullah, S.; Saqib, Z.A.; Abbas, A.; Ali, A.; Iqbal, M.S.; Hussain, K.; Shakir, M.; Shah, M.; Butt, M.U. Estimation and Forecasting of Rice Yield Using Phenology-Based Algorithm and Linear Regression Model on Sentinel-II Satellite Data. *Agriculture* **2021**, *11*, 1026. [CrossRef]
22. Xu, B.; Wang, W.; Guo, L.; Chen, G.; Wang, Y.; Zhang, W.; Li, Y. Evaluation of Deep Learning for Automatic Multi-View Face Detection in Cattle. *Agriculture* **2021**, *11*, 1062. [CrossRef]

23. Rokhafrouz, M.; Latifi, H.; Abkar, A.A.; Wojciechowski, T.; Czechlowski, M.; Naieni, A.S.; Maghsoudi, Y.; Niedbała, G. Simplified and Hybrid Remote Sensing-Based Delineation of Management Zones for Nitrogen Variable Rate Application in Wheat. *Agriculture* **2021**, *11*, 1104. [CrossRef]
24. Peng, Y.; Wang, L.; Zhao, L.; Liu, Z.; Lin, C.; Hu, Y.; Liu, L. Estimation of Soil Nutrient Content Using Hyperspectral Data. *Agriculture* **2021**, *11*, 1129. [CrossRef]
25. Sabzi-Nojadeh, M.; Niedbała, G.; Younessi-Hamzekhanlu, M.; Aharizad, S.; Esmaeilpour, M.; Abdipour, M.; Kujawa, S.; Niazian, M. Modeling the Essential Oil and Trans-Anethole Yield of Fennel (*Foeniculum vulgare* Mill. Var. *vulgare*) by Application Artificial Neural Network and Multiple Linear Regression Methods. *Agriculture* **2021**, *11*, 1191. [CrossRef]
26. Ahmad, M.N.; Shariff, A.R.M.; Aris, I.; Abdul Halin, I. A Four Stage Image Processing Algorithm for Detecting and Counting of Bagworm, Metisa Plana Walker (Lepidoptera: Psychidae). *Agriculture* **2021**, *11*, 1265. [CrossRef]
27. Lei, K.; Zong, C.; Yang, T.; Peng, S.; Zhu, P.; Wang, H.; Teng, G.; Du, X. Detection and Analysis of Sow Targets Based on Image Vision. *Agriculture* **2022**, *12*, 73. [CrossRef]
28. Lei, T.C.; Wan, S.; Wu, Y.C.; Wang, H.-P.; Hsieh, C.-W. Multi-Temporal Data Fusion in MS and SAR Images Using the Dynamic Time Warping Method for Paddy Rice Classification. *Agriculture* **2022**, *12*, 77. [CrossRef]
29. Gorzelany, J.; Belcar, J.; Kuźniar, P.; Niedbała, G.; Pentoś, K. Modelling of Mechanical Properties of Fresh and Stored Fruit of Large Cranberry Using Multiple Linear Regression and Machine Learning. *Agriculture* **2022**, *12*, 200. [CrossRef]
30. Liu, C.; Su, J.; Wang, L.; Lu, S.; Li, L. LA-DeepLab V3+: A Novel Counting Network for Pigs. *Agriculture* **2022**, *12*, 284. [CrossRef]
31. Niedbała, G.; Kurasiak-Popowska, D.; Piekutowska, M.; Wojciechowski, T.; Kwiatek, M.; Nawracała, J. Application of Artificial Neural Network Sensitivity Analysis to Identify Key Determinants of Harvesting Date and Yield of Soybean (*Glycine max* [L.] Merrill) Cultivar Augusta. *Agriculture* **2022**, *12*, 754. [CrossRef]
32. Yascaribay, G.; Huerta, M.; Silva, M.; Clotet, R. Performance Evaluation of Communication Systems Used for Internet of Things in Agriculture. *Agriculture* **2022**, *12*, 786. [CrossRef]
33. Zhou, J.; Hu, W.; Zou, A.; Zhai, S.; Liu, T.; Yang, W.; Jiang, P. Lightweight Detection Algorithm of Kiwifruit Based on Improved YOLOX-S. *Agriculture* **2022**, *12*, 993. [CrossRef]
34. Cui, Q.; Yang, B.; Liu, B.; Li, Y.; Ning, J. Tea Category Identification Using Wavelet Signal Reconstruction of Hyperspectral Imagery and Machine Learning. *Agriculture* **2022**, *12*, 1085. [CrossRef]
35. Zhao, H.; Gan, S.; Yuan, X.; Hu, L.; Wang, J.; Liu, S. Application of a Fractional Order Differential to the Hyperspectral Inversion of Soil Iron Oxide. *Agriculture* **2022**, *12*, 1163. [CrossRef]
36. Alfatni, M.S.M.; Khairunniza-Bejo, S.; Marhaban, M.H.B.; Saaed, O.M.B.; Mustapha, A.; Shariff, A.R.M. Towards a Real-Time Oil Palm Fruit Maturity System Using Supervised Classifiers Based on Feature Analysis. *Agriculture* **2022**, *12*, 1461. [CrossRef]
37. Dang, Y.; Ma, H.; Wang, J.; Zhou, Z.; Xu, Z. An Improved Multi-Objective Optimization Decision Method Using NSGA-III for a Bivariate Precision Fertilizer Applicator. *Agriculture* **2022**, *12*, 1492. [CrossRef]
38. Choi, J.-H.; Park, S.H.; Jung, D.-H.; Park, Y.J.; Yang, J.-S.; Park, J.-E.; Lee, H.; Kim, S.M. Hyperspectral Imaging-Based Multiple Predicting Models for Functional Component Contents in Brassica Juncea. *Agriculture* **2022**, *12*, 1515. [CrossRef]
39. Jajja, A.I.; Abbas, A.; Khattak, H.A.; Niedbała, G.; Khalid, A.; Rauf, H.T.; Kujawa, S. Compact Convolutional Transformer (CCT)-Based Approach for Whitefly Attack Detection in Cotton Crops. *Agriculture* **2022**, *12*, 1529. [CrossRef]
40. Simionescu, C.S.; Plenovici, C.P.; Augustin, C.L.; Rahoveanu, M.M.T.; Rahoveanu, A.T.; Zugravu, G.A. Fuzzy Quality Certification of Wheat. *Agriculture* **2022**, *12*, 1640. [CrossRef]
41. Sun, J.; Jiang, T.; Song, Y.; Guo, H.; Zhang, Y. Research on the Optimization of Fresh Agricultural Products Trade Distribution Path Based on Genetic Algorithm. *Agriculture* **2022**, *12*, 1669. [CrossRef]
42. Ließ, M. Modeling the Agricultural Soil Landscape of Germany—A Data Science Approach Involving Spatially Allocated Functional Soil Process Units. *Agriculture* **2022**, *12*, 1784. [CrossRef]
43. Antanaitis, R.; Malašauskienė, D.; Televičius, M.; Urbutis, M.; Rutkauskas, A.; Šertvytytė, G.; Anskienė, L.; Baumgartner, W. Associations of Automatically Recorded Body Condition Scores with Measures of Production, Health, and Reproduction. *Agriculture* **2022**, *12*, 1834. [CrossRef]
44. Jin, H.; Meng, G.; Pan, Y.; Zhang, X.; Wang, C. An Improved Intelligent Control System for Temperature and Humidity in a Pig House. *Agriculture* **2022**, *12*, 1987. [CrossRef]
45. Chen, Y.; Chen, X.; Lin, J.; Pan, R.; Cao, T.; Cai, J.; Yu, D.; Cernava, T.; Zhang, X. DFCANet: A Novel Lightweight Convolutional Neural Network Model for Corn Disease Identification. *Agriculture* **2022**, *12*, 2047. [CrossRef]
46. Niedbała, G.; Kurek, J.; Świderski, B.; Wojciechowski, T.; Antoniuk, I.; Bobran, K. Prediction of Blueberry (*Vaccinium corymbosum* L.) Yield Based on Artificial Intelligence Methods. *Agriculture* **2022**, *12*, 2089. [CrossRef]
47. Dang, Y.; Yang, G.; Wang, J.; Zhou, Z.; Xu, Z. A Decision-Making Capability Optimization Scheme of Control Combination and PID Controller Parameters for Bivariate Fertilizer Applicator Improved by Using EDEM. *Agriculture* **2022**, *12*, 2100. [CrossRef]
48. Hara, P.; Piekutowska, M.; Niedbała, G. Prediction of Protein Content in Pea (*Pisum sativum* L.) Seeds Using Artificial Neural Networks. *Agriculture* **2022**, *13*, 29. [CrossRef]
49. Yong, L.Z.; Khairunniza-Bejo, S.; Jahari, M.; Muharam, F.M. Automatic Disease Detection of Basal Stem Rot Using Deep Learning and Hyperspectral Imaging. *Agriculture* **2022**, *13*, 69. [CrossRef]
50. Johari, S.N.A.M.; Khairunniza-Bejo, S.; Shariff, A.R.M.; Husin, N.A.; Masri, M.M.M.; Kamarudin, N. Automatic Classification of Bagworm, Metisa Plana (Walker) Instar Stages Using a Transfer Learning-Based Framework. *Agriculture* **2023**, *13*, 442. [CrossRef]

51. Khalid, M.; Sarfraz, M.S.; Iqbal, U.; Aftab, M.U.; Niedbała, G.; Rauf, H.T. Real-Time Plant Health Detection Using Deep Convolutional Neural Networks. *Agriculture* **2023**, *13*, 510. [CrossRef]
52. Liang, C.; Pan, K.; Zhao, M.; Lu, M. Multi-Node Path Planning of Electric Tractor Based on Improved Whale Optimization Algorithm and Ant Colony Algorithm. *Agriculture* **2023**, *13*, 586. [CrossRef]
53. Tende, I.G.; Aburada, K.; Yamaba, H.; Katayama, T.; Okazaki, N. Development and Evaluation of a Deep Learning Based System to Predict District-Level Maize Yields in Tanzania. *Agriculture* **2023**, *13*, 627. [CrossRef]
54. Hara, P.; Piekutowska, M.; Niedbała, G. Prediction of Pea (*Pisum sativum* L.) Seeds Yield Using Artificial Neural Networks. *Agriculture* **2023**, *13*, 661. [CrossRef]
55. Åström, O.; Hedlund, H.; Sopasakis, A. Machine-Learning Approach to Non-Destructive Biomass and Relative Growth Rate Estimation in Aeroponic Cultivation. *Agriculture* **2023**, *13*, 801. [CrossRef]
56. Ribeiro, A.L.A.; Maciel, G.M.; Siquieroli, A.C.S.; Luz, J.M.Q.; Gallis, R.B.d.A.; Assis, P.H.d.S.; Catão, H.C.R.M.; Yada, R.Y. Vegetation Indices for Predicting the Growth and Harvest Rate of Lettuce. *Agriculture* **2023**, *13*, 1091. [CrossRef]
57. Ibrahim, M.F.; Khairunniza-Bejo, S.; Hanafi, M.; Jahari, M.; Ahmad Saad, F.S.; Mhd Bookeri, M.A. Deep CNN-Based Planthopper Classification Using a High-Density Image Dataset. *Agriculture* **2023**, *13*, 1155. [CrossRef]

agriculture

Article

Can Milk Flow Traits Act as Biomarkers of Lameness in Dairy Cows?

Vida Juozaitienė [1,*], Ramūnas Antanaitis [2,*], Gediminas Urbonavičius [1], Mingaudas Urbutis [2], Saulius Tušas [3] and Walter Baumgartner [4]

1 Department of Animal Breeding, Veterinary Academy, Lithuanian University of Health Sciences, Tilžės St. 18, LT-47181 Kaunas, Lithuania; gediminas.urbonavicius@lsmuni.lt
2 Large Animal Clinic, Veterinary Academy, Lithuanian University of Health Sciences, Tilžės St. 18, LT-47181 Kaunas, Lithuania; mingaudas.urbutis@lsmuni.lt
3 Institute of Animal Rearing Technologies, Veterinary Academy, Lithuanian University of Health Sciences, Tilžės St. 18, LT-47181 Kaunas, Lithuania; saulius.tusas@lsmuni.lt
4 University Clinic for Ruminants, University of Veterinary Medicine, Veterinaerplatz 1, A-1210 Vienna, Austria; walter.baumgartner@vetmeduni.ac.at
* Correspondence: vida.juozaitiene@lsmuni.lt (V.J.); ramunas.antanaitis@lsmuni.lt (R.A.); Tel.: +370-67425663 (V.J.)

Abstract: We hypothesized that lameness has an impact on milk flow traits. The aim of the current study was therefore to investigate the relation between lameness and milk flow traits in dairy cows. For this study 73 healthy and 55 cows with lameness were selected. Lameness was diagnosed by a local specialized veterinarian, according to the standard procedure. The blood samples were collected during clinical examination. The milking properties of cows were evaluated twice in a row—during evening and morning milking. The selected cows in the current lactation did not receive veterinary treatment, and correct hoof trimming was not performed at least four weeks before the experiment. The measurements were taken by two electronic mobile milk flow meters (Lactocorder®®, WMB AG, Balgache, Switzerland). Milk flow data were processed using LactoPro 5.2.0 software (Biomelktechnik Swiss). Cortisol concentration was measured with the automated analyzer TOSOH®® AIA-360 (South San Francisco, CA, USA). We found out that milk flow traits can act as biomarkers of lameness in dairy cows. We determined that the milk yield in the first minute of healthy dairy cows was 1.77 kg higher than that of lame cows. The electrical conductivity during the initial time of milking of healthy cows was 0.24 mS/cm lower than that of the lame group. The milking duration of LA cows was 1.07 min shorter and the time of incline in milk flow from 0.5 kg/min till the reach of the plateau phase was longer. The risk of lameness was most clearly indicated by an increase in blood cortisol concentration; if its blood level in cows exceeds 1 μg/dL, the risk of identifying lameness increases 4.9 times.

Keywords: lameness; biomarkers; dairy cows; precision dairy farming

Citation: Juozaitienė, V.; Antanaitis, R.; Urbonavičius, G.; Urbutis, M.; Tušas, S.; Baumgartner, W. Can Milk Flow Traits Act as Biomarkers of Lameness in Dairy Cows? *Agriculture* **2021**, *11*, 227. https://doi.org/10.3390/agriculture11030227

Academic Editors: Gniewko Niedbała and Sebastian Kujawa

Received: 12 February 2021
Accepted: 5 March 2021
Published: 9 March 2021

1. Introduction

Lameness is one of the most severe health problems in dairy cows and is described as a disease that presents itself with a walking dysfunction, which is becoming a frequent and serious problem of animal welfare, herd productivity, and herd management [1]. The clinically severe lameness ranges from 26 to 54% [2], but foot lesions, that cause lameness in cattle, are multifactorial and trauma, infection, and metabolic disorders are contributing factors [3]. In order to identify healthy and lame cows, the information of the herd health reports can be used [4]. Higher surveillance rates may increase the possibility of an earlier detection of lameness, since most of the time farmers tend to underestimate the actual lameness prevalence in their herd [5].

Although, measurement of pain in animals is challenging, there are established methods of estimating stress, caused by pain or discomfort. Thus, the assessment of stress in farm animals can be used to evaluate animal welfare [6]. Lameness can be defined as an abnormal behavior (presenting itself with reduced velocity and altered stride, lowered head, and curved back). It negatively affects animal production and welfare, such as a reduced ability of the cow to engage in social interactions and with its environment [7]. Cortisol concentration of the animal can be a valuable biomarker of chronic stress. Significant differences were found when comparing cortisol concentrations in hair of healthy and diseased cows (i.e., retained placenta, displaced abomasum, clinical hypocalcemia, metritis, clinical mastitis, and surgical procedures) [8]. Pain experienced from lameness could act as a stressor in dairy cattle [9]. Adverse situations trigger responses of the adrenal glands, which in turn increases glucocorticoid concentrations [10]. Cortisol has been suggested as a stress biomarker in lame cattle [11]. Our past results indicate that lameness influenced the frequency of visits to the automatic milking system (AMS), the productivity of the cows, and the interval between milking. The sum of all the negative consequences has a major impact on herd profitability, as well as on the health and welfare status of the cows. It is highly recommended to perform a comprehensive analysis of AMS variables in order to guarantee an effective management of performance and hoof health of dairy cows [12]. Miguel-Pacheco et al. [13] stated that further studies are required to investigate the potential use of capabilities and maximal benefits of the technologies available in AMS as a tool to measure and monitor the health status of cows. According Van Hertem et al. [14] in future research image-processing techniques should be tested for improving the lameness detection accuracy. To increase the prediction accuracy in automatic lameness detection, associations between the independent variables should be included [5].

The aim of the current study was therefore to evaluate the relation between lameness and milk flow traits in dairy cows.

2. Materials and Methods

2.1. Animal Selection

The research was carried out in accordance with the standards set by the Animal Welfare and Protection of the Republic of Lithuania (No. 108-2728; 2012, No. 122-6126). The study approval number is PK016965.

Lithuanian black and white dairy cows without any reproductive or other disorders were monitored carefully for claw health status on a commercial dairy farm. On the farm 128 cows (on average 2.8 ± 0.2 lactations and 60.1 ± 3.1 days postpartum) were selected for the experiment: 73 healthy (HL group) and 55 cows with lameness (LA group). Lameness was diagnosed by a local veterinarian who specializes in hoof care, according to the standard procedure described by Sprecher et al. [15]: 1 = normal, 2 = presence of a slightly asymmetric gait, 3 = the cow clearly favored one or more limbs (moderately lame), 4 = severely lame, and 5 = extremely lame (non-weight-bearing lame). Visual locomotion scoring was conducted once weekly for four consecutive weeks by the same observer.

The blood samples were collected during the exact general clinical examination by using a tube without anticoagulant (BD Vacutainer, Crawley, UK) and were centrifuged at 3.500 RPM, 20 °C for 10–15 min. Samples were delivered to the Large Animal Clinic's Laboratory of Clinical Tests at the Lithuanian University of Health Sciences Veterinary Academy.

All cows with lameness were treated with Naxcel (100 mg ceftiofur/mL; Zoetis Canada, Kirkland, QC, USA) administered subcutaneously at the dosage of 2.2 mg/kg of body weight. Treatment was repeated at 24-h intervals. At the same time, Rimadyl Cattle®® solution (50 mg carprofen/mL; Zoetis, Belgium) was delivered by subcutaneous injection at a dose of 1.4 mg per 1 kg body weight only once.

We hypothesized that lameness has an impact on milk flow traits in dairy cows such as: total milk yield, milk yield during the first minute, milk output during the first two minutes, milk output during the first three minutes, time to milk flow of 0.5 kg/min, time of main milking phase, time of incline in milk flow from 0.5 kg/min till the reach of the plateau phase, time at plateau phase, time at decline phase, time at stripping, milking speed traits (kg/min), highest milk flow, average milk flow on main milking phase, maximum milk yield per minute, and electrical conductivity (mS/cm). The latter includes electrical conductivity at highest milk flow, mS/cm, electrical conductivity during the initial time, mS/cm of milking (beginning peak level of the electrical conductivity), and maximum electrical conductivity after reaching the highest milking speed, mS/cm (Table 1).

Table 1. Description of evaluated traits with Lactocorder®®.

Indicator	Description
MGG	Total milk yield (kg/milking)
MG1	Milk yield during first min (kg)
MG2	Milk yield during first 2 min (kg)
MG3	Milk yield during first 3 min (kg)
tMGG	Time at MGG (min)
tS500	Time to milk flow of 0.5 kg/min
tMHG	Time of main milking phase (min)
tAN	Time of incline in milk flow from 0.5 kg/min till the reach of the plateau phase (min)
tPL	Time at plateau phase (min)
tAB	Time at decline phase (min)
tMNG	Time at stripping (min)
HMF	Highest milk flow (kg/min)
DMHG	Average milk flow on main milking phase (kg/min)
HMG	Maximum milk yield (kg) per min
ELHMF	Electrical conductivity at highest milk flow (mS/cm)
ELAP	Electrical conductivity during the initial time of milking (beginning peak level of the electrical conductivity) (mS/cm)
ELMAX	Maximum electrical conductivity after reaching the highest milking speed (mS/cm)
BIMO	Absence or presence of bimodality

Cows on the farm are milked twice a day with 24 places of parallel milking parlors (DeLaval VMS; DeLaval International AB Tumba, Botkyrka, Sweden). The milking properties of cows were evaluated twice in a row—during evening and morning milking. The selected cows in the current lactation did not receive veterinary treatment, and correct hoof trimming was not performed at least four weeks before the experiment.

The cows were studied during the winter period of the year 2019. Diets were formulated according to requirements of the seventh revised edition of the Nutrient Requirements of Dairy Cattle by National Research Council (NRC 2001) that meet or exceed the energy needs for 550 kg lactating Holstein dairy cows producing 35 kg/day. TMR for cows composed of 35% corn silage, 10% grass silage, 5% grass hay, and 50% grain concentrate mash (50% barely and 50% wheat). Composition of ration—DM (%)—48.8; 83 NDF (% of DM)—28.2; ADF (% of DM)—19.8; NFC (% of DM)—38.7; CP (% of DM)—15.8; NEL 84 (Mcal/kg)—1.6. TMR was fed to the cows twice per day at 10:00 a.m. and 08:00 p.m.

2.2. Measurements

Lameness in cows was assessed on the visual locomotion scale (VLS) from 1 to 5. Based on these results, the cows were divided into two groups: HL group—healthy cows (VLS score = 1–2 for all hooves) and LA group—lame cows (VLS $\geq$ 3 for at least one hoof) [15].

The milk flow of cows was recorded twice in a row during the morning and the evening milking. The measurements were taken by two electronic mobile milk flow meters (Lactocorder®®, WMB AG, Balgache, Switzerland). Milk flow data were processed using LactoPro 5.2.0 software (Biomelktechnik Swiss). Table 1 provides a detailed description of the milk flow traits of the cows studied in this experiment.

The milk flow curves were grouped according to their shape: the normal curve (absence of bimodality, BIMO = 0) and the bimodal curve (BIMO = 1; a flow pattern with two rises separated by a clear drop in milk flow below 0.2 kg/min shortly after the start of milking).

Cortisol concentration was measured using the automated analyzer TOSOH®® AIA-360 (South San Francisco, CA, USA), which uses a competitive fluorescent enzyme immunoassay, which runs in small, single-use test cups that contain all necessary reagents. Daily checks, calibration curves, and maintenance procedures were performed as described in the system operator's manual.

2.3. Statistical Analysis

Data were analyzed using the IBM SPSS Statistics software (version 25.0, IBM, Munich, Germany). Prior to analyses, the normality of all the data recorded in the study was assessed by the Shapiro–Wilk normality test. The descriptive statistics of data are presented as mean ± standard error of the mean of a sample (M ± SEM). Differences between the groups of cows were inspected with the independent T test ($p < 0.05$). Pearson's chi-square test ($\chi 2$) for independence was used to determine if there was a statistically significant relationship between cow health and the bimodality of their milk flow curve.

Using a binary multivariable logistic regression, we investigated the relationship of lameness with blood cortisol concentration, the bimodality of the milk flow curve, and those milk flow variables that showed statistically significant differences between the groups of cows. Lameness for logistic regression analysis was defined as the dependent variable (where 1 means lameness and 0 means no lameness), all explanatory variables were divided into two classes based on their arithmetic mean (class 0 < M, class 1 ≥ M). In the backward stepwise model, all insignificant variables were excluded in accordance with the significance of the Wald test and the Hosmer and Lemeshow method. Wald's test was used to determine whether the effect on the dependent variable was statistically significant, and the Hosmer–Lemeshow test—to confirm that the logistic regression model fits the data and Nagelkerke coefficient R^2—was used to assess the suitability of the model. Three statistically significant indicators were used in the final multivariable binary regression model. The continuous explanatory variables of the model (based on their arithmetic mean) were divided into two categorical classes: MGG < 16.5 kg or ≥16.5 kg; blood cortisol concentration < 1.00 μg/dL or ≥1.00 μg/dL. The results of the logistic regression are presented in terms of the odds ratio (OR) and 95% confidence interval (PI).

3. Results

3.1. Relationship of Lameness with Milk Yield and Electrical Conductivity Traits of Dairy Cows

The data presented in Table 2 demonstrate that the milk yield of healthy dairy cows was 1.77 kg higher than that of lame cows, as well as the milk yield in the first minute (0.34 kg more; $p < 0.01$). Higher milk yields in healthy cows were also observed during the second and third minutes of milking, but these differences between the groups were not statistically significant.

Table 2. Milk yield (kg) traits of cows ($n = 128$).

Variable	Group	M	SE
MGG	HL	17.14	0.366
	LA	15.37 **	0.422
MG1	HL	2.28	0.085
	LA	1.94 **	0.098
MG2	HL	5.38	0.177
	LA	4.96	0.204
MG3	HL	8.33	0.256
	LA	7.84	0.295

** $p < 0.01$. MGG—total milk yield (kg/milking); MG1—milk yield during first minute (kg); MG2—milk yield during first 2 min (kg); MG3—milk yield during first 3 min (kg); HL—healthy group; LA—lameness group. M—mean; SE—standard error of the mean of a sample.

MGG—total milk yield (kg/milking); MG1—milk yield during first min (kg); MG2—milk yield during first 2 min (kg); MG3—milk yield during first 3 min (kg); HL—healthy group; LA—lameness group.

MGG was higher in the group of healthy cows (0.3 kg) in the morning and in the group of LA cows. In contrast, it was higher in the evening (1.0 kg). The data are presented in Figure 1.

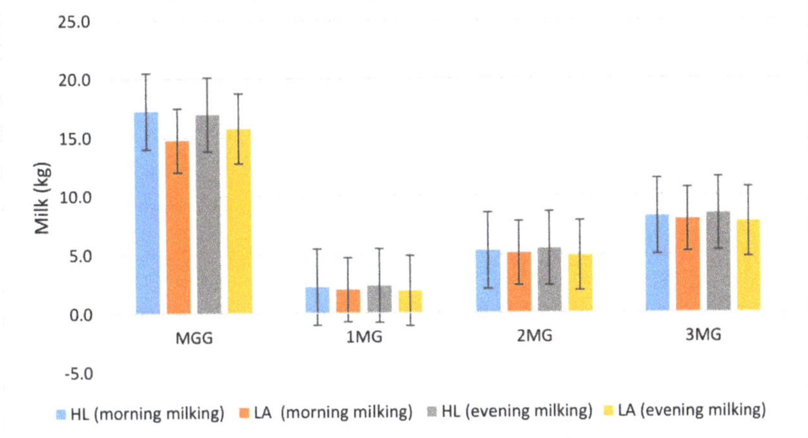

Figure 1. Milk yield (kg) traits of cows by milking. MGG—total milk yield (kg/milking); MG1—milk yield during first minute (kg); 2MG—milk yield during first 2 min (kg); 3MG—milk yield during first 3 min (kg).

The ELAP of healthy cows was statistically significantly lower (−0.24 mS/cm, $p < 0.01$) than that of the LA group. It should also be noted that all the studied indicators of milk electrical conductivity were higher in cows with signs of lameness compared with healthy cows (Table 3).

Table 3. Electrical conductivity traits (mS/cm) in milk of cows, (n = 128).

Variable	Group	M	SE
ELHMF	HL	5.72	0.042
	LA	5.73	0.048
ELAP	HL	6.10	0.052
	LA	6.34 **	0.060
ELMAX	HL	5.98	0.054
	LA	5.99	0.062

** $p < 0.01$; ELHMF—electrical conductivity at highest milk flow; ELAP—electrical conductivity during the initial time of milking; ELMAX—maximum electrical conductivity after reaching the highest milking speed; HL—healthy group; LA—lameness group. M—mean; SE—standard error of the mean of a sample.

We did not find statistically significant differences between the electrical conductivity of evening and morning milking (Figure 2).

ELHMF—electrical conductivity at highest milk flow; ELAP—electrical conductivity during the initial time of milking; ELMAX—maximum electrical conductivity after reaching the highest milking speed; HL—healthy group; LA—lameness group.

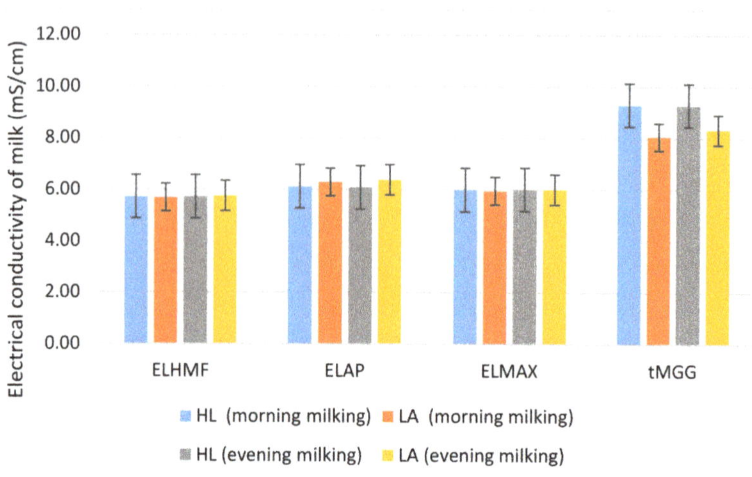

Figure 2. Electrical conductivity traits (mS/cm) in milk of cows by milking.

3.2. Relationship of Lameness with Milking Time and Speed Traits

The milking duration (tMGG) of LA cows was shorter (−1.07 min, $p < 0.05$), as was the value of the tMNG (−0.41 min, $p < 0.01$), but the tAN duration was longer (0.26 min, $p < 0.001$) compared to lame cows (Table 4). Similar trends were observed between the groups of cows when comparing morning milking and evening milking data (Figure 3).

Table 4. Milking time (min) traits of cows by health status and lameness level (n = 0.128).

Variable	Group	M	SE
tMGG	HL	9.29	0.304
	LA	8.22 *	0.350
tS500	HL	0.18	0.012
	LA	0.10 ***	0.014
tMHG	HL	7.62	0.260
	LA	7.11	0.300
tAN	HL	0.57	0.035
	LA	0.83 ***	0.041
tPL	HL	3.65	0.194
	LA	3.13	0.224
tAB	HL	3.40	0.215
	LA	3.14	0.247
tMNG	HL	0.61	0.104
	LA	0.20 **	0.120

* $p < 0.05$, ** $p < 0.01$, *** $p < 0.01$; tMGG—time at total milk yield; tS500—time to milk flow of 0.5 kg/min; tMHG—time of main milking phase (min.); tAN—time of incline in milk flow from 0.5 kg/min till the reach of the plateau phase (min); tPL—time at plateau phase (min); tAB—time at decline phase (min); tMNG—time at stripping (min); HL—healthy group; LA—lameness group. M—mean; SE—standard error of the mean of a sample.

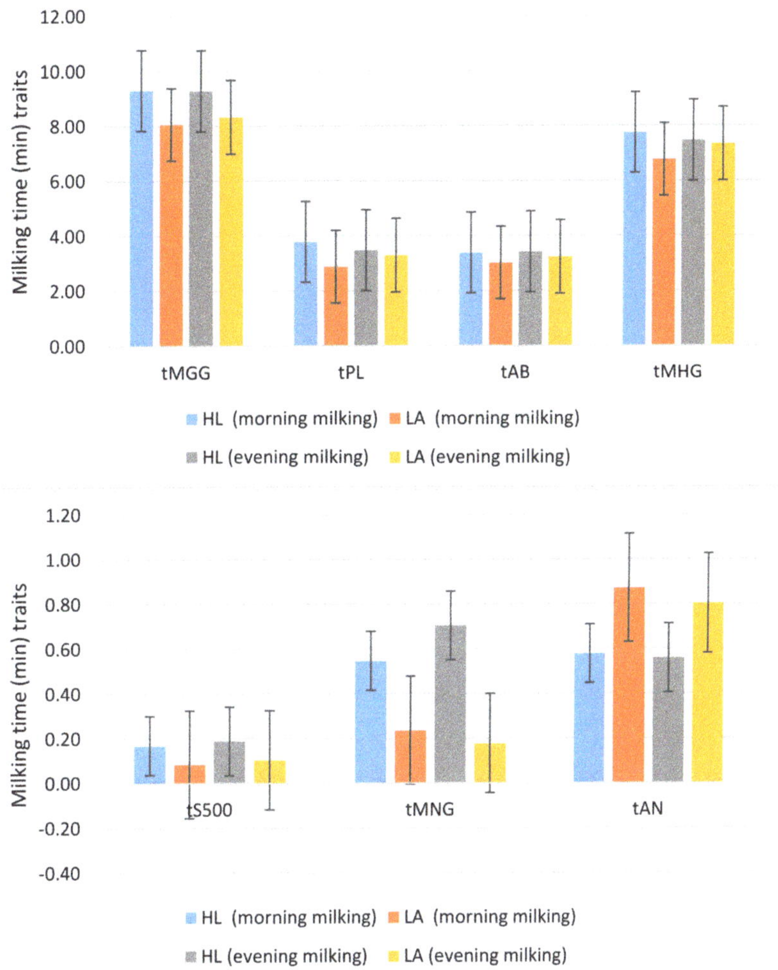

Figure 3. Milking time (min) traits of cows by milking. tS500—time to milk flow of 0.5 kg/min; tMNG—time at stripping (min); tAN—time of incline in milk flow from 0.5 kg/min till the reach of the plateau phase (min).

All means of the milking duration indices in healthy cows were higher during morning and evening milking, except for tAN, which was higher in the LA cow group during both milking periods (0.24–0.29 min, $p < 0.05$).

tMGG—time at total milk yield; tS500—time to milk flow of 0.5 kg/min; tMHG—time of main milking phase (min); tAN—time of incline in milk flow from 0.5 kg/min till the reach of the plateau phase (min); tPL—time at plateau phase (min); tAB—time at decline phase (min); tMNG—time at stripping (min); HL—healthy group; LA—lameness group.

Lame cows had higher HMF and HMG values, but the differences with the healthy group were not significant. The data are presented in Table 5.

Table 5. Milking speed traits (kg/min) of cows by health status and lameness level (n = 0.128).

Dependent Variable	Group	M	SEM
HMF	HL	3.19	0.099
	LA	3.22	0.115
DMHG	HL	2.26	0.063
	LA	2.24	0.073
HMG	HL	3.12	0.097
	LA	3.15	0.112

HMF—highest milk flow (kg/min); DMHG—average milk flow during main milking phase (kg/min); HMG—maximum milk yield (kg) per minute; HL—healthy group; LA—lameness group. M—mean; SEM—standard error of the mean of a sample.

HMF—highest milk flow (kg/min); DMHG—average milk flow during main milking phase (kg/min); HMG—maximum milk yield (kg) per minute; HL—healthy group; LA—lameness group.

The average values of all milking speed traits were slightly (0.02–0.07 kg/min) higher during the morning milking of healthy cows and, conversely, during the evening milking (0.01–0.05 kg/min) higher in the group of lame cows (Figure 4).

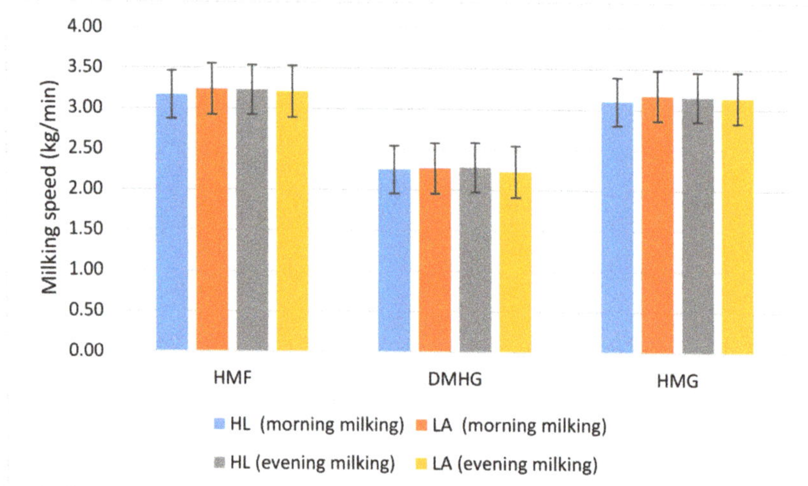

Figure 4. Milking speed traits (kg/min) of cows by milking.

3.3. Relationship of Lameness with Bimodality in Milk Flow of Cows

After evaluating the milk flow curves of cows, we determined 37.5% of the total bimodal curves. The analysis showed that the bimodality of the milk flow curve was statistically significantly associated with the health status of cows ($p < 0.001$). As a result, 23.3% of HL cows were found to have bimodal curves, whereas for 56.4% of LA cows bimodal curves were obtained.

In healthy cows (Figure 5) more bimodal milk flow curves were found during evening milking (1.94 times), while in sick cows, on the contrary, during morning milking (1.24 times).

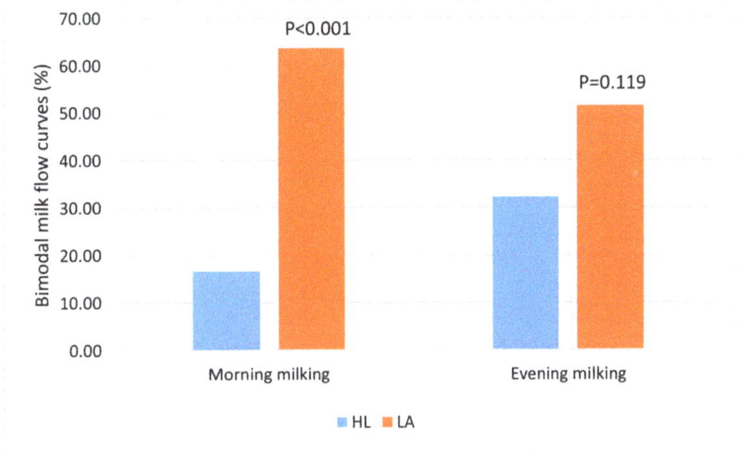

Figure 5. Prevalence of bimodality in milk flow curves by health status of cows and milking. HL—healthy group; LA—lameness group.

3.4. *Relationship of Lameness with Blood Cortisol Concentration and Milk Flow Traits in Cows*

The average blood cortisol concentration of lame cows (1.40 ± 0.082 µg/dL) was 2.1 times higher ($p < 0.001$) than in healthy cows (0.68 ± 0.071 µg/dL).

As can be seen from the data in Figure 6, healthy cows showed a higher blood cortisol concentration before evening milking (1.93 times, $p < 0.001$), while lame cows, in contrast, before morning milking (1.14 times).

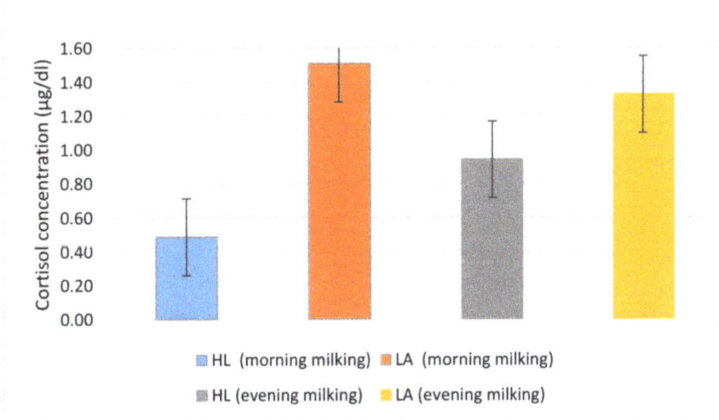

Figure 6. Cortisol concentration (µg/dL) in blood of cows by health status and milking. HL—healthy group; LA—lameness group.

We also found that the concentration of cortisol was 1.9 times higher ($p < 0.001$) in cows with a bimodal milk flow curve (1.42 ± 0.090 µg/dL) compared to cows with a normal milk flow curve (0.73 ± 0.070 µg/dL). Blood cortisol concentration was positively associated with lameness as well as the bimodality of the milk flow curve.

Multivariable binary logistic regression analysis (Table 6) showed that the blood cortisol concentration ($p = 0.001$), the total cow's milk yield ($p = 0.021$), and the bimodality of the milk flow curve ($p = 0.015$) can be used to predict lameness early. Of all the parameters studied, the risk of lameness was most clearly indicated by an increase in blood cortisol concentration. If its blood level in cows exceeds 1 μg/dL, the risk of identifying lameness increases 4.9 times (95% CI = 1.943–12.328).

Table 6. Binary multivariable regressions factors indicating a statistically significant risk in predicting lameness of cows (n = 0.128).

Categorical Variables	Classes	p	OR	95% CI. for OR	
				Lower	Upper
BIMO	0 1	0.015	2.860	1.222	6.690
Cortisol (μg/dl)	<1 ≥1	0.001	4.895	1.943	12.328
MGG (kg)	<16.5 ≥16.5	0.021	0.382	0.169	0.865

OR—odds ratio, PI—95% confidence interval. BIMO—absence (BIMO = 0) or presence of bimodality (BIMO = 1); MGG—total milk yield (kg/milking).

4. Discussion

Lameness is the biggest challenge for dairy farms to overcome [14]. Associations were found between lameness and many parameters measured automatically by sensors, revealing the high potential of detecting lameness by analyzing automatically measured performance and behavior data [16]. This study showed that the level of blood cortisol in the HL group was lower than that in the LA group. Significantly higher levels of plasma cortisol concentration in the LA group can be associated with stress experienced by the animals in this group, and more pronounced reactions to the milking process. Cortisol concentration can be a valuable biomarker of chronic stress. On the day of diagnosis, elevated serum cortisol concentrations have been registered in cows diagnosed with lameness [17].

The measurement of cortisol concentrations could detect clearly defined differences between healthy and clinically diseased (i.e., retained placenta, clinical hypocalcaemia, clinical mastitis, displaced abomasum, metritis, and surgical procedures) lactating cows for evaluation of the effect of a stressor, though a comparison of cortisol concentration with or without stressor would be necessary [18]. As previously noted, on the day of diagnosis, increased concentrations of serum cortisol were determined in cows diagnosed with lameness [19]. In a previous study the hair cortisol concentration of clinically sick (i.e., laminitis, metritis, mastitis) and cows with a compromised physiological state (parturition) was higher than that of clinically healthy cows [19]. Redbo [20] reported that chronic stress diminished the sensitivity of the adrenal cortex. Growing bulls responded to a long stress period due to tethering with a lower plasma cortisol concentration after receiving ACTH stimulation in comparison to bulls in the control group [21]. Hair cortisol concentrations have been shown to correlate with adrenocortical activity after ACTH challenge during the 14-d interval before the collection of hair [22], indicating its value as a biomarker for painful and acute events in this period for animals kept in the same environment [8]. Fischer-Tenhagen et al. [6] claim that cortisol concentration acts as a valuable biomarker for chronic lameness in dairy cows. Cortisol has been employed as a stress biomarker in lame cattle [11]. On the day of diagnosis, serum cortisol concentrations were elevated in cows diagnosed with lameness [23]. The study of O'Driscoll et al. [17] demonstrated, that on the day of diagnosis, the cortisol concentration in serum was elevated in cows with sole ulcers. Our data did not show any influence of the lameness score on the cortisol concentration, which is in agreement with O'Driscoll et al. [17] who reported the cortisol concentration in cows with sole hemorrhages.

In our study, the negative relationship between milk yield and blood cortisol concentration in cows of the LA group could be attributed to the deteriorating welfare of animals due to lameness. The results of this study are confirmed by the results obtained by Miyazawa [24], where a negative association between milk yield and plasma cortisol concentration and hormone release in response to the process of milking has been established [25]. This study concluded that lameness is a stress factor in Lithuanian black and white cows, which leads to elevated cortisol concentrations. The increase of stress in lame cows affected the milking process—weaker stimulation of milk ejection reflex and consequently lower milk yield in lame cows. During milking, the milk flow was recorded with a specially rebuilt set of Lactocorders (Werkzeug und Maschinenbau Berneck AG, Balgach, Switzerland). Over-milking has been defined as milking that has been continued after the milk flow has suddenly dropped below 0.2 kg/min. Milk within the udder can be divided into two fractions: cisternal milk, which is immediately extracted by the machine and does not require oxytocin release, and alveolar milk, which can only be removed by the active involvement of the animal, when oxytocin release affects the milk ejection [26]. When the animals are physiologically stressed, suffering from acute events, or long-term bad conditions of milking [27], we can record physiological responses such as high levels of cortisol and diminished sensitivity to ACTH [28]. Redbo et al. [20] documented that chronic stress leads to a lowered sensitivity of the adrenal cortex. Reduced secretion of ACTH and cortisol during continuous stressful situations, such as chronic lameness, is a clear example of the hormonal regulation to prevent prolonged exposure to elevated concentrations of cortisol as described by Knights and Smith [29]. It is often described that this leads to a clear partial or total inhibition of the milk ejection reflex, a delay in milk ejection, and/or a reduced milk flow [30]. Such a delayed milk ejection presents itself with bimodal milk flow curves [31]. De Mol et al. [32] and Kamphuis et al. [33] also showed this effect in their models. Milk yield may have an effect on prediction accuracy for lameness, but Pavlenko et al. [34] even excluded it from their models because it had no association with lameness in their data.

5. Conclusions

Concerning the important findings of our present study, we conclude that milk flow traits can act as biomarkers of lameness in dairy cows. We found that milk yield and yield in the first minute (0.34 kg) of healthy dairy cows was 1.77 kg higher than that of lame cows. The electrical conductivity during the initial time of milking of health cows was 0.24 mS/cm lower than that of the LA group. The milking duration of LA cows was 1.07 min shorter, the time of incline in milk flow from 0.5 kg/min till the reach of the plateau phase was longer. The risk of lameness was most clearly indicated by an increase in blood cortisol concentration. Cows with blood cortisol exceeding 1 ug/dL are 4.9 times more likely to be lame. From practical point of view the results of our study suggest that lameness of cows had negative impact on milk flow traits. This can help to detect lame cows via their milk flow traits.

Author Contributions: V.J.: assisted in designing and setting up field data collection activities and developed the software and algorithm for data analysis. The entire process was revised by the co-authors. R.A.: overall research study process including literature search, carrying out research experiments, and compiling the final manuscript; G.U.: sampling of the experimental animals; M.U.: aided in fieldwork set up, data collection, and sampling of the experimental animals; W.B. and S.T.: provided criticism and revised the manuscript with major support in processing of data in the study. All authors have read and agreed to the published version of the manuscript.

Funding: This research received no external funding.

Institutional Review Board Statement: The study was conducted according to the guidelines of the Declaration of Helsinki, and ap-proved by Ethics Committee (The study approval number is PK016965, 2017.06.06).

Informed Consent Statement: Informed consent was obtained from all subjects involved in the study.

Data Availability Statement: The data presented in this study are available within the article.

Conflicts of Interest: The authors declare no conflict of interest.

References

1. Booth, C.J.; Warnick, L.D.; Gröhn, Y.T.; Maizon, D.O.; Guard, C.L.; Janssen, D. Effect of lameness on culling in dairy cows. *J. Dairy Sci.* **2004**, *87*, 4115–4122. [CrossRef]
2. Bicalho, R.C.; Machado, V.S.; Caixeta, L.S. Lameness in dairy cattle: A debilitating disease or a disease of debilitated cattle? A cross-sectional study of lameness prevalence and thickness of the digital cushion. *J. Dairy Sci.* **2009**, *92*, 3175–3184. [CrossRef] [PubMed]
3. Walker, S.L.; Smith, R.F.; Jones, D.N.; Routly, J.E.; Morris, M.J.; Dobson, H. The effect of a chronic stressor, lameness, on detailed sexual behaviour and hormonal profiles in milk and plasma of dairy cattle. *Reprod. Domest. Anim.* **2010**, *45*, 109–117. [CrossRef]
4. Van Hertem, T.; Maltz, E.; Antler, A.; Romanini, C.E.B.; Viazzi, S.; Bahr, C.; Schlageter-Tello, A.; Lokhorst, C.; Berckmans, D.; Halachmi, I. Lameness detection based on multivariate continuous sensing of milk yield, rumination, and neck activity. *J. Dairy Sci.* **2013**, *96*, 4286–4298. [CrossRef] [PubMed]
5. Grimm, K.; Haidn, B.; Erhard, M.; Tremblay, M.; Döpfer, D. New insights into the association between lameness, behavior, and performance in Simmental cows. *J. Dairy Sci.* **2019**, *102*, 2453–2468. [CrossRef] [PubMed]
6. Fischer-Tenhagen, C.; Ladwig-Wiegard, M.; Heuwieser, W.; Thöne-Reineke, C. Is hair cortisol a potential indicator for stress caused by chronic lameness in dairy cows? *J. Dairy Sci.* **2018**, *101*, 5439–5443. [CrossRef] [PubMed]
7. Scott, G.B. Changes in limb loading with lameness for a number of Friesian cattle. *Br. Vet. J.* **1989**, *145*, 28–38. [CrossRef]
8. Burnett, T.A.; Madureira, A.M.; Silper, B.F.; Tahmasbi, A.; Nadalin, A.; Veira, D.M.; Cerri, R.L. Relationship of concentrations of cortisol in hair with health, biomarkers in blood, and reproductive status in dairy cows. *J. Dairy Sci.* **2015**, *98*, 4414–4426. [CrossRef]
9. Underwood, W.J. Pain and distress in agricultural animals. *J. Am. Vet. Med. Assoc.* **2002**, *221*, 208–211. [CrossRef] [PubMed]
10. Möstl, E.; Palme, R. Hormones as indicators of stress. *Domest. Anim. Endocrinol.* **2002**, *23*, 67–74. [CrossRef]
11. Bustamante, H.A.; Rodríguez, A.R.; Herzberg, D.E.; Werner, M.P. Stress and pain response after oligofructose induced-lameness in dairy heifers. *J. Vet. Sci.* **2015**, *16*, 405. [CrossRef]
12. Urbonavicius, G.; Antanaitis, R.; Zilaitis, V.; Tusas, S.; Kajokiene, L.; Zymantiene, J.; Spancerniene, U.; Gavelis, A.; Juskiene, V.; Juozaitiene, V. The influence of lameness on several automatic milking system variablesand reproductive performance indicators in dairy cows. *Pol. J. Vet. Sci.* **2020**, 383–390. [CrossRef]
13. Miguel-Pacheco, G.G.; Kaler, J.; Remnant, J.; Cheyne, L.; Abbott, C.; French, A.P.; Pridmore, T.P.; Huxley, J.N. Behavioural changes in dairy cows with lameness in an automatic milking system. *Appl. Anim. Behav. Sci.* **2014**, *150*, 1–8. [CrossRef]
14. Bicalho, R.C.; Vokey, F.; Erb, H.N.; Guard, C.L. Visual locomotion scoring in the first seventy days in milk: Impact on pregnancy and survival. *J. Dairy Sci.* **2007**, *90*, 4586–4591. [CrossRef]
15. Sprecher, D.E.A.; Hostetler, D.E.; Kaneene, J.B. A lameness scoring system that uses posture and gait to predict reproductive performance. *Theriogenology* **1997**, *47*, 1179–1187. [CrossRef]
16. Thorup, V.M.; Nielsen, B.L.; Robert, P.E.; Giger-Reverdin, S.; Konka, J.; Michie, C.; Friggens, N.C. Lameness affects cow feeding but not rumination behavior as characterized from sensor data. *Front. Vet. Sci.* **2016**, *3*, 37. [CrossRef] [PubMed]
17. O'Driscoll, K.; Gleeson, D.; O'Brien, B.; Boyle, L. Effect of milking frequency and nutritional level on hoof health, locomotion score and lying behaviour of dairy cows. *Livest. Sci.* **2010**, *127*, 248–256. [CrossRef]
18. Bertulat, S.; Fischer-Tenhagen, C.; Suthar, V.; Möstl, E.; Isaka, N.; Heuwieser, W. Measurement of fecal glucocorticoid metabolites and evaluation of udder characteristics to estimate stress after sudden dry-off in dairy cows with different milk yields. *J. Dairy Sci.* **2013**, *96*, 3774–3787. [CrossRef]
19. O'Driscoll, K.; McCabe, M.; Earley, B. Leukocyte profile, gene expression, acute phase response, and metabolite status of cows with sole hemorrhages. *J. Dairy Sci.* **2017**, *100*, 9382–9391. [CrossRef]
20. Redbo, I. Relations between oral stereotypies, open-field behavior, and pituitary–adrenal system in growing dairy cattle. *Physiol. Behav.* **1998**, *64*, 273–278. [CrossRef]
21. Ladewig, J.; Smidt, D. Behavior, episodic secretion of cortisol, and adrenocortical reactivity in bulls subjected to tethering. *Horm. Behav.* **1989**, *23*, 344–360. [CrossRef]
22. Del Rosario, G.D.L.V.; Valdez, R.A.; Lemus-Ramirez, V.; Vázquez-Chagoyán, J.C.; Villa-Godoy, A.; Romano, M.C. Effects of adrenocorticotropic hormone challenge and age on hair cortisol concentrations in dairy cattle. *Can. J. Vet. Res.* **2011**, *75*, 216–221.
23. O'Driscoll, K.; McCabe, M.; Earley, B. Differences in leukocyte profile, gene expression, and metabolite status of dairy cows with or without sole ulcers. *J. Dairy Sci.* **2015**, *98*, 1685–1695. [CrossRef]
24. Miyazawa, K. Changes in plasma cortisol and glucose level after applying stimulus to the jugular vein to collect blood, stimulus on the teat and the stimulus of milking in cows. *Res. Bull. Obihiro Univ.* **1983**, *13*, 1–7.
25. Stelwagen, K.; Knight, C.H. Effect of unilateral once twice milking of cows on milk yield and udder characteristics in early and late lactation. *J. Dairy Res.* **1997**, *64*, 487–494. [CrossRef] [PubMed]
26. Bruckmaier, R.M.; Wellnitz, O. Induction of milk ejection and milk removal in different production systems. *J. Anim. Sci.* **2007**, *86*, 15–20. [CrossRef] [PubMed]
27. Tancin, V.; Urhincat, M.; Mačuhovà, J.; Bruckmaier, R.M. Effect of pre-stimulation on milk flow pattern and distribution of milk constituents at a quarter level. *Czech J. Anim. Sci.* **2007**, *52*, 117–121. [CrossRef]
28. Bruckmaier, R.M.; Blum, J.W. Simultaneous recording of oxytocin release, milk ejection and milk flow during milking in dairy cows with and without prestimulation. *J. Dairy Res.* **1996**, *63*, 201–208. [CrossRef]

29. Knights, M.; Smith, G.W. Decreased ACTH secretion during prolonged transportation stress is associated with reduced pituitary responsiveness to tropic hormone stimulation in cattle. *Domest. Anim. Endocrinol.* **2007**, *33*, 442–450. [CrossRef]
30. Wellnitz, O.; Bruckmaier, R.M. Central and peripheral inhibition of milk ejection. *Livest. Prod. Sci.* **2001**, *70*, 135–140. [CrossRef]
31. Tancin, V.; Bruckmaier, R.M. Factors affecting milk ejection and removal during milking and suckling of dairy cows. *Vet. Med.* **2001**, *46*, 108–118. [CrossRef]
32. De Mol, R.M.; Ouweltjes, W.; Kroeze, G.H.; Hendriks, M.M.W.B. Detection of estrus and mastitis: Field performance of a model. *Appl. Eng. Agric.* **2001**, *17*, 399. [CrossRef]
33. Kamphuis, C.; DelaRue, B.; Burke, C.R.; Jago, J. Field evaluation of 2 collar-mounted activity meters for detecting cows in estrus on a large pasture-grazed dairy farm. *J. Dairy Res.* **2012**, *95*, 3045–3056. [CrossRef] [PubMed]
34. Pavlenko, A.; Bergsten, C.; Ekesbo, I.; Kaart, T.; Aland, A.; Lidfors, L. Influence of digital dermatitis and sole ulcer on dairy cow behaviour and milk production. *Animal* **2011**, *5*, 1259–1269. [CrossRef] [PubMed]

agriculture

MDPI

Article

Simulation of Fuel Consumption Based on Engine Load Level of a 95 kW Partial Power-Shift Transmission Tractor

Md. Abu Ayub Siddique [1], Seung-Min Baek [2], Seung-Yun Baek [2], Wan-Soo Kim [1], Yeon-Soo Kim [1,3], Yong-Joo Kim [1,2,*], Dae-Hyun Lee [1,*], Kwan-Ho Lee [4] and Joon-Yeal Hwang [5]

[1] Department of Biosystems Machinery Engineering, Chungnam National University, Daejeon 34134, Korea; ayub@cnu.ac.kr (M.A.A.S.); ws6602@cnu.ac.kr (W.-S.K.); kimtech612@kitech.re.kr (Y.-S.K.)
[2] Department of Smart Agricultural Systems, Chungnam National University, Daejeon 34134, Korea; bsm1104@o.cnu.ac.kr (S.-M.B.); kelpie0037@o.cnu.ac.kr (S.-Y.B.)
[3] Smart Agricultural Machinery R&D Group, Korea Institute of Industrial Technology (KITECH), Gimje 54325, Korea
[4] CAE Solution Team, TYM ICT Inc., Gongju 32530, Korea; lkh@tymict.com
[5] Smart Solution Team, TYM ICT Inc., Gongju 32530, Korea; fing0828@tymict.com
* Correspondence: babina@cnu.ac.kr (Y.-J.K.); leedh7@cnu.ac.kr (D.-H.L.); Tel.: +82-42-821-6716 (Y.-J.K.)

Citation: Siddique, M..A.A.; Baek, S.-M.; Baek, S.-Y.; Kim, W.-S.; Kim, Y.-S.; Kim, Y.-J.; Lee, D.-H.; Lee, K.-H.; Hwang, J.-Y. Simulation of Fuel Consumption Based on Engine Load Level of a 95 kW Partial Power-Shift Transmission Tractor. *Agriculture* **2021**, *11*, 276. https://doi.org/10.3390/agriculture11030276

Academic Editors: Gniewko Niedbała and Sebastian Kujawa

Received: 26 February 2021
Accepted: 20 March 2021
Published: 23 March 2021

Publisher's Note: MDPI stays neutral with regard to jurisdictional claims in published maps and institutional affiliations.

Abstract: This study is focused on the estimation of fuel consumption of the power-shift transmission (PST) tractor based on PTO (power take-off) dynamometer test. The simulation model of PST tractor was developed using the configurations and powertrain of the real PST tractor. The PTO dynamometer was installed to measure the engine load and fuel consumption at various engine load levels (40, 50, 60, 70, 80, and 90%), and verify the simulation model. The axle load was also predicted using tractor's specifications as an input parameter of the simulation model. The simulation and measured results were analyzed and compared statistically. It was observed that the engine load, as well as fuel consumption, were directly proportional to the engine load levels. However, it was statistically proved that there was no significant difference between the simulation and measured engine torque and fuel consumption at each load level. The regression equations show that there was an exponential relationship between the fuel consumption and engine load levels. However, the specific fuel consumptions (SFC) for both simulation and measured were linear relationships and had no significant difference between them at each engine load level. The results were statistically proved that the simulation and measured SFCs were similar trends. The plow tillage operation could be performed at the gear stage of 7.65 km/h with higher working efficiency at low fuel consumption. The drawback of this study is to use a constant axle load instead of dynamic load. This study can provide useful information for both researchers and manufacturers related to the automated transmission of an agricultural tractor, especially PST tractor for digital farming solutions. Finally, it could contribute to the manufacturers developing a new agricultural tractor with higher fuel efficiency.

Keywords: tractor; powershift transmission; fuel consumption; load level; simulation model

1. Introduction

Tractors, which are machines, deal with various works including agricultural, construction, and forestry [1]. Specifically, agricultural tractors perform various agricultural works such as plow tillage [2], subsurface drainage operation [3], rotary [4], and baler [5]. According to the Mordor Intelligence statistics [6], the global market of agricultural tractors is expected to have 4.02% of annual growth rate in 2025 than that in 2020. Among them, approximately 50% of the tractor global market is in the Asia–Pacific region. In Korea, farmers aged over 65 years old account for 6.7% of the total in 2010; this is expected to be 11.3% in 2050 [7,8]. To compete with the largest tractor manufacturing companies in the global market, and to fulfill the consumers' demands, advanced technologies along with the highest facilities should be introduced.

Advanced technologies such as autonomous, artificial intelligence are applied for agricultural tractors considering driving comfort for aged farmers, higher working efficiency due to lack of labor, and precision farming for increasing production [9]. To confirm the users' flexibility, precise work, and higher efficiencies during operations, several researchers and manufacturers are developing numerous powertrain systems, these include manual transmission (MT), automatic transmission (AT), dual-clutch transmission (DCT), continuously variable transmission (CVT), and power-shift transmission (PST) for agricultural tractors [10,11].

The PST is one of the standard modern tractor systems that is comparatively easier and convenient for the user to control and maintain the vehicle on-field [12]. The PST has two types: the partial PST, which can shift two or more speeds without clutch, having clutch to shift gears, and cannot control machine itself; and the full PST, which can shift all gears without clutching, and the machine can control itself. The PST allows changing gear stages precisely on the run under load conditions of the vehicles [13]. The PST is equipped with a wide range of gear stages without a loss (or minimum) of power during power delivery from engine to driving axles [9]. The PST can apply from low power vehicles to high power vehicles and has become popular as a precise technology. The performance of agricultural machinery varies from nation to nation based on the working environment. Therefore, it is important to conduct the efficiency of the PST tractor.

The fuel consumption of the tractor depends on workloads because the working loads are varied by operation conditions such as engine speed, soil properties of the field, operation types based on implements, and transmission gear stages. Among them, the engine speed and transmission gear stages mostly affect the fuel efficiency of the tractor [14]. The fuel efficiency of the tractor can be optimized by adjusting engine load conditions. The engine load varies on throttle opening of the engine [15]. Therefore, the engine load level of an engine was considered as mostly affected factor to estimate fuel consumption in this study.

Numerous approaches have been proposed to optimize engine speed for improving the fuel efficiency of the tractors. A passive eco-driving system derived for optimal engine speed considering workloads [16]. Jiang [17] developed a PID (proportional-integral-derivative) controller to maintain engine speed for estimating fuel consumption. The performance was reported as better than the existing mechanical governor. Lee [14] developed a model-based controller for fuel consumption based on working loads during plow tillage operation. Therefore, the engine load level is an important factor to estimate the optimal fuel consumption of the agricultural tractors.

The above literature revealed that fuel consumption can be optimized by controlling engine load levels. However, the development of theoretical or model-based control algorithms to control the engine speed is comparatively difficult for highly sensible various tuning methods and has limitations due to time delay, and feedback control process [18]. Moreover, the model-based controllers are needed to validate by field experiment, which requires a tractor installed measurement system that is highly time-consuming and expensive. The only simulation can be an alternative method to minimize cost and experimental time. Saunders [19] developed the discrete element method (DEM) simulation to improve and verify the moldboard plow performance. They believed the simulation method is much more convenient, easier, high accuracy, as well as time-saving. Therefore, a comprehensive, easier, and highly adaptable and reliable simulation method is applied to estimate the fuel consumption in this study.

In addition, digital farming has recently been one of the interesting and value-added topics in automation or unmanned agricultural machinery as well as precision agriculture. As the engine load greatly affects fuel consumption, power transmission efficiency, the decision-making support system of an automatic transmission system or unmanned agricultural vehicles, is one of the key issues for smart digital farming. To address the current issue for digital farming, PTO (power-take-off) dynamometer, which is an indoor test bench, is commercially used as a simple engine load control device in this study. PTO dynamometer, which does not require a long period to install, and comparatively low cost than that of measurement tractor for measuring the fuel consumption according to the engine load conditions. The engine load is adjusted, controlled, and optimized by the PTO dynamometer load.

This study is a basic study to develop an efficient PST tractor based on different field conditions, implement type, and crop types according to the required hydraulic power, as the PST tractor transmission is performed by the hydraulic pressures. The novelty of this study is that the fuel consumption of the PST tractor was estimated by the simulation method and verified for optimal engine load conditions based on the measured PTO load by PTO dynamometer. Therefore, the objective of this study is to estimate the fuel consumption of PST tractor based on engine load levels using the measured PTO loads by PTO dynamometer. The specific objectives are as follows: (i) to measure engine loads by PTO dynamometer for verifying the simulation model, and (ii) to estimate and analyze the specific fuel consumption of PST tractor.

2. Materials and Methods

2.1. Tractor Transmission Configurations

In this study, a 95 kW PST tractor (TS130, TYM Co., Ltd., Gongju, Korea) was used to estimate the fuel consumption. The dimension of the tractor (Length × Width × Height) were 4490 × 2360 × 2940 mm. The rated engine torque is 415 Nm at the rated rotational speed of 2200 rpm. The transmission is power-shift with a combination of 18 × 18 gear stages for both forward and reverse directions. The weight distribution of the PST tractor is 40.3 and 59.7% of the front and rear axle, respectively, whereas the gross weight is 44,587 N. The specifications of the PST tractor were listed in Table 1.

Table 1. The specifications of the power-shift transmission (PST) tractor used in this study.

Parameters	Specifications	
Model	TS130, TYM, Korea	
Weight (N)	Gross weight (N) Weight distribution (%)	44,587 40.3 and 59.7
Engine	Type Rated power (kW) Rated torque (Nm) Rated speed (rpm) Shifting method	Tier 4 95 415 2200 Power-shift
Transmission	Gear stages Sub-shifting stages Combinations (forward × reverse)	6 (1, 2, 3, 4, 5, 6) 3 (L, M, H) 18 × 18
Tire	Model (front and rear) Diameter (front and rear) (mm)	380/85R24 and 460/85R38 1256 and 1770

The powertrain of the PST tractor is composed of 2 power shifts (high and low), main clutch: 6 driving shifts (1, 2, 3, 4, 5, and 6) of hydraulic power-shift type and sub-shifting: 3 range shifts (L, M, and H) of mechanical type. The engine power is transmitted to the main clutch dealing with the forward-reverse and high-low shifts. Sequentially the driving shaft is connected to driving shift gear, range shaft gear, and PTO shift gear to drive the rear PTO. This is the partial PST because the sub-shift (Range shift) is performed by a mechanical gear system. The schematic diagram of the powertrain of the partial PST tractor is shown in Figure 1.

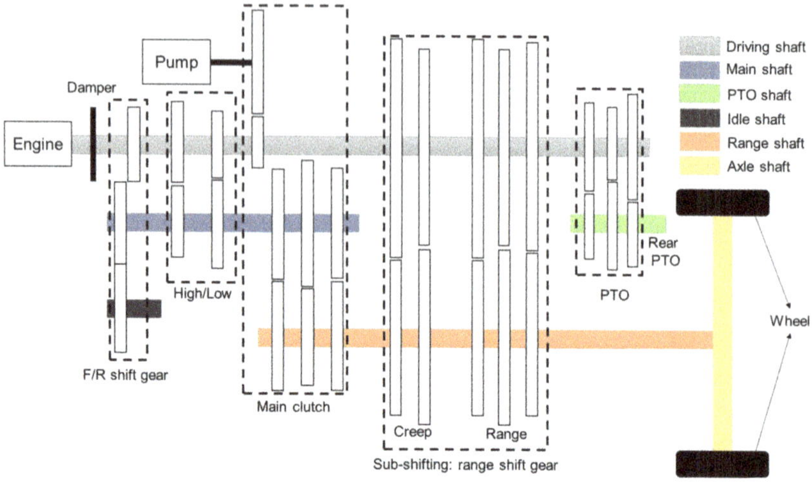

Figure 1. The schematic diagram of the PST tractor transmission.

2.2. *Tractor Dynamic Model*

2.2.1. Axle Load Prediction

To estimate the fuel consumption of an agricultural tractor, load conditions should be considered. As the main clutch of PST is performed by the hydraulic power, the axle torque cannot predict using the gear ratio of the transmission. Therefore, the axle torque was predicted using tractor specifications. Tractor based prediction model is defined as the theoretical axle torque. Theoretical axle torque of a tractor is calculated under ideal conditions using weight and engine specifications [20]. Both front and rear axle torques of the PST tractor can be calculated using Equations (1)–(4) based on the gross weight of tractor, weight distribution ratio, traction coefficient, and wheel radius.

$$T_f = W_f \times \mu \times r_f, \quad (1)$$

$$T_r = W_r \times \mu \times r_r, \quad (2)$$

$$W_f = W \times \omega_f, \quad (3)$$

$$Wr = W \times \omega_r, \quad (4)$$

where T_f and T_r are the front and rear axles torque (Nm), respectively; W, W_f, and W_r are the gross, front, and rear axle weight of the tractor (N), respectively; r_f and r_r are the front and rear wheel tires radius (m), respectively; μ is the coefficient of traction (0.8) [21], ω_f, and ω_r are the weight distribution ratio of both front and rear axles (%), respectively.

2.2.2. Specific Fuel Consumption (SFC)

To analyze the actual fuel consumption, the engine loads, which were measured by PTO dynamometer were used to calculate the specific fuel consumption (SFC) [22]. The engine performance of an agricultural tractor is highly affected by fuel efficiency due to the working load variation. It does not mean that fuel consumption is the only factor of fuel efficiency. The SFC is the index of the fuel efficiency that is the work done by an engine per unit horsepower and per unit time. The SFC of an agricultural tractor was calculated using the following Equations (5) and (6) [23].

$$SFC = \frac{FC}{P_{engine}}, \quad (5)$$

$$P_{engine} = \frac{2\pi \times T_{engine} \times N_{engine}}{60,000}, \quad (6)$$

where FC is the fuel consumption (kg/h); SFC is the specific fuel consumption (g/kWh); P_{engine} is the engine power (kW); T_{engine} is the engine torque (Nm), and N_{engine} is the engine speed (rpm).

2.3. Simulation Model of PST Tractor

In this study, the simulation model of PST was developed based on the powertrain of the PST tractor. The simulation model of the PST was developed using the commercial simulation software namely LMS AMESim (version 16, SIEMENS AG, Munich, Germany), which is operated by the 95 kW engine. The engine power is generally divided into the main transmission and rear PTO. The engine power of the main transmission is supplied to the driving axle through the high-low, hydraulic clutch pack (main clutch), and mechanical sub-shifting (range shifts: creep and range). In the PST, the constructions of three main clutch packs are the same. Therefore, the simulation model was simplified and conducted simulation using one main clutch pack to estimate the fuel consumption precisely. In this study, the measured data and engine characteristics map were applied to conduct and verify the simulation model of the PST. The predicted axle load was also used as an input parameter to characterize the simulation model as a real PST tractor used in this study. The simulation model of the entire PST tractor was shown in Figure 2.

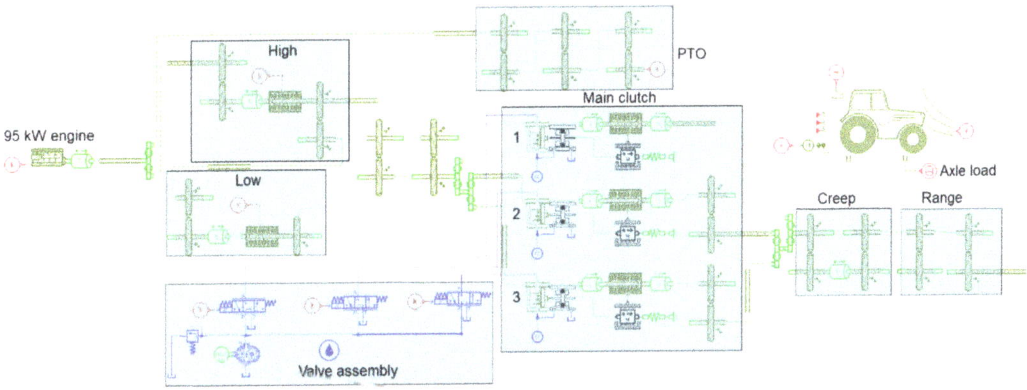

Figure 2. The simulation model of the entire PST tractor used in this study.

2.4. Dynamometer Test Bench and Specifications

In this study, a PTO dynamometer, which is an indoor test bench was installed to measure the fuel consumption of the PST tractor according to the engine speed. The fuel consumption measurement device (REO-CFMT, 3R Co., Ltd., Siheung, Korea), is used in this study. The PTO dynamometer was connected to the rear PTO of the PST tractor and the engine load was adjusted by controlling the dynamometer load. Quantum X (HBM: MX840B) data acquisition system (DAQ) was used to measure the dynamometer load and fuel consumption of the PST tractor. Using the measured fuel consumption, the SFC of the PST tractor was determined at various throttle levels of engine. The experiment of the PTO dynamometer test was carried out by [24]. The PTO dynamometer test bench was shown in Figure 3.

In this study, the engine load levels were divided into six levels like 40, 50, 60, 70 80, and 90%. The full load (100%) condition was applied to develop the engine characteristics map to conduct the simulation because the fuel consumption of tractor can be optimized by operating the engine at full load condition [25,26]. Farias et al. [27] also considered not less than 30% of the engine load level. According to the Nebraska tractor test, the SFC increases for more than 30% of engine loads [28]. Because the engine is required to have very low speed and longer gears; also, the tractor is required to apply withstanding its load.

Figure 3. The PTO dynamometer test bench used in this study.

The eddy current type PTO dynamometer (SE 500, SAJ, Pune, India) was used in this study. This type of dynamometer offers a wide range of capacities from 5 to 720 kW for engine test. In this study, a 500 kW PTO dynamometer was installed and the load cell type was U4000. It has the precision strain gauge load cell torque measurement system that provides high accuracy torque measurement for engine test, where was maximum torque of 3000 Nm at 1800 rpm. The accuracy of this dynamometer: ±0.25% of the rated torque and ±1% of speed, where was the speed range of 1600~4500 rpm. The detailed specifications of the PTO dynamometer used in this study were listed in Table 2.

Table 2. The specifications of the dynamometer used in this study.

Parameters	Specifications
Model	SE 500, SAJ, India
Type	EDDY Current
Maximum power (kW)	500
Maximum torque (Nm)	3000 @1800 rpm
Speed (rpm)	1600~4500
Inertia (kgm^2)	2.196
Load cell type	U4000
Weight (kg)	1500

2.5. Simulation Procedures

2.5.1. Model Verification and Generalization

The raw data, which were measured using PTO dynamometer were preprocessed by data filtering. The data filtering was used to remove the observations that contain the errors or undesirable observations for analysis and applied in the simulation model for verifying the model. The entire measured data were 12 sets, which were divided into 7 and 5 sets for the calibration and validation, respectively, in this study. The calibration datasets were applied to the simulation model of the PST tractor to determine the hourly fuel consumption and finally, the specific fuel consumptions were calculated. The validation datasets were used to verify the simulation model.

In this study, the plow tillage load was also applied to generalize the verified simulation model. The representative plow tillage load data was calculated from literature [14]. To select the suitable gear stage at low fuel consumption, the representative load data was plotted in the engine performance curve that generalized the simulation model for agricultural operations. Because, the engine performance curve has two regions: one is governed region which was controlled by governor, and another is the ungoverned region which was controlled by the engine load. Generally, the fuel consumption is comparatively low before ungoverned region but the engine might be shut down at a low speed due to the working load fluctuations. Moreover, the fuel consumption in the governed region was higher whereas was the engine torque is lower. Tractor might be unable to perform field operations at high load conditions. Therefore, the plow load data should be plotted on the engine performance curve to select a suitable gear stage for plow operation. The procedure of the model verification and generalization is shown in Figure 4.

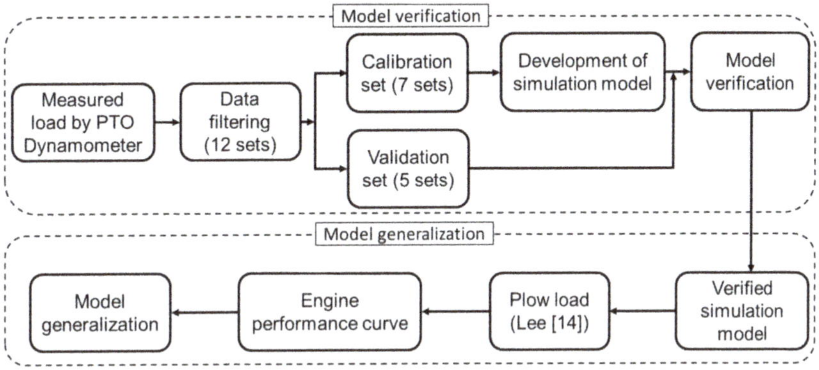

Figure 4. The block diagram of model verification and generalization used in this study.

2.5.2. Engine Characteristics Map

The engine static torque is a function of engine speed and throttle level [10,29]. The mathematical model of engine torque is in Equation (7) for a 95 kW engine.

$$T_{engine} = f(N_{engine}, a), \tag{7}$$

where T_{engine} is the engine torque (Nm); N_{engine} is the engine speed (rpm), and a is the throttle level (%).

Throttle level means to control the engine power by regulating the fuel amount to enter into the engine. Engine throttle opening greatly depends on throttle angles [30] In this study, the engine characteristics map was developed using AMESim 3D graphical platform. The engine characteristic map was developed for engine full load (100%) condition from engine test. The engine torque was calculated at each throttle level (%) using Equation (7), which was applied to develop the static engine map used in this study, whereas the maximum torque was 500 Nm at 1600 rpm for 100% of throttle opening. The simulation was conducted using the developed engine map, shown in Figure 5.

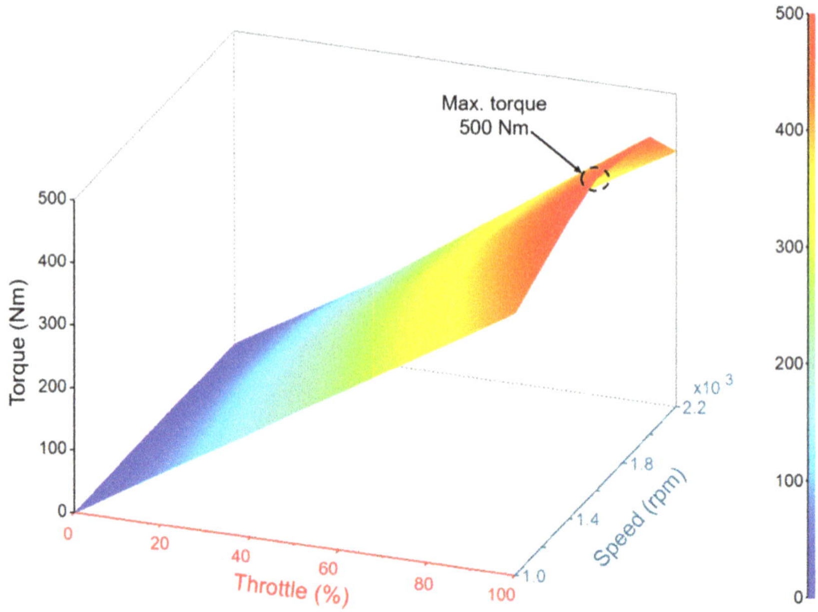

Figure 5. The 95 kW engine map used in this study.

2.5.3. Simulation Parameters

The block diagram of the simulation procedure was shown in Figure 6. The engine power is divided into the transmission and rear PTO of the tractor. The transmission power supplied by the forward-reversed shaft was delivered to the driving axle through the high-low, the main clutch operated hydraulically, and mechanical sub-shifting. In this study, the measured load was applied in the rear PTO to conduct the simulation. The axle load was also predicted using a theoretical model, where the specifications of a real PST tractor were used, and a 95 kW engine was applied in the simulation model. The engine characteristics map (Figure 5) was applied to the engine.

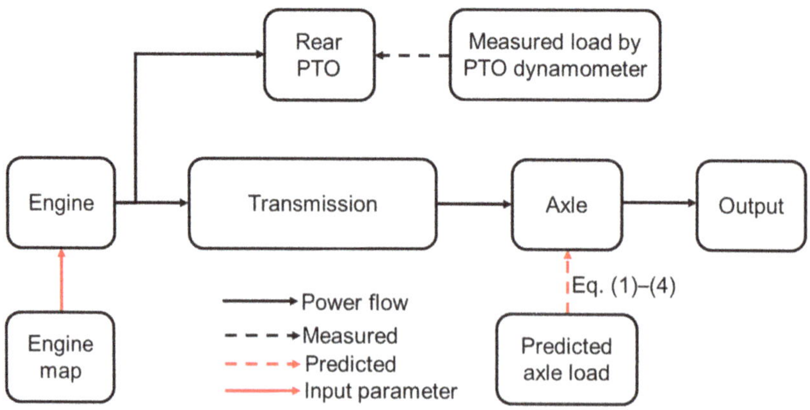

Figure 6. The block diagram for applied simulation parameters in this study.

2.6. Analysis Method

In this study, several statistical approaches were used in this study to evaluate the estimated fuel consumptions for both simulation and measured experiment. One-way

ANOVA (Analysis of variance), and Duncan's multiple range test (DMRT) were performed to analyze the significance of the engine torque for both simulation and experimental methods with respect to the throttle level of the engine. The fuel consumption was also analyzed statistically. The software used for the analysis was IBM SPSS Statistics (SPSS 25, SPSS Inc., New York, NY, USA). The error and accuracy of the SFC for various throttle levels were analyzed for both simulation and experimental methods by regression methods. The coefficient of determination and p-value were also determined. The R-square value, which is over 0.90 is considered reliable for comparison between two variables [31]. The R-squared can be obtained by the following Equation (8).

$$R^2 = 1 - \frac{\sum_i (y_i - \hat{y}_i)^2}{\sum_i (y_i - \acute{y})^2}, \quad (8)$$

where R^2 is the regression coefficients of the SFC for both methods; y_i is the ith measured SFC (g/kWh); $\hat{y}_i$ is the ith simulation SFC (g/kWh), and $\acute{y}$ is the mean of the measured SCF (g/kWh).

The simulation and experimental methods by PTO dynamometer test to measure the SFC of the tractor were also compared statistically and determined the accuracy and error by the root mean square error (RMSE) and relative deviation (RD) along with the R-squared value for both methods. The RMSE and RD can be obtained using the following Equations (9) and (10).

$$\text{RMSE} = \sqrt{\frac{1}{N} \sum_i (\hat{y}_i - y_i)^2}, \quad (9)$$

$$\text{RD} = \frac{\text{RMSE}}{\text{Mean}} \times 100, \quad (10)$$

where N is the number of the total SFC data; RMSE is the root mean square error of the SFC (%), and RD is the relative deviation of the SFC (%), which was calculated by the ratio of the RMSE to the mean of the SFC of the tractor.

3. Results

3.1. Engine Torque

In this study, engine torque was measured based on the engine load levels (40, 50, 60, 70, 80, and 90%) by installing PTO dynamometer and the simulation model was calibrated to conduct the simulation of the model. It was observed that the engine torques were gradually increased and reached the maximum engine torques for each load level at 1600 rpm of the engine speed. After then, the engine torques decreased gradually until the engine speed of 2200 rpm. It was noticed that there was a similar trend for both simulation and measured engine torques. Even there was almost the same increasing rate of the engine torque for each load level.

It was observed that the highest maximum engine torques for both simulation and measured were found at 90% of load level and the lowest maximum torques for both methods were at 40% of load level. It means that the engine torques were increasing with an increasing rate of the engine load levels. The results indicate that the engine torques were directly proportional to the engine load levels. Both simulation and measured engine torques were shown in Figure 7.

After analyzing, it was found that the highest maximum engine torques for both simulation and measured were 455 and 450 Nm at 90% of load level, respectively. The lowest maximum torques for both methods were also found 200 Nm at 40% of load level. The simulation and measured engine torques were also compared statistically using ANOVA along with DMRT test. It was noticed that there was no significant difference among the simulation and measured engine torques at each load level, where the p-value was less than the significant value of 5%. The ANOVA along with DMRT test results of both simulation and measured engine torques were listed in Table 3.

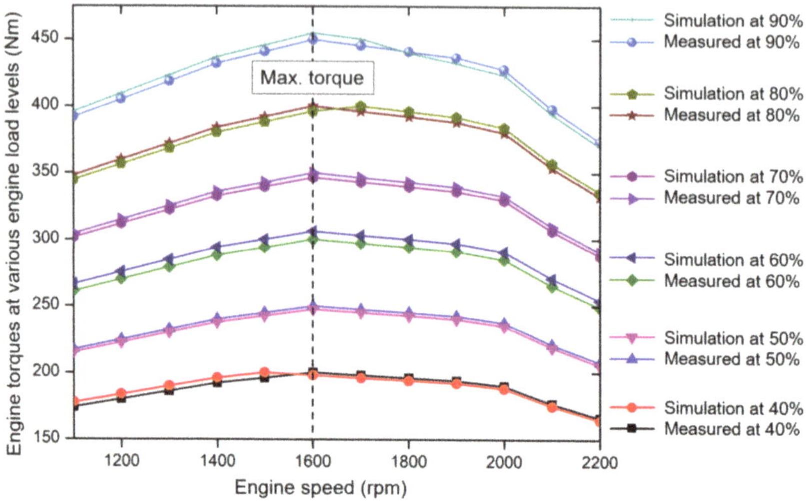

Figure 7. The measured and simulation engine torques at various load levels.

Table 3. The statistical analysis of both simulation and measured engine torques at various engine load levels.

Method	Simulation Engine Torques (Nm)						*p*-Value
Load Levels (%)	40	50	60	70	80	90	
Maximum	200	247.52	306.12	346.53	400	455	
Minimum	164.36	205.45	254.08	287.62	335	369.80	
Avg. ± S.D.*	186.62 ± 11.30 [a]	230.08 ± 14.49 [b]	284.55 ± 17.92 [c]	322.11 ± 20.28 [d]	371.76 ± 23.70 [e]	419.41 ± 28.07 [f]	0.000
Method	**Measured Engine Torques (Nm)**						
Maximum	200	250	300	350	400	450	
Minimum	166	207.50	249	290.50	332	373.50	
Avg. ± S.D.*	185.91 ± 11.71 [a]	232.38 ± 14.63 [b]	278.86 ± 17.56 [c]	325.34 ± 20.48 [d]	371.81 ± 23.41 [e]	418.29 ± 26.34 [f]	

[a,b,c,d,e,f] Means within each column with the same lettering are not significantly different at $p < 0.05$ according to Duncan's multiple range test. * Avg. ± S.D. is the Average ± Standard Deviation.

3.2. *Specific Fuel Consumption (SFC)*

In this study, the fuel consumption was measured to estimate the SFC. Figure 8 shows the simulation and measured hourly fuel consumption at six levels (40, 50, 60, 70, 80, and 90) of engine load. It was observed that fuel consumption of both simulation and measured were also increasing sharply with an increase of the engine load levels. For load levels of 40, 50, and 60%, the increasing rate of fuel consumption was almost parallel with each other. In the case of 70% of engine load level, the fuel consumption was dramatically increased after 2000 rpm of engine speed. For 80 and 90% of engine load levels, the fuel consumptions were also parallel between them but there were comparatively higher mean differences from the fuel consumption of 70% of engine load level.

The highest and lowest maximum hourly fuel consumptions for both simulation and measured were found around 20.22 and 19.95, and 6.42 and 6.55 kg/h at 90 and 40% of engine loads, respectively. It was observed that the average increasing rate for both simulation and measured fuel consumption were comparatively higher at 70% of engine load than that of other load levels, accounting for 8.45 and 8.60 kg/h, whereas the average fuel consumptions for both simulation and measured at 60 and 80% of engine loads were 6.14 and 6.24, and 12.93 and 11.56 kg/h, respectively. The statistical analysis results show that there was no significant difference between simulation and measured hourly fuel consumption at each load level because the statistical analysis proved that the p-value between the simulation and measured fuel consumption was less than the significance level of 5%. The statistical analysis results of hourly fuel consumptions for both methods were listed in Table 4.

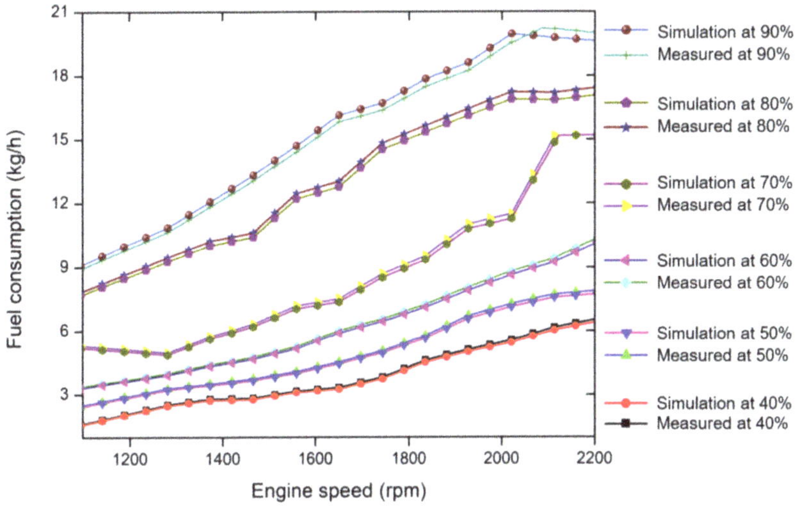

Figure 8. The simulation and measured hourly fuel consumption at various load levels.

Table 4. The statistical analysis of both simulation and measured hourly fuel consumptions at various engine load levels.

Method	Simulation Fuel Consumption (kg/h)						*p*-Value
Load Levels (%)	40	50	60	70	80	90	
Max. at 2200 rpm	6.42	7.72	10.14	15.19	17.08	20.22	
Min. at 1100 rpm	1.57	2.42	2.42	4.88	7.67	8.90	
Avg. ± S.D.	3.76 ± 1.41 [a]	4.88 ± 1.70 [b]	6.14 ± 2.04 [c]	8.45 ± 3.23 [d]	12.93 ± 3.18 [e]	15.15 ± 3.65 [f]	0.000
Method	Measured Fuel Consumption (kg/h)						
Max. at 2200 rpm	6.55	7.88	10.35	15.19	17.43	19.95	
Min. at 1100 rpm	1.61	2.47	3.36	4.98	7.83	9.08	
Avg. ± S.D.*	3.84 ± 1.44 [a]	4.98 ± 1.73 [b]	6.27 ± 2.09 [c]	8.60 ± 3.23 [d]	11.56 ± 3.25 [e]	15.35 ± 3.58 [f]	

[a,b,c,d,e,f] Means within each column with the same lettering are not significant different at $p < 0.05$ according to Duncan's multiple range test.
* Avg. ± S.D. is the Average ± Standard Deviation.

In this study, both simulation and measured hourly fuel consumptions (FC) were used to calculate the SFC of the engine. Further, the engine power (P_{engine}) was calculated from the measured engine torque and speed to determine the SFC (g/kWh). Figure 9 shows the average SFC of the engine for both simulation and measured at different load levels. The regression equations (y) of the SFC of both simulation and measured with respect to the engine load levels represented that the SFC of both simulation and measured were increased almost 13.66 and 13.94 g/kWh, respectively for each 10% increase of engine loads. It indicates the engine load levels have a significant effect on the SFC of the PST tractor. From regression equations, it was also observed that the SFC of both methods were an exponential relationship with different engine load levels. The R-squared of both simulation and measured SFC were found almost 0.9831 and 0.9835, respectively.

The ANOVA and DMRT test results of the SFC for both simulation and measured at various load levels were listed in Table 5. The analysis results show that there was no significant difference between the simulation and measured SFC at the significance level of 5%, whereas the standard error (SE) was 1.14. The maximum average (Avg. ± S.D.) SFC of both simulation and measured were 239.91 ± 51.37 and 244.81 ± 52.41 g/kWh, respectively at 90% of engine load level, whereas the minimum average (Avg. ± S.D.) SFCs were calculated at 40% of engine load, accounting for approximately 51.35 ± 9.81 and 52.40 ± 10.01 g/kWh for both simulation and measured methods, respectively.

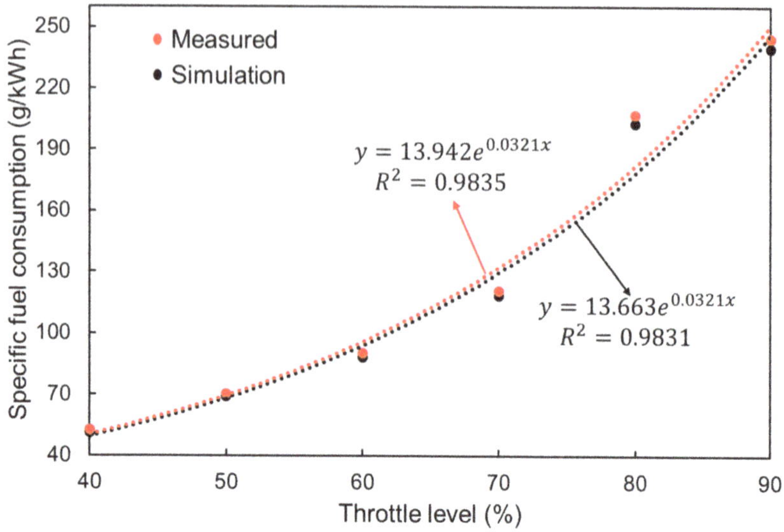

Figure 9. The regression of the average measured and simulation Specific Fuel Consumption (SFC) at various load levels.

Table 5. The statistical analysis of both simulation and measured specific fuel consumptions (SFC) at various engine load levels.

Method	Specific Fuel Consumption (g/kWh)						*p*-Value	SE
Load Levels (%)	**40**	**50**	**60**	**70**	**80**	**90**		
Simulation	51.35 ± 9.81 [a] *	68.81 ± 14.22 [b]	18.71 ± 18.71 [c]	118.46 ± 30.24 [d]	202.92 ± 44.86 [e]	239.91 ± 51.37 [f]	0.000	1.14
Measured	52.40 ± 10.01 [a]	70.21 ± 14.51 [b]	90.14 ± 19.09 [c]	120.88 ± 30.24 [d]	207.06 ± 45.78 [e]	244.81 ± 52.41 [f]		

[a,b,c,d,e,f] Means within each column with the same lettering are not significant different at $p < 0.05$ according to Duncan's multiple range test.
* Average ± Standard Deviation.

To identify the similarity, accuracy, tendency, and error between the simulation and measured SFC, linear regression, R^2, RMSE, and RD were calculated. Figure 10 shows the regression analysis of SFC for both methods. The R-squared value was 0.99, whereas the RMSE and RD were found approximately 1.89% and 2.54%. From the regression equation (y), it is clear that the simulation and measured SFCs were a linear relationship between each other. The results indicate that the SFC for both simulation and measured were a similar trend.

It is important to generalize the simulation model at low fuel consumption and high engine torque for agricultural operations. Therefore, the representative plow tillage load data was plotted in the engine performance curve (Figure 11). It was observed that the SFC was steadily increasing after 1500 rpm of the engine. It means that the engine should operate at a low speed to reduce fuel consumption. However, the working efficiency would be lower at low engine speed. It might suddenly turn off the engine due to the fluctuation of loads. Therefore, the engine speed should be adjusted within the ungoverned region where the working efficiency was comparatively higher but the fuel consumption was relatively lower.

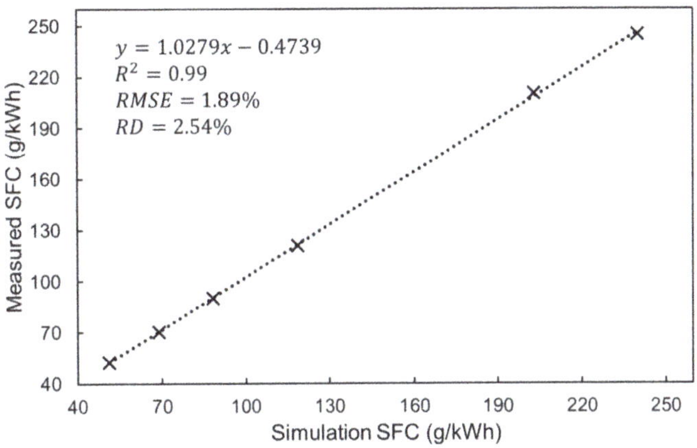

Figure 10. The linear regression of the average simulation and measured SFC.

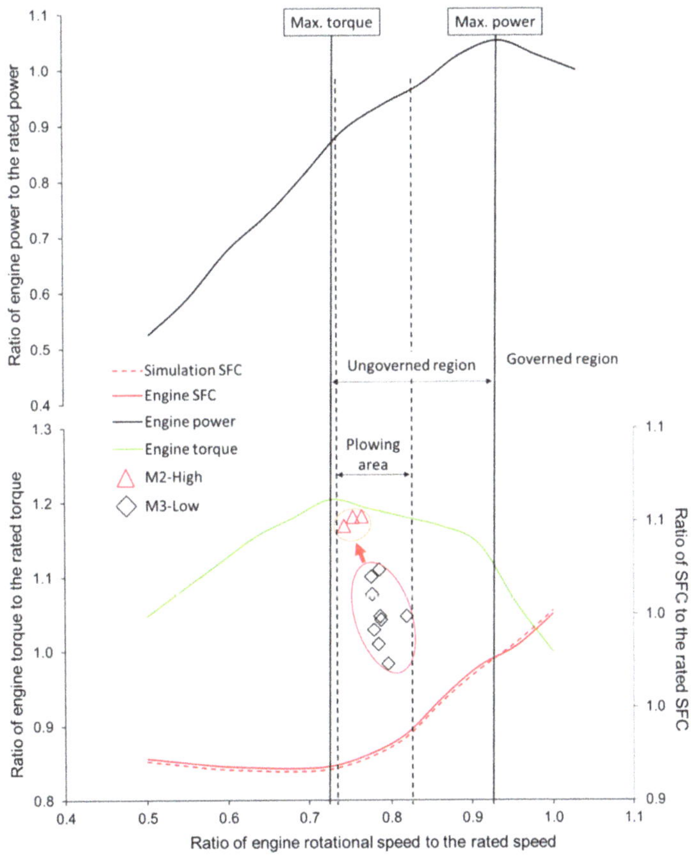

Figure 11. The engine performance curve at engine full load condition.

The plow operation loads, which were calculated comparing with the literature [14], were performed for two gear selections. M2-High and M3-Low were 7.56 km/h and 9.81 km/h, respectively, which were selected to perform plow tillage at the same agricultural field. The ratios of the calculated torque to the rated torque were presented in Figure 11. It showed that the plow tillage operation loads belonged to the ungoverned region of the engine performance curve, where the plowing at the gear stage of 7.56 km/h was relatively low fuel consumption and high working load for full engine load conditions. In the case of the 9.81 km/h gear stage, the fuel consumption was comparatively higher with low working load.

4. Discussion

In this study, the engine torque and fuel consumption of the PST tractor were measured at various engine load levels (40, 50, 60, 70, 80, and 90%); also, the simulation model of the PST tractor was verified by the measured PTO load data. The results of this study were discussed as below:

(1) It was noted that the engine torques for both simulation and measured were directly proportional to the engine load level. The statistical analysis (DMRT) proved that there was no significant difference between the simulation and measured engine torque at each load level. Kolator and Bialobrzewski [32] reported that engine load condition has a highly significant effect on tractor performance.
(2) It was observed that the increasing rates of the fuel consumptions were parallel with each other and sharply increased with respect to the engine speed. Only for 70% of engine load level, the fuel consumption was dramatically increased after 2000 rpm of engine speed. It might be caused by an excessive flow that is the main reason for fuel losses [33]. Those results indicate that the hourly fuel consumptions were also directly proportional to the engine load levels. The SFCs at various engine load levels were analyzed statistically. The statistical analysis showed that there was no significant difference between the simulation and measured SFC as the significant level of 5%. The regression analysis with respect to throttle level showed that the increasing rate for both the simulation and measured SFCs at each load level. The results indicated that the engine load levels have a highly significant effect on the SFC of the PST tractor. Shafaei [31] reported over 0.90 of R-squared value will be reliable to verify the relationship. From the regression equation, it is clear that there was an exponential relationship between the SFC and engine load levels.
(3) To distinguish the similarity, accuracy, and error between the simulation and experimental methods to estimate the SFC, statistical tests (R-squared, RMSE, and RD) were also conducted. From the regression equation, it was clear that there was a linear relationship between the simulation and measured SFC. The statistical results proved that the simulation SFC was similar to the measured SFC.
(4) It was observed that the engine generated power was higher at 9.81 km/h (M3-Low) than the soil strength, which indicated the loss of power. It was believed that the power loss had occurred by the travel reduction ratio (slip) [34]. The results indicated that the gear selection of 7.56 km/h (M2-High) was highly beneficial to conduct plow tillage operation using the 95 kW partial PST tractor considering high engine torque. To optimize the fuel consumption, the gear selection should be shifted to the M2-High instead of M3-Low because M3-Low generates low engine torque that might be suddenly turn-off the engine due to high soil strength [14].

In addition, several researchers proved the reliability of the simulation method based on comfortability, easier, and convenience. The simulation method, which has also been reliable to conduct the performance, design, estimation, and evaluation [35] of the agricultural machinery, reduce labor cost and time-consuming instead of field experiment [36]. The results revealed that the PTO dynamometer can easily control the engine load to estimate fuel consumption. As the simulation results of SFC represent the PTO dynamometer measured SFC results, the simulation results can be commercially applied to improve the fuel efficiency of the PST tractor. Finally, this study can contribute to the manufacturers to develop a new agricultural tractor with high fuel efficiency.

5. Conclusions

This study was emphasized to simulate the fuel consumption of the PST tractor. The simulation of the PST tractor was developed using the configurations and powertrain of the real PST tractor manufactured by the Korean company. The PTO dynamometer, an indoor test bench, was installed to measure the engine load and fuel consumption at various throttle levels (40, 50, 60, 70, 80, and 90%) of the PST tractor according to the engine speed. The tractor axle load was predicted using the specifications of the tractor as an input parameter of the simulation model. The major findings of this study were listed as below:

(1) It was observed that the highest maximum engine torques for both simulation and measured were 455 and 450 Nm at 90% of engine load, respectively. The lowest maximum torques for both methods were also found at 200 Nm at 40% of engine load. It was also observed that the maximum engine torques for both simulation and measured were at 1600 rpm of engine speed for all engine throttle levels. The statistical analysis (DMRT) proved that there was no significant difference between the simulation and measured engine torque at each throttle level. However, it was noticed that the engine torques for both simulation and measured were directly proportional to the engine throttle level.

(2) The highest and lowest maximum hourly fuel consumptions for both simulation and measured were found around 20.22 and 19.95, and 6.42 and 6.55 kg/h at 90 and 40% of engine load, respectively. It was observed that the average increasing rate for both simulation and measured fuel consumption were comparatively higher at 70% of engine load than that of other engine load levels, accounting for 8.45 and 8.60 kg/h. The regression equations of the SFC of both simulation and measured with respect to the engine throttle levels represented that the SFC of both simulation and measured were increased almost 13.66 and 13.94 g/kWh, respectively for each 10% increase of engine load. The R-squared of both simulation and measured SFC were found almost 0.9831 and 0.9835, respectively. The analysis results show that there was no significant difference between the simulation and measured SFC, whereas the standard error (SE) was 1.14. The R-squared value was 0.99, whereas the RMSE and RD were approximately 1.89% and 2.54%, respectively.

In summary, the engine torques were directly proportional to the engine load levels. The statistical analysis (DMRT) proved that there was no significant difference between the simulation and measured engine torques. The simulation and measured SFCs were an exponential relationship with various engine load levels. However, both simulation and measured SFCs were linearly relationship and have statistically no significant difference between them. It was also found a similar trend for both methods.

The drawback of this study is the prediction load using tractor and engine specifications, which were applied to the axle, were constant. Howard [37] conducted the fuel efficiency of the continuously variable transmission (CVT) tractor with respect to drawbar power that was a dynamic load [38]. Gui [28] suggested applying the engine speed controller to estimate the optimal fuel efficiency of the agricultural tractor. Lee [14] also developed the engine speed control system to improve fuel efficiency and verify by tillage operation, which was also a dynamic workload. Finally, it can be said that the field operation is needed to estimate the fuel efficiency for commercialization. However, it is planned to conduct field operations considering various types of implements based on major agricultural operations.

In conclusion, it can be said that PTO dynamometer, which can control the engine loads to estimate the fuel consumption, minimize labor cost and time rather than field operations, and this study can be helpful to the manufacturers to develop and improve a new agricultural tractor with higher fuel efficiency. It could be recommended that the users should perform the plow tillage at the gear stage of 7.65 km/h for low fuel consumption with respect to higher working efficiency. This study could also contribute to digital farming by improving fuel efficiency, which was the key issue for the automated transmission of an agricultural tractor. In the future, it is planned to conduct field operations based on implements type and crops for dynamic load data, and field data will be stored in a server system, which is a core goal of smart digital farming for an agricultural machinery sector.

Author Contributions: Conceptualization, M.A.A.S. and Y.-J.K.; methodology, M.A.A.S., W.-S.K., and Y.-J.K.; software, M.A.A.S.; validation, M.A.A.S., Y.-S.K., S.-M.B.; S.-Y.B.; K.-H.L., and J.-Y.H. formal analysis, M.A.A.S.; investigation, W.-S.K., Y.-S.K., S.-M.B., S.-Y.B.; K.-H.L.; J.-Y.H., and Y.-J.K., and D.-H.L.; writing—original draft preparation, M.A.A.S.; writing—review and editing, M.A.A.S., and D.-H.L.; visualization, M.A.A.S., D.-H.L., and Y.-J.K.; supervision, Y.-J.K.; project administration, Y.-J.K.; funding acquisition, Y.-J.K. All authors have read and agreed to the published version of the manuscript.

Funding: This work was carried out with the support of "Cooperative Research Program for Agriculture Science and Technology Development (Project No. PJ01498102)" Rural Development Administration, Republic of Korea.

Institutional Review Board Statement: Not applicable.

Informed Consent Statement: Not applicable.

Data Availability Statement: The data presented in this study are available within the article.

Conflicts of Interest: The authors declare no conflict of interest.

References

1. Kim, W.S.; Kim, Y.J.; Kim, Y.S.; Baek, S.Y.; Baek, S.M.; Lee, D.H.; Nam, K.C.; Kim, T.B.; Lee, H.J. Development of control system for automated manual transmission of 45-kW agricultural tractor. *Appl. Sci.* **2020**, *10*, 2930. [CrossRef]
2. Shafaei, S.M.; Loghavi, M.; Kamgar, S. Fundamental realization of longitudinal slip efficiency of tractor wheels in a tillage practice. *Soil Tillage Res.* **2021**, *205*, 104765. [CrossRef]
3. Islam, M.N.; Iqbal, M.Z.; Kabir, M.S.N.; Jung, K.Y.; Mun, D.H.; Chung, S.O. Performance Evaluation of Trenchless Subsurface Drainage Piping Machine. *J. Biosyst. Eng.* **2019**, *44*, 218–225. [CrossRef]
4. Mairghany, M.; Yahya, A.; Adam, N.M.; Mat Su, A.S.; Aimrun, W.; Elsoragaby, S. Rotary tillage effects on some selected physical properties of fine textured soil in wetland rice cultivation in Malaysia. *Soil Tillage Res.* **2019**, *194*, 104318. [CrossRef]
5. Kim, W.S.; Kim, Y.J.; Baek, S.M.; Moon, S.P.; Lee, N.G.; Kim, Y.S.; Park, S.U.; Choi, Y.; Kim, Y.K.; Choi, I.S.; et al. Fatigue life simulation of tractor spiral bevel gear according to major agricultural operations. *Appl. Sci.* **2020**, *10*, 8898. [CrossRef]
6. Mordor Intelligence. Available online: https://www.mordorintelligence.com/industry-reports/india-agricultural-tractormachinery-%0Amarket (accessed on 16 March 2021).
7. Kim, Y.S.; Lee, P.U.; Kim, W.S.; Kwon, O.W.; Kim, C.W.; Lee, K.H.; Kim, Y.J. Strength analysis of a PTO (Power Take-Off) gear-train of a multi-purpose cultivator during a rotary ditching operation. *Energies* **2019**, *12*, 1100. [CrossRef]
8. Siddique, M.A.A.; Kim, W.S.; Kim, Y.S.; Kim, T.J.; Choi, C.H.; Lee, H.J.; Chung, S.O.; Kim, Y.J. Effects of temperatures and viscosity of the hydraulic oils on the proportional valve for a rice transplanter based on PID control algorithm. *Agriculture* **2020**, *10*, 73. [CrossRef]
9. Tanelli, M.; Panzani, G.; Savaresi, S.M.; Pirola, C. Transmission control for power-shift agricultural tractors: Design and end-of-line automatic tuning. *Mechatronics* **2011**, *21*, 285–297. [CrossRef]
10. Siddique, M.A.A.; Kim, W.S.; Kim, Y.S.; Baek, S.Y.; Baek, S.M.; Kim, Y.J.; Park, S.U.; Choi, C.H. Simulation of Design Factors of a Clutch Pack for Power-Shift Transmission for an Agricultural Tractor. *Sensors* **2020**, *20*, 7293. [CrossRef]
11. Liyou, X.; Yihao, Z.; Jinzhong, S.; Xianghai, Y. Optimization of power shift tractor clutch based on ahp and improved genetic algorithm. *Acta Tech.* **2017**, *62*, 373–384.
12. Siddique, M.A.A.; Kim, T.J.; Kim, Y.J. Technical Trend of the Power Shift Transmission (PST) of Agricultural Tractor. *J. Drive Control* **2020**, *17*, 68–75.
13. Liang, J.; Yang, H.; Wu, J.; Zhang, N.; Walker, P.D. Power-on shifting in dual input clutchless power-shifting transmission for electric vehicles. *Mech. Mach. Theory* **2018**, *121*, 487–501. [CrossRef]
14. Lee, J.W.; Kim, S.C.; Oh, J.; Chung, W.J.; Han, H.W.; Kim, J.T.; Park, Y.J. Engine speed control system for improving the fuel efficiency of agricultural tractors for plowing operations. *Appl. Sci.* **2019**, *9*, 3898. [CrossRef]
15. Grisso, R.; Pitman, R.; Perumpral, J.V.; Roberson, G.T. Gear up and Throttle down to Save Fuel. In *Virginia Cooperative Extension*; Virginia Tech.: Blacksburg, VA, USA, 2010; pp. 442–450.
16. Park, S.H.; Kim, Y.J.; Im, D.H.; Kim, C.K.; Jung, S.C.; Kim, H.J.; Jang, Y.; Kim, S.S. Development of Eco Driving System for Agricultural Tractor. *J. Biosyst. Eng.* **2010**, *35*, 77–84. [CrossRef]
17. Jiang, J. Optimal Gain Scheduling Controller for a Diesel Engine. *IEEE Control Syst.* **1994**, *14*, 42–48. [CrossRef]
18. Sung, S.W.; Lee, I.B. Limitations and countermeasures of PID controllers. *Ind. Eng. Chem. Res.* **1996**, *35*, 2596–2610. [CrossRef]
19. Saunders, C.; Ucgul, M.; Godwin, R.J. Discrete element method (DEM) simulation to improve performance of a mouldboard skimmer. *Soil Tillage Res.* **2021**, *205*, 104764. [CrossRef]

20. Kim, W.S.; Kim, Y.J.; Baek, S.Y.; Baek, S.M.; Kim, Y.S.; Park, S.U. Development of a prediction model for tractor axle torque during tillage operation. *Appl. Sci.* **2020**, *10*, 4195. [CrossRef]
21. Kim, W.-S.; Kim, Y.-S.; Kim, Y.-J. Development of Prediction Model for Axle Torque of Agricultural Tractors. *Trans. ASABE* **2020**, *63*, 1773–1786. [CrossRef]
22. Kim, Y.S.; Kim, W.S.; Baek, S.Y.; Baek, S.M.; Kim, Y.J.; Lee, S.D.; Kim, Y.J. Analysis of tillage depth and gear selection for mechanical load and fuel efficiency of an agricultural tractor using an agricultural field measuring system. *Sensors* **2020**, *20*, 2450. [CrossRef]
23. Goering, C.E. *Engine and Tractor Power*; American Society of Agricultural Engineers (ASAE): St. Joseph, MI, USA, 1992.
24. Bietresato, M.; Friso, D.; Sartori, L. Assessment of the efficiency of tractor transmissions using acceleration tests. *Biosyst. Eng.* **2012**, *112*, 171–180. [CrossRef]
25. Chancellor, W.J.; Thai, N.C. Automatic Control of Tractor Transmission Ratio and Engine Speed. *Trans. ASAE* **1984**, *27*, 642–0646. [CrossRef]
26. Zhang, N.; Perumpral, J.; Byler, R.K. Automatic control system for optimizing diesel engine performance. *Comput. Electron. Agric.* **1987**, *2*, 31–46. [CrossRef]
27. De Farias, M.S.; Schlosser, J.F.; Linares, P.; Bertollo, G.M.; Martini, A.T. Reduction of fuel consumption using driving strategy in agricultural tractor. *Rev. Bras. Eng. Agric. E Ambient.* **2019**, *23*, 144–149. [CrossRef]
28. Gui, X.Q.; Goering, C.E.; Buck, N.L. Simulation of a fuel-efficient augmented engine. *Trans. Am. Soc. Agric. Eng.* **1989**, *32*, 1875–1881. [CrossRef]
29. Li, B.; Sun, D.; Hu, M.; Liu, J. Research on economic comprehensive control strategies of tractor-planter combinations in planting, including gear-shift and cruise control. *Energies* **2018**, *11*, 686. [CrossRef]
30. Bera, P. Torque characteristic of SI engine in dynamic operating states. *Combust. Engines* **2017**, *171*, 175–180. [CrossRef]
31. Shafaei, S.M.; Kamgar, S. A comprehensive investigation on static and dynamic friction coefficients of wheat grain with the adoption of statistical analysis. *J. Adv. Res.* **2017**, *8*, 351–361. [CrossRef]
32. Kolator, B.; Białobrzewski, I. A simulation model of 2WD tractor performance. *Comput. Electron. Agric.* **2011**, *76*, 231–239. [CrossRef]
33. Knauder, C.; Allmaier, H.; Sander, D.E.; Sams, T. Investigations of the friction losses of different engine concepts. Part 1: A combined approach for applying subassembly-resolved friction loss analysis on a modern passenger-car diesel engine. *Lubricants* **2019**, *7*, 39. [CrossRef]
34. Renius, K.T. *Fundamentals of Tractor Design*, 1st ed.; Springer International Publishing: Cham, Switzerland, 2019; ISBN 978-3-030-32804-7.
35. Iqbal, M.Z.; Islam, M.N.; Chowdhury, M.; Islam, S.; Park, T.; Kim, Y.-J.; Chung, S.-O. Working Speed Analysis of the Gear-Driven Dibbling Mechanism of a 2.6 kW Walking-Type Automatic Pepper Transplanter. *Machines* **2021**, *9*, 6. [CrossRef]
36. Baek, S.M.; Kim, W.S.; Kim, Y.S.; Baek, S.Y.; Kim, Y.J. Development of a simulation model for HMT of a 50 kW class agricultural tractor. *Appl. Sci.* **2020**, *10*, 4064. [CrossRef]
37. Howard, C.N.; Kocher, M.F.; Hoy, R.M.; Blankenship, E.E. Testing the fuel efficiency of tractors with continuously variable and standard geared transmissions. *Trans. ASABE* **2013**, *56*, 869–879. [CrossRef]
38. Kim, Y.S.; Kim, W.S.; Siddique, M.A.A.; Baek, S.Y.; Baek, S.M.; Cheon, S.H.; Lee, S.D.; Lee, K.H.; Hong, D.H.; Park, S.U.; et al. Power transmission efficiency analysis of 42 kW power agricultural tractor according to tillage depth during moldboard plowing. *Agronomy* **2020**, *10*, 1263. [CrossRef]

agriculture

MDPI

Article

Prediction of Reproductive Success in Multiparous First Service Dairy Cows by Parameters from In-Line Sensors

Ramūnas Antanaitis [1,*], Vida Juozaitienė [2], Dovilė Malašauskienė [1], Mindaugas Televičius [1], Mingaudas Urbutis [1], Gintaras Zamokas [3] and Walter Baumgartner [4]

1 Large Animal Clinic, Veterinary Academy, Lithuanian University of Health Sciences, Tilžės str 18, LT-47181 Kaunas, Lithuania; dovile.malasauskiene@lsmuni.lt (D.M.); mindaugas.televicius@lsmuni.lt (M.T.); mingaudas.urbutis@lsmuni.lt (M.U.)
2 Department of Animal Breeding, Veterinary Academy, Lithuanian University of Health Sciences, Tilžės str 18, LT-47181 Kaunas, Lithuania; vida.juozaitiene@lsmuni.lt
3 Dr. Leonas Kriaučeliūnas Small Animal Clinic, Veterinary Academy, Lithuanian University of Health Sciences, Tilžės str 18, LT-47181 Kaunas, Lithuania; gintaras.zamokas@lsmuni.lt
4 University Clinic for Ruminants, University of Veterinary Medicine, Veterinaerplatz 1, A-1210 Vienna, Austria; walter.baumgartner@vetmeduni.ac.at
* Correspondence: ramunas.antanaitis@lsmuni.lt; Tel.: +370-67349064

Abstract: The aim of the current study was to evaluate the relationship of different parameters from an automatic milking system (AMS) with the pregnancy status of multiparous cows at first service and to assess the accuracy of such a follow-up with regard to blood parameters. Before the insemination of cows, blood samples for measuring biochemical indices were taken from the coccygeal vessels and the concentrations of blood serum albumin (ALB), cortisol, non-esterified fatty acids (NEFA) and the activities of aspartate aminotransferase (AST) and gamma glutamyltransferase (GGT) were determined. From oestrus day to seven days after oestrus, the following parameters were registered: milk yield (MY), electric milk conductivity, lactate dehydrogenase (LDH) and β-hydroxybutyric acid (BHB). The pregnancy status was evaluated using ultrasound "Easy scan" 30–35 days after insemination. Cows were grouped by reproductive status: PG− (non-pregnant; n = 48) and PG+ (pregnant; n = 44). The BHB level in PG− cows was 1.2 times higher ($p < 0.005$). The electrical conductivity of milk was statistically significantly higher in all quarters of PG− cows (1.07 times) than of PG+ cows ($p < 0.05$). The arithmetic mean of blood GGT was 1.61 times higher in PG− cows and the NEFA value 1.23 times higher ($p < 0.05$) compared with the PG+ group. The liver function was affected, the average ALB of PG− cows was 1.19 times lower ($p < 0.05$) and the AST activity was 1.16 times lower ($p < 0.05$) compared with PG+ cows. The non-pregnant group had a negative energy balance demonstrated by high in-line milk BHB and high blood NEFA concentrations. We found a greater number of cows with cortisol >0.0.75 mg/dL in the non-pregnant group. A higher milk electrical conductivity in the non-pregnant cows pointed towards a greater risk of mastitis while higher GGT activities together with lower albumin concentrations indicated that the cows were more affected by oxidative stress.

Keywords: automatic milking system; reproduction; blood; metabolic profile

Citation: Antanaitis, R.; Juozaitienė, V.; Malašauskienė, D.; Televičius, M.; Urbutis, M.; Zamokas, G.; Baumgartner, W. Prediction of Reproductive Success in Multiparous First Service Dairy Cows by Parameters from In-Line Sensors. *Agriculture* **2021**, *11*, 334. https://doi.org/10.3390/agriculture11040334

Academic Editors: Gniewko Niedbała and Sebastian Kujawa

Received: 21 February 2021
Accepted: 6 April 2021
Published: 8 April 2021

1. Introduction

As milk production per cow has increased, fertility in dairy cows has decreased. This can be explained by genetics, physiology, nutrition and management issues and these variables have been examined at the animal, organ and cellular level at critical time points during the production life of dairy cows [1]. An increased adoption of technologies will enable farmers to have access to rich data sources that can aid in further improving animal health and welfare [2].

The system "Herd Navigator" (Lattec I/S, Hillerød, Denmark), according to Yu and Maeda [3], works autonomously and delivers real-time physiological information about lactating cows and at the same time also helps with farm management decisions. This system alerts and advises dairy farmers of each cow's condition and health. The system contains unique biological models that consider measured parameters, cow information and additional risk factors to keep the herd healthy. In this way, it prevents costly treatments and large production losses. Significant improvement results on reproduction, mastitis and ketosis have been proven on the farms that are running the system [3]. The use of Herd Navigator technology enhances our understanding of the factors influencing the modern milk cow's reproductive physiology and informed decision-making can enhance fertility in dairy cows. In addition to helping with reproductive management decisions, Herd Navigator generated data offers a new opportunity to evaluate luteal activity parameters [4]. Inferior cattle performance can be monitored on a real-time basis with Herd Navigator and factors that lead to ketosis and reproductive management can be recognized and corrected to increase farm performance [5]. Although it requires a considerable investment during the installation stage, which would be costly for small farms, the tool is profitable for large farms compared with the frequent manual collection of progesterone information [6]. Additional assays augmented by the analyzer (urea nitrogen, lactate dehydrogenase and β-hydroxybutyrate) can be used in real-time to classify metabolic diseases and animal mastitis for more detailed monitoring [7]. According to Yu and Maeda [3], an in-line Herd Navigator system automatically provides real-time physiological information on lactating dairy cows for farm management decisions. This is not only a new tool for scientific research but also may enhance production, food safety, animal wellbeing and the environment. According to our past study, we found that the real-time measured β-hydroxybutyrate levels and changes in their dynamics correlated with different reproductive statuses, productivity and number of lactations [8]. Previous results highlighted that lactate dehydrogenase changes in the automatic milking system (AMS) indicators of cows may be considered an additional tool for improving reproductive management in dairy herds but further research-based studies are necessary before a practical application [9]. The Herd Navigator system has led to promising preliminary results but further work is required to validate it [2]. The aim of this study was to determine the prediction of reproductive success in multiparous first service dairy cows by parameters from in-line sensors and blood metabolic profiles. We hypothesized that in-line measurements could be predictive of the fertility status in dairy cows. To test our hypothesis, we investigated the relationship between the AMS and blood biochemical parameters with cow reproductive success.

2. Materials and Methods

2.1. Location and Animals

The research was conducted during the period from 1 February to 30 October in 2019 (272 days). The experiment was carried out on a commercial dairy farm (in total around 1300 cows) located in the southern part of Lithuania (54.9753923° N, 23.7662303° E). A total of 92 Lithuanian Holstein cows were selected based on the following criteria: having had two or more lactations (2.9 ± 0.13 lactation), from 60–90 (on average 78 ± 6) days in milk (DIM) and being clinically healthy. The cows were kept in a free housing system, milked with 15 DeLaval milking robots (DeLaval Inc. Tumba, Sweden) (average number of milkings per day = 3.5) and were fed a total mixed ration (TMR) two times per day at a set time, balanced according to their physiological requirements. The TMR was composed of 25% corn silage, 15% grass silage, 4% grass hay and 50% grain concentrate. The main nutritional characteristics were dry matter (DM) (%) 48.8%, neutral detergent fiber (% of DM) 28.2% DM, acid detergent fiber (% of DM) 19.8% DM, non-fiber carbohydrates (% of DM) 38.7% DM and crude protein (% of DM) 15.8% DM; the net energy for lactation (Mcal/kg DM) was 1.6. The feed ration was composed to fit the requirements of a 550 kg Holstein cow producing on average 35 kg/day of milk (during the experiment). The average of the body condition score in a 5-point scale was 2.70 (±0.15). Oestrus was identified when a cow exhibited one or more of the following signs: progesterone alarm (registered by the Herd Navigator system), an increase on the walking activity of cows (registered by the AMS), mucosal discharge, restlessness, vigilance, tail lift, the occlusion of the vulvar mucosa and a strong uterine tone by clinical examination. Twelve hours after the onset of oestrus, the cows were inseminated. The pregnancy status was evaluated using ultrasound "Easy scan" (IVM imaging, Scotland) 30–35 days after insemination. Cows were grouped by reproductive status: PG− (non-pregnant; n = 48) and PG+ (pregnant; n = 44).

Welfare and Protection of the Republic of Lithuania (study approval number PK016965).

2.2. Measurements

From oestrus day to seven days after oestrus, the following parameters from the AMS were registered: milk yield (MY), electric milk conductivity (EC) front right (FR), front left (FL), back left (BL), back right (BR), lactate dehydrogenase (LDH), β-hydroxybutyric acid (BHB) and blood parameters: albumin (ALB), aspartate aminotransferase (AST), gamma glutamyltransferase (GGT), blood cortisol concentration and non-esterified fatty acids (NEFA) (Table 1).

Table 1. Measured parameters and their source.

Measured Parameters	Source
Milk yield (MY)	Herd Navigator
Electric milk conductivity (EC) front right (FR), front left (FL), back left (BL), back right (BR)	Herd Navigator
Lactate dehydrogenase (LDH)	Herd Navigator
β-hydroxybutyric acid (BHB)	Herd Navigator
Albumin (ALB)	Hitachi 705 analyzer
Aspartate aminotransferase (AST)	Hitachi 705 analyzer
Gamma glutamyltransferase (GGT)	Hitachi 705 analyzer
Cortisol	Analyzer AIA-360
Non-esterified fatty acids (NEFA)	Rx Daytona, Randox Laboratories

The research was carried out in accordance with the provisions of the Law on Animal.

The blood samples were collected using an evacuated tube without an anticoagulant (BD Vacutainer®, Eysin, Switzerland). The blood samples were centrifuged at 3500 RPM for 10–15 min. Samples were delivered to the Large Animal Clinic's Laboratory of Clinical Tests at the Lithuanian University of Health Sciences Veterinary Academy. The obtained blood serum was examined using the Hitachi 705 analyzer (Hitachi, Tokyo, Japan)) and DiaSys reagents (Diagnostic Systems GmbH, Berlin, Germany) to determine the concentrations of blood serum albumin (ALB) and the activities of aspartate aminotransferase (AST) and gamma glutamyltransferase (GGT). The blood cortisol concentration levels were determined with automated immunoassay analyzer AIA-360 (Tosoh Bioscience, USA) using the fluorescence enzyme immunoassay method. The samples for NEFA were analyzed using an automated wet chemistry analyzer (Rx Daytona, Randox Laboratories Ltd., London, UK) using reagents from Rx Daytona (Randox Laboratories Ltd., London, UK).

An automated real-time analyzer, Herd Navigator, was used in combination with a DeLaval milking robot for milk BHB and LDH detection (DeLaval Inc, Tumba, Sweden). An in-line sampler in the milking robot automatically took a representative sample of several milliliters of milk from an individual cow during milking and determined the concentration of those parameters. In the Herd Navigator system, the LDH activity and BHB values were measured using dry stick technology. The raw measurements were corrected according to company-determined methods to account for variations in the surrounding humidity and differences between sets of dry sticks. The most extreme outliers were then taken out of the equation. Measurements exceeding 200 μmol/min per liter were set to a 200 maximum value and all negative values were removed because they did not comply with the usual range of measurements obtained from the Herd Navigator system. This editing of data is usual for data from the Herd Navigator system.

2.3. Data Analysis and Statistics

The statistical analysis of the AMS and Herd Navigator variables was carried out using SPSS 25.0 software (IBM Corp, released 2017, IBM SPSS Statistics for Windows, Version 25.0. Armonk, NY, USA: IBM Corp.). The data (according to a Shapiro–Wilk test) were normally distributed. The results were expressed as a standard error of the mean of a sample (M and SEM). Data of parameters from in-line sensors were tested for significance ($p < 0.05$) using the repeated measures analysis. A general linear model was used to evaluate the differences between groups (PG− and PG+) of cows and variables with repeated (seven days) measurements (MY, FL, FR, BL, BR, LDH, BHB and cortisol, ALB, AST, GGT, NEFA). The difference in times of the variables with repeated measurements was tested using a Bonferroni post-hoc test.

The relationship between the level of indicators of the metabolic blood profile, the AMS variables and the group of cows according to their reproductive status was determined using the chi-squared ($\chi 2$) statistic. A multivariable binary logistic regression technique was carried out using pregnancy as the dependent variable (where 1 denoted pregnancy and 0 denoted absence) to investigate the relationship between the predicted variable and variables from the AMS and the blood metabolic profile indicators of cows. The predictors for logistic regression were considered class variables in the analyses. The values of ALB were classified: <21 g/L, 21–36 g/L and > 36 g/L [10], AST: <78 U/L and 78–132 U/L [10], GGT: <8.11 U/L and 8.11–27.79 U/L [11] and NEFA: <0.320 mmol/L and ≥0.320 mmol/L [12]. The cortisol of cows was recorded as < 0.45 mg/dL, 0.47–0.75 mg/dL and > 0.0.75 mg/dL [10]. The values of milk LDH were grouped as being ≤ 25 µmol/min or > 25 µmol/min, MY ≤ 44 kg/d and >44 kg/d, BHB < 0.06 mmol/L and >0.06 mmol/L and were classified on the basis of the arithmetic mean. Based on the average of all udder quarters, cows were divided into three EC levels: (1) EC = 3.5–3.9 mS/cm, (2) EC = 4.0–4.4 mS/cm and (3) EC = 4.5–5.0 mS/cm [13]. For each cow, we calculated the difference in udder quarters EC (DifEC) between the maximum and minimum EC values. With this in mind, we grouped the cow data into classes: DifEC < 0.5 mS/cm and DifEC ≥ 0.5 mS/cm. The factors that contributed to the possibility of pregnancy were analyzed with multivariable logistic regression models by applying a backward stepwise logistic model to eliminate all non-significant explanatory variables. The results of the logistic regression are presented in terms of the odds ratio (OR) and its 95% confidence interval (PI), which indicates a 95% probability that the true OR is likely to be within the specified range.

3. Results

3.1. Relation of Cows' Pregnancy Success with Automatic Milking System (AMS) Parameters

The analysis revealed a statistically significant relationship of changes of the AMS parameters on the pregnancy success of cows (Table 2). The BHB level in PG− cows was 1.2 times higher ($p = 0.003$). The electrical conductivity of milk was statistically significantly higher in all quarters of PG− cows (1.06–1.07 times) than that of PG+ cows ($p < 0.05$). Although LDH was 1.28 times higher in the PG− group, the difference with the PG+ group was not statistically significant.

Table 2. Automatic milking system (AMS) variables by groups of cows.

Parameter	PG− (n = 48)		PG+ (n = 44)		p (Between Days)	p (Between Groups
	M	SEM	M	SEM		
MY (kg/d)	44.56	2.252	43.99	2.203	0.009	0.231
FL (mS/cm)	4.41	0.06	4.17	0.048	0.423	0.002
FR (mS/cm)	4.45	0.05	4.17	0.043	0.192	<0.001
BL (mS/cm)	4.37	0.043	4.15	0.05	0.521	0.002
BR (mS/cm)	4.42	0.047	4.14	0.044	0.458	<0.001
LDH (μmol/min)	29.23	3.288	22.81	2.310	0.560	0.366
BHB (mmol/L)	0.06	0.001	0.05	0.001	0.078	0.003

PG+ = pregnant cows; PG− = non-pregnant cows; M = mean; SEM = standard error of the mean of a sample; MY = milk yield; FL = electric milk conductivity of front left; FR = electric milk conductivity of front right; BL = electric milk conductivity of back left; BR = electric milk conductivity of back right; LDH = lactate dehydrogenase; BHB = β-hydroxybutyric acid.

3.2. Relation Between Cows' Pregnancy Success and Biochemical Parameters in Serum

The arithmetic mean of blood GGT activity was 1.61 times higher in PG− cows compared with the PG+ group (p = 0.017). The NEFA value was 1.23 times higher compared with PG+ cows (p = 0.049). On the other hand, the average ALB level of PG− cows was 1.19 times lower (p = 0.017) and the AST activity was1.15 times lower (p = 0.048). The average cortisol value was 1.66 times lower in PG+ cows but this difference was not statistically significant (Table 3).

Table 3. Metabolic profile indicators of blood.

Parameter	PG−		PG+		p (Between Days)	p (Between Groups)
	M	SEM	M	SEM		
Cortisol (mmol/L)	1.37	0.358	0.83	0.119	0.612	0.227
ALB (g/L)	20.85	1.257	24.85	1.046	0.201	0.017
AST (IU/L)	49.51	2.666	57.37	2.755	0.543	0.048
GGT (IU/L)	10.46	1.555	6.50	0.519	0.487	0.017
NEFA (mmol/L)	0.16	0.008	0.14	0.006	0.199	0.049

PG+ = pregnant cows; PG− = non-pregnant cows; M = mean; SEM = standard error of the mean of a sample; ALB = albumin; AST = aspartate aminotransferase; GGT = gamma glutamyltransferase; NEFA = non-esterified fatty acids.

3.3. The Relationship of Cow Pregnancy with Milk and Blood Parameters

There were 1.7-fold more cows with cortisol > 0.0.75 mg/dL in the PG− group compared with the PG+ group but also 1.7-fold fewer cows with cortisol < 0.45 mg/dL (Figure 1A). PG+ cows tended to have higher blood ALB levels compared with the PG− group (p = 0.008). Cows with ALB < 21 g/L accounted for 27.3% in the PG+ group (there were 2.1 times fewer in the PG− group). The recommended ALB level was set for a 1.3 times higher number of cows in the PG+ group compared with the PG− group. On the other hand, 9.1% of cows with ALB > 36 g/L were found in the PG+ group while there were no such cows in the PG− group (Figure 1B). There were 1.2 times more cows with AST < 78 U/L (p = 0.117) in the PG+ group compared with the PG− group. The chi-squared independence test (Figure 1), in which a question was asked about the relationship between the pregnancy of cows and the level of AST, did not show a statistically significant relationship (Figure 1C). PG+ cows tended to have a lower GGT (Figure 1D). If there were 50% of such cows in the PG− group, then in the PG+ group there were twice as many. There were four times more cows with NEFA > 0.320 mmol/L (p = 0.031) in the PG− group compared with the PG+ group (Figure 1E).

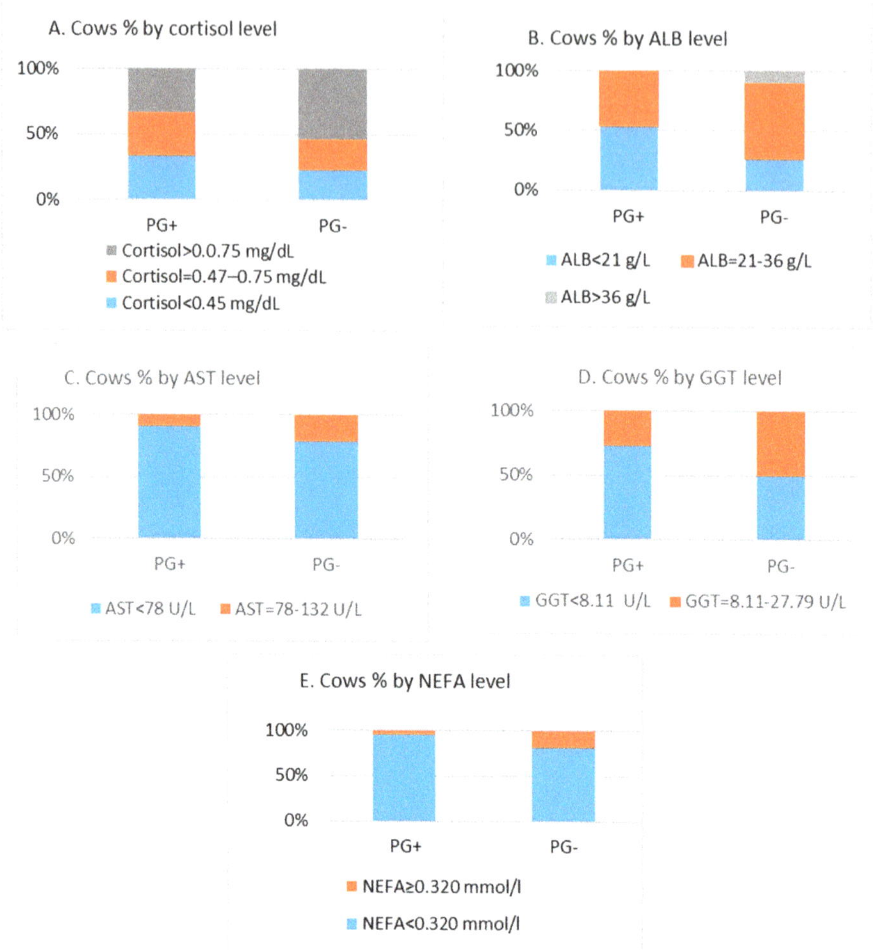

Figure 1. Frequency of PG+ and PG− cows by level of metabolic blood profile indicators: (**A**)—cows % by cortisol level; (**B**)—cows % by albumin level; (**C**)—cows % by aspartate aminotransferase level; (**D**)—cows % by gamma glutamyltransferase level; (**E**)—cows % by non-esterified fatty acids level; ALB = albumin; AST = aspartate aminotransferase; GGT = gamma glutamyltransferase; NEFA = non-esterified fatty acids; PG+ = pregnant cows; PG− = non-pregnant cows.

When analyzing the dependence of cow pregnancy success and studied milk indicators (Figure 2), a statistically significant relationship was found only with milk BHB and EC ($p < 0.001$). The number of cows with BHB $\leq$ 0.06 mmol in the PG+ group was 2.6 times higher than in the PG− group (Figure 2B). The analysis showed that relatively low EC levels in milk (3.5–5.0 mS/cm) were found in the studied cows (Figure 2B). The PG+ group had a 1.6-fold higher number of cows ($p < 0.001$) with a low EC (3.5–3.9 mS/cm) compared with the PG− group. In the PG− group, more (1.4–1.5-fold) cows were found with a larger EC difference between udder quarters in milk (Figure 2E) and also a larger MY (Figure 2A).

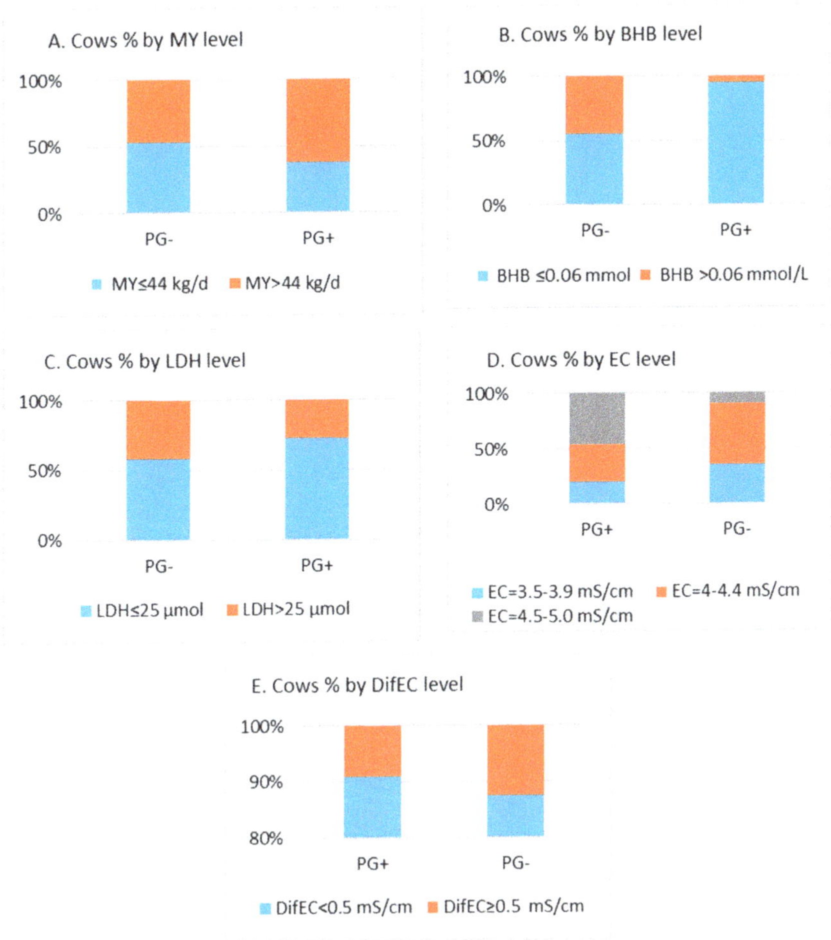

Figure 2. Frequency of PG+ and PG− cows by level of AMS variables. (**A**)—cows % by milk yield level; (**B**)—cows % by β-hydroxybutyric acid level; (**C**)—cows % by lactate dehydrogenase level; (**D**)—cows % by electric milk conductivity level; (**E**)—cows % by difference between electric milk conductivity values in udder quarters level; MY = milk yield; BHB = β-hydroxybutyric acid; LDH = lactate dehydrogenase; EC = electric milk conductivity; DifEC = difference between EC values in udder quarters. PG+ = pregnant cows; PG− = non-pregnant cows.

A multivariable logistic regression analysis showed that cows with a higher ALB activity in the blood were less likely (OR = 4.230; 95% CI = 1.635–10.946, p = 0.003) to become pregnant (p = 0.049). We also found that at relatively low EC levels in milk, cows were more likely to become pregnant (OR = 4.230; 95% CI = 1.635–10.946, p = 0.01) when their milk ECs were higher and met recommended levels [14].

4. Discussion

We found that in-line milk BHB levels in non-pregnant multiparous dairy cows at first service was 1.2 times higher than in pregnant cows. Animals with an elevated blood BHB had lower pregnancy success at first artificial insemination than healthy cows [3]. However, a decrease of pregnancy success within 70 days post-voluntary waiting period was reported by Ospina et al. [14]. Moreover, a greater number of inseminations per pregnancy (2.8 vs. 2.0, respectively), a lower peak activity (35% less activity), a shorter activity at oestrus (14% less hours) and a longer interval from calving to first observed oestrus also prolonged days open for multiparous cows in cows with ketosis than in healthy cows [15]. The mean predicted blood BHB of cows with ketosis was higher in early [16] lactation but decreased rapidly with the stage of lactation toward the level found in more healthy cows [8]. Subsequent studies revealed that there was an association between excessive negative energy balance (NEB) and decreased reproductive performance [17] According to the milk BHB results we can state that non-pregnant cows had a higher risk of ketosis.

According to our results, the in-line electrical milk conductivity of non-pregnant cows was 1.06–1.07 higher. Norberg et al. [18] aimed to evaluate the accuracy of electric milk conductivity (EMC) for predicting infection status; it also has some potential in terms of detecting mastitis. EMCs are parameters for the detection of mastitis in dairy farms equipped with in-line sensors [19]. During mammary tissue inflammation, the osmotic balance is maintained by an increase of $Na+$ and $Cl-$; in particular, $Na+$ derived from the highly $Na+$ concentrated extracellular environment is the main ion responsible for the increase of the electrical conductivity [20]. Based on that, we can suspect that non-pregnant cows have a greater risk of mastitis.

Our results of the blood biochemical parameters of non-pregnant cows showed that the GGT activities and NEFA values were higher and the albumin concentration was lower. We found that pregnant cows tended to have a higher blood albumin concentration. In dairy cows in the puerperium, a certain degree of fatty liver is noticed, which leads to the dysfunction of organs, releasing the enzymes of hepatocytes. Thus, the GGT activity in the blood increases significantly [21]. The strong association between NEFA concentrations and reproductive performance is likely because of the more direct physiological relationship between NEFA concentrations and a NEB [22]. The concentration of NEFA increases because of lipolysis, which is positively stimulated by glucagon [23]. According to Yang et al. [24], increased serum GGT may be a marker of oxidative stress, which is strongly associated with hypertension, dyslipidemia and an abnormal glucose tolerance. Albumin can be considered to be a negative acute-phase protein [25] with subnormal concentrations indicating an impaired liver function following a diverted synthesis to positive acute-phase proteins [26]. A low level of serum albumin may be associated with hepatocyte dysfunction, fat infiltration, degeneration and damage to liver tissue [27]. Albumins are also negative acute-phase proteins and their low serum levels can be found in various acute or chronic inflammations [28]. In the cows tested, slightly reduced serum albumin levels may have been caused by testing and a slightly reduced feed intake, which only confirmed the need to monitor the internal homeostasis of high-yielding animals [29].

Considering the data of our study, the AST was significantly lower in PG− compared with PG+ but the GGT activity was higher. This can be explained as the AST acts as a non-specific liver enzyme and is not inevitably accompanied by liver damage but often appears in connection with puerperal disorders. Increased AST activities were also measured shortly before and after normal parturition in cows and can be explained by the caruncle transformation and their degradation. Higher levels of AST enzymes were also recorded with a higher rate of muscle cell damage caused by the mobilization of body reserves [22].

Our results described in this study showed that the non-pregnant group had a negative energy balance demonstrated by high in-line milk BHB and high blood NEFA concentrations. Several studies have reported that a NEB as measured by NEFA or BHB has a strong association with reproductive performance. NEFA and BHB have a detrimental effect on reproductive performance [8]. In the early postpartum period, dairy cows resynthesize fat and muscle to support lactation and this in turn alters the blood metabolic and hormone profiles, which affects milk yield and fertility [16].

In our study we found a greater number of cows with cortisol > 0.0.75 mg/dL in the non-pregnant group. According to Burnett et al. [30] chronic and acute stress can trigger disruptions in reproductive function. Chronic stress is linked to disruptions in the pulsatile pattern of gonadotropin-releasing hormone (GnRH) and the consequent impairment of estrous behavior expression [31]. Stressors such as isolation, transport and restraint (acute causes) have been found to interfere with the endocrine events preceding ovulation and thus results in ovulation failure [32].

According to our results we found a positive correlation between LDH and milk EC at udder quarter level and BHB. In the AMS, the sensors that measure electrical conductivity are the most commonly used to detect mastitis. These sensors can continuously measure the concentration of ions in milk during the milk harvesting process, albeit with variable results [33]. Although lactate dehydrogenase testing at present is only commercially available in Herd Navigator, it is currently used for the automatic detection of mastitis [34]. According to Khatun et al. [33] LDH activity was strongly associated with mastitis. Subclinical ketosis is a risk factor for subsequent diseases and has been associated with mastitis. It has been demonstrated that BHB concentrations exceeding thresholds of 1.1 and 1.6 mmol/L are associated with a decreased probability of pregnancy and an increased culling risk, respectively [35]. According to our results we found that the Herd Navigator system does not just take a role as a novel tool for scientific research but could also be used in managing cow reproduction from a practical point. An in-line system automatically provides real-time physiological information for farm management decisions [3]. We confirmed the hypothesis that that in-line measurements could be predictive of the fertility status in dairy cows. By monitoring in-line parameters from the Herd Navigator such as BHB and EMC we could predict reproductive success in multiparous first service dairy cows. Our results suggest paying attention to non-pregnant cows having a higher risk of a negative energy balance and mastitis greatly affected by oxidative stress.

5. Conclusions

According to the results of the current study, we can conclude that the prediction of reproductive success in multiparous first service dairy cows by parameters from in-line sensors can be achieved through in-line biomarkers such as BHB and EMC. BHB levels were 1.2 times higher and electrical milk conductivity was 1.06–1.07 times higher in non-pregnant cows compared with pregnant cows. We also found that cows in the non-pregnant group had a negative energy balance indicated by high milk BHB and high blood NEFA concentrations. Higher milk EMC in the non-pregnant cows pointed towards a greater risk of mastitis while higher GGT activities together with lower albumin concentrations indicated that the cows were more affected by oxidative stress. More cows with a higher cortisol concentration in the non-pregnant cow group indicated the negative impact of stress on pregnancy success.

Author Contributions: R.A.—overall research study process including literature search, carrying out research experiments and compiling the final manuscript. The entire process was revised by the co-authors. V.J.—assisted in designing and setting up field data collection activities and developed the software and algorithm for data analysis. D.M. and M.T.—aided in fieldwork set up, data collection and sampling of the experimental animals. M.U. and G.Z.—design of the field experiment and data collection. W.B.—major support in processing of data in the study. All authors have read and agreed to the published version of the manuscript.

Funding: This research received no external funding.

Institutional Review Board Statement: The study was conducted according to the guidelines of the Declaration of Helsinki, and ap-proved by Ethics Committee (The study approval number is PK016965, 6 June 2017).

Informed Consent Statement: Informed consent was obtained from all subjects involved in the study.

Data Availability Statement: The data presented in this study are available within the article.

Conflicts of Interest: The authors declare no conflict of interest.

References

1. Walsh, S.W.; Williams, E.J.; Evans, A.C.O. A review of the causes of poor fertility in high milk producing dairy cows. *Animal Reprod. Sci.* **2011**, *123*, 127–138. [CrossRef] [PubMed]

2. Barkema, H.W.; Von Keyserlingk, M.A.; Kastelic, J.P.; Lam, T.J.; Luby, C.; Roy, J.P.; LeBlanc, S.J.; Keefe, G.P.; Kelton, D.F. Invited review: Changes in the dairy industry affecting dairy cattle health and welfare. *J. Dairy Sci.* **2015**, *98*, 7426–7445. [CrossRef] [PubMed]
3. Yu, G.M.; Maeda, T. Inline progesterone monitoring in the dairy industry. *Trends Biotechnol.* **2017**, *35*, 579–582. [CrossRef]
4. Larson, S.F.; Butler, W.R.; Currie, W.B. Reduced fertility associated with low progesterone postbreeding and increased milk urea nitrogen in lactating cows. *J. Dairy Sci.* **1997**, *80*, 1288–1295. [CrossRef]
5. Blom, J.Y.; Christensen, J.M.; Ridder, C. Real-time analyses of BHB in milk can monitor ketosis and its impact on reproduction in dairy cows. In *Precision Livestock Farming Applications: Making Sense of Sensors to Support Farm Management*; Wageningen Academic Publishers: Wageningen, The Netherlands, 2015; pp. 2788–2796.
6. Nyman, S.; Johansson, K.; de Koning, D.J.; Berry, D.P.; Veerkamp, R.F.; Wall, E.; Berglund, B. Genetic analysis of atypical progesterone profiles in Holstein-Friesian cows from experimental research herds. *J. Dairy Sci.* **2014**, *97*, 7230–7239. [CrossRef]
7. Saint-Dizier, M.; Chastant-Maillard, S. Methods and on-farm devices to predict calving time in cattle. *Vet. J.* **2015**, *205*, 349–356. [CrossRef]
8. Antanaitis, R.; Juozaitienė, V.; Televičius, M.; Malašauskienė, D. Changes in the real-time registration of milk β-hydroxybutyrate according to stage and lactation number, milk yield, and status of reproduction in dairy cows. *Pol. J. Vet. Sci.* **2018**, *21*, 763–768.
9. Antanaitis, R.; Malašauskienė, D.; Televičius, M.; Juozaitienė, V.; Rutkauskas, A.; Palubinskas, G. Inline changes in lactate dehydrogenase, milk concentration according to the stage and number of lactation periods, including the status of reproduction and milk yield in dairy cows. *Pol. J. Vet. Sci.* **2020**, *23*, 153–156.
10. Jackson, P.G.G.; Cockcroft, P.D. *Clinical Examination of Farm Animals*; Copyright© Blackwell Science Ltd.: Oxford, UK, 2002.
11. Stojević, Z.; Piršljin, J.; Milinković-Tur, S.; Zdelar-Tuk, M.; Ljubić, B.B. Activities of AST, ALT and GGT in clinically healthy dairy cows during lactation and in the dry period. *Veterinarski Arhiv* **2005**, *75*, 67–73.
12. Petrera, F.; Napolitano, F.; Dal Prà, A.; Abeni, F. Plasma parameters related to energy and lipid metabolism in periparturient M odenese and Italian Friesian cows. *J. Animal Physiol. Animal Nutr.* **2015**, *99*, 962–973. [CrossRef] [PubMed]
13. Hillerton, J.E.; Walton, A.W. Identification of subclinical mastitis with a hand-held electrical conductivity meter. *Vet. Record* **1991**, *128*, 513–515. [CrossRef]
14. Ospina, P.A.; Nydam, D.V.; Stokol, T.; Overton, T.R. Associations of elevated nonesterified fatty acids and β-hydroxybutyrate concentrations with early lactation reproductive performance and milk production in transition dairy cattle in the northeastern United States. *J. Dairy Sci.* **2010**, *93*, 1596–1603. [CrossRef]
15. Rutherford, A.J.; Oikonomou, G.; Smith, R.F. The effect of subclinical ketosis on activity at estrus and reproductive performance in dairy cattle. *J. Dairy Sci.* **2016**, *99*, 4808–4815. [CrossRef] [PubMed]
16. Ospina, P.A.; Nydam, D.V.; Stokol, T.; Overton, T.R. Association between the proportion of sampled transition cows with increased nonesterified fatty acids and β-hydroxybutyrate and disease incidence, pregnancy rate, and milk production at the herd level. *J. Dairy Sci.* **2010**, *93*, 3595–3601. [CrossRef]
17. Wathes, D.C.; Fenwick, M.; Cheng, Z.; Bourne, N.; Lewellyn, S.; Morris, D.G.; Fitzpatrick, R. Influence of negative energy balance on cyclicity and fertility in the high producing dairy cow. *Theriogenology* **2007**, *68*, S232–S241. [CrossRef] [PubMed]
18. Norberg, E.; Hogeveen, H.; Korsgaard, I.R.; Friggens, N.C.; Sloth, K.N.; Løvendahl, P. Electrical conductivity of milk: Ability to predict mastitis status. *J. Dairy Sci.* **2004**, *87*, 1099–1107. [CrossRef]
19. Televičius, M.; Juozaitiene, V.; Malašauskienė, D.; Antanaitis, R.; Rutkauskas, A.; Urbutis, M.; Baumgartner, W. Inline Milk Lactose Concentration as Biomarker of the Health Status and Reproductive Success in Dairy Cows. *Agriculture* **2021**, *11*, 38. [CrossRef]
20. Costa, A.; Egger-Danner, C.; Mészáros, G.; Fuerst, C.; Penasa, M.; Sölkner, J.; Fuerst-Waltl, B. Genetic associations of lactose and its ratios to other milk solids with health traits in Austrian Fleckvieh cows. *J. Dairy Sci.* **2019**, *102*, 4238–4248. [CrossRef] [PubMed]
21. Ježek, J.; Cincović, M.R.; Nemec, M.; Belić, B.; Djoković, R.; Klinkon, M.; Starič, J. Beta-hydroxybutyrate in milk as screening test for subclinical ketosis in dairy cows. *Pol. J. Vet. Sci.* **2017**, *20*, 507–512. [CrossRef]
22. Bastin, C.; Abdelsayed, M.; Gengler, N.; Bradley, A.J. Invited review: Opportunities for genetic improvement of metabolic diseases. *J. Dairy Sci.* **2016**, *99*, 6855–6873.
23. Tatone, E.H.; Duffield, T.F.; LeBlanc, S.J.; DeVries, T.J.; Gordon, J.L. Investigating the within-herd prevalence and risk factors for ketosis in dairy cattle in Ontario as diagnosed by the test-day concentration of β-hydroxybutyrate in milk. *J. Dairy Sci.* **2017**, *100*, 1308–1318. [CrossRef]
24. Yang, D.A.; Johnson, W.O.; Müller, K.R.; Gates, M.C.; Laven, R.A. Estimating the herd and cow level prevalence of bovine digital dermatitis on New Zealand dairy farms: A Bayesian superpopulation approach. *Prev. Vet. Med.* **2019**, *165*, 76–84. [CrossRef]
25. Fleck, A. Clinical and nutritional aspects of changes in acutephase proteins during inflammation. *Proc. Nutr. Soc.* **1989**, *48*, 347–354. [CrossRef] [PubMed]
26. Mohebbi-Fani, M.; Omidi, A.; Mirzaei, A.; Nazifi, S.; Nowroozi, K.H. A field study on glucose, non-esterified fatty acids, betahydroxybutyrate and thyroid hormones in dairy cows during the breeding period in Fars province, Iran. *Iran J. Vet. Res.* **2018**, *1*, 55.
27. Farid, A.S.; Honkawa, K.; Fath, E.M.; Nonaka, N.; Horii, Y. Serum paraoxonase-1 as biomarker for improved diagnosis of fatty liver in dairy cows. *BMC Vet. Res.* **2013**, *9*, 73. [CrossRef] [PubMed]
28. Ametaj, B.N.; Bradford, B.J.; Bobe, G.; Lu, Y.; Nafikov, R.; Sonon, R.N.; Young, J.W.; Beitz, D.C. Acute phase response indicates inflammatory conditions may play a role in the pathogenesis of fatty liver in dairy cows. *J. Dairy Sci.* **2002**, *85*, 189.

29. Mordak, R.; Kupczyński, R.; Kuczaj, M.; Niżański, W. Analysis of correlations between selected blood markers of liver function and milk composition in cows during late lactation period. *Ann. Anim. Sci.* **2020**, *20*, 871–886. [CrossRef]
30. Burnett, T.A.; Madureira, A.M.; Silper, B.F.; Tahmasbi, A.; Nadalin, A.; Veira, D.M.; Cerri, R.L. Relationship of concentrations of cortisol in hair with health, biomarkers in blood, and reproductive status in dairy cows. *J. Dairy Sci.* **2015**, *98*, 4414–4426. [CrossRef]
31. Walker, S.L.; Smith, R.F.; Jones, D.N.; Routly, J.E.; Dobson, H. Chronic stress, hormone profiles and estrus intensity in dairy cattle. *Horm. Behav.* **2008**, *53*, 493–501. [CrossRef] [PubMed]
32. Moberg, G.P. Biological response to stress: Implications for animal welfare. In *The Biology of Animal Stress: Basic Principles and Implications for Animal Welfare 1*; University of California: Davis, CA, USA, 2000; p. 21.
33. Khatun, M.; Bruckmaier, R.M.; Thomson, P.C.; House, J.; García, S.C. Suitability of somatic cell count, electrical conductivity, and lactate dehydrogenase activity in foremilk before versus after alveolar milk ejection for mastitis detection. *J. Dairy Sci.* **2019**, *102*, 9200–9212. [CrossRef] [PubMed]
34. Mollenhorst, H.; Rijkaart, L.J.; Hogeveen, H. Mastitis alert preferences of farmers milking with automatic milking systems. *J. Dairy Sci.* **2012**, *95*, 2523–2530. [CrossRef] [PubMed]
35. Roberts, T.; Chapinal, N.; LeBlanc, S.J.; Kelton, D.F.; Dubuc, J.; Duffield, T.F. Metabolic parameters in transition cows as indicators for early-lactation culling risk. *J. Dairy Sci.* **2012**, *95*, 3057–3063. [CrossRef] [PubMed]

agriculture

Article

Extraction of Arecanut Planting Distribution Based on the Feature Space Optimization of PlanetScope Imagery

Yu Jin [1,2], Jiawei Guo [1,3,4,†], Huichun Ye [1,3,*], Jinling Zhao [2], Wenjiang Huang [1,2,3] and Bei Cui [1,3]

1 Key Laboratory of Digital Earth, Aerospace Information Research Institute, Chinese Academy of Sciences, Beijing 100094, China; ahuJingy@163.com (Y.J.); guojw@jou.edu.cn (J.G.); huangwj@aircas.ac.cn (W.H.); cuibei@aircas.ac.cn (B.C.)
2 National Engineering Research Center for Agro-Ecological Big Data Analysis & Application, Anhui University, Hefei 230601, China; zhaojl@ahu.edu.cn
3 Hainan Key Laboratory of Earth Observation, Hainan Institute of Aerospace Information Research Institute, Chinese Academy of Sciences, Sanya 572029, China
4 School of Marine Technology and Geomatics, Jiangsu Ocean University, Lianyungang 222005, China
* Correspondence: yehc@aircas.ac.cn
† Co-first author.

Abstract: The remote sensing extraction of large areas of arecanut (*Areca catechu* L.) planting plays an important role in investigating the distribution of arecanut planting area and the subsequent adjustment and optimization of regional planting structures. Satellite imagery has previously been used to investigate and monitor the agricultural and forestry vegetation in Hainan. However, the monitoring accuracy is affected by the cloudy and rainy climate of this region, as well as the high level of land fragmentation. In this paper, we used PlanetScope imagery at a 3 m spatial resolution over the Hainan arecanut planting area to investigate the high-precision extraction of the arecanut planting distribution based on feature space optimization. First, spectral and textural feature variables were selected to form the initial feature space, followed by the implementation of the random forest algorithm to optimize the feature space. Arecanut planting area extraction models based on the support vector machine (SVM), BP neural network (BPNN), and random forest (RF) classification algorithms were then constructed. The overall classification accuracies of the SVM, BPNN, and RF models optimized by the RF features were determined as 74.82%, 83.67%, and 88.30%, with Kappa coefficients of 0.680, 0.795, and 0.853, respectively. The RF model with optimized features exhibited the highest overall classification accuracy and kappa coefficient. The overall accuracy of the SVM, BPNN, and RF models following feature optimization was improved by 3.90%, 7.77%, and 7.45%, respectively, compared with the corresponding unoptimized classification model. The kappa coefficient also improved. The results demonstrate the ability of PlanetScope satellite imagery to extract the planting distribution of arecanut. Furthermore, the RF is proven to effectively optimize the initial feature space, composed of spectral and textural feature variables, further improving the extraction accuracy of the arecanut planting distribution. This work can act as a theoretical and technical reference for the agricultural and forestry industries.

Keywords: arecanut; PlanetScope satellite image; random forest algorithm; feature optimization; area extraction

Citation: Jin, Y.; Guo, J.; Ye, H.; Zhao, J.; Huang, W.; Cui, B. Extraction of Arecanut Planting Distribution Based on the Feature Space Optimization of PlanetScope Imagery. *Agriculture* **2021**, *11*, 371. https://doi.org/10.3390/agriculture11040371

Academic Editors: Gniewko Niedbała and Sebastian Kujawa

Received: 16 March 2021
Accepted: 16 April 2021
Published: 19 April 2021

1. Introduction

Arecanut (*Areca catechu* L.) is a perennial evergreen tree of the palm family and an important Chinese medicinal plant. It is common in some areas of southern Asia to chew the fruit; however, it is currently listed as a class 1 carcinogen by the World Health Organization International Agency for Research on Cancer. At present, arecanut is principally distributed in the Asian countries of India, Indonesia, Bangladesh, China, Myanmar, Thailand, the Philippines, Vietnam, and Cambodia [1]. It is a key economic crop in the tropical and subtropical regions of China, with a planting history of more than 1000 years. Principle sowing locations include tropical regions such as Hainan and Taiwan, with smaller distributions in Guangxi, Yunnan, Hunan, and Fujian. The production of arecanut in Hainan Province currently accounts for more than 90% of the domestic total. Furthermore, in 2019, the planting area, harvest area, and total output reached 115,171 ha, 83,318 ha, and 272,200 t, respectively [2]. This identifies arecanut as one of the largest tropical cash crops in Hainan Province, playing a crucial role in the industry and farmers' income within the province.

The negative impact of diseases (i.e., yellow leaf disease) has resulted in the recent reduction in the arecanut planting area and yield. The area of areca affected by yellow leaf disease has reached 533.3 km^2, with an increase of 20–30 km^2 per year; moreover, the annual loss caused by yellow leaf disease is estimated to exceed 2 billion yuan [3]. As a major pillar of industry in Hainan Province, the reduction in the arecanut output has generated huge economic losses to arecanut growers in the province. Therefore, there is an urgent need for the timely and accurate extraction of the planting area of arecanut in Hainan in order to grasp the dynamic changes of this crop and to effectively manage the development of the arecanut industry in Hainan.

Remote sensing technology holds numerous advantages, such as high efficiency, dynamic applications, wide spatial coverage, and fast data acquisition, allowing for the rapid, accurate, and dynamic monitoring of crop planting areas [4–6]. Current research on the monitoring of crop areas typically employs remote sensing technology to classify crops and extract planting information. In small-scale areas, unmanned aerial vehicle (UAV) aerial remote sensing platforms are often used for the extraction of crop planting areas. For example, Zheng et al. [7] used RGB, NIR-GB, and multispectral images from unmanned aerial vehicle (UAV) to extract rice plants information at the early growth stages. Shen et al. [8] integrated UAV technology with moderate spatial resolution (MSR) data to estimate crop planting areas using random stratified sampling, making the crop area estimation accuracy more than 95% with a 95% confidence interval. Based on the extraction of crop area, more scholars use UAV remote sensing to monitor growth and predict yield [9]. However, UAV remote sensing has limitations in its endurance time and flight radius, and it is not suitable for large-scale crop surveys.

Satellite imagery is associated with a high and wide field of view, fast data collection, repeatable coverage, and continuous observations [10], and is frequently applied for the large-scale extraction of crop planting areas. Based on moderate resolution imaging spectroradiometer (MODIS) time series data, Pan et al. [11] established the crop proportion phenology index (CPPI) for estimating wheat area, with the root mean square error (RMSE) in fractional crop area predictions ranging roughly from 15% in the individual pixels to 5% above 6.25 km^2. Zhang and Lin [12] fused Landsat-8 OLI time series with phenological parameters for the extraction of rice planting area in cloudy areas based on object-oriented algorithms, providing high-precision rice distribution maps with an overall accuracy of 92.38%. Liu et al. [13] constructed a decision tree model based on multitemporal HJ-1 CCD images to accurately extract corn planting area in Zhecheng County, Henan Province, China. However, these satellite data are mostly limited by low spatial resolution or short revisit period, which are not suitable for crop monitoring in some regions (i.e., the tropical and subtropical regions) with fragmented plots and cloudy and rainy weather.

With the development of remote sensing technology, the high-resolution PlanetScope satellite cluster can achieve daily global coverage with a 3 m spatial resolution, providing an effective data source for the extraction of agricultural and forestry planting information in tropical and subtropical regions. Arecanut is a tropical palm typically reaching 10–20 m tall with a straight and slender trunk. Its dark green leaves can spread 2 m across. These morphological features of arecanut present its distinctive spectral and texture features from the high-resolution imagery that differentiate arecanut land from other lands. The objective of this research was to (i) establish a high-precision arecanut information extraction method based on feature space optimization, which is composed of spectral and texture features extracted from PlanetScope satellite images, and (ii) evaluate the performance of three machine learning algorithms with support vector machine (SVM), BP neural network (BPNN), and random forest (RF) algorithms combined with the optimized feature space in an attempt to extract the arecanut information. The results provide theoretical and technical references for the remote sensing extraction of agricultural and forestry information.

2. Materials and Method

2.1. Study Area

The study area is located in Beida Town, Wanning City, Hainan Province, China (110°23′–110°40′ E, 18°86′–19°01′ N) with an area of 276.09 km^2 (Figure 1). The area has a tropical monsoon climate, with an average annual temperature of 23.6 °C, a monthly average temperature of 18.7–28.5 °C, annual precipitation of approximately 2200 cm, and average annual sunshine hours over 1800. Beida Town is located in a hilly mountainous area. The soil type is Ferralsols according to IUSS Working Group WRB [14].

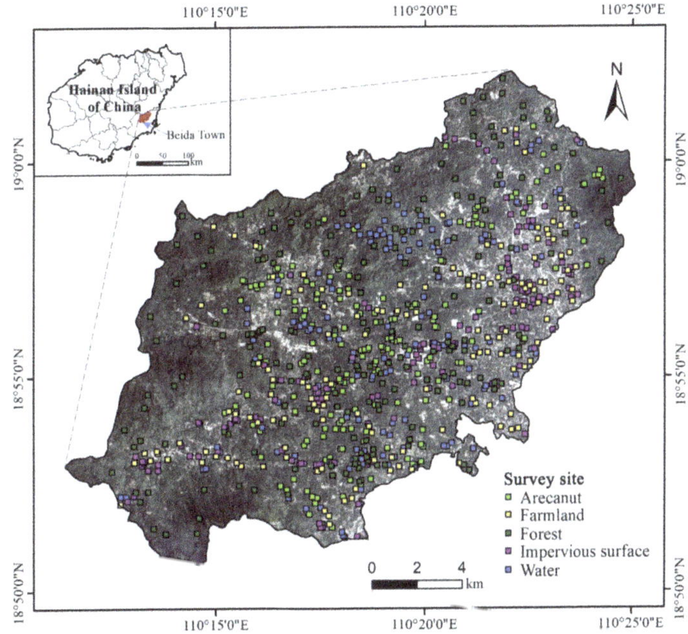

Figure 1. Geographic location of the study area with the spatial distribution of land use/cover type survey sites.

Hainan Province contains the largest arecanut production area in China, with the greatest planting area located in Wanning City. In 2018, the planting area reached 18,138 hm^2, accounting for 16.5% of the total planting area in Hainan [2]. Beida Town is the principal planting area of arecanut in Wanning City. The town also grows cash crops such as rubber, pineapple, and lychee.

2.2. Data Acquisition and Processing

2.2.1. PlanetScope Satellite Image Acquisition and Preprocessing

The PlanetScope small satellite constellation currently has more than 170 satellites in orbit, surpassing all current satellites in terms of resolution (3–4 m), frequency (daily), and global coverage [15]. In the current paper, we selected a high-quality clear and cloudless PlanetScope satellite image collected on 21 March 2019. The PlanetScope image used is an orthographic data product (3B) that has undergone sensor and radiometric calibration, as well as orthorectification and atmospheric correction. The satellite image has a spatial resolution of 3 m and contains four spectral bands in the blue, green, red, and near-infrared regions. Table 1 lists the PlanetScope satellite parameters.

Table 1. Specifications of PlanetScope satellite.

Parameter	Parameter Value
Track	International Space Station OrbitSun-synchronous orbit
Orbital inclination	52°, 98°
Spatial resolution	3–4 m
Spectral band	Band1: Blue (455–515 nm) Band2: Green (500–590 nm) Band3: Red (590–670 nm) Band4: NIR (780–860 nm)
Track height	400 km, 475 km
Sensor type	Bayer filter CCD camera
Width	24.6 km × 16.4 km

2.2.2. Ground Sample Data Collection

The principal land use/cover types in the study area are farmland, forest, impervious surface (urban and rural areas; industrial and mining, water conservancy construction, and transportation land), water (rivers, lakes, ponds, etc.), and arecanut grove. Table 2 lists the visual interpretation characteristics of the main features in the study area. Ground sample data were obtained through field surveys with a GPS receiver on 19–21 March 2019. The coverage size of the field should be more than 10 m × 10 m. According to the location of the survey sites, the field boundaries were then drawn based on Google Earth Pro (version 7.3.2.5776). Finally, a total of 850 field polygon samples were determined. There are 150 samples for water, 150 for impervious surface, 200 for forest, 150 for farmland, and 200 for arecanut grove, with 70% and 30% of the samples used for training and verification, respectively.

Table 2. Visual interpretation signs of features in the study area.

Feature Category	Image Characteristics	Feature Description
Water		Light green, the larger the water body, the darker the color. Pit ponds are small in area, with clear boundaries and irregular shapes; rivers are in regular curved strips; lakes have large water areas, darker colors, and irregular shapes.
Impervious surface		Light purple and brown with irregular shapes, bare soil, and less vegetation coverage.
Forest		Dark green, the plots are irregularly distributed, with uniform tone and clear texture.

Table 2. *Cont.*

Feature Category	Image Characteristics	Feature Description
Farmland		Light green, with clear stripes, regular continuous distribution, and uniform texture.
Arecanut		Light green, granular canopy distributed in a large area, irregular plot shape, uniform texture, and small amount of soil exposure.

2.3. Feature Variable Selection

2.3.1. Primary Selection of Characteristic Variables

- Primary selection of spectral characteristic variables

The spectral characteristic variables initially selected included four original spectral bands and five widely used vegetation indices (Table 3). A spectral band can act as an important indicator for the extraction of ground feature information from remote sensing images. Here, we used the blue, green, red, and near-infrared reflectance bands of the PlanetScope image as the primary selection variables for spectral features. The blue band is susceptible to factors such as soil background, and plays a key role in the distinction between soil and vegetation; the green band is more sensitive to different types of plants and can be used to distinguish between vegetation types [16]; the red band is the principal absorption band of chlorophyll and is an important indicator of plant vitality [17]; and the near-infrared band can remove the influence of the atmosphere (e.g., aerosols and thin clouds) and can reflect the vegetation growth and coverage [18].

Table 3. Description of the spectral characteristic variables selected in this study.

Spectral Characteristic	Formula [1]	Reference
Blue band	R_B	[16]
Green band	R_G	[16]
Red band	R_R	[17]
Near-infrared band	R_{NIR}	[18]
Difference Vegetation Index (DVI)	$R_{NIR} - R_R$	[19]
Modified Soil Adjusted Vegetation Index (MSAVI)	$\frac{1}{2}[(2R_{NIR}+1) - \sqrt{(2R_{NIR}+1)^2 - 8(R_{NIR} - R_R)}]$	[20]
Normalized Difference Vegetation Index (NDVI)	$(R_{NIR} - R_R)/(R_{NIR} + R_R)$	[21]
Ratio Vegetation Index (RVI)	R_{NIR}/R_R	[22]
Soil Brightness Index (SBI)	$\sqrt{R_{NIR}^2 + R_R^2}$	[23]

[1] R_R, R_G, R_B, and R_{NIR} denote the red, green, blue, and near-infrared band.

Based on the principal feature types in the study area, the Difference Vegetation Index (DVI), Modified Soil Adjusted Vegetation Index (MSAVI), Normalized Difference Vegetation Index (NDVI), Ratio Vegetation Index (RVI), and Soil Brightness Index (SBI) were selected (Table 3). DVI is extremely sensitive to changes in the soil background and can better identify vegetation and water bodies [19]; MSAVI can reflect the soil and vegetation coverage information on the ground under the influence of soil background factors, and can accurately identify low vegetation coverage [20]; NDVI is sensitive to green vegetation and can reflect vegetation growth status and coverage [21]; RVI enhances the radiation difference between vegetation and soil, and can characterize biomass information under different vegetation coverage [22]; and SBI is sensitive to the soil background and can effectively extract construction and bare land in the absence of vegetation cover [23].

- Primary selection of texture feature variables

Texture is a visual feature that reflects the homogeneity of the image and describes the grayscale spatial distribution of the image pixel neighborhood [24]. Each feature contained in remote sensing images has its own unique texture structure. However, the spectrum of an object may vary between features, while different objects may also have the same spectrum, resulting in difficulties in identifying objects. The stem of the arecanut tree is upright, arborous, up to 30 m tall, and with obvious circular leaf marks. Moreover, the leaves are clustered at the top of the stem and with a length of 1.3–2 m. The tree bears fruit to many pines of a long and narrow lanceolate shape, a 30–60 cm length, and 2.5–4 cm width. The upper pinna is connate and the tip has irregular teeth. Furthermore, the arecanut has a row spacing of 2.5 to 3.0 m and a plant spacing of 2.0 to 2.5 m. Thus, the forests of this crop present textural features that are obviously distinct from other ground objects within the high spatial resolution remote sensing image.

We selected the Gray-Level Co-Occurrence Matrix (GLCM) method to select eight texture feature indicators: Mean (Me), Variance (Var), Homogeneity (Hom), Contrast (Con), Dissimilarity (Dis), Entropy (Ent), Second moment (SM), and Correlation (Cor), Table 4 shows their formulas [25]. Based on the four spectral bands of the PlanetScope satellite image, a total of 32 texture features were subsequently extracted.

Table 4. Description of the texture features selected in this study.

Texture Feature	Formula [1]	Description
Mean	$\sum_i \sum_j P(i,j) \times i$	Reflects the regular degree of texture.
Variance	$\sum_i \sum_j (i - Mean)^2 \times P(i,j)$	Reflects the deviation between the pixel and mean values; the larger the grayscale change, the larger the value.
Homogeneity	$\sum_i \sum_j P(i,j) \times \frac{1}{1+(i-j)^2}$	Reflects the local gray uniformity of the image; the more uniform the local, the larger the value.
Contrast	$\sum_i \sum_j P(i,j) \times (i-j)^2$	Reflects the sharpness of the image and the depth of the texture.
Dissimilarity	$\sum_i \sum_j P(i,j) \times \lvert i-j \rvert$	Similar to contrast, with greater linearity; the higher the local contrast, the higher the dissimilarity.
Entropy	$-\sum_i \sum_j P(i,j) \times log\ P(i,j)$	Reflects the texture complexity; the larger the value, the more complex the texture.
Second Moment	$\sum_i \sum_j P(i,j)^2$	Reflects the uniformity of the image distribution and texture thickness.
Correlation	$\sum_i \sum_j \frac{(i-Mean) \times (j-Mean) \times P(i,j)^2}{Variance}$	Reflects the image local relevance.

[1] $P(i, j)$ is the element value of the image at point (i, j).

2.3.2. Feature Variable Optimization Method

The random forest (RF) algorithm, proposed by Breiman and AdeleCulter in 2001, integrates multiple trees based on ensemble learning, with a single decision tree taken as the basic unit [26,27]. Due to its strong noise tolerance, avoidance of overfitting, and ability to handle high-dimensional data, RF is not only applied to classification tasks, but can also calculate the importance of a single feature variable. In particular, RF performs feature screening using a feature importance evaluation, whereby the contribution value of each feature on each decision tree is determined, and the average values are compared between features. The out-of-bag (OOB) error rate is typically used as the evaluation index to measure the feature contribution, denoting the variable importance (*VI*) of different feature variables. The feature optimization is then realized by ranking features by importance. The feature variable importance, *VI*, is calculated as follows:

$$VI = \frac{\sum_{k=1}^{N} B_{n_k}^{M} - B_{o_k}^{M}}{M} \tag{1}$$

where N is the number of generated decision trees, M is the number of feature variables, $B_{n_k}^{M}$ is the out-of-bag error of the k-th decision tree when feature M is added to the noise interference, and $B_{o_k}^{M}$ is the out-of-bag error of the k-th decision tree without noise interference. The addition of feature M with random noise dramatically reduces the accuracy rate of the out-of-bag data, indicating the strong influence of this feature on the prediction results of the sample, and thus its importance is relatively high [28].

2.4. Classification Model Building Method

The proposed classification framework initially calculates the training sample feature variables and subsequently constructs three classification models based on the BP neural network (BPNN), RF, and support vector machine (SVM) algorithms. These models are applied to extract the arecanut and other surrounding ground features. The extraction results are then verified using the verification samples and compared to determine the most accurate classification model.

2.4.1. BP Neural Network Algorithm

BPNN is a multilayer feedforward neural network that adopts the error back propagation algorithm to train the model. As the most widely used neural network to date [29], it uses the gradient descent method to minimize the mean square error between the actual and expected output values. BPNN is able to perform both signal forward propagation and error backward propagation. The input signal of BPNN propagates forward through the input layer and each hidden layer, and finally reaches the output layer, where the actual output value is obtained and compared with the expected output value. If the two output values are not equal, the error will enter back propagation, where the output error is adjusted with a threshold and weight at each layer via gradient descent. This results in a neural network model that has an expected output value within the error tolerance range. The BPNN is composed of the input, output, and hidden layers, and is trained by constantly adjusting the threshold and weight. The specific implementation process is as follows:

- Dataset entry: define randomly divided training set P_train, validation set T_test, training label P class and verification label T class.
- Data normalization: the mapminmax function is used to normalize and map the data to the range of 0–1 to avoid significant differences between the input and output data.
- A neural network is established and the network parameters are set.
- The training parameters are defined and network training is performed. The number of iterations, learning rate, training error target, and maximum number of failures are set to 200, 0.001, 0.0001, and 10, respectively. The train (net, P, T) function is used for network training.
- Network simulation is performed using the sim (net, test matrix) function and the overall recognition accuracy of BPNN is obtained based on the predicted and expected values.

2.4.2. Random Forest Algorithm

The application of RF in classification tasks centers around the bootstrap method to randomly extract and return s samples from the sample set. Following n sample iterations, n training sets are obtained with n decision tree models. The generated n decision trees are then integrated into a random forest with multiple tree classifiers to determine the final prediction result [28]. Multiple decision trees are constructed during the training phase and the final class output is the pattern of a single decision tree class. The number of decision trees ntree is set to 500, while all other parameters are taken as the default values.

2.4.3. Support Vector Machine

SVM, first proposed by Vapnik [30], is based on statistical learning theory, and in particular, the principle of structural risk minimization [31]. Under linear separability, SVM aims to determine the optimal classification hyperplane of the two types of samples in the original space. For linear inseparability, the relaxation variable is added for analysis, and the samples in the low dimensional input space are mapped to the high dimensional feature space via nonlinear mapping, resulting in linearity and allowing for the determination of the optimal hyperplane in the feature space. Optimizing the segmentation hyperplane separates the sample types and minimizes the error, resulting in the accurate classification of data. More details of the SVM calculation process can be found in the relevant literature [32]. SVM has a simple structure and strong adaptability and robustness, and is thus applicable to a wide range of linear, nonlinear, classification, and regression problems. We employed SVM to build a monitoring model for arecanut yellowing disease using the mapminmax function to normalize the training and validation sets and to scale the data within [0,1]. The svmtrain and svmpredict commands were then implemented in LIBSVM 3.23 to train the samples and test the validation set, respectively. SVM system default values were used for the linear kernel function and parameters such as penalty factor c and kernel parameter g.

3. Results

3.1. Feature Space Optimization

The initial feature space contains a total of 41 feature variables (9 spectral and 32 textural feature variables). A large number of feature variables will generate redundant data, increasing the model complexity and affecting the classification accuracy. We employed the RF algorithm to evaluate the importance of the 41 feature variables in the initial feature space, ranking their importance based on the feature variable weights. Figure 2 presents the importance rankings of the target feature variables, whereby the first 12 feature weights are greater than 1, the middle 12 feature weights range between 0.5 and 1, and the latter 17 feature weights are less than 0.5. Then, according to the order of feature importance, the first k ($i = 1, 2, \ldots\ldots, 41$) feature variables were selected to construct the random forest classification model of arecanut, and the overall classification accuracy was subsequently calculated. The overall classification accuracy is maximized to 88.3% when the number of feature variables equals 14. Therefore, the first 14 feature variables (CorNIR, VarNIR, MeNIR, MeR, RB, EntNIR, RR, RNIR, ConNIR, MeB, RG, NDVI, SBI, and HomNIR) were selected to construct the optimized feature space.

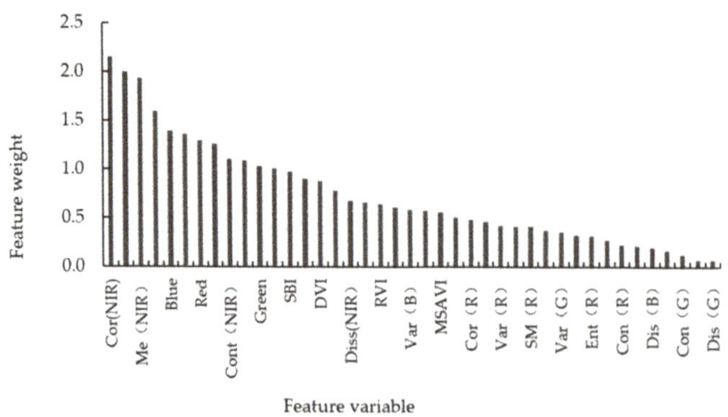

Figure 2. Importance ranking of the first 14 feature variables for constructing the optimized feature space.

3.2. *Extraction of Arecanut Planting Information*

In this study, two feature spaces (i.e., initial feature space and optimized feature space) were used as the input of three machine learning algorithms (i.e., SVM, BPNN, and RF) to extract the arecanut planting information, respectively. A total of six classification models were constructed based on SVM, BPNN, and RF to extract the arecanut planting area in the study region, denoted as SVM-1, BPNN-1, and RF-1 for the initial feature space input and SVM-2, BPNN-2, and RF-2 for the optimized feature space as input. Ground survey data were used to evaluate the classification accuracy of the initial and optimized feature space inputs, and the impact of feature optimization on the extraction accuracy of arecanut planting area was then analyzed.

Table 5 reports the classification accuracy of arecanut based on the different classification models. Following RF feature optimization, the user's and producer's accuracy of SVM-2 are observed to exceed those of SVM-1 by 10.35% and 7.54%, respectively. The producer's accuracy for BPNN-2 remained changed, while the user's accuracy increased from 81.86% to 87.50%. In addition, the user's and producer's accuracy of the RF-2 model is 0.60% and 7.58% higher than those of the SVM-1 and RF-1 models, respectively. The overall accuracy of SVM-2, BPNN-2, and RF-2 is determined as 74.82%, 83.67%, and 88.30%, respectively, which is 3.90%, 7.77%, and 7.45% higher than that of SVM-1, BPNN-1, and RF-1. Moreover, the Kappa coefficients of SVM-2, BPNN-2, and RF-2 are 0.680, 0.795, and 0.853, respectively, exceeding those of SVM-1, BPNN-1, and RF-1.

Table 5. The classification accuracy of arecanut based on different classification models with different feature subsets.

Model	Omission Error/%	Commission Error/%	User's Accuracy/%	Producer's Accuracy/%	Overall Accuracy/%	Kappa Coefficient
SVM-1	24.24	27.54	72.46	75.76	70.92	0.630
BPNN-1	15.15	18.84	81.16	84.85	75.90	0.698
RF-1	13.64	8.06	91.94	86.36	80.85	0.760
SVM-2	19.70	17.19	82.81	83.30	74.82	0.680
BPNN-2	15.15	12.50	87.50	84.85	83.67	0.795
RF-2	6.06	7.46	92.54	93.94	88.30	0.853

In order to compare the extraction effects of SVM, BPNN, and RF on the arecanut planting area, we further compared and analyzed the classification results of SVM-2, BPNN-2, and RF-2 following feature space optimization (Table 5). RF-2 is observed to have the highest overall accuracy, improving on those of BPNN-2 and SVM-2 by 5.53% and 18.02%, respectively. In summary, the classification model following feature space optimization has the ability to improve the extraction accuracy of the arecanut planting area, with the feature optimized RF 2 model identified as the most suitable for arecanut planting information extraction, effectively improving the extraction accuracy of arecanut.

In order to verify the influence of different classification algorithms on the extraction accuracy of arecanut, we further constructed a confusion matrix for the classification results of SVM-2, BPNN-2, and RF-2 (Table 6) and investigated the omission and misclassification of arecanut. The SVM-2 results reveal that 17.19% of the identified arecanut are misclassified forest land and farmland, while 19.70% are misclassified as forest land. The BPNN-2 and RF-2 models reduced the omission and commission errors of arecanut compared to SVM-2, with RF-2 exhibiting the lowest omission and commission errors. Thus, the optimized RF-2 model is identified to have the greatest separability for arecanut, forest, and farmland.

Table 6. Confusion matrix of the classification results based on SVM-2, BPNN-2 and RF-2 models.

Model.	Land Use Type	Water	Impervious Surface	Forest	Farmland	Arecanut	Total
SVM-2	Water	49	0	0	0	0	49
	Impervious surface	0	50	0	0	0	50
	Forest	1	0	59	46	13	119
	Farmland	0	0	0	0	0	0
	Arecanut	0	0	7	4	53	64
	Total	50	50	66	50	66	282
BPNN-2	Water	49	0	0	0	0	49
	Impervious surface	0	50	0	0	0	50
	Forest	0	0	50	15	4	69
	Farmland	1	0	12	31	6	50
	Arecanut	0	0	4	4	56	64
	Total	50	50	66	50	66	282
RF-2	Water	49	0	0	0	0	49
	Impervious surface	0	50	0	0	0	50
	Forest	0	0	57	17	1	75
	Farmland	1	0	6	31	3	41
	Arecanut	0	0	3	2	62	67
	Total	50	50	66	50	66	282

3.3. *Regional Application*

In order to visually compare the classification effect of the study area images under different methods, we selected the central area of the study area image and employed SVM-2, BPNN-2, and RF-2 to determine the distribution map of the arecanut extraction results on a regional scale (Figure 3). Table 6 and Figure 3 reveal that the SVM-2 model has a serious leakage of cultivated land in the study area, almost all of which is classified as forest and arecanut. Furthermore, a large extent of forest is wrongly divided into arecanut. The BPNN model effectively overcomes the mixed separation of farmland and other vegetation, and the distinction between arecanut and forest is more obvious. The classification results of the RF model are generally consistent with those of the BPNN model, while the former improves on the misclassification of farmland in the northern region. In summary, based on the feature variables following RF feature optimization, the application of the RF method can extract the arecanut planting area in the study region more effectively compared to the BPNN and SVM models.

Arecanut planting area extraction models based on the support vector machine (SVM), BP neural network (BPNN), and random forest (RF) classification algorithms were then constructed. The overall classification accuracies of the SVM, BPNN, and RF models optimized by the RF features were determined as 74.82%, 83.67%, and 88.30%, with Kappa coefficients of 0.680, 0.795, and 0.853, respectively.

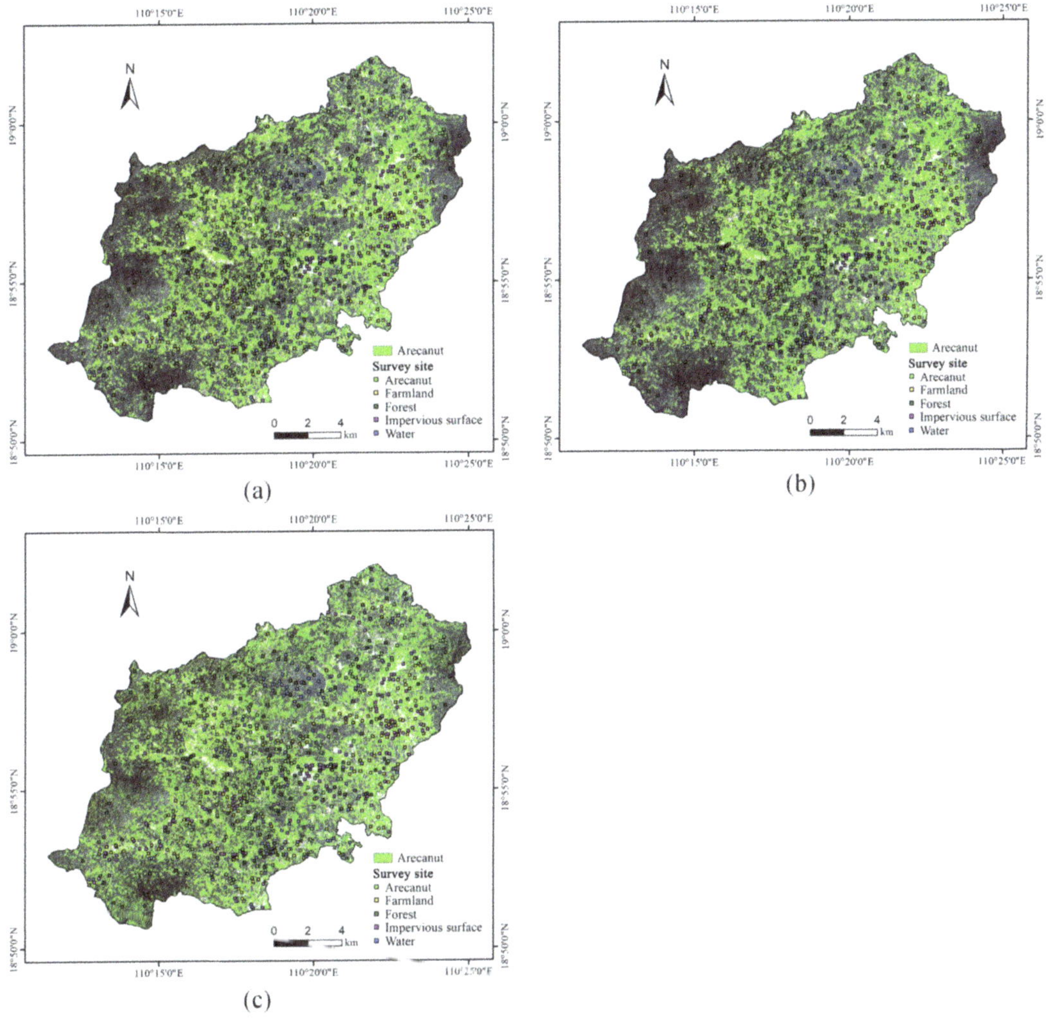

Figure 3. The distribution map of arecanut extracted based on different classification models with the feature space optimization of PlanetScope imagery. (**a**) SVM-2, (**b**) BPNN-2, and (**c**) RF-2.

4. Discussion

The fusion of spectral and textural features to construct a classification model presented in the current paper has been demonstrated to achieve promising results in the extraction of arecanut planting area. However, numerous spectral features have not been considered. Follow-up research should include additional spectral features based on the extraction method of arecanut planting area employed in this paper. In addition, more data sources should be adopted to construct a more accurate method for the extraction of arecanut planting area.

The selection of characteristic variables is crucial for the construction of a classification model. In particular, the presence of irrelevant, weakly related, or redundant features in the primary selected features will directly affect the classification accuracy and generalization ability of the model [33]. Therefore, feature selection is required to remove such features. The RF feature variable optimization algorithm can determine each feature variable weight, reducing the redundancy of the feature variables and improving the classification accuracy and generalization ability. However, this optimization approach does not consider the correlation between various features, thus further improvement is required. Future research will focus on determining the feature correlations while removing the redundancy between the features. In addition, simpler and more efficient feature selection algorithms are required for the model input selection.

The choice of modeling method affects the accuracy of the classification model. Although SVM, BPNN, and RF have strong applications in research, they are associated with several limitations. For example, although the SVM method is able to deal with various nonlinear problems through the selection of the kernel function, determining the kernel function and related parameters proves to be a difficult task, thus restricting its application [34]. The work presented in the current paper is not based on certain theoretical standards and only the linear kernel function is selected. The next step should consider other kernel functions in order to select the optimal function, which can then allow for the further optimization of the model parameters to obtain a higher precision model. The BPNN method has strong nonlinear fitting and generalization abilities, and the established network model is stable. However, the BPNN method also faces limitations, for example, the accurate determination of the number of hidden layer nodes. In particular, the network fails to converge for small node numbers and the fault tolerance is poor, while for a large number of nodes, the network has a long learning time and is prone to overfitting. Although RF has a strong tolerance to noise and is not prone to overfitting, its parameters are more complicated and features with more value divisions are likely to have a greater impact on RF decision-making, thereby affecting the accuracy of the model. Determining how to improve these methods is reserved for future work.

5. Conclusions

Current methods based on low- and medium-resolution satellite images are not able to meet the demand for the high-precision extraction of arecanut area in Hainan due to the cloudy and rainy climate and severe land fragmentation. In the current paper, a high-precision extraction method for arecanut planting area was proposed based on image feature space optimization using PlanetScope satellite imagery. Results demonstrate the ability of the spectral and texture features of PlanetScope satellite data to effectively extract the planting distribution of arecanut. The Kappa coefficients of the SVM, BPNN, and RF models following the RF feature optimization were determined as 0.680, 0.795, and 0.853, with overall classification accuracies of 74.82%, 83.67%, and 88.30%, respectively. The application of feature optimization improves the overall accuracy by 3.90%, 7.77% and 7.45%, respectively. This indicates the strong applicability of feature space optimization based on PlanetScope satellite imagery for the extraction of arecanut planting area. The research results provide theoretical and technical references for the remote sensing extraction of agricultural and forestry information.

Author Contributions: Conceptualization, information analysis, and writing of original draft, Y.J. and J.G.; methodology, W.H. and J.Z.; software, B.C.; investigation, writing—review and editing and funding acquisition, H.Y. All authors have read and agreed to the published version of the manuscript.

Funding: This research was funded by Hainan Provincial Major Science and Technology Program of China (ZDKJ2019006); Youth Innovation Promotion Association CAS (2021119); Future Star Talent Program of Aerospace Information Research Institute, Chinese Academy of Sciences (2020KTY-WLZX08); National special support program for high-level personnel recruitment (Wenjiang Huang).

Institutional Review Board Statement: Not applicable.

Informed Consent Statement: Not applicable.

Data Availability Statement: The data are available from the corresponding author, at reasonable request.

Conflicts of Interest: The authors declare no conflict of interest.

References

1. Graham, P. Traditional medical treatments III: Betel nut (Areca catechu). *Ann. ACTM Int. J. Trop. Travel Med.* **2005**, *6*, 13–14. [CrossRef]
2. Hainan Provincial Bureau of Statistics, Survey Office of National Bureau of Statistics in Hainan. *Hainan Statistical Yearbook 2020*; China Statistical Publishing House: Beijing, China, 2020. (In Chinese)
3. Sun, H.; Gong, M. Current development status and countermeasures of arecanut planting and processing industry in Hainan. *Chin. J. Trop. Agric.* **2019**, *39*, 91–94, (In Chinese with English abstract). [CrossRef]
4. Noguchi, N.; O'Brien, J.G. Remote sensing technology for precision agriculture. *Environ. Control Biol.* **2003**, *41*, 107–120. [CrossRef]
5. Hayes, M.J.; Decker, W.L. Using satellite and real-time weather data to predict maize production. *Int. J. Biometeorol.* **1998**, *42*, 10–15. [CrossRef]
6. Onojeghuo, A.O.; Blackburna, G.A.; Huang, J.; Kindredc, D.; Huang, W. Applications of satellite 'hyper-sensing' in Chinese agriculture: Challenges and opportunities. *Int. J. Appl. Earth Obs. Geoinf.* **2018**, *64*, 62–86. [CrossRef]
7. Zheng, H.; Zhou, X.; He, J.; Yao, X.; Tian, Y. Early season detection of rice plants using rgb, nir-g-b and multispectral images from unmanned aerial vehicle (UAV). *Comput. Electron. Agric.* **2020**, *169*, 105223. [CrossRef]
8. Shen, K.; Li, W.; Pei, Z.; Fei, W.; Sun, G.; Zhang, X.; Chen, X.; Ma, S. Crop area estimation from UAV transect and MSR image data using spatial sampling method. *Procedia Environ. Sci.* **2015**, *26*, 95–100. [CrossRef]
9. Fu, Z.; Jiang, J.; Gao, Y.; Krienke, B.; Liu, X. Wheat growth monitoring and yield estimation based on multi-rotor unmanned aerial vehicle. *Remote Sens.* **2020**, *12*, 508. [CrossRef]
10. Ládai, A.D.; Barsi, Á. Analysing automatic satellite image classification in the desert of Sudan. *Period. Polytech. Civ. Eng.* **2008**, *52*, 23–27. [CrossRef]
11. Pan, Y.; Li, L.; Zhang, J.; Liang, S.; Zhu, X.; Sulla-Menashe, D. Winter wheat area estimation from modis-evi time series data using the crop proportion phenology index. *Remote Sens. Environ.* **2012**, *119*, 232–242. [CrossRef]
12. Zhang, M.; Lin, H. Object-based rice mapping using time-series and phenological data. *Adv. Space Res.* **2019**, *63*, 190–202. [CrossRef]
13. Liu, J.; Tian, Q.; Huang, Y.; Du, L.; Wang, L. Extraction of the corn planting area based on multi-temporal HJ-1 satellite data. In Proceedings of the 19th International Conference on Geoinformatics, Shanghai, China, 24–26 June 2011; IEEE: Piscataway, NJ, USA, 2011. [CrossRef]
14. IUSS Working Group WRB. *World Reference Base for Soil Resources 2006*; World Soil Resources Reports No. 103; FAO: Rome, Italy, 2006.
15. Planet Team. *Planet Application Program Interface: In Space for Life on Earth*; Planet Labs Inc.: San Francisco, CA, USA, 2018. Available online: https://api.planet.com (accessed on 2 March 2018).
16. Huang, Z.; Cao, C.; Chen, W.; Xu, M.; Dang, Y.; Singh, R.P.; Bashir, B.; Xie, B.; Lin, X. Remote sensing monitoring of vegetation dynamic changes after fire in the Greater Hinggan Mountain Area: The algorithm and application for eliminating phenological impacts. *Remote Sens.* **2020**, *12*, 156. [CrossRef]
17. Zhang, J.; Pu, R.; Yuan, L.; Huang, W.; Nie, C.; Yang, G. Integrating remotely sensed and meteorological observations to forecast wheat powdery mildew at a regional scale. *IEEE J. Sel. Top. Appl. Earth Observ. Remote Sens.* **2017**, *7*, 4328–4339. [CrossRef]
18. Zeng, C.; Binding, C. The effect of mineral sediments on satellite chlorophyll-a retrievals from line-height algorithms using red and near-infrared bands. *Remote Sens.* **2019**, *11*, 2306. [CrossRef]
19. Jordan, C.F. Derivation of leaf area index from quality of light on the forest floor. *Ecology* **1969**, *50*, 663–666. [CrossRef]
20. Qi, J.; Chehbouni, A.; Huete, A.R.; Kerr, Y.H.; Sorooshian, S. A modified soil adjusted vegetation index. *Remote Sens. Environ.* **1994**, *48*, 119–126. [CrossRef]
21. Rouse, J.W.; Haas, R.H.; Deering, D.W.; Schell, J.A. *Monitoring the Vernal Advancement of Retrogradation of Natural Vegetation*; NASA/GSFC, Type III, Final Report; NASA/GSFC: Greenbelt, MD, USA, 1974. Available online: https://ntrs.nasa.gov/api/citations/19740004927/downloads/19740004927.pdf (accessed on 20 December 2020).
22. Huete, A.R.; Jackson, R.D. Suitability of spectral indices for evaluating vegetation characteristics on arid rangelands. *Remote Sens. Environ.* **1987**, *23*, 213–232. [CrossRef]
23. Ren, C. *Study on Extraction of Mango Forest with High Resolution Remote Sensing Image*; Institute of Remote Sensing and Digital Earth, Chinese Academic of Sciences: Beijing, China, 2017. (In Chinese with English Abstract)
24. Soares, J.V.; Rennó, C.D.; Formaggio, A.R.; Yanasse, C.C.F.; CesarFrery, A. An investigation of the selection of texture features for crop discrimination using SAR imagery. *Remote Sens. Environ.* **1997**, *59*, 234–247. [CrossRef]
25. Haralick, R.M.; Shanmugam, K.; Dinstein, I. Textural features for image classification. *Stud. Media Commun.* **1973**, *3*, 610–621. [CrossRef]
26. Kwok, S.W.; Carter, C. Multiple decision trees. *Mach. Intell.* **2013**, *9*, 327–335. [CrossRef]
27. Pavlov, Y.L. Limit distributions of the height of a random forest. *Theor. Probab. Appl.* **2006**, *28*, 471–480. [CrossRef]
28. Rodriguez-Galiano, V.F.; Chica-Olmo, M.; Abarca-Hernandez, F.; Atkinson, P.M.; Jeganathan, C. Random forest classification of mediterranean land cover using multi-seasonal imagery and multi-seasonal texture. *Remote Sens. Environ.* **2012**, *121*, 93–107. [CrossRef]
29. Wang, S.; Zhang, N.; Wu, L.; Wang, Y. Wind speed forecasting based on the hybrid ensemble empirical mode decomposition and GA-BP neural network method. *Renew. Energy* **2016**, *94*, 629–636. [CrossRef]

30. Mavroforakis, M.E.; Theodoridis, S. A geometric approach to support vector machine (SVM) classification. *IEEE Trans. Neural Networks* **2006**, *17*, 671–682. [CrossRef] [PubMed]
31. Lesser, B.; Mücke, M.; Gansterer, W.N. Effects of reduced precision on floating-point SVM classification accuracy. *Procedia Comput. Sci.* **2011**, *4*, 508–517. [CrossRef]
32. He, Q.; Xie, Z.; Hu, Q.; Wu, C. Neighborhood based sample and feature selection for SVM classification learning. *Neurocomputing* **2011**, *74*, 1585–1594. [CrossRef]
33. Dash, M.; Liu, H. Feature selection for classification. *Intell. Data Anal.* **1997**, *1*, 131–156. [CrossRef]
34. Wang, L.; Zhang, S. Incorporation of texture information in a SVM method for classifying salt cedar in western China. *Remote Sens. Lett.* **2014**, *5*, 501–510. [CrossRef]

agriculture

MDPI

Article

Automated Chicken Counting in Surveillance Camera Environments Based on the Point Supervision Algorithm: LC-DenseFCN

Liangben Cao [1,†], Zihan Xiao [1,†], Xianghui Liao [1], Yuanzhou Yao [1], Kangjie Wu [1], Jiong Mu [1,2], Jun Li [1,2] and Haibo Pu [1,2,*]

[1] College of Information Engineering, Sichuan Agricultural University, Ya'an 625000, China; caoliangben@stu.sicau.edu.cn (L.C.); xiaozihan@stu.sicau.edu.cn (Z.X.); liaoxianghui@stu.sicau.edu.cn (X.L.); yaoyuanzhou@stu.sicau.edu.cn (Y.Y.); wukangjie@stu.sicau.edu.cn (K.W.); jmu@sicau.edu.cn (J.M.); lijun@sicau.edu.cn (J.L.)

[2] Sichuan Key Laboratory of Agricultural Information Engineering, Ya'an 625000, China

* Correspondence: puhb@sicau.edu.cn; Tel.: +86-182-2758-9988

† These authors contributed to the work equally and should be regarded as co-first authors.

Abstract: The density of a chicken population has a great influence on the health and growth of the chickens. For free-range chicken producers, an appropriate population density can increase their economic benefit and be utilized for estimating the economic value of the flock. However, it is very difficult to calculate the density of chickens quickly and accurately because of the complicated environmental background and the dynamic number of chickens. Therefore, we propose an automated method for quickly and accurately counting the number of chickens on a chicken farm, rather than doing so manually. The contributions of this paper are twofold: (1) we innovatively designed a full convolutional network—DenseFCN—and counted the chickens in an image using the method of point supervision, which achieved an accuracy of 93.84% and 9.27 frames per second (FPS); (2) the point supervision method was used to detect the density of chickens. Compared with the current mainstream object detection method, the higher effectiveness of this method was proven. From the performance evaluation of the algorithm, the proposed method is practical for measuring the density statistics of chickens in a farm environment and provides a new feasible tool for the density estimation of farm poultry breeding.

Keywords: deep learning; aquaculture automation; computer vision; chicken detection

Citation: Cao, L.; Xiao, Z.; Liao, X.; Yao, Y.; Wu, K.; Mu, J.; Li, J.; Pu, H. Automated Chicken Counting in Surveillance Camera Environments Based on the Point Supervision Algorithm: LC-DenseFCN. *Agriculture* **2021**, *11*, 493. https://doi.org/10.3390/agriculture11060493

Academic Editors: Gniewko Niedbała and Sebastian Kujawa

Received: 12 April 2021
Accepted: 20 May 2021
Published: 26 May 2021

1. Introduction

Refined agriculture is a significant trend of agricultural development for the future, among which agricultural informatization is a development direction vigorously advocated for at present [1]. Realizing the informatization of the agricultural industry is helpful to promote the intellectualization of agricultural management, increase the output of agricultural products, and obtain greater economic benefits [2,3].

The aquaculture industry also has great prospects in the trend of agricultural informatization. As with other production technologies in the agricultural field, the main goal of intelligent farming is to increase productivity and take operational measures related to the environment to reduce costs [4–7]. In intelligent aquaculture, some of the main parameters concerned include temperature, humidity, light intensity, and population density. Welfare considerations affect the sale of poultry products, and breeding density is seen as a priority for animal welfare [8]. For broiler breeding, compared with cage rearing, group rearing is more reliable [9]. An appropriate breeding density will improve the growth performance of chickens as well as their immunity and carcass yield [10–12]. However, there are few studies on the rapid monitoring of chicken population density. Most methods of monitoring chicken populations involve studying the morphology of the chickens and observing their physiological behaviors. For example, Yao Y. et al. [13] designed a classifier to determine the sex of chickens. A. Aydin [14] used 3D vision cameras to observe chickens' walking posture to assess whether they had lameness or other behavioral problems. Tu X. et al. [9] developed a real-time automatic feed weight monitoring system which could automatically detect the intake and weight of individual turkeys. Chakchai so-in et al. [15] used traditional image processing methods and four algorithms, such as K-means, to classify images of chickens taken by cameras arranged in a chicken farm; the accuracy of the logic filter algorithm was the highest, achieving an accuracy of 80%. Similarly to Chakchai so-in, we also monitored the density of chickens and counted them from images taken with a camera in a chicken farm. The difference is that we chose to use a deep learning method to process the images. In free-range chicken farms, the traditional method of counting chickens is often manual, which has the following challenges:

1. The motility of the flock makes it difficult to perform a complete count of the flock;
2. The process is time consuming.

Deep learning has been widely used in different environments of many fields and has proven to be a very efficient method. Of course, it is also widely used in various agriculture-related fields [16], such as animal density detection and animal counting. Beibei Xu et al. [17] used drones to collect images of cattle raised on a large scale and used the method of instance segmentation to detect and count the photos of these cattle, with the highest accuracy of 94%. Hung Nguyen et al. [18] used object detection to identify and count wild animals, and they were able to identify the animals with 96% accuracy by setting up a fixed camera position to photograph the path. Mengxiao Tian et al. [19] used the density map method to count pigs in a pigsty and finally achieved a mean absolute error (MAE) of 1.67. According to a recent survey of deep learning in agriculture [20], deep learning has a wide application value in poultry breeding. This article focuses on the detection and counting of free-range chickens on a poultry farm to achieve rapid management of the poultry farm.

In this study, we used a deep learning method to automate the counting of free-range chickens on a poultry farm. The chicken farm was indoors and covered by greenhouses; therefore, we used surveillance cameras to collect data from the chicken farm. Due to the high density of the chicken population, there was a high degree of overlap between the individual chickens, and the overlapping chicken population made it difficult to distinguish the individuals. Therefore, the object detection and density map methods were not suitable for this task; as such, we chose to use the method of point supervision for processing. We innovatively combined the segmentation loss function designed by Issam H. Laradji et al. [21] and DenseNet [22] to create a new semantic segmentation model, LC-DenseFCN, to meet the actual requirements.

2. Related Work

2.1. DenseNet

Since ResNet [23] was put forward, a variety of ResNet networks have emerged in an endless stream, and each has its own characteristics; the network performance has also been improved to some extent. With its excellent performance, the Dense Convolutional Network (DenseNet) won the best paper of IEEE Conference on Computer Vision and Pattern Recognition (CVPR) 2017. The DenseNet (Dense Convolutional Network) proposed in the paper is mainly a comparison with ResNet and Inception Network, which has some reference in thought but is a new structure. The network structure is not complicated, but it is effective and comprehensively outperforms ResNet in the CIFAR index. It can be said that DenseNet has absorbed the best part of ResNet and has had more innovative work conducted on it, further improving the network performance.

2.2. FCN

The fully convolutional network (FCN) [24] is the pioneering work that applied the convolutional neural network (CNN) structure to the field of image semantic segmentation and achieved outstanding results; accordingly, it was awarded the Best Paper Honorable

Mention of CVPR in 2015. CNNs have been driving progress in the field of image recognition in recent years. Whether it is whole-picture classification or object detection, key point detection has been greatly developed with the help of CNNs. However, image semantic segmentation is different from the above task, which is a space-intensive prediction task. In other words, it needs to predict the categories of all pixels in an image. FCN trains an end-to-end, point-to-point network for semantic segmentation and represents the state of the art. This is the first time that an end-to-end FCN has been trained for pixel-level prediction.

2.3. Localization-Based Counting Algorithm

Issam H. Laradji et al. proposed a positioning counting method based on point supervision, i.e., an object counting model in which the dataset labels are all point labels of unit pixels [21]. In this model, a compound loss function (LC-Loss) is designed for monitoring counting. The structure of the model itself can use most of the semantic segmentation models of FCN, so there is only a need to modify the function to achieve a good counting effect. This model is suitable for object counting tasks in the order of 10 or 100, and the annotation of datasets is relatively simple. Its backbone network structure is not complicated and has great value in migration applications. In the localization-based counting algorithm model, the backbone networks used are LC-FCN [21] and LC-ResFCN [21]. The reasons for choosing to use both networks are as follows: (1) FCN is relatively simple and has a fast segmentation result in semantic segmentation; (2) based on FCN, combined with ResNET-50 network, the accuracy can be relatively improved, and due to the unique residual structure of ResNet, the computing speed will not be greatly affected.

2.4. Object Detection Algorithms for Contrast Experiment

In recent years, great progress has been made in object detection algorithms, and current object detection algorithms are mainly divided into two categories. The first are two-stage algorithms, such as R-CNN [25], Fast R-CNN [26], Faster R-CNN [27], and Mask R-CNN [28], which rely on a CNN to generate a region proposal network first, and then classify and regress on the region proposal network. The other are one-stage algorithms, which can directly predict the bounding box and class probability from the input image by using only a convolutional neural network structure. The three algorithms used in this paper are Mask R-CNN, YOLOv3 [29], and EfficientDet [30].

The early YOLOv1 [31] has several shortcomings: (1) the input size is fixed, and the output layer is a fully connected layer; (2) it is not suitable for detecting small objects—although each grid can predict many bounding boxes, only the bounding box with the highest IoU (intersection over union) is selected as the object detection output. To address these problems, YOLOv2 [32] was proposed. In YOLOv2, by using a convolution layer, the output layer replaces the fully connected layer of YOLOv1. YOLOv2 cancels all dropout and uses batch normalization in the convolution layer. Compared with the previous YOLOv1 and YOLOv2, YOLOv3 shows great improvements. The backbone network of YOLOv3 is Darknet53, and its most important feature is the use of the residual network structure. Furthermore, the DarknetConv2D structure is used in the convolution part. As a result, YOLOv3 largely improves the detection accuracy while maintaining a high detection speed. Based on the excellent performance of Yolov3, Yao Y. et al. [13] used the YOLOv3 algorithm to detect the images of chickens, to achieve the purpose of counting chickens and segmenting single chickens.

Mask R-CNN is a compact, versatile object detection framework. It can not only detect the objects in an image, but also give a high-quality detection result for each object. Beibei Xu et al. [17] used Mask R-CNN to detect and count cattle, and achieved good results in the field of animal detection and counting. Mask R-CNN extends the Faster R-CNN by parallel adding a new branch to the bounding box recognition branch for predicting the object mask. Mask R-CNN (an extension of Faster R-CNN) which also allows for instance segmentation (associating specific image pixels to the detected object) is selected for further study. Instance segmentation allows not only the detection of each animal, but also the delineation of its boundaries within the image, thereby allowing further potential applications for livestock welfare monitoring. The benefits provided by instance segmentation allow for diverse future applications including estimations of animal pose and direction of travel. In this work, however, we constrain interest to the object detector capabilities of Mask R-CNN.

The EfficientDet algorithm was proposed by the Google brain team. By means of improving the multiple dimensioned feature fusion structure of FPN and borrowing ideas from the EfficientNet [33] model scaling method for reference, its contribution mainly includes two points: (1) EfficientDet proposed the BiFPN network, which allows simple and fast multi-scale feature fusion; (2) a compound scaling method was proposed, which can uniformly scale the resolution, depth, and width of all the backbone, characteristic and predictive networks. Based on these improvements, EfficientDet can provide high-accuracy and rapid testing with fewer parameters.

3. Materials and Methods

3.1. Overview of Our Framework

This section describes the pipeline which is proposed for processing RGB images that are captured by camaras to detect and count chickens using a deep learning algorithm. The structure of chicken detection and counting in camera images is illustrated in Figure 1. The RGB images acquired by the camera are used to extract features from the full image using DenseNet (our DenseFCN network is a fully convolutional network; therefore, we removed the full connection layer for processing pixel-level predictions). Then, a deconvolution operation is carried out to up-sample the feature map of the last convolutional layer and restore it to the same size as the input image; in this way, a prediction can be generated for each pixel while retaining the spatial information in the original input image. Finally, the feature map of the up-sampled is classified, pixel by pixel. The ground truth was annotated manually for every chicken in the training sets (we used LC-Loss as the loss function for the network; therefore, we only needed to mark one pixel of the chicken); then, network training was performed after labeling for parameter optimization, followed by chicken detection and counting in testing sets.

3.2. Dataset Preparation and Pre-Processing

The provision of sufficient and diverse chicken population datasets was a pre-requisite for this work. As the research area is at an advanced level in the industry, there was a lack of appropriate publicly available datasets; therefore, we completed the whole process of data collection by ourselves. The images used in this research were taken at Sichuan Meishan City Song's chicken farm. Part of the chicken farm is indoors and covered by greenhouses; therefore, we used surveillance cameras to collect data on the environment of the free-range chicken farm. We installed multiple Hikvision DS-IPC-B12-I cameras in the chicken farm and filmed it from different locations. The shooting angle was mainly from top to bottom, and the collection time was in March 2020. The collected videos were pre-processed, the monitoring videos were framed, the invalid fragments were deleted, and the key frames were then extracted. Some images that were not easy to identify in morphology, did not have obvious features, or were too blurred were abandoned, and most of the images obtained were clear. However, due to the movement characteristics of chickens, a small number of images were relatively fuzzy. In order to improve the robustness of the model, we added some relatively fuzzy images into the dataset. In addition, the dataset also contained some daily behaviors of chickens, such as eating, drinking, and jumping, which improved the richness and diversity of the dataset. We finally selected 1200 valid images of chickens with a resolution of 1080P as the dataset, including images of chickens with different densities. The detection dataset was divided in a ratio of 5:1 to build the training set and the test set. For data annotation, we used the deep learning annotation tool LabelMe to manually label the images and marked the position of a single chicken in the image with RGB (255.0.0) pixels.

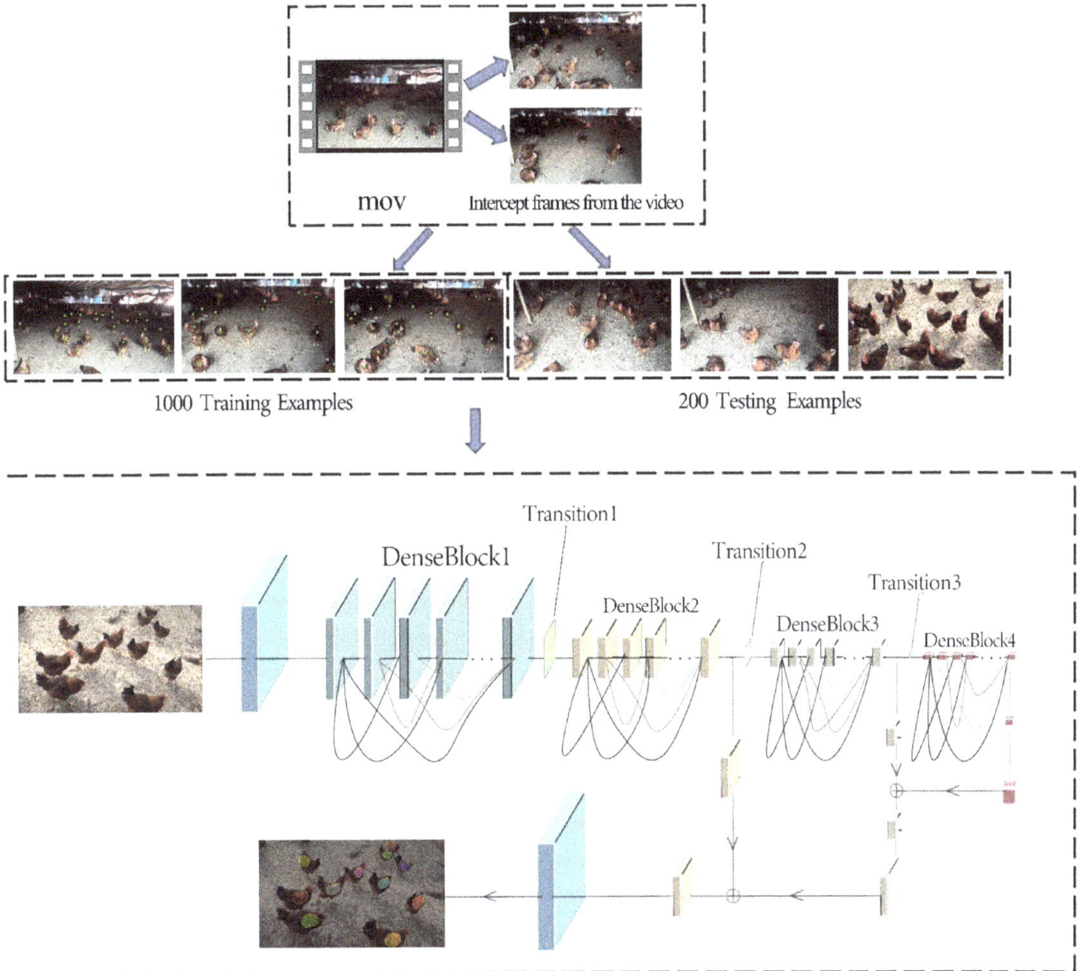

Figure 1. The structure of the chicken detection and counting algorithm.

The environment in which chickens are raised is complex and sometimes affected by inclement weather, such as low temperatures and heavy rain. In the process of digital image acquisition, coding, transmission, and processing, noise always exists [34]. Due to the advanced equipment we used, and the good weather conditions during the data collection stage, there was less noise in the collected dataset. In an actual scene, the equipment is often affected by aging circuits and the environment, and the noise level is very high. In order to improve the robustness of the model, we randomly selected 200 images from the dataset and added salt and pepper noise to simulate a more realistic shooting environment. Noise brings a lot of difficulties to image processing, which has a direct impact on image segmentation, feature extraction and image recognition. Therefore, it was necessary to filter the collected images. Median filtering is based on the theory of order statistics of a nonlinear signal processing technology and can effectively restrain noise. Its basic principle is to replace the value of a point in a digital image or sequence with the median value of each point in a neighborhood of the point, so that it becomes the true value close to the surrounding pixel value, to eliminate the isolated noise points, which is particularly useful for speckle noise and salt and pepper noise, because it does not depend on the surroundings of those values with a typical value difference that is very large. Figure 2 shows some of the filtered images with salt and pepper noise.

Figure 2. Adding salt and pepper noise and median filtering process.

3.3. The Detection and Counting Algorithm

LC-DenseFCN is an extension of LC-FCN, which takes DenseNet as the backbone network, removes the full connection layer, and fuses and deconvolves the feature map with rich semantic information, eventually forming a fully convolutional network that can perform pixel-level prediction. Like FCN, LC-DenseFCN is divided into two stages: (1) the convolution process of feature extraction from the image; and (2) the deconvolution process of fusion and deconvolution of the extracted feature maps of different layers.

3.3.1. Convolution

The process of feature extraction mainly consists of four DenseBlocks and three Transition layers. The DenseBlock is composed of several composite functions, including batch normalization, ReLU activation function, and a convolutional layer. DenseNet is a convolutional neural network with tight connections. Any two layers in this neural network are directly connected, i.e., the input of each layer in the network is the union of the output of all previous layers, and the features learned by this layer will be directly transmitted to all subsequent layers as inputs. This tight connection only exists in the same DenseBlock, but there is no such tight connection in different DenseBlocks. This structure can reduce the network parameters, reduce gradient disappearance, and improve feature utilization.

3.3.2. Deconvolution

In general CNN structures, a pooling layer is used to reduce the size of the output image. The input image of VGG16 [35] was shrunk 32 times after pooling five times; in ResNet, some convolutional layers are also involved in the process of reducing the image size. What we needed to obtain was a segmented image with the same size as the original image; therefore, we needed to deconvolve the last layer. In the process of DenseNet's feature extraction of the image, the size of the feature image is gradually reduced. Firstly, we deconvolved the feature graph of DenseBlock4 to make its size the same as that of the feature graph of DenseBlock3 and fused them. Then, deconvolution was performed on the feature map obtained after fusion to make its size the same as that of DenseBlock2, and we then fused it with the feature map of DenseBlock2. Finally, deconvolution with a stride of 8 was performed on the fused feature image to obtain the detection results.

3.3.3. Loss Function

LC-Loss enables the model to produce an area block where each object is and finally counts the number of areas; the required monitoring signal is a position point for each object, not a bounding box. The loss function is called location-based counting loss, and it has four items. The first two require the model to give the semantic label of each pixel of the graph, while the last two require the model to learn to separate the areas with multiple objects and remove the areas without objects.

$$\mathcal{L}(S,T) = \mathcal{L}_I(S,T) + \mathcal{L}_P(S,T) + \mathcal{L}_S(S,T) + \mathcal{L}_F(S,T) \tag{1}$$

Here, S is the figure given by the ground truth (GT), and T is the output of the network. On each pixel is a softmax vector, which represents the probability that this pixel belongs to each kind of object. Finally, Argmax is taken on each output pixel to divide the region.

Image-level loss: C_e is given by the GT of this figure and the set of object classes that exist in the figure; $C_{\neg e}$ represents the set of object classes that do not exist. $S_{t_c c}$ is the maximum probability of category c of S per pixel. That is, the model is encouraged to predict c if there is class c in the GT and is penalized to predict c if there is.

$$\mathcal{L}_I(S,T) = -\frac{1}{|C_e|}\sum_{c\in C_e} log(S_{t_c c}) - \frac{1}{|C_{\neg e}|}\sum_{c\in C_{\neg e}} log(1 - S_{t_c c}) \tag{2}$$

Point-level loss: This loss is normal softmax cross-entropy loss; thus, the background category needs to be ignored, and this item will only be calculated for the marked position points.

$$\mathcal{L}_P(S,T) = -\sum_{i\in I_s} log(S_{iT_i}) \tag{3}$$

Split-level loss: The loss contains two implementation methods—the line split method and the watershed split method.

1. Line split method.

$$z(E) = \frac{1}{|E|}\sum_{i\in E} S_{i0} \tag{4}$$

Each blob in the blob set B is formed by a series of point sets around the central coordinate p of the blob. For any point in the point set BP, it will form a pairing with its surrounding points (P_i, P_j). Additionally, a line can then be used to separate P_i and P_j. The position where the lines separate will be the background with the highest probability of learning. This separates one blob from the surrounding blobs.

2. Watershed split method.

$$\mathcal{L}_S(S,T) = -\sum_{i\in T_b} \alpha_i log(S_{i0}) \tag{5}$$

Here, S_{i0} represents the probability that pixel I belongs to background 0, and a_i represents the number of pixels in each blob that belong to that blob. This loss lets the model learn so that there is a clear dividing line between the two adjacent blobs.

False positive loss: $\mathcal{L}_F$ discourages the model from predicting a blob with no point annotations in order to reduce the number of false positive predictions (FP). The loss function is defined as follows:

$$\mathcal{L}_F(S,T) = -\sum_{i\in B_{fp}} log(S_{i0})B_{fp} \tag{6}$$

where B_{fp} represents the pixels that predict which category the blob belongs to, excluding the background category, which is a circle of pixels around a point; and S_{i0} represents the probability that the category i belongs to the background category. The probability of category i belonging to the background category in the whole process optimization is 1, to achieve the purpose of removing false positive predictions, where the loss is optimized only when there is an false positive prediction; otherwise, the loss is not optimized.

3.4. Experimental Environment and Evaluation Protocol

The operating system of the experiment was Ubuntu 16.04, the deep learning framework used in all experiment was Pytorch 0.4, and all experimental results were obtained on an NVIDIA Geforce RTX 2080 super GPU (Santa Clara, CA, USA), with a video memory of 8 GB.

In this paper, there are some differences between the evaluation criteria of point supervision algorithms and object detection algorithms. Mean Absolute Error (MAE), Mean Square Error (MSE), Root-Mean-Square Error (RMSE), Mean Relative Error (MRE), and accuracy are used as the evaluation metrics of the point supervision algorithms, including LC-FCN, LC-ResFCN and LC-DenseFCN. MAE is the most commonly used index to measure accuracy. It is the average error in a more general form and also an important ruler for evaluating models in machine learning. MSE is the most commonly used error in regression loss function. It is the mean of the sum of squares of the difference between the predicted value, $f(x)$, and the target value, y. RMSE reflects the square root of the mean of the square variance between the predicted value and the actual observed value. MRE is generally used to analyze the accuracy of results. Accuracy represents the ratio of correctly predicted samples to the total number of predicted samples; the definition formula is as follows:

$$\text{Accuracy} = \frac{TP + TN}{TP + TN + FP + FN} \tag{7}$$

where *TP*, *FP*, *TN* and *FN* represent true positive, false positive, true negative and false negative, respectively.

Moreover, the precision, recall and average precision (AP) and frames per second (FPS) are utilized as the evaluation metrics of point supervision algorithms and object detection algorithms, including LC-DenseFCN, YOLOv3, Mask R-CNN and EfficientDet. The precision reflects the proportion of true predicted positive in all the predicted positive, but the recall reflects the proportion of true predicted positive in all of the positives. For the precision–recall curve [36], the larger the area enclosed by the curve, the better the performance. Another important performance index is speed; only fast speed can realize real-time detection, which was extremely important for our application scenario. A common measure of speed is FPS, i.e., the number of images that can be processed per second.

4. Results

This section presents the performance evaluation of the proposed method to detect and count chickens in different experimental settings. First, we conducted a performance comparison of the segmentation model under different backbones; then, we conducted a loss function analysis; and finally, we performed a comparison with the state-of-the-art object detection algorithms.

4.1. Performance Comparison of Segmentation Model under Different Backbones

We compared the proposed LC-DenseFCN model with two typical existing methods: LC-FCN and LC-ResFCN. We evaluated the two competing methods on the same test images collected from the chicken farm; the training process of these competing methods is shown in Figure 3. The following can be noted: (1) the loss of LC-FCN converged slowly and fluctuated greatly, while Mean Absolute Error (MAE) jittered greatly and even showed an upward trend with the increase in epoch; (2) the loss of LC-ResFCN converged faster, but the MAE was not stable enough and the jitter was obvious; (3) the MAE of the early training process oscillated, but the late convergence effect was better. The convergence speed of loss was a little slower than that of LC-ResFCN in the early stage, but the overall convergence was more stable without large fluctuations. After analysis, the deep network framework was not suitable for such a small dataset. LC-DenseFCN uses DenseNet as the backbone network, and its smaller number of parameters makes the network easier to train; therefore, LC-DenseFCN had better convergence. Moreover, we have summarized the performances in terms of error and accuracy in Table 1. As can be seen from the table, LC-DenseFCN, in this paper, achieved the best accuracy of 93.84%, and all errors were the lowest in test datasets. It is obvious that the dense connection structure and feature reuse method of DenseNet achieved better results than the other two methods in our experiment.

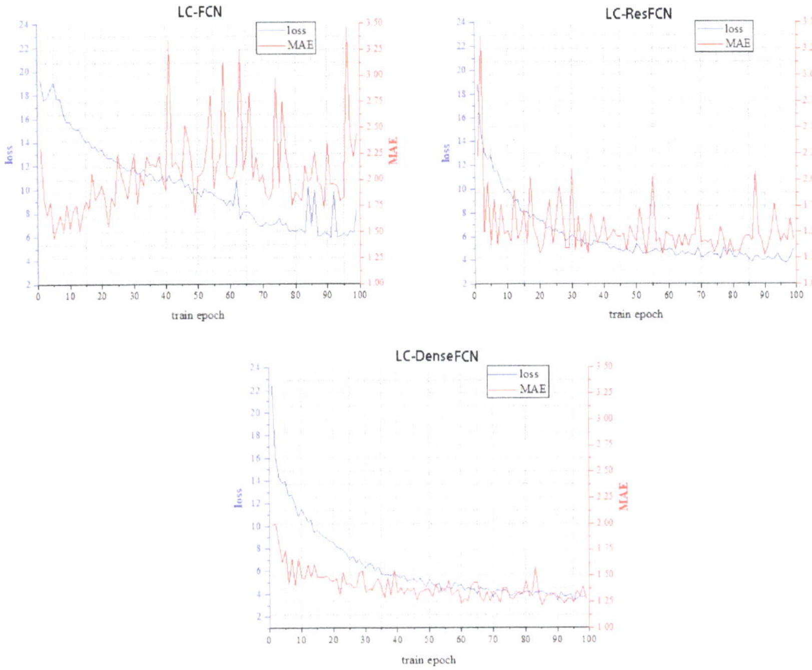

Figure 3. The point supervision algorithm training progress through 100 epochs (MAE means Mean Absolute Error).

Table 1. Accuracy and error values of three point supervision methods.

	Accuracy (%)	Mean Absolute Error	Mean Square Error	Root-Mean-Square Error	Mean Relative Error
LC-FCN	91.40	1.60	5.14	2.27	0.86
LC-ResFCN	92.32	1.38	3.71	1.93	0.77
LC-DenseFCN	93.84	1.30	2.25	1.51	0.67

4.2. Loss Function Analysis

In this section, we present assessments of the effect of each term of the loss function on the counting and localization results. As can be seen from Figure 4b, the model using only two terms (the image-level loss $\mathcal{L}_I$ and the point-level loss $\mathcal{L}_P$) resulted in a single blob that grouped many object instances together. From Figure 5, we can see that this performed poorly in terms of the MAE and counting accuracy. Then, we introduced the split-level loss function $\mathcal{L}_S$ to encourage the model to predict blobs that did not contain more than one point-annotation. As shown in Figure 4c, the model after adding $\mathcal{L}_S$ predicted several blobs as object instances rather than one large single blob. However, because $\mathcal{L}_I + \mathcal{L}_P + \mathcal{L}_S$ did not penalize the model for predicting blobs with no point annotations, it caused the model to make false predictions, which also resulted in a model counting accuracy of only 0.68 (see Figure 5b). Finally, we introduced the false positive loss $\mathcal{L}_F$ which discouraged the model from predicting blobs with no point annotations. By adding this loss term to the optimization, LC-DenseFCN achieved significant improvements, as seen in the qualitative and quantitative results, which are shown in Figures 4d and 5. Furthermore, the results of Figures 4 and 5 also verify the role played by each part of LC-Loss on the whole network. Each term of the loss function plays a corresponding role, which is well adapted to the model that we proposed, and a complete LC-Loss can make our network achieve the best performance.

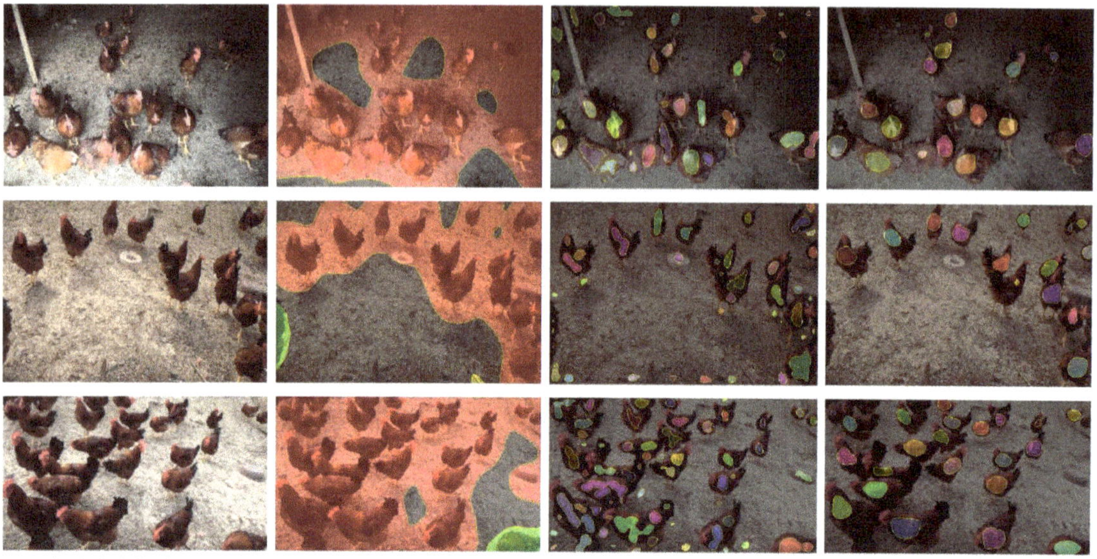

Figure 4. Qualitative results of LC-DenseFCN trained with different terms of the proposed loss function. (**a**) Test images obtained from our test dataset. (**b**) Prediction results using only image-level and point-level loss terms. (**c**) Prediction results using image-level, point-level and split-level loss terms. (**d**) Prediction results trained with the full proposed loss function.

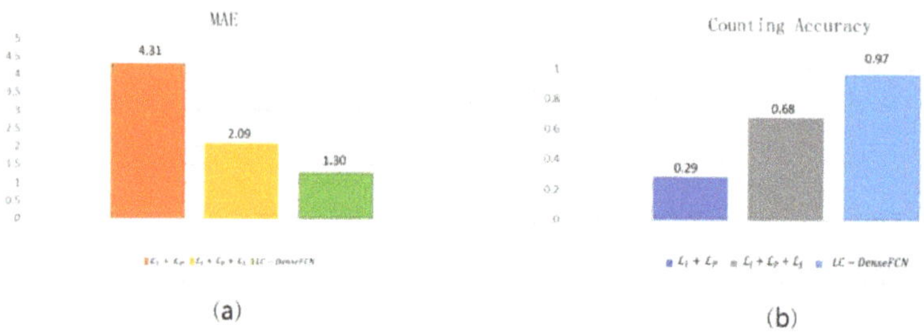

Figure 5. Comparison of different parts of the proposed loss function for counting and localization performance. (**a**) The mean absolute error (MAE) performance of the three loss functions. (**b**) The counting accuracy of the three loss functions.

4.3. Comparison with State-of-the-Art Object Detection Algorithms

Object counting is a computer vision task that can be applied to surveillance and vehicle counting. In the past, object detection has been regarded as the mainstream method for counting tasks. We compared LC-DenseFCN with three state-of-the-art object detection algorithms, namely, YOLOv3, EfficientDet and Mask R-CNN. To allow the readers to visually compare the results of the different methods, the predictions processed by the competing methods are visualized in Figure 6. As shown, all four methods used in the experiment achieved a good performance for the less dense image of chickens. However, object detection methods did not perform well for dense or heavily overlapped parts of the image. After a preliminary analysis, the purpose of the object detection method is to frame the object out; as such, it requires more complete characteristics of the object, resulting in a poor detection effect for multiple overlapping objects. As a point supervision method, LC-DenseFCN does not need the full features of the object; hence, it can perform well in processing images where multiple objects overlap.

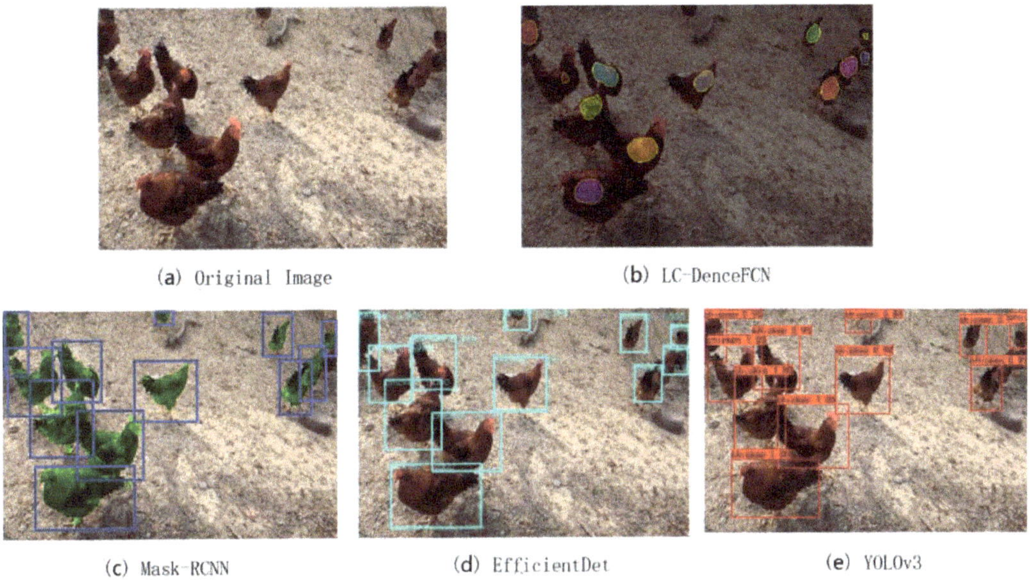

Figure 6. Predictions on the test images for four detection algorithms.

It can be seen from Figure 7 there is an inverse relationship between the precision and recall, which means the higher the precision, the lower the recall. However, we expected to detect all the target objects, which means higher recall rates, and also expected higher precision rates of the detected objects. At around recall = 0.82, 0.93, 0.95, 0.96, the inflection point appeared in all four curves known as balance points where the precision and recall achieved the best values, and then the precision dropped sharply.

The counting experiment results of the test image set are shown in Table 2. It can be seen from the table that the counting accuracy of LC-DenseFCN reached 0.97, which was the highest among the four methods, and the values of FPS and AP were also higher than those of the other three object detection methods. Among the three object detection algorithms, the counting accuracy of EfficientDet reached 0.96, which was only 0.01% lower than that of LC-DenseFCN. Moreover, the AP of EfficientDet was also very close to that of LC-DenseFCN, but the FPS value was only 4.33. The counting accuracy and AP of Mask R-CNN were not good enough, because compared with other object detection algorithms, it also needed to segment the objects, which increased the difficulty of detection and resulted in a low FPS. If using cameras to observe a chicken farm in real time, a high detection speed is very necessary. The result indicates that LC-DenseFCN is most effective in real-world datasets, because such datasets consist of data from different, complex scenes with different density distributions and different degrees of occlusion.

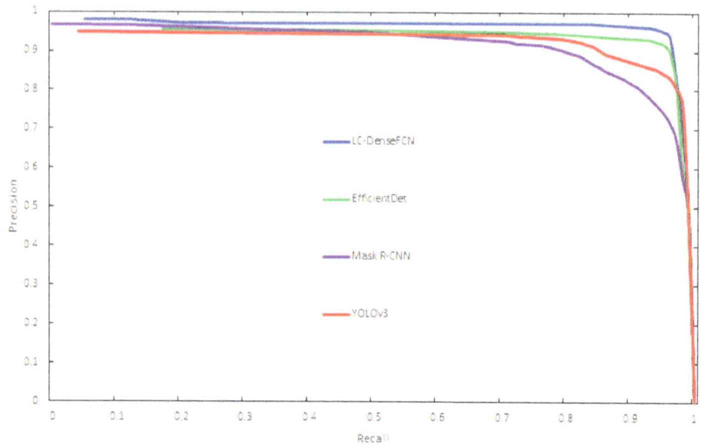

Figure 7. Comparisons of precision–recall curves for four detection cases.

Table 2. Comparison of counting results with three competing methods.

	Ground Truth	Detected	Counting Accuracy	Counting Error	Average Precision	Frames Per Second
LC-DenseFCN	4872	4748	0.97	0.03	0.94	9.27
YOLOv3	4872	4396	0.90	0.10	0.89	5.11
EfficientDet	4872	4697	0.96	0.04	0.92	4.33
Mask R-CNN	4872	4213	0.86	0.14	0.86	2.04

5. Discussion

In this paper, we have proposed LC-DenseFCN, a deep learning algorithm, for the detection and counting of chickens from camera images. The key novelty of the work is the presentation of the LC-DenseFCN algorithm and the demonstration of its effectiveness for this important poultry monitoring task. As for the feasibility of this approach, we discuss the following:

(1) In terms of processing accuracy, the source of detection was images collected by only two cameras on the chicken farm. Due to the fixed collection angle, the information of the overall environment could not be obtained; as a result, the information acquired was limited. Meanwhile, the movement characteristics of the chickens could also lead to a large error. In this regard, multiple cameras can be installed at different angles of the chicken farm, and the images collected by multiple cameras can be integrated to assess the overall situation of the chicken farm. Moreover, the overlapping situation of multiple chickens can be improved by installing cameras on top of the farm, which helps to effectively reduce counting errors. In general, although some individuals may have been missing in the test process, we finally achieved real-time monitoring of the number of chickens. The number of chickens should, theoretically, be in a dynamic range, and the final data can be obtained by statistical methods; thus, it will not have a significant impact on the overall accuracy;

(2) In terms of processing speed, in order to meet a farmer's need to obtain information on their chickens at any time, real-time processing of images collected by the camera is required. This is why we considered using the point supervision method. Compared to object detection methods, the point supervision method does not need to know the exact size of the object; therefore, the processing method is much simpler, and for counting tasks, the size of an object is of no practical use. The current semantic segmentation and instance segmentation methods were not considered due to

their slow processing speed and the large amount of time spent on data annotation. The experimental results indicate that the LC-DenseFCN algorithm performs well; the detection speed was 9.27 FPS, which is sufficient for real-time detection.

Based on the above discussion, we believe that the proposed method is effective for automatic poultry management and the development of refined agriculture. Furthermore, we have provided a new idea for the application of agricultural intelligence in the breeding industry.

6. Conclusions

In this study, deep learning technology was applied to the detection and counting of chickens, and a high-precision algorithm for chicken detection and counting was proposed. Firstly, we obtained video data from a camera installed on a chicken farm. Then, we took frames from the video and created a dataset of 1200 images. For meeting the needs of practical applications, we designed a point supervision model (LC-DenseFCN) and compared it with other point supervision models (LC-FCN and LC-ResFCN) and object detection models (EfficientDet, YOLOv3 and SSD). The experimental results indicate that the LC-DenseFCN algorithm performs well; the detection accuracy was 93.84%, and the counting accuracy reached up to 97%. Moreover, LC-DenseFCN could process images at 9.27 FPS, which was faster than any other model; therefore, its performance meets the requirements of practical applications. In order to prove the effectiveness of the combination of LC-Loss and model, we conducted a split experiment on LC-Loss, and the results showed that each loss of LC-Loss had a corresponding effect, and our model was matched with LC-Loss. As for the experimental results, we discussed the realization method of the automatic estimation of the quantity of chickens on the farm, which provided the train of thought for the automatic management of chicken farms. In future work, we will concentrate on assessing the performance of LC-DenseFCN in counting poultry species and further explore the impact of stocking density on animal welfare.

Author Contributions: Conceptualization, L.C. and X.L.; methodology, H.P.; software, Z.X.; validation, K.W.; formal analysis, Y.Y.; investigation, J.M.; resources, H.P.; data curation, J.M.; project administration, J.L.; writing—original draft preparation, L.C. and Z.X.; writing—review and editing, H.P. All authors have read and agreed to the published version of the manuscript.

Funding: This work was supported by the Sichuan Province Department of Education (Grant No. JG2018-348) and the Sichuan Agricultural University (Grant No.2021993069).

Institutional Review Board Statement: Not applicable.

Informed Consent Statement: Not applicable.

Acknowledgments: Thanks to Song's chicken farm for their help in collecting the dataset and Haoyang Yu for his advice on methods. Thanks to Xiao Chen and Yuanjiang Luo for their advice on drawing.

Conflicts of Interest: The authors declare no conflict of interest.

References

1. Abbasi, A.Z.; Islam, N.; Shaikh, Z.A. A review of wireless sensors and networks' applications in agriculture. *Comput. Stand. Interfaces* **2014**, *36*, 263–270.
2. Ruiz-Garcia, L.; Lunadei, L.; Barreiro, P.; Robla, I. A Review of Wireless Sensor Technologies and Applications in Agriculture and Food Industry: State of the Art and Current Trends. *Sensors* **2009**, *9*, 4728–4750. [CrossRef] [PubMed]
3. Ning, W.; Zhang, N.; Wang, M. Wireless sensors in agriculture and food industry—Recent development and future perspective. *Comput. Electron. Agric.* **2006**, *50*, 1–14.
4. Neethirajan, S.; Tuteja, S.K.; Huang, S.T.; Kelton, D. Recent advancement in biosensors technology for animal and livestock health management. *Biosens. Bioelectron.* **2017**, *98*, 398–407. [CrossRef] [PubMed]
5. Handcock, R.N.; Swain, D.L.; Bishop-Hurley, G.J.; Patison, K.P.; Wark, T.; Valencia, P.; Corke, P.; O'Neill, C. Monitoring Animal Behaviour and Environmental Interactions Using Wireless Sensor Networks, GPS Collars and Satellite Remote Sensing. *Sensors* **2009**, *9*, 3586–3603. [CrossRef]

6. Neethirajan, S. Recent advances in wearable sensors for animal health management. *Sens. Bio-Sens. Res.* **2017**, *12*, 15–29. [CrossRef]
7. Frost, A.R.; Schofield, C.P.; Beaulah, S.A.; Mottram, T.T.; Lines, J.A.; Wathes, C.M. A review of livestock monitoring and the need for integrated systems. *Comput. Electron. Agric.* **1997**, *17*, 139–159. [CrossRef]
8. Brown, K.H.; Hollingsworth, J. The Food Marketing Institute and the National Council of Chain Restaurants: Animal welfare and the retail food industry in the United States of America. *Rev. Sci. Tech.* **2005**, *24*, 655–663. [CrossRef]
9. Tu, X.; Du, S.; Tang, L.; Xin, H.; Wood, B. A real-time automated system for monitoring individual feed intake and body weight of group housed turkeys. *Comput. Electron. Agric.* **2011**, *75*, 313–320. [CrossRef]
10. Thaxton, J.P.; Dozier, W.A., III; Branton, S.L.; Morgan, G.W. Environment, Well-Being, and behavior Stocking Density and Physiological Adaptive Responses of Broilers. *Poult. Sci.* **2006**. [CrossRef]
11. Tong, H.B.; Lu, J.; Zou, J.M.; Wang, Q.; Shi, S.R. Effects of stocking density on growth performance, carcass yield, and immune status of a local chicken breed. *Poult. Sci.* **2012**, *91*, 667. [CrossRef]
12. Dozier, W.A.; Thaxton, J.P.; Purswell, J.L.; Olanrewaju, H.A.; Roush, W.B. Stocking density effects on male broilers grown to 1.8 kilograms of body weight. *Poult. Sci.* **2006**, *85*, 344–351. [CrossRef]
13. Yao, Y.; Yu, H.; Mu, J.; Li, J.; Pu, H. Estimation of the Gender Ratio of Chickens Based on Computer Vision: Dataset and Exploration. *Entropy* **2020**, *22*, 719. [CrossRef] [PubMed]
14. Aydin, A. Using 3D vision camera system to automatically assess the level of inactivity in broiler chickens. *Comput. Electron. Agric.* **2017**, *135*, 4–10. [CrossRef]
15. So-In, C.; Poolsanguan, S.; Rujirakul, K. A hybrid mobile environmental and population density management system for smart poultry farms. *Comput. Electron. Agric.* **2014**, *109*, 287–301. [CrossRef]
16. Alahmari, S.S.; Goldgof, D.; Hall, L.; Phoulady, H.A.; Patel, R.H.; Mouton, P.R. Automated Cell Counts on Tissue Sections by Deep Learning and Unbiased Stereology. *J. Chem. Neuroanat.* **2019**, *96*, 94–101. [CrossRef]
17. Xu, B.; Wang, W.; Falzon, G.; Kwan, P.; Schneider, D. Automated cattle counting using Mask R-CNN in quadcopter vision system. *Comput. Electron. Agric.* **2020**, *171*, 12. [CrossRef]
18. Nguyen, H.; Maclagan, S.J.; Tu, D.N.; Nguyen, T.; Phung, D. Animal Recognition and Identification with Deep Convolutional Neural Networks for Automated Wildlife Monitoring. In Proceedings of the 2017 IEEE International Conference on Data Science and Advanced Analytics (DSAA), Tokyo, Japan, 19–21 October 2017; pp. 40–49.
19. Tian, M.; Guo, H.; Chen, H.; Wang, Q.; Long, C.; Ma, Y. Automated pig counting using deep learning. *Comput. Electron. Agric.* **2019**, *163*, 104840. [CrossRef]
20. Zhang, Q.; Liu, Y.; Gong, C.; Chen, Y.; Yu, H. Applications of Deep Learning for Dense Scenes Analysis in Agriculture: A Review. *Sensors* **2020**, *20*, 1520. [CrossRef] [PubMed]
21. Laradji, I.H.; Rostamzadeh, N.; Pinheiro, P.O.; Vazquez, D.; Schmidt, M. Where are the blobs: Counting by localization with point supervision. In Proceedings of the European Conference on Computer Vision (ECCV), Munich, Germany, 8–14 September 2018; pp. 547–562.
22. Huang, G.; Liu, Z.; Van Der Maaten, L.; Weinberger, K.Q. Densely connected convolutional networks. In Proceedings of the IEEE conference on computer vision and pattern recognition, Honolulu, HI, USA, 21–26 July 2017; pp. 4700–4708.
23. He, K.; Zhang, X.; Ren, S.; Sun, J. Deep residual learning for image recognition. In Proceedings of the IEEE conference on computer vision and pattern recognition, Las Vegas, NV, USA, 27–30 June 2016; pp. 770–778.
24. Long, J.; Shelhamer, E.; Darrell, T. Fully Convolutional Networks for Semantic Segmentation. *IEEE Trans. Pattern Anal. Mach. Intell.* **2015**, *39*, 640–651.
25. Girshick, R.; Donahue, J.; Darrell, T.; Malik, J. Rich feature hierarchies for accurate object detection and semantic segmentation. In Proceedings of the IEEE Conference on Computer Vision and Pattern Recognition, Columbus, OH, USA, 23–28 June 2014; pp. 580–587.
26. Girshick, R. Fast R-Cnn. In Proceedings of the IEEE International Conference on Computer Vision, Santiago, Chile, 7–13 December 2015; pp. 1440–1448.
27. Ren, S.; He, K.; Girshick, R.; Sun, J. Faster R-CNN: Towards Real-Time Object Detection with Region Proposal Networks. *IEEE Trans. Pattern Anal. Mach. Intell.* **2017**, *39*, 1137–1149. [CrossRef] [PubMed]
28. He, K.; Gkioxari, G.; Dollár, P.; Girshick, R. Mask R-Cnn. In Proceedings of the IEEE International Conference on Computer Vision, Venice, Italy, 22–29 October 2017; pp. 2961–2969.
29. Redmon, J.; Farhadi, A. YOLOv3: An Incremental Improvement. In Proceedings of the CVPR 2018: IEEE Conference on Computer Vision and Pattern Recognition, Salt Lake, UT, USA, 18–22 June 2018.
30. Tan, M.; Pang, R.; Le, Q.V. Efficientdet: Scalable and efficient object detection. In Proceedings of the IEEE/CVF Conference on Computer Vision and Pattern Recognition, Seattle, WA, USA, 13–19 June 2020; pp. 10781–10790.
31. Redmon, J.; Divvala, S.; Girshick, R.; Farhadi, A. In You only look once: Unified, real-time object detection. In Proceedings of the IEEE Conference on Computer Vision and Pattern Recognition, Las Vegas, NV, USA, 27–30 June 2016; pp. 779–788.
32. Redmon, J.; Farhadi, A. In YOLO9000: Better, faster, stronger. In Proceedings of the IEEE Conference on Computer Vision and Pattern Recognition, Honolulu, HI, USA, 21–26 July 2017; pp. 7263–7271.
33. Tan, M.; Le, Q. In Efficientnet: Rethinking model scaling for convolutional neural networks. In Proceedings of the International Conference on Machine Learning, Long Beach, CA, USA, 9–15 June 2019; pp. 6105–6114.

34. Boyat, A.K.; Joshi, B.K. A Review Paper: Noise Models in Digital Image Processing. *Signal Image Process. Int. J.* **2015**, *6*, 63–75. [CrossRef]
35. Simonyan, K.; Zisserman, A. Very deep convolutional networks for large-scale image recognition. *arXiv* **2014**, arXiv:1409.1556.
36. Kuznetsova, A.A. Statistical Precision-Recall Curves for Object Detection Algorithms Performance Measurement. In *Cyber-Physical Systems Modelling and Intelligent Control*; Springer: Berlin/Heidelberg, Germany, 2021; pp. 335–348.

agriculture

MDPI

Article

Identification of Risk Factors for Lameness Detection with Help of Biosensors

Ramūnas Antanaitis [1,*], Vida Juozaitienė [2,*], Gediminas Urbonavičius [3], Dovilė Malašauskienė [1], Mindaugas Televičius [1], Mingaudas Urbutis [1], Karina Džermeikaitė [1] and Walter Baumgartner [4]

1 Large Animal Clinic, Veterinary Academy, Lithuanian University of Health Sciences, Tilžės Street 18, LT-47181 Kaunas, Lithuania; dovile.malasauskiene@lsmuni.lt (D.M.); mindaugas.televicius@lsmuni.lt (M.T.); mingaudas.urbutis@lsmuni.lt (M.U.); k.dzermeikaite@gmail.com (K.D.)
2 Department of Biology, Faculty of Natural Sciences, Vytautas Magnus University, K. Donelaičio 58, LT-44248 Kaunas, Lithuania
3 Department of Animal Breeding, Veterinary Academy, Lithuanian University of Health Sciences, Tilžės Street 18, LT-47181 Kaunas, Lithuania; gediminas.urbonavicius@lsmuni.lt
4 University Clinic for Ruminants, University of Veterinary Medicine, Veterinaerplatz 1, A-1210 Vienna, Austria; walter.baumgartner@vetmeduni.ac.at
* Correspondence: ramunas.antanaitis@lsmuni.lt (R.A.); vida.juozaitiene@vdu.lt (V.J.); Tel.: +370-6734-9064 (R.A.)

Abstract: In this study we hypothesized that the lameness of early lactation dairy cows would have an impact on inline biomarkers, such as rumination time (RT), milk fat (%), milk protein (%), milk fat/protein ratio (F/P), milk lactose (L, %), milk electrical conductivity of all udder quarters, body weight (BW), temperature of reticulorumen content (TRR), pH of reticulorumen content (pH), and walking activity (activity). All 30 lame cows (LCs) used in this experiment had a score of 3–4, identified according to the standard procedure of Sprecher et al. The 30 healthy cows (HC) showed a lameness score of one. RT, milk fat, MY, milk protein, F/P, L, milk electrical conductivity of all udder quarters, and BW were registered using Lely Astronaut® A3 milking robots each time the cow was being milked. The TRR, cow activity, and pH of the contents of each cow's reticulorumen were registered using specific smaXtec boluses. The study lasted a total of 28 days. Days "−14" to "−1" denote the days of the experimental period before the onset of clinical signs of lameness (day "0"), and days "1" to "13" indicate the period after the start of treatment. We found that from the ninth day before the diagnosis of laminitis until the end of our study, LCs had higher milk electrical conductivity in all udder quarters, and higher milk fat to protein ratios. On the 3rd day before the onset of clinical signs of the disease until the day of diagnosis, the milk fat of the LC group was reduced. The activity of the LCs decreased sharply from the second day to the first day after treatment. RT in the HC group tended to decrease during the experiment. pH in LCs also increased on the day of the appearance of clinical signs.

Keywords: lameness; inline biomarkers; fresh dairy cows

Citation: Antanaitis, R.; Juozaitienė, V.; Urbonavičius, G.; Malašauskienė, D.; Televičius, M.; Urbutis, M.; Džermeikaitė, K.; Baumgartner, W. Identification of Risk Factors for Lameness Detection with Help of Biosensors. *Agriculture* **2021**, *11*, 610. https://doi.org/10.3390/agriculture11070610

Academic Editors: Gniewko Niedbała and Sebastian Kujawa

Received: 17 May 2021
Accepted: 27 June 2021
Published: 29 June 2021

1. Introduction

Lameness is an important health and welfare concern in dairy farming. Producers are aware of how painful this condition is to the animal. They are more than willing to put effort into controlling lameness, even if these control measures seem inconvenient; however, the prevalence of lameness in the herd remains underestimated [1]. Many cases go untreated for several weeks, and those that are treated often develop repeated cases, requiring further treatment [2]. In any herd, lameness has negative implications on productivity and behavior [3], and this may be heightened in farms where milking robots are used. In problematic herds with a high incidence of lameness, there is a huge economic loss due to the reduction in milk yield and weight loss [4–6]. Lame cows have lower milk yields, shorter rumination times, lower milk temperatures, and a lower intake of supplements, and exhibit greater refusal to go to milking [7]. These changes in milk composition can be used to determine lameness of infectious and non-infectious origins; this observation is supported by Bonfatti et al. (2020) [8]. Bramley et al. (2013) [9] found that herds that are at higher risk for lameness were also at much higher risk for acidosis. Acidosis systematically impacts the physiology of animals, including laminitis and a diffuse aseptic inflammation of the laminae [10,11]. Vasoactive substances (histamine and endotoxins) are released during a decline in ruminal pH and as the result of bacteriolysis and tissue degradation. These substances cause vasoconstriction and dilation, which ultimately destroy the microvasculature of the corium. The effects of rumen acidosis in fresh dairy cows are mainly manifested by decreased milk fat content, decreased fiber digestibility, and diarrhea [11]. AlZahal (2008) [12] also observed a decrease in pH below 5.6 and a rumen temperature exceeding 40 °C. According to AlZahal et al. (2008) [12], rumen temperature negatively correlates with rumen pH. Therefore, changes in rumen temperature can be used to determine acidosis, and this can potentially be achieved using radio-telemetric devices [13]. According to Zhang et al. (2015) [14], lameness also decreases milk yield and days in milk (DIM) as well as the milk fat and fat-to-protein ratio. Milk fat depression has been associated with a decreased acetate-to-propionate ratio [15]. Analyzing the fluctuations in the ratio of milk fat to protein and their relationship to the pH value of the rumen content, it was found that the ratio of milk fat to protein, showing sub-acute ruminal acidosis (SARA), is <1 [15]. Miekley et al. (2012) [16] investigated the use of milk electrical conductivity and cow activity for the early diagnosis of mastitis and laminitis. Lukas et al. (2009) [17] reported that the milk electrical conductivity is a reliable indicator of mastitis, but Khatun et al. (2019) [18] argue that the electrical conductivity of milk is not a very sensitive indicator for the detection of mastitis, even if is the most widely used method for the early diagnosis of mastitis recorded by automatic milking systems. However, Lukas et al. (2009) [17] described that other metabolic and digestive problems such as ketosis, left displaced abomasum, a retained placenta, and lameness were also associated with a clear increase in milk electrical conductivity. Walker et al. (2008) [19] observed that lame cows spend more time lying down, less time standing, and less time walking. Mazrier et al. (2006) [20] confirmed that the use of electronic devices to record walking time of cows can detect lameness 7–10 days before the onset of clinical signs, which is associated with decreased activity in cows. Cows with lameness have also been found to spend less time eating and to be less active compared with non-lame cows [21,22].

Demonstrating the negative changes associated with lameness, especially when accounting for other factors, would help dairy producers to better evaluate the negative effects of even simple cases of lameness and would likely lead to more improved lameness monitoring and treatment methods. The expanding use of automatic milking systems (AMSs) provides many challenges and opportunities to dairy producers. The use of AMSs also has the advantage of monitoring cow-level milking frequency and quarter-level production and milk quality, which can support illness detection tools [23]. King et al. (2018) [23] also found that some of the markers registered by automatic milking systems change in association with lameness in cows. Nonetheless, not all health disorders can be detected electronically, and producers must still physically assess and fetch cows for milking if their milking interval is too long [24–26].

Based on the information in the literature, we hypothesized that the lameness of fresh dairy cows would have an impact on inline biomarkers, such as rumination time (RT), milk yield (MY), milk fat (%), milk protein (%), milk fat/protein ratio (F/P), milk lactose (L, %), milk electrical conductivity of all udder quarters, body weight (BW), temperature of reticulorumen content (TRR), pH of reticulorumen content (pH), and walking activity (activity). With this study we aimed to investigate the identification of risk factors for the detection of lameness with the help of biosensors.

2. Materials and Methods

2.1. Location and Animals

The investigation was conducted at the Lithuanian University of Health Sciences, and at one farm containing dairy cows (54.9754° N, 23.7684° E) in the period from 1 July 2020 to 15 December 2020.

All animal experimental procedures were approved; the approval number is PK016965. Sixty Lithuanian black and white breed cows (on average 5 years old) were selected from a herd of 500 dairy cows, which were being kept in a loose system. A lameness diagnosis was performed by trained staff (by the same one person) according to the standard procedure described by Sprecher et al. [26]: 1 = normal, 2 = presence of a slightly asymmetric gait, 3 = the cow clearly protects one or more limbs (moderately lame), 4 = severely lame, and 5 = extremely lame (non-weight-bearing lame). Visual locomotion scoring was conducted once weekly for four consecutive weeks by the same observer. Based on Warnick et al., [27], the causes of lameness were categorized as sole ulcer, abscess, and foot rot. Sole ulcer (pododermatitis circumscripta) included degenerative or necrotic defects in the sole near the sole heel junction. Abscess (white line abscess, sole abscess) was defined as a pus-filled cavity of the white line or sole of the foot. Foot rot (interdigital phlegmon) was swelling of the soft tissues of the foot, resulting in symmetrical swelling above the coronary band and spreading of the toes, in some cases with necrosis of the tissue between the claws.

All 30 LCs used in this experiment had a score of 3–4, presenting severe lameness with pronounced arching of the back, they were reluctant to move, and were unwilling to complete weight transfer off the affected limb. The 30 HCs had a lameness score of 1. The study lasted for a total of 28 days. Days "−14" to "−1" denote the days of the experimental period before the onset of clinical signs of lameness (day "0"), and days "1" to "13" indicate the period after the start of treatment. Average DIM was 60 (±10) days. Treatment was repeated at 24-h intervals for a total of three days. Rimadyl Cattle® (50 mg/mL) s.c. injection (Zoetis, Belgium) at a dose of 1.4 mg/kg body weight was performed once as well.

Cows were provided with a total mixed ration (TMR) consisting of 20% corn silage, 20% grass silage, 50% flaked grain concentrate, 5% grass hay, and 5% of mineral mixture. The ration was composed as to exceed or meet the requirements of a 550-kg Holstein cow producing 40 kg of milk per day. Composition of ration: dry matter (DM) (%) 49.00; neutral detergent fiber (% of DM) 28.00; acid detergent fiber (% of DM) 20.00; crude protein (% of DM) 16.00; non-fiber carbohydrates (% of DM) 39.00; net energy for lactation (Mcal/kg). Feeding was carried out every day at 05:00 and 17:00. The milking process was done with five Lely Astronaut® A3 milking robots with free traffic.

2.2. Measurements

Lely Astronaut® A3 milking robots with free traffic were used to milk the cows and they collected information about the cows: MY, BW, RT, L, F/P, milk fat, and milk protein milk electrical conductivity (mS/cm) of all quarters of the udders (front left (MCFL), front right (MCFR), rear left (MCBL), rear right (MCBR)).

With the help of smaXtec boluses (smaXtec animal care technology®), the parameters pH, TRR, and cow activity were monitored in real-time and were registered every 10 min each day. Data were measured with the help of specific antennas (smaXtec animal care technology®). For the monitoring of pH, TRR, and activity an indwelling and wireless data transmitting system (smaXtec animal care GmbH, Graz, Austria) was used. The system was controlled by a microprocessor. Data (pH, TRR) were collected using an analogue-to-digital converter (A/D converter) and stored in an external memory chip. Calibration of the pH-probes was performed using pH 4 and pH 7 buffer solutions at the beginning of the experiment. All data were obtained using smaXtec messenger® computer software.

2.3. Data Analysis and Statistics

Data were analyzed with the SPSS 26.0 (SPSS Inc., Chicago, IL, USA) program package. The normal distribution of variables was assessed using the Kolmogorov–Smirnov test (inline variables were not normally distributed). To assess the differences between the compared LC and HC groups over a 28-day period, we used the general linear model repeated measures and Fisher's standard deviation (SD) criterion. Differences were considered statistically significant at $p \leq 0.05$. Linear regression was also used to analyze changes in the studied inline biomarkers.

3. Results

The average weight of the cows was 700 kg. Average milk yield during 2020 was 12,000 kg per cow and year.

3.1. Relationship of Lameness in Fresh Dairy Cows with Milk Traits

The lame cows were less productive (on average—9.4 kg/d; $p < 0.001$ of milk yield), had a higher milk fat concentration (on average 0.74 percentage points $p < 0.001$), lower milk protein content (on average—0.243 percentage points; $p = 0.031$), and had a higher milk fat/protein ratio (on average—0.297; $p < 0.001$), compared to healthy cows. Lame cows' milk was found to have a lower lactose content (on average–0.141 percentage points; $p < 0.001$) and all udder quarters had higher milk electrical conductivity (on average—2.1–7.6 mS/cm; $p < 0.001$) (Table 1).

Table 1. Milk traits by group of cows (LSM ± SD).

Variable	HC	LC	*p*
MY (kg)	43.39 ± 7.65	34.00 ± 9.59	<0.001
Fat (%)	4.15 ± 0.54	4.89 ± 0.67	<0.001
Protein (%)	3.74 ± 0.35	3.50 ± 0.44	0.031
F/P	1.11 ± 0.09	1.41 ± 0.11	<0.001
Lactose (%)	4.66 ± 0.05	4.52 ± 0.07	<0.001
MCFL (mS/cm)	67.1 ± 2.24	74.6 ± 2.81	<0.001
MCFR (mS/cm)	66.3 ± 2.98	73.9 ± 3.74	<0.001
MCBL (mS/cm)	67.5 ± 2.61	71.7 ± 3.27	<0.001
MCBR (mS/cm)	69.5 ± 1.96	71.5 ± 0.05	<0.001

MY—milk yield; Fat—milk fat; Protein—milk protein; F/P—milk fat to protein ratio; Lactose—milk lactose; MCFL—milk electrical conductivity of front left udder quarter; MCFR—milk electrical conductivity of front right udder quarter; MCRL—milk electrical conductivity of rear left udder quarter; MCRR—milk electrical conductivity of rear right udder quarter; HC—healthy cow group; LC—lame cow group; data are presented as least square means (LSM), standard deviation (SD); $p \leq 0.05$—means in the row differed significantly.

In 90.48% of the milk samples of the LC group (Figure 1A), the F/P ratio exceeded 1.2, and only 9.52% of the samples met the reference norm ($p < 0.001$). The lactose level in milk exceeded 4.6% in 90.91% of milk samples from HC cows and only 4.76% in milk samples from LC cows ($p < 0.001$). Results are presented in Figure 1B.

Milk productivity of HC increased linearly from 38.7 ± 0.15 kg to 57.90 ± 0.18 kg ($p < y = 0.5504x + 39.869$; $R^2 = 0.9322$, $p < 0.001$). From the ninth day before the diagnosis of laminitis until the end of the experiment, the milk production of the LC group of cows was statistically significantly lower compared to the group of healthy cows ($p < 0.05$); and only on the eighth day after the start of treatment did the milk yield of LC cows begin to increase slightly (Figure 2A).

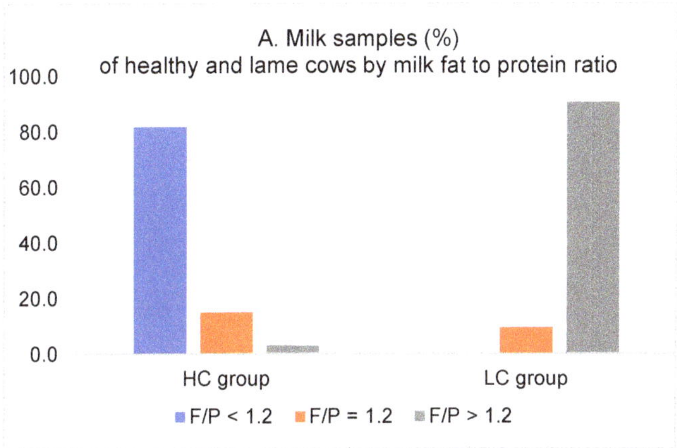

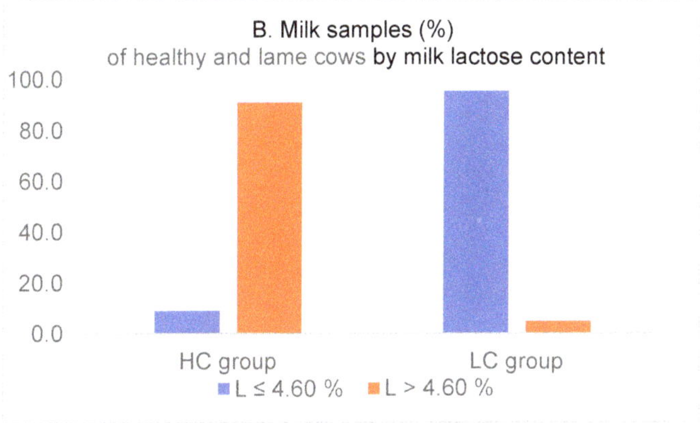

Figure 1. Relation of laminitis with milk fat-to-protein ratio (**A**) and lactose level (**B**). F/P—milk fat to protein ratio; L—milk lactose (%); HC—healthy cow group; LC—lame cow group.

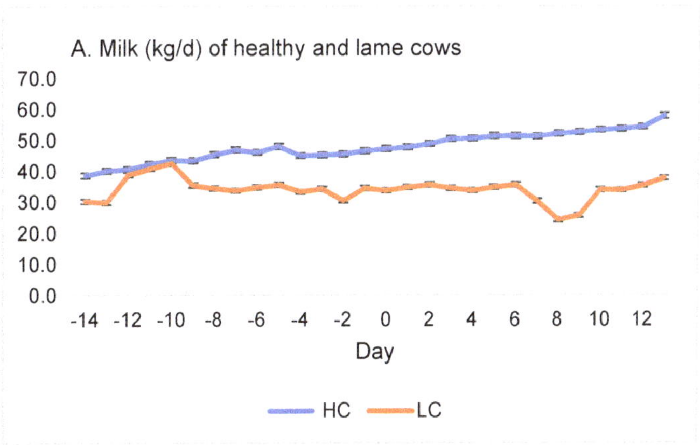

Figure 2. *Cont.*

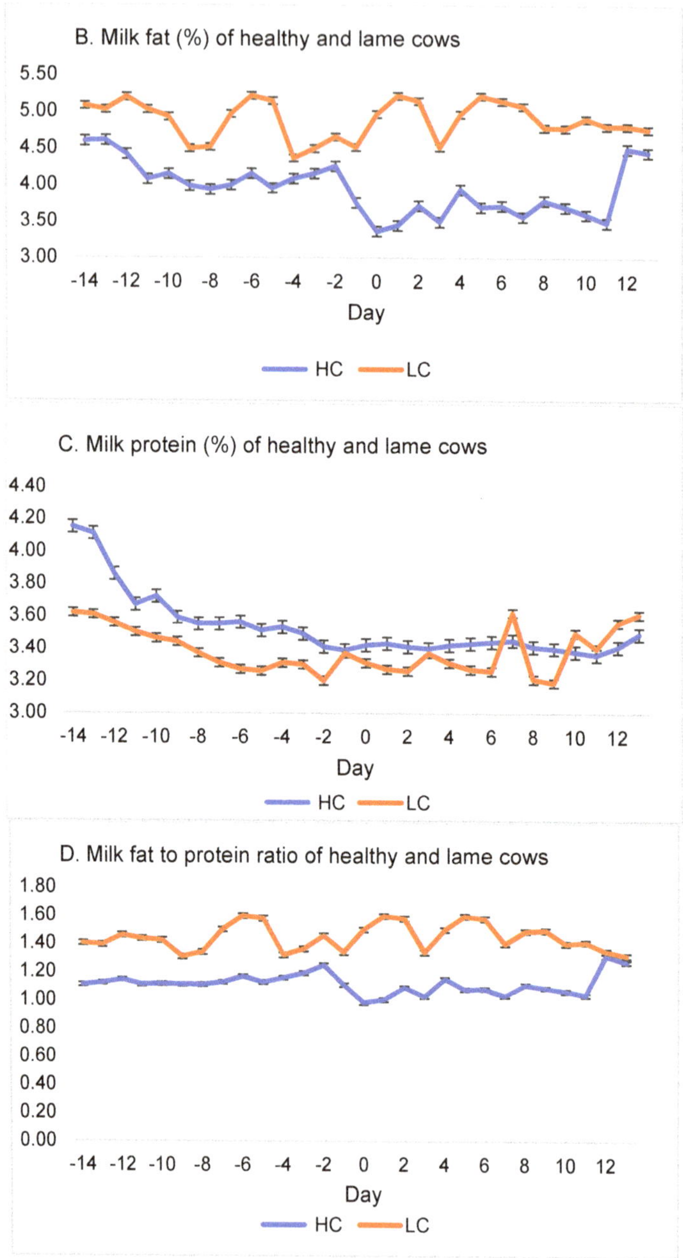

Figure 2. *Cont.*

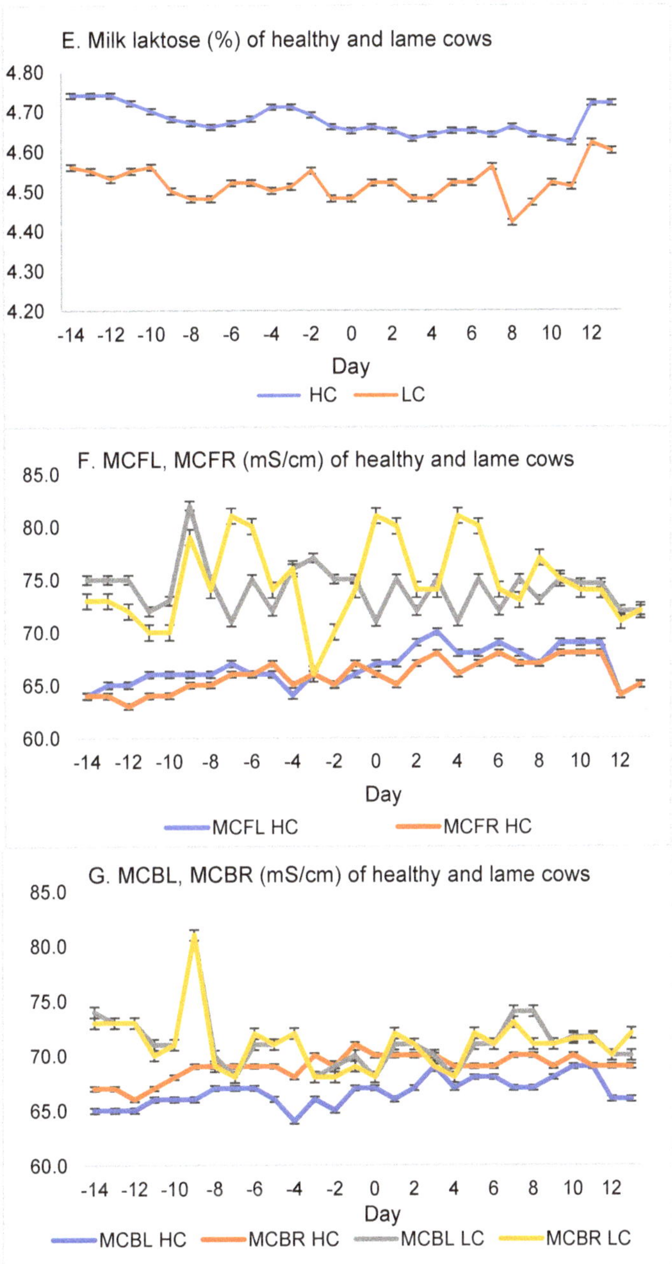

Figure 2. Changes in milk milk yield (**A**), fat (**B**), protein (**C**), fat to protein ratio (**D**), lactose (**E**), milk electrical conductivity of front left and right udder quarter (**F**), milk electrical conductivity of rear left and right udder quarter (**G**). MCFL—milk electrical conductivity of front left udder quarter; MCFR—milk electrical conductivity of front right udder quarter; MCBL—milk electrical conductivity of rear left udder quarter; MCBR—milk electrical conductivity of rear right udder quarter; HC—healthy cow group; LC—lame cow group.

Milk fat (y = −0.0202x + 4.2357, R^2 = 0.2156, p = 0.013) and protein content in HC decreased during the experiment (y = −0.0187x + 3.8052, R^2 = 0.5748, p < 0.001). On the 3rd day before the onset of clinical signs of the disease until the day of diagnosis, the milk fat of LC was significantly reduced (p < 0.001). In these cows, milk fat (Figure 2B) began to stabilize at day 8 and milk protein began to increase at day 9 after treatment (Figure 2C).

The milk fat-to-protein ratio in LC was higher compared to healthy cows during the first 25 days of the experiment (p < 0.001). The average value of this indicator in both groups of cows at the end of the experiment was the same (Figure 2D).

Milk lactose in HC tended to decrease during the experimental period (y = −0.0025x + 4.7129, R^2 = 0.3256, p = 0.002) and was higher than in LC (p < 0.001). A steady increase in this indicator in LC was observed from the 8th day from the beginning of treatment to the end of the experiment (Figure 2E).

The electrical conductivity of HC tended to increase according to linear regression analysis (R^2 = 0.26–0.40, p = 0.005, and p < 0.001). In LC this indicator was higher compared to HC and fluctuated during the experiment (Figure 2F,G).

3.2. *Relationship of Lameness in Cows with Reticulorumen Indicators, Activity, and Body Weight*

In the LC group, we found lower mean RT values (−133.42 min/d, p < 0.001) and higher activity values (+1.38 steps/h, p = 0.027) and body weight (+5.69 kg, p < 0.001) compared to the HC group (Table 2).

Table 2. Reticulorumen indicators, activity, and body weight by group of cows (LSM ± SD).

Variable	HC	LC	p
RT (min/day)	517.6 ± 21.80	384.2 ± 52.41	<0.001
pH	6.43 ± 0.17	6.46 ± 0.22	0.231
TRR °C	39.16 ± 0.87	39.33 ± 1.17	0.059
Activity (steps/h)	8.55 ± 0.45	9.93 ± 0.37	0.027
Body weight (kg)	734.5 ± 15.44	728.8 ± 20.89	<0.001

RT—rumination time; pH—reticulorumen pH; HC—healthy cow group; LC—lame cow group; data are presented as least square means (LSM), standard deviation (SD); $p \leq 0.05$—means in the row differed significantly.

The RT in HC tended to decrease during the experiment (y = −1.9354x + 553.78. R^2 = 0.3647, p < 0.001) and was significantly higher than in LCs from the start of the experiment until the 11th day after treatment (p < 0.001). For LCs we could not apply linear regression in the analysis of their RT change (Figure 3A).

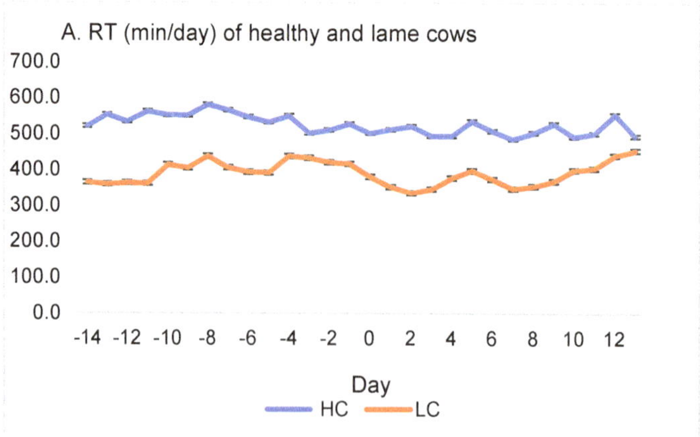

Figure 3. *Cont.*

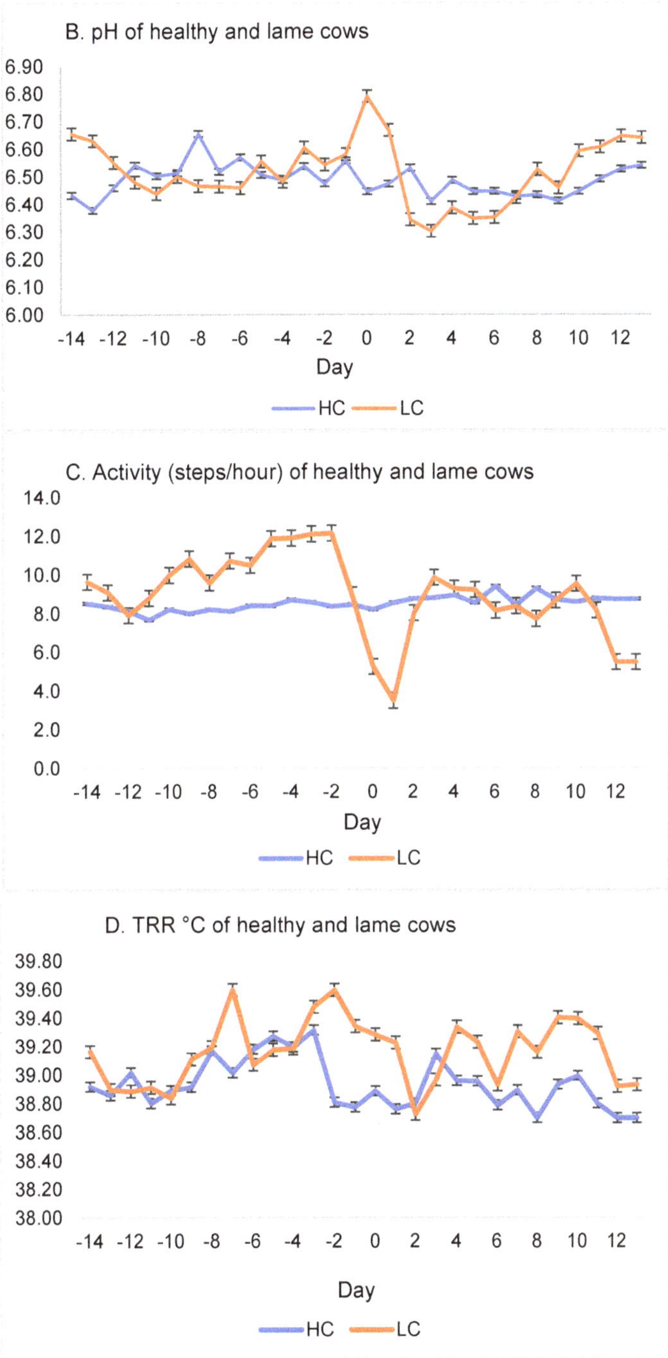

Figure 3. *Cont.*

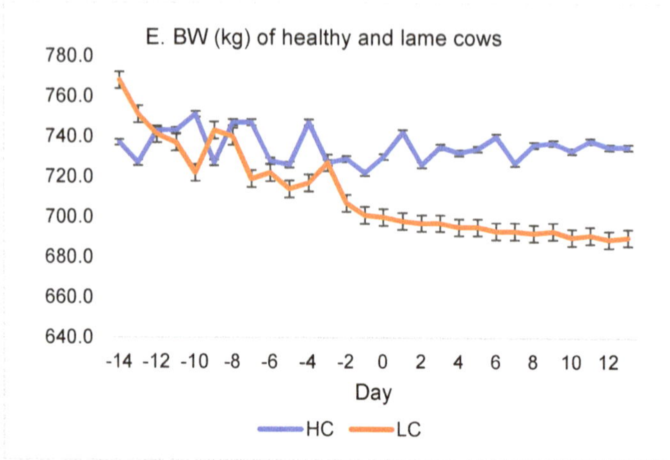

Figure 3. Change in rumination time (**A**), reticulorumen pH (**B**), activity (**C**), reticulorumen temperature (**D**), and body weight (**E**) of cows by group during the experiment. RT—rumination time; Activity—cow activity; TRR—reticulorumen temperature; BW—body weight; HC—healthy cow group; LC—lame cow group.

As seen in Figure 3B, the pH in LC cows increased significantly from day −2 to day 0 ($p < 0.001$), then decreased until the 3rd day after treatment ($p < 0.001$), and increased until the end of the experiment. The pH of the reticulorumen in HCs was more stable throughout the observation period.

The activity of cows in the LC group decreased sharply from day "−2" to the first day after treatment, then increased, but began to decrease again at 10 days after the start of treatment. In HC this indicator changed less throughout the experiment (Figure 3C).

The highest TRR (Figure 3D) in cows of the LC group was recorded from "−3" to "0" days (40.00 °C–40.16 °C). At the end of the experiment, the TRR of this group of cows dropped to 39.32 °C ± 0.229 °C ($p < 0.001$).

The BW of LC decreased throughout the experimental period ($y = -2.474x + 747.37$, $R^2 = 0.8409$, $p < 0.001$), whereas in HC it was more stable (Figure 3E).

4. Discussion

According to the results of our study, we found that lame cows were less productive; the milk production of lame cows was lower than in healthy cows. Huxley et al. (2013) reported milk yield losses per case of lameness, with most losses ranging from 270 to 574 kg/lactation [3]. Cow-level studies of lameness in AMS herds have reported reduced milk yield, reduced total and voluntary milking frequency, and greater daily lying time for lame cows [28–30]. Lameness had a detrimental effect on most outcome variables analyzed, especially to parameters related to milk production and AMS visits. Lame cows produced 1.6 kg/d less milk during the 6-d data collection period in some studies. Furthermore, a higher milking order index was registered for moderately lame cows [31].

We found that lame cows in all udder quarters had higher milk electrical conductivity. The monitoring of affected milk electrical conductivity has promise as an indirect and rapid method for the detection of subclinical mastitis [32]. The technology is based on measuring potassium, sodium, and other free ions, especially chloride, which is directly proportional to electrical conductivity. In normal milk, the concentration of Cl^- is around 75–130 mg/100 mL; however, due to inflammation, the amount of free Cl^- can increase up to 111–198 mg/100 mL. These changes can appear rapidly and sporadically, depending on the type of mastitis developing [32]. Lameness, as a problem affecting crossbred dairy cattle herds, has tremendous effects on animal health, production, welfare, and reproduction [33].

Severe pain in lame cows modifies the normal rising and lying behavior of cattle. The chances of mastitis increased the longer the cows remained lying down [33]. A positive relationship was observed between poor claw health and an increased incidence of clinical mastitis [34].

Our study shows that lame cows had a lower milk protein content and a higher milk fat-to-protein ratio. On the 3rd day before the onset of clinical signs of the disease until the day of diagnosis, the milk fat of lame cows was significantly reduced. In the milk of lame cows, lower concentrations of milk lactose were registered. Lameness in crossbred dairy cattle may affect the milk composition. Olechnowicz and Jacekowski (2010) detected a significant decrease in the average milk yield and in milk components of fat, protein, and lactose content in cows with clinical lameness, as compared to non-lame cows [35]; due to lameness, the cows exhibited poor absorption and assimilation of nutrients, since they were under more stress and pain and had increased oxidative agents, which in turn resulted in significantly lower average monthly protein, fat, and lactose contents and production compared to non-lame cows. In their subsequent trial, Olechnowicz and Jacekowski (2012) [36] observed that clinical lame cows had significantly lower milk protein content than healthy cows.

The activity of cows in the LC group decreased sharply from day "−2" to the first day after treatment, then increased but began to decrease again at the end of the experiment. According to literature, alterations in limb movements develop due to pain from claw disorders [36]. In multiple studies, however, it has become evident that it is difficult for farmers to reliably detect lameness. For this reason, a number of attempts have been made to incorporate technical equipment into gait analysis and to develop automatic detection systems to warn farmers about lameness incidences in their herds [31,37,38]. Currently, two main types of systems are used on farms [38–41]: permanently installed in the animal's environment (e.g., weighing platforms, 3-dimensional cameras, video analysis), or systems attached to the cow (e.g., pedometer). However, according to Rutten et al. (2013) [38], the aforementioned systems are mainly used to detect severe lameness, which is already easily detected visually. Lameness leads to several changes for cows, both in their gait and behavior. However, in most studies the cows with different lameness scores have been grouped together, whereas little is known about lameness-induced behavioral changes in moderately lame cows specifically.

Moderately lame cows, compared with non-lame cows, demonstrated a lower average locomotor activity, which can be explained by the reduced time of being elevated on their feet [19] and is in agreement with observations in cows with lameness of differing severity [22,41]. Furthermore, the locomotory activity registered during the first hour after feed delivery or push-ups was lower in moderately lame cows than in non-lame cows. This result suggests that non-lame cows are quicker to react to the feed push-ups or delivery [42–44] and provides proof for the hypothesis that lame cows have trouble getting up and walking to the feed alley for freshly delivered feed. Furthermore, a greater difference in locomotion activity 1 h after milking has been suspected between lame and non-lame cows. Juarez et al. (2003) [45] found that the percentage of cows lying down after returning from the milking parlor increased with the growing severity of lameness. This presumably occurs due to a painful weight-bearing process (while waiting for and during milking), which increases with the severity of lameness. The average locomotor activity and the locomotor activity during the hour after feed delivery or push-ups were lower in moderately lame cows compared to non-lame cows. The daily lying duration of moderately lame cows in the present study ranged between that of non-lame and previously observed severely lame cows, which points to a positive relationship between lying duration and the degree of lameness. The prolonged lying duration in combination with an increased number of lying bouts in moderately lame cows resulted in a longer average lying duration than in non-lame cows. The changes in lying behavior are associated with the reduced time that lame cows spend on their feet [19], possibly as a way to avoid bearing weight on damaged limbs [45]. Furthermore, moderately lame cows also showed a lower average level of neck locomotion than non-lame cows [29].

During our study, RT in healthy cows tended to decrease during the experiment and was higher than in lame cows from the start of the experiment until the 11th day after treatment. According to Weigele et al. (2018) [30], the number of eating chews and eating time per 24 h were shorter in moderately lame than in non-lame cows. Moreover, no effect of moderate lameness was evident on the number of ruminating chews, ruminating time, number of boluses, average ruminating speed, and the average number of ruminating chews per bolus [30]. Moderately lame cows also displayed less frequent visits to the concentrate feeder than non-lame cows. In contrast, both lame and non-lame cows showed a similar probability of concentrate leftovers [30]. A reduced number of eating chews and eating time were found in moderately lame cows, which was also found by Beer et al. (2016) [46]. In accordance with Palmer et al. (2012) [47], due to painful limbs or feet, lame cows spent less time standing and eating. Bareille et al. (2003) [48] observed a lower total feed intake in lame cows compared to non-lame cows, which could be related to their poorer BCS [19,49]. On the other hand, several studies have reported a higher feeding rate (feed intake per min) in lame cows and thus came to a conclusion that they eat faster than their non-lame counterparts [21,50,51]. Taking these results (no difference in average mastication speed but reduced eating time and number of eating chews) into account, Weigele et al. (2018) [30] assumed that moderately lame cows had a poorer breakdown of feed. Moderate lameness did not affect rumination duration, number of ruminating chews, number of boluses, or the average ruminating speed. Although cows can ruminate while standing, the greater part of rumination usually takes place while lying down [52,53], whereas eating is performed while standing. Therefore, the effect of lameness on rumination might not be as significant as its effect on eating. This notion is supported by Thorup et al. (2016) [51], who also observed reduced eating behavior but similar rumination behavior when comparing lame and non-lame cows.

We found that pH in lame cows increased on the day of the appearance of clinical signs, then decreased until the 3rd day after treatment, and increased until the end of the experiment. Changes in the rumen initiate several systemic changes. Increased organic acid, particularly lactic acid, and thus a reduced pH value, result in decreased ruminal motility, predisposing cows to ruminitis and hyperkeratosis [54]. In many dairy operations, the biggest concern is not acute acidosis but subclinical acidosis, whereby a minute accumulation of lactic acid is detectable in the rumen; however, pH still decreases. Daily episodes of pH < 5.5 for given periods ultimately lead cattle to low-grade, subclinical acidosis. Symptoms include an erratic appetite, bodyweight loss, diarrhea, and lameness [55]. Laminitis occurs in acute, chronic, and subclinical forms [56]. Acidosis and laminitis are linked with alterations in the hemodynamics of the peripheral microvasculature [57]. Various theories have been developed to explain the pathogenesis of laminitis. Vasoactive substances such as histamine and endotoxins are released during ruminal acidosis and as the end-products of bacteriolysis and tissue degradation. These substances influence vasodilation and vasoconstriction, ultimately destroying the corium's microvasculature [58]. Ischemia results in reduced oxygen and nutrient availability for the extremities of the corium. This degrades the junctions between tissues that are critical for locomotion [57]. The insidious rotation of the distal phalanx (pedal bone) often results in permanent damage. Signs of subclinical laminitis are yellowish discolorations and hemorrhages of the sole [56]. Other clinical manifestations include dorsal wall concavity, double soles, heel erosion, and ridging of the dorsal wall [59]. Pododermatitis aseptica diffusa is the scientific name for laminitis, an aseptic inflammation of the foot's dermal layers. Studies have concluded that body condition score, body weight, rear lateral digital variations, non-tarsal rear limb superficial swelling, abnormal hoof overgrowth, and limb laceration were risk factors for clinical lameness.

5. Conclusions

The risk factors identified for lameness with the help of sensors were lower milk yield, milk protein and lactose contents, bodyweight, activity, and rumination time, and higher milk electrical conductivity and milk fat-to-protein ratio.

From the ninth day before the diagnosis of laminitis until the end of our experiment, the milk production of lame cows was lower than that of healthy cows. LCs exhibited a higher milk electrical conductivity in all udder quarters, lower milk protein and lactose contents, as well as a higher milk fat-to-protein ratio. On the 9th day before the onset of clinical signs of the disease until the day of diagnosis, the milk fat of the lame cows was reduced. The bodyweight of LCs decreased throughout our experimental period. The activity of lame cows decreased sharply from the first day to the second day after treatment. RT in healthy cows tended to decrease during the experiment and was higher than in lame cows from the start of the experiment until the 11th day after treatment. The reticulorumen pH in lame cows also increased on the day of the appearance of clinical signs.

From a practical point of view, we recommend the use of the following biomarkers for the early diagnosis of lameness in fresh dairy cows: milk yield, milk electrical conductivity, milk composition (protein, lactose, and fat), walking activity, rumination time, and reticulorumen pH. In future studies, we recommend extending the study period to 30 days or more before the onset of clinical signs of lameness.

Author Contributions: R.A.: supervision of the whole study; V.J.: software and algorithm development, design, and setup of field experiments; G.U.: data collection and analysis; D.M.: setup of the field experiment and data collection, selection, and management of the experimental group of animals; M.T.: research consultancy; M.U.: design of field experiment and data collection and analysis; and W.B.: intensive support in processing of data in the manuscript. The manuscript was written by R.A. and revised by all co-authors. All authors have read and agreed to the published version of the manuscript.

Funding: This research received no external funding.

Institutional Review Board Statement: The study was conducted according to the guidelines of the Declaration of Helsinki, and approved by Ethics Committee (The study approval number is PK016965, 2017.06.06).

Informed Consent Statement: Not applicable.

Data Availability Statement: The data presented in this study are available within the article.

Conflicts of Interest: The authors declare no conflict of interest.

References

1. Bennett, R.; Barker, Z.; Main, D.; Whay, H.; Leach, K. Investigating the value dairy farmers place on a reduction of lameness in their herds using a willingness to pay approach. *Vet. J.* **2014**, *199*, 72–75. [CrossRef] [PubMed]
2. Green, L.; Huxley, J.; Banks, C.; Green, M. Temporal associations between low body condition, lameness and milk yield in a UK dairy herd. *Prev. Vet. Med.* **2014**, *113*, 63–71. [CrossRef] [PubMed]
3. Huxley, J. Impact of lameness and claw lesions in cows on health and production. *Livest. Sci.* **2013**, *156*, 64–70. [CrossRef]
4. Zhang, X.; Ding, J.; Li, Y.; Song, Q.; Li, S.; Hayat, M.A.; Zhang, J.; Wang, H. The changes of inflammatory mediators and vasoactive substances in dairy cows' plasma with pasture-associated laminitis. *BMC Vet. Res.* **2020**, *16*, 119. [CrossRef]
5. Donovan, G.A.; Risco, C.A.; Temple, G.D.; Tran, T.Q.; Van Horn, H.H. Influence of transition diets on occurrence of sub-clinical laminitis in Holstein dairy cows. *J. Dairy Sci.* **2004**, *87*, 73–84. [CrossRef]
6. Shearer, J.K. *Lameness of Dairy Cattle: Consequences and Causes*; Department of Large Animal Clinical Sciences College of Veterinary Medicine University of Florida Gainesville: Gainesville, FL, USA, 1997; pp. 38–50.
7. King, M.; Leblanc, S.; Pajor, E.; Devries, T. Cow-level associations of lameness, behavior, and milk yield of cows milked in automated systems. *J. Dairy Sci.* **2017**, *100*, 4818–4828. [CrossRef] [PubMed]
8. Bonfatti, V.; Ho, P.; Pryce, J. Usefulness of milk mid-infrared spectroscopy for predicting lameness score in dairy cows. *J. Dairy Sci.* **2020**, *103*, 2534–2544. [CrossRef] [PubMed]
9. Bramley, E.; Costa, N.; Fulkerson, W.; Lean, I. Associations between body condition, rumen fill, diarrhoea and lameness and ruminal acidosis in Australian dairy herds. *N. Z. Vet. J.* **2013**, *61*, 323–329. [CrossRef]
10. Nocek, J.E. Bovine Acidosis: Implications on Laminitis. *J. Dairy Sci.* **1997**, *80*, 1005–1028. [CrossRef]

11. Plaizier, J.; Krause, D.; Gozho, G.; McBride, B. Subacute ruminal acidosis in dairy cows: The physiological causes, incidence and consequences. *Vet. J.* **2008**, *176*, 21–31. [CrossRef]
12. AlZahal, O.; Kebreab, E.; France, J.; Froetschel, M.; McBride, B. Ruminal Temperature May Aid in the Detection of Subacute Ruminal Acidosis. *J. Dairy Sci.* **2008**, *91*, 202–207. [CrossRef] [PubMed]
13. AlZahal, O.; Steele, M.; Van Schaik, M.; Kyriazakis, I.; Duffield, T.; McBride, B.; AlZahal, H. The use of a radiotelemetric ruminal bolus to detect body temperature changes in lactating dairy cattle. *J. Dairy Sci.* **2011**, *94*, 3568–3574. [CrossRef] [PubMed]
14. Zhang, G.; Hailemariam, D.; Dervishi, E.; Deng, Q.; Goldansaz, S.A.; Dunn, S.M.; Ametaj, B.N. Alterations of Innate Immunity Reactants in Transition Dairy Cows before Clinical Signs of Lameness. *Animals* **2015**, *5*, 717–747. [CrossRef] [PubMed]
15. Enemark, D.J.M.; Jørgensen, R.J.; Enemark, S.S. Rumen acidosis with special emphasis on diagnostic aspects of subclinical rumen acidosis: A review. *Vet. Zootech.* **2002**, *20*, 42.
16. Miekley, B.; Traulsen, I.; Krieter, J. Detection of mastitis and lameness in dairy cows using wavelet analysis. *Livest. Sci.* **2012**, *148*, 227–236. [CrossRef]
17. Lukas, J.; Reneau, J.; Wallace, R.; Hawkins, D.; Munoz-Zanzi, C. A novel method of analyzing daily milk production and electrical conductivity to predict disease onset. *J. Dairy Sci.* **2009**, *92*, 5964–5976. [CrossRef] [PubMed]
18. Khatun, M. Data-Based Approaches to Improve Accuracy and Timing of Mastitis Detection in Automatic Milking Systems. Ph.D. Thesis, The University of Sydney, Sydney, Australia, 2019.
19. Walker, S.L.; Smith, R.; Routly, J.E.; Jones, D.N.; Morris, M.J.; Dobson, H. Lameness, Activity Time-Budgets, and Estrus Expression in Dairy Cattle. *J. Dairy Sci.* **2008**, *91*, 4552–4559. [CrossRef]
20. Mazrier, H.; Tal, S.; Aizinbud, E.; Bargai, U. A field investigation of the use of the pedometer for the early detection of lameness in cattle. *Can. Vet. J.* **2006**, *47*, 883–886.
21. Gonzalez, L.A.; Tolkamp, B.J.; Coffey, M.; Ferret, A.; Kyriazakis, I. Changes in Feeding Behavior as Possible Indicators for the Automatic Monitoring of Health Disorders in Dairy Cows. *J. Dairy Sci.* **2008**, *91*, 1017–1028. [CrossRef]
22. O'Callaghan, K.A.; Cripps, P.J.; Downham, D.Y.; Murray, R.D. Subjective and objective assessment of pain and discomfort due to lameness in dairy cattle. *Anim. Welf.* **2003**, *12*, 605–610.
23. King, M.; Pajor, E.; Leblanc, S.; Devries, T. Associations of herd-level housing, management, and lameness prevalence with productivity and cow behavior in herds with automated milking systems. *J. Dairy Sci.* **2016**, *99*, 9069–9079. [CrossRef] [PubMed]
24. Jacobs, J.A.; Siegford, J.M. Invited review: The impact of automatic milking systems on dairy cow management, behavior, health, and welfare. *J. Dairy Sci.* **2012**, *95*, 2227–2247. [CrossRef] [PubMed]
25. King, M.; Leblanc, S.; Pajor, E.; Wright, T.; Devries, T. Behavior and productivity of cows milked in automated systems before diagnosis of health disorders in early lactation. *J. Dairy Sci.* **2018**, *101*, 4343–4356. [CrossRef] [PubMed]
26. Sprecher, D.; Hostetler, D.; Kaneene, J. A lameness scoring system that uses posture and gait to predict dairy cattle reproductive performance. *Theriogenology* **1997**, *47*, 1179–1187. [CrossRef]
27. Warnick, L.; Janssen, D.; Guard, C.; Gröhn, Y. The Effect of Lameness on Milk Production in Dairy Cows. *J. Dairy Sci.* **2001**, *84*, 1988–1997. [CrossRef]
28. Bach, A.; Dinarés, M.; Devant, M.; Carré, X. Associations between lameness and production, feeding and milking attendance of Holstein cows milked with an automatic milking system. *J. Dairy Res.* **2006**, *74*, 40–46. [CrossRef]
29. Deming, J.A.; Bergeron, R.; Leslie, K.E.; Devries, T.J. Associations of cow-level factors, frequency of feed delivery, and standing and lying behaviour of dairy cows milked in an automatic system. *Can. J. Anim. Sci.* **2013**, *93*, 427–433. [CrossRef]
30. Weigele, H.C.; Gygax, L.; Steiner, A.; Wechsler, B.; Burla, J.B. Moderate lameness leads to marked behavioral changes in dairy cows. *J. Dairy Sci.* **2018**, *101*, 2370–2382. [CrossRef]
31. Gáspárdy, A.; Ismach, G.; Bajcsy, Á.C.; Veress, G.; Márkus, S.; Komlósi, I. Evaluation of the on-line electrical conductivity of milk in mastitic dairy cows. *Acta Vet. Hung.* **2012**, *60*, 145–155. [CrossRef] [PubMed]
32. Ito, K.; von Keyserlingk, M.; Leblanc, S.; Weary, D. Lying behavior as an indicator of lameness in dairy cows. *J. Dairy Sci.* **2010**, *93*, 3553–3560. [CrossRef] [PubMed]
33. Singh, A.; Singh, S.; Gupta, D.K.; Bansal, B.K. Relationship of lameness to body condition score, udder health and milk quality in crossbred dairy cattle. *Vet. Arh.* **2018**, *88*, 179–190. [CrossRef]
34. Olechnowicz, J.; Jaskowski, J.M. Impact of clinical lameness, calving season, parity, and month of lactation on milk, fat, protein, and lactose yields during early lactation of dairy cows. *Bull. Vet. Inst. Pulawy* **2010**, *54*, 605–610.
35. Olechnowicz, J.; Jaskowski, M.J. Relationship between clinical lameness and somatic cell counts, and fat and protein contents in the milk of dairy cows. *Med. Weter* **2012**, *68*, 12.
36. Randall, L.; Green, M.; Chagunda, M.; Mason, C.; Archer, S.; Green, L.; Huxley, J. Low body condition predisposes cattle to lameness: An 8-year study of one dairy herd. *J. Dairy Sci.* **2015**, *98*, 3766–3777. [CrossRef]
37. Dyer, R.; Neerchal, N.; Tasch, U.; Wu, Y.; Dyer, P.; Rajkondawar, P. Objective Determination of Claw Pain and Its Relationship to Limb Locomotion Score in Dairy Cattle. *J. Dairy Sci.* **2007**, *90*, 4592–4602. [CrossRef] [PubMed]
38. Rutten, C.; Velthuis, A.; Steeneveld, W.; Hogeveen, H. Invited review: Sensors to support health management on dairy farms. *J. Dairy Sci.* **2013**, *96*, 1928–1952. [CrossRef]
39. Nechanitzky, K.; Starke, A.; Vidondo, B.; Müller, H.; Reckardt, M.; Friedli, K.; Steiner, A. Analysis of behavioral changes in dairy cows associated with claw horn lesions. *J. Dairy Sci.* **2016**, *99*, 2904–2914. [CrossRef] [PubMed]

40. Jabbar, K.A.; Hansen, M.F.; Smith, M.L.; Smith, L. Early and non-intrusive lameness detection in dairy cows using 3-dimensional video. *Biosyst. Eng.* **2017**, *153*, 63–69. [CrossRef]
41. Thorup, V.; Munksgaard, L.; Robert, P.-E.; Erhard, H.; Thomsen, P.T.; Friggens, N. Lameness detection via leg-mounted accelerometers on dairy cows on four commercial farms. *Animals* **2015**, *9*, 1704–1712. [CrossRef]
42. Solano, L.; Barkema, H.; Pajor, E.; Mason, S.; Leblanc, S.; Nash, C.; Haley, D.; Pellerin, D.; Rushen, J.; De Passillé, A.; et al. Associations between lying behavior and lameness in Canadian Holstein-Friesian cows housed in freestall barns. *J. Dairy Sci.* **2016**, *99*, 2086–2101. [CrossRef] [PubMed]
43. Mandel, R.; Harazy, H.; Gygax, L.; Nicol, C.; Ben-David, A.; Whay, H.; Klement, E. Short communication: Detection of lameness in dairy cows using a grooming device. *J. Dairy Sci.* **2018**, *101*, 1511–1517. [CrossRef]
44. Blackie, N.; Amory, J.; Bleach, E.; Scaife, J. The effect of lameness on lying behaviour of zero grazed Holstein dairy cattle. *Appl. Anim. Behav. Sci.* **2011**, *134*, 85–91. [CrossRef]
45. Juarez, S.T.; Robinson, P.H.; DePeters, E.J.; Price, E.O. Impact of lameness on behavior and productivity of lactating Holstein cows. *Appl. Anim. Behav. Sci.* **2003**, *83*, 1–14. [CrossRef]
46. Beer, G.; Alsaaod, M.; Starke, A.; Schüpbach-Regula, G.; Müller, H.; Kohler, P.; Steiner, A. Use of Extended Characteristics of Locomotion and Feeding Behavior for Automated Identification of Lame Dairy Cows. *PLoS ONE* **2016**, *11*, e0155796. [CrossRef]
47. Palmer, M.A.; Law, R.; O'Connell, N. Relationships between lameness and feeding behaviour in cubicle-housed Holstein–Friesian dairy cows. *Appl. Anim. Behav. Sci.* **2012**, *140*, 121–127. [CrossRef]
48. Bareille, N.; Beaudeau, F.; Billon, S.; Robert, A.; Faverdin, P. Effects of health disorders on feed intake and milk production in dairy cows. *Livest. Prod. Sci.* **2003**, *83*, 53–62. [CrossRef]
49. Bicalho, R.; Machado, V.; Caixeta, L. Lameness in dairy cattle: A debilitating disease or a disease of debilitated cattle? A cross-sectional study of lameness prevalence and thickness of the digital cushion. *J. Dairy Sci.* **2009**, *92*, 3175–3184. [CrossRef] [PubMed]
50. Norring, M.; Haggman, J.; Simojoki, H.; Tamminen, P.; Winckler, C.; Pastell, M. Short communication: Lameness impairs feeding behavior of dairy cows. *J. Dairy Sci.* **2014**, *97*, 4317–4321. [CrossRef]
51. Thorup, V.M.; Nielsen, B.L.; Robert, P.-E.; Giger-Reverdin, S.; Konka, J.; Michie, W.; Friggens, N.C. Lameness Affects Cow Feeding but Not Rumination Behavior as Characterized from Sensor Data. *Front. Vet. Sci.* **2016**, *3*, 37. [CrossRef] [PubMed]
52. Kilgour, R.J. In pursuit of "normal": A review of the behaviour of cattle at pasture. *Appl. Anim. Behav. Sci.* **2012**, *138*, 1–11. [CrossRef]
53. Schirmann, K.; Chapinal, N.; Weary, D.; Heuwieser, W.; von Keyserlingk, M. Rumination and its relationship to feeding and lying behavior in Holstein dairy cows. *J. Dairy Sci.* **2012**, *95*, 3212–3217. [CrossRef] [PubMed]
54. Dirksen, G. Rumen Function and Disorders Related to Production Disease. In Proceedings of the VII International Conference Distance Farm Animal; Cornell University: Ithaca, NY, USA, 1989; p. 350.
55. Gentile, G.; Cinotti, S.; Ferri, G.; Famigli-Bergamini, P. Nutritional acidosis and technological characteristics of milk in high producing dairy cows. In Proceedings of the XIV World Congress on Diseases of Cattle, Dublin, Ireland, 26–29 August 1986; pp. 823–829.
56. Bergsten, C. Hemorrhages of the sole horn of dairy cows as a retrospective indicator of laminitis: An epidemiological study. *Acta Vet. Scand.* **1994**, *35*, 55. [CrossRef] [PubMed]
57. Boosman, R.; Németh, F.; Gruys, E.; Klarenbeek, A. Arteriographical and pathological changes in chronic laminitis in dairy cattle. *Vet. Q.* **1989**, *11*, 144–155. [CrossRef] [PubMed]
58. Brent, B.E. Relationship of Acidosis to Other Feedlot Ailments. *J. Anim. Sci.* **1976**, *43*, 930–935. [CrossRef] [PubMed]
59. Ossent, P.; Lischer, C.J. Theories on the pathogenesis of bovine laminitis. In *Proceedings of the 8th International Symposium Disorder Ruminant Digit International Conference*; Greenough, P.R., Ed.; Bovine Lameness: Saskatoon, SK, Canada; Banff, AB, Canada, 1994; p. 207.

agriculture

MDPI

Article

Impact of Lameness on Attributes of Feeding Registered with Noseband Sensor in Fresh Dairy Cows

Ramūnas Antanaitis [1,*], Vida Juozaitienė [2], Gediminas Urbonavičius [3], Dovilė Malašauskienė [1], Mindaugas Televičius [1], Mingaudas Urbutis [1] and Walter Baumgartner [4]

1 Large Animal Clinic, Veterinary Academy, Lithuanian University of Health Sciences, Tilžės str 18, LT-47181 Kaunas, Lithuania; dovile.malasauskiene@lsmuni.lt (D.M.); mindaugas.televicius@lsmuni.lt (M.T.); mingaudas.urbutis@lsmuni.lt (M.U.)
2 Department of Biology, Faculty of Natural Sciences, Vytautas Magnus University, K. Donelaičio 58, LT-44248 Kaunas, Lithuania; vida.juozaitiene@vdu.lt
3 Department of Animal Breeding, Veterinary Academy, Lithuanian University of Health Sciences, Tilžės str 18, LT-47181 Kaunas, Lithuania; gediminas.urbonavicius@lsmuni.lt
4 University Clinic for Ruminants, University of Veterinary Medicine, Veterinaerplatz 1, A-1210 Vienna, Austria; walter.baumgartner@vetmeduni.ac.at
* Correspondence: ramunas.antanaitis@lsmuni.lt; Tel.: +370-67349064

Abstract: We hypothesized that lameness in fresh dairy cows (1–30 days after calving) has an impact on attributes of feeding registered with a noseband sensor. The aim of this study was to investigate the impact of lameness in fresh dairy cows on attributes of feeding (registered with the RumiWatch noseband sensor): rumination time (RT), drinking time (DT), eating time (ET), rumination chews (RC), eating chews (EC), chews per minute (CM), drinking gulps (DG), bolus count (B), and chews per bolus (CB). The measurement registration was started at the first day after calving and continued until 30 days after calving. There were 20 Lithuanian black and white breed cows selected. Lameness diagnosis was performed by trained staff based on a locomotion score system and it was diagnosed on average on the 15th day after calving. The causes of lameness were categorized as sole ulcer, abscess and foot rot. Special attention was paid to attributes of feeding registered 14 days before and 13 days after diagnosis. The 10 lame cows (LG) used in this experiment had a lameness score of 3–4 presented with severe lameness: they were reluctant to move and unwilling to complete weight transfer off the affected limb. The 10 healthy cows (HG) were given a lameness score of 1. We found that lameness of fresh dairy cows has an impact on inline registered ingestive behaviors biomarkers—the mean RT of HG cows was as much as 2.19 times higher than that of LG cows on the day of diagnosis of lameness, later this difference between the groups decreased to the sixth day of treatment, then increased again and decreased at the end of the experiment. The lowest eating time was found on diagnosis day and the highest on the ninth day before determination of lameness. Drinking time was higher in the HG group, with the exception of 10 and 9 days prior to clinical signs of disease in LG cows. A downward trend in rumination chews was observed in LG cows from day 7 until the onset of clinical symptoms. The bolus count decreased from day 3 before diagnosis to day 1 after diagnosis in LG cows. The largest difference in this indicator between groups was found on day of diagnosis. Analysing the pattern of CM values in the LG group, we found a decrease from 10 days before to 2 days after diagnosis. The CB value was almost the same in both groups of cows at the end of the experiment, but largest difference between the groups was found on day 7 after clinical sings of lameness.

Keywords: rumination; locomotion; precision dairy farming

Citation: Antanaitis, R.; Juozaitienė, V.; Urbonavičius, G.; Malašauskienė, D.; Televičius, M.; Urbutis, M.; Baumgartner, W. Impact of Lameness on Attributes of Feeding Registered with Noseband Sensor in Fresh Dairy Cows. *Agriculture* **2021**, *11*, 851. https://doi.org/10.3390/agriculture11090851

Academic Editors: Gniewko Niedbała, Sebastian Kujawa and Francesco Marinello

Received: 19 July 2021
Accepted: 4 September 2021
Published: 6 September 2021

1. Introduction

In dairy farming, lameness is one of the major welfare and health problems [1]. Traditionally, the detection of lameness has relied on visual inspection to detect animals with abnormal gait [2]. The early detection of health disorders is an important point in dairy cattle farming, since early detection of the disease has a major influence on the severity of diseases, the animal welfare and the economics of dairy production [1]. Identification of healthy and lame cows was entirely based on the herd health reports of the farm. The accuracy of these herd health reports depended on the correct identification of lameness by the herdsman, the amount of reported cases of lameness and the expertise of the veterinarian [3]. The monitoring of physiological parameters may allow to detect changes before the appearance of clinical signs and earlier diagnosis would greatly benefit cows by preventing disease progression [4]. Sensors have become an integral part of modern dairy farming, which allows for continuous monitoring of the individual in the herd [3]. Accelerometer sensors are used in lame animal detection in livestock industries [2]. The automation of lameness detection or gait scoring is an important topic in research and several different approaches have been developed [3]. Studies focusing on ground reaction forces, pressure sensitive walkways and accelerometers measured the asymmetry of the gait when walking, other studies used computer vision to analyze the gait automatically. In these studies, the main focus was put on gait parameters-step overlap, hoof release angle and back arch curvature [5–7]. Raw data of Benaissa et al. (2019) [8]. illustrate that the widely used collar-mounted accelerometer classifies ruminating and feeding behaviors with accuracy closely comparable to the RumiWatch noseband sensor [8]. In our past studies, we found that cows with subclinical ketosis had shorter rumination and drinking times and lesser rumination chews, chews per minute, boluses, and chews per bolus [9]. Van Hertem et al. (2013) reported that for the automated detection of lameness a combination of farm data on the milk yield, rumination time and neck activity can be used [3]. During the nighttime, a significantly lower rumination time and milk yield were registered in lame animals. Lame animals showed a higher night to day ratio of neck activity. In a 10-fold cross-validation procedure, a logistic regression model for automated clinical lameness detection according to behavioral and performance data achieved a correct classification rate of 0.86, a specificity of 0.85, and a sensitivity of 0.89. An ear-tag sensor, a type of on-animal sensor, was used to precisely detect walking, standing, grazing, and lame walking activities [3]. Furthermore, strong correlations with lameness are suggested with unique animal-based parameters—such as leg hygiene, body condition scoring (BCS), and hock condition—that have been used in animal welfare determination practices [10]. With more research and use, this classification algorithm could be integrated into an automatic livestock monitoring system to provide real time information on an individual's health status, that is problematic under current extensive livestock production systems [2]. A tri-axial accelerometer fitted in the ear-tag can successfully differentiate lame walking activity from normal walking, standing and grazing behaviors. In one research paper, it was noted that the leg and collar deployed accelerometers failed to successfully classify both normal and lame walking activity. However, this could have been affected by sensor placement and the simulation not being a true representation of lameness [2]. Miguel-Pacheco et al. (2014) stated that further studies are needed in order to expand the use and benefits of the technologies available as a tool to measure and monitor the health status of cows [11].

The RumiWatch Sensor (RWS) built into the cow nose-halter, uses data gathered by a pressure sensor in combination with data gathered by a triaxial accelerometer to detect specific behavioral characteristics in dairy cows [12]. The RWS is able to record and store individual animal behavior over several days with high precision [13,14]. The RWS is a useful tool for research as it records bolus counts, number of rumination chews, total number of eating chews, and the time spent ruminating for grazing cows and supplemented grazing cows [12].

We hypothesized that lameness of fresh dairy cows (1–30 days after calving) has an impact on the attributes of feeding registered by the noseband sensor. The aim of this study was to investigate the impact of lameness on attributes of feeding registered with the RumiWatch noseband sensor: rumination time, drinking time, eating time, rumination chews, eating chews, drinking gulps, bolus count, chews per minute, and chews per bolus before and after diagnosis of lameness fresh dairy cows.

2. Materials and Methods

2.1. Animals, Location, and Experimental Design

This study was conducted at the Lithuanian University of health sciences, and one commercial dairy farm (54.9754° N, 23.7684° E) during 1 July 2020–15 December 2020. All animal experimental procedures were approved, approval number is PK016965. During investigation time, the total incidence of lameness in the herd was about 10%. Fifty from 1550 Lithuanian Black and White breed multiparous dairy cows coming up to parturition were selected (on average 5 years of age), which were being kept in a loose-housing system. The recording of the feeding behavioral data was started at the first day after calving and continued until 30 days after calving by applying the RumiWatch noseband halter.

The selected cows were monitored daily for signs of lameness. Lameness diagnosis was performed by trained staff according to the standard procedure described by Sprecher et al. [15]: 1 = normal; 2 = presence of a slightly asymmetric gait; 3 = the cow clearly favored one or more limbs (moderately lame); 4 = severely lame; and 5 = extremely lame (non-weight-bearing lame). Of cows monitored in this experiment, 10 out of 50 had a lameness score of 3–4, represented by severe lameness and they were diagnosed on average on the 15th day post partum. They were reluctant to move and unwilling to complete weight transfer off the affected limb. The 10 lame cows comprised the lame group (LG) and were examined by a claw health professional to determine the cause of lameness. Six of the cows were diagnosed with sole ulcers, three with abscess, and one with foot rot. These results match the findings of Warnick et al. [16], where the causes for lameness in a herd were categorized as sole ulcer (60%), abscess (20%), and foot rot (20%). According Warnick et al. [16], sole ulcer was defined as degenerative or necrotic defects in the sole near the sole–heel junction. Abscess was defined as a pus-filled cavity of the white line or sole of the foot. Foot rot was swelling of the soft tissues of the foot resulting in symmetrical swelling above the coronary band and spreading of the toes, in some cases with necrosis of the tissue between the claws.

Immediately after identification of lameness, all lame cows had their claws treated and received subcutaneous administrations of Naxcel (100 mg/mL; Zoetis, Kirkland, QC, Canada), at the dosage of 2.2 mg/kg of body weight. Treatment was repeated at 24-h intervals for a total of three consecutive days and a subcutaneous injection of Rimadyl Cattle® (50 mg/mL) solution (Zoetis, Belgium) at a dose of 1.4 mg per 1 kg body weight once. After treatment lameness was identified by the same trained staff according to the standard procedure described by Sprecher et al. [15]. After treatment, the lameness score of the cows improved.

Based on the principle of analogues (Table 1), 10 healthy cows (HG) were selected-same breed, same production, and lactation number. Cows in the healthy group had maintained a lameness score of 1 throughout the study period. Data from the RumiWatch noseband halter from 14 days before and 13 days after the diagnosis of lameness was used to compare with non-lame cow data during the same period. The healthy cows were monitored during the same days as the lame cows.

Table 1. Characteristics of experimental groups.

Group	Breed	Average of DIM	Average of Milk Yield of Past Lactation (kg/Year)	Partulation	Average of Number of Lactation
HG	Lithuanian Black and White	15	12500 (±500)	Multiparous	3
LG	Lithuanian Black and White	15	12100 (±350)	Multiparous	3

HG—healthy cows; LG—lame group; DIM—days in milk; kg/year—kilograms per year.

Cows were provided with a total mix ration (TMR) consisting of 20% corn silage, 5% grass hay, 20% grass silage, 50% grain concentrate mash, and 5% of mineral mixture. TMR was formulated accordingly to meet or exceed the requirements of a 550 kg Holstein cow producing 40 kg/d of milk. Diet was delivered every day at 05.00 and 17.00 h. Composition of ration are presented in Table 2. Cows were milked with DeLaval milking robot (DeLaval Inc., Tumba, Sweden).

Table 2. Composition of ration.

Parameters	Values	Units
Dry matter (DM)	49	%
Neutral detergent fiber	28	% of DM
Acid detergent fiber	20	% of DM
Non-fiber carbohydrates	39	% of DM
Crude protein	16	% of DM
Net energy for lactation	1.7	Mcal/kg

DM—dry matter; % of DM—percent of dry matter; Mcal/kg—megacalorie per kilogram.

2.2. Measurements

Experimental days from '−14' to '−1' denote the period before the onset of clinical signs of lameness (day '0') in the LG (lame) group of cows and days '1' through '13' indicate the period from the start of lameness treatment and recovery period.

Throughout the investigation time, the RumiWatch Sensor (RWS; Itin + Hoch GmbH, Liestal, Switzerland), for the measure of the ruminative behavior, was used on the 50 cows selected for the study (Figure 1). The RWS is comprised of a noseband halter that has a built-in pressure detector and a liquid-filled pressure tube.

Figure 1. Cow with the RumiWatch Sensor.

The pressure sensor and the tube are placed in a belt on the nose bridge of the cow. The triaxial accelerometer, the data logger, and a secure digital memory card are located on the right belt ring, mounted in a waterproof plastic box. In another box on the left belt ring, the batteries are incorporated. Data collected by the pressure sensor and the triaxial accelerometer were recorded with a frequency of 10 Hz and saved on the secure digital memory card. To ensure the best possible detection of the jaw movements by the pressure sensor, the belt around the nose and the lower jaw left 3 to 5 cm of movement space to the nose bridge and was located between 11 and 16 cm behind the nasal tip. The recorded data were read out by the RumiWatch Manager (version 2.1.0.0, Itin and Hoch GmbH) and processed through the evaluation software, called the RumiWatch Converter (C2, version 0.7.3.2, Itin and Hoch GmbH).

With the help of RWS the following attributes of feeding were registered: Drinking time (DT)—time spent for drinking, including interruptions between drinking gulps up to 5 s.; Rumination time (RT)—time spent for rumination chews including chewing interruptions up to 5 s.; Eating time (ET)—time spent for eating chews, including interruptions between eating chews up to 5 s.; Rumination chews (RC)—number of chews during rumination for breakdown of the regurgitated materials into smaller particles using the molars; Eating chews (EC)—total number of prehension bites and breakdown chews while eating; Drinking gulps (DG)—total number of drinking gulps while drinking; Bolus (B)—number of boluses during rumination; Chews per minute during rumination (CM)—chews for one minute; Chews per bolus (CB)—chews performed during rumination between the regurgitation and swallowing of one bolus.

2.3. *Data Analysis and Statistics*

The statistical analysis of RumiWatch data was performed with the SPSS 25.0 (SPSS Inc., Chicago, IL, USA) program package. The normal distribution of variables was assessed using the Kolmogorov–Smirnov test. After that, we used the general linear repeated measures model and Fisher's standard deviation criterion to compare the indicators between the study (healthy and lame) groups.

The results of RumiWatch indicators were given as least square means (LSM) and standard deviation (SD). A linear regression was applied to determine the change in the estimated RumiWatch variables during the two experiment periods: (1) period until the onset of clinical signs (days from '−14' to '−1') and (2) treatment period (days from '0' to '13'). A probability of less than 0.05 was considered significant ($p < 0.05$).

3. Results

3.1. *Impact of Fresh Dairy Cows Lameness on Ingestive Behaviors Registered with RumiWatch Noseband Sensor*

As can be seen from the data in in the Table 3, the average values of all indicators recorded by the RumiWatch sensor were significantly higher in healthy cows. The average indicators of HG cows were higher than the average indicators of lame cows from 1.1 times ($p \leq 0.05$; drinking gulps and chews per minute) to 3.1 times ($p < 0.001$; chews per bolus).

3.2. *Changes in Ingestive Behaviors Registered with RumiWatch Noseband Sensor in Lame and Non-Lame Cows during the Experiment*

The rumination time, eating time, rumination chews and bolus count indicators in healthy cows were higher ($p < 0.001$) than in lame cows throughout the experiment (Figure 1).

In the lame group, there was a linear trend towards a decrease in rumination time ($R^2 = 0.30$), rumination chews ($R^2 = 0.31$), and bolus ($R^2 = 0.25$) indicators from the beginning of the experiment to lameness identification day. In the healthy group, a similar trend was observed only for drinking gulps values ($R^2 = 0.26$). For lame cows, an increase in drinking gulps ($R^2 = 0.54$) and chews per minute ($R^2 = 0.37$) was observed from the start of lameness treatment to the end of the experiment. Other indicators changed less (Table 4).

Table 3. Results of RumiWatch noseband sensor data of dairy cows that were eventually diagnosed with subclinical lameness and cows that remained healthy. Least square means (LSM) and standard deviation (SD) by groups of cows.

Indicator	Healthy	Lame	*p*
RT (min/h)	24.56 ± 0.44	15.94 ± 0.43	<0.001
ET (min/h)	6.44 ± 0.2	3.71 ± 0.12	<0.001
DT (min/h)	0.70 ± 0.03	0.28 ± 0.03	<0.001
RC (n/h)	1558.57 ± 33.12	1053.9 ± 34.88	<0.001
EC (n/h)	430.96 ± 13.25	242.69 ± 13.35	<0.001
DG (n/h)	152.96 ± 5	135.61 ± 4.21	0.048
B (n/rumination)	24.55 ± 0.52	16.32 ± 0.54	<0.001
CM (n/min)	61.66 ± 1.1	56.75 ± 1.02	0.050
CB (n/rumination)	11.10 ± 0.4	3.61 ± 0.42	<0.001

RT—Rumination time (time in minutes spent for rumination chews); ET—Eating time (time in minutes spent for eating chews); DT—Drinking time (time in minutes spent for drinking); RC—Rumination chews (chews during rumination for mechanical breakdown of the regurgitated materials); EC—Eating chews (number of prehension bites); DG—Drinking gulp (total number of drinking gulps while drinking); B—number of boluses per rumination) CM—Chews per minute (chews for one minute); CB—Chews per bolus (chews performed during rumination).

Table 4. Evaluation of changes in ingestive-related behaviors Indicators by period of the experiment.

Indicator	Period until the Onset of Clinical Signs		Treatment Period	
	Healthy	Lame	Healthy	Lame
RT	$y = -0.14x + 24.61$, $R^2 = 0.1$, $p = 0.75$	$y = -0.33x + 13.99$, $R^2 = 0.3$, $p < 0.001$	$y = 0.1x + 22.69$, $R^2 = 0.05$, $p = 0.45$	$y = 0.18x + 14.28$, $R^2 = 0.07$, $p = 0.39$
ET	$y = -0.1x + 5.93$, $R^2 = 0.16$, $p = 0.16$	$y = -0.05x + 3.47$, $R^2 = 0.05$, $p = 0.39$	$y = -0.04x + 6.48$, $R^2 = 0.05$, $p = 0.41$	$y = -0.002x + 3.62$, $R^2 = 0.0007$, $p = 0.89$
DT	$y = 0.01x + 0.69$, $R^2 = 0.17$, $p = 0.15$	$y = -0.008x + 0.22$, $R^2 = 0.15$, $p = 0.18$	$y = -0.02x + 0.95$, $R^2 = 0.3$, $p < 0.001$	$y = 0.005x + 0.23$, $R^2 = 0.05$, $p = 0.5$
RC	$y = -1.68x + 1595.3$, $R^2 = 0.002$, $p = 0.67$	$y = -24.56x + 923.8$ $R^2 = 0.31$, $p < 0.001$	$y = 3.67x + 1470$, $R^2 = 0.007$, $p = 0.73$	$y = 20.85x + 866.65$, $R^2 = 0.15$, $p = 0.17$
EC	$y = -8.83x + 382.05$, $R^2 = 0.22$, $p = 0.082$	$y = -4.43x + 225.11$, $R^2 = 0.11$, $p = 0.18$	$y = -0.49x + 419.57$, $R^2 = 0.001$, $p = 0.86$	$y = 2.94x + 206.64$, $R^2 = 0.11$, $p = 0.19$
D	$y = -4.52x + 145.63$, $R^2 = 0.25$, $p = 0.049$	$y = -5.04x + 133.83$, $R^2 = 0.21$, $p = 0.083$	$y = 3.34x + 84.59$, $R^2 = 0.15$, $p = 0.163$	$y = 5.95x + 92.65$, $R^2 = 0.53$, $p < 0.001$
B	$y = 0.11x + 26.53$, $R^2 = 0.07$, $p = 0.32$	$y = -0.25x + 14.75$, $R^2 = 0.24$, $p = 0.05$	$y = 3.34x + 84.59$, $R^2 = 0.15$, $p = 0.16$	$y = 0.2x + 14.65$, $R^2 = 0.09$, $p = 0.22$
CM	$y = 0.14x + 64.18$, $R^2 = 0.01$, $p = 0.62$	$y = -0.19x + 59.14$, $R^2 = 0.01$, $p = 0.68$	$y = -0.42x + 62.89$, $R^2 = 0.10$, $p = 0.27$	$y = 1.67x + 42.88$, $R^2 = 0.37$, $p < 0.001$
CB	$y = 0.19x + 10.66$, $R^2 = 0.18$, $p = 0.09$	$y = -0.11x + 2.76$, $R^2 = 0.23$, $p = 0.06$	$y = -0.45x + 16.37$, $R^2 = 0.25$, $p = 0.05$	$y = 0.15x + 2.56$, $R^2 = 0.13$, $p = 0.22$

RT—Rumination time (time in minutes spent for rumination chews); ET—Eating time (time in minutes spent for eating chews); DT—Drinking time (time in minutes spent for drinking); RC—Rumination chews (chews during rumination for mechanical breakdown of the regurgitated materials); EC—Eating chews (number of prehension bites); DG—Drinking gulp (total number of drinking gulps while drinking); B—number of boluses per rumination); CM—Chews per minute (chews for one minute); CB—Chews per bolus (chews performed during rumination).

The mean rumination time of healthy cows was as much as 2.19 times higher ($p < 0.001$) than that of lame cows on the lameness identification day, later this difference between the groups decreased to the sixth day of treatment, then increased again and decreased at the end of the experiment (Figure 2A). The changes in rumination time of lame cows could be described by linear regression, showing a downward trend in this indicator from the beginning to the end of the experiment. In the healthy group, a decreasing trend of this indicator was also observed during the experiment, but the value of $\mathbf{R}^2$ was small (Table 2).

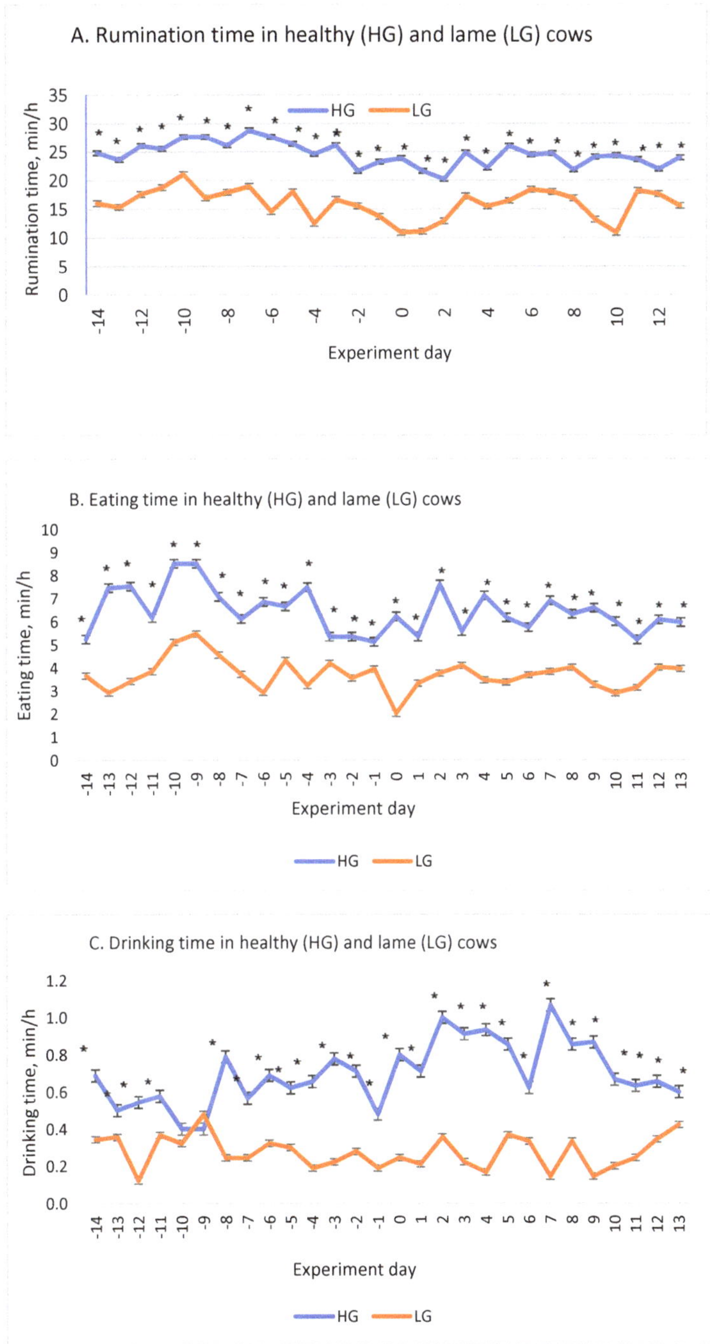

Figure 2. *Cont.*

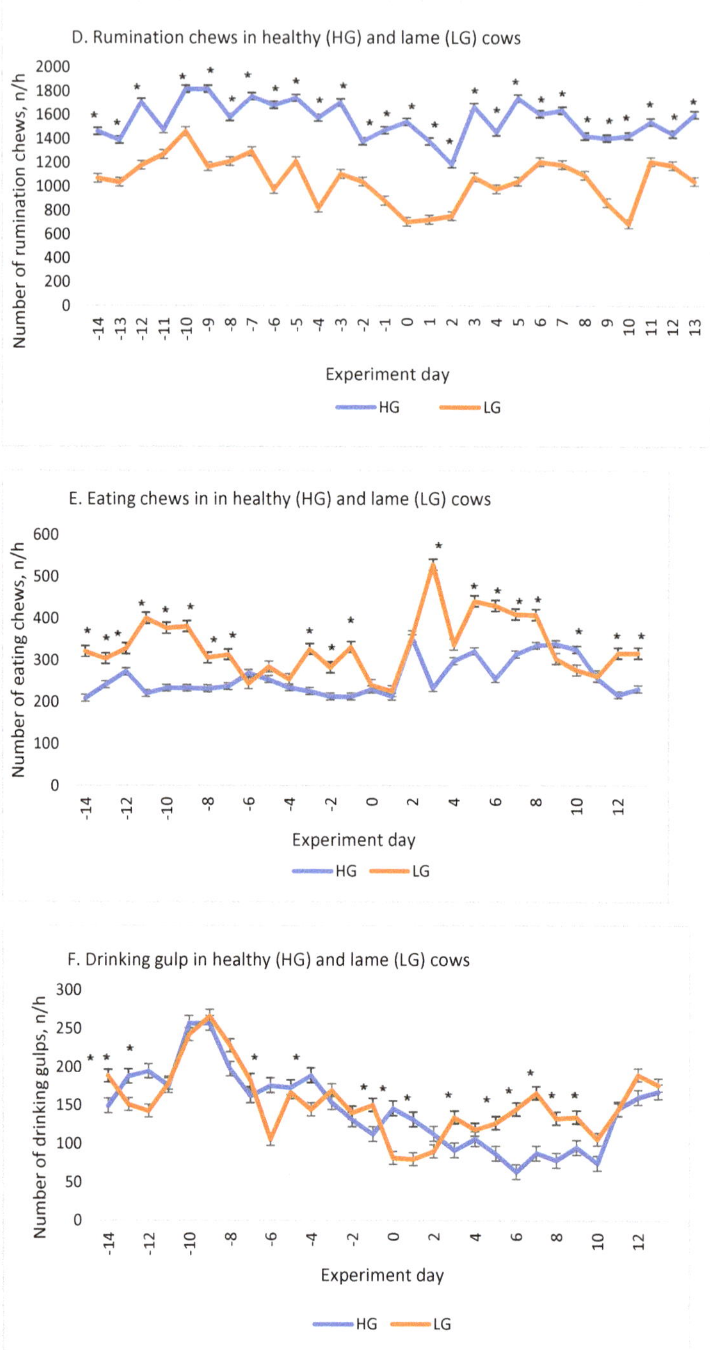

Figure 2. *Cont.*

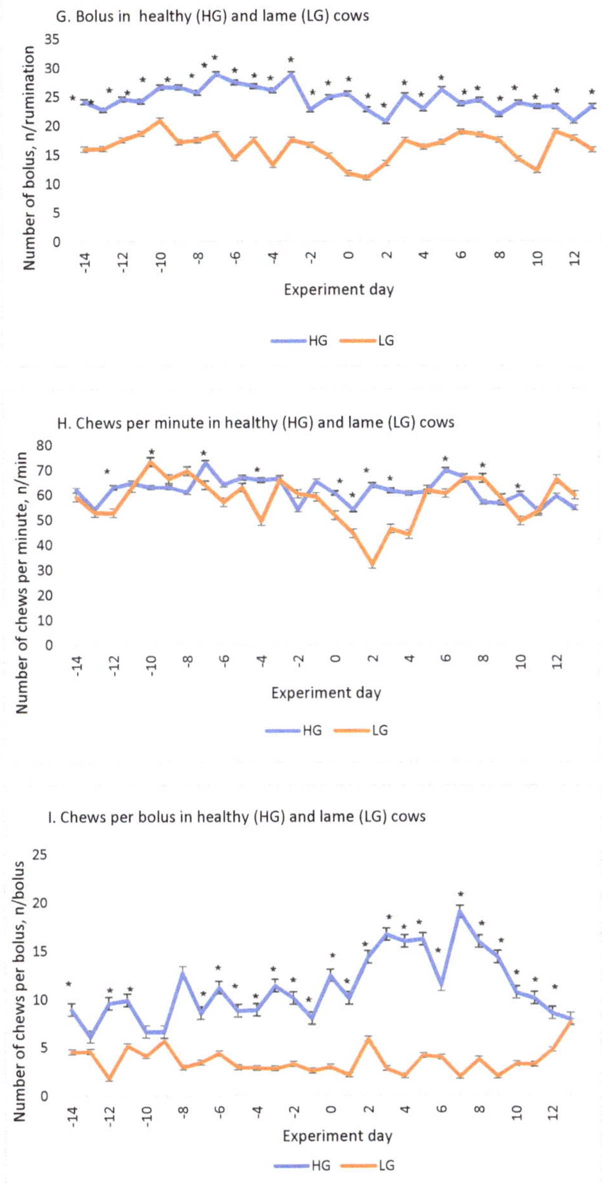

Figure 2. Changes in ingestive-related behaviors parameters by day of the experiment. * shows that the values of the HG and LG groups differ significantly at $p < 0.05$. RT—Rumination time (time in minutes spent for rumination chews); ET—Eating time (time in minutes spent for eating chews); DT—Drinking time (time in minutes spent for drinking); RC—Rumination chews (chews during rumination for mechanical breakdown of the regurgitated materials); EC—Eating chews (number of prehension bites); DG—Drinking gulp (total number of drinking gulps while drinking); B—number of boluses per rumination) CM—Chews per minute (chews for one minute); CB—Chews per bolus (chews performed during rumination). LG—lame cows' group, HG—healthy group. Days '−14' to '−1' denote the experimental period before the onset of clinical signs of lameness (day '0') in the LG group, and days '1' to '13'—the period after the start of treatment.

The lowest eating time (2.01 ± 0.1) was found on lameness identification day and the highest (2.72 times higher, $p < 0.001$) on day '−9' before the onset of clinical signs of the disease (Figure 2B).

Drinking time was significantly ($p < 0.001$) higher in the healthy group, with the exception of '−10' and '−9' days prior to clinical signs of disease in lame cows. The largest difference between groups (7.38-fold) was found on day 7 of treatment (Figure 2C).

A downward trend in rumination chews (Figure 2D) was observed in lame cows from day 7 before until the identification of lameness. During the mentioned period, the value of the indicator decreased 2.07 times ($p < 0.001$). On the last day of the experiment, the mean rumination chews differed between the groups 1.53-fold ($p < 0.001$).

The mean eating chews value of lame cows decreased 1.50 times from the beginning of the experiment to lameness identification day ($p < 0.001$), while in the healthy group, on the contrary, it increased 1.13 times. Eating chews was 1.55-fold higher in the healthy cow group ($p < 0.001$) on the last day of the experiment compared to the lame group (Figure 2E).

Eating chews (Figure 2E) in the healthy group during the experiment increased ($R^2 = 0.4718$) from the second day to the end of the experiment (1.45 times, $p < 0.001$). The highest eating chews value were found in lame cows on the third day after identification of lameness, and the lowest (2.19 times lower) on the first day after identification of lameness ($p < 0.001$).

During the experimental period, drinking gulps decreased in the healthy group ($y = -4.0578x + 203.98$, $R^2 = 0.434$) (Table 2). In lame cows, the lowest drinking gulps (176.00 ± 4.351) (total number of drinking gulps while drinking) was set on lameness identification day, and the highest value (1.59 times higher, $p < 0.001$) was on the 12th day after identification of lameness (Figure 2F).

The value of indicator bolus (Figure 2G) decreased (1.59 times $p < 0.001$) from day '−3' to day '+1' in lame cows. The largest difference in this indicator between groups was found on day '0' (2.17 times, $p < 0.001$). Twelve days after the identification of lameness, the difference between the groups decreased.

Analyzing the change in chews per minute (Figure 2H) in the lame group, we found a decrease (2.27 times, $p < 0.001$) in this indicator from '−10' to '2' days of the experiment. After that, we noticed an increase up to day '+8'. Chews per minute values in the healthy group fluctuated less during the experimental period.

The chews per bolus value were almost the same in both groups of cows at the end of the experiment, but the largest difference between the groups (9.70 times) was found on day 7 (Figure 2I).

4. Discussion

Lameness has a large impact on dairy farming industry development [17]. Therefore, in the past few years a growing number of electronic techniques have been incorporated into the dairy industry with the aim to detect lameness quicker and more accurate [7]. Automated lameness detection could provide useful cow- and herd-level information to address an information gap, particularly regarding mild and moderately lame cows [18]. The indirect methods mainly used an accelerometer to detect the behavior and activity of cows [19]. In this study, we investigated the impact of lameness on attributes of feeding registered with the RumiWatch noseband sensor: rumination time, drinking time, eating time, rumination chews, eating chews, drinking gulps, bolus count, chews per minute, and chews per bolus before and after diagnosis of lameness fresh dairy cows. According to our results, lame cows demonstrated a reduction in the rumination time (on average 8.62 min/h less), eating time (on average 2.73 min/h less), rumination chews (on average 504.67 n/h less), drinking time (on average 0.42 min/h) and bolus count (on average 8.23 n/rumination) less that non lame cow throughout the experiment).

Monitoring of rumination time alone, or when combined with other variables, has been described as a means of detecting illness in dairy cattle and it was closely associated with subclinical and clinical health disorders [20].

Past studies found that lameness has affected a number of rumination behavior characteristics, in terms of decreased daily feed intake, feeding time [21–23], rumination time [21,23], and increased feeding rate [21,22]. However, none of these studies investigated lameness, rumination, and feeding behavior at the same time [24]. According to Thorup et al. [25] when compared to not lame cows, lame cows are likely to exhibit a different feeding behavior such as increased feeding rate and decreased feeding time, whereas rumination time seems much less affected by lameness. Thorup et al. [25] found that lameness problems would require measurement of feeding as well as rumination to be registered automatically. Using a combination of two or more feeding behaviors is likely to increase the accuracy of detecting problems. Also, Soriani et al. [26] found that cows with a shorter rumination time had a greater incidence of clinical disease (including ketosis and lameness).

Rumination can also be affected by the social hierarchy of cows, where a lower rank implies less eating and therefore less ruminating, less lying, and more standing, which in turn leads to an increased chance of lameness [3].

Also, the lowest eating time (2.01 $\pm$ 0.1 min/h) was found on lameness identification day for lame cows). According literature, lame cows may compensate for the reduction in total feeding time by increasing their feed intake rate [22]. Alternatively, the change in feeding behavior observed in lame animals may adversely affect rumen function—e.g., consuming the total daily dry matter intake over fewer meals, at an increased rate—may decrease rumination. Finally, the discomfort or stress associated with lameness may directly affect rumen function via central depression of the centers controlling rumination. A previous work has demonstrated that rumination is negatively associated with higher levels of cortisol [26]. The time dairy cattle spend eating varies greatly when combined across experimental conditions and there are signs of a compensatory mechanism between ruminating and eating time [27]. According to our results, the highest eating chews value was found in lame cows on the third day after identification of lameness, and the lowest (2.19 times lower) on the first day after identification of lameness According to literature, eating duration and the number of eating chews decreased in moderately lame cows [28]. Lame cows spent less time eating to reduce the time spent standing on painful feet or limbs [29]. Bareille et al. [30] state that lame cows have reduced total feed intake than non-lame cows. Thorup et al. found weaker eating behavior but similar rumination behavior in lame cows compared with non-lame cows [25]. According to the measurement of reduced eating time and number of eating chews, we can assume that moderately lame cows in our study potentially had a poorer comminution of feed [28]. Weigele et al. [29] found that the number of boluses, ruminating time, number of ruminating chews, and average ruminating speed, however, were not affected by moderate lameness. Lameness reduced feeding frequency by 44% [24]. Constrained feeding time forces cows to eat faster, and in this respect, lameness may be viewed as a feeding time constraint by cows [30].

The lowest number of drinking gulps was found on lameness identification day for lame cows (176 $\pm$ 4.35 n/h) and the highest (1.59 times higher) was found on the 12 day after identification of lameness). Many factors can put a strain on water consumption: dry matter of the diets, changes in ambient temperatures, increased loss of water due to increased milk production, amount of feed ingested, consumption of sodium and potassium, not to mention physiological factors and diseases [31].

To our knowledge, the present study is the first study to investigate impact of lameness in fresh dairy cows on attributes of feeding registered with on-line sensors (such as rumination time, drinking time, eating time, rumination chews, eating chews, drinking gulps, bolus count, chews per minute, chews per bolus before, and after diagnosis of lameness fresh dairy cows). This study showed that the attributes of feeding registered with on-line sensors could also be used for an automatic detection of lameness.

5. Conclusions

In conclusion, lameness seemed to impair attributes of feeding registered with noseband sensor in fresh dairy cows and confirming our hypothesis, results of the present study could be integrated for early identification of lameness in fresh dairy cows. According to our results, we found that lameness of fresh dairy cows affects the inline registered ingestive behaviors biomarkers. In addition, reduction in the rumination time, eating time, rumination chews, drinking time and bolus count in lame group on lameness identification day, may serve as lameness indicators. Also, eating behavior changes as early as 10 days before the visual identification of lameness.

For practicians, we recommend the use of the following biomarkers for the early diagnosis of lameness in fresh dairy cows such as rumination time, eating time, rumination chews, drinking time and bolus count.

Further studies are required for the early automatic detection of lameness, combining the parameters from current study with other in-line biomarkers (such as reticulorumen pH, temperature and etc.), also, with including also the social hierarchy of cows.

Author Contributions: R.A.: overall research study process including literature search, carrying out research experiments, and compiling the final manuscript. The entire process was revised by the co-authors. V.J.: Assisted in designing and setting up field data collection activities and developed the software and algorithm for data analysis. G.U.: sampling of the experimental animals; D.M. and M.T.: aided in fieldwork set up, data collection, and sampling of the experimental animals. M.U.: literature search. W.B.: major support in processing of data in the study. All authors have read and agreed to the published version of the manuscript.

Funding: This research received no external funding.

Institutional Review Board Statement: The study was conducted according to the guidelines of the Declaration of Helsinki, and ap-proved by Ethics Committee (The study approval number is PK016965, 6 June 2017).

Informed Consent Statement: Not applicable.

Data Availability Statement: The data presented in this study are available within the article.

Conflicts of Interest: The authors declare no conflict of interest.

References

1. Bruijnis, M.; Beerda, B.; Hogeveen, H.; Stassen, E.N. Assessing the welfare impact of foot disorders in dairy cattle by a modeling approach. *Animal* **2012**, *6*, 962–970. [CrossRef] [PubMed]
2. Barwick, J.; Lamb, D.; Dobos, R.; Schneider, D.; Welch, M.; Trotter, M. Predicting lameness in sheep activity using tri-axial acceleration signals. *Animals* **2018**, *8*, 12. [CrossRef]
3. Van Hertem, T.; Maltz, E.; Antler, A.; Romanini, C.; Viazzi, S.; Bahr, C.; Schlageter-Tello, A.; Lokhorst, C.; Berckmans, D.; Halachmi, I. Lameness detection based on multivariate continuous sensing of milk yield, rumination, and neck activity. *J. Dairy Sci.* **2013**, *96*, 4286–4298. [CrossRef] [PubMed]
4. Stangaferro, M.L.; Wijma, R.; Caixeta, L.S.; Al-Abri, M.A.; Giordano, J.O. Use of rumination and activity monitoring for the identification of dairy cows with health disorders: Part III. Metritis. *J. Dairy Sci.* **2016**, *99*, 7422–7433. [CrossRef]
5. Poursaberi, A.; Bahr, C.; Pluk, A.; Van Nuffel, A.; Berckmans, D. Real-time automatic lameness detection based on back posture extraction in dairy cattle: Shape analysis of cow with image processing techniques. *Comput. Electron. Agric.* **2010**, *74*, 110–119. [CrossRef]
6. Pluk, A.; Bahr, C.; Leroy, T.; Poursaberi, A.; Song, X.; Vranken, E.; Maertens, W.; Van Nuffel, A.; Berckmans, D. Evaluation of step overlap as an automatic measure in dairy cow locomotion. *Trans. ASABE* **2010**, *53*, 1305–1312. [CrossRef]
7. Pluk, A.; Bahr, C.; Poursaberi, A.; Maertens, W.; Van Nuffel, A.; Berckmans, D. Automatic measurement of touch and release angles of the fetlock joint for lameness detection in dairy cattle using vision techniques. *J. Dairy Sci.* **2012**, *95*, 1738–1748. [CrossRef]
8. Benaissa, S.; Tuyttens, F.A.; Plets, D.; Cattrysse, H.; Martens, L.; Vandaele, L.; Joseph, W.; Sonck, B. Classification of ingestive-related cow behaviours using RumiWatch halter and neck-mounted accelerometers. *Appl. Anim. Behav. Sci.* **2019**, *211*, 9–16. [CrossRef]
9. Antanaitis, R.; Juozaitienė, V.; Televičius, M.; Malašauskienė, D.; Urbutis, M.; Baumgartner, W. Influence of Subclinical Ketosis in Dairy Cows on Ingestive-Related Behaviours Registered with a Real-Time System. *Animals* **2020**, *10*, 2288. [CrossRef]

10. Sadiq, M.B.; Ramanoon, S.Z.; Shaik Mossadeq, W.M.; Mansor, R.; Syed-Hussain, S.S. Association between Lameness and Indicators of Dairy Cow Welfare Based on Locomotion Scoring, Body and Hock Condition, Leg Hygiene and Lying Behavior. *Animals* **2017**, *7*, 79. [CrossRef]
11. Miguel-Pacheco, G.G.; Kaler, J.; Remnant, J.; Cheyne, L.; Abbott, C.; French, A.P.; Pridmore, T.P.; Huxley, J.N. Behavioural changes in dairy cows with lameness in an automatic milking system. *Appl. Anim. Behav. Sci.* **2014**, *150*, 1–8. [CrossRef]
12. Rombach, M.; Münger, A.; Niederhauser, J.; Südekum, K.; Schori, F. Evaluation and validation of an automatic jaw movement recorder (RumiWatch) for ingestive and rumination behaviors of dairy cows during grazing and supplementation. *J. Dairy Sci.* **2018**, *101*, 2463–2475. [CrossRef]
13. Büchel, S. Sensor-Based Control of Chewing and Rumination Behavior of Dairy Cows. Ph.D. Thesis, University of Kassel, Kassel, Germany, 2014.
14. Zehner, N.; Huerlimann, M.; Hoch, M. User Guide RumiWatch Converter Version 0.7. 3.2 and higher. In *User Guide*; CIGR-AgEng: Bennwil, Switzerland, 2014.
15. Sprecher, D.E.A.; Hostetler, D.E.; Kaneene, J.B. A lameness scoring system that uses posture and gait to predict dairy cattle reproductive performance. *Theriogenology* **1997**, *47*, 1179–1187. [CrossRef]
16. Warnick, L.D.; Janssen, D.; Guard, C.L.; Gröhn, Y.T. The effect of lameness on milk production in dairy cows. *J. Dairy Sci.* **2001**, *84*, 1988–1997. [CrossRef]
17. Van Nuffel, A.; Zwertvaegher, I.; Pluym, L.; Van Weyenberg, S.; Thorup, V.M.; Pastell, M.; Sonck, B.; Saeys, W. Lameness detection in dairy cows: Part 1. How to distinguish between non-lame and lame cows based on differences in locomotion or behavior. *Animals* **2015**, *5*, 838–860. [CrossRef]
18. O'Leary, N.W.; Byrne, D.T.; O'Connor, A.H.; Shalloo, L. Invited review: Cattle lameness detection with accelerometers. *J. Dairy Sci.* **2020**, *103*, 3895–3911. [CrossRef] [PubMed]
19. Alsaaod, M.; Fadul, M.; Steiner, A. Automatic lameness detection in cattle. *Vet. J.* **2019**, *246*, 35–44. [CrossRef] [PubMed]
20. Liboreiro, D.N.; Machado, K.S.; Silva, P.R.; Maturana, M.M.; Nishimura, T.K.; Brandão, A.P.; Endres, M.I.; Chebel, R.C. Characterization of peripartum rumination and activity of cows diagnosed with metabolic and uterine diseases. *J. Dairy Sci.* **2015**, *98*, 6812–6827. [CrossRef] [PubMed]
21. Almeida, P.E.; Weber, P.S.D.; Burton, J.L.; Zanella, A.J. Depressed DHEA and increased sickness response behaviors in lame dairy cows with inflammatory foot lesions. *Domest. Anim. Endocrinol.* **2008**, *34*, 89–99. [CrossRef]
22. Gonzàlez, L.A.; Tolkamp, B.J.; Coffey, M.P.; Ferret, A.; Kyriazakis, I. Changes in feeding behavior as possible indicators for the automatic monitoring of health disorders in dairy cows. *J. Dairy Sci.* **2008**, *91*, 1017–1028. [CrossRef]
23. Norring, M.; Häggman, J.; Simojoki, H.; Tamminen, P.; Winckler, C.; Pastell, M. Short communication: Lameness impairs feeding behavior of dairy cows. *J. Dairy Sci.* **2014**, *97*, 4317–4321. [CrossRef] [PubMed]
24. Thorup, V.M.; Nielsen, B.L.; Robert, P.; Giger-Reverdin, S.; Konka, J.; Michie, C.; Friggens, N.C. Lameness affects cow feeding but not rumination behavior as characterized from sensor data. *Front. Vet. Sci.* **2016**, *3*, 37. [CrossRef] [PubMed]
25. Soriani, N.; Trevisi, E.; Calamari, L. Relationships between rumination time, metabolic conditions, and health status in dairy cows during the transition period. *J. Anim. Sci.* **2012**, *90*, 4544–4554. [CrossRef] [PubMed]
26. Bristow, D.J.; Holmes, D.S. Cortisol levels and anxiety-related behaviors in cattle. *Physiol. Behav.* **2007**, *90*, 626–628. [CrossRef] [PubMed]
27. White, R.R.; Hall, M.B.; Firkins, J.L.; Kononoff, P.J. Physically adjusted neutral detergent fiber system for lactating dairy cow rations. I: Deriving equations that identify factors that influence effectiveness of fiber. *J. Dairy Sci.* **2017**, *100*, 9551–9568. [CrossRef] [PubMed]
28. Weigele, H.C.; Gygax, L.; Steiner, A.; Wechsler, B.; Burla, J. Moderate lameness leads to marked behavioral changes in dairy cows. *J. Dairy Sci.* **2018**, *101*, 2370–2382. [CrossRef]
29. Bareille, N.; Beaudeau, F.; Billon, S.; Robert, A.; Faverdin, P. Effects of health disorders on feed intake and milk production in dairy cows. *Livest. Prod. Sci.* **2003**, *83*, 53–62. [CrossRef]
30. Munksgaard, L.; Jensen, M.B.; Pedersen, L.J.; Hansen, S.W.; Matthews, L. Quantifying behavioural priorities—Effects of time constraints on behaviour of dairy cows, Bos taurus. *Appl. Anim. Behav. Sci.* **2005**, *92*, 3–14. [CrossRef]
31. Brandstetter, V.; Neubauer, V.; Humer, E.; Kröger, I.; Zebeli, Q. Chewing and drinking activity during transition period and lactation in dairy cows fed partial mixed rations. *Animals* **2019**, *9*, 1088. [CrossRef]

agriculture

MDPI

Article

Lettuce Growth Pattern Analysis Using U-Net Pre-Trained with *Arabidopsis*

Sungyul Chang [1], Unseok Lee [2], Min Jeong Hong [1], Yeong Deuk Jo [1] and Jin-Baek Kim [1,*]

[1] Radiation Breeding Research Team, Advanced Radiation Technology Institute (ARTI), Korea Atomic Energy Research Institute (KAERI), 29 Geumgu-gil, Jeongeup-si 56212, Jeollabuk-do, Korea
[2] Smart Farm Research Center, Korea Institute of Science and Technology (KIST), 679 Saimdang-ro, Gangneung 210-340, Gangwon-do, Korea
* Correspondence: jbkim74@kaeri.re.kr

Abstract: To overcome the challenges related to food security, digital farming has been proposed, wherein the status of a plant using various sensors could be determined in real time. The high-throughput phenotyping platform (HTPP) and analysis with deep learning (DL) are increasingly being used but require a lot of resources. For botanists who have no prior knowledge of DL, the image analysis method is relatively easy to use. Hence, we aimed to explore a pre-trained *Arabidopsis* DL model to extract the projected area (PA) for lettuce growth pattern analysis. The accuracies of the extract PA of the lettuce cultivar "Nul-chung" with a pre-trained model was measured using the Jaccard Index, and the median value was 0.88 and 0.87 in two environments. Moreover, the growth pattern of green lettuce showed reproducible results in the same environment ($p < 0.05$). The pre-trained model successfully extracted the time-series PA of lettuce under two lighting conditions ($p < 0.05$), showing the potential application of a pre-trained DL model of target species in the study of traits in non-target species under various environmental conditions. Botanists and farmers would benefit from fewer challenges when applying up-to-date DL in crop analysis when few resources are available for image analysis of a target crop.

Keywords: digital farming; deep learning; image analysis; plant area; growth pattern

Citation: Chang, S.; Lee, U.; Hong, M.J.; Jo, Y.D.; Kim, J.-B. Lettuce Growth Pattern Analysis Using U-Net Pre-Trained with *Arabidopsis*. *Agriculture* **2021**, *11*, 890. https://doi.org/10.3390/agriculture11090890

Academic Editors: Gniewko Niedbała and Sebastian Kujawa

Received: 16 August 2021
Accepted: 13 September 2021
Published: 16 September 2021

1. Introduction

Food security is a big challenge for many geographic areas, and present agricultural production practices could not support the present food demand worldwide [1]. To overcome this problem, agricultural data are utilized, which is called digital agriculture, to more effectively cultivate plants in real time [2]. Phenotyping data is a key process in digital agriculture as it can reveal the status of plant extracts from image-based data [3].

The amount of phenomics data extracted from a high-throughput phenotyping platform (HTPP) has been increasing, and more diverse image-based sensor data are expected from the platform [4]. However, extracting features of interesting traits from an image is challenging because there are currently no general analysis tools available. Projected area (PA), defined as the measured whole plant area based on images, is considered as the most utilized feature in plants. The semantic segmentation of plants based on legacy methods such as adjusting the contrast with the region of interest (ROI) was applied, but the result was not successful because the legacy method is very sensitive to lighting conditions. Machine learning (ML)-based image analysis has shown superior performance over the legacy method [5,6]. Deep learning (DL)-based image analysis methods even further reduced the error rates [7–9]. A previous study indicated that U-Net showed superior performance in the semantic segmentation of *Arabidopsis* [8]. In this study, U-Net successfully distinguished the subtle differences among plants exposed to various types of gamma radiation [8].

The extraction of target phenotypes in target plant species using image analysis showed a very effective process with DL [10,11]. However, DL training for crop segmentation requires a large amount of resources and time, as shown in lettuce [12]. The main reason was due to labeling images that include plant and background information usually generated by manual laborers [13]. In addition, trained label images for the DL model need to be embedded in various steps for the pre-and post-processing of the original images. As a result, the built image analysis pipeline for a crop requires time and collaboration with other scientists such as an image scientist [14]. The *Arabidopsis* pre-trained model already has the necessary information for separate plant areas over the background area that was defined as PA, but this has not been explored in other plant species. Investigating the possibility of utilizing a pre-trained model has a notable advantage over building a new model for a target plant. In addition to the minimal effort needed for building a new model, the manually intense annotation of a large number of plant images and training DL is not necessary. Moreover, botanists could utilize the pre-trained model with little knowledge of DL models for a target plant. A previous study on plant disease detection suggested that the pre-training of VGG-16 that was used for various general tasks could be applied in plant disease detection [15]. However, there has been no report that utilizes a model pre-trained in a model plant for different species in a growth pattern analysis.

In this study, we aimed to explore the application of a pre-trained *Arabidopsis* DL model called U-Net for the segmentation of green lettuce cultivars under the same and different environmental conditions.

2. Materials and Methods

2.1. Plant Growth Condition, Image Acquisition, and Analysis

Green lettuce cultivars (Danong seeds, Andong, Kangwon, Korea) were placed in a soil mixture (Hueng-nong Bio, Pyeontaek-Si, Republic of Korea) and covered with a translucent plastic dome in an environmentally controlled room. Two plant-to-sensor type HTPPs were utilized, and each HTPP had different lighting conditions. In HTPP one, the environmental conditions were identical to those of the pre-trained DL model in *Arabidopsis*. In HTPP two, all environmental conditions were identical to those of the platform, except that the lighting intensity was 400 μmol ms^{-1} s^{-1}, which was two-fold higher than that of platform one (200 umol ms^{-1} s^{-1}). In each platform, two trays were randomly placed for technical replication. After 3 days of planting, the plastic dome was removed, and images were obtained with HTTP [8] every hour from 7:00 to 20:00 for 23 days. In HTPP two, a black plastic dome was placed around the camera to remove excessive lighting.

A schematic of the image analysis pipeline was available [8] and the image process was executed with a 4×2 image cropping Python line option within the pipeline.

2.2. Evalution of Image Analysis Result

Thirty-five images of lettuce were randomly selected in HTPP one in various plants and growth stages. In addition, the same number of images were selected at HTPP two. Seventy images that were selected from the HTTP one and two were manually annotated (Figure 1A) with an annotation tool [8]. Binary transformation was applied on the annotated images using a polygon drawing function in the Python image library (PIL), and this binary image was defined as ground truth (GT) data for further analysis. Errors at the edge of the binary images from the pre-trained DL model were removed (Figure 1C,D) with post-processing. Figure 1B shows the comparison of the GT image and post-processed pre-trained image (Figure 1D) with intersection over union evaluation (IoU) in lettuce [16]. The IoU is known as the Jaccard index and is utilized for segmentation of images [17] and plant images [18]. The IoU scores for HTPPs one and two were calculated.

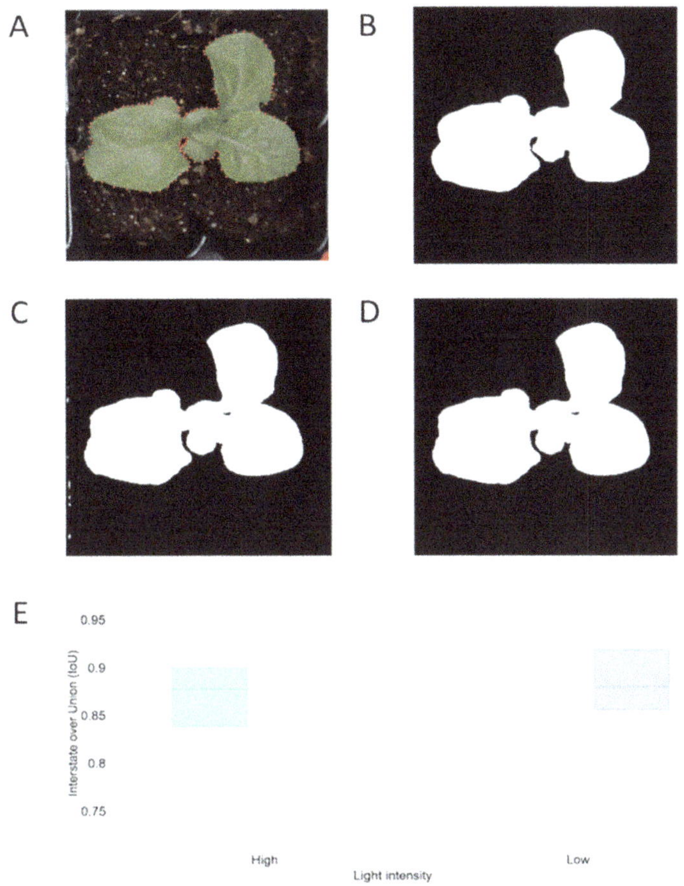

Figure 1. Summary of extract projected area (PA) between ground truth (GT) and U-Net model at two lighting intensities. Thirty-five images were selected at various plant and growth stages in a random manner. (**A**) Manual labeling of the leaf area of a selected lettuce image. (**B**) Binary transformation (PA) of GT for the lettuce image. (**C**) Binary transformation (PA) of the selected image with U-Net. (**D**) Binary transformation (PA) with U-Net using the error correction method. (**E**) Comparing accuracies of extract PA between the GT and U-Net models using the Jaccard index.

2.3. *Growth Analysis and Statistical Analysis*

The average PA with standard deviation was calculated using the native function in R (R Foundation for statically computing, Vienna, Austria) [19] and eight samples per 14 time-steps per day were visualized with Plotly [20]. For technical replication in the same environment, two randomly placed trays were compared. We then compared lettuce growth under different lighting conditions using the daily changes graph in Plotly (Plotly Technologies Inc., Montréal, QC, Canada) [20].

Statistical analyses were performed using R [19] for the selected time, 13:00, on 9, 12, and 18 days after sowing (DAS) to determine whether there were differences between the technical replications using *t*-test. The entire process was repeated for the effect of environmental variation for different lighting intensities at the same time on 14, 18, and 21 DAS using Duncan's multiple range test, based on a significance level of $p < 0.05$.

3. Results

3.1. Evaluation of PA Extraction in Green Lettuce

In a previous study, the U-Net DL model accurately extracted the PA for *Arabidopsis* growth patterns [8]. This DL model was built for *Arabidopsis*, and in this model, the accuracies in the lettuce PA were needed to be checked before further analysis. The evaluation matrix had an IoU score that ranged from 0.8 to 0.97, with a median value of 0.88 in the HTPP one (Figure 1E). In a previous study, a crop segmentation study using an ML model obtained a yield of approximately 0.85 [21]. Therefore, although no lettuce image was used for the DL model, we obtained results that were comparable to those of the previous crop segmentation study. This showed the possibility of detecting seedling and mature lettuce even when pre-trained in different plant species. Furthermore, these results indicate that the extraction of the growth pattern of green lettuce in the time-series data was possible.

3.2. Lettuce Growth Analysis

The growth pattern analyzed with PA showed different growth rates on different dates in green lettuce (Figure 2). Previous lettuce studies suggested that initial growth was relatively slow, followed by a very rapid growth phase in various studies, including exposure to different lighting sources [22] and carbon dioxide levels [23]. In these studies, the PA or biomass of lettuce was measured days apart and showed very similar growth patterns. In this study, the growth patterns in hourly intervals and a previous study were compared [24]. In a previous study, a rapid growth phase 12 days after emergence (DAE) was found as well as a rapid growth phase 16 DAS or 13 DAE (Figure 2). The slight differences might be due to the different cultivars used in each study. Overall, our results indicate that image analysis using a pre-trained DL model is suitable for analyzing the growth pattern of lettuce in an environmentally controlled setting. This information is helpful for investigating interesting traits in lettuce given that it could detect drastic growth phases and changes in hours rather than days. The technical replication study showed reproducible results ($p < 0.05$) in lettuce placed in a slightly different location in HTPP one (Table 1).

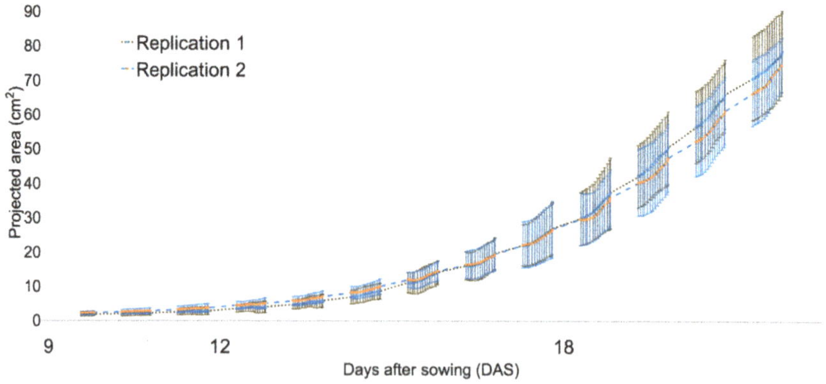

Figure 2. Comparing the growth pattern of the green lettuce cultivar between technical replications. Images were acquired between 07:00 to 20:00 in a 1-h interval. The results are presented as the mean of each time point from replication one (n = 8) and replication two (n = 8).

Table 1. Comparing projected area (PA) of green lettuce at fixed lighting condition (200 μmol ms^{-1} s^{-1}) on multiple days after sowing (DAS) and time. Values in the same column followed by an asterix are significantly different ($p < 0.05$).

DAS	Time (24 h)	Replication	PA (cm^2)
9	13:00	One	1.8394
	13:00	Two	2.2356
12	13:00	One	3.9754
	13:00	Two	4.6765
18	13:00	One	32.3245
	13:00	Two	30.7201

3.3. Lettuce Growth at Differernt Lighting Intensities

The HTPP one, which had the same lighting conditions as the pre-trained model, produced very uniform images of lettuce (Figure 3A) and accurate image analysis results (Figure 3C). At HTPP two, a black plastic dome was placed over the camera to remove excess light, and it showed lettuce images with various backgrounds (Figure 3B). Surprisingly, the pre-trained model accurately extracted the PA, even in various background images (Figure 3D).

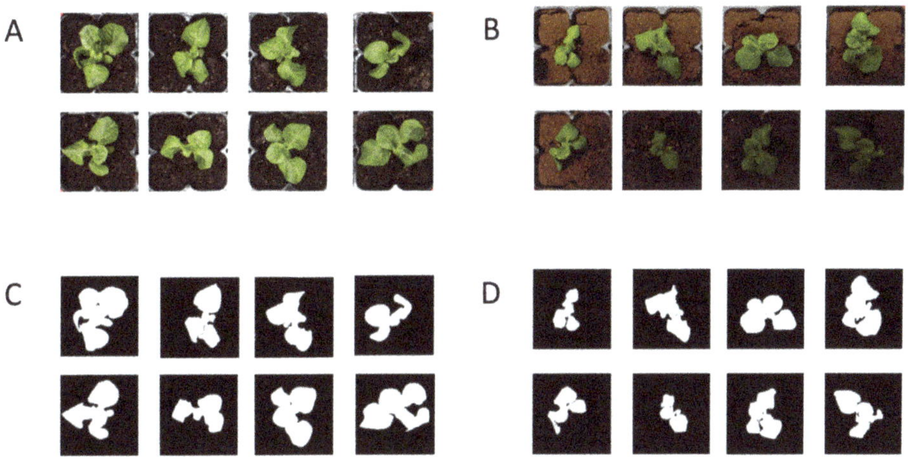

Figure 3. Visualized of cropped and processed images of green lettuce at 19 days after sowing (DAS). (**A**) Cropped individual lettuce images at 200 μmol ms^{-1} s^{-1}. (**B**) Cropped individual lettuce images at 400 μmol ms^{-1} s^{-1}. (**C**) Visualized image analysis result of lettuce at 200 μmol ms^{-1} s^{-1}. (**D**) Visualized image analysis result of lettuce at 200 μmol ms^{-1} s^{-1}.

In summary, the technical replication study showed that green cultivars grown in different trays did not differ from each other (Table 2) at 14, 18, and 21 DAS ($p < 0.05$). In addition, drastic changes were observed under different lighting conditions with technical replication in each environment (Supplementary Figure S1 and Table S2). The results indicate that the pre-trained DL model could be applied for the comparison of a large number of samples from different environments.

Table 2. Comparing projected area (PA) of green lettuce at two lighting conditions on multiple days after sowing (DAS). Values in the same column followed by a different letter are significantly different ($p < 0.05$).

DAS	Light Condition	Replication	PA (cm^2)
14	200 μmol ms^{-1} s^{-1}	One	7.8281a
	200 μmol ms^{-1} s^{-1}	Two	8.6423a
	400 μmol ms^{-1} s^{-1}	One	6.3605b
	400 μmol ms^{-1} s^{-1}	Two	5.5361b
18	200 μmol ms^{-1} s^{-1}	One	33.1341a
	200 μmol ms^{-1} s^{-1}	Two	31.8327a
	400 μmol ms^{-1} s^{-1}	One	21.4811b
	400 μmol ms^{-1} s^{-1}	Two	19.9635b
21	200 μmol ms^{-1} s^{-1}	One	74.2587a
	200 μmol ms^{-1} s^{-1}	Two	69.8884a
	400 μmol ms^{-1} s^{-1}	One	45.7692b
	400 μmol ms^{-1} s^{-1}	Two	46.1067b

4. Discussion

Digital agriculture requires a lot of data for cultivation to maximize yield and select the optimal harvest time [2]. Even though the conventional practice utilized a large amount of data, only a small fraction of the data was quantified. The main reason for this is that a farmer accesses the current status of a plant with experience and intuition, but little plant data are recorded. The emerging inexpensive image sensors [4] lead to the acquisition of more quantitative data for each crop in different environments. In addition, more uniform data acquired from the HTPP is expected [25]. This would enable the accumulation of image data to study agronomical traits in the future [26].

The quality of image processed output for agriculture data has improved significantly since the process is utilized using ML and DL [9]. Machine learning- and DL-based methods have been applied in various crops and model plants, and they demonstrated the effectiveness of the process [9]. However, DL models constructed for crops require significant amounts of time and resources [12], given that each model requires training data, and the data is generated by manual labeling of individual images for specific traits [27]. In addition, the pre-and post-processing steps to acquire quality data require effort. Thus, the construction of an entire image analysis pipeline might not be achievable for a small research group to analyze images [14]. Therefore, it is necessary for small researchers to utilize pre-trained DL models in non-target species for interesting traits in their target species.

A pre-trained general DL model called the VGG-16 model was applied for plant disease detection studies [28] but the VGG-16 model was not built for a plant. A previous study utilized generally built pre-trained DL for disease detection, and there have been no known studies for the agronomical traits in crops. To the best of our knowledge, this is the first study to utilize pre-trained DL for target plants in non-target plants. *Arabidopsis* and lettuce are completely different species but share a similar genetic leaf shape controlling mechanism; thus, the leaf shape characteristics are similar. The lettuce seedlings and mature plants were accurately detected with the *Arabidopsis* pre-trained model (Figure 1E). In addition, this study showed that a very fine-scale growth pattern analysis was possible with reproducible results. The U-Net could have analyzed the leaf shape information in order to separate it from the soil or background information, even though it is almost impossible to understand learned information (features) from a pre-trained DL model. Future research on whether learned information from a pre-trained DL model transfer information into a new DL model [27], also known as "transfer learning", for interesting traits in different plant species could be performed. The method significantly reduced the efforts needed to train and utilize the DL model for agronomical traits because relatively little information is needed to construct new DL models [27]. The phenotypic effect of lighting in lettuce has been well documented in multiple studies [22,29]. However, time-series responses in different lighting conditions are difficult to find, given that the image analysis pipeline utilized limited images in a few environment settings. As a result, the image analysis pipeline performed very well in a specific dataset. Even in DL-based models, limited annotated images perform very well in a specific environment [15]. Recently, fully documented time-series data have been available [12] but they were tested in two growing seasons in a greenhouse for various environmental factors, including lighting. At this point, fully documented time-series data on the effects of lighting are not available.

The increased volume of images that require up-to-date image analysis tools [9] and DL models [12] could provide solutions future applications. This study showed the possibility of applying pre-trained DL to study interesting traits in different species with less effort. This could be a very cost-effective process when groups of botanists and image scientists develop generalized tools for botanists who have little or no prior knowledge of ML or DL models.

5. Conclusions

The pre-trained model showed accurate results, expressed as an IoU score as high as 0.97, in extracting lettuce growth patterns under the same environmental conditions as *Arabidopsis*. The reproducible result confirmed by statistical test (t-test) between replication measurement on 9, 12, and 18 DAS. Moreover, lettuce grown under the two lighting conditions showed significant separation on 14, 18 and 21 DAS between the two environments ($p < 0.05$). This study clearly indicates the feasibility of applying a pre-trained DL model to analyze the growth patterns of another crop under various environmental conditions.

Supplementary Materials: The following are available online at https://www.mdpi.com/article/10.3390/agriculture11090890/s1, Figure S1: Comparing the growth pattern of a green lettuce between the two lighting conditions. The results are presented as the mean of eight samples for each tray at different conditions. Each tray had a lighting condition of 200 μmol ms^{-1} s^{-1} (A1) with replication (A2) and a lighting condition at 400 μmol L ms^{-1} s^{-1} (B1) with replication (B2).

Author Contributions: Conceptualization, S.C. and J.-B.K.; methodology, S.C.; software, U.L.; validation, S.C.; formal analysis, S.C.; investigation, S.C., M.J.H. and J.-B.K.; data curation, S.C. and U.L.; writing—original draft preparation, S.C., U.L., M.J.H., Y.D.J. and J.-B.K.; writing—review and editing, S.C., M.J.H., Y.D.J. and J.-B.K.; supervision, J.-B.K.; project administration, J.-B.K.; funding acquisition, J.-B.K. All authors have read and agreed to the published version of the manuscript.

Funding: This work was supported by grants from the Nuclear R & D programs of the Ministry of Science and ICT (MSIT) and the research program of KAERI, Republic of Korea.

Data Availability Statement: Data available upon request.

Conflicts of Interest: The authors declare no conflict of interest.

References

1. Godfray, H.C.J.; Beddington, J.R.; Crute, I.R.; Haddad, L.; Lawrence, D.; Muir, J.F.; Pretty, J.; Robinson, S.; Thomas, S.M.; Toulmin, C. Food security: The challenge of feeding 9 billion people. *Science* **2010**, *327*, 812. [CrossRef]
2. Klerkx, L.; Jakku, E.; Labarthe, P. A review of social science on digital agriculture, smart farming and agriculture 4.0: New contributions and a future research agenda. *NJAS-Wagening. J. Life Sci.* **2019**, *90–91*, 100315. [CrossRef]
3. Furbank, R.T.; Tester, M. Phenomics—Technologies to relieve the phenotyping bottleneck. *Trends Plant Sci.* **2011**, *16*, 635–644. [CrossRef]
4. Pieruschka, R.; Schurr, U. Plant phenotyping: Past, present, and future. *Plant Phenomics* **2019**, *26*, 7507131. [CrossRef]
5. Lee, U.; Chang, S.; Putra, G.A.; Kim, H.; Kim, D.H. An automated, high-throughput plant phenotyping system using machine learning-based plant segmentation and image analysis. *PLoS ONE* **2018**, *13*, e0196615. [CrossRef] [PubMed]
6. Nagano, S.; Moriyuki, S.; Wakamori, K.; Mineno, H.; Fukuda, H. Leaf-movement-based growth prediction model using optical flow analysis and machine learning in plant factory. *Front. Plant Sci.* **2019**, *10*, 227. [CrossRef] [PubMed]
7. Jiang, H. The analysis of plants image recognition based on deep learning and artificial neural network. *IEEE Access* **2020**, *8*, 68828–68841.

8. Chang, S.; Lee, U.; Hong, M.J.; Jo, Y.D.; Kim, J.-B. High-throughput phenotyping (htp) data reveal dosage effect at growth stages in *arabidopsis thaliana* irradiated by gamma rays. *Plants* **2020**, *9*, 557. [CrossRef]
9. Jiang, Y.; Li, C. Convolutional neural networks for image-based high-throughput plant phenotyping: A review. *Plant Phenomics* **2020**, *9*, 4152816. [CrossRef] [PubMed]
10. Toda, Y.; Okura, F.; Ito, J.; Okada, S.; Kinoshita, T.; Tsuji, H.; Saisho, D. Training instance segmentation neural network with synthetic datasets for crop seed phenotyping. *Commun. Biol.* **2020**, *3*, 173. [CrossRef]
11. Gao, J.; French, A.P.; Pound, M.P.; He, Y.; Pridmore, T.P.; Pieters, J.G. Deep convolutional neural networks for image-based convolvulus sepium detection in sugar beet fields. *Plant Methods* **2020**, *16*, 29. [CrossRef]
12. Zhang, L.; Xu, Z.; Xu, D.; Ma, J.; Chen, Y.; Fu, Z. Growth monitoring of greenhouse lettuce based on a convolutional neural network. *Hortic. Res.* **2020**, *7*, 124. [CrossRef]
13. Ronneberger, O.; Fischer, P.; Brox, T. U-net: Convolutional networks for biomedical image segmentation. In *International Conference on Medical Image Computing and Computer-Assisted Intervention*; Springer: Cham, Switzerland, 2015; pp. 234–241.
14. Gehan, M.A.; Fahlgren, N.; Abbasi, A.; Berry, J.C.; Callen, S.T.; Chavez, L.; Doust, A.N.; Feldman, M.J.; Gilbert, K.B.; Hodge, J.G.; et al. Plantcv v2: Image analysis software for high-throughput plant phenotyping. *PeerJ* **2017**, *5*, e4088. [CrossRef]
15. Sladojevic, S.; Arsenovic, M.; Anderla, A.; Culibrk, D.; Stefanovic, D. Deep neural networks based recognition of plant diseases by leaf image classification. *Comput. Intell. Neurosci.* **2016**, *2016*, 3289801. [CrossRef] [PubMed]
16. Du, J.; Lu, X.; Fan, J.; Qin, Y.; Yang, X.; Guo, X. Image-based high-throughput detection and phenotype evaluation method for multiple lettuce varieties. *Front. Plant Sci.* **2020**, *11*, 56386. [CrossRef] [PubMed]
17. Everingham, M.; Van Gool, L.; Williams, C.K.I.; Winn, J.; Zisserman, A. The pascal visual object classes (voc) challenge. *Int. J. Comput. Vis.* **2010**, *88*, 303–338. [CrossRef]
18. Feng, X.; Zhan, Y.; Wang, Q.; Yang, X.; Yu, C.; Wang, H.; Tang, Z.; Jiang, D.; Peng, C.; He, Y. Hyperspectral imaging combined with machine learning as a tool to obtain high-throughput plant salt-stress phenotyping. *Plant J.* **2020**, *101*, 1448–1461. [CrossRef]
19. R Core Team. *R: A Language and Environment for Statistical Computing, Version 3.2.2*; R Foundation for Statistical Computing: Vienna, Austria, 2019.
20. Plotly Technologies. *Collaborative Data Science*; Plotly Technologies Inc.: Montréal, QC, Canada, 2015.
21. Haug, S.; Ostermann, J. A crop/weed field image dataset for the evaluation of computer vision based precision agriculture tasks. In *European Conference on Computer Vision*; Springer: Cham, Switzerland, 2014; pp. 105–116.
22. Han, T.; Vaganov, V.; Cao, S.; Li, Q.; Ling, L.; Cheng, X.; Peng, L.; Zhang, C.; Yakovlev, A.N.; Zhong, Y.; et al. Improving "color rendering" of led lighting for the growth of lettuce. *Sci. Rep.* **2017**, *7*, 45944. [CrossRef]
23. Jung, D.; Kim, D.; Yoon, H.; Moon, T.; Park, K.; Son, J. Modeling the canopy photosynthetic rate of romaine lettuce (*Lactuca sativa* L.) grown in a plant factory at varying CO_2 concentrations and growth stages. *Hortic. Environ. Biotechnol.* **2016**, *57*, 487–492. [CrossRef]
24. Klassen, S.; Ritchie, G.; Frantz, J.; Pinnock, D.; Bugbee, B. Real-time imaging of ground cover: Relationships with radiation capture, canopy photosynthesis, and daily growth rate. *Digit. Imaging Spectr. Tech. Appl. Precis. Agric. Crop. Physiol.* **2004**, *66*, 1–14. [CrossRef]
25. Tardieu, F.; Cabrera-Bosquet, L.; Pridmore, T.; Bennett, M. Plant phenomics, from sensors to knowledge. *Curr. Biol.* **2017**, *27*, R770–R783. [CrossRef] [PubMed]
26. Weersink, A.; Fraser, E.; Pannell, D.; Duncan, E.; Rotz, S. Opportunities and challenges for big data in agricultural and environmental analysis. *Annu. Rev. Resour. Econ.* **2018**, *10*, 19–37. [CrossRef]
27. Yuan, Y.; Fang, S.; Chen, L. Crop disease image classification based on transfer learning with dcnns. In *Chinese Conference on Pattern Recognition and Computer Vision (PRCV)*; Springer: Cham, Switzerland, 2018; pp. 457–468.
28. Krishnaswamy Rangarajan, A.; Purushothaman, R. Disease classification in eggplant using pre-trained vgg16 and msvm. *Sci. Rep.* **2020**, *10*, 2322. [CrossRef] [PubMed]
29. Van Henten, E.J. Validation of a dynamic lettuce growth model for greenhouse climate control. *Agric. Syst.* **1994**, *45*, 55–72. [CrossRef]

agriculture

MDPI

Article

Estimation and Forecasting of Rice Yield Using Phenology-Based Algorithm and Linear Regression Model on Sentinel-II Satellite Data

Abid Nazir [1,*], Saleem Ullah [1], Zulfiqar Ahmad Saqib [2,3], Azhar Abbas [4,*], Asad Ali [5], Muhammad Shahid Iqbal [1], Khalid Hussain [6], Muhammad Shakir [1], Munawar Shah [1] and Muhammad Usman Butt [7]

1 Department of Space Science, Institute of Space Technology, P.O. Box 2750, Islamabad 44000, Pakistan; saleem.ullah@mail.ist.edu.pk (S.U.); shahid.iqbal@mail.ist.edu.pk (M.S.I.); m.shakir@mail.ist.edu.pk (M.S.); munawar.shah@mail.ist.edu.pk (M.S.)
2 Institute of Soil and Environmental Sciences, University of Agriculture, Faisalabad 38040, Pakistan; zulfiqar.dasti@uaf.edu.pk
3 Agricultural Remote Sensing Laboratory (ARSL), National Centre of GIS and Space Application (NCGSA), University of Agriculture, Faisalabad 38040, Pakistan
4 Institute of Agriculture and Resource Economics, University of Agriculture, Faisalabad 38040, Pakistan
5 Department of Applied Mathematics and Statistics, Institute of Space Technology, P.O. Box 2750, Islamabad 44000, Pakistan; asad.ali@mail.ist.edu.pk
6 Department of Agronomy, Faculty of Agriculture, University of Agriculture, Faisalabad 38040, Pakistan; khalidkhanuaf@gmail.com
7 Sustainable Rice Production, Galaxy Rice Mills Pvt Ltd., Gujranwala 52230, Pakistan; usmanbutt9200@gmail.com
* Correspondence: gee2018abid@gmail.com (A.N.); azhar.abbas@uaf.edu.pk (A.A.)

Citation: Nazir, A.; Ullah, S.; Saqib, Z.A.; Abbas, A.; Ali, A.; Iqbal, M.S.; Hussain, K.; Shakir, M.; Shah, M.; Butt, M.U. Estimation and Forecasting of Rice Yield Using Phenology-Based Algorithm and Linear Regression Model on Sentinel-II Satellite Data. *Agriculture* **2021**, *11*, 1026. https://doi.org/10.3390/agriculture11101026

Academic Editors: Gniewko Niedbała and Sebastian Kujawa

Received: 11 September 2021
Accepted: 12 October 2021
Published: 19 October 2021

Abstract: Rice is a primary food for more than three billion people worldwide and cultivated on about 12% of the world's arable land. However, more than 88% production is observed in Asian countries, including Pakistan. Due to higher population growth and recent climate change scenarios, it is crucial to get timely and accurate rice yield estimates and production forecast of the growing season for governments, planners, and decision makers in formulating policies regarding import/export in the event of shortfall and/or surplus. This study aims to quantify the rice yield at various phenological stages from hyper-temporal satellite-derived-vegetation indices computed from time series Sentinel-II images. Different vegetation indices (viz. NDVI, EVI, SAVI, and REP) were used to predict paddy yield. The predicted yield was validated through RMSE and ME statistical techniques. The integration of PLSR and sequential time stamped vegetation indices accurately predicted rice yield (i.e., maximum R^2 = 0.84 and minimum RMSE = 0.12 ton ha^{-1} equal to 3% of the mean rice yield). Moreover, our results also established that optimal time spans for predicting rice yield are late vegetative and reproductive (flowering) stages. The output would be useful for the farmer and decision makers in addressing food security.

Keywords: rice yield; vegetation indices; hyper-temporal data; PLSR

1. Introduction

The rapid increase in the world population exerts pressure on the agriculture sector and threatening the food security of the world [1]. Among cereals, rice is one of the prime sources of food with high nutritive value (i.e., containing carbohydrate, vitamins (B, E, thiamine), and minerals (Ca, Mg, Fe). Rice is widely grown, consumed globally (i.e., daily food of 3.5 billion people worldwide), and accounts for 19% of the dietary energy [2]. Globally, 90% of the rice comes from Asia, which is approximately 640 million tons per annum [3,4]. Pakistan ranks 11th at the global rice production list and contributes 8% to the world's total rice trade [3]. Pakistan produced seven (7) million tons of rice in the year

2017–2018 and earned a foreign exchange of two-billion dollars ($USD) from rice export [5]. The high-quality nutritious rice (e.g., basmati) produced in the country is available at affordable price in the international market and thus contributing in food security for the increasing global population [6].

Timely and accurate predictions of crop yield before harvest at a large scale is critical for food security and administrative planning, especially in the current continually changing global environment and international situation [7,8]. Different approaches have been adopted for precise yield estimation and each method has its own strengths and limitations. For instance, the traditional field surveys and crop statistics are useful for precisely estimating crop yield; however, when crop yield prediction of the large region is desired, the surveys prove inadequate due to budget, time, and large skilled manpower constraints [9]. The use of Earth observation data (remote sensing) offers an effective system for monitoring agriculture and quantifying crop yield at large spatial extent. The remotely sensed solution is fast, cost-efficient, and non-destructive [10,11]. In addition, the repetitive data acquisition capability of remote sensing sensors makes them an ideal choice for retrieving temporal information of crop phenology, plants health (stress), response to weather and soil nutrients (i.e., manure and fertilizer), variation in plant biomass, and ultimately its effect on yield production [12,13].

Satellite remote sensing also enables crop yield estimation at field, landscape, and regional scales for making policies and ensuring food security [14,15]). Yield estimation of various crops, such as wheat [11], corn [16], and sugar beet [17] is done successfully using assimilation algorithms on RS data. In recent research, two approaches are commonly used for this purpose: one is canopy reflectance data, and the other is based on the spectral indices. The free availability of optical remote sensing data of Sentinel-2 satellite with multiple spectral bands in the red, red edge, and near infrared (NIR) is making RS an ideal choice for monitoring agricultural crops, vegetation phenology [18] (Caballero et al., 2020), temporal variability in cropping [19], as well as environmental monitoring and land cover mapping [20].

Different vegetation indices (VIs) derived from satellite images are effective indicators of vegetation status and are positively correlated with crop yield. The Normalized Difference Vegetation Index (NDVI) has been widely used for predicting crop yield and identifying growth stages [20–22]. Similarly, other variants of NDVI such as Soil Adjusted Vegetation Index (SAVI) and Enhanced Vegetation Index (EVI) have been found effective for crop growth monitoring in the initial filling stages of the crop [21,22]. While using canopy reflectance data to directly estimate crop yield, multivariable analysis methods are commonly introduced to support the dataset analysis. Partial least squares regression (PLSR), stepwise multiple linear regression (SMLR), artificial neural network (ANN), etc. are helpful to construct and validate the multivariate remote sensing models of estimating the yield and improve the accuracy of crop yield estimation through satellite remote sensing, specifically when analyzing the quantitative relationship between RS variables obtained from satellite images and crop yield [11].

The impacts of different phenological and growth stages (i.e., vegetative, reproductive, and ripening) on yield production are rarely explored. Very few researchers quantified the impact of growth stages on crop yields and assigned different weightage to different growth stages [21,23]. Some studies conclude that physiological status (e.g., crop growth) and biochemical contents (e.g., nitrogen) of pre-heading stage is more crucial [24,25], while other found that high rice biomass at post-heading stages is essential for optimum production [23]. Similarly, few studies highlight the relationship between late reproductive growth period and rice yield [26,27]. The overall objective of this study is to forecast rice yield and investigate the relationship between remote sensing derived VIs at different phenological stages of rice crop and its yield. The study also aims to identify the most critical growth period (phenological stage) for quantification of rice yield with hyper-temporal sentinel-II and derived-indices using Partial Least Square Regression (PLSR) model to improve the estimation accuracy of rice yield by remote sensing.

2. Materials and Methods

2.1. Study Area

This study was carried out at Sheikhupora district (Latitude 31°30′00″ to 31°65′30″ N and Longitude 73°40′00″ to 74°23′00″ E) of Punjab, Pakistan (Figure 1). Sheikhupora district is located between Ravi and Chenab rivers and irrigated by river water from two canals (Upper Chenab and Khanpur Canals). Climatically the region is dominated by the wet monsoon, thus making it favorable for rice crop. The annual precipitation ranges from 120 to 720 mm, which mainly occurs between July and August [28]. The study area is dominated by alluvial clay and loamy soil rich in humus and mineral composition. The mineralogical, chemical and geotechnical compositions of soil (pH = 8, EC = 1.1−4.5 dS m^{-1} with soil field capacity from 45–71%) make the region ideal for rice cultivation [29]. Due to the favorable conditions, Sheikhupora district is the second largest rice-producing district in Pakistan with an average production of 2–2.5 million tons annually [5,30].

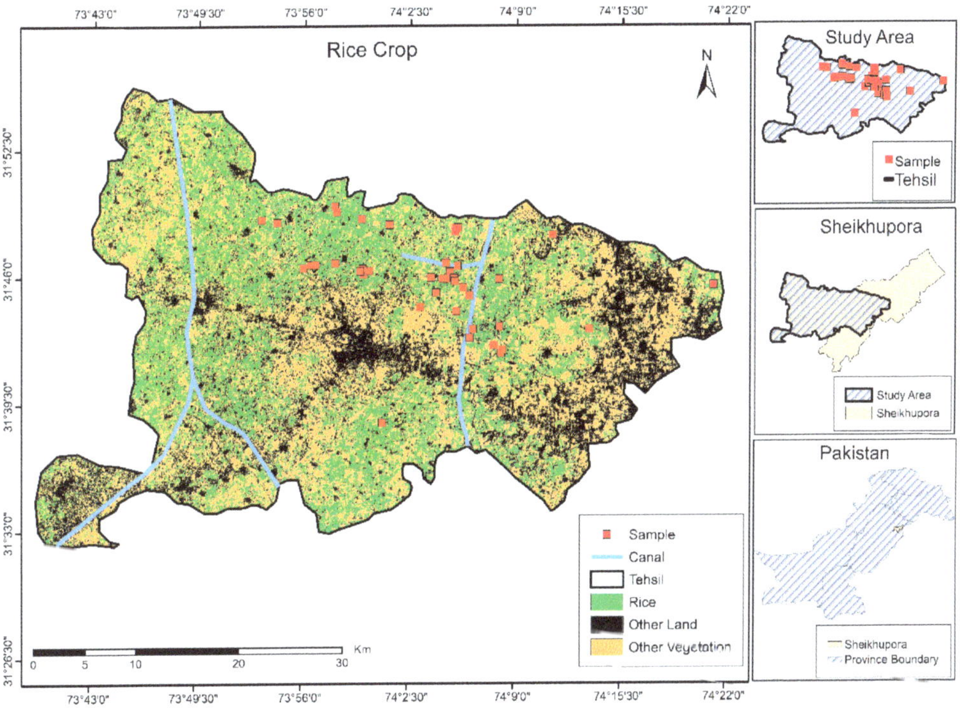

Figure 1. Map of the study area (sampling sites are marked as red square dots).

2.2. Data Collection

2.2.1. Field Data

Stratified-random-sampling procedure was used to collect data from 137 plots well distributed in the study area. The rice fields with minimum size of 60 × 60 m (i.e., corresponding to the coarse pixel size of satellite images used in this study) were considered for the purpose of sampling, monitoring, and analysis. In most of the paddy fields, the rice crop was transplanted in start of July (after seed sown in the nursery at the start of June). These sampled plots were carefully monitored from transplanting till harvesting period. The rice produced in each plot (Figure 2) was carefully measured and recorded for further analysis.

Figure 2. The pictorial view of different phenological or growth stages of rice in field conditions.

2.2.2. Satellite Data Preprocessing and Vegetation Index

The optical remote sensing data of Sentinel-II satellite for year 2016 was used in this study. The availability of multiple spectral bands in the red, red edge, and near infrared (NIR) part of the EM spectrum with 10–20 m resolution makes Sentinel-II an ideal choice for studying vegetation phenology and monitoring agricultural crops for stress level, nutrient contents, pest attack, and yield estimation [31–33]. The cloud free Sentinel-2 time-series images (i.e., spanning from growing to harvesting phase) were pre-processed for atmospheric correction using Sen2Cor processing. The atmospherically corrected images were then used for computing vegetation indices (Table 1) from times series satellite images spanning from sowing till harvesting period of the rice crop. Vegetation indices are mathematical transformations using two or more spectral bands devised to enhance certain characteristics of vegetation [34]. Several images to cover the entire growth period of rice crop were used and vegetation indices were computed using Google Earth Engine (GEE).

Table 1. Mathematical formulas of vegetation indices (NDVI, SAVI, EVI, and Red Edge Position).

Vegetation Indices (VIs)	References
$NDVI = \frac{\rho(NIR)-\rho(Red)}{\rho(NIR)+\rho(Red)}$	Rouse et al. (1974) [35]
$SAVI = 1 + L\frac{\rho(NIR)-\rho(Red)}{\rho(NIR)+\rho(Red)+L}$ where L = 0.5, to minimize the brightness effect of soil	Huete (1988) [36]
$EVI = G\frac{\rho(NIR)-\rho(Red)}{\rho(NIR)+C1\times\rho(Red)-C2\times\rho(Blue)+L}$ where G = 2.5; L = 0.5 (Soil adjusted factor); *C1* and *C2* are constants to reduce aerosols effects.	Liu and Huete (1995) [37]
$Red\ Edge = \frac{\rho(Red)+\rho(Red\ Edge3)}{2}$ $REP = 704 + 35\left[\frac{Red\ Edge-\rho(Red\ Edge1)}{\rho(Red\ Edge2)-\rho(Red\ Edge1)}\right]$ where 704 and 35 represent interpolation constants that can be adjusted according to available band's wavelength	Filella and Penuelas (1994) [38]

2.3. Geo-Statistical Analysis

The PLSR is an established multivariate analysis technique commonly used in chemomatric and hyperspectral data analysis [39,40]. The Partial Least Square Regression (PLSR) analyses were performed to all time-series vegetation indices (explanatory variables) with rice yields (response variable). In PLSR model development, the selection of optimum number of latent variables (LVs) is more critical, as increase of the number of LVs would improve the accuracy of the model, while selection of too many variables can lead to the over fitting and the error would increase [41]. To minimize this over fitting problem, the optimal number of LVs was selected based on achieving a combination of a high R^2 and a low root mean squared error of the prediction (RMSEP) (Figure 3). The PLSR model was evaluated by plotting the 1:1 relationship graph between the predicted and measured values of the yield (Figure 3). The evaluation indices were the R^2 and the RMSE. A larger R^2 shows that the model is better, while smaller RMSE values indicate the stronger estimation ability of the model. To evaluate the performances of the prediction models, leave-one-out cross-validation [42] was used, in which the model was iteratively trained on multiple time series data and then used to predict yield. PLSR can be mathematically expressed as:

$$Y = a + b_1X_1 + b_2X_2 + \cdots + b_nX_n \quad (1)$$

where Y is response variable (rice yield), X_1, X_2–X_n are the selected latent variables (LVs), which are the time series images, in this case a is the intercept and b_1–b_n represent the regression coefficients (also known as β-coefficients) for different predictors.

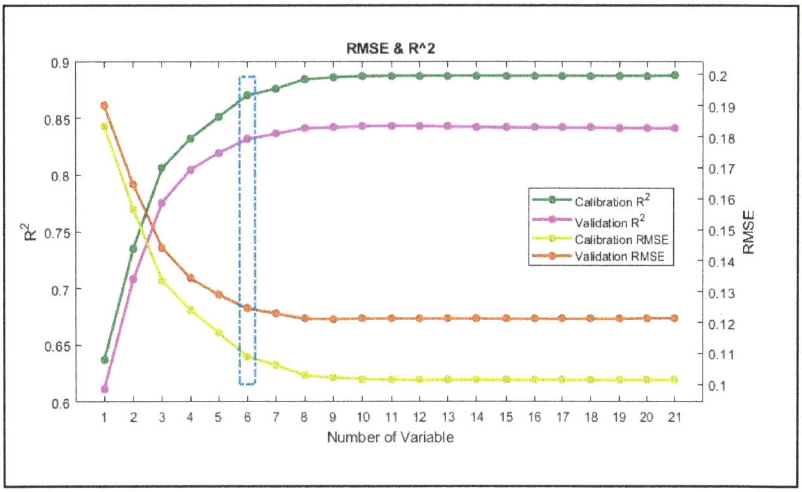

Figure 3. The RMSE and R^2 plot. The RMSE is decreasing as the number of latent variable increases. After a certain number of latent variables, the decrease in RMSE was negligible and that was taken as the optimum number of variables for PLSR model development.

2.4. Spatial Distribution and Mapping of Rice Yield

To model the spatial distribution of rice yields, a two-step procedure was adopted. The thematic map of rice crop was developed using phenological based mapping algorithms (Figure 4). In this routine, the phenology profiles (or signatures) serves as numerical key for discerning different crop types grown in the region of interest [43,44]. The time series vegetation indices (e.g., NDVI profiles computed from optical data for the entire growth span of rice crop: 130 days) were used to demarcate the rice crop and to compute the rice cultivated area. The phenological-mapping-routine takes into account the entire range (minimum–maximum) of vegetation index values (e.g., NDVI in this case) at each time stamp (e.g., from transplanting till harvesting) and can be mathematically expressed as:

$$R = (NDVI^1 min \text{ and } NDVI^1 max) \text{ and } (NDVI^2 min \text{ and } NDVI^2 max) \; NDVI^n min \text{ and } NDVI^n max) \quad (2)$$

where R represents response variable (rice), and $NDVI^1$, $NDVI^2$, and $NDVI^n$ represent NDVIs derived from RS data at different crop phenological stages starting from transplanting till harvesting.

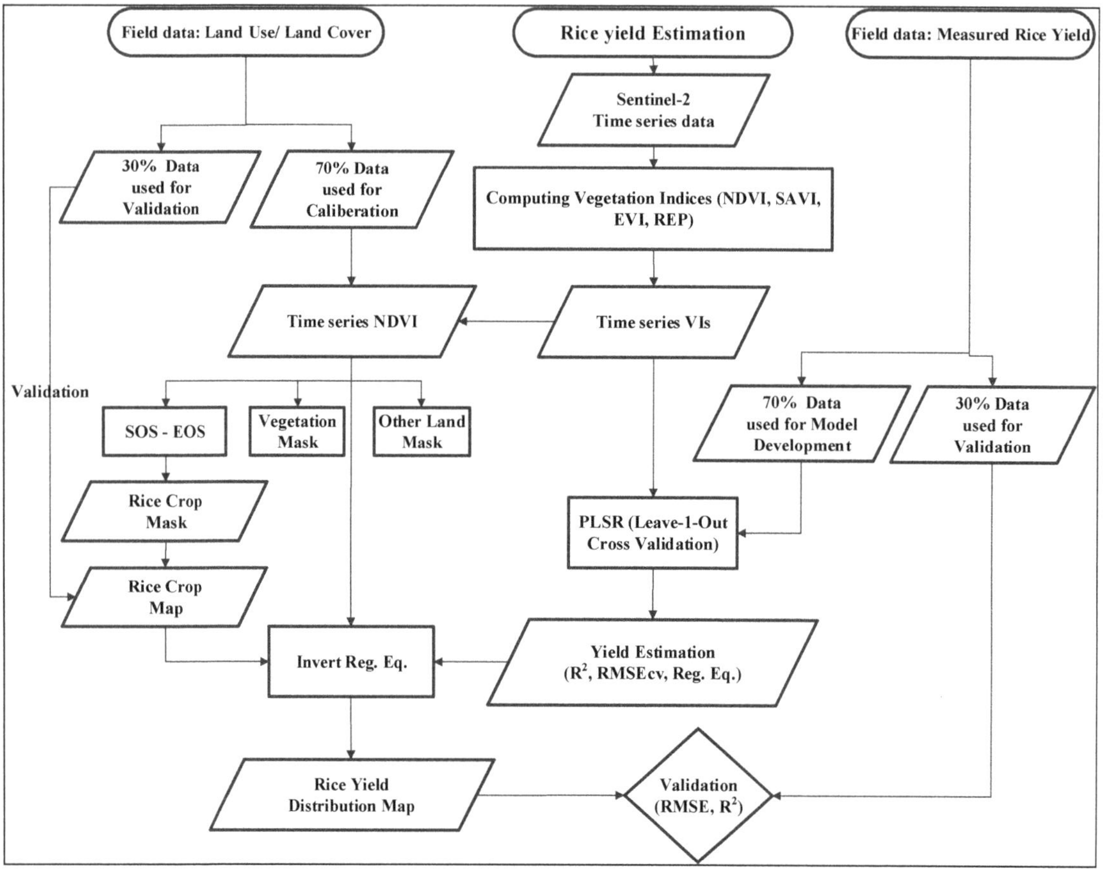

Figure 4. The framework of methodology followed for phenology-based rice mapping and yield estimation.

To develop the spatial distribution maps of rice yield, the models of statistical analysis (the regression equations of PLSR analysis) were inverted to the time series vegetation indices of rice masked areas (developed from phenological based mapping) and were validated with an independent dataset.

3. Results

3.1. Rice Yield Estimation (Field Level Data)

The statistical descriptions of the measured rice yield are summarized in Table 2. In calibration datasets (with sample size n = 96), the rice yield varies between 3.06 ton/ha and 4.15 ton/ha with mean equal to 3.70 ton/ha. The standard deviation (SD) was ±0.31 ton/ha and coefficient of covariance (CV) equal to 0.083 ton/ha. The graphical display reflects that the calibration dataset is near normally distributed.

Table 2. Statistical description of the field measured rice yield with graphical displays of calibration and validation datasets.

Dataset Type	Sample Size (n)	Minimum (ton/ha)	Maximum (ton/ha)	Mean (ton/ha)	SD (ton/ha)	CV (ton/ha)	Graphical Distribution
Calibration	96	3.06	4.15	3.70	0.31	0.083	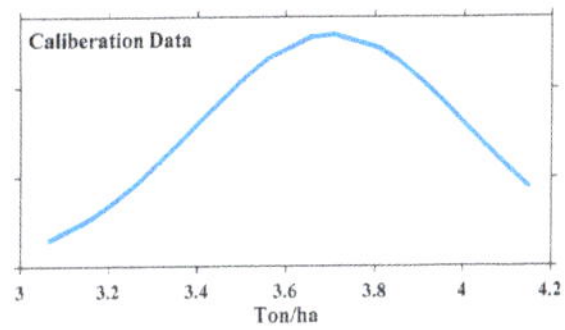
Validation	41	3.16	4.15	3.71	0.29	0.078	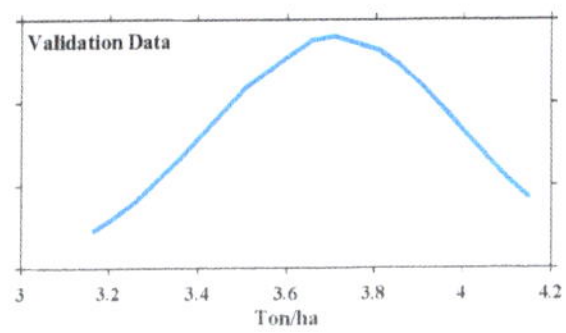

The summary statistics of validation sets (n = 41) shows that rice yield ranges between 3.16 ton/ha and 4.15 ton/ha with average equal to 3.71 ton/ha. The standard deviation (SD) was ±0.29 and coefficient of covariance (CV) equal to 0.078 ton/ha. The validation data are also normally distributed (Table 2: see the graphical display at last column).

3.2. Variation in Temporal Profiles of Vegetation Indices with Rice Phenology

The temporal profiles of vegetation index spanning across the full growth period of rice crop are shown in Figure 5. The vegetation index (e.g., NDVI) values were least (minimum) at transplanting phase and showed gradual increase with increase in vegetative parts (Figure 5). The vegetation indices reached at peak in the late vegetative phase and continually maintained high values (e.g., spectral plateau) till flowering phase. At post flowering phase (e.g., ripening phase), the vegetation index values started declining and reached its minimum at fully ripened harvesting phase (see Figure 5).

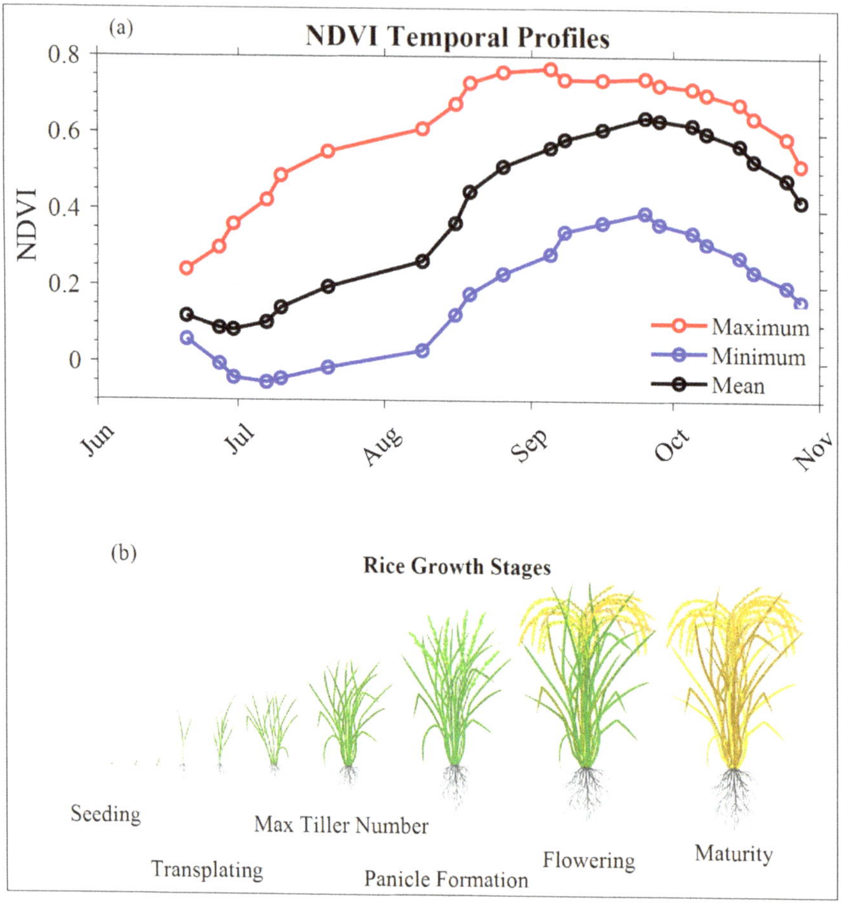

Figure 5. (**a**) The temporal profiles of vegetation index (NDVI) and (**b**) different growth stages of rice crop. The points are not equally spaced (subplot **a**) due to the unavailability of cloud free images of the study area (i.e., monsoon season; late July and early August).

3.3. Prediction of Rice Yield and the Performance of Vegetation Indices

The prediction of rice yields using PLSR and time series vegetation indices are summarized Table 3. The integration of PLSR and sequential-time-stamped-vegetation indices were found effective for quantifying rice yields. The time series NDVI yielded the highest $R^2 = 0.83$ (lowest $RMSE_{cv} = 0.12$ ton/ha) followed by EVI ($R^2 = 0.80$, $RMSE_{cv} = 0.14$ ton/ha), SAVI ($R^2 = 0.79$, $RMSE_{cv} = 0.14$ ton/ha), and REP ($R^2 = 0.64$, $RMSE_{cv} = 0.17$ ton/ha). The performance of different indices (NDVI, SAVI, EVI, REP) for rice yield estimation were consistent for both calibration and validation datasets (Table 3, Figure 6c,f,i,l).

Table 3. Results of the PLSR applied to time series vegetation indices (NDVI, SAVI, EVI, and REP). Number of latent variables (in PLSR model), calibration R^2, validation R^2, calibration RMSE ($RMSE_C$), and cross validation RMSE ($RMSE_{CV}$) are summarized.

Indices	No. of Latent Variables in PLSR Model	Calibration R^2	RMSEC (ton/ha)	Validation R^2	RMSECV (ton/ha)
NDVI	6	0.87	0.11	0.83	0.12
EVI	6	0.85	0.12	0.80	0.14
SAVI	6	0.84	0.12	0.79	0.14
REP	5	0.70	0.16	0.62	0.17

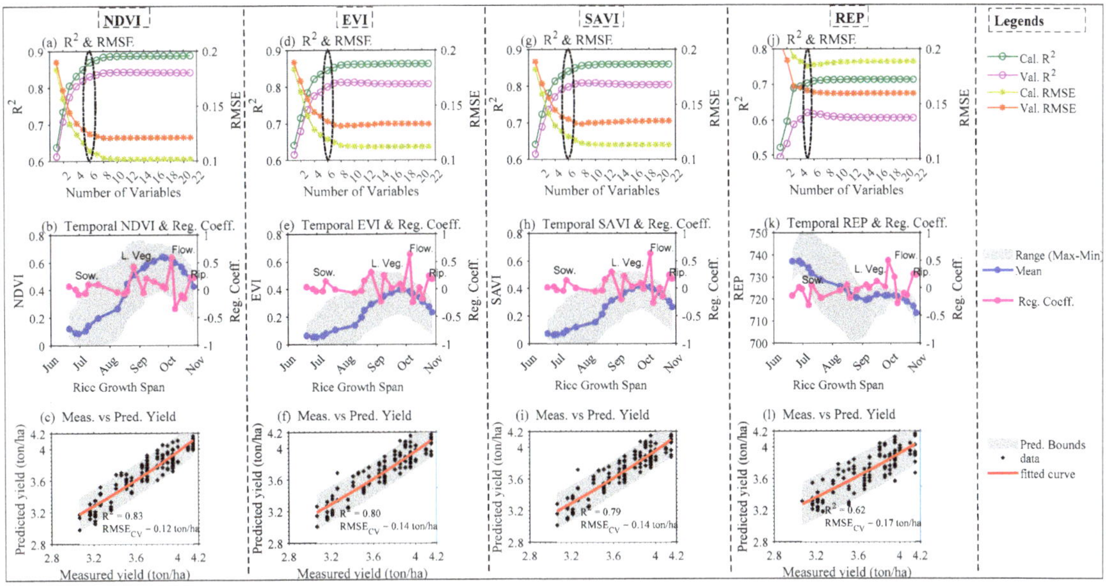

Figure 6. The analysis based on temporal NDVI, EVI, SAVI, and REP. The R^2 increases and RMSE decreases with augmenting the number of variables unless it reaches saturation (dashed elliptical in Panels **a**,**d**,**g**,**j**). After certain number of latent variables, the decrease in RMSE was negligible and that was taken as the optimum number of variables for PLSR model development. The temporal profiles of vegetation indices are similar in shape except REP (Panels **b**,**e**,**h**,**k**). The regression coefficients lines show that sowing, late vegetative, flowering and ripening are important phases for predicting rice yields using PLSR. The measured vs. predicted (Panels **c**,**f**,**i**,**l**) manifest that yield was best predicted using temporal NDVI data (yielded highest R^2 and lowest RMSE).

Using the time series profiles of NDVI, EVI and SAVI, the PLSR models selected six latent variables (Figure 6, Table 3). The selected six latent variables explained most of the variance (e.g., as the case of NDVI where R^2 = 0.83) and the addition of further variables hardly improve the model performance (e.g., the total 21 variables yield maximum R^2 of 0.84) (Figure 6a,d,g,j). The important latent variable (in this study, time stamped vegetation indices) was located at late vegetative, reproductive (flowering), and ripening phases of rice growth (Figure 6b,e,h,k). Using time series Red Edge Position (REP) data, the number of selected latent variables were five (05) with high regression coefficient (or *B*-coefficient) values at flowering, ripening, and late vegetative phases (Figure 6c).

3.4. Spatial Varability in Rice Yield Potential

The distribution map (developed from best predicting PLSR model, time series vegetation indices and map of rice grown area) reflects that rice yield distribution varies in space and ranges between 1.5 to 4.2 ton/ha (Figure 7a). The upper limit of rice yields (i.e., 4.20 ton/ha) in the distribution map were closely matching with the ground measured yield (4.15 ton/ha). The minimum limit was underestimated in the spatial distribution map of rice yield (1.5 ton/ha) compared to the ground measured minimum yield (3.06 ton/ha). The validation of spatial distribution maps against independent ground measured yield data (30% of the total samples) confirms that yield was predicted with high accuracy (see Figure 7b). The high R^2 (0.83) and low RMSE (0.14 tons/ha) manifested a close match between measured and predicted rice yields (Figure 7b).

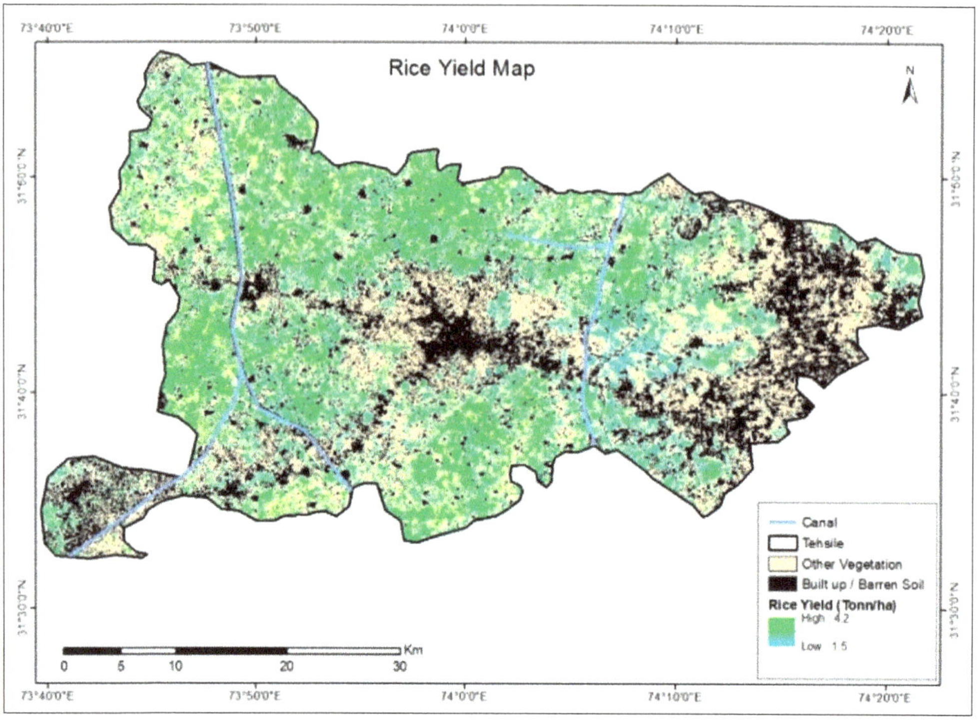

(**a**)

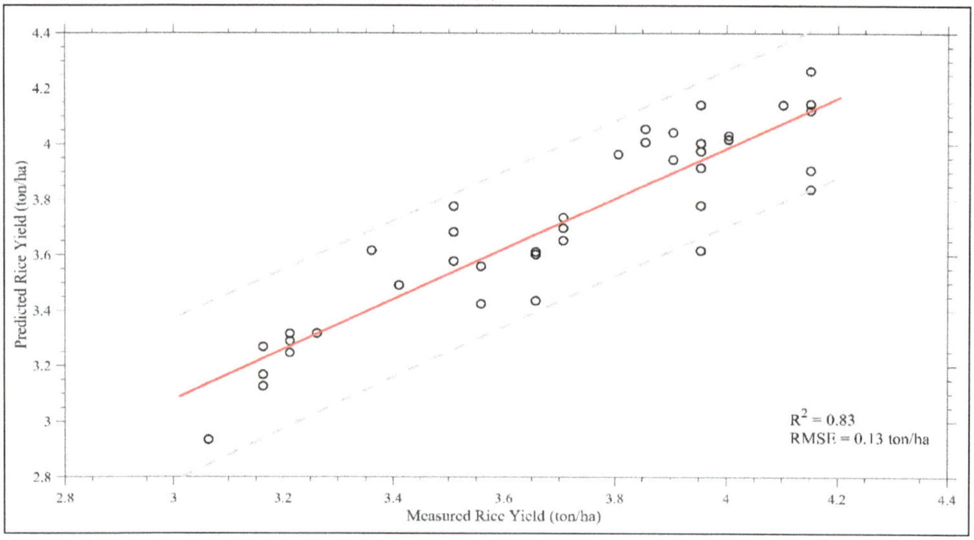

(**b**)

Figure 7. The distribution of rice yield varies spatially (ranging from 1.5 to 4.2 ton/ha) in the study area (**a**). The predicted yield (based on distribution map) shows strong relation with measured yield (**b**).

4. Discussion

The field data (Table 2) reflect that the measured rice yield production (minimum (3.06 ton/ha), mean (3.7 ton/ha), maximum 4.15 (ton/ha) in the study area is within the limits of rice crop statistics within the country (i.e., ranging from 2.4 ton/ha to 10 ton/ ha). These numbers are far less than the statistics of the neighboring countries (i.e., China, Vietnam, Bangladesh) and could be attributed to the variety of rice [45], uneven water usage, weeds and pest attacks, and post-harvest loses (e.g., shattering and improper drying and storing etc.). The super basmati grown in the study area produces high quality rice and is famous for its aromatic fragrance; however, it is less productive compared to other hybrid varieties.

The time series profiles show that vegetation indices (Figure 8a) are minimum at the transplantation stage and gradually increase in vegetative phase (tillering, panicle, and flowering stages). A decline was observed in the vegetation indices after post flowering phases (e.g., dough, ripening). This initial increase in vegetation index values could be associated with increase in leaf area coverage (LAI, biomass) and the post flowering phase decline could be attributed to the senescing of rice crop. The temporal variation in vegetation indices (in this study) are in line with the findings of previous studies [46,47], where the peak greenness is achieved at flowering/milk phases and decline was observed in dough stages and reaching minimum at ripening phase (Figure 8).

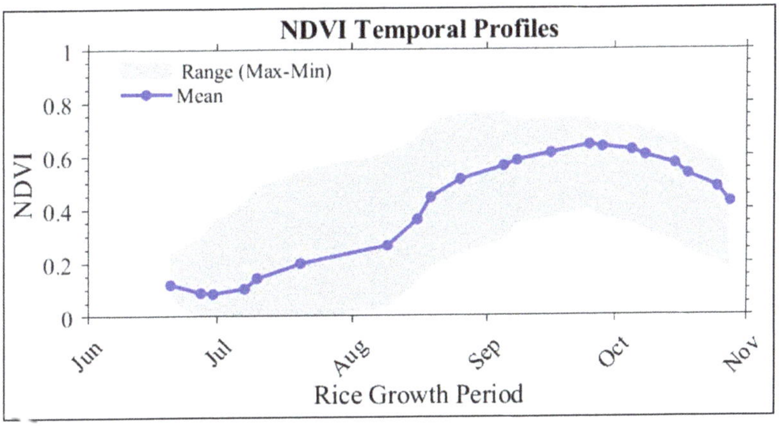

(**a**)

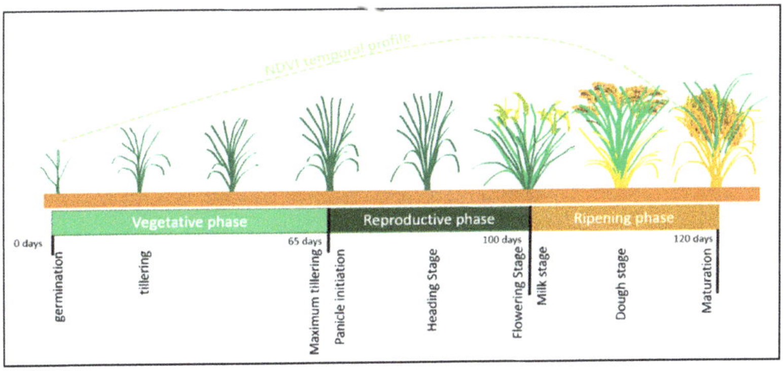

(**b**)

Figure 8. (**a**) The range and mean temporal profiles of Normalized Difference Vegetation Index (NDVI) spanning the entire growth period of rice crop. (**b**) The temporal profile of vegetation index (NDVI) is closely in-line with the findings of [47] (shown in plot "**b**").

The integration of PLSR and time series vegetation indices accurately predicted rice yields (the maximum $R^2 = 0.83$ and least RMSE = 0.12 ton/ha (3.12% of the mean yield). The slightly better performance of NDVI (relative to SAVI, EVI, and REP) could be attributed to the canopy characteristics (e.g., structural, vegetation percent cover) of rice crop. The abundance of stem and leaf blades of rice crop obscures the background soil visibility and leads to enhanced vegetation (rice) reflectance signals, thus allowing slope-based-index (such as NDVI) to perform more precise estimates of rice yield [48,49].

Using PLSR regression, the prediction accuracy enhances with augmenting the number of variables (NDVI, EVI, SAVI, and REP) and leads to high R^2 and low RMSE until the model stabilized at a certain point. In this study, the PLSR model saturate at the addition of six latent variables (Figure 6a,d,g,j) and captured maximum variance present in dataset. The addition of further variables hardly improves the prediction of rice yield and displays a flat line [50]. The selected latent variables belong to late vegetative, reproductive (panicle, flowering, milky), and ripening phases (Figure 6b,e,h,k) and reflects the critical importance of these growth stages in rice yield production. The picking of late vegetative phase may be associated with an increase in biomass and leaf covered area (leaves and shoots fully develop at this stage and gain maximum crop height). The high correlation at reproductive phases could be associated with the formation panicle, flowering, milk, dough, and maturity of grain, which directly influence the crop yields. The results of this study are also consistent with the findings of existing literature, where booting stage highly influences the rice yield production [51]. The outcomes of this study help in estimating the rice yield and highlight the critical phases in the life cycle (of rice crop) where monitoring and human intervention (such as usage of water and agrochemicals) can enhance the yield production.

5. Conclusions

This study aims to accurately quantify rice yield and identify the critical growth stages that influence the rice yield production. The integration of PLSR and time series vegetation indices (i.e., spanning across the entire rice crop growth period) results in accurate predictions of rice yield. Among vegetation indices, NDVI performs the best (yield high R^2 and low RMSE) followed by EVI, SAVI, and REP. Using the time stamped vegetation indices, the PLSR coefficients identified the growth stages that influence the rice yield. The growth stages (selected latent variables) belong to late vegetative, reproductive (panicle, flowering, milky), and ripening phases. The selected critical growth stages were common in all four types of vegetation indices (i.e., NDVI, EVI, SAVI, and REP) used. This study concludes that PLSR can effectively be used for rice yield estimation and identifying critical stages of the rice growth cycle. The precise yield estimation (rice in this case) allows decision makers to strategize policy regarding yield import and export. The outcome of this study can also help the farmers to monitor rice at critical time spans and allow them to intervene (e.g., usage of water, fertilizer, pesticides etc.) in a timely manner. The timely interventions thus help in producing more yields which in turn is essential for minimizing hunger (SDG 2), alleviating poverty (SDG 1), ensuring land (SDG 15) and food security.

However, the present study did not compare the accuracy of PLS algorithm with artificial neural networks, support vector machines, other geo-statistics, etc., for yield forecasting. These would be interesting directions for future study.

Author Contributions: The study was conceptualization by A.N. and S.U.; methodology, A.A. (Azhar Abbas), M.S.I.; data collection and validation, Z.A.S. and K.H.; formal analysis, M.S. (Muhammad Shakir) and M.S. (Munawar Shah); Planning and Execution, M.U.B.; writing—review and editing, A.N. and A.A. (Asad Ali). All authors have read and agreed to the published version of the manuscript.

Funding: We are thankful for the financial support of University of Agriculture Faisalabad (APC funding) and Galaxy Rice Mills Pvt Ltd. Gujranwala for logistics during field data collection.

Institutional Review Board Statement: This study is 'Not applicable' for Institutional Review Board and Ethical review as this research work does not involve any humans or animals or biological material.

Informed Consent Statement: Not applicable.

Data Availability Statement: Data will be made available for research purpose upon request.

Acknowledgments: We acknowledge the Copernicus Open Access Hub for providing freely available Sentinel-2 data. We also acknowledge M. Hamid Choudhary, GIS Centre The University of Punjab, Lahore Pakistan for helping in software and Field survey.

Conflicts of Interest: The authors declare no conflict of interest.

References

1. Godfray, H.C.J.; Beddington, J.R.; Crute, I.R.; Haddad, L.; Lawrence, D.; Muir, J.F.; Pretty, J.; Robinson, S.; Thomas, S.M.; Toulmin, C. Food Security: The Challenge of Feeding 9 Billion People. *Science* **2010**, *327*, 812–818. [CrossRef]
2. Dawe, D. The contribution of rice research to poverty alleviation. In *Redesigning Rice Photosynthesis to Increase Yield, Proceedings of the Workshop on the Quest to Reduce Hunger: Redesigning Rice Photosynthesis, Los Baños, Philippines, 30 November–3 December 1999*; Sheehy, J.E., Mitchell, P.L., Hardy, B., Eds.; Elsevier: Amsterdam, The Netherlands, 2000; pp. 3–12.
3. FAO. *The State of Food and Agriculture (SOFA): Climate Change, Agriculture and Food Security*; Food and Agriculture Organization of the United Nations (FAO): Rome, Italy, 2016.
4. Bandumula, N. Rice Production in Asia: Key to Global Food Security. *Proc. Natl. Acad. Sci. USA India Sect. B Biol. Sci.* **2018**, *88*, 1323–1328. [CrossRef]
5. PBS. *Prices of Agricultural Commodities*; Agriculture Statistics; Pakistan Bureau of Statistics (PBS): Islamabad, Pakistan, 2019; pp. 124–126.
6. Wang, F.; Wang, F.; Zhang, Y.; Hu, J.; Huang, J.; Xie, J. Rice Yield Estimation Using Parcel-Level Relative Spectral Variables From UAV-Based Hyperspectral Imagery. *Front. Plant Sci.* **2019**, *10*, 453. [CrossRef]
7. Gebbers, R.; Adamchuk, V.I. Precision Agriculture and Food Security. *Science* **2010**, *327*, 828–831. [CrossRef]
8. Ji, Z.; Pan, Y.; Zhu, X.; Wang, J.; Li, Q. Prediction of crop yield using phenological information extracted from remote sensing vegetation index. *Sensors* **2021**, *21*, 1406. [CrossRef]
9. Fang, H.; Liang, S.; Hoogenboom, G.; Teasdale, J.; Cavigelli, M. Corn-yield estimation through assimilation of remotely sensed data into the CSM-CERES-Maize model. *Int. J. Remote Sens.* **2008**, *29*, 3011–3032. [CrossRef]
10. Lobell, D.B. The use of satellite data for crop yield gap analysis. *Field Crop. Res.* **2013**, *143*, 56–64. [CrossRef]
11. Zhang, P.P.; Zhou, X.X.; Wang, Z.X.; Mao, W.; Li, W.X.; Yun, F.; Guo, W.S.; Tan, C.W. Using HJ-CCD image and PLS algorithm to estimate the yield of field-grown winter wheat. *Sci. Rep.* **2020**, *10*, 5173. [CrossRef] [PubMed]
12. Savin, I.Y.; Isaev, V.A. Rice yield forecast based on satellite and meteorological data. *Russ. Agric. Sci.* **2010**, *36*, 424–427. [CrossRef]
13. Shiu, Y.-S.; Chuang, Y.-C. Yield Estimation of Paddy Rice Based on Satellite Imagery: Comparison of Global and Local Regression Models. *Remote Sens.* **2019**, *11*, 111. [CrossRef]
14. Peng, D.; Huang, J.; Li, C.; Liu, L.; Huang, W.; Wang, F.; Yang, X. Modelling paddy rice yield using MODIS data. *Agric. For. Meteorol.* **2014**, *184*, 107–116. [CrossRef]
15. Gilardelli, C.; Stella, T.; Confalonieri, R.; Ranghetti, L.; Campos-Taberner, M.; García-Haro, F.J.; Boschetti, M. Downscaling rice yield simulation at sub-field scale using remotely sensed LAI data. *Eur. J. Agron.* **2019**, *103*, 108–116. [CrossRef]
16. Ines, A.V.M.; Das, N.N.; Hansen, J.W.; Njoku, E.G. Assimilation of remotely sensed soil moisture and vegetation with a crop simulation model for maize yield prediction. *Remote Sens. Environ.* **2013**, *138*, 149–164. [CrossRef]
17. Launay, M.; Guerif, M. Assimilating remote sensing data into a crop model to improve predictive performance for spatial applications. *Agric. Ecosyst. Environ.* **2005**, *111*, 321–339. [CrossRef]
18. Caballero, I.; Fernández, R.; Escalante, O.M.; Mamán, L.; Navarro, G. New capabilities of Sentinel-2A/B satellites combined with in situ data for monitoring small harmful algal blooms in complex coastal waters. *Sci. Rep.* **2020**, *10*, 8743. [CrossRef]
19. Basterrechea, D.A.; Parra, L.; Parra, M.; Lloret, J. A Proposal for Monitoring Grass Coverage in Citrus Crops Applying Time Series Analysis in Sentinel-2 Bands. In Proceedings of the 4th International Conference on Industrial IoT Technologies and Applications, Virtual Event; Springer: Cham, Switzerland, 11 December 2020; pp. 193–206. Available online: https://link.springer.com/chapter/10.1007/978-3-030-71061-3_12 (accessed on 30 September 2021).
20. Phiri, D.; Simwanda, M.; Salekin, S.; Nyirenda, V.R.; Murayama, Y.; Ranagalage, M. Sentinel-2 data for land cover/use mapping: A review. *Remote Sens.* **2020**, *12*, 2291. [CrossRef]
21. Zheng, H.; Cheng, T.; Yao, X.; Deng, X.; Tian, Y.; Cao, W.; Zhu, Y. Detection of rice phenology through time series analysis of ground-based spectral index data. *Field Crop. Res.* **2016**, *198*, 131–139. [CrossRef]
22. Guo, C.; Zhang, L.; Zhou, X.; Zhu, Y.; Cao, W.; Qiu, X.; Cheng, T.; Tian, Y. Integrating remote sensing information with crop model to monitor wheat growth and yield based on simulation zone partitioning. *Precis. Agric.* **2018**, *19*, 55–78. [CrossRef]
23. Murchie, E.H.; Yang, J.; Hubbart, S.; Horton, P.; Peng, S. Are there associations between grain-filling rate and photosynthesis in the flag leaves of field-grown rice? *J. Exp. Bot.* **2002**, *53*, 2217–2224. [CrossRef] [PubMed]
24. Ntanos, D.A.; Koutroubas, S.D. Dry matter and N accumulation and translocation for Indica and Japonica rice under Mediterranean conditions. *Field Crop. Res.* **2002**, *74*, 93–101. [CrossRef]
25. Ryu, C.; Suguri, M.; Umeda, M. Multivariate analysis of nitrogen content for rice at the heading stage using reflectance of airborne hyperspectral remote sensing. *Field Crop. Res.* **2011**, *122*, 214–224. [CrossRef]

26. Horie, T.; Ohnishi, M.; Angus, J.F.; Lewin, L.G.; Tsukaguchi, T.; Matano, T. Physiological characteristics of high-yielding rice inferred from cross-location experiments. *Field Crop. Res.* **1997**, *52*, 55–67. [CrossRef]
27. Yang, Q.; Shi, L.; Han, J.; Zha, Y.; Zhu, P. Deep convolutional neural networks for rice grain yield estimation at the ripening stage using UAV-based remotely sensed images. *Field Crop. Res.* **2019**, *235*, 142–153. [CrossRef]
28. PMD-Pakistan Meteorological Department. *Decadal Agromet Bulletin of Pakistan*; National Agromet Centre (NAMC) Pakistan Meteorological Department: Islamabad, Pakistan, 2019; p. 8.
29. Khaliq, T.; Gaydon, D.S.; Ahmad, M.-D.; Cheema, M.J.M.; Gull, U. Analyzing crop yield gaps and their causes using cropping systems modelling–A case study of the Punjab rice-wheat system, Pakistan. *Field Crop. Res.* **2019**, *232*, 119–130. [CrossRef]
30. Qamar, W.; Younas, M.; Waseem, M. Price Fluctuations of Rice Crop in District Sheikhupura. *J. Agric. Stud.* **2019**, *7*, 227. [CrossRef]
31. Transon, J.; D'Andrimont, R.; Maugnard, A.; Defourny, P. Survey of Hyperspectral Earth Observation Applications from Space in the Sentinel-2 context. *Remote Sens.* **2018**, *10*, 157. [CrossRef]
32. Cohrs, C.W.; Cook, R.L.; Gray, J.M.; Albaugh, T.J. Sentinel-2 Leaf Area Index Estimation for Pine Plantations in the Southeastern United States. *Remote Sens.* **2020**, *12*, 1406. [CrossRef]
33. Narin, O.G.; Abdikan, S. Monitoring of phenological stage and yield estimation of sunflower plant using Sentinel-2 satellite images. *Geocarto. Int.* 2020. [CrossRef]
34. Thenkabail, P.S.; Smith, R.B.; De Pauw, E. Hyperspectral Vegetation Indices and Their Relationships with Agricultural Crop Characteristics. *Remote Sens. Environ.* **2000**, *71*, 158–182. [CrossRef]
35. Rouse, J.W., Jr.; Haas, R.H.; Schell, J.A.; Deering, D.W. *Monitoring Vegetation Systems in the Great Plains with ERTS*; NASA Special Pub-351; NASA: Washington, DC, USA, 1974; pp. 309–317.
36. Huete, A.R. A soil-adjusted vegetation index (SAVI). *Remote Sens. Environ.* **1988**, *25*, 295–309. [CrossRef]
37. Liu, H.Q.; Huete, A. A feedback based modification of the NDVI to minimize canopy background and atmospheric noise. *IEEE Trans. Geosci. Remote Sens.* **1995**, *33*, 457–465. [CrossRef]
38. Filella, I.; Penuelas, J. The red edge position and shape as indicators of plant chlorophyll content, biomass and hydric status. *Int. J. Remote Sens.* **1994**, *15*, 1459–1470. [CrossRef]
39. Kawamura, K.; Watanabe, N.; Sakanoue, S.; Lee, H.-J.; Inoue, Y.; Odagawa, S. Testing genetic algorithm as a tool to select relevant wavebands from field hyperspectral data for estimating pasture mass and quality in a mixed sown pasture using partial least squares regression. *Grassl. Sci.* **2010**, *56*, 205–216. [CrossRef]
40. Ullah, S.; Skidmore, A.K.; Ramoelo, A.; Groen, T.A.; Naeem, M.; Ali, A. Retrieval of leaf water content spanning the visible to thermal infrared spectra. *SPRS J. Photogramm. Remote Sens.* **2014**, *93*, 56–64. [CrossRef]
41. Denham, M.C. Choosing the number of factors in partial least squares regression: Estimating and minimizing the mean squared error of prediction. *J. Chemom.* **2000**, *14*, 351–361. [CrossRef]
42. Li, Y.; Guan, K.; Yu, A.; Peng, B.; Zhao, L.; Li, B.; Peng, J. Toward building a transparent statistical model for improving crop yield prediction: Modeling rainfed corn in the U.S. *Field Crop. Res.* **2019**, *234*, 55–65. [CrossRef]
43. Gong, Z.; Kawamura, K.; Ishikawa, N.; Goto, M.; Wulan, T.; Alateng, D.; Yin, T.; Ito, Y. MODIS normalized difference vegetation index (NDVI) and vegetation phenology dynamics in the Inner Mongolia grassland. *Solid Earth* **2015**, *6*, 1185–1194. [CrossRef]
44. Cai, Z.; Jönsson, P.; Jin, H.; Eklundh, L. Performance of Smoothing Methods for Reconstructing NDVI Time-Series and Estimating Vegetation Phenology from MODIS Data. *Remote Sens.* **2017**, *9*, 1271. [CrossRef]
45. Laenoi, S.; Rerkasem, B.; Lordkaew, S.; Prom-u-thai, C. Seasonal variation in grain yield and quality in different rice varieties. *Field Crop. Res.* **2018**, *221*, 350–357. [CrossRef]
46. Kuenzer, C.; Knauer, K. Remote sensing of rice crop areas. *Int. J. Remote Sens.* **2013**, *34*, 2101–2139. [CrossRef]
47. Mosleh, M.K.; Hassan, Q.K.; Chowdhury, E.H. Application of Remote Sensors in Mapping Rice Area and Forecasting its Production: A Rev. *Sensors* **2015**, *15*, 769–791. [CrossRef]
48. Panda, S.S.; Ames, D.P.; Panigrahi, S. Application of Vegetation Indices for Agricultural Crop Yield Prediction Using Neural Network Techniques. *Remote Sens.* **2010**, *2*, 673–696. [CrossRef]
49. Mirasi, A.; Mahmoudi, A.; Navid, H.; Valizadeh Kamran, K.; Asoodar, M.A. Evaluation of sum-NDVI values to estimate wheat grain yields using multi-temporal Landsat OLI data. *Geocarto. Int.* **2019**, *36*, 1309–1324. [CrossRef]
50. Arshad, M.; Ullah, S.; Khurshid, K.; Ali, A. Estimation of leaf water content from mid- and thermal-infrared spectra by coupling genetic algorithm and partial least squares regression. *J. Appl. Remote Sens.* **2018**, *12*, 022203. [CrossRef]
51. Chang, K.-W.; Shen, Y.; Lo, J.-C. Predicting Rice Yield Using Canopy Reflectance Measured at Booting Stage. *Agron. J.* **2005**, *97*, 872–878. [CrossRef]

agriculture

MDPI

Article

Evaluation of Deep Learning for Automatic Multi-View Face Detection in Cattle

Beibei Xu [1], Wensheng Wang [1,2,3,*], Leifeng Guo [1,3], Guipeng Chen [4], Yaowu Wang [5], Wenju Zhang [1] and Yongfeng Li [1]

1 Agricultural Information Institute, Chinese Academy of Agriculture Sciences, Beijing 100086, China; xuxiaobei224@163.com (B.X.); guoleifeng@caas.cn (L.G.); zhangwenju@caas.cn (W.Z.); liyongfeng_1116@163.com (Y.L.)
2 Information Centre, Ministry of Agriculture and Rural Affairs, Beijing 100125, China
3 Key Laboratory of Agricultural Big Data, Ministry of Agriculture and Rural Affairs, Beijing 100086, China
4 Agricultural Economics and Information Institute, Jiangxi Academy of Agriculture Sciences, Nanchang 330200, China; chenguipeng1983@163.com
5 Laboratory of Geo-Information Science and Remote Sensing, Wageningen University, 6708 PB Wageningen, The Netherlands; wangyaowu@caas.cn
* Correspondence: wangwensheng@caas.cn

Citation: Xu, B.; Wang, W.; Guo, L.; Chen, G.; Wang, Y.; Zhang, W.; Li, Y. Evaluation of Deep Learning for Automatic Multi-View Face Detection in Cattle. *Agriculture* **2021**, *11*, 1062. https://doi.org/10.3390/agriculture11111062

Academic Editors: Gniewko Niedbała and Sebastian Kujawa

Received: 23 September 2021
Accepted: 23 October 2021
Published: 28 October 2021

Abstract: Individual identification plays an important part in disease prevention and control, traceability of meat products, and improvement of agricultural false insurance claims. Automatic and accurate detection of cattle face is prior to individual identification and facial expression recognition based on image analysis technology. This paper evaluated the possibility of the cutting-edge object detection algorithm, RetinaNet, performing multi-view cattle face detection in housing farms with fluctuating illumination, overlapping, and occlusion. Seven different pretrained CNN models (ResNet 50, ResNet 101, ResNet 152, VGG 16, VGG 19, Densenet 121 and Densenet 169) were fine-tuned by transfer learning and re-trained on the dataset in the paper. Experimental results showed that RetinaNet incorporating the ResNet 50 was superior in accuracy and speed through performance evaluation, which yielded an average precision score of 99.8% and an average processing time of 0.0438 s per image. Compared with the typical competing algorithms, the proposed method was preferable for cattle face detection, especially in particularly challenging scenarios. This research work demonstrated the potential of artificial intelligence towards the incorporation of computer vision systems for individual identification and other animal welfare improvements.

Keywords: cattle face detection; RetinaNet; deep learning; precision livestock

1. Introduction

Animal husbandry is undergoing a transition from extensive farming to precision livestock farming and welfare breeding. However, the farming facilities and technologies play crucial parts in affecting the economic benefits of large-scale pastures. Inadequate management probably directly damages the health of livestock and is adverse to the food quality and safety, and the development of the livestock industry [1]. Therefore, there is an urgent need for cost-effective technology methods to address these challenges in animal agricultural systems, such as lack of labor and difficulties in real-time monitoring. Precision farming has aroused more interest recently due to the increasing concern over sustainable livestock and production efficiency [1–5]. Precision farming takes advantage of modern information technologies as an enabler of more efficient, productive, and profitable farming enterprises. For example, Internet of Things (IoT) are used for collecting data on the whole lifecycle of livestock, including breeding, slaughtering, meat processing, and marketing; Big Data and Artificial Intelligence (AI) can provide accurate analysis and real-time physical dynamics of each animal species as for a scientific basis for decision-making and analysis of farm managers. Among these, recognition of individual livestock is an indispensable and significant task in precision livestock management since it has a tremendous breadth of applications. Quick and accurate recognition of individual farm animals is of major importance for illness prevention and control [6], genetic enhancement of varieties [7], quality security of dairy products [8], and reduction of agricultural fake insurance claims [9].

Classical livestock identification techniques, such as ear notching [10], ear tattooing [11], hot iron branding [12], and ear tags [13–16], is subject to equipment loss, duplication, fraud, animal welfare security, monitoring cost, and distance challenges. Instead, based on biometric traits, non-contact identification is a new trend in livestock identification due to its uniqueness, invariance, low cost and easy operation, and high animal welfare. The non-contact identification methods, such as retinal vascular patterns [17,18], iris patterns, and muzzle print patterns [19,20], utilize computer vision and pattern recognition to extract biological features of livestock for individual identification. However, as an individual's most direct external visual information, the difference in facial features allows the livestock's face to be used more extensively to identify the individual. From the perspective of farm practice, compared with the biometric recognition methods noted, face identification is more intuitive and compatible with habits. There is also no need for cooperation of livestock fixed postures. In addition, face identification has great advantages in terms of anti-interference and scalability, which is analogous to human face identification.

Detection of livestock face is often conducted prior to individual identification and tracking in biometric and surveillance systems. Many approaches have been put forward in the literature for animal face detection. Mukai et al. employed the Haar and HOG (Histogram of Oriented Gradients) feature to build the classifiers for pet faces and proved the effectiveness for detecting the cat and dog faces [21]. Local Binary Pattern (LBP) features have been used to extract local texture features from different levels of Gaussian filtered images of cattle faces for face detection [22]. Clark presented pigs' face detection by identifying the features utilizing the Viola–Jones approach for cascade classifiers and basic likelihood functions [23]. Mohammed et al. aimed to detect multi-view faces in cattle with accuracy enhancement using three classifiers and temperature thresholding. Cattle face detection was established in thermal imaging by adopting HOG as a feature and Support Vector Machine (SVM) as a classifier [24]. Akihiko et al. combined face detection with digital cameras to automatically find dogs and cats in the images with acceptable speed performance by integrating edge-based features with multi-layer classifiers [25].

However, the heavy involvement of handcrafted features prevents these approaches from application to complex scenarios in terms of speed and accuracy. The use of the Convolutional Neural Network (CNN) to detect livestock has been demonstrated as successful and promising for further research with regard to variable inputs, processing speed, and accuracy for object detection in images [26]. Alžběta et al. dealt with a reliable dog face detection approach in the images by adopting the two-step technique using the cascade of regressors [27]. The recent advances in deep learning [28–34] have shown their great potential in object detection and classification of thousands of global images due to higher accuracy, precision, and quicker processing speed. Faster R-CNN has been directly used for face detection combined with PANSNet-5 in the cow face recognition framework [35]. Considering the practical scenario of multi-face detection task of livestock cattle identity authentication, Gou et al. improved Faster R-CNN by substituting ZF network for Inception v2 as the basic network [36].

Despite these advances in livestock face detection, the subtle changes in lighting, severe pose variation, false acceptances because of complex background, color similarity between livestock and background, shape deformation, and occlusion present serious challenges to face detection in an actual setting such as a cattle feedlot. Consequently, it is highly necessary to perform a wider assessment of face detection algorithm performance across a range of livestock production settings. The rapid development of object detection with deep learning provides promising techniques for face detection. RetinaNet, a recently proposed powerful object detection framework, which surpasses the detection performance of cutting-edge, two-stage R-CNN family object detectors and matches the speed of one-stage object detectors, appears to be the most prospective for livestock face detection. In the previous research, RetinaNet was used to explore for detection of road damages [37], automated detection of firearms in cargo X-ray images [38], and the task of indoor assistance navigation for blind and visually impaired persons [39]. Despite the general appeal of RetinaNet, it has not been evaluated in great detail for precision livestock monitoring practices. Given the urgent need to develop technologies that can assist with livestock production and welfare management, it is timely to assess the application of a state-of-the-art machine learning algorithm for precision livestock monitoring. Due to their great significance concerning animal husbandry, cattle were chosen as the case study to explore the performance of RetinaNet-based object detection for multi-view face detection.

2. Related Work

Face detection is a particular application of object detection that accurately finds the target face and its location in images. Object detection is currently a very active research field in computer vision that facilitates high-level tasks such as automatic individual identification and intelligent image recognition. The early object detection methods, including Viola–Jones detectors, HOG detector, and deformable part-based model were built based on handcrafted features, which render the time complexity high and many of the windows redundant [40]. In addition, manually designed features in the traditional object detection are not sufficiently robust to deal with the wide diversity of image changes encountered in practice; thereby, CNN was introduced into the object detection community. Due to its relatively superior performance of learning for robust and high-level feature representations of an image, CNN-based object detection prevents extracting complicated features and their reconstruction process in traditional object detection. Therefore, after R. Girshick et al. took the lead to propose the region-based CNN features for object detection in 2014, the object detection algorithms evolved from R-CNN at an unprecedented speed and have made much progress in recent years. Current state-of-the-art CNN-based object detectors can be grouped into two-stage algorithms and one-stage detection algorithms.

The two-stage detectors start with the extraction of object proposals through selective search or Region Proposal Network (RPN), and then the candidate regions are classified and regressed for precise coordinates. Regression-based algorithms such as Yolo and SSD require the sampling densely at various positions with different aspect ratios first, then provide the direct prediction of object categorization and a bounding box using CNN. Although the end-to-end procedure of the regression-based detectors outperforms the region-based detectors in processing speed, they achieve lower mean average precision because of example imbalance between object and background. As a result, T.-Y. Lin et al. designed a novel one-stage detector called RetinaNet in 2017 to address the class imbalance and increase the importance of hard examples [41]. "Focal loss" was used in RetinaNet to redefine the standard cross-entropy loss, so the training could automatically downweight the simple examples and center more on hard and misclassified examples. Focal loss enables RetinaNet to achieve comparable accuracy of two-stage algorithms and also maintains relatively high processing speed [41].

Considering the aspects of operating speed and accuracy in farming practice, RetinaNet was selected in this paper for further study. For face detection, unlike the human face, consideration should be given to changes in cattle's face and body orientation due to their random roaming. Therefore, this paper will explore the effectiveness of RetinaNet for multi-view cattle face detection. Advancements in deep learning networks present an opportunity to extend the research to the empirical comparisons of the typical CNN backbones for RetinaNet in the task of detecting multi-view cattle face.

3. Materials and Methods

3.1. Overview of the Proposed Framework

Figure 1 shows the overall workflow proposed for processing RGB images that are captured by 2-D cameras to detect multi-view cattle faces based on RetinaNet. The RGB images acquired by 2-D cameras are used as input images after image preprocessing, including image partitioning and image resize. The backbone, including ResNet, VGG, and Densenet, is selected for feature extraction, and then the Feature Pyramid Network (FPN) strengthens the multi-scale features formed in the former convolutional network to obtain more expressive feature maps, which contain a rich and multi-scale feature pyramid. The feature map selects two Fully Convolutional Network (FCN) sub-networks with the same structure but without sharing parameters for cattle face classification prediction and bounding-box prediction. Ground truth was annotated manually for every cattle face in the training sets and then network training was performed after labeling for forming the cattle face detector, followed by the output of multi-view cattle face detection in testing sets.

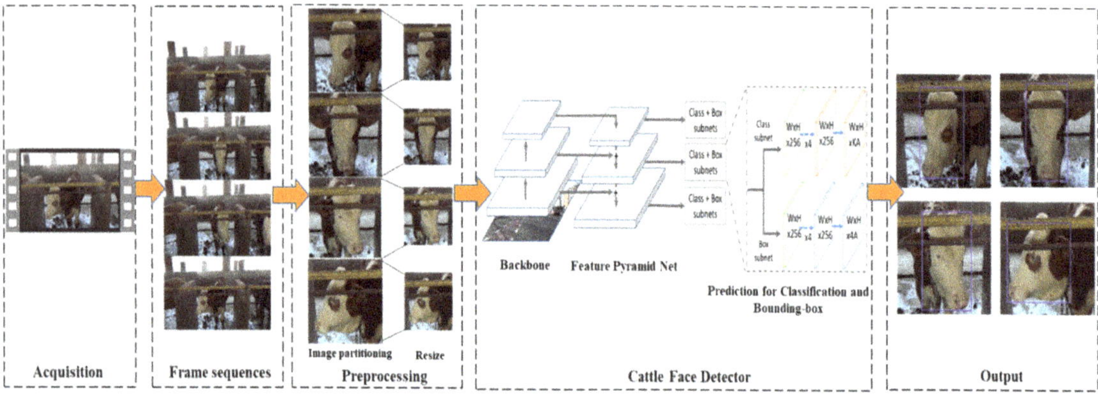

Figure 1. The proposed framework for the detection of multi-view cattle face based on RetinaNet.

3.2. RetinaNet-Based Object Detection

The name of RetinaNet comes from its dense sampling on the input image. RetinaNet is designed to evaluate the proposed focal loss for class imbalance in regression-based algorithms. The framework consists of three parts: (i) the front backbone network for feature extraction, (ii) FPN for constructing the multi-scale feature pyramids, and (iii) two subnetworks for object classification and bounding box regression. Focal loss is a newly high-sufficient loss function that replaces the training with the sampling heuristics and two-stage cascade while dealing with class imbalance. The details for backbones and FCN sub-networks, commonly used in R-CNN-like detectors, are expounded in the original papers, and this section mainly describes FPN and focal loss of the algorithm.

3.2.1. Feature Pyramid Networks

FPN is adopted to strengthen the feature extraction of backbone for weak semantic features using a top-down pyramid and lateral connections (see Figure 2). As indicated in the blue blocks, the bottom-up path is the feed-forward calculation for the main convolutional network, which calculates the feature hierarchy with different proportions. For the feature pyramid, the pyramid level is defined for each stage and the output of the last layer in each stage is chosen as the feature map because the deepest layer of each stage should have the strongest characteristics. Specifically, for the ResNet101 used in the RetinaNet, the outputs of these final residual blocks for conv2_x, conv3_x, conv4_x, and conv5_x are denoted as {C2, C3, C4, C5}. Since conv1 will occupy plenty of memory, it is not included in the pyramid.

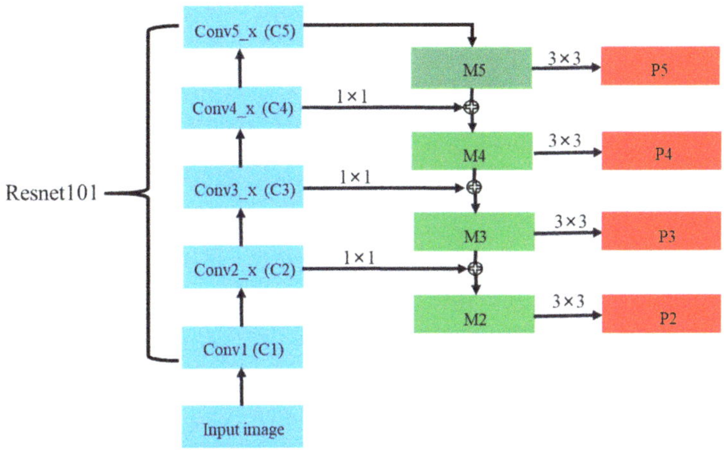

Figure 2. The architecture of FPN.

The top-down flow marked in green obtains high-resolution features by upsampling the feature maps with coarser space but stronger semantics from higher pyramid levels. Later, the bottom-up path is connected laterally to reinforce these features. Specifically, the weak feature map is upsampled twice, and then the upsampling map is merged with the corresponding bottom-up map. This cycle is repeated until the final resolution map is produced. We only need to combine a 1×1 convolutional layer with C5 to produce low-resolution images to run the iteration. Next, we append a 3×3 convolution to perform on each merged image so as to diminish the aliasing effect of upsampling. The same applies to other layers and the final feature map set is called {P2, P3, P4, P5} for object classification and bounding box regression, corresponding to {C2, C3, C4, C5}, respectively.

3.2.2. Focal Loss

The box regression sub-net and classification sub-net in the RetinaNet are implemented using the standard Smooth L_1 loss (Formula (1)) and the Focal loss (Formula (3)), respectively, as the loss functions. Focal loss is a cross-entropy loss that can be dynamically scaled. A weighting factor is added for the traditional cross-entropy function, which can automatically drop the weight of the loss contributed by simple examples and center more on hard samples to solve the class imbalance.

$$SmoothL_1(x) = \begin{cases} 0.5\,x^2 & if\ |x| < 1 \\ |x| - 0.5 & otherwise \end{cases} \quad (1)$$

$$x = f(x_i) - y_i \quad (2)$$

$$FL(P) = -\partial_t (1 - P)^{\gamma} \log(P) \quad (3)$$

$$P = \begin{cases} p & if\ z = 1 \\ 1 - p & otherwise \end{cases} \quad (4)$$

Here, x is the error value between the estimated value $f(x_i)$ and ground truth y_i; ∂_t and γ are two tunable focusing hypermeters and they function as the role of balancing the ratio between simple and difficult examples; p is the estimated possibility for the given label class. Thus, if the figure of math is 1, it specifies the label class and P is the same as the p in this situation.

3.3. Datasets Preparation and Preprocessing

To address the scarce dataset for cattle face detection and recognition using deep learning, datasets were collected from two housing farms located in Jiangxi Province, China, and there were 85 healthy scalpers and Simmental ranging in age from 6 to 20 months. The experiment was conducted under various scenes such as different illumination, overlapping, and postures without human intervention, and it took three days to complete this data collection. Examples of multi-view cattle face in different scenes are displayed in Figure 3. This work aims to simulate and facilitate the detection and identification of cattle face by future mobile devices instead of surveillance cameras, and it is common to collect the images where the cattle faces occupy large areas. The cattle were filmed using a Sony FDR-AX 40 camera with MOV video format (3840 × 2160 pixels) at 25 frames per second. The camera on a tripod was fronted straight to the standing cow with a view of 3 cow's face width and 1.5 cow's face length. The original images cropped from videos were in JPG format at 3840 by 2160 pixels. After extracting valuable data frames of every video in MATLAB, the selected images were clipped using MATLAB and then be resized to 224 × 224 pixels. Notably, to ensure the effectiveness of detection performance, during the image selection, different situations of cattle faces for each cow were selected and highly similar faces, especially in consecutive frames, were avoided. The datasets contained a total of 3000 images (1000 negative images included) that were split into training and testing in the proportion 2:1.

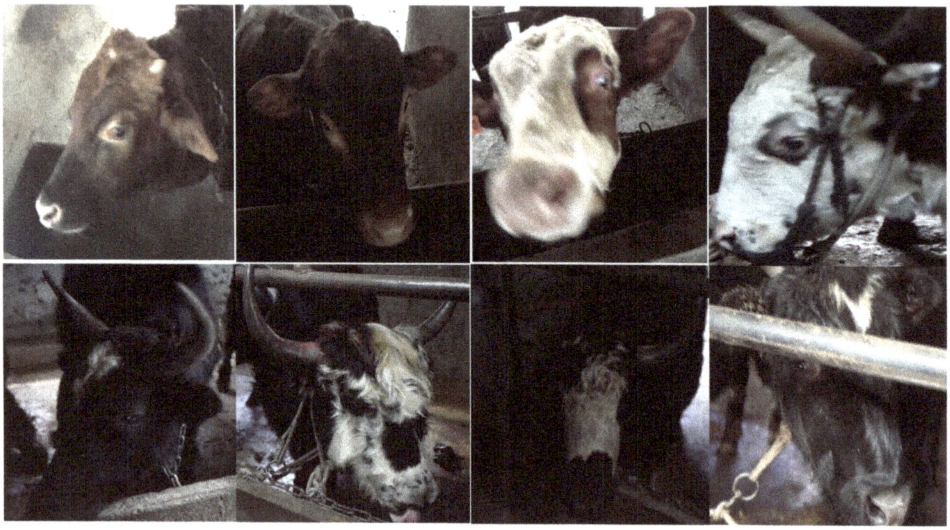

Figure 3. Examples of cattle faces in different scenes.

LabelImg is the annotation tool that was used to label the ground truth for cattle faces using RectBox for training datasets. For labeling, the region of every cattle face was selected and annotated using the RectBox in the image. Then, the class label named cattle face needed to be marked on the bubble pop up on the screen. The details of data annotation include object name, box location, and image size, as shown in Figure 4.

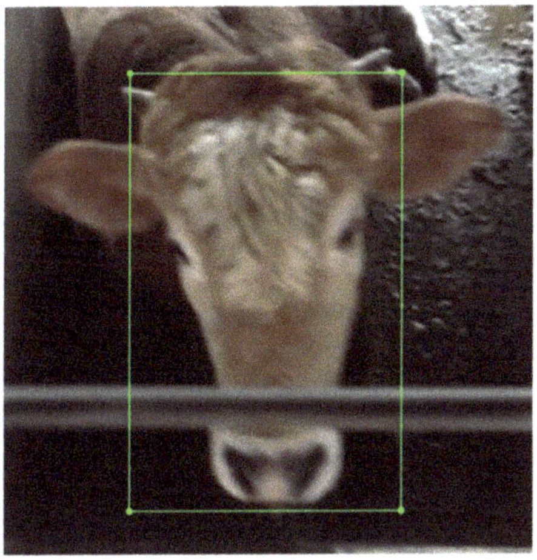


```
<annotation>
    <folder>Desktop</folder>
    <filename>00797.jpg</filename>
    <path>C:/Users/Desktop/00797.jpg</path>
    <source>
        <database>Unknown</database>
    </source>
    <size>
        <width>224</width>
        <height>224</height>
        <depth>3</depth>
    </size>
    <segmented>0</segmented>
    <object>
        <name>cattle face</name>
        <pose>Unspecified</pose>
        <truncated>0</truncated>
        <difficult>0</difficult>
        <bndbox>
            <xmin>55</xmin>
            <ymin>22</ymin>
            <xmax>161</xmax>
            <ymax>203</ymax>
        </bndbox>
    </object>
</annotation>
```


Figure 4. Example of annotation (green) for cattle face and details of a labeled cattle face.

4. Results

4.1. Implementation Details

The experiment was conducted on a desktop computer equipped with Windows 10 64-bit and an NVIDIA GeForce GTX 1080 graphics card. The proposed framework was written employing available libraries including numpy 1.16.5 and scikit-learn 0.21.3 in Python3.6. Keras 2.31 combined with tensorflow-gpu-2.1.0 was installed to provide a deep neural network framework for Python that was compatible with the Python version.

Transfer learning was adopted because of the limited computing resources and datasets for training. Transfer learning was to fine-tune a particular model for the intended task based on existing models. The backbones used in the proposed framework were initialized by ResNet-pretrained model using COCO datasets and VGG-pretrained model using ImageNet datasets and Densenet-pretrained model using ImageNet datasets. All 200,000 training iterations took approximately 17 h, and the best performing epoch for the model was chosen on testing data after the training loss converged. The threshold was set at 0.5 for the Intersection-over-Union (IoU) of confidence and bounding-box in all network models.

4.2. Performance Analysis with Different Backbones

As referred in Section 3.1., the original ResNet 50 backbone model of RetinaNet can be replaced with ResNet 101, ResNet 152, VGG 16, VGG 19, Densenet 121, and Densenet 169. The experiment compared the RetinaNet with ResNet 50 with these various backbone CNNs. The results in Figure 5 demonstrate the comparison Average Precision (AP) and Average Processing Time (Atime) between different backbones using 1000 images, including 500 positive samples with cattle face and 500 negative samples without cattle face. In addition, to better assess the performance of various models on cattle face detection in detail, we also computed True Positive (TP), False Positive (FP), and False Negatives (FN) of seven backbones and then calculated the corresponding precision, recall, and F1 score, as presented in Table 1.

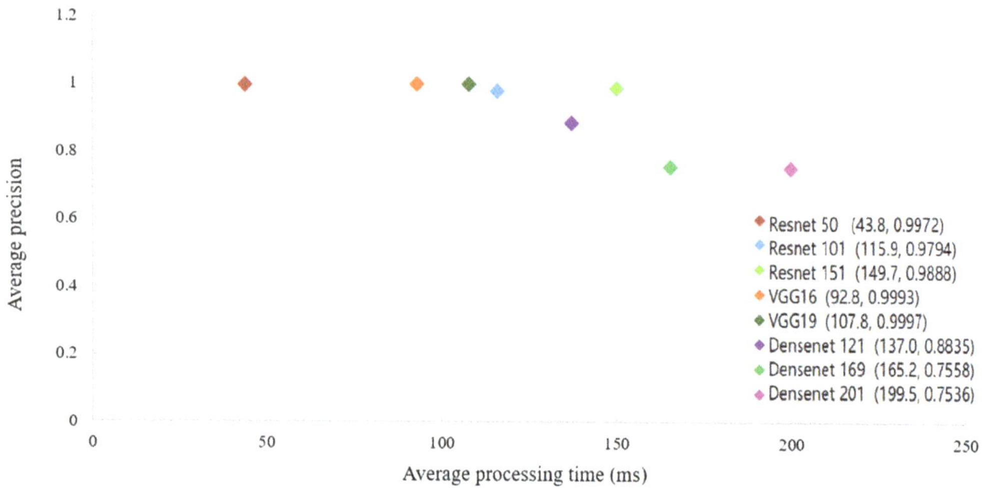

Figure 5. Average precision and average processing time of cattle face detection using different backbones.

Table 1. Comparison of detection results with different backbones.

Backbone	Precision	Recall	F1	TP	FP	FN
ResNet 50	0.9980	1.0000	0.9990	500	1	0
ResNet 101	0.9840	1.0000	0.9920	500	8	0
ResNet 152	0.9840	1.0000	0.9920	500	8	0
VGG16	0.8040	1.0000	0.8910	500	122	0
VGG19	0.8800	1.0000	0.9390	500	65	0
Densenet 121	0.3850	0.4220	0.4030	211	337	289
Densenet 169	0.6270	0.2760	0.3830	138	82	362

It can be seen from Figure 5 that the average precision of VGG 16 and VGG 19 are slightly higher than the value of ResNet 50 and achieve the best average precision, but the average processing time of ResNet 50 outperforms other backbones. As for cattle face detection, Densenet has a poor detection effect with the best average precision of 88.35% and the fastest processing time of 0.1370 s. AP and Atime are both significant metrics in the matter of how practical the system might be in actual use. Therefore, considering processing time and accuracy, the detection algorithm with ResNet 50 as the feature extraction model is regarded as having the best performance, whose AP reaches 99.8% and Atime is 0.0438 s per image.

As observed in Table 1, the cattle face detection model using ResNet 50 yields a precision of 99.8%, a 100% of recall and an F1 score of 0.9990, which are higher than other backbones. Moreover, the results concerning cattle face detection errors depict that the model achieves the lowest FP and FN rates with only 1 in 500 cattle faces potentially being misclassified in the case of ResNet 50. In contrast, although deeper ResNet including ResNet101, ResNet 152, and VGG network architectures obtain better performance on FP, they are reported to receive more falsely detected cattle face, especially using VGG. As with the results shown in Figure 5, the lowest scores on precision, recall, and F1 score are reported by employing Densenet due to the superior FP and FN rates but the lowest TP rate. Some representative examples for the prediction on the test image processed by seven different backbones is visualized in Figure 6.

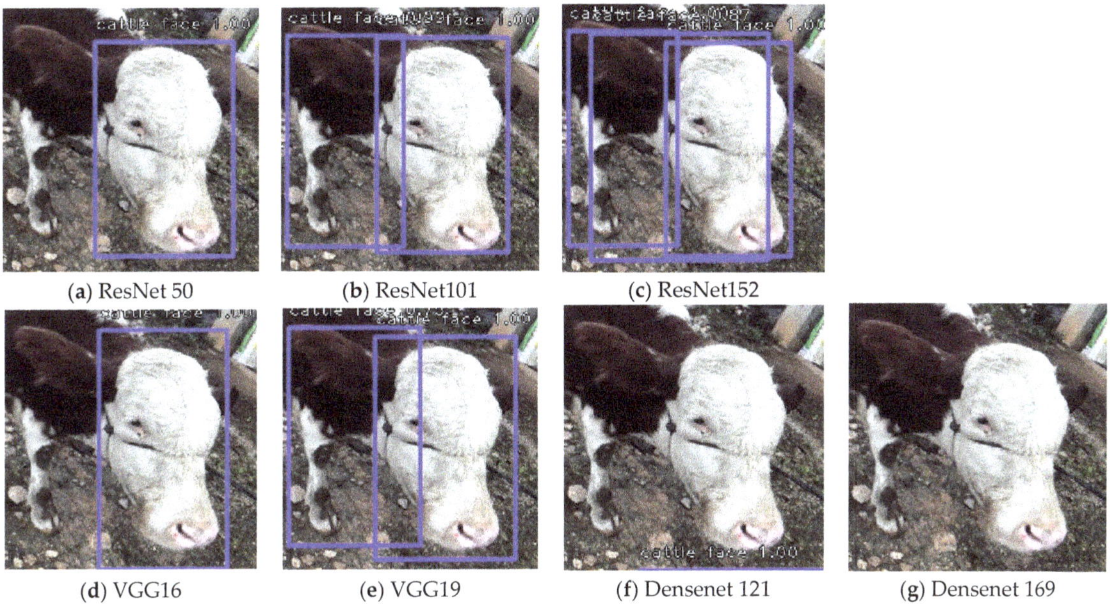

Figure 6. Comparisons of cattle face detection processed by seven different backbones.

4.3. Comparison with Other State-of-the-Art Object Detection Algorithms

The proposed RetinaNet based multi-view cattle face detection is also compared to show its advantages over the typical existing object detection approaches. Yolov3 and Faster R-CNN are the typical works of object detectors in practice. For instance, Faster R-CNN has been attempted to explore the multi-class fruit detection [42–44], livestock detection [45], posture detection of pigs [46], and cattle face detection [35]. Yolov3 has also been applied to fruit and fruit disease detection [47–50], plant and plant disease and pest detection [51–53], livestock behavior detection [47,54], and fish detection [55]. Therefore, experiments in this paper are conducted to compare the testing results of these competing methods with the ground truth information, and the results are summarized in Table 2.

Table 2. Comparison of detection results with three competing methods.

Methods	AP	Atime	Precision	Recall	F1	TP	FP	FN
Yolov3	0.9968	0.1368	0.8700	1	0.9300	498	72	2
Faster R-CNN	0.9857	0.1526	0.9940	1	0.9970	500	3	0
RetinaNet + ResNet 50	0.9980	0.0438	0.9980	1	0.9990	500	1	0

It is observed from Table 2 that RetinaNet with ResNet 50 show better detection performance than Yolov3 and Faster R-CNN in both detection accuracy and calculation requirement for future online detection (AP of 99.8% and Atime of 0.0438 s). The results indicate that RetinaNet is most competent in real-world practice as the datasets are in different complex scenes with severe face-pose variation and different degrees of occlusion. Yolov3 and Faster R-CNN achieved nearly similar performance with RetinaNet in AP (99.68% for Yolov3 and 99.8% for RetinaNet) and F1 score (0.9970 for Faster R-CNN and 0.9990 for RetinaNet), respectively, but the F1 score is preferable as the metric for "true positive detection" whilst average precision is preferable for "boundary extraction" of cattle face. Therefore, Yolov3 and Faster R-CNN are not sufficiently reliable in complex multi-view cattle face detection.

4.4. *Evaluation of Multi-View Cattle Face Detection Results*

The major misdetections of the abovementioned algorithms concern multi-view cattle face in complex conditions. To clearly observe the comparisons of results for multi-view cattle face detection in different scenes, 100 images were selected from 500 positive samples for three scenes of partial occlusion, light change, and posture change, and then the detection AP values and F1 scores were calculated separately for these competing detection models, as shown in Table 3.

Table 3. Comparison of detection results under different conditions.

Methods	Partial Occlusion		Light Variation		Posture Change	
	AP	F1	AP	F1	AP	F1
Yolov3	0.9980	1.0000	1.0000	1.0000	0.9720	0.9980
Faster R-CNN	0.9910	0.9990	1.0000	1.0000	0.9840	0.9980
RetinaNet + ResNet 50	1.0000	1.0000	1.0000	1.0000	0.9980	0.9990

As seen in Table 3, RetinaNet with ResNet 50 outperforms Yolov3 and Faster R-CNN under three particularly challenging situations. Three detection models all present very accurate detection results with AP of 100% and F1 score of 1.0000 in the situation with light changes, which implies that CNN-based deep learning algorithms are robust to illumination variations. However, as observed, there are inaccurate detection boundaries using Yolov3 and false cattle face detections using Faster R-CNN while the performance of RetinaNet remains relatively high in partial occlusion situation. Although three detection models do not present good detection results in posture change situations, RetinaNet achieves better performance in detection accuracy and boundary accuracy owing to the structure of FPN and focal loss in the model. Faster R-CNN presents the advantage of RPN, which is commonly used in two-stage detectors, and thus the boundary precision is higher than Yolov3. To facilitate the readers to visually observe the comparisons of results, this paper compares the predictions processed by the above-competing methods under partial occlusion and posture change situations, as shown in Figure 7.

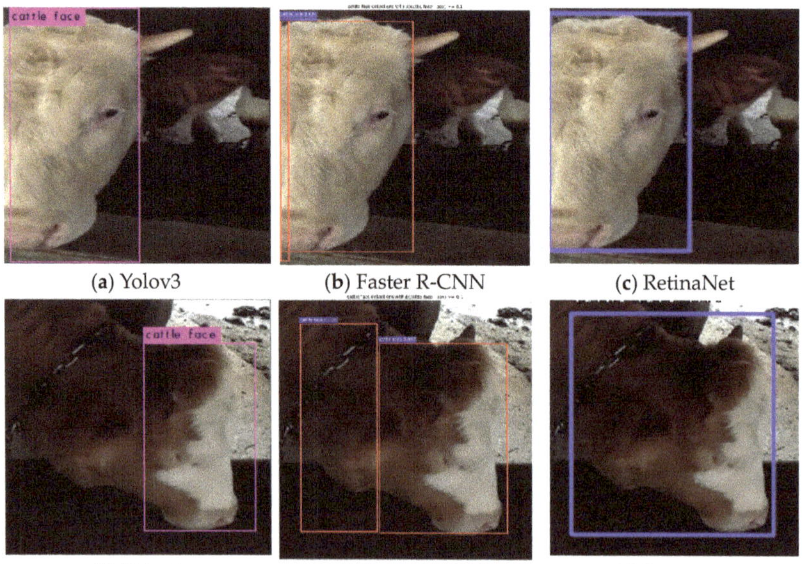

Figure 7. Comparisons of cattle face detection processed by three object detection algorithms in partial occlusion and posture change situations.

5. Discussion

This paper evaluated an up-to-date object detector, RetinaNet, to automate the face detection process for a livestock identification vision system in the farmland. The key novelty of the study is the application evaluation of the RetinaNet algorithm with various backbones and comparisons with typical competing detection models for multi-view cattle face detection in complex and relevant cattle production scenarios. The essence of the detection in this paper is bounding-box location and classification with confidence. Previous studies in cattle face suffered the deviation of the bounding-box [56] and the challenge for dataset collection from complex scenarios [35]. The strong point of the RetinaNet is the capability to perform both relatively high detection accuracy and fast processing time of cattle face within the imagery. This allows for the development of further algorithms to perform tasks such as facial expression assessment from the imagery for welfare monitoring. Cattle face detection in the paper is the first step toward real-time individual livestock identification in farming environments that have different applications, such as the cattle insurance industry, meat products traceability [57], and other animal welfare improvements.

Transfer learning is an essential part of machine learning as pretrained CNN models can be fine-tuned and re-trained to perform new tasks when limited annotated data exists for training. However, the generalization capabilities of various deep networks on different datasets might change due to their architecture [43,58,59]. Therefore, this study compared the performance quantitatively of ResNet, VGG, and Densenet with different depth to select the optimal backbone in this detection task. The results indicate that RetinaNet with ResNet 50 achieves the best performance with an average precision of 99.8%, F1 score of 0.9990, and average processing time of 0.0438 s. Since backbones with better performance can improve the accuracy of detection, and there is no agreed pretrained CNN model in object detection algorithms, this backbone could be properly adjusted and optimized depending on the circumstances and applications. For instance, Yolov3 incorporating the DenseNet for apple detection in various growth periods [49] was considered to perform well. Still, ResNet may be better for fruit detection and instance segmentation [43], and plant disease detection achieves better results using VGG architecture [60].

For demonstrating the feasibility of the proposed framework further, this study made the performance comparisons with two competitive algorithms of object detection on the same datasets. The detection results presented illustrate that the AP and Atime provided by the RetinaNet with ResNet 50 model are significantly better than the other two models, reflecting the superiority of the proposed cattle face detection model. Considering the multi-view face caused by various unstructured scenes in actual cattle production scenarios, such as overlapping, occlusion, and illumination changes, the cattle face detection accuracy could be reduced to some extent. The F1 scores and average precision metrics were assessed over unstructured scenes in the study, and it is worth mentioning that the performance of RetinaNet was better than other algorithms. Some detection results of cattle faces are shown in Figure 8. Especially for partial occlusion and light variation situations, the accuracy of cattle face detection using RetinaNet reaches 100%, but the posture change situation is particularly challenging, even using RetinaNet and computer vision in general. The suggested main reason for this performance discrepancy of posture change situation can be attributed to multiple behaviors, such as leaning over to graze or drink and lying on the side to rest, which then bring difficulties to cattle face detection.

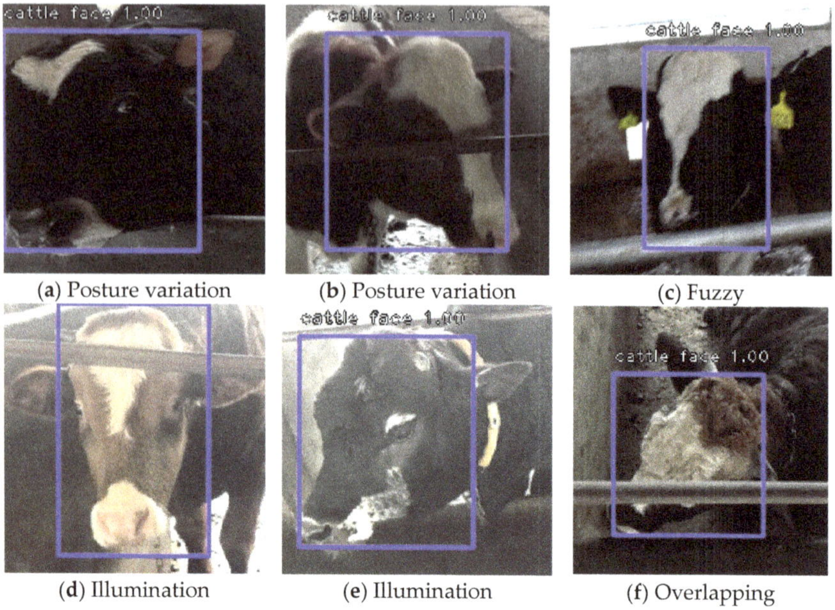

Figure 8. Detection results of cattle faces in various unstructured scenes.

6. Conclusions

Developing deep learning for object detection and image processing is crucial to the livestock identification system, which substitutes for wearable devices such as RFID ear tags, thus reducing the damage to animals. To establish the livestock machine vision system capable of monitoring individuals, this paper focused on cattle face detection, which is an important component of envisaged future technology. The state-of-art RetinaNet detection model proposed in this study was assessed on various unstructured scenes. The compared metrics performed successfully across a range of scenarios with an average precision score of 99.8% and an average processing time of 0.0438 s. The results presented indicate that the proposed model was particularly effective for the detection of cattle faces with illumination changes, overlapping, and occlusion. Compared to the existing algorithms, the proposed model has better universality and robustness both in accuracy and speed, which makes it generally more applicable for actual scenes. However, the conditions of training and testing are the same in this work, and the robustness of the system may be questioned; thus, further experiments are needed.

This work has potential for computer vision system integration into mobile apps to perform not only livestock detection and counting and individual identification, but also facial expression recognition for animal welfare. Despite the significantly high success of the proposed method, it is still far from being a generic tool that could be used in actual livestock production scenarios. Future work will focus on a lightweight neural network to improve the running speed of cattle face detection. In addition, future work will also concentrate on building an autonomous livestock individual identification system using facial features.

Author Contributions: Conceptualization, W.W. and B.X.; Methodology, B.X., L.G. and G.C.; Software, B.X. and Y.W.; Validation, B.X., W.Z. and Y.L.; Formal Analysis, Y.L.; Investigation, B.X. and Y.W.; Resources, L.G. and G.C.; Data Curation, W.Z. and Y.L.; Writing—Original Draft Preparation, B.X. and. W.W.; Writing—Review & Editing, W.W.; Visualization, B.X.; Supervision, W.W.; Project Administration, W.W., L.G.; Funding Acquisition, L.G. and G.C. All authors have read and agreed to the published version of the manuscript.

Funding: This research was supported by China Scholarship Council (202003250122) and was funded by Inner Mongolia Autonomous Region Science and Technology Major Project (2020ZD0004), National Natural Science Foundation of China (32060776), Youth Science Foundation of Jiangxi Province (20192ACBL21023), and Hebei Province Key Research and Development Plan (20327202D, 20327401D).

Acknowledgments: We are grateful to two private housing farms in Jiangxi Province in China for their kindly support with data collection.

Conflicts of Interest: The authors declare no conflict of interest.

References

1. Liaghat, S.; Balasundram, S.K. A Review: The Role of Remote Sensing in Precision Agriculture. *Am. J. Agric. Biol. Sci.* **2010**, *5*, 553–564. [CrossRef]
2. Auernhammer, H. Precision farming—The environmental challenge. *Comput. Electron. Agric.* **2001**, *30*, 31–43. [CrossRef]
3. Vranken, E.; Berckmans, D. Precision livestock farming for pigs. *Anim. Front.* **2017**, *7*, 32–37. [CrossRef]
4. Andonovic, I.; Michie, C.; Cousin, P.; Janati, A.; Pham, C.; Diop, M. Precision Livestock Farming Technologies. In Proceedings of the 2018 Global Internet of Things Summit (GIoTS), Bilbao, Spain, 4–7 June 2018; pp. 1–6.
5. Xu, B.; Wang, W.; Falzon, G.; Kwan, P.; Guo, L.; Chen, G.; Tait, A.; Schneider, D. Automated cattle counting using Mask R-CNN in quadcopter vision system. *Comput. Electron. Agric.* **2020**, *171*, 105300. [CrossRef]
6. Disney, W.T.; Green, J.W.; Forsythe, K.W.; Wiemers, J.F.; Weber, S.J.R.T. Benefit-cost analysis of animal identification for disease prevention and control. *Rev. Sci. Tech. l'OIE* **2001**, *20*, 385–405. [CrossRef]
7. Gwaza, D.; Gambo, D. Application of Radio Frequency Identification to Selection for Genetic improvement of Rural Livestock Breeds in Developing Countries. *J. Anim. Husb. Dairy Sci.* **2017**, *1*, 38–52.
8. Yordanov, D.; Angelova, G. Identification and Traceability of Meat and Meat Products. *Biotechnol. Biotechnol. Equip.* **2006**, *20*, 3–8. [CrossRef]
9. Awad, A.I. From classical methods to animal biometrics: A review on cattle identification and tracking. *Comput. Electron. Agric.* **2016**, *123*, 423–435. [CrossRef]
10. Leslie, E.; Hernández-Jover, M.; Newman, R.; Holyoake, P. Assessment of acute pain experienced by piglets from ear tagging, ear notching and intraperitoneal injectable transponders. *Appl. Anim. Behav. Sci.* **2010**, *127*, 86–95. [CrossRef]
11. Jones, S.M. *Tattooing of Cattle and Goats*; University of Arkansas System: Little Rock, AR, USA, 2014.
12. Adcock, S.J.J.; Tucker, C.B.; Weerasinghe, G.; Rajapaksha, E. Branding Practices on Four Dairies in Kantale, Sri Lanka. *Animals* **2018**, *8*, 137. [CrossRef] [PubMed]
13. Stanford, K.; Stitt, J.; Kellar, J.A.; McAllister, T. Traceability in cattle and small ruminants in Canada. *Rev. Sci. Et Tech. Int. Off. Epizoot.* **2001**, *20*, 510–522. [CrossRef] [PubMed]
14. Yang, L.; Liu, X.Y.; Kim, J.S. Cloud-based Livestock Monitoring System Using RFID and Blockchain Technology. In Proceedings of the 2020 7th IEEE International Conference on Cyber Security and Cloud Computing (CSCloud)/2020 6th IEEE International Conference on Edge Computing and Scalable Cloud (EdgeCom), New York, NY, USA, 1–3 August 2020; pp. 240–245.
15. Klindtworth, M.; Wendl, G.; Klindtworth, K.; Pirkelmann, H. Electronic identification of cattle with injectable transponders. *Comput. Electron. Agric.* **1999**, *24*, 65–79. [CrossRef]
16. Zaiqiong, W.; Zetian, F.; Wei, C.; Jinyou, H. A RFID-based traceability system for cattle breeding in China. In Proceedings of the 2010 International Conference on Computer Application and System Modeling (ICCASM 2010), Taiyuan, China, 22–24 October 2010; pp. V2-567–V2-571.
17. Whittier, J.C.; Shadduck, J.A.; Golden, B.L. *Secure Identification, Source Verification of Livestock—The Value of Retinal Images and GPS*; Wageningen Academic Publishers: Wageningen, The Netherlands, 2003; pp. 167–172.
18. Gonzales Barron, U.; Corkery, G.; Barry, B.; Butler, F.; McDonnell, K.; Ward, S. Assessment of retinal recognition technology as a biometric method for sheep identification. *Comput. Electron. Agric.* **2008**, *60*, 156–166. [CrossRef]
19. Kumar, S.; Pandey, A.; Sai Ram Satwik, K.; Kumar, S.; Singh, S.K.; Singh, A.K.; Mohan, A. Deep learning framework for recognition of cattle using muzzle point image pattern. *Measurement* **2018**, *116*, 1–17. [CrossRef]
20. Kumar, S.; Singh, S.K.; Abidi, A.I.; Datta, D.; Sangaiah, A.K. Group Sparse Representation Approach for Recognition of Cattle on Muzzle Point Images. *Int. J. Parallel Program.* **2018**, *46*, 812–837. [CrossRef]
21. Mukai, N.; Zhang, Y.; Chang, Y. Pet Face Detection. In Proceedings of the 2018 Nicograph International (NicoInt), Tainan, Taiwan, 28–29 June 2018; pp. 52–57.
22. Kumar, S.; Tiwari, S.; Singh, S.K. Face recognition for cattle. In Proceedings of the 2015 Third International Conference on Image Information Processing (ICIIP), Waknaghat, India, 21–24 December 2015; pp. 65–72.
23. Clark, A.W. Calculating the Weight of a Pig through Facial Geometry Using 2-Dimensional Image Processing. Master's Thesis, Texas Tech University, Lubbock, TX, USA, 2015.
24. Jaddoa, M.; Gonzalez, L.; Cuthbertson, H.; Al-Jumaily, A. Multi View Face Detection in Cattle Using Infrared Thermography. In *Proceedings of the Applied Computing to Support Industry: Innovation and Technology*; Ramadi, Iraq, 15–16 September 2019, Springer: Cham, Switzerland, 2020; pp. 223–236.

25. Yamada, A.; Kojima, K.; Kiyama, J.; Okamoto, M.; Murata, H. Directional edge-based dog and cat face detection method for digital camera. In Proceedings of the 2011 IEEE International Conference on Consumer Electronics (ICCE), Las Vegas, NV, USA, 9–12 January 2011; pp. 87–88.
26. Lecun, Y.; Bottou, L.; Bengio, Y.; Haffner, P. Gradient-based learning applied to document recognition. *Proc. IEEE* **1998**, *86*, 2278–2324. [CrossRef]
27. Vlachynska, A.; Oplatkova, Z.K.; Turecek, T. Dogface Detection and Localization of Dogface's Landmarks. In *Artificial Intelligence and Algorithms in Intelligent Systems*; Springer: Cham, Switzerland, 2019; pp. 465–476.
28. Girshick, R.; Donahue, J.; Darrell, T.; Malik, J. Rich Feature Hierarchies for Accurate Object Detection and Semantic Segmentation. In Proceedings of the 2014 IEEE Conference on Computer Vision and Pattern Recognition, Columbus, OH, USA, 23–28 June 2014; pp. 580–587.
29. Girshick, R. Fast R-CNN. In Proceedings of the 2015 IEEE International Conference on Computer Vision (ICCV), Santiago, Chile, 7–13 December 2015; pp. 1440–1448.
30. Ren, S.; He, K.; Girshick, R.; Sun, J. Faster R-CNN: Towards Real-Time Object Detection with Region Proposal Networks. *IEEE Trans. Pattern Anal. Mach. Intell.* **2017**, *39*, 1137–1149. [CrossRef]
31. Redmon, J.; Divvala, S.; Girshick, R.; Farhadi, A. You Only Look Once: Unified, Real-Time Object Detection. In Proceedings of the 2016 IEEE Conference on Computer Vision and Pattern Recognition (CVPR), Las Vegas, NV, USA, 27–30 June 2016; pp. 779–788.
32. Redmon, J.; Farhadi, A. YOLO9000: Better, Faster, Stronger. In Proceedings of the 2017 IEEE Conference on Computer Vision and Pattern Recognition (CVPR), Honolulu, HI, USA, 21–26 July 2017; pp. 6517–6525.
33. Liu, W.; Anguelov, D.; Erhan, D.; Szegedy, C.; Reed, S.; Fu, C.-Y.; Berg, A.C. SSD: Single Shot MultiBox Detector. In Proceedings of the Computer Vision—ECCV 2016, Amsterdam, The Netherlands, 11–14 October 2016; Springer: Cham, Switzerland, 2016; pp. 21–37.
34. Redmon, J.; Farhadi, A. Yolov3: An incremental improvement. *arXiv* **2018**, arXiv:1804.02767.
35. Yao, L.; Hu, Z.; Liu, C.; Liu, H.; Kuang, Y.; Gao, Y. Cow face detection and recognition based on automatic feature extraction algorithm. In Proceedings of the ACM Turing Celebration Conference—China, Chengdu, China, 17–19 May 2019.
36. Gou, X.; Huang, W.; Liu, Q. A Cattle Face Detection Method Based on Improved NMS. *Comput. Modernization* **2019**, *7*, 43–46.
37. Ochoa-Ruiz, G.; Angulo-Murillo, A.A.; Ochoa-Zezzatti, A.; Aguilar-Lobo, L.M.; Vega-Fernández, J.A.; Natraj, S. An Asphalt Damage Dataset and Detection System Based on RetinaNet for Road Conditions Assessment. *Appl. Sci.* **2020**, *10*, 3974. [CrossRef]
38. Yunqi, C.; Basak, O. *Automated Firearms Detection in Cargo X-Ray Images using RetinaNet*; International Society for Optics and Photonics: Bellingham, WA, USA, 2019; Volume 10999. [CrossRef]
39. Afif, M.; Ayachi, R.; Said, Y.; Pissaloux, E.; Atri, M. An Evaluation of RetinaNet on Indoor Object Detection for Blind and Visually Impaired Persons Assistance Navigation. *Neural Process. Lett.* **2020**, *51*, 2265–2279. [CrossRef]
40. Zou, Z.; Shi, Z.; Guo, Y.; Ye, J. Object detection in 20 years: A survey. *arXiv* **2019**, arXiv:1905.05055.
41. Lin, T.; Goyal, P.; Girshick, R.; He, K.; Dollár, P. Focal Loss for Dense Object Detection. *IEEE Trans. Pattern Anal. Mach. Intell.* **2020**, *42*, 318–327. [CrossRef] [PubMed]
42. Tu, S.; Pang, J.; Liu, H.; Zhuang, N.; Chen, Y.; Zheng, C.; Wan, H.; Xue, Y. Passion fruit detection and counting based on multiple scale faster R-CNN using RGB-D images. *Precis. Agric.* **2020**, *21*, 1072–1091. [CrossRef]
43. Kang, H.; Chen, C. Fruit detection, segmentation and 3D visualisation of environments in apple orchards. *Comput. Electron. Agric.* **2020**, *171*, 105302. [CrossRef]
44. Wan, S.; Goudos, S. Faster R-CNN for multi-class fruit detection using a robotic vision system. *Comput. Netw.* **2020**, *168*, 107036. [CrossRef]
45. Xu, B.; Wang, W.; Falzon, G.; Kwan, P.; Guo, L.; Sun, Z.; Li, C. Livestock classification and counting in quadcopter aerial images using Mask R-CNN. *Int. J. Remote Sens.* **2020**, *41*, 8121–8142. [CrossRef]
46. Nasirahmadi, A.; Sturm, B.; Edwards, S.; Jeppsson, K.-H.; Olsson, A.-C.; Müller, S.; Hensel, O. Deep Learning and Machine Vision Approaches for Posture Detection of Individual Pigs. *Sensors* **2019**, *19*, 3738. [CrossRef] [PubMed]
47. Fu, L.; Feng, Y.; Wu, J.; Liu, Z.; Gao, F.; Majeed, Y.; Al-Mallahi, A.; Zhang, Q.; Li, R.; Cui, Y. Fast and accurate detection of kiwifruit in orchard using improved YOLOv3-tiny model. *Precis. Agric.* **2020**, *22*, 754–776. [CrossRef]
48. Koirala, A.; Walsh, K.B.; Wang, Z.; McCarthy, C. Deep learning for real-time fruit detection and orchard fruit load estimation: Benchmarking of 'MangoYOLO'. *Precis. Agric.* **2019**, *20*, 1107–1135. [CrossRef]
49. Tian, Y.; Yang, G.; Wang, Z.; Li, E.; Liang, Z. Detection of Apple Lesions in Orchards Based on Deep Learning Methods of CycleGAN and YOLOV3-Dense. *J. Sens.* **2019**, *2019*, 7630926. [CrossRef]
50. Kuznetsova, A.; Maleva, T.; Soloviev, V. Detecting Apples in Orchards Using YOLOv3. In Proceedings of the Computational Science and Its Applications (ICCSA) 2020, Cagliari, Italy, 1–4 July 2020; Springer: Cham, Switzerland, 2020; pp. 923–934.
51. Zhou, J.; Tian, Y.; Yuan, C.; Yin, K.; Yang, G.; Wen, M. Improved UAV Opium Poppy Detection Using an Updated YOLOv3 Model. *Sensors* **2019**, *19*, 4851. [CrossRef]
52. Liu, J.; Wang, X. Tomato Diseases and Pests Detection Based on Improved Yolo V3 Convolutional Neural Network. *Front. Plant Sci.* **2020**, *11*, 898. [CrossRef]
53. Liu, G.; Nouaze, J.C.; Touko Mbouembe, P.L.; Kim, J.H. YOLO-Tomato: A Robust Algorithm for Tomato Detection Based on YOLOv3. *Sensors* **2020**, *20*, 2145. [CrossRef]

54. Wang, J.; Wang, N.; Li, L.; Ren, Z. Real-time behavior detection and judgment of egg breeders based on YOLO v3. *Neural Comput. Appl.* **2020**, *32*, 5471–5481. [CrossRef]
55. Raza, K.; Hong, S. Fast and Accurate Fish Detection Design with Improved YOLO-v3 Model and Transfer Learning. *Int. J. Adv. Comput. Sci. Appl.* **2020**, *11*, 7–16. [CrossRef]
56. Wang, H.; Qin, J.; Hou, Q.; Gong, S. Cattle Face Recognition Method Based on Parameter Transfer and Deep Learning. *J. Phys. Conf. Ser.* **2020**, *1453*, 012054. [CrossRef]
57. Zhao, J.; Li, A.; Jin, X.; Pan, L. Technologies in individual animal identification and meat products traceability. *Biotechnol. Biotechnol. Equip.* **2020**, *34*, 48–57. [CrossRef]
58. Wang, G.; Sun, Y.; Wang, J. Automatic Image-Based Plant Disease Severity Estimation Using Deep Learning. *Comput. Intell. Neurosci.* **2017**, *2017*, 2917536. [CrossRef] [PubMed]
59. Ayan, E.; Erbay, H.; Varçın, F. Crop pest classification with a genetic algorithm-based weighted ensemble of deep convolutional neural networks. *Comput. Electron. Agric.* **2020**, *179*, 105809. [CrossRef]
60. Ferentinos, K.P. Deep learning models for plant disease detection and diagnosis. *Comput. Electron. Agric.* **2018**, *145*, 311–318. [CrossRef]

agriculture

MDPI

Article

Simplified and Hybrid Remote Sensing-Based Delineation of Management Zones for Nitrogen Variable Rate Application in Wheat

Mohammad Rokhafrouz [1], Hooman Latifi [1,2,*], Ali A. Abkar [3], Tomasz Wojciechowski [4], Mirosław Czechlowski [4], Ali Sadeghi Naieni [5], Yasser Maghsoudi [1] and Gniewko Niedbała [4]

1 Department of Photogrammetry and Remote Sensing, Faculty of Geodesy and Geomatics Engineering, K. N. Toosi University of Technology, Tehran 15433-19967, Iran; m.rokhafrooz@email.kntu.ac.ir (M.R.); ymaghsoudi@kntu.ac.ir (Y.M.)
2 Department of Remote Sensing, University of Würzburg, Oswald Külpe Weg 86, 97074 Würzburg, Germany
3 AgriWatch BV, World Trade Center Twente, Industrieplein 2, 7553 LL Hengelo, The Netherlands; ali.abkar@agriwatch.nl
4 Department of Biosystems Engineering, Faculty of Environmental and Mechanical Engineering, Poznań University of Life Sciences, Wojska Polskiego 50, 60-627 Poznań, Poland; tomasz.wojciechowski@up.poznan.pl (T.W.); miroslaw.czechlowski@up.poznan.pl (M.C.); gniewko.niedbala@up.poznan.pl (G.N.)
5 Photogrammetry and Remote Sensing Department, National Cartographic Center, Tehran 1684-13185, Iran; naeini29857@alumni.itc.nl
* Correspondence: hooman.latifi@kntu.ac.ir; Tel.: +98-21-8887070-3 (ext. 312)

Citation: Rokhafrouz, M.; Latifi, H.; Abkar, A.A.; Wojciechowski, T.; Czechlowski, M.; Naieni, A.S.; Maghsoudi, Y.; Niedbała, G. Simplified and Hybrid Remote Sensing-Based Delineation of Management Zones for Nitrogen Variable Rate Application in Wheat. *Agriculture* **2021**, *11*, 1104. https://doi.org/10.3390/agriculture11111104

Academic Editor: Yongchao Tian

Received: 20 September 2021
Accepted: 2 November 2021
Published: 5 November 2021

Publisher's Note: MDPI stays neutral with regard to jurisdictional claims in published maps and institutional affiliations.

Abstract: Enhancing digital and precision agriculture is currently inevitable to overcome the economic and environmental challenges of the agriculture in the 21st century. The purpose of this study was to generate and compare management zones (MZ) based on the Sentinel-2 satellite data for variable rate application of mineral nitrogen in wheat production, calculated using different remote sensing (RS)-based models under varied soil, yield and crop data availability. Three models were applied, including (1) a modified "RS- and threshold-based clustering", (2) a "hybrid-based, unsupervised clustering", in which data from different sources were combined for MZ delineation, and (3) a "RS-based, unsupervised clustering". Various data processing methods including machine learning were used in the model development. Statistical tests such as the Paired Sample *T*-test, Kruskal–Wallis H-test and Wilcoxon signed-rank test were applied to evaluate the final delineated MZ maps. Additionally, a procedure for improving models based on information about phenological phases and the occurrence of agricultural drought was implemented. The results showed that information on agronomy and climate enables improving and optimizing MZ delineation. The integration of prior knowledge on new climate conditions (drought) in image selection was tested for effective use of the models. Lack of this information led to the infeasibility of obtaining optimal results. Models that solely rely on remote sensing information are comparatively less expensive than hybrid models. Additionally, remote sensing-based models enable delineating MZ for fertilizer recommendations that are temporally closer to fertilization times.

Keywords: precision agriculture; management zones; remote sensing; Sentinel-2; clustering; winter wheat; drought; digital agriculture

1. Introduction

In recent years, there has been an intense growth in the world population, which is projected to reach 9.7 billion by 2050 [1]. Population growth puts enormous pressure on agricultural productivity growth, but also on the increasing environmental impact of the agri-food sector [2,3]. In many regions of the world, small farms are the main food producers, and this group will be under pressure to increase production efficiency [4].

Important elements in the process of agricultural production growth are biological and technological progress [5]. In the scope of technological progress, the most promising developments are considered to be precision agriculture (PA) (agriculture 3.0) and digital agriculture (agriculture 4.0) [6–8].

Precision agriculture or site-specific management can provide food security and sustainable development [3,6,9–11]. It is based on innovative system approaches which comprise several technologies, such as global navigation satellite system (GNSS), geographic information system (GIS), proximal sensing (PS) and remote sensing (RS), artificial intelligence (AI), machine learning (ML), automatic guidance, section control, variable rate technology (VRT) and advanced information processing for timely within- and between-season crop managements [12–15]. The main purpose of PA is to optimize crop management concerning spatial and temporal variabilities, which results in optimized utilization of farm inputs such as fertilizers, pesticides, herbicides and seeds [16]. All of this is aimed at increasing farm profitability and achieving the Sustainable Development Goals (SDGs) such as No Poverty, Zero Hunger and Reduced Inequalities. For such a purpose, a wide range of data and information from field inventory, crop growth and yield patterns must be analysed [17]. With this correctly processed information, agricultural inputs such as fertilizers, water or energy can be applied in a spatially variable manner using homogeneous production zones, i.e., management zones (MZ) [16].

The RS and PS are the most researched technology in PA [18]. Their effective use requires the delivery of information and communication technologies (ICT) tools, including algorithms for RS, that are useful to the user: farmers with limited knowledge capital, which is currently indicated as one of the main barriers to the adaptation and spread of PA [18]. The need to support and develop simplified, low-tech precision farming methods seems therefore justified.

MZ are defined as sub-units of farm fields with a relatively homogeneous combination of yield-limiting factors [16,19]. Each zone can be managed with a different but specific single-rate management practice to maximize the efficiency of farm inputs [16,19]. Methods to create MZ have been developed for almost 3 decades, and the evolution of methods to create them is widely reported in the literature [2]. Generally, MZ delineation approaches can be categorized based on the provided data and information from different sources [2]. These methods are generally based on farmers' knowledge [20], soil physical and chemical attributes [21–24], geomorphology [25], yield [26–31], electrical conductivity (EC) [32,33] and RS [17,34,35] data, and also hybrid models that combine information from different data sources [36–47].

Each of those approaches has its pros and cons. Although the hybrid method is theoretically comprehensive and more accurate, additional field measurements such as soil sampling or proximal soil sensing are expensive, labor-intensive and time-consuming, and require seasonal sampling to specify nutrient level due to temporal variability of soil properties [22,48–50]. Besides, large commercial agricultural fields reportedly do not completely represent spatial variability [51]. RS methods deliver key components of precision farming and provide valuable data from crop coverage and actual crop growth patterns to delineate MZ [52]. With spatiotemporal continuity along with cost-effectiveness, RS has a capacity for time-series analysis [17]. However, optical satellite imagery is associated with the main limitation of being affected by atmospheric haze or cloud, which often occurs in temperate and rainy regions [17].

A broad range of active/passive satellite RS data is currently available with various properties, such as spatial resolution, temporal resolution, spectral range and viewing geometry [53]. The launch of Sentinel-2A (2015) and Sentinel-2B (2017) satellites by the European Space Agency (ESA) boosted the PA applications since the data are freely available [54]. Sentinel-2 satellites are equipped with a passive multispectral instrument (MSI), including 13 spectral channels, 4 bands at 10 m, 6 bands at 20 m and 3 bands at 60 m spatial resolution [55]. These satellites have a high revisit time of ten days with one satellite and five days as a constellation (2–3 days at mid-latitudes) [56]. Currently, ESA provides the Level-2A products of this mission as bottom of atmosphere (BOA) reflectance and atmospherically corrected images [57].

Remote sensing of vegetation is mainly based on the green (495–570 nm), red (620–750 nm), red-edge (680–730 nm), near- and mid-infrared bands (850–1700 nm) regions of the spectrum [35]. In order to obtain information from vegetation status, these bands can be used to derive vegetation indices [35]. The normalized difference vegetation index (NDVI) [58], a normalized difference between the reflectance of red and near-infrared (NIR) spectral bands, is the most common crop parameter used in MZ delineation, because of its ease of calculation and interpretation [2,35]. Along with NDVI, some biophysical variables such as leaf area index (LAI), fraction of absorbed photosynthetically active radiation (FAPAR) and the fraction of vegetation cover (FVC) were also used in this research [59]. LAI is defined as half the developed area of photosynthetically active elements of the vegetation per unit of horizontal ground area [59], FAPAR corresponds to the fraction of photosynthetically active radiation absorbed by the canopy [59] and FVC is the ratio of the vertically projected area of vegetation to the total surface area [60].

Several studies for MZ delineation on multispectral satellite images exist in the literature. Song et al. [37] delineated and compared MZ based on soil data, yield data, and crop RS information from one multispectral satellite scene, as well as their combination. Additionally, Martins et al. [47] generated a MZ map with a combination of soil attributes, EC, yield maps and a vegetation index (VI) of one multispectral satellite image. Georgi et al. [17] developed a segmentation algorithm for generating MZ from within-field crop patterns using only multi-temporal, multi-spectral satellite images.

The above examples [17,37,47] reveal that several studies conducted MZ delineation with various methods and different data types. However, none of them considered agronomy (e.g., BBCH stage) and especially climate information (e.g., soil moisture conditions, drought), which could negatively affect the implemented models. Additionally, comparative approaches need to be tested to improve the MZ delineation considering data and/or knowledge availability, time and cost-benefit analysis and accuracy.

Above all, the extent of available agricultural spatial data for farmers is also an important issue in the development of homogeneous MZ and homogeneous productivity zones. The ideal scenario is when the producer/farmer has multi-years spatial data on yield, crop vegetation, soil and climate, but literature reports indicate that the level of farm datafication varies [61]. Practice shows that the ideal scenario exists in regions with high adaptation of PA technologies, e.g., where the use of combine harvester yield monitors is common practice [18]. In most cases, these data are less available. This also applies to smaller farms with lower adaptation of PA technologies. Proposing new digital technologies for such users, such as Decision Support Systems (DSS) and Farm Management Information Systems (FMIS), there is a need to develop simplified algorithms/methods for the creation of MZ, based on one or two seasons' data. The problem of determining the MZ of simplified algorithms that work in practice, i.e., in FMIS, is presented in research by Santaga et al. [62].

The overall aim of the study is to generate and compare MZ maps prepared using different models. The first model was termed as "RS- and threshold-based clustering" and was adopted from Georgi et al. [17] with partial modifications. The second model was a "hybrid-based and unsupervised clustering" model, a hybrid model in which a combination of data from different sources was utilized for MZ delineation. The last approach was called "RS-based unsupervised clustering", which was similar to the first model but with a different classification. The secondary objectives were to statistically analyse the models' outputs, as well as to improve the accuracy of MZ delineation by incorporating agronomy and climate information in the applied models. Finally, a MZ map was presented to guide spring mineral nitrogen fertilization of winter wheat in specific phenological phases based on fertilization dates.

2. Materials and Methods

2.1. Study Area

The research was conducted on the 50.2 ha field on the experimental farm of the Poznań University of Life Sciences, RGD Brody, located in Brody (52.43 N, 16.29 E according to WGS84), Wielkopolskie Voivodeship, Poland (Figure 1). The field dedicated to the research was covered with winter wheat crop (*Triticum aestivum* cv. 'RGT Reform'), carried out in the reduced soil tillage system and non-irrigated.

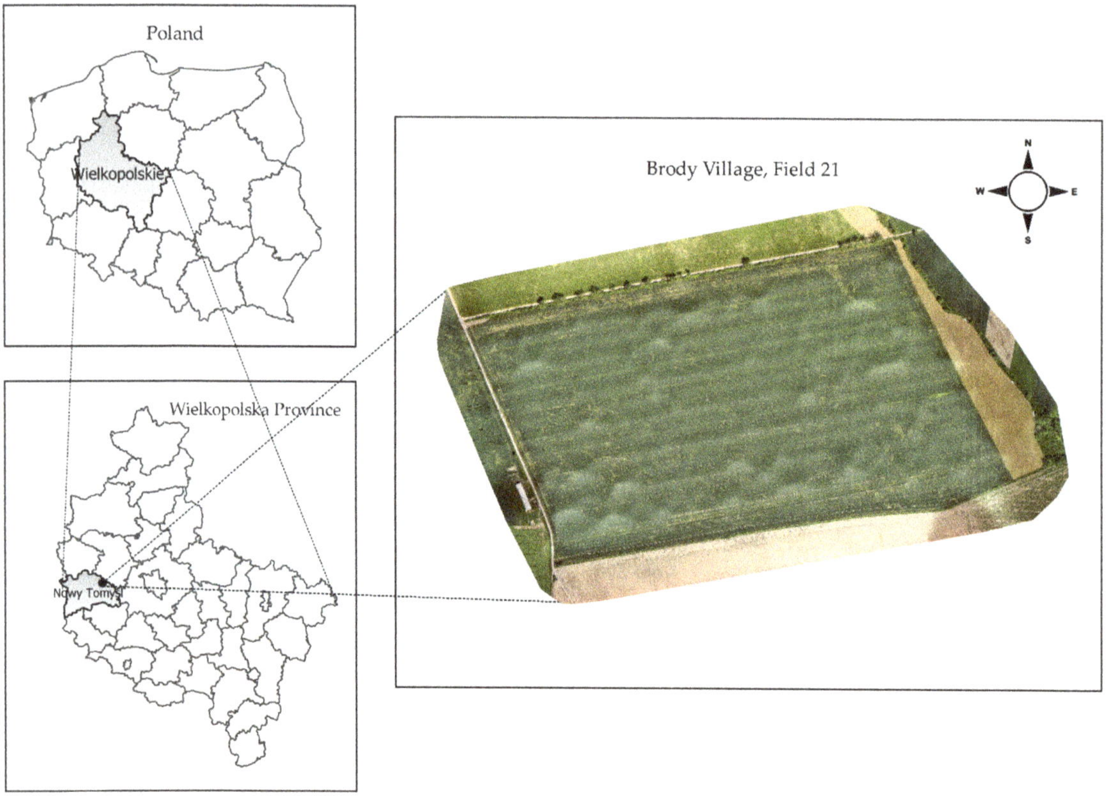

Figure 1. The geographic location of the study area in Brody, Wielkopolska province, Poland.

The average annual precipitation sum of the study area (1960–2019) was 599 mm, and the annual mean air temperature was 8.5 °C, while in 2019 and 2020, the annual precipitation sum was 462 and 520 mm respectively, with the mean air temperature of the study at 10.8 and 10.6 °C, respectively. Meteorological data were obtained from the meteorological station of Brody Experimental Station and were recorded according to the World Meteorological Organization guidelines.

The farm soils in the study area are light, loamy sands, developed on loamy sands overlying loamy material, and are classified as Albic Luvisols according to World Reference Base nomenclature [63,64].

2.2. Data

2.2.1. Soil Sampling

Soil physical and chemical properties were determined by infield sampling and laboratory analysis. Soil samples were taken mechanically, in a semi-automatic operation, from the 0–30 cm field layer in a 4 ha grid, where one average, mixed sample consisted of 16 primary samples. Sampling was carried out in spring 2020. In the present study, analyses were performed to determine pH_{KCl}, phosphorous (P_2O_5), potassium (K_2O) and magnesium (Mg) contents. These are the most commonly used parameters determined by farmers in agrotechnical practice, especially small-scale farming, because of the low implementation costs. Additionally, an analysis of soil organic matter (OM) content was performed as one of the important factors determining sorption and water properties of arable soils.

Soil physico-chemical parameters were determined as follows: pH in 1M KCl according to PN-ISO 10390-1997, available phosphorus and potassium by the Egner-Riehm method [65] according to PN-R-04023 and PN-R-04022 respectively, and magnesium by the Schachtschabel method [66] according to PN-R-04020. OM content was determined by the Tiurin method (T methode) [67] with the Van Bemmelen coefficient of 1.724.

2.2.2. Yield Data

Winter wheat yield data were recorded automatically during harvest in 2019 and 2020 with a modified Claas Lexion 480 combine harvester. Data were recorded at a temporal resolution of 1 Hz for each of the harvester passes. The recorded raw yield data were post-processed and filtered to mitigate lag times and exclude outliers. A detailed description of the combine harvester prototype equipped with a system for monitoring qualitative and quantitative grain parameters is described by Czechlowski and Wojciechowski [68,69].

2.2.3. Elevation Data

The basic hypsometric data were obtained using the measuring system of the modified combine harvester described by Czechlowski and Wojciechowski [64]. The combine was equipped with a Novatel RT2 PROPAK V3 GNSS receiver with a GPS-702-GG: a dual-frequency (L1/L2) antenna and a SmallTRIP 3.2 GPRS/NTRIP modem with automatic connection to the Real-Time Kinematic (RTK) NAWGEO service of the ASG-Eupos network. The possibility of using this type of data to create digital elevation models of agricultural field surfaces was reported by Czechlowski et al. [70]. The slope map was generated based on the interpolated elevation map (DEM) as a percent slope (See Preprocessing Section).

2.2.4. RS Data

A time series of Sentinel-2 L2A images from 1 January 2018 to 1 July 2020 was downloaded from The Copernicus Open Access Hub [71]. Sentinel-2 L2A data are atmospherically and geometrically corrected. Additionally, layers such as a scene classification layer, cloud mask, cloud shadow mask and snow mask were provided along with Sentinel-2 L2A raw data. During the downloading stage, it was attempted to download the data with minimum possible cloud cover by checking the quick layer of every data point. The quick layer of each data point is accessible in the Copernicus Open Access Hub which is embedded in the detailed information of each Sentinel-2 scene. Finally, 119 scenes of Sentinel-2 L2A were downloaded. Table 1 shows the number of downloaded data per year and month.

Table 1. Monthly distribution of 119 Sentinel-2 L2A scenes available from 1 January 2018 to 1 July 2020.

Year \ Month	Jan	Feb	Mar	Apr	May	June	July	Aug	Sep	Oct	Nov	Dec
2018	1	2	1	5	8	2	6	7	5	8	2	1
2019	0	4	2	6	5	8	3	3	5	5	2	4
2020	3	3	4	7	4	3	- *	- *	- *	- *	- *	- *

* Not analysed.

2.3. Models

Three models were separately implemented to delineate MZ. In the following sections, each model is described in detail. First, a brief introduction of each model is provided, then its flowchart is presented and related parts are explained.

2.3.1. Model-1 (RS- and Threshold-Based Clustering)

The first model was fundamentally adopted from Georgi et al. [17], who developed a segmentation algorithm for generating MZ of within-field crop patterns by solely using multi-temporal and multi-spectral satellite images. Thus, the input to the model was the time series of RS data. However, the RapidEye data used in [17] was replaced with Sentinel-2 L2A data because of its free and open data policy. Moreover, a different approach was applied for selecting cloud-free and cloud shadow-free data using mask layers embedded in Sentinel-2 L2A products. The workflow of model-1 is summarized in Figure 2 and the whole process of this model was subdivided into 4 parts (Sentinel-2 data processing, Data selection, Processing of NIR bands, and Segmentation and classification), and a detailed description of each part is provided in the following sections.

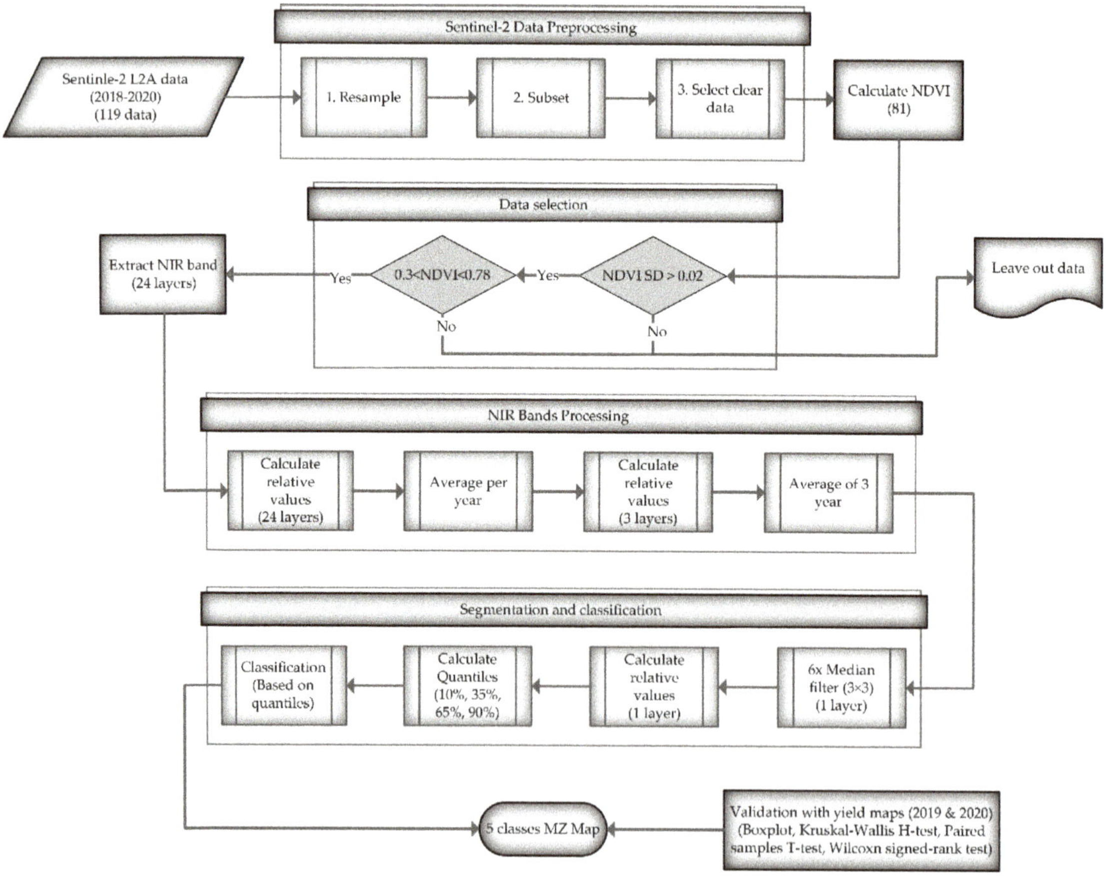

Figure 2. The workflow of model-1 (RS- and threshold-based clustering).

Sentinel-2 Data Processing

As inputs to the model, 119 scenes of the Sentinel-2 L2A were imported. First, all bands were resampled to 10 m pixel size, which was inevitable due to differences in spatial resolutions. The cloud-free and cloud shadow-free data were extracted by investigating the quality scene classification layer, quality cloud confidence, cloud probability mask, cloud mask and shadow mask layers embedded in Sentinel-2 L2A products. For each scene, the mentioned layers were investigated by visual quality control. Finally, NDVI was calculated for all images. The entire workflow was conducted in SNAP V7 software [72].

Data Selection

The data were selected with two constraints. Standard deviation (SD) was calculated for each NDVI data point, and the SD values <0.02 were dropped to exclude the images of dense vegetation cover with no spatial patterns. Then, the mean of each NDVI data point was calculated, and the values between 0.3 and 0.78 were selected since the values <0.3 and >0.78 depict the corresponding images of vegetation canopy, which is too sparse (bare soil background) or too dense, respectively [73]. When the canopy becomes too dense, NDVI saturates because red reflectance does not change much, but near-IR reflectance increases [73]. This stage was implemented in Python 3.8 by using 'rasterio', 'unidip' and 'numpy' packages. For selected dates, NIR bands of Sentinel-2 data were extracted, on which the consequent steps were implemented. This model was conducted based on NIR bands, while NDVI data were used only for data selection since ratio indices such as NDVI cause noise patterns and artifacts that challenge the MZ delineation [17,74]. This resulted in 24 out of 119 raster data, as summarized in Table 2.

Table 2. Acquisition dates of final selected Sentinel-2 data.

Data	Data	Data
25 February 2018	8 February 2019	20 February 2020
6 April 2018	18 February 2019	11 March 2020
9 April 2018	25 February 2019	5 April 2020
31 May 2018	28 February 2019	8 April 2020
3 June 2018	18 June 2019	22 June 2020
8 June 2018	20 June 2019	
3 July 2018	25 June 2019	
31 October 2018	24 August 2019	
7 November 2018	27 August 2019	
5 December 2018		
Total in 2018: 10	Total in 2019: 9	Total in 2020: 5

Processing of NIR Bands

Each image was converted to relative values by Equation (1), i.e., a normalization to a percentage, where 100% was equal to the average NIR value of each image.

$$Normed\ pixel\ value = \left(\frac{Pixel\ value - Minimum}{Mean - Minimum}\right) \times 100 \quad (1)$$

where *Minimum* is the minimum value of the whole scene and *Mean* is the mean value of the whole scene. Then, an average of NIR bands for each year was calculated, thus the NIR time series of each year formed a raster stack. Then, normalization and averaging for the generated data were applied as well.

Segmentation and Classification

A 3 × 3 median filter was applied to eliminate the small zones and smooth the class boundaries, and the result was normalized as previously mentioned. Eventually, a thresholding method was implemented on the classification, whereby the 10%, 35%, 65%

and 90% quantiles were calculated, and the final raster was classified into five classes. These quantile values were empirically chosen [17]. The final five classes were termed 'very low' (1), 'low' (2), 'average' (3), 'high' (4) and 'very high' (5), which correspond to yield expectancy. The processing of NIR bands, classification and mapping was conducted in QGIS V3.10 [75].

2.3.2. Model-2 (Hybrid-Based, Unsupervised Clustering)

The Second approach is based on a hybrid model for MZ delineation that combined data from different sources (see Figure 3). This method was recently applied by researchers in several studies [44–47]. However, in these studies [44–47], principal component analysis (PCA) was utilized to reduce data dimensionality and minimize the dependencies among variables. In this study, PCA was replaced with machine learning-based feature selection (random forest (RF) feature importance). The flowchart of model-2 is shown in Figure 3 and the whole workflow of this model was subdivided into 5 parts (Input data, Preprocessing, Processing, Output (MZ map) and Validation), and a detailed description of each part is provided in the following sections.

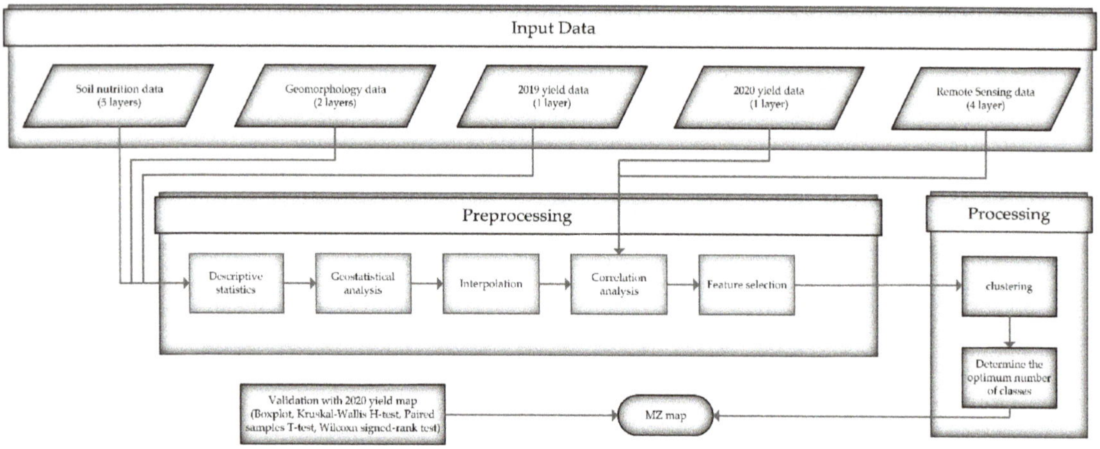

Figure 3. The workflow of model-2 (hybrid-based, unsupervised clustering).

Input Data

For this method, data from 4 different sources were integrated, including (1) soil nutrition data comprising soil pH_{KCl}, P_2O_5, K_2O, Mg and OM, (2) topographical data comprising elevation and slope of the study area, (3) yield data from 2019 and 2020 and (4) RS data comprising NDVI and biophysical variables of LAI, FAPAR and FVC of Sentinel-2 data at the heading stage of wheat (15 May 2020 for this study area) [37]. The NDVI and biophysical variables were derived in SNAP V7 [72].

Preprocessing

Descriptive statistics (minimum, maximum, mean, SD, standard error (SE), coefficient of variation (CV), skewness and kurtosis) of soil, elevation and yield samples were calculated. Since the locations of soil, yield and elevation data are different, no geostatistical analysis was possible prior to correlation analysis and feature selection. Thus, semi-variogram parameters (nugget (C0), sill (C + C0) and range) were estimated to represent the spatial distribution of soil, yield and elevation data [76]. Several semi-variogram models, including circular, spherical, tetraspherical, pentaspherical, exponential, Gaussian, rational quadratic, hole effect, k-Bessel, J-Bessel and stable, were evaluated. The best-fit model with the lowest root-mean-square (RMS) error was selected for each data point. Then, the data were interpolated using the best-fit model and ordinary kriging (OK) procedure. Since the resolution of the satellite data is 10 m, all the variables (e.g., soil, yield and elevation data) were interpolated to 10 m spatial resolution. Based on Martins et al. [44], variables that are temporally stable and correlated with crop yield were selected, which are crucially significant to delineate MZ [47]. Therefore, a correlation matrix was generated using Spearman's correlation to specify the relationship among variables on interpolated data with the spatial resolution of 10 m. Unlike other recent studies [44–47], in this study, RF with variance reduction criterion was utilized instead of PCA to rank the features, reduce data dimensionality and minimize the dependencies amongst variables. The final variables were selected by considering Spearman's correlation matrix (using correlation coefficient criterion) and RF feature importance (using mean squared error criterion). Deriving descriptive statistics, correlation analysis and feature selection was performed in Python 3.8 using 'pandas', 'scipy', 'rasterio', 'numpy', 'matplotlib', 'seaborn' and 'sklearn' packages. The geostatistical analysis and interpolation were conducted in ArcGIS V10.7 [77].

Processing

MZ delineation was performed by the fuzzy c-means algorithm using Management Zone Analyst (MZA) software V1.0. [36,78]. Furthermore, two types of cluster validity functions, fuzzy performance index (FPI) [79,80] and normalized classification entropy (NCE) [81], were used to determine the optimum number of zones. The FPI is a measure of the degree of separation, i.e., fuzziness, between classes, with values ranging from 0 to 1 [36]. Additionally, the NCE measures the degree of disorganization between classes [36]. The minimum values of these indices suggest the optimum number of clusters since it represents the least membership sharing (FPI) or the highest amount of organization (NCE), as shown in Equations (2) and (3) [36]. The settings used in the MZA software included similarity measure = Mahalanobis distance, fuzziness exponent = 1.3, the maximum number of iterations = 300, convergence criteria = 0.0001, minimum number of zones = 2 and maximum number of zones = 8. Finally, the ideal number of zones was selected by considering the lowest values of FPI and NCE:

$$FPI = 1 - \frac{c}{(c-1)}\left[1 - \frac{1}{n}\sum_{k=1}^{n}\sum_{i=1}^{c}(u_{ik})^2\right] \quad (2)$$

$$NCE = \frac{c}{(n-c)}\left[-\frac{1}{n}\sum_{k=1}^{n}\sum_{i=1}^{c}u_{ik}log_a(u_{ik})\right] \quad (3)$$

where c is the number of clusters, n is the number of observations, u_{ik} is the fuzzy membership and log_a is the natural logarithm. Following clustering, mapping was conducted in QGIS V3.10 [75].

2.3.3. Model-3 (RS-Based, Unsupervised Clustering)

The approach applied here differed from model-1, in that a K-means clustering algorithm was conducted to compare the classification of this model (threshold-based clustering) with a simple clustering procedure. Other components of model-3 were similar to model-1 (Figure 2). As can be seen from Figure 4, the whole workflow of model-3 was subdivided into 3 sections (Sentinel-2 data processing, Data selection and Classification) and the description of each part was provided in Section 2.3.1.

The K-means algorithm was performed with 5 classes and 100 iterations in SNAP V7 [72]. Moreover, mapping was conducted in QGIS V3.10 [75].

2.4. Model Improvement

To improve the result of the final MZ maps, the RS-based models (model-1 and model-3) were enriched with climate and agronomy information. The agronomic information used is the date of sowing and the dates of subsequent spring mineral nitrogen fertilization treatments, which are carried out three times, as is typical for this region. The timing of the three application rates is linked to the phenological phases of plant development and to climate and soil conditions (beginning of vegetation, stem-shooting phase, earing). Agronomic and drought information were used to select RS images. This is a knowledge-based selection of input RS data under new climate conditions, such as drought. Additionally, both RS-based models were performed concerning phenological phases, which starts from seeding to harvesting time using expert knowledge. To overcome this, the models should be run with single-year RS data and considering phenological phases in a specific year. The abnormal climate condition, such as drought in the analysed period, can impact the final yield map, which affects the performance of the models. The RS-based models were performed only for 2020 RS data to avoid this, by selecting data after 19 September 2019 (seeding date) in the 2019–2020 season. This resulted in 5 RS datasets that passed the data selection constraints (Table 2). Since this analysis was conducted for single-year data, the generated MZ maps can guide fertilization before the fertilization time approaches considering the growth stages of wheat. In this study area, fertilization was performed at four dates in the 2019–2020 season (23 January 2020, 17 February 2020, 25 March 2020 and 21 April 2020). With regard to fertilization dates and available selected RS data, two MZ maps for fertilization were generated for 25 March 2020 and 21 April 2020.

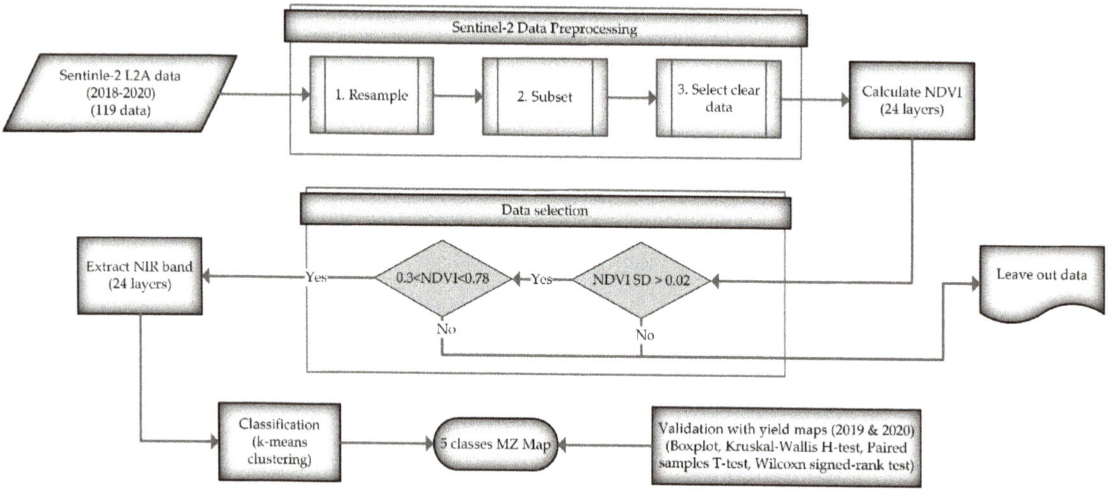

Figure 4. The workflow of model-3 (RS-based, unsupervised clustering).

2.5. *Sampling for Validation*

A stratified random sampling procedure was conducted for drawing validation data. Yield maps were converted to relative values (see the Processing of NIR Bands Section) and then averaged based on the models over the available years [17]. For each sample, the relative yield value and corresponding class ID were sampled based on the MZ map [17]. The size and conjectured SD of each class were considered to determine the sample size of each class. The number of samples was computed by Equation (4) [82]:

$$N = \left(\frac{\sum_{i=1} W_i S_i}{S_o}\right)^2 \quad (4)$$

where N is the number of samples, W_i is the proportion of mapped area for class i, S_i is the SD of stratum i, S_o is the expected SE of overall accuracy and C is the total number of classes. The S_i of each class was conjectured since no specific data for samples were available to determine the S_i of each class by the assumption that S_i is higher for classes with low area proportion. Afterwards, S_o was assumed equal to 0.01 [82]. Finally, equal distribution (ED_i) (Equation (5)) and weighted distribution (WD_i) (Equation (6)) of each class' samples were calculated to determine the sample size of each class. The final number of samples for each class (N_i) (Equation (7)) was assessed by averaging these distributions. The sampling procedure was performed in QGIS V3.10 [75].

$$ED_i = N \times W_i \tag{5}$$

$$WD_i = \frac{N}{C} \tag{6}$$

$$N_i = \frac{(ED_i + WD_i)}{2} \tag{7}$$

2.6. Validation

To validate and evaluate the final MZ, a variety of statistical tests were performed. The Paired Sample *T*-test, Kruskal–Wallis H-test and Wilcoxon signed-rank test were applied to samples of each class and compared with each other to explore whether there were statistically significant differences. The Paired Sample *T*-test requires normally distributed data, so it was performed on logarithmically transformed sample values for the second time after the test was performed without normalization. The purpose of MZ delineation is to classify the wheat parcel into homogeneous zones, thus the separability of final zones was tested. Finally, boxplots for each model were used along with fitted lines through the medians of each class. The validation was conducted in Python 3.8 by using 'pandas', 'scipy', 'numpy' and 'matplotlib' packages.

3. Results

3.1. Model-1

Figure 5a shows the delineated, 5-class MZ map based on model-1. The higher the number of zones, the better the crop vitality and yield expectancy will be. Based on this map, the north and south-west parts of the field (Zones 5 and 4) show a more productive crop pattern compared with the west and east parts. The statistical tests (Table S1 see the (Supplementary Material)) indicated inseparability between classes for values with $p > 0.05$, and these values are highlighted in red in Table S1. According to Table S1, although the result of the Kruskal–Wallis H-Test, which compares the separabilities of the zones all at once, showed that all classes are separable with the *p*-value of 2.9×10^{-54} in all other tests, the pairs of the zones 3–4, 3–5 and 4–5 are inseparable. The results of the Paired Samples *T*-test for the pairs of the zones 3–4, 3–5 and 4–5 were 0.60, 0.63 and 0.95, respectively. The *p*-values of the Paired Samples *T*-test (log of data) for the mentioned zones were equal to 0.86, 0.80 and 0.70, respectively. Finally, the results of the Wilcoxon signed-rank test for these zones were 0.34, 0.08 and 0.77. Moreover, in the Wilcoxon signed-rank test, the pair of the zones 1–2 did not support the separability hypothesis, with a *p*-value of 0.08. On the other hand, values with $p < 0.05$ indicate separable zones. Additionally, the boxplot shown in Figure 6a confirms the result of the statistical tests, with overlaps observed for the pairs of the zones 3–5 and 1–2.

3.2. Model-2

The results of descriptive statistics for soil, yield and elevation samples are summarized in Table 3. Despite the sufficient number of samples for elevation (DEM), OM and yields for 2020 and 2019, the numbers of soil pH_{KCl}, P_2O_5, K_2O and Mg samples were fewer than the expected number for interpolation as there were just 14 samples in the 50 ha area. The authors of this study are well aware of the small sample size, but this is the typical soil sampling density used in state public advisory practice. As shown in Table 3, the average yield in 2020 dropped by approximately 400 kg ha^{-1} compared with that of 2019. This lower productivity is attributed to the severe spring drought that occurred in major parts of Poland, including this study area [83]. The CV of the analysed attributes can be categorized from low (CV < 12%) to moderate (12% ≤ CV < 60%) based on the classification suggested by Warrick and Nielsen [84].

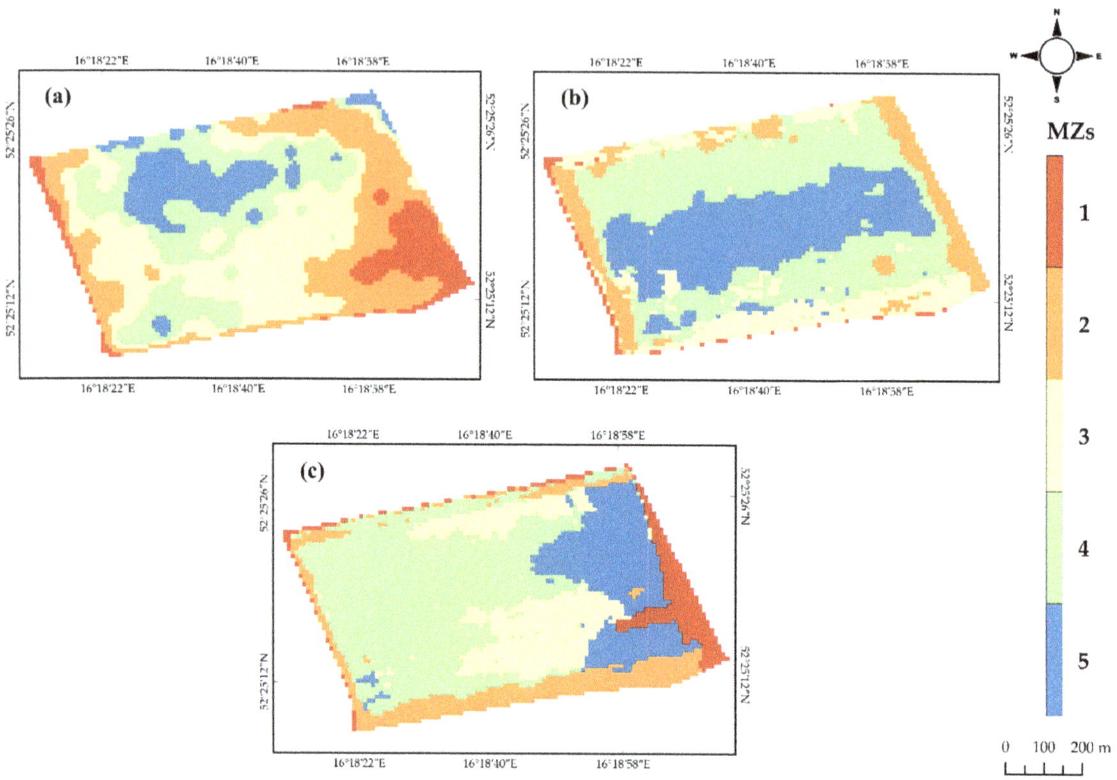

Figure 5. Delineated MZ maps of model-1 (a), model-2 (b) and model-3 (c).

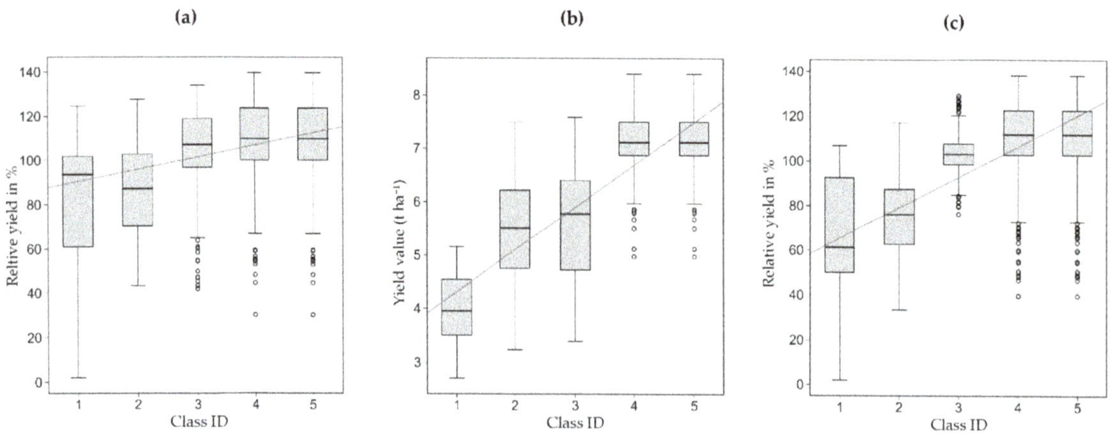

Figure 6. Boxplots of stratified sampling for model-1 (a), model-2 (b) and model-3 (c). The white circles out of the whiskers show outliers.

Table 3. Descriptive statistics of DEM, soil attributes and wheat yield.

Attributes	n	Min	Max	Mean	SD	SE	CV	Skewness	Kurtosis
DEM 2020 (m)	67,158	92.42	100.12	96.21	2.36	0.01	0.02	0.06	−1.45
OM 2011 (%)	52	1.14	2.64	1.66	0.44	0.06	0.26	0.92	−0.33
pH_{KCl} 2020	14	6.00	7.1	6.44	0.32	0.08	0.05	0.44	−0.45
P_2O 2020 (mg 100 g soil^{-1})	14	17.20	36.6	25.71	5.92	1.58	0.22	0.20	−1.07
K_2O 2020 (mg 100 g soil^{-1})	14	23.00	34.0	26.86	3.55	0.95	0.13	0.64	−0.80
Mg 2020 (mg 100 g soil^{-1})	14	8.50	12.4	9.89	1.41	0.38	0.14	0.70	−1.04
Yield 2019 (t ha^{-1})	9613	0.39	15.18	7.19	1.56	0.02	0.22	−1.25	5.01
Yield 2020 (t ha^{-1})	8520	0.36	13.12	6.82	1.58	0.02	0.23	−1.30	3.11

Table 4 represents the results of the geostatistical analysis. It suggested the best-fit models to be exponential (DEM, yield 2019, yield 2020) J-Bessel (OM, pH_{KCl}), Gaussian (K_2O, Mg) and Hole Effect (P_2O_5), based on the minimum RMSE. The values of Nugget/Sill could be used to determine the degree of the spatial autocorrelation, in which the values < 25%, 25–75% and >75% suggest strong, moderate and weak spatial dependencies, respectively [85]. DEM and OM showed strong spatial dependence, while other parameters showed moderate degrees. The range of the semi-variogram was the distance over which the samples are correlated with each other [86]. A low value of Nugget/Sill and a high range of an attribute generally indicate that high precision can be obtained by kriging [85].

Table 4. Semi-variogram parameters of DEM, soil attributes and wheat yields.

Variables	Model	Nugget (C_0)	Partial Sill (C_1)	Sill ($C_0 + C_1$)	Nugget/Sill $C_0/(C_0 + C_1)$	Range (m)	RMSE
DEM	Exponential	0	0.0005	0.0005	0	1.4042	0.0209
OM	J-Bessel	0.0266	0.2443	0.2709	0.0982	1247.5	0.1665
pH_{KCl}	J-Bessel	0.0299	0.0778	0.1077	0.2776	925.03	0.2391
P_2O_5	Hole Effect	12.883	28.331	41.214	0.3126	915.83	4.2782
K_2O	Gaussian	6.4560	16.516	22.972	0.2810	1147.2	2.8914
Mg	Gaussian	1.0110	2.7338	3.7448	0.2970	1147.2	1.0427
Yield 2019	Exponential	1.6169	0.7382	2.3551	0.6865	490.05	1.3108
Yield 2020	Exponential	1.2226	2.2208	3.4434	0.3550	1301.7	1.0925

Figure 7 shows the maps using the best-fit model, including interpolated soil, elevation and yield, along with RS data.

The maps suggested a high correlation of RS data with the yield map of 2020. Moreover, high values were generally observed in the central part of the field. Soil pH_{KCl} and P_2O_5 maps were consistent since the values of P_2O_5 were high and neutral in the west and east parts of the field. However, the values were low and somewhat acidic in the southern part. In terms of K_2O, Mg and OM, higher values were observed in the Eastern part. Besides, DEM and slope values were higher in the western part. Figures 8 and 9 show the results of correlation analysis and selection of features with high correlation with yield 2020 data. The RS data (NDVI, LAI, FCV and FAPAR) were highly and positively correlated with yield 2020. The feature selection also showed that yield 2019 and NDVI are appropriate features for clustering.

Besides, the optimum number of classes was found by computing two cluster validity indices (FPI, NCE). Figure 10 shows the plotted values of FPI and NCE against the number of clusters, with the optimum number being the value at which FPI and NCE are minimum, i.e., 5 clusters. This was consistent with model-1 results in terms of the number of zones.

The result of MZ in 5 classes by fuzzy c-means clustering (Figure 5b) showed a better condition in the central part of the field from west to east (Zone 5). This map was consistent with the features (NDVI and yield 2019 data) that were used in the MZ delineation process. As reported in Table S2, the *p*-value of the Kruskal–Wallis H-Test for this model was 3.2×10^{-117}, which means that all of the zones support the separability hypothesis. However, the results of other statistical tests showed that the pair of the zones 2–3 did not support the separability hypothesis, with *p*-values of 0.37 (Paired Samples *T*-test), 0.46 (Paired Samples *T*-test (log of data)) and 0.55 (Wilcoxon signed-rank test). This can also be observed in the boxplot (Figure 6b).

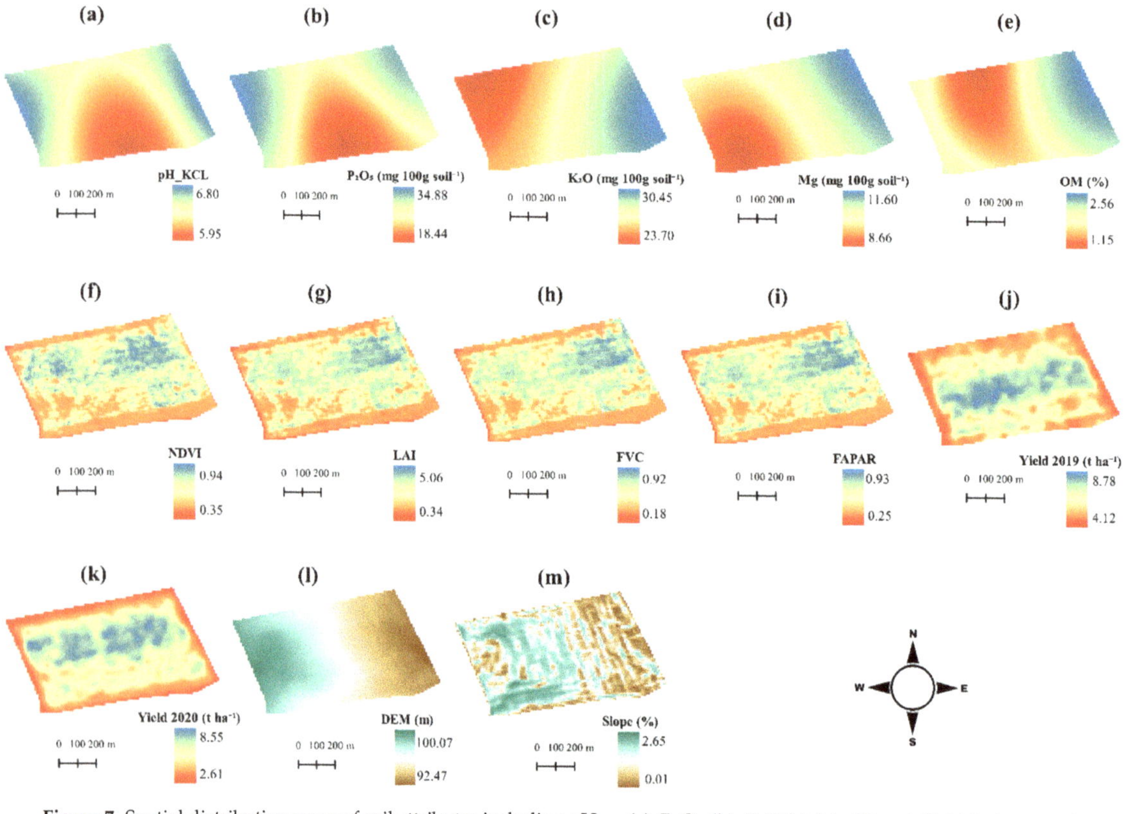

Figure 7. Spatial distribution maps of soil attributes including pH_{KCl} (**a**), P_2O_5 (**b**), K_2O (**c**), Mg (**d**) and OM (**e**), the spatial distribution maps of RS data including NDVI (**f**), LAI (**g**), FCV (**h**) and FAPAR (**i**), the spatial distribution maps of yield for 2019 (**j**) and 2020 (**k**) and the spatial distribution maps of geomorphology data including DEM (**l**) and Slope (**m**).

3.3. Model-3

Figure 5c shows the delineated MZ map for model-3, which suggests that the central, eastern and western parts of the field have higher yield expectancy. However, a low yield pattern was observed at the edge of the field. According to Table S3, the *p*-value of the Kruskal–Wallis H-Test for this model was 1.2×10^{-102}. Nevertheless, the pair of the zones 4–5 did not support the separability hypothesis, with *p*-values of 0.37 (Paired Samples *T*-test), 0.18 (Paired Samples *T*-test (log of data)) and 0.99 (Wilcoxon signed-rank test), as they are also overlapping along with the pair of the zones 1–2 (Figure 6c).

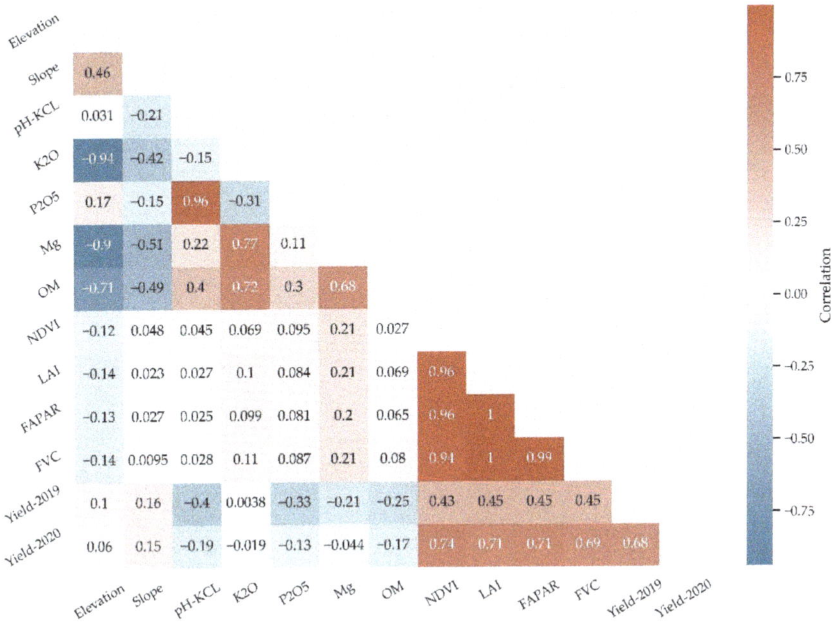

Figure 8. Spearman's correlation matrix amongst soil attributes, RS data, yield data and geomorphology data.

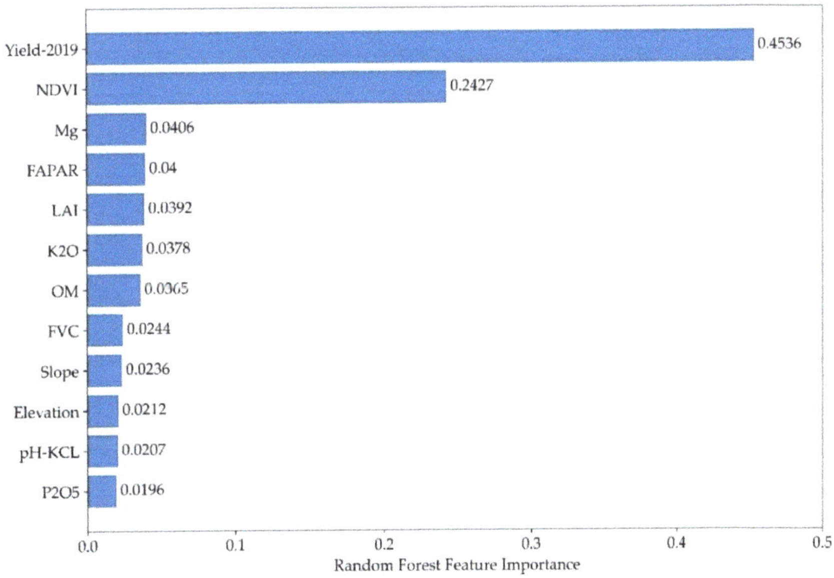

Figure 9. Result of RF feature importance.

3.4. Improvement of Model Results

3.4.1. Model-1

Figure 11 shows the delineated MZ maps in 5 classes for model-1 considering agronomy and climate information. Figure 11a shows the MZ map that is based on the RS data before 25 March 2020 (see Table 2). Thus, it can be used as a fertilization recommendation map at 25 March 2020. Figure 11b shows the result of model-1 with the RS data before 21 April 2020. Additionally, this map can be utilized as a fertilizer recommendation map at 21 April 2020. As can be seen in Tables S4 and S5, all zones passed the separability hypothesis (p-value < 0.5), thus all zones are separable. The boxplots of these maps also confirmed the results of statistical tests (Figure 12). Zones 4 and 5 passed the test, though they showed partial overlap.

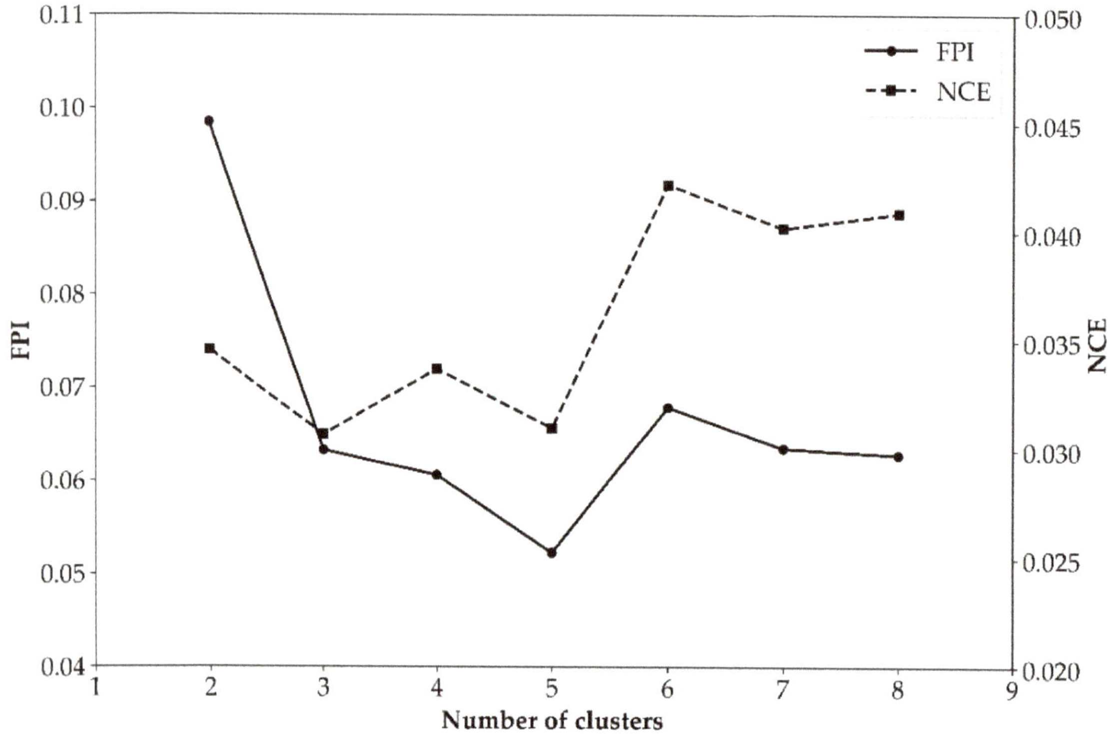

Figure 10. Fuzzy performance index (FPI) and normalized classification entropy (NCE) calculated to determine the optimum number of clusters for the study area.

3.4.2. Model-3

Likewise, MZ delineation was conducted for model-3 with the hypothesis (agronomy and climate information) that was considered for improving the MZ results. Figure 13 shows the MZ maps for two dates before fertilization. As shown in Table S6, all zones were separable (p-value < 0.5). However, zones 3 and 4 did not pass the statistical test, and the p-values of statistical tests were 0.78 (Paired Samples T-test), 0.73 (Paired Samples T-test (log of data)) and 0.08 (Wilcoxon signed-rank test) (Table S7). The boxplot of both maps confirms the statistical tests (Figure 14). Similar to model-1, zones 4 and 5 were overlapping.

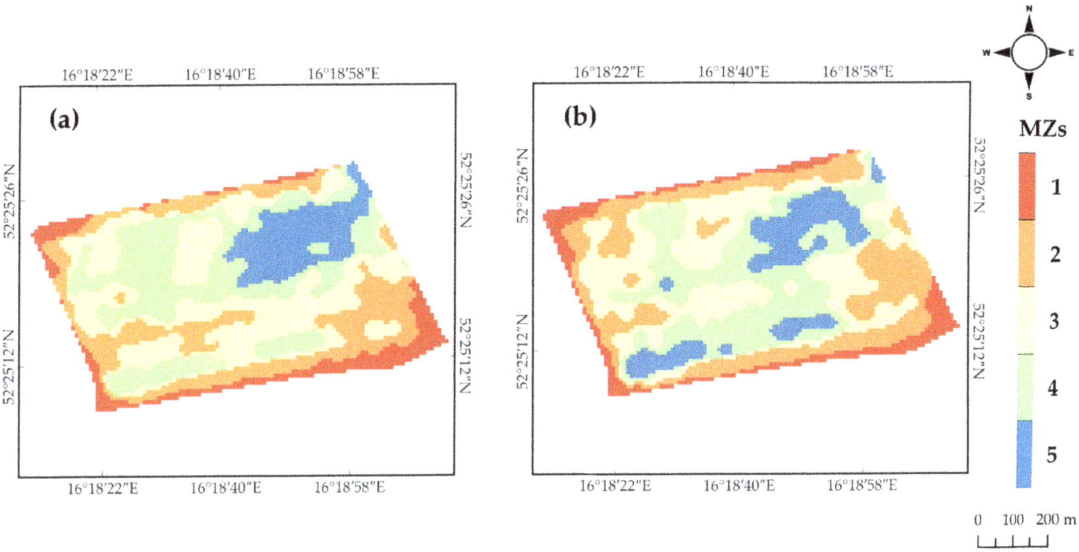

Figure 11. Delineated MZ maps of model-1 for fertilization at 25 March 2020 (**a**) and 21 April 2020 (**b**).

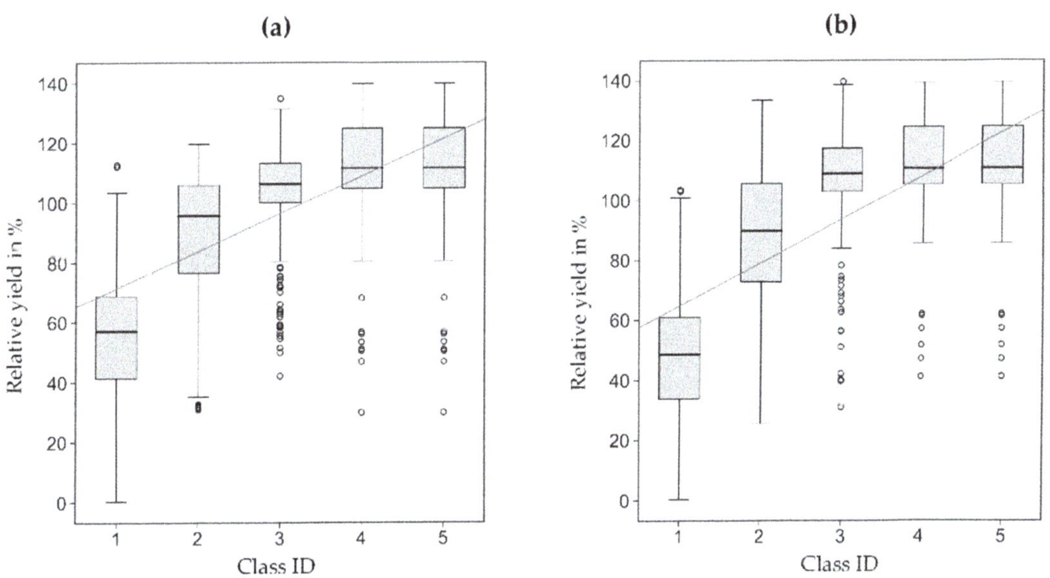

Figure 12. Boxplots of stratified sampling for model-1 for fertilizations at 25 March 2020 (**a**) and 21 April 2020 (**b**). The white circles out of the whiskers show outliers.

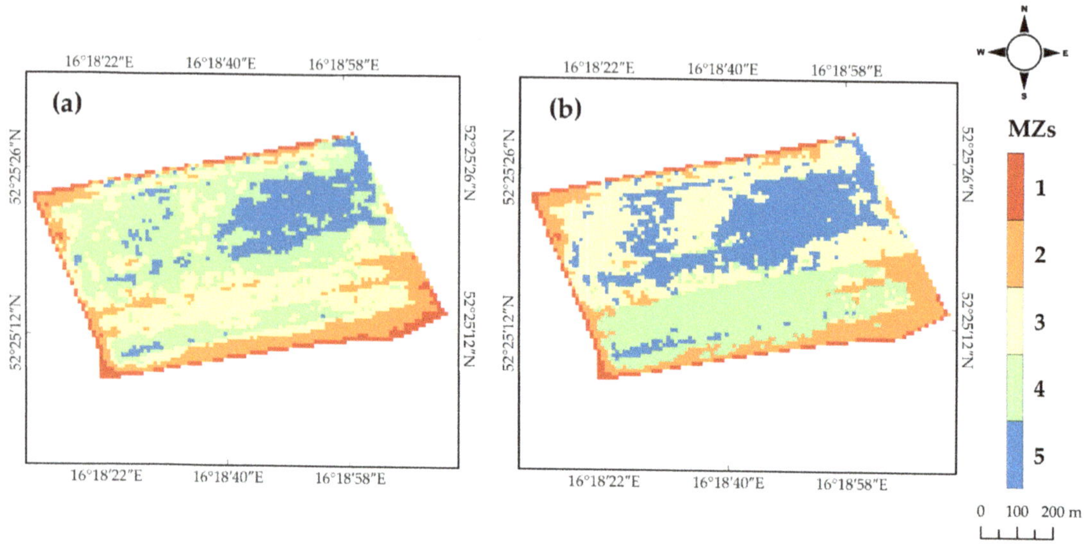

Figure 13. Delineated MZ maps of model-3 for fertilization at 25 March 2020 (**a**) and 21 April 2020 (**b**).

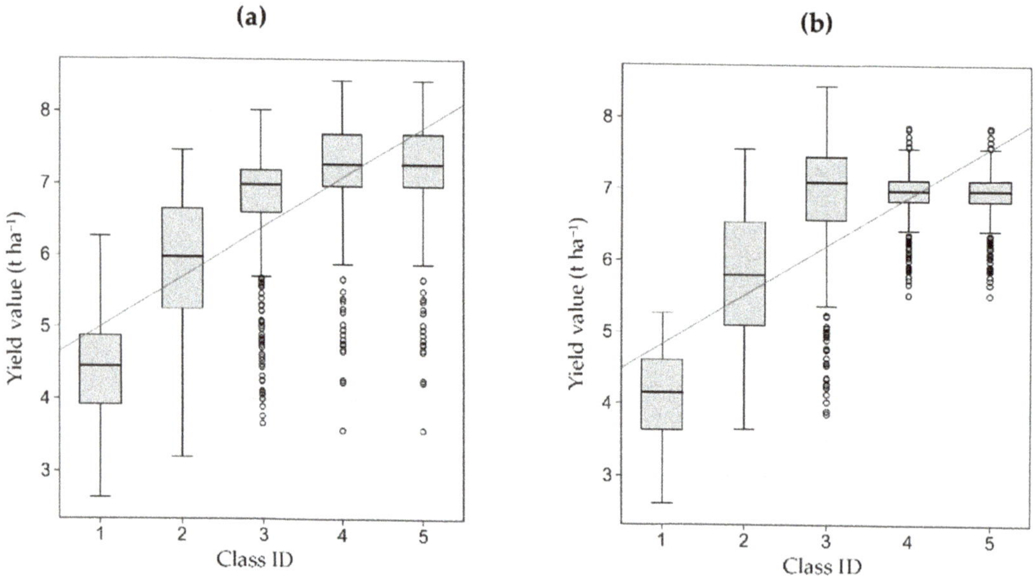

Figure 14. Boxplots of stratified sampling for model-3 for fertilization at 25 March 2020 (**a**) and 21 April 2020 (**b**). The white circles out of the whiskers show outliers.

4. Discussion

This study followed a comparative analysis for MZ mapping that was aimed at optimizing fertilization in the context of PA. The comparative analysis has been tested by experiments based on the Sentinel-2 satellite data. The results showed only marginal consistency among the mapping outputs without any additional agronomy and climate information, which confirms the hypothesized challenges in the best model selection. Although statistical tests were considered to be appropriate validation tools, it is also suggested that the quality of the final results be eventually confirmed by a farmer and those engaged in the ongoing cultivation process.

In terms of the applied models, model-1 was partially adopted from Georgi et al. [17], and solely relied on RS data without any additional source of information. One of the advantages of this model is the utilization of NIR bands that are less noisy and more physical than band ratio vegetation indices such as NDVI. Therefore, NIR bands were used in model-3 as well. Besides, labelling of the zones from low to high was convenient compared with using a thresholding procedure for MZ delineation in other methods. On the other hand, labelling and sorting the zones in a clustering method (e.g., k-means) is a demanding task. All in all, the main disadvantage of model-1 was the lack of agronomy and climate information that can negatively impact the result.

Model-2 integrated data from multiple sources that potentially provide additional information for MZ delineation. This resulted in an enhanced accuracy of the final results. Nevertheless, one may note that such a workflow increases the costs, whereas this model is considered impractical without the incorporated sample data (soil, yield, elevation, etc.). Here, the optimum number of soil sampling data was not accessible (14 samples in 50 ha) (Table 3). Furthermore, the soil sampling data which were applied here comprised only chemical attributes and lacked soil physical attributes, such as soil compaction, texture and electric conductivity, providing further influential information on soil condition. Additionally, the certainty of utilizing RS data was confirmed in this model, as previously claimed by Martins et al. [47] and Song et al. [37]. Besides, the results of this study showed the high correlation of RS data with the yield map in correlation analysis and feature selection (See Figures 8 and 9). The k-means clustering method was used in model-3 to eliminate the NIR bands' preprocessing part (see Figure 2) and reduce the data computation of model-1. However, the labelling and sorting of the zones remain an issue in this method, as previously mentioned.

Although Georgi et al. [17] indicated that a three-year timeframe would be sufficient to depict the crop pattern for MZ delineation, the results of this study witnessed an inferior performance considering the years 2018, 2019 and 2020. Instead, focusing on one crop growing season in 2019–2020 yielded the best results. Climate change is to blame for this, as it influences crop patterns and results in year-to-year variations in crop yield. For example, in 2020, the average yield had dropped by 400 kg/ha in comparison to 2019 (see Table 3). As a result, one of the finest data sources for MZ delineation would be high-resolution remote sensing data, such as unmanned aerial vehicle (UAV) data, as it contains detailed spatial information and one UAV data point before the fertilization date can be used to create MZ and apply the generated MZ map for fertilization. Additionally, Nawar et al. [2] asserted that in terms of performance and cost, remote sensing data such as UAV or satellite data are more suitable for informing variable rate nitrogen fertilizer application than soil characteristics' data, such as electric conductivity and soil texture.

Several studies [45,47] recently applied Cohen's Kappa coefficient to evaluate delineated MZ maps with yield maps. However, this method was avoided due to some rationales. First, the classification of yield maps entails highly expert agronomic knowledge by which elements such as climate and agronomy information should be considered to classify the yield maps. Further, yield maps inherently comprise continuous values, subject to bias and information loss when converted and interpreted in categorical/integer data values. Last but not least, statistical tests from Georgi et al. [17] were adopted, which we think will provide an appropriate platform for validating the MZ map.

Further analyses for improvement of MZ delineation showed superior performance of model-1 and model-3 when agronomy and climate information were considered. The output MZ maps of both performed better and were similar and consistent, unlike other delineated MZ maps that lacked agronomy and climate information (see Figures 5, 11 and 13). This supports the importance of this information.

As an additional analysis, RS-based models were calibrated by solely considering agronomy information. Thus, the data within the phenological phases from 2018 to 2020

(2 seasons) were selected, i.e., leaving out data before seeding and after harvesting in two seasons. However, the results showed inferior performance of the RS-based models compared to the model considering both agronomy and climate information. This is parallel to suggestions presented in [4] indicating that information on crops, soil or phenological phases increases the quality of the model. Thus, incorporating both agronomy and climate information was strongly suggested in MZ delineation for PA. This issue is inevitable under new climate conditions with more frequent drought and other extreme events. As such, the requirement of considering this information is to conduct a model in a single year. Accordingly, the generated MZ map can be used as a guide for fertilization.

In a similar study, Santaga et al. [45] investigated simplified and advanced models for creating variable nitrogen fertilization maps using satellite imagery, yield maps and protein maps embedded in FMIS architecture. In contrast to the results of [45], in this study, in addition to simplified models, hybrid models were tested using non-advanced soil data, which should be considered more accessible to agricultural producers than yield or protein maps. In both studies, both simplified and hybrid or advanced models can be effectively implemented in the expanding FMIS [87].

5. Conclusions

This study compared three models based on the Sentinel-2 satellite data for management zones' (MZ) delineation in the context of precision agriculture (PA) on an example of a winter wheat field. The remote sensing (RS)-based models 1 and 3 did not require any additional data and were thus considered less expensive than model-2. Additionally, agronomy (wheat phenological phases) and climate information (drought) were considered to improve MZ maps when applying RS-based models. The results also showed that MZ delineation is prone to uncertain results, particularly under new and rapidly changing climate conditions, if no agronomy expert knowledge and climate information were incorporated. These findings enhance the understanding of the role of new climate conditions for RS-based PA algorithms in rainfed farming with the potential agricultural drought and its impact in applying fertilizer. As drought stress is on the rise and had a significant impact on the growth of most arable crops in central Europe, it is suggested to perform low-cost RS-based techniques only for the current season. This information provides additional, reliable and current information to improve MZ delineation for optimizing fertilization in a single-year context.

In conclusion, an algorithm has been designed and its use has been evaluated for a few test cases for the integration of soil, crop and yield information, together with knowledge about agronomy and climate information. This will improve the results of MZ delineation and generate a guiding map to prescribe variable rates of fertilization before the necessary fertilizer application dates.

It is recommended for future research to use remote sensing data with high spatial resolution, such as satellite images with higher resolution and drone images, in the delineation of MZ maps since delineation of MZ maps with single-year data is feasible using the proposed method. It is also suggested to use cost-benefit analysis to evaluate the implemented MZ maps, e.g., in the context of variable rate fertilization. However, it was concluded that the quality of the final MZ maps should be eventually calibrated in collaboration with farmers and all these steps help farmers set up their fertilization operation to address problem areas and maximize yield.

This study has compared three models for creating MZ, focused on implementation in cloud-based farm management information systems (FMIS). The results of this work, as well as those of Santaga et al. [45], indicate that for FMIS, it is necessary to develop not only advanced models for creating MZ or variable nitrogen fertilization strategies, but also simplified and hybrid models for those users who do not have multiyear crop vegetation data or simplified soil data. Further work in this area is recommended.

Supplementary Materials: The following are available online at https://www.mdpi.com/article/10.3390/agriculture11111104/s1, Table S1: The results of statistical tests for model-1. The red-shaded cells show inseparability between classes ($p > 0.05$), Table S2: The results of statistical teste for model-2. The red-shaded colour follows the description from Table S1, Table S3: The results of statistical tests for model-3. The red-shaded colour follows the description from Table S1, Table S4: The results of statistical tests for model-1 for fertilization at 25 March 2020, Table S5: The results of statistical tests for model-1 for fertilization at 21 April 2020, Table S6: The results of statistical tests for model-3 for fertilization at 25 March 2020, Table S7: The results of statistical tests for model-3 for fertilization at 21 April 2020. The red-shaded colour follows the description from Table S1.

Author Contributions: Conceptualization, H.L., A.A.A. and M.R.; methodology, M.R., A.A.A., H.L., A.S.N. and T.W.; software, M.R.; validation, M.R., T.W., H.L. and G.N.; formal analysis, M.R., A.A.A., T.W., H.L. and G.N.; investigation, M.R., A.A.A., T.W., M.C. and H.L.; data curation, M.C., T.W., M.R., A.A.A. and G.N.; writing—original draft preparation, M.R. and H.L.; writing—review and editing, M.R., H.L., A.A.A., Y.M., T.W., A.S.N. and G.N.; visualization, M.R.; supervision, H.L., A.A.A. and Y.M. All authors have read and agreed to the published version of the manuscript.

Funding: This research has received funding from the European Union's Horizon 2020 research and innovation programme, performed as part of the project "SmartAgriHubs—Connecting the dots to unleash the innovation potential for digital transformation of the European agri-food sector", under grant agreement No. 818182.

Institutional Review Board Statement: Not applicable.

Informed Consent Statement: Not applicable.

Data Availability Statement: Not applicable.

Acknowledgments: The authors would like to thank KNTU for technical support. The research was conducted within the Research Laboratory "Remote Sensing for Ecology and Ecosystem Conservation (RSEEC)" of the KNTU. We would like to thank RGB Brody Poznań University of Life Sciences for providing field data. We also thank ESA for providing open access Sentinel-2 data.

Conflicts of Interest: The authors declare no conflict of interest.

References

1. Assembly, G. *United Nations: Transforming Our World: The 2030 Agenda for Sustainable Development*; UN: New York, NY, USA, 2015.
2. Nawar, S.; Corstanje, R.; Halcro, G.; Mulla, D.; Mouazen, A.M. Delineation of Soil Management Zones for Variable-Rate Fertilization. *Adv. Agron.* **2017**, *143*, 175–245. [CrossRef]
3. Foley, J.A.; Ramankutty, N.; Brauman, K.A.; Cassidy, E.S.; Gerber, J.S.; Johnston, M.; Mueller, N.D.; O'Connell, C.; Ray, D.K.; West, P.C., et al. Solutions for a cultivated planet. *Nature* **2011**, *478*, 337–342. [CrossRef] [PubMed]
4. Sims, B.; Kienzle, J. Sustainable Agricultural Mechanization for Smallholders: What Is It and How Can We Implement It? *Agriculture* **2017**, *7*, 50. [CrossRef]
5. Larkin, D.L.; Lozada, D.N.; Mason, R.E. Genomic Selection—Considerations for Successful Implementation in Wheat Breeding Programs. *Agronomy* **2019**, *9*, 479. [CrossRef]
6. Gebbers, R.; Adamchuk, V.I. Precision agriculture and food security. *Science* **2010**, *327*, 828–831. [CrossRef]
7. Lee, C.-L.; Strong, R.; Dooley, K.E. Analyzing Precision Agriculture Adoption across the Globe: A Systematic Review of Scholarship from 1999–2020. *Sustainability* **2021**, *13*, 10295. [CrossRef]
8. Lajoie-O'Malley, A.; Bronson, K.; van der Burg, S.; Klerkx, L. The future(s) of digital agriculture and sustainable food systems: An analysis of high-level policy documents. *Ecosyst. Serv.* **2020**, *45*, 101183. [CrossRef]
9. Pierce, F.J.; Nowak, P. Aspects of precision agriculture. *Adv. Agron.* **1999**, *67*, 1–85.
10. Bongiovanni, R.; Lowenberg-DeBoer, J. Precision agriculture and sustainability. *Precis. Agric.* **2004**, *5*, 359–387. [CrossRef]
11. Mueller, N.D.; Gerber, J.S.; Johnston, M.; Ray, D.K.; Ramankutty, N.; Foley, J.A. Closing yield gaps through nutrient and water management. *Nature* **2012**, *490*, 254–257. [CrossRef]
12. Liaghat, S.; Balasundram, S.K. A review: The role of remote sensing in precision agriculture. *Am. J. Agric. Biol. Sci.* **2010**, *5*, 50–55.
13. Wojciechowski, T.; Niedbała, G.; Czechlowski, M.; Nawrocka, J.R.; Piechnik, L.; Niemann, J. Rapeseed seeds quality classification with usage of VIS-NIR fiber optic probe and artificial neural networks. In Proceedings of the 2016 International Conference on Optoelectronics and Image Processing (ICOIP), Warsaw, Poland, 10–12 June 2016; pp. 44–48.
14. Niazian, M.; Niedbała, G. Machine Learning for Plant Breeding and Biotechnology. *Agriculture* **2020**, *10*, 436. [CrossRef]
15. Szwedziak, K.; Polańczyk, E.; Grzywacz, Ż.; Niedbała, G.; Wojtkiewicz, W. Neural Modeling of the Distribution of Protein, Water and Gluten in Wheat Grains during Storage. *Sustainability* **2020**, *12*, 5050. [CrossRef]
16. Mulla, D.J. Twenty five years of remote sensing in precision agriculture: Key advances and remaining knowledge gaps. *Biosyst. Eng.* **2013**, *114*, 358–371. [CrossRef]

17. Georgi, C.; Spengler, D.; Itzerott, S.; Kleinschmit, B. Automatic delineation algorithm for site-specific management zones based on satellite remote sensing data. *Precis. Agric.* **2017**, *19*, 684–707. [CrossRef]
18. Cisternas, I.; Velásquez, I.; Caro, A.; Rodríguez, A. Systematic literature review of implementations of precision agriculture. *Comput. Electron. Agric.* **2020**, *176*, 105626.
19. Vrindts, E.; Mouazen, A.M.; Reyniers, M.; Maertens, K.; Maleki, M.R.; Ramon, H.; De Baerdemaeker, J. Management Zones based on Correlation between Soil Compaction, Yield and Crop Data. *Biosyst. Eng.* **2005**, *92*, 419–428. [CrossRef]
20. Fleming, K.; Heermann, D.; Westfall, D. Evaluating soil color with farmer input and apparent soil electrical conductivity for management zone delineation. *Agron. J.* **2004**, *96*, 1581–1587. [CrossRef]
21. Van Alphen, B.; Stoorvogel, J. A Methodology to Define Management Units in Support of an Integrated, Model-Based Approach to Precision Agriculture. In Proceedings of the Fourth International Conference on Precision Agriculture, Saint Paul, MN, USA, 19–22 July 1998; pp. 1265–1278.
22. Nolan, S.; Goddard, T.; Lohstraeter, G.; Coen, G. Assessing managements units on rolling topography. In Proceedings of the 5th International Conference on Precision Agriculture, Bloomington, MN, USA, 16–19 July, 2000; pp. 1–12.
23. Fraisse, C.; Sudduth, K.; Kitchen, N. Delineation of site-specific management zones by unsupervised classification of topographic attributes and soil electrical conductivity. *Trans. ASAE* **2001**, *44*, 155. [CrossRef]
24. Fu, Q.; Wang, Z.; Jiang, Q. Delineating soil nutrient management zones based on fuzzy clustering optimized by PSO. *Math. Comput. Model.* **2010**, *51*, 1299–1305. [CrossRef]
25. MacMillan, R.; Pettapiece, W.; Watson, L.; Goddard, T. A landform segmentation model for precision farming. In Proceedings of the Fourth International Conference on Precision Agriculture, Saint Paul, MN, USA, 19–22 July 1998; pp. 1335–1346.
26. Taylor, R.; Kluitenberg, G.; Schrock, M.; Zhang, N.; Schmidt, J.; Havlin, J. Using Yield Monitor Data To Determine Spati Al Crop Production Potential. *Trans. ASAE* **2001**, *44*, 1409.
27. Diker, K.; Heermann, D.; Brodahl, M. Frequency analysis of yield for delineating yield response zones. *Precis. Agric.* **2004**, *5*, 435–444. [CrossRef]
28. Milne, A.E.; Webster, R.; Ginsburg, D.; Kindred, D. Spatial multivariate classification of an arable field into compact management zones based on past crop yields. *Comput. Electron. Agric.* **2012**, *80*, 17–30. [CrossRef]
29. Niedbała, G.; Kozlowski, J. Application of artificial neural networks for multi-criteria yield prediction of winter wheat. *J. Agric. Sci. Technol.* **2019**, *21*, 51–61.
30. Niedbała, G.; Nowakowski, K.; Rudowicz-Nawrocka, J.; Piekutowska, M.; Weres, J.; Tomczak, R.J.; Tyksiński, T.; Álvarez Pinto, A. Multicriteria prediction and simulation of winter wheat yield using extended qualitative and quantitative data based on artificial neural networks. *Appl. Sci.* **2019**, *9*, 2773. [CrossRef]
31. Hara, P.; Piekutowska, M.; Niedbała, G. Selection of Independent Variables for Crop Yield Prediction Using Artificial Neural Network Models with Remote Sensing Data. *Land* **2021**, *10*, 609. [CrossRef]
32. Kitchen, N.R.; Sudduth, K.A.; Myers, D.B.; Drummond, S.T.; Hong, S.Y. Delineating productivity zones on claypan soil fields using apparent soil electrical conductivity. *Comput. Electron. Agric.* **2005**, *46*, 285–308. [CrossRef]
33. Cambouris, A.; Nolin, M.; Zebarth, B.; Laverdière, M. Soil management zones delineated by electrical conductivity to characterize spatial and temporal variations in potato yield and in soil properties. *Am. J. Potato Res.* **2006**, *83*, 381–395.
34. Ahn, C.W.; Baumgardner, M.; Biehl, L. Delineation of soil variability using geostatistics and fuzzy clustering analyses of hyperspectral data. *Soil Sci. Soc. Am. J.* **1999**, *63*, 142–150.
35. Vizzari, M.; Santaga, F.; Benincasa, P. Sentinel 2-Based Nitrogen VRT Fertilization in Wheat: Comparison between Traditional and Simple Precision Practices. *Agronomy* **2019**, *9*, 278. [CrossRef]
36. Fridgen, J.J.; Kitchen, N.R.; Sudduth, K.A.; Drummond, S.T.; Wiebold, W.J.; Fraisse, C.W. Management Zone Analyst (MZA) Software for Subfield Management Zone Delineation. *Agron. J.* **2004**, *96*, 100–108.
37. Song, X.; Wang, J.; Huang, W.; Liu, L.; Yan, G.; Pu, R. The delineation of agricultural management zones with high resolution remotely sensed data. *Precis. Agric.* **2009**, *10*, 471–487. [CrossRef]
38. De Benedetto, D.; Castrignanò, A.; Rinaldi, M.; Ruggieri, S.; Santoro, F.; Figorito, B.; Gualano, S.; Diacono, M.; Tamborrino, R. An approach for delineating homogeneous zones by using multi-sensor data. *Geoderma* **2013**, *199*, 117–127. [CrossRef]
39. Yao, R.-J.; Yang, J.-S.; Zhang, T.-J.; Gao, P.; Wang, X.-P.; Hong, L.-Z.; Wang, M.-W. Determination of site-specific management zones using soil physico-chemical properties and crop yields in coastal reclaimed farmland. *Geoderma* **2014**, *232–234*, 381–393. [CrossRef]
40. Farid, H.U.; Bakhsh, A.; Ahmad, N.; Ahmad, A.; Mahmood-Khan, Z. Delineating site-specific management zones for precision agriculture. *J. Agric. Sci.* **2015**, *154*, 273–286. [CrossRef]
41. Peralta, N.R.; Costa, J.L.; Balzarini, M.; Castro Franco, M.; Córdoba, M.; Bullock, D. Delineation of management zones to improve nitrogen management of wheat. *Comput. Electron. Agric.* **2015**, *110*, 103–113. [CrossRef]
42. Basso, B.; Dumont, B.; Cammarano, D.; Pezzuolo, A.; Marinello, F.; Sartori, L. Environmental and economic benefits of variable rate nitrogen fertilization in a nitrate vulnerable zone. *Sci. Total Environ.* **2016**, *545–546*, 227–235. [CrossRef]
43. Gavioli, A.; de Souza, E.G.; Bazzi, C.L.; Guedes, L.P.C.; Schenatto, K. Optimization of management zone delineation by using spatial principal components. *Comput. Electron. Agric.* **2016**, *127*, 302–310. [CrossRef]
44. Oldoni, H.; Silva Terra, V.S.; Timm, L.C.; Júnior, C.R.; Monteiro, A.B. Delineation of management zones in a peach orchard using multivariate and geostatistical analyses. *Soil Tillage Res.* **2019**, *191*, 1–10. [CrossRef]

45. Gavioli, A.; de Souza, E.G.; Bazzi, C.L.; Schenatto, K.; Betzek, N.M. Identification of management zones in precision agriculture: An evaluation of alternative cluster analysis methods. *Biosyst. Eng.* **2019**, *181*, 86–102. [CrossRef]
46. Moharana, P.C.; Jena, R.K.; Pradhan, U.K.; Nogiya, M.; Tailor, B.L.; Singh, R.S.; Singh, S.K. Geostatistical and fuzzy clustering approach for delineation of site-specific management zones and yield-limiting factors in irrigated hot arid environment of India. *Precis. Agric.* **2019**, *21*, 426–448. [CrossRef]
47. Nogueira Martins, R.; Magalhães Valente, D.S.; Fim Rosas, J.T.; Souza Santos, F.; Lima Dos Santos, F.F.; Nascimento, M.; Campana Nascimento, A.C. Site-specific Nutrient Management Zones in Soybean Field Using Multivariate Analysis: An Approach Based on Variable Rate Fertilization. *Commun. Soil Sci. Plant. Anal.* **2020**, *51*, 687–700. [CrossRef]
48. Gotway, C.; Ferguson, R.; Hergert, G. The Effects of Mapping and Scale on Variable-Rate Fertilizer Recommendations for Corn. In Proceedings of the Third International Conference on Precision Agriculture, Minneapolis, MN, USA, 23–26 June 1996; pp. 321–330.
49. Fleming, K.; Westfall, D.; Bausch, W. Evaluating management zone technology and grid soil sampling for variable rate nitrogen application. In Proceedings of the 5th International Conference on Precision Agriculture, Bloomington, MN, USA, 16–19 July 2000; pp. 16–19.
50. Koch, B.; Khosla, R. The Role of Precision Agriculture in Cropping Systems. *J. Crop. Prod.* **2003**, *9*, 361–381. [CrossRef]
51. Cohen, S.; Cohen, Y.; Alchanatis, V.; Levi, O. Combining spectral and spatial information from aerial hyperspectral images for delineating homogenous management zones. *Biosyst. Eng.* **2013**, *114*, 435–443. [CrossRef]
52. Basnyat, P.; McConkey, B.G.; Selles, F.; Meinert, L.B. Effectiveness of using vegetation index to delineate zones of different soil and crop grain production characteristics. *Can. J. Soil Sci.* **2005**, *85*, 319–328. [CrossRef]
53. Oza, S.R.; Panigrahy, S.; Parihar, J.S. Concurrent use of active and passive microwave remote sensing data for monitoring of rice crop. *Int. J. Appl. Earth Obs. Geoinf.* **2008**, *10*, 296–304. [CrossRef]
54. Segarra, J.; Buchaillot, M.L.; Araus, J.L.; Kefauver, S.C. Remote Sensing for Precision Agriculture: Sentinel-2 Improved Features and Applications. *Agronomy* **2020**, *10*, 641. [CrossRef]
55. European Space Agency (ESA). MultiSpectral Instrument (MSI) Overview. Available online: https://sentinels.copernicus.eu/web/sentinel/technical-guides/sentinel-2-msi/msi-instrument (accessed on 20 October 2021).
56. European Space Agency (ESA). Sentinel-2 (MSI) Overview. Available online: https://sentinel.esa.int/web/sentinel/missions/sentinel-2 (accessed on 20 October 2021).
57. European Space Agency (ESA). Data Products (MSI) Overview. Available online: https://sentinels.copernicus.eu/web/sentinel/missions/sentinel-2/data-products (accessed on 20 October 2021).
58. Rouse, J.W.; Haas, R.H.; Schell, J.A.; Deering, D.W. Monitoring vegetation systems in the Great Plains with ERTS. *NASA Spec. Publ.* **1974**, *351*, 309.
59. Weiss, M.; Baret, F. S2ToolBox Level 2 Products: LAI, FAPAR, FCOVER, Version 1.1. Available online: https://step.esa.int/docs/extra/ATBD_S2ToolBox_L2B_V1.1.pdf (accessed on 20 October 2021).
60. Song, W.; Mu, X.; Ruan, G.; Gao, Z.; Li, L.; Yan, G. Estimating fractional vegetation cover and the vegetation index of bare soil and highly dense vegetation with a physically based method. *Int. J. Appl. Earth Obs. Geoinf.* **2017**, *58*, 168–176.
61. Kosior, K. Łańcuch wartości dużych zbiorów danych (big data) w rolnictwie–problemy i wyzwania regulacyjne. *Studia BAS* **2020**, *3*, 101–125. [CrossRef]
62. Santaga, F.S.; Benincasa, P.; Toscano, P.; Antognelli, S.; Ranieri, E.; Vizzari, M. Simplified and Advanced Sentinel-2-Based Precision Nitrogen Management of Wheat. *Agronomy* **2021**, *11*, 1156.
63. Majchrzak, L.; Sawinska, Z.; Natywa, M.; Skrzypczak, G.; Głowicka-Wołoszyn, R. Impact of different tillage systems on soil dehydrogenase activity and spring wheat infection. *J. Agric. Sci. Technol.* **2016**, *18*, 1871–1881.
64. Tryjanowski, P.; Sparks, T.H.; Blecharczyk, A.; Małecka-Jankowiak, I.; Switek, S.; Sawinska, Z. Changing Phenology of Potato and of the Treatment for its Major Pest (Colorado Potato Beetle)—A Long-term Analysis. *Am. J. Potato Res.* **2017**, *95*, 26–32. [CrossRef]
65. Egnér, H.; Riehm, H.; Domingo, W. Untersuchungen über die chemische Bodenanalyse als Grundlage für die Beurteilung des Nährstoffzustandes der Böden. II. *Chem. Extra Ktionsmethoden Zur Phosphor-Und Kaliumbestimmung K. Lantbr. Ann.* **1960**, *26*, 199–215.
66. Schachtschabel, P.V. Das pflanzenverfügbare Magnesium des Boden und seine Bestimmung. *Z. Für Pflanz. Düngung Bodenkd.* **1954**, *67*, 9–23. [CrossRef]
67. Tiurin, I. Kmietodikie analiza dla srawnitielnogo izuczenia sostawa poczwiennogo gumusa. *Tr. Poczw. Inst. AN SSSR* **1951**, *38*, 1–250.
68. Czechlowski, M.; Wojciechowski, T. The utilization of information about local variable environmental conditions to predict the quality of wheat grain during the harvest. *J. Res. Appl. Agric. Eng.* **2013**, *58*, 31–34.
69. Czechlowski, M.; Wojciechowski, T. System mechatroniczny do selektywnego zbioru ziarna zbóż. *Zeszyty Naukowe Instytutu Pojazdów* **2013**, *4*, 95.
70. Czechlowski, M.; Wojciechowski, T.; Adamski, M.; Niedbała, G.; Piekutowska, M. Application of ASG-EUPOS high precision positioning system for cereal harvester monitoring. *J. Res. Appl. Agric. Eng.* **2018**, *63*, 44–50.
71. European Space Agency (ESA). Copernicus Open Access Hub. Available online: https://scihub.copernicus.eu/ (accessed on 20 October 2021).

72. European Space Agency (ESA). STEP-Scientific Toolbox Exploitation Platform Ver 7.0. Available online: https://step.esa.int/main/ (accessed on 20 October 2021).
73. Liang, S. *Quantitative Remote Sensing of Land Surfaces*; John Wiley & Sons Inc.: Hoboken, NJ, USA, 2005; Volume 30.
74. Lillesand, T.; Kiefer, R.W.; Chipman, J. *Remote Sensing and Image Interpretation*; John Wiley & Sons Inc.: Hoboken, NJ, USA, 2015.
75. QGIS, Version 3.10.8. Available online: https://www.qgis.org/en/site/ (accessed on 8 February 2021).
76. Webster, R.; Oliver, M.A. *Geostatistics for Environmental Scientists*; John Wiley & Sons Inc.: Hoboken, NJ, USA, 2007.
77. ArcGIS, Version 10.7. Available online: https://desktop.arcgis.com/en/arcmap/ (accessed on 8 February 2021).
78. Management Zone Analyst (MZA), Version 1.0. Available online: https://www.ars.usda.gov/research/software/download/?softwareid=24&modecode=50-70-10-00 (accessed on 8 February 2021).
79. Odeh, I.; McBratney, A.; Chittleborough, D. Soil pattern recognition with fuzzy-c-means: Application to classification and soil-landform interrelationships. *Soil Sci. Soc. Am. J.* **1992**, *56*, 505–516. [CrossRef]
80. Boydell, B.; McBratney, A. Identifying potential within-field management zones from cotton-yield estimates. *Precis. Agric.* **2002**, *3*, 9–23.
81. Bezdek, J.C. *Pattern Recognition with Fuzzy Objective Function Algorithms*; Springer: Boston, MA, USA, 1981; pp. 43–93.
82. Olofsson, P.; Foody, G.M.; Herold, M.; Stehman, S.V.; Woodcock, C.E.; Wulder, M.A. Good practices for estimating area and assessing accuracy of land change. *Remote. Sens. Environ.* **2014**, *148*, 42–57. [CrossRef]
83. Pińskwar, I.; Choryński, A.; Kundzewicz, Z.W. Severe Drought in the Spring of 2020 in Poland—More of the Same? *Agronomy* **2020**, *10*, 1646. [CrossRef]
84. Warrick, A. Spatial variability of soil physical properties in the field. *Appl. Soil Phys.* **1980**, 319–344. [CrossRef]
85. Cambardella, C.A.; Moorman, T.B.; Novak, J.; Parkin, T.; Karlen, D.; Turco, R.; Konopka, A. Field-scale variability of soil properties in central Iowa soils. *Soil Sci. Soc. Am. J.* **1994**, *58*, 1501–1511. [CrossRef]
86. Reza, S.K.; Nayak, D.C.; Mukhopadhyay, S.; Chattopadhyay, T.; Singh, S.K. Characterizing spatial variability of soil properties in alluvial soils of India using geostatistics and geographical information system. *Arch. Agron. Soil Sci.* **2017**, *63*, 1489–1498. [CrossRef]
87. Giua, C.; Materia, V.C.; Camanzi, L. Management information system adoption at the farm level: Evidence from the literature. *Br. Food J.* **2020**, *123*, 884–909. [CrossRef]

agriculture

MDPI

Article

Estimation of Soil Nutrient Content Using Hyperspectral Data

Yiping Peng [1,†], Lu Wang [1,2,3,4,5,†], Li Zhao [1], Zhenhua Liu [1,3,4,5], Chenjie Lin [1], Yueming Hu [1,2,3,4,5,*] and Luo Liu [1,3,4,5]

1 College of Natural Resources and Environment, South China Agricultural University, Guangzhou 510642, China; pengyp@stu.scau.edu.cn (Y.P.); selinapple@scau.edu.cn (L.W.); zhaoli@stu.scau.edu.cn (L.Z.); zhenhua@scau.edu.cn (Z.L.); linchenjie@stu.scau.edu.cn (C.L.); liuluo@scau.edu.cn (L.L.)
2 College of Tropical Crops, Hainan University, Haikou 570228, China
3 Guangdong Provincial Key Laboratory of Land Use and Consolidation, South China Agricultural University, Guangzhou 510642, China
4 Guangdong Province Engineering Research Center for Land Information Technology, South China Agricultural University, Guangzhou 510642, China
5 Key Laboratory of Construction Land Transformation, Ministry of Land and Resources, South China Agricultural University, Guangzhou 510642, China
* Correspondence: ymhu@scau.edu.cn; Tel.: +86-020-852-88307
† These authors contributed equally to this work.

Abstract: Soil nutrients play a vital role in plant growth and thus the rapid acquisition of soil nutrient content is of great significance for agricultural sustainable development. Hyperspectral remote-sensing techniques allow for the quick monitoring of soil nutrients. However, at present, obtaining accurate estimates proves to be difficult due to the weak spectral features of soil nutrients and the low accuracy of soil nutrient estimation models. This study proposed a new method to improve soil nutrient estimation. Firstly, for obtaining characteristic variables, we employed partial least squares regression (PLSR) fit degree to select an optimal screening algorithm from three algorithms (Pearson correlation coefficient, PCC; least absolute shrinkage and selection operator, LASSO; and gradient boosting decision tree, GBDT). Secondly, linear (multi-linear regression, MLR; ridge regression, RR) and nonlinear (support vector machine, SVM; and back propagation neural network with genetic algorithm optimization, GABP) algorithms with 10-fold cross-validation were implemented to determine the most accurate model for estimating soil total nitrogen (TN), total phosphorus (TP), and total potassium (TK) contents. Finally, the new method was used to map the soil TK content at a regional scale using the soil component spectral variables retrieved by the fully constrained least squares (FCLS) method based on an image from the HuanJing-1A Hyperspectral Imager (HJ-1A HSI) of the Conghua District of Guangzhou, China. The results identified the GBDT-GABP was observed as the most accurate estimation method of soil TN (R^2_{cv} of 0.69, the root mean square error of cross-validation (RMSECV) of 0.35 g kg^{-1} and ratio of performance to interquartile range (RPIQ) of 2.03) and TP (R^2_{cv} of 0.73, RMSECV of 0.30 g kg^{-1} and RPIQ = 2.10), and the LASSO-GABP proved to be optimal for soil TK estimations (R^2_{cv} of 0.82, RMSECV of 3.39 g kg^{-1} and RPIQ = 3.57). Additionally, the highly accurate LASSO-GABP-estimated soil TK (R^2 = 0.79) reveals the feasibility of the LASSO-GABP method to retrieve soil TK content at the regional scale.

Keywords: VIS-NIR spectroscopy; screening algorithm; estimation model; HJ-1A imagery

Citation: Peng, Y.; Wang, L.; Zhao, L.; Liu, Z.; Lin, C.; Hu, Y.; Liu, L. Estimation of Soil Nutrient Content Using Hyperspectral Data. *Agriculture* **2021**, *11*, 1129. https://doi.org/10.3390/agriculture11111129

Academic Editors: Gniewko Niedbała and Sebastian Kujawa

Received: 9 October 2021
Accepted: 9 November 2021
Published: 11 November 2021

1. Introduction

The rapid and efficient monitoring of soil nutrients has become an important prerequisite for agricultural production management and ensuring the healthy development of crops. However, current soil nutrient estimations are often obtained using field sampling and laboratory analysis, which is time-consuming and costly. The monitoring of soil nutrients via hyperspectral remote-sensing techniques is rapid and efficient, and numerous related studies have been performed within the past 30 years [1–4].

Current research on the retrieval of soil nutrients via hyperspectral remote-sensing technology typically focuses on two factors: the determination of characteristic variables and the construction of the estimation model. The determination of suitable characteristic variables ensures high-precision estimations. Statistical analyses (e.g., Pearson correlation coefficient (PCC) and partial least squares regression (PLSR)) are frequently employed to determine these variables [5,6]. For example, Liu et al. (2007) obtained the 620–810 nm characteristic variables of soil organic matter by correlation and multiple regression analyses [5]. Vibhute et al. (2019) determined the characteristic variables of soil nitrogen at 480, 511, 653, 997, 1472, 1795, 2210, and 2296 nm based on correlation analysis [6]. However, statistical variable selection methods in high-dimensional space can fail due to a lack of significance testing and parameter estimations in the model [7]. Following the development of data mining technology, several machine learning algorithms have been introduced to determine the characteristic variables of soil nutrients. Zhang et al. (2019) proposed a method combining mutual information and ant colony optimization to select soil total nitrogen (TN) characteristic bands at 943, 1004, 1097, 1351, 1550, 1710, 2123, and 2254 nm [8]. Despite the great progress made by these studies, determining the characteristic variables remains to be difficult due to the weak spectral responses to soil nutrients. Therefore, additional screening algorithms, particularly machine learning approaches, are required in order to accurately determine the characteristic variables.

Existing relationship models between spectral variables and soil nutrient contents can be classified into two categories; linear and nonlinear models. Linear estimation methods build linear mathematical relationships between spectral variables and soil nutrient contents. Multiple linear regression (MLR) and partial least squares regression (PLSR) are the most commonly used linear estimation methods for soil nutrients [9–11]. However, correlations between spectral variables and soil nutrients are rarely linear in nature [12]. Thus, machine learning models were introduced to solve this problem. The random forest (RF), support vector machine (SVM), and back propagation neural network (BPNN) algorithms are frequently employed to estimate soil nutrients [13–15]. Compared with linear models, nonlinear methods improve on the explanatory power of the spectral changes related to soil nutrients. However, large-scale training samples for SVM approaches are difficult to obtain and implement due to their complexity, huge memory requirements, and extensive computational time in quadratic programming routines [16]. In addition, RF is prone to overfitting in regression models when learning specific details and noise in the training data [17,18]. BPNN is associated with large weights and threshold uncertainties, affecting the estimation accuracy [19,20]. Therefore, there is a great need to determine an optimal algorithm for high accuracy soil nutrient content estimations.

This study has the aim of developing a new method to accurately estimate soil nutrient contents. In order to achieve this aim, we set the following objectives: (1) to determine the optimal screening algorithm from three algorithms (Pearson correlation coefficient, PCC; least absolute shrinkage and selection operator, LASSO; and gradient boosting decision tree, GBDT) for the accurate selection of soil nutrient characteristic variables; (2) to implement the MLR, ridge regression (RR), back propagation neural network with genetic algorithm optimization (GABP), and SVM to determine a high-accuracy model for the estimation of soil nutrient contents; and (3) to apply a high-accuracy method to map the soil nutrient contents at the regional scale using HuanJing-1A Hyperspectral Imager (HJ-1A HSI) imagery. Both the hyperspectral data and HJ-1A HSI images were collected in the Guangdong province and Conghua District of Guangzhou, China.

2. Materials and Methods

2.1. Study Area

Guangdong province, China was selected as the study area in order to build the optimal hyperspectral estimation model of soil nutrients (Figure 1a), while Conghua District within Guangdong was selected to map the soil nutrient contents (Figure 1b). The East-West and North-South spans of Guangdong province are approximately 800 and

600 km, respectively. The province belongs to the East Asian monsoon region, with middle subtropical, south subtropical, and north tropical zone climate types from the north to south. Mean annual temperature and precipitation of the area are 21.8 °C and 1789.3 mm, respectively. Guangdong is an important grain production region, with a crop planting area of 4.28 × 10^4 km^2 in 2019 and total grain yield of 1.19 × 10^{10} kg.

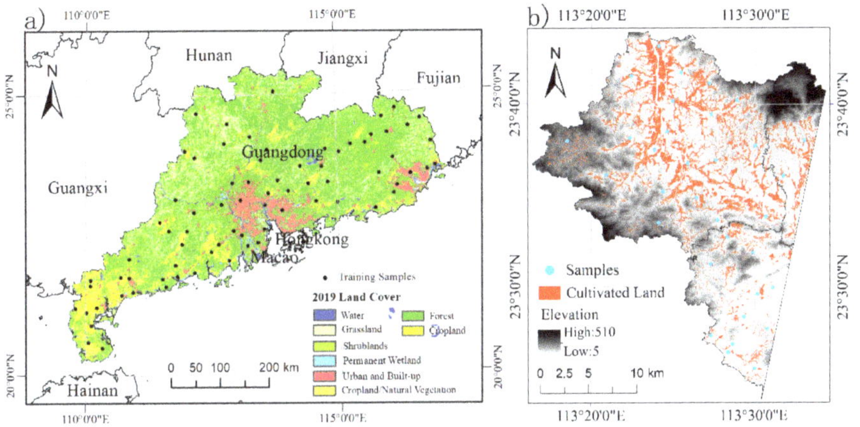

Figure 1. (**a**) MODIS land cover map from the MCD12 product of the study area with a spatial distribution of 75 soil samples; (**b**) the test study area determined from the 2016 Cultivated Land Map planted with rice of the Conghua National Land Department, with a spatial distribution of 33 soil samples used to assess the accuracy of the estimated soil nutrient content.

2.2. *Data and Pre-Processing*

2.2.1. Soil Sampling and Chemical Analysis

A total of 75 soil samples were gathered for constructing hyperspectral estimate models of soil nutrients contents based on a 50 × 50 km sampling grid within Guangdong province and field actual conditions to ensure uniform distribution of the soil samples [21,22]. Surface soil samples (0–20 cm) were collected at five sampling locations at each site. To remove stones and other large debris, the samples were air-dried and sieved through a 2 mm polyethylene sieve. After that, the samples were pulverized into fine powder. The soil nutrient content and soil spectral reflectance were then determined by dividing each sample into two parts. Soil TN was measured using the semi-micro Kjeldahl method described by Walkley and Black [23]. Soil TP and TK were determined via an ultraviolet spectrophotometer (UV-2600, Shimadzu CO, LTD., Kyoto, Japan) and a flame photometer (FP640, INESA Analytical Instrument CO, LTD., Shanghai, China), respectively. The soil nutrient content statistics from the 75 soil samples are presented in Table 1.

Table 1. Statistic information for soil nutrient contents in the study area.

Soil Nutrients	Min	Q1	Median	Q3	Max	Mean	SD	Skewness	Kurtosis	CV
TN	0.21	0.99	1.33	1.70	2.79	1.36	0.57	0.43	0.21	41.91
TP	0.13	0.37	0.59	1.00	3.15	0.75	0.55	1.90	5.21	73.33
TK	0.62	4.75	9.66	16.84	30.39	10.55	7.61	0.61	−0.23	72.13

Note: soil total nitrogen, TN; total phosphorus, TP; and total potassium, TK; unit: g kg^{-1}. Q1, first quartile; Q3, third quartile; SD, standard deviation; CV, coefficient of variation (%).

Moreover, the mapping of Guangdong Province needs multiple HJ-1A images with 100 m spatial resolution. In addition, it is very difficult to obtain multiple high-quality satellite images of the whole province on the same day. Therefore, in this study, Conghua district was selected for conducting the soil nutrient mapping experiment. A total of

33 soil samples were collected in Conghua District (Figure 1b) to verify the feasibility of mapping soil nutrient content at the regional scale. The acquisition time of the HJ-1A HSI image coincided with the collection of the samples, which are evenly distributed in the whole image. The soil sample collection principle and pretreatment are consistent with the Guangdong province samples.

2.2.2. Spectral Measurements and Pre-Processing of Soil Samples

Soil spectral measurements were performed on 75 soil samples collected across the province. An AvaField portable spectrometer (Avantes, Inc., Apeldoorn, Holland) was used to measure soil spectral reflectance, which has a spectral range and resolution of 340–2511 and 0.6 nm, respectively. The spectral measurements were carried out in a dark room to regulate the lighting environment and minimize the influence of stray light. The soil spectral reflectance values were measured using a 50 W halogen lamp with a 10° field of view in vertical contact with the soil sample. Each sample was uniformly tiled on a black cloth and measured five times. The average spectrum was calculated and used in further processing. Prior to the collection of the reflectance readings, the spectrometer was calibrated every three samples with a white Spectralon. To decrease signal noise, we used Savitzky–Golay smoothing with a window size of 10. In addition, the smoothed spectral data (raw spectral, R) were processed with the first derivative (FD), second derivative (SD), and reciprocal logarithmic (RL) to eliminate or reduce the effect of background noise and account for signal intensity fluctuations induced by soil surface spectral scattering and absorption. The outcomes of the processing are shown in Figure 2.

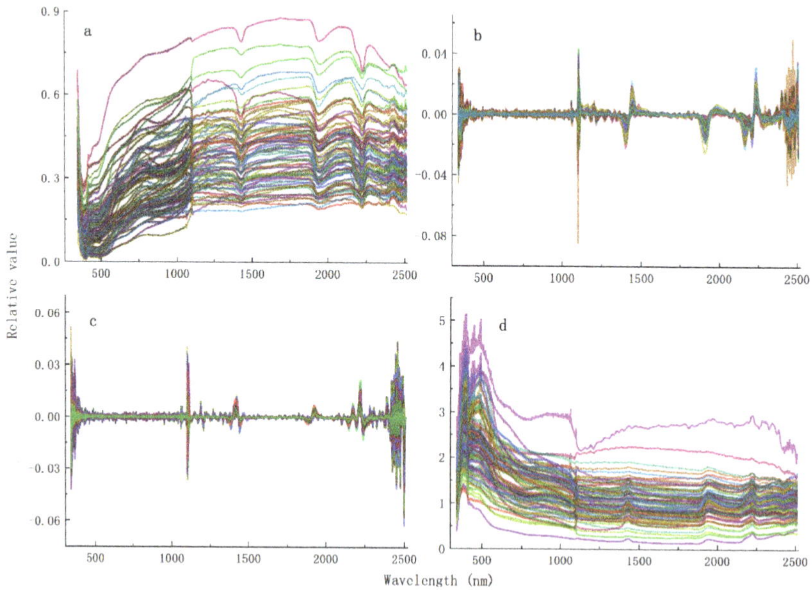

Figure 2. Transformed spectral indices of soil samples: (**a**) raw spectral curves; (**b**) first derivative spectral curves; (**c**) second derivative spectral curves; and (**d**) reciprocal logarithmic spectral curves.

2.2.3. Image Acquisition and Pre-Processing

In order to extend the application of the established model at the regional scale, a HJ-1A image acquired on 30 October 2017 with a 100 m spatial resolution and 115 bands (459–956 nm) was used to map the soil nutrient contents. The image was subjected to radiometric correction, atmospheric correction, geometric precision correction, and stripe noise reduction (Figure 3) using ENVI 5.3 (Exelis Visual Information Solutions, Inc., Boulder,

CO, USA). The image's spectral resolution was 5 nm, which was substantially coarser than the AvaField portable spectrometer's measured spectral interval of 0.6 nm. ENVI's spectral resampling technique was utilized to spectrally resample the measured soil spectral data gathered with the AvaField portable spectrometer in order to match the spectral resolution of the HJ-1A HSI data.

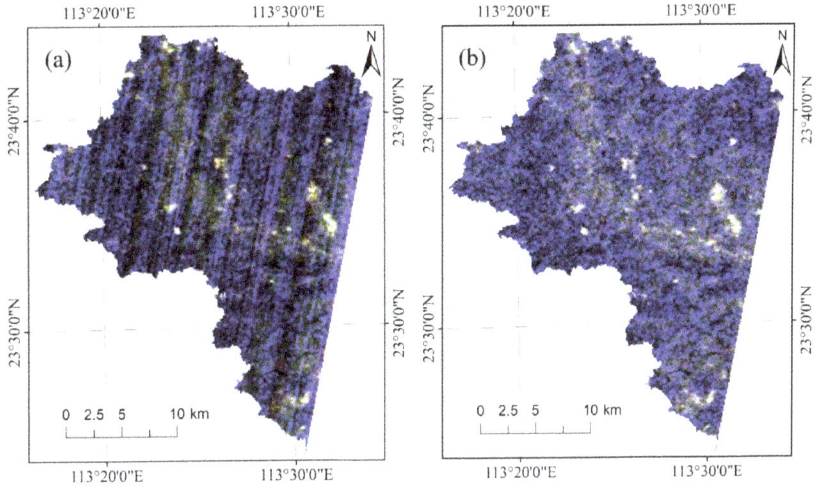

Figure 3. The HJ-1A image: (**a**) untreated and (**b**) stripe noise reduction.

2.3. Methods

This section is organized into four parts. In Section 2.3.1, we describe how the optimal algorithm of screening the soil nutrient characteristic variables can be determined using the PLSR fit degrees. The second section explains how the optimal prediction algorithm for soil nutrients can be screened from four algorithms by their prediction accuracy. In Section 2.3.3, we detail the mapping of soil nutrient base on HJ-1A image data using the above the optimal screening and predicting algorithms. Section 2.3.4 describes accuracy validation methods for the predicting models and mapping.

2.3.1. Determining the Optimal Screening Algorithm of the Characteristic Variables

One of the most important steps in the development of the optimal hyperspectral estimation method of the soil nutrient contents was the determination of the characteristic variables [8,24,25]. Additionally, the determination of the accurate screening algorithms is key for characteristic variables of the soil nutrient content. In order to determine the optimal screening algorithm of the characteristic variables, we compared traditional linear screening algorithm (PCC) and nonlinear screening algorithms (LASSO and GBDT) based on two evaluation steps. First, the characteristic variables were screened using PCC, GBDT and LASSO might be correlated with each other. That is, there are collinearities among the variables. Therefore, the variance inflation factor (VIF) of a stepwise regression was applied to eliminate the collinearity of the selected characteristic variables. The set of variables having a VIF lower than 10 [26] was retained. The three screening algorithms are described in detail as follows:

LASSO: The least absolute shrinkage and selection operator, proposed by Tibshirani (1996), minimizes the sum of squares of residuals under the constraint that the sum of the absolute values of the regression coefficients (penalty coefficient) is less than a pre-defined constant. This produces regression coefficients (RC) strictly equal to 0 and removes low-weight variables and can therefore effectively deal with complex high-dimensional data problems [27]. LASSO can be defined as follows:

$$\arg\min_{B}\left\{\sum_{j=1}^{n} y_i - \sum_{j=1}^{p} x_{ij}B_j\right\}, \quad \text{subject to } \sum_{j=1}^{p}\left|B_j\right| \leq t, \tag{1}$$

where y_i represents the measured spectral data in the ith band; n is the spectral dimensionality; B_j denotes the input weight in the jth spectral sample; x_{ij} is the covariate vector of the ith measured spectral data and j spectral sample; and p is the spectral sample number.

GBDT: The gradient boosting decision tree is a boosting algorithm that calculates the information gain during the branching of the decision tree to determine the spectral variable to be split and the corresponding split value. Once all decision trees are constructed, the feature importance (FI) is obtained by calculating the information gain of the decision tree feature and dividing by the total frequency of the feature in all trees of the GBDT strong learner [28]:

$$\text{FI} = \frac{\sum I(a, D)}{N_a}, \tag{2}$$

where $I(a, D)$ denotes the feature (spectral variable) information gain; a is the feature; D is the soil sample; and N_a is the total frequency of feature a in all trees.

PCC: The Pearson correlation coefficient is commonly employed to screen characteristic variables. Here, the PCC was implemented between the spectral variables and soil nutrient content to determine characteristic variables with the largest correlation coefficient ($p \leq 0.05$ significance level). The Pearson correlation coefficient can be expressed as:

$$\text{r}_i = \frac{\sum_{n=1}^{N}\left(R_{ni} - \overline{R_i}\right)\left(y_n - \overline{y}\right)}{\sqrt{\sum_{n=1}^{N}\left(R_{ni} - \overline{R_i}\right)^2 \sum_{n=1}^{N}\left(y_n - \overline{y}\right)^2}}, \tag{3}$$

where R_{ni} is the spectral value of the ith spectral variable of the nth soil sample point; $\overline{R_i}$ is the average spectral value of the ith spectral variable; y_n is the soil nutrient content of the nth soil sample point; and $\overline{y}$ is the average value of the soil nutrient content.

Once the characteristic variables were selected by each algorithm, PLSR fit degrees (R^2) [29–31] between the measured soil nutrient contents and characteristic variables were compared. The screening algorithm with the maximum fit degree was determined as the optimal.

2.3.2. Determining the Accurate Model for Estimating Soil Nutrients

In this study, the screened characteristic variables were used as independent variables and each of the soil nutrient (TN, TP and TK) contents were used as the dependent variable. Additionally, four different algorithms were applied to build the relationship models between characteristic variables and soil nutrients: MLR, RR, SVM, and GABP. The four algorithms are described in detail as follows:

(1) Multi-Linear Regression

Multi-linear regression is a type of regression analysis for multiple independent variables. The optimal combination of these independent variables is taken to estimate the dependent variables. This model can describe the influence of each variable on the soil properties and is widely used in soil property estimations [32–34]. We adopted MLR to estimate soil nutrient content using the following formula:

$$Z_{MLR} = a_0 + a_1x_1 + a_2x_2 + \cdots + a_nx_n, \tag{4}$$

where Z_{MLR} is the dependent variable (soil nutrient content); x_i $(i = 1, 2, \ldots, n)$ is the independent variable (spectral variables); a_i $(i = 1, 2, \ldots, n)$ represents the regression fitting coefficient; and a_0 is the intercept.

(2) Ridge Regression

Ridge regression (RR) is a least square estimation method that improves on its predecessors. In particular, it abandons the unbiasedness of the least square method, thus losing part of the information and reducing the accuracy and making the regression coefficient more realistic and reliable [35]. The existence of multiple collinear relations between independent variables magnifies the mean square error. This error is reduced by using RR estimation rather than the standard least square estimation [36,37]. The RR is expressed as:

$$\hat{\beta}(k) = (X'X + kI)^{-1}X'Y, \tag{5}$$

where $\hat{\beta}(k)$ is the ridge regression estimate of β; and k is the ridge parameter. When $k = 0$, the least square estimate of β is equal to $\hat{\beta}(0)$.

(3) Support Vector Machine

SVM, proposed by Cortes and Vapnik (1995), is a robust supervised learning model with a capacity for solving practical problems (e.g., nonlinearity and high dimensionality). SVM greatly simplifies the traditional regression process through efficient "transduction inference" from training samples to predictions [38]. The SVM model can be expressed as:

$$\mathrm{f(x)} = w_i \cdot \varnothing_i(x) + b, \tag{6}$$

where $\mathrm{f(x)}$ is the soil nutrient estimate; x is the characteristic variable; w_i is the weight coefficient; b is the error term; $\varnothing_i$ denotes a nonlinear transfer function; and ω and b are calculated by the following convex optimization problem with an e-insensitivity loss function [39]:

$$\min: \frac{1}{2}||w||^2 + \mathrm{C}\sum_i^N (\xi_i + \xi_i^*), \tag{7}$$

$$\text{s.t.} \begin{cases} y_i - w\varnothing(x) - b \le \varepsilon + \xi_i^* \\ w\varnothing(x) + b - y_i \le \varepsilon + \xi_i^* \\ \xi_i, \xi_i^* \ge 0 \end{cases}, (i = 1, \cdots, n), \tag{8}$$

where $||w||^2$ represents the flatness of the m-dimensional space; ε is a parameter that indicates the maximum allowed error between the measured and estimated values; ξ_i and ξ_i^* are slack variables and C is the penalty factor. Equations (7) and (8) belong to the convex quadratic programming problem with inequality constraints. In order to obtain the Lagrangian multipliers, the equations are converted into a dual problem via the Lagrange multiplier method. The constrained original objective function (Equation (8)) is then transformed into the unconstrained Lagrangian objective function:

$$\min: \frac{1}{2}\sum_{\mathrm{i,j=1}}^{\mathrm{n}}(\alpha_i - \alpha_i^*)\left(\alpha_j - \alpha_j^*\right)(\varnothing(x_i)\varnothing(x_j)) + \varepsilon\sum_{\mathrm{i=1}}^{\mathrm{n}}(\alpha_i^* + \alpha_i) - \sum_{\mathrm{i=1}}^{\mathrm{n}} y_i(\alpha_i^* - \alpha_i), \tag{9}$$

$$\text{s.t.} \begin{cases} \sum_{\mathrm{i=1}}^{\mathrm{n}}(\alpha_i - \alpha_i^*) = 0 \\ 0 \le \alpha_i^* \le \mathrm{C}, i = 1, \cdots, n \end{cases}, \tag{10}$$

where $\alpha_i - \alpha_i^*$ is the transformation of w. The SVM function is expressed as:

$$\mathrm{f(x)} = w_i \cdot \varnothing_i(x) + b = \sum_{\mathrm{i=1}}^{\mathrm{n}}(\alpha_i - \alpha_i^*)\mathrm{K}(\mathrm{x_i}, \mathrm{x}) + \mathrm{b}, \tag{11}$$

where $\mathrm{K}\ (\mathrm{x_i}, \mathrm{x}) = \varnothing(x_i)\varnothing(x_j)$ is the kernel function. The radial basis function was selected as the kernel function.

(4) Genetic Algorithm-Back Propagation Neural Network

The GABP algorithm optimizes the structure and connection weight of the back propagation neural network using the parallel random search ability of the genetic algorithm, effectively avoiding a local optimal solution [40]. We adopted the population search method to optimize the weights and thresholds of the neural network (Figure 4).

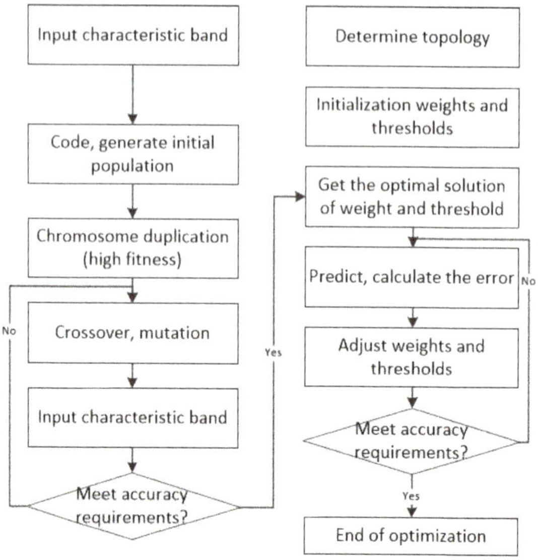

Figure 4. Flow chart for the genetic algorithm-back propagation neural network.

2.3.3. Estimating Regional-Scale Soil Nutrient Contents Using HJ-1A Hyperspectral Data

Once the optimal variable screening and predictor models were selected, the method was applied to mapping the contents of the soil nutrient using HJ-1A image with 115 bands (459–956 nm) and 5 nm spectral resolution, which will not provide the above characteristic variables with beyond 956 nm wavelength. Thus, the characteristic variables should be re-screened from the resampling measured soil spectral data with 5 nm spectral resolution using the above optimal screening algorithm and to develop the corresponding estimation models. Then, the model was applied to mapping the contents of the soil TK using the HJ-1A HSI image for the Conghua district at the regional scale.

Moreover, in order to apply the methods to Conghua district, the HJ-1A image was considered to contain pure pixels. However, the coarse image spatial resolution of 100 × 100 m generally prevents the existence of pure pixels, with mixed pixels (including crop and soil) typically dominating the study area. Thus, the fully constrained least squares (FCLS) method [41] was used to obtained pure pixels and spectral reflectance of soil (Figure 5).

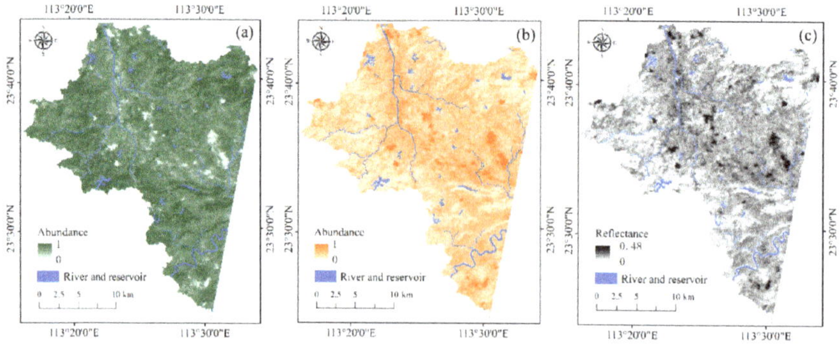

Figure 5. Component decomposition maps of the mixed pixels using FCLS: (**a**) vegetation abundance maps, (**b**) soil abundance maps, and (**c**) spectral reflectance of soil at the 900 nm band of the HJ-1A image.

2.3.4. Accuracy Validation

The coefficient of determination (R^2), concordance correlation coefficient (CCC), ratio of performance to interquartile range (RPIQ), the root mean square error of calibration (RMSEC), and cross-validation (RMSECV) were used as statistical measures to assess the performance of estimation models. The RPIQ is defined as the ratio of IQ to RMSECV [42]. IQ is the interquartile range (IQ = Q3 − Q1) of the observed values. Q1 and Q3 denote the first and third quartile, respectively.

3. Results

3.1. Optimal Algorithm for the Screening of the Characteristic Variables

In order to determine the characteristic variables, the three choosing algorithms (PCC, LASSO, and GBDT) were implemented on 6272 spectral data of the R, FD, SD, and RL (Figure 2) and soil nutrient contents in the 75 sample points collected across the province. Figure 6 illustrates the correlation coefficients of the spectral variables. Stepwise regression with VIF analysis was further used to eliminate the collinearity among the spectral variables screened by the PCC algorithms (Table 2).

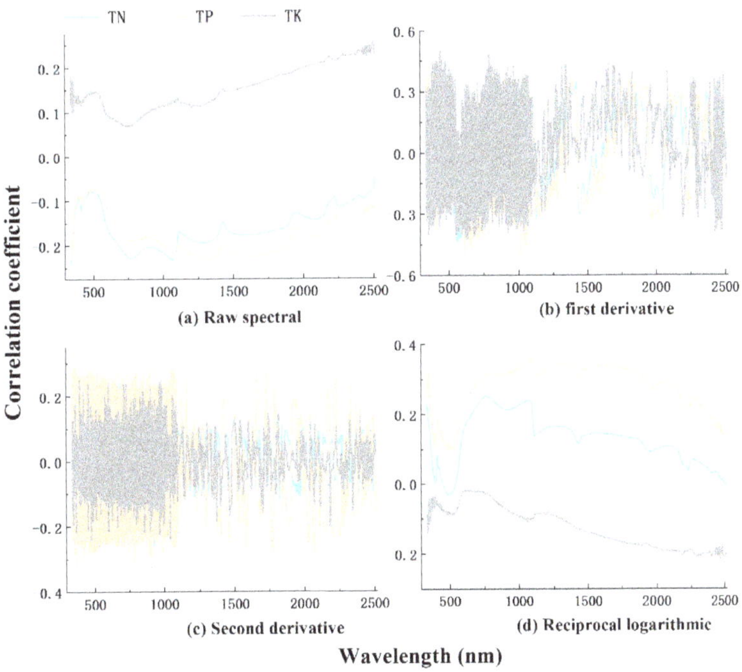

Figure 6. Correlation coefficients between the soil total nitrogen (TN), total phosphorus (TP), and total potassium (TK) concentrations and the various spectral variables.

Table 2. PCC-determined characteristic variables of the three soil nutrients.

Soil Nutrient	Spectral Variables	Correlation Coefficients	VIF
TN	FD_{562}, SD_{714}	−0.44, −0.26	1.70, 1.51
TP	FD_{1009}, FD_{356}, SD_{905}	−0.50, 0.45, −0.32	2.65, 1.32, 1.42
TK	R_{2498}, FD_{442}	0.20, 0.50	1.08, 4.14

Considering the possible existence of a nonlinear relationship between the spectral variables and soil nutrient contents, we introduced the GBDT and LASSO algorithms for the screening task. Numerous experiments were performed, identifying the prediction error of the GBDT algorithm to tend towards stability for screening criteria of soil TN, TP, and TK in the GBDT algorithm equal to FI > 0.015, FI > 0.015, and FI > 0.01, respectively. Figure 7 depicts the feature importance and regression coefficient of the screened spectral variables. For the LASSO algorithm we employed RC $\neq 0$ as the screening criteria. Stepwise regression with VIF analysis was further employed to eliminate the collinearity among the spectral variables screened by the GBDT and LASSO algorithms. Table 3 reports the final results of the screening characteristic variables for the three soil nutrients (TN, TP, and TK).

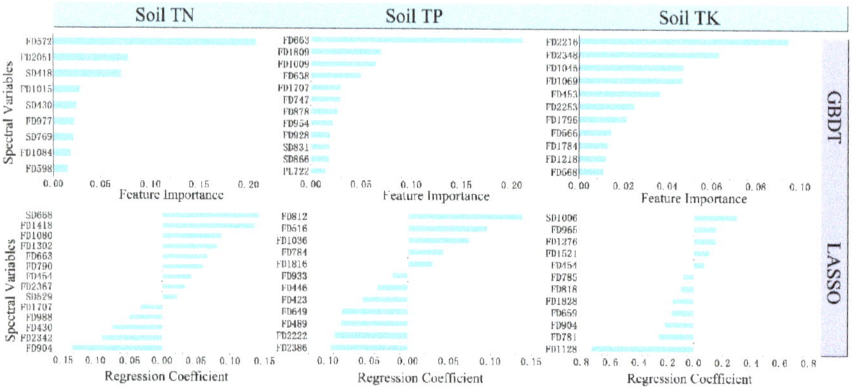

Figure 7. GBDT feature importance and LASSO regression coefficient.

Table 3. GBDT- and LASSO-determined characteristic variables of the three soil nutrients.

Models	Soil Nutrient	Spectral Variables	VIF
LASSO	TN	FD_{454}, FD_{904}, FD_{1302}, FD_{1418}, FD_{1707}, FD_{2342}, FD_{2367}, SD_{529}, SD_{668}	3.14, 2.85, 4.42, 6.35, 3.48, 3.68, 6.44, 3.99, 3.78
	TP	FD_{423}, FD_{489}, FD_{516}, FD_{649}, FD_{1816}, FD_{2222}, FD_{2386}	2.34, 2.48, 2.28, 2.13, 3.01, 1.66, 6.07
	TK	FD_{659}, FD_{904}, FD_{965}, FD_{1128}, FD_{1521}, SD_{1006}	2.89, 4.21, 2.78, 4.40, 3.10, 1.48
GBDT	TN	FD_{572}, FD_{977}, FD_{1084}, FD_{1015}, FD_{2051}, SD_{418}	8.97, 4.56, 1.45, 3.87, 1.37, 3.42
	TP	FD_{663}, FD_{747}, FD_{1009}, SD_{831}	2.68, 3.00, 3.42, 7.45
	TK	FD_{1045}, FD_{1069}, FD_{1784}, FD_{1796}, FD_{2348}	4.42, 2.52, 5.57, 8.53, 6.17

In order determine the most accurate screening algorithm, the PLSR approach was selected to construct the model between soil nutrients and the characteristic variables from the three algorithms based on 75 soil samples from the province. The PLSR relationship models are described as follows:

$$\text{PCC} - \text{PLSP}\begin{cases} Y_N = 1.505 - 224 \times FD_{562} - 9717 \times SD_{714} & \left(R^2 = 0.17\right) \\ Y_P = 1.698 - 1091 \times FD_{1009} + 46 \times FD_{356} + 224 \times SD_{905} & \left(R^2 = 0.35\right) \\ Y_K = 12.91 + 9 \times R_{2498} + 6173 \times FD_{442} & \left(R^2 = 0.40\right) \end{cases}$$

$$\text{LASSO}-\text{PLSP}\begin{cases} Y_N = 1.530 + 4928 \times SD_{668} + 38 \times FD_{1418} + 53 \times FD_{1302} + 786 \times FD_{454} + 26 \\ \quad \times FD_{2367} - 4079 \times SD_{529} - 37 \times FD_{1707} + 32 \times FD_{2342} - 157 \times FD_{904} & \left(R^2 = 0.15\right) \\ Y_P = 0.646 + 1362 \times FD_{516} + 159 \times FD_{1816} - 414 \times FD_{423} - 57 \times FD_{649} \\ \quad - 121 \times FD_{489} - 50 \times FD_{2222} + 67 \times FD_{2386} & \left(R^2 = 0.15\right) \\ Y_K = 8.871 + 290,321 \times SD_{1006} + 15,335 \times FD_{965} + 2839 \times FD_{1521} - 11,052 \\ \quad \times FD_{659} - 15,362 \times FD_{904} - 1372 \times FD_{1128} & \left(R^2 = 0.47\right) \end{cases}$$

$$\text{GBDT}-\text{PLSP}\begin{cases} Y_N = 1.908 - 366 \times FD_{572} - 265 \times FD_{2051} + 237 \times FD_{1084} - 1086 \times SD_{418} \\ \quad -61 \times FD_{977} + 364 \times FD_{1015} & \left(R^2 = 0.26\right) \\ Y_P = 1.657 - 504 \times FD_{663} - 1232 \times FD_{1009} + 915 \times FD_{747} - 4946 \times SD_{831} & \left(R^2 = 0.37\right) \\ Y_K = 19.188 - 2601 \times FD_{2348} - 7380 \times FD_{1045} - 14,742 \times FD_{1069} \\ \quad +1605 \times FD_{1796} + 1207 \times FD_{1784} & \left(R^2 = 0.24\right) \end{cases}$$

The results identify the optimal algorithms of soil TN, TP, and TK as GBDT, GBDT, and LASSO, with R^2 values of 0.26, 0.37, and 0.47, respectively. Among three nutrients, the LASSO-PLSR showed the best estimation of soil TK.

3.2. Determining the Optimal Model for Soil Nutrient Content Estimations

The MLR, RR, SVM, and GABP models were adopted to determine the relationship between the characteristic variables and soil nutrients (Figure 8). The GABP model exhibited the highest predicative capability for the three soil nutrients, with scatter plots closer to the 1:1 line compared to MLR, RR, and SVM. Additionally, it offered the most accurate estimates in cross-validation with R^2_{cv} of 0.69, RMSECV of 0.35, and RPIQ = 2.03 for TN; R^2_{cv} of 0.73, RMSECV of 0.30 and RPIQ = 2.10 for TP, R^2_{cv} of 0.82, RMSECV of 3.39, and RPIQ = 3.57 for TK, respectively (Table 4). The prediction effect of soil TK is obviously better than that of TN and TP, which may be due to potassium being a metal element, with a spectral response sensitivity that exceeds other non-metal elements (e.g., nitrogen and phosphorus).

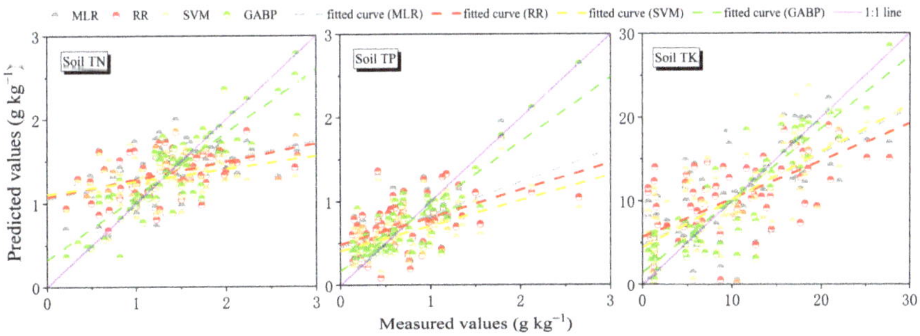

Figure 8. Scatter plots of measured and estimated values.

Table 4. Accuracy assessment of estimated soil nutrient contents (unit: g kg^{-1}).

Soil Nutrients	Model	R^2 (C)	CCC	RMSEC	R^2_{cv}	RMSECV	RPIQ
TN	MLR	0.22	0.37	0.50	0.17	0.51	1.39
	RR	0.21	0.35	0.50	0.18	0.51	1.39
	SVM	0.13	0.26	0.53	0.11	0.57	1.25
	GABP	0.76	0.86	0.28	0.69	0.35	2.03
TP	MLR	0.36	0.55	0.40	0.32	0.47	1.34
	RR	0.34	0.47	0.43	0.33	0.44	1.43
	SVM	0.36	0.49	0.41	0.35	0.41	1.54
	GABP	0.77	0.87	0.26	0.73	0.30	2.10
TK	MLR	0.48	0.67	5.30	0.42	5.52	2.19
	RR	0.44	0.61	5.32	0.43	5.33	2.27
	SVM	0.54	0.72	5.17	0.52	5.31	2.28
	GABP	0.86	0.92	2.88	0.82	3.39	3.57

3.3. Mapping Soil Nutrient Contents Using the Proposed Method

Table 4 demonstrates the soil TK estimation accuracy exceeding that of TN and TP. Therefore, we applied the proposed method to map soil TK contents in Conghua District at the regional scale using HJ-1A imagery because the spectral wavelength of HJ-1A data ranged from 459 to 956 nm, which had different range and spectral bands of wavelengths from those of the spectral variables involved in the above estimation models. The model based on 75 sample points collected across the province could not be utilized for the HJ-1A images. We employed the LASSO-GABP method to re-screen the optimal spectral variables from the resample measured soil spectral data with 5 nm spectral resolution and to develop the corresponding estimation models. The screened spectral variables were determined as band462, band464, band466, band470, band477, band484, band574, and band652. The soil TK was estimated with reliable accuracy (R^2 of 0.82, RMSEC of 3.28 g kg^{-1}; Figure 9).

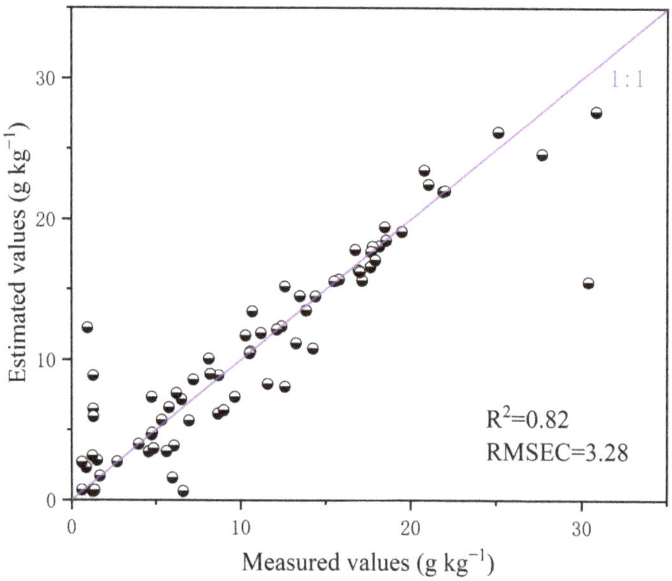

Figure 9. Scatter plots of the measured and estimated values based on resample measured soil spectral data.

Figure 10 demonstrates the spatial distribution of the soil TK contents obtained using the estimation model. The soil TK content is generally concentrated within 10–20 g kg^{-1}, with flat areas exhibiting a higher content and areas with high slopes and close proximity to rivers associated with lower content. This may be linked to soil erosion, which is consistent with the actual situation.

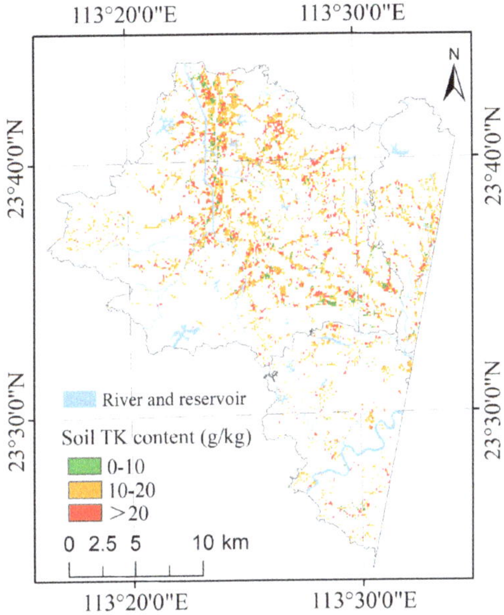

Figure 10. Spatial distribution of the soil total potassium content for the study area.

The 33 sample plots (Figure 1b) were used to verify the feasibility of mapping soil nutrient content by calculating the R^2, RMSE, and RPIQ values (Table 5). The estimation accuracy of soil TK content was relatively high, with an R^2 of 0.79 and RMSE of 4.01 g kg^{-1}. This indicates that the GABP model is capable of mapping the soil TK content. However, the estimation accuracy of the regional-scale retrievals is lower than that of the point-scale. This may be due to the limitation of the narrow spectral region of the HJ-1A HSI data (450–960 nm).

Table 5. Estimation accuracy of soil total potassium content using the GABP model based on the 33 validation sample plots (unit: g kg^{-1}).

	Dataset	Mean	Max	Min	SD	R^2	RMSE	RPIQ
Soil TK	Measured Value	18.35	30.57	2.64	6.67	0.79	4.01	1.86
	Estimated Value	20.01	36.42	1.36	8.86			

4. Discussion

In the current paper we compared three algorithms (PCC, LASSO, and GBDT) and four models (MLR, RR, SVM, and GABP) in terms of soil nutrient estimations in order to determine a method for the prediction of high-accuracy soil nutrients.

In this method, to the best of our knowledge, this is the first attempt to use the LASSO and GBDT algorithms to determine the characteristic variables for soil nutrient estimations. LASSO with PLSR fit degree (R^2) of 0.47 was determined as optimal for the accurate selection of soil TK characteristic variables, and GBDT for TN and TP with R^2 of 0.26 and 0.37. This indicates the significant nonlinear spectral response mechanism of the soil nutrients. The result found that 16 characteristic variables obtained using the optimal screening algorithms are sensitive to soil nutrients: FD_{572}, FD_{977}, FD_{1084}, FD_{1015}, FD_{2051}, SD_{418} for TN, FD_{663}, FD_{747}, FD_{1009}, SD_{831} for TP and FD_{659}, FD_{904}, FD_{965}, FD_{1128}, FD_{1521}, SD_{1006} for TK. Some selected wavelengths are in general agreement with previous research [43–45].

Previous studies generally employ linear models to estimate soil nutrients [2,46–48]. In order to improve the estimation accuracy of soil nutrients, we adopted linear (MLR and RR) and nonlinear (SVM and GABP) algorithms to construct the soil nutrient estimation models based on the determined spectral characteristic variables (Table 3). The validation results (Table 4) revealed the GBDT-GABP algorithm to perform the best in soil TN (R^2_{cv} of 0.69, RMSECV of 0.35, and RPIQ = 2.03) and TP (R^2_{cv} of 0.73, RMSECV of 0.30, and RPIQ = 2.10) estimations, while LASSO-GABP was optimal for soil TK (R^2_{cv} of 0.82, RMSECV of 3.39, and RPIQ = 3.57), which are in general agreement with previous research results with R^2 from 0.56 to 0.84 (TN), 0.65 to 0.81 (TP), and 0.67 to 0.82 (TK) [43,49–54]. The proposed model constructed using machine learning algorithms outperformed the linear models. This indicates the existence of a significant nonlinear relationship between the soil nutrients and spectral characteristic variables.

In order to validate the regional-scale applicability of the new method, HJ-1A image data obtained from pure pixels using the fully constrained least squares (FCLS) method was used to map soil TK with the best estimation accuracy (R^2 = 0.86) on point scale. Results using the 33 validation sample plots demonstrate the screened spectral characteristic variables to explain 79% of the variance in the TK content, with an RMSE of 4.01 g kg^{-1} for the mapping of TK content. This indicates the great potential of GABP to map the soil TK content at a large scale. However, the point-scale estimation accuracy is higher than that of the regional-scale due to the narrow spectral range of the HJ-1A HSI data. Future research will map the TK contents using satellite hyperspectral images covering a wider spectral region (350–2500 nm).

The prediction effect of soil TK is obviously better than that of TN and TP (Table 3). This may be because potassium is a metal element, with a spectral response sensitivity that exceeds other non-metal elements (e.g., nitrogen and phosphorus). The introduction of additional soil elements (including metals and nonmetals) to explore this phenomenon will be the focus of further work.

We employed 75 soil samples to develop the models and validate the method for the whole Guangdong province, while 33 sample plots were used to verify the feasibility of mapping soil nutrient content in Conghua District. Although the sampling design was conducted based on different soil characteristics and soil types, the sample sizes were relatively small. Future studies will employ larger sample sizes to further develop and validate the proposed method.

5. Conclusions

The determination of characteristic variables is key for accurate hyperspectral estimation models of the soil nutrient content. This paper introduced the LASSO and GBDT algorithms to screen the optimal relevant characteristic variables of soil TN, TP, and TK. The estimation models of soil nutrient content were subsequently developed using the selected characteristic variables and field observations of soil nutrient content. The most accurate estimation model was then adopted to explore the possibility of spatially mapping the soil nutrient content using HJ-1A data. The results demonstrated that compared with the statistical analysis method, the machine learning method effectively screened the characteristic variables. In addition, based on the RMSECV values, the GABP models of the soil nutrient contents determined the most accurate estimates at the soil sample point level. The new method provides the potential for soil nutrient mapping at the regional scale with a reasonable accuracy using hyperspectral imagery. Results indicate the ability of the LASSO and GBDT algorithms to improve the estimation accuracy of soil TN, TP, and TK, which are crucial for agricultural management. The proposed machine learning method has the potential to effectively select the spectral characteristic indices of soil nutrients, increasing the accuracy of the results.

Author Contributions: Conceptualization, Y.P. and L.W.; methodology, Y.P.; software, C.L.; validation, Y.P., L.Z. and L.L.; investigation, Y.P.; resources, L.W. and Y.H.; data curation, Y.P., L.W. and L.Z.; writing—original draft preparation, Y.P.; writing—review and editing, Y.P., L.W. and Z.L.; funding acquisition, Y.H. and L.W. All authors have read and agreed to the published version of the manuscript.

Funding: This research was funded by the National Key Research and Development Program of China (No. 2020YFD1100203), National Natural Science Foundation of China (No. U1901601), and Guangdong Province Agricultural Science and Technology Innovation and Promotion Project (No. 2021KJ102).

Data Availability Statement: Not applicable.

Acknowledgments: We gratefully acknowledge the paper writing assistance of Mingbang Zhu as well as the experimental assistance of Ziqing Xia.

Conflicts of Interest: The authors declare no conflict of interest.

References

1. Jacquemoud, S.; Baret, F.; Hanocq, J.F. Modeling spectral and bidirectional soil reflectance. *Remote Sens. Environ.* **1992**, *41*, 123–132. [CrossRef]
2. Yu, X.; Liu, Q.; Wang, Y.B.; Liu, X.Y.; Liu, X. Evaluation of MLSR and PLSR for estimating soil element contents using visible/near-infrared spectroscopy in apple orchards on the Jiaodong peninsula. *Catena* **2016**, *137*, 340–349. [CrossRef]
3. Rossel, R.A.V.; Walvoort, D.J.J.; Mcbratney, A.B.; Janik, L.J.; Skjemstad, J.O. Visible, near infrared, mid infrared or combined diffuse reflectance spectroscopy for simultaneous assessment of various soil properties. *Geoderma* **2006**, *131*, 59–75. [CrossRef]
4. An, X.F.; Li, M.Z.; Zheng, L.H.; Liu, Y.M.; Sun, H. A portable soil nitrogen detector based on NIRS. *Precis. Agric.* **2014**, *15*, 3–16. [CrossRef]
5. Liu, H.J.; Zhang, B.; Zhao, J.; Zhang, X.Y.; Song, K.S.; Wang, Z.M.; Duan, H.T. Spectral models for prediction of organic matter in black soil. *Acta Pedol. Sin.* **2007**, *44*, 27–32.
6. Vibhute, A.D.; Kale, K.V.; Gaikwad, S.V.; Dhumal, R.K. Estimation of soil nitrogen in agricultural regions by VNIR reflectance spectroscopy. *SN Appl. Sci.* **2020**, *2*, 1523. [CrossRef]
7. Liu, Q.; Qin, Z.G.; Luo, X.C.; Cheng, H.R. *Summary of Feature Selection Methods in Statistical Machine Learning*; China National Computer Congress: Tianjin, China, 2009.
8. Zhang, Y.; Li, M.Z.; Zheng, L.H.; Qin, Q.M.; Lee, W.S. Spectral features extraction for estimation of soil total nitrogen content based on modified ant colony optimization algorithm. *Geoderma* **2019**, *333*, 23–34. [CrossRef]
9. Casa, R.; Castaldi, F.; Pascucci, S.; Basso, B.; Pignatti, S. Geophysical and Hyperspectral Data Fusion Techniques for In-Field Estimation of Soil Properties. *Vadose Zone J.* **2013**, *12*, vzj2012.0201. [CrossRef]
10. Cao, F.X.; Yang, Z.J.; Ren, J.C.; Jiang, M.Y.; Ling, W.K. Linear vs Nonlinear Extreme Learning Machine for Spectral-Spatial Classification of Hyperspectral Image. *Sensors* **2017**, *17*, 2603. [CrossRef]
11. Leone, A.P.; Viscarra-Rossel, R.A.; Amenta, P.; Buondonno, A. Prediction of Soil Properties with PLSR and vis-NIR Spectroscopy: Application to Mediterranean Soils from Southern Italy. *Curr. Anal. Chem.* **2012**, *8*, 283–299. [CrossRef]
12. Song, Y.Q.; Xin, Z.; Su, H.Y.; Li, B.; Hu, Y.M.; Cui, X.S. Predicting Spatial Variations in Soil Nutrients with Hyperspectral Remote Sensing at Regional Scale. *Sensors* **2018**, *18*, 3086. [CrossRef]
13. Mouazen, A.M.; Kuang, B.; De Baerdemaeker, J.; Ramon, H. Comparison among principal component, partial least squares and back propagation neural network analyses for accuracy of measurement of selectedsoil properties with visible and near infrared spectroscopy. *Geoderma* **2010**, *158*, 23–31. [CrossRef]
14. Peng, X.T.; Shi, T.Z.; Song, A.H.; Chen, Y.Y.; Gao, W.X. Estimating soil organic carbon using VIS/NIR spectroscopy with SVMR and SPA methods. *Remote Sens.* **2014**, *6*, 2699–2717. [CrossRef]
15. Moura-Bueno, J.M.; Dalmolin, R.S.D.; Caten, A.T.; Dotto, A.C.; Demattê, J.A.M. Stratification of a local VIS-NIR-SWIR spectral library by homogeneity criteria yields more accurate soil organic carbon predictions. *Geoderma* **2019**, *337*, 565–581. [CrossRef]
16. Tang, F.; Chen, M.; Wang, Z. New approach to training support vector machine. *J. Syst. Eng. Electron* **2006**, *17*, 200–219.
17. Cutler, D.R.; Edwards, T.C.; Beard, K.H.; Cutler, A.; Hess, K.T.; Gibson, J.; Lawler, J.J. Random forests for classification in ecology. *Ecology* **2007**, *88*, 2783–2793. [CrossRef] [PubMed]
18. Ma, L.; Chen, C.; Shen, Y.; Wu, L.H.; Huang, Z.L.; Gao, H.L. Determinants of tree survival at local scale in a sub-tropical forest. *Ecol. Res.* **2014**, *29*, 69–80. [CrossRef]
19. Wen, J. *Research of Neural Network Theory and Application*; Southwest Jiaotong University Press: Chengdu, China, 1996.
20. Wang, F.; Gao, J.; Zha, Y. Hyperspectral sensing of heavy metals in soil and vegetation: Feasibility and challenges. *ISPRS J. Photogramm. Remote Sens.* **2018**, *136*, 73–84. [CrossRef]

21. Zhang, S.M.; Xu, M.X.; Zhang, Z.X.; Li, B.B. Methods of sampling soil organic carbon in farmlands with different landform types on the Loess Plateau. *J. Nat. Resour.* **2018**, *33*, 634–643.
22. Yang, J.Y.; Tang, S.; Yun, W.J.; Zhang, C.; Zhu, D.H.; Chen, Y.Q. Sampling method for monitoring classification of cultivated land in county area based on Kriging estimation error. *Trans. CSAE* **2013**, *29*, 223–230.
23. Walkley, A.J.; Black, C.A. An estimation of the Degtjareff method for determining soil organic matter and a proposed modification of the chromic acid titration method. *Soil Sci.* **1934**, *37*, 29–38. [CrossRef]
24. Petropoulos, G.P.; Arvanitis, K.; Sigrimis, N. Hyperion hyperspectral imagery analysis combined with machine learning classifiers for land use/cover mapping. *Expert Syst. Appl.* **2012**, *39*, 3800–3809. [CrossRef]
25. Sorol, N.; Arancibia, E.; Bortolato, S.A.; Olivieri, A.C. Visible/near infrared-partial least-squares analysis of Brix in sugar cane juice: A test field for variable selection methods. *Chemom. Intell. Lab. Syst.* **2010**, *102*, 100–109. [CrossRef]
26. Allouis, T.; Durrieu, S.; Véga, V.; Couteron, P. Stem Volume and Above-Ground Biomass Estimation of Individual Pine Trees from LiDAR Data: Contribution of Full-Waveform Signals. *IEEE J. Sel. Top. Appl. Earth Obs. Remote Sens.* **2013**, *6*, 924–934. [CrossRef]
27. Tibshirani, R.J. Regression shrinkage and selection via the lasso. *J. R. Stat. Soc. Ser. B* **1996**, *58*, 267–288. [CrossRef]
28. Xu, W.Q.; Ning, L.K.; Luo, Y. Wind Speed Forecast Based on Post-Processing of Numerical Weather Predictions Using a Gradient Boosting Decision Tree Algorithm. *Atmosphere* **2020**, *11*, 738. [CrossRef]
29. Wold, H. Nonlinear Estimation by Iterative Least Squares Procedure. *Res. Pap. Stat.* **1966**, 441–444.
30. Wold, S.; Sjöström, M.; Eriksson, L. PLS-regression: A basic tool of chemometrics. *Chemometr. Intell. Lab.* **2001**, *58*, 109–130. [CrossRef]
31. Ni, W.C. Discussion on the definition of R^2 equivalence. *Stat. Decis.* **2009**, *09*, 141–142.
32. Zhang, H.M.; Liu, W.; Han, W.T.; Liu, Q.Z.; Song, R.J.; Hou, G.H. Inversion of Summer Maize Leaf Area Index Based on Gradient Boosting Decision Tree Algorithm. *Trans. Chin. Soc. Agric. Mach.* **2019**, *50*, 251–259.
33. Wang, F.; Shi, Z.; Biswas, A.; Yang, S.T.; Ding, J.L. Multi-algorithm comparison for predicting soil salinity. *Geoderma* **2020**, *365*, 114211. [CrossRef]
34. Fathololoumi, S.; Vaezi, A.R.; Alavipanah, S.K.; Ghorbani, A.; Biswas, A. Comparison of spectral and spatial-based approaches for mapping the local variation of soil moisture in a semi-arid mountainous area. *Sci. Total Environ.* **2020**, *724*, 138319. [CrossRef]
35. Zhang, Z.T.; Wang, H.F.; Karnieli, A.; Chen, J.Y.; Han, W.T. Inversion of Soil Moisture Content from Hyperspectra Based on Ridge Regression. *Trans. Chin. Soc. Agric. Mach.* **2018**, *49*, 240–248.
36. Kennard, H.R.W. Ridge Regression: Applications to Nonorthogonal Problems. *Technometrics* **1970**, *12*, 69–82.
37. Hernandez, J.; Lobos, G.A.; Matus, I.; Del Pozo, A.; Silva, P.; Galleguillos, M. Using Ridge Regression Models to Estimate Grain Yield from Field Spectral Data in Bread Wheat (Triticum Aestivum L.) Grown under Three Water Regimes. *Remote Sens.* **2015**, *7*, 2109–2126. [CrossRef]
38. Wang, X.; Wang, Z.Q.; Jin, G.; Yang, J. Land reserve prediction using different kernel based support vector regression. *Trans. Chin. Soc. Agric. Eng.* **2014**, *30*, 204–211.
39. Cortes, C.; Vapnik, V. Support-vector networks. *Mach. Learn.* **1995**, *20*, 273–297. [CrossRef]
40. Saleh, S.M.; Ibrahim, K.H.; Magdi Eiteba, M.B. Study of genetic algorithm performance through design of multi-step LC compensator for time-varying nonlinear loads. *Appl. Soft Comput.* **2016**, *48*, 535–545. [CrossRef]
41. Xie, H.; Luo, X.; Xu, X.; Pan, H.; Tong, X. Automated Subpixel Surface Water Mapping from Heterogeneous Urban Environments Using Landsat 8 OLI Imagery. *Remote Sens.* **2016**, *8*, 584. [CrossRef]
42. Hermansen, C.; Norgaard, T.; Jonge, L.; Moldrup, P.; Müller, K.; Knadel, M. Predicting glyphosate sorption across New Zealand pastoral soils using basic soil properties or Vis–NIR spectroscopy. *Geoderma* **2020**, *360*, 114009. [CrossRef]
43. Jia, S.; Li, H.; Wang, Y.; Tong, R.; Li, Q. Hyperspectral Imaging Analysis for the Classification of Soil Types and the Determination of Soil Total Nitrogen. *Sensors* **2017**, *17*, 2252. [CrossRef]
44. Kawamura, K.; Tsujimoto, Y.; Rabenarivo, M.; Asai, H.; Andriamananjara, A.; Rakotoson, T. Vis-NIR Spectroscopy and PLS Regression with Waveband Selection for Estimating the Total C and N of Paddy Soils in Madagascar. *Remote Sens.* **2017**, *9*, 1081. [CrossRef]
45. Lin, C.; Ma, R.H.; Zhu, Q.; Li, J.T. Using hyper-spectral indices to detect soil phosphorus concentration for various land use patterns. *Environ. Monit. Assess.* **2015**, *187*, 4130. [CrossRef]
46. Yang, M.H.; Mouazen, A.; Zhao, X.M.; Guo, X. Assessment of a soil fertility index using visible and near-infrared spectroscopy in the rice paddy region of southern China. *Eur. J. Soil Sci.* **2020**, *71*, 615–626. [CrossRef]
47. Baldock, J.A.; Beare, M.H.; Curtin, D.; Hawke, B. Stocks, composition and vulnerability to loss of soil organic carbon predicted using mid-infrared spectroscopy. *Soil Res.* **2018**, *56*, 468–480. [CrossRef]
48. Munnaf, M.A.; Guerrero, A.; Nawar, S.; Haesaert, G.; Meirvenne, M.V.; Mouazen, A.M. A combined data mining approach for on-line prediction of key soil quality indicators by Vis-NIR spectroscopy. *Soil Tillage Res.* **2021**, *205*, 104808. [CrossRef]
49. An, X.F.; Zheng, L.H.; Li, M.Z. Real-Time Analysis of Soil Total Nitrogen and Soil Total Phosphorus with NIR Spectroscopy. *Sens. Lett.* **2010**, *8*, 163–166. [CrossRef]
50. Udelhoven, T.; Emmerling, C.; Jarmer, T. Quantitative analysis of soil chemical properties with diffuse reflectance spectrometry and partial least-square regression: A feasibility study. *Plant Soil* **2003**, *251*, 319–329. [CrossRef]
51. Shi, T.Z.; Cui, L.J.; Wang, J.J.; Fei, T.; Chen, Y.Y.; Wu, G.F. Comparison of multivariate methods for estimating soil total nitrogen with visible/near-infrared spectroscopy. *Plant Soil* **2013**, *366*, 363–375. [CrossRef]

According to the literature review, there is no report on forecasting fennel EOY% and *trans*-anethole yield (TAY%) using ANN methods, and predicting these traits based on the characteristics and parameters affecting them using an effective ANN method seems necessary to facilitate the process of breeding such complex traits. Therefore, the aims of the current project were to (1) model and predict the most important ingredient in the essential oil of fennel using an artificial neural network, (2) compare the predicted results with the results of conventional regression-based method and (3) determine the most important selection criteria for EOY% and TAY% in different fennel populations using sensitivity analysis. The developed model could be helpful to improve the TAY% and EOY% of different fennel subspecies/varieties.

2. Materials and Methods

2.1. Plant Material Source and Recorded Traits

In the present study, the seeds of 16 populations of fennel (*Foeniculum vulgare* Mill. var. *vulgare*) from Iran, 2 populations from Turkey and 2 populations from Germany (Table 1) were collected and cultured in the experimental field section at the Faculty of Agriculture, University of Tabriz, Iran (46°17′ N, 37°5′ E), with an altitude of 710 m during 2017–2018. Plant authentication was performed by using the voucher specimens (Table 1) available at the herbarium of the University of Tabriz, Tabriz, Iran. The soil texture was loamy clay with a pH value of 7.5 and less than 1% organic matter. The field was not under cultivation for any plant during the past year. The seeds of each population were sown in a plot (4 m × 0.5 m) as a randomized complete block design (RCBD) with three replications. The fertilizers used in this study were 70 kg N, 40 kg P and 25 kg K per hectare. Essential oil yield (EOY%), *trans*-anethole yield (TAY%), and different agro-morphological traits including number of days to germination (NDG), number of days to 50% flowering (NDF50%), number of days to 100% flowering (NDF100%), number of days to maturity (NDM), initial plant height (IPH) (plant height at the time of the first inflorescence emergence), final plant height (FPH) (plant height at harvest time), number of stems (NS), stem diameter (SD), number of internodes (NI), length of the first internode (LFI), length of the longest internode (LLOI), length of the last internode (LLAI), length of the peduncle (LP), number of umbels (NU), biomass per square meter (B/m^2), 1000-seed weight (TSW), seed yield per plant (SY/P), seed yield per square meter (SY/m^2), number of seeds per plant (NS/P), number of seeds per umbel (NS/U) and harvest index (HI) were randomly recorded from 15 plants per plot. An analysis of variance (ANOVA) was conducted to assess the significant statistical differences among evaluated fennel populations for the studied characteristics. A normality test was conducted with SAS software before the analysis of variance. The means of significant differences of traits (average of two years) were used for the statistical analysis (Table 2).

2.2. Isolation of Essential Oils and GC/MS Analysis

For the isolation of essential oils, 100 g of mature seeds of each population were subjected to hydro-distillation using a Clevenger-type apparatus for 3 h and the collected essential oil was dried over anhydrous sodium sulfate and kept at 4 °C until analysis. Some physical characteristics of mature seed including moisture content and average length, thickness and density of selected seeds were equal to 8%, 5.5 mm, 1.6 mm and 410 Kg/m^3, respectively. Chemical compositions of essential oils were analyzed by an Agilent 7890A Network GC system pooled with Agilent 5975C Network with Triple-Axis mass detector. The GC analysis was carried out on the Agilent 7890A Network GC system equipped with a splitless model injector (with 1.0 µm volume and 250 °C temperature). The carrier gas was helium with a flow rate of 1.1 mL/min and the capillary column used was HP 5-MS (30 m × 0.25 mm, film thickness 0.25 µm). The column pressure was fixed to 56,054.38 Pa. The oven temperature was initially kept at 50 °C for 2 min after injection and then increased to 250 °C with a rate of 6 °C/min heating ramp and kept constant at 250 °C for 4 min. The ionization voltage and mass range were 70 eV and 34–500 m/z, respectively. The temperatures 280 °C and 250 °C were used as anion source and interface temperatures, respectively. Constituents of the essential oils were recognized based on their retention time and mass spectra pattern with related available data or with the Wiley library and literature. Percentages of each compound were calculated from the given GC peak area and these data were used for quantification purposes.

Table 1. Locality and average of essential oil yield and *trans*-anethole yield of studied fennel populations.

No	Population	Variety	Locality	Voucher Number	Latitude (N)	Longitude (E)	Essential Oil Yield (%)	*trans*-Anethole Yield (%)
1	Salzland	*Vulgare*	Germany	Ah123	51°78′	11°77′	2.29 ± 0.85	1.46 ± 0.59
2	Gotha	*Vulgare*	Germany	Ah115	51°07′	10°87′	2.14 ± 0.73	1.77 ± 0.57
3	Gazianetp	*Vulgare*	Turkey	Ah114	37°05′	37°37′	2.67 ± 0.79	2.30 ± 0.68
4	Izmir	*Vulgare*	Turkey	Ah113	38°35′	27°07′	1.63 ± 0.41	1.17 ± 0.32
5	Bonab	*Vulgare*	Iran	Ah111	37°35′	46°03′	2.89 ± 1.03	2.24 ± 0.91
6	Birjand	*Vulgare*	Iran	Ah110	32°84′	59°18′	0.73 ± 0.24	0.54 ± 0.16
7	Tatmaj	*Vulgare*	Iran	Ah126	33°69′	51°62′	1.88 ± 0.79	1.50 ± 0.54
8	Torbatejam	*Vulgare*	Iran	Ah127	35°23′	60°66′	2.90 ± 0.92	2.45 ± 0.76
9	Meshkinshahr	*Vulgare*	Iran	Ah120	38°37′	47°69′	2.30 ± 0.61	1.70 ± 0.62
10	Khorobiabanak	*Vulgare*	Iran	Ah118	33°89′	54°87′	0.99 ± 0.41	0.73 ± 0.51
11	Moghan	*Vulgare*	Iran	Ah121	39°62′	47°87′	4.12 ± 1.32	2.68 ± 0.68
12	Ziar	*Vulgare*	Iran	Ah129	32°50′	51°94′	1.66 ± 0.69	1.14 ± 0.42
13	Shirvan	*Vulgare*	Iran	Ah124	37°39′	57°96′	2.42 ± 0.86	1.77 ± 0.68
14	Karaj	*Vulgare*	Iran	Ah116	35°77′	51°06′	1.54 ± 0.64	1.07 ± 0.44
15	Kerman	*Vulgare*	Iran	Ah117	30°30′	57°13′	0.82 ± 0.33	0.55 ± 0.19
16	Khorramabad	*Vulgare*	Iran	Ah119	33°48′	48°44′	2.89 ± 0.67	2.10 ± 0.71
17	Neishabour	*Vulgare*	Iran	Ah122	36°19′	58°83′	2.02 ± 0.77	0.33 ± 0.15
18	Varamin	*Vulgare*	Iran	Ah128	35°34′	51°62′	3.77 ± 0.94	3.12 ± 0.67
19	Hamedan	*Vulgare*	Iran	Ah112	34°81′	48°48′	2.92 ± 0.91	2.16 ± 0.56
20	Tabriz	*Vulgare*	Iran	Ah125	38°07′	46°08′	2.50 ± 0.76	0.50 ± 0.12

Table 2. Descriptive statistics of morphological, phonological and yield-related characteristics in the fennel populations.

Characteristic	Abbreviation	Min	Max	Mean	Standard Deviation
Number of days to germination	NDG	7	18	12.45	4.21
Number of days to 50% flowering	NDF50%	59	102	79.18	16.56
Number of days to 100% flowering	NDF100%	82	114	92.36	24.98
Number of days to maturity	NDM	126	180	145.68	36.15
Initial plant height (cm)	IPH	39.47	82.89	58.29	17.14
Final plant height (cm)	FPH	68.56	198	107.25	36.84
Number of stems	NS	1	4	2.58	1.25
Stem diameter (cm)	SD	2.75	15.85	8.54	3.62
Number of internodes	NI	6	14	9.35	3.48
Length of the first internode (cm)	LFI	3.11	9.32	6.13	2.04
Length of the longest internode (cm)	LLOI	5.48	19.14	14.59	5.74
Length of the last internode (cm)	LLAI	2.36	11.71	7.64	1.91
Length of the peduncle (cm)	LP	4.85	14.29	9.25	4.89
Number of umbels	NU	12	58	36.25	15.70
Biomass (g/m^2)	B/m^2	654.25	1457.83	124.91	62.25
Thousand seed weight (g)	TSW	2.85	7.65	5.16	3.11
Seed yield per plant (g)	SY/P	12.35	86.54	32.67	10.29
Seed yield (g/m^2)	SY/m^2	115.12	542.28	315.21	82.27
Number of seeds per plant	NS/P	985	9153	7859	2141
Number of seeds per umbel	NS/U	112	276	192.51	78.29
Harvest index (%)	HI	12.11	46.82	37.26	16.28

2.3. Data Processing and Statistical Analysis

2.3.1. Input Variables Selection

To ensure a realistic model, only a portion of independent variables was carefully chosen. At the beginning of this process, the relationship between the dependent and independent variables was found using Pearson's correlation. The important input variables were selected based on the stepwise regression (SWR) analysis results. Pearson correlation and SWR analyses were conducted using SAS® software.

2.3.2. Multiple Linear Regression

Multiple linear regression (MLR) (stepwise method) was used and for each dependent variable (EOY% and TAY%), the desired models were fixed. All values for the independent variables X (Table 2) are related to the dependent value of the variable Y. The general equation is as follows (Equation (1)):

$$y_i = \beta_0 + \beta_{1\times 1} + \beta_{2\times 2} + \ldots + \beta_n x_n + \varepsilon_i \tag{1}$$

where y_i is EOY% or TAY%, β_0 + β_n are coefficients of regression, $x_1 - x_n$ are input variables and ε is an error associated with the ith observation. Stepwise regression was applied to estimate the MLR coefficients. The MLR analysis was carried out using SAS® software.

2.3.3. Artificial Neural Network

The computation of ANN was conducted using Neuro-Solutions software version 5.07 software package from NeuroDimension Inc. (http://www.Neurosolutions.com, accessed on 22 November 2021). The variables including EOY% and TAY% were used as the dependent and the other traits were defined as independent variables. Two different ANN models were established for each EOY% and TAY%. The training and testing of ANN and MLR were carried out based on the recorded traits from 15 samples of each of the 20 fennel populations over two years. Therefore, the field experiment dataset was based on the 15 samples of 20 fennel populations. All the data were randomly divided into three subsets: 65% for training, 20% for network test and 15% for validation. This classification was based on (i) the results of previous studies with the same number of data, and (ii) trial and error and comparing the modeling results with different ratios. Descriptive statistical analysis of the measured traits in two years is shown in Table 2.

For the efficient ANN analysis and to avoid bias estimation due to differences in units of input variables, all data were normalized and transferred into values between −1 and +1 for hyperbolic tangent and 0 and 1 for sigmoid transfer functions [25] using Equation (2).

$$x_{norm} = \left[\left(\left(\frac{x_i - x_{min}}{x_{max} - x_{min}}\right)\right) \times 0.8\right] + 1 \tag{2}$$

where x_i is the original data, x_{norm} is the normalized input or output values and x_{max} and x_{min} are the maximum and minimum values of the resultant variable, respectively.

The three main input, hidden and output layers are essential for building the topology of a neural network system. In this study, the output of the network is assumed by Equation (3).

$$y_t = \alpha_0 + \sum_{j=1}^{n} \alpha_j f\left(\sum_{i=1}^{m} \beta_{ij} y_{t-1} + \beta_{0j}\right) + \varepsilon_t \tag{3}$$

where y_t is the network output (essential oil), n and m are the number of hidden nodes and number of input nodes, respectively, and f shows the transfer function. β_{ij} $\{i = 1, 2, \ldots, m;$ $j = 0, 1, \ldots, n\}$ are the weights from the input to hidden nodes, α_j $\{j = 0, 1, \ldots, n\}$ are the vectors of weights from the hidden to the output nodes and α_0 and β_{0j} denote the weights of arcs leading from the bias terms, which always are equal to 1.

The feed-forward multilayer perceptron (MLP) architecture with three layers and a Back-Propagation (BP) training algorithm along with the Levenberg–Marquardt, Momentum and Conjugate Gradient learning algorithms were applied in the present study. The most appropriate topology in various numbers of hidden layers (1–4) and neurons related to each layer was determined by trial and error tests. Several activation functions including Tangent Hyperbolic Axon, Linear Tangent Hyperbolic Axon, Sigmoid Axon and Linear Sigmoid Axon were tested with the aim of finding the equation with high capability in both hidden and output layers.

2.4. Performance and Sensitivity Analysis

Three statistical quality parameters, mean absolute error (MAE), root mean square error (RMSE) and coefficient of determination (R^2), were used to compare the performance of the developed ANN with different transfer functions and hidden layers and MLR models for estimating the desired output of EOY% and TAY% according to Equations (4)–(6), respectively.

$$\mathrm{MAE} = \frac{1}{n}\sum_{i,j=1}^{n}|y_i - y_j| \tag{4}$$

$$\mathrm{RMSE} = \sqrt{\frac{\sum_{i,j=1}^{n}(y_i - y_j)^2}{n}} \tag{5}$$

$$\mathrm{R}^2 = \frac{\sum_{i,j=1}^{n}(y_i - \overline{y_i})(y_j - \overline{y_j})}{\sqrt{\sum_{i=1}^{n}(y_i - \overline{y_i})^2 \sum_{j=1}^{n}(y_j - \overline{y_j})^2}} \tag{6}$$

where n is the number of data, y_i is the observed values, y_j represents the predicted values and the bars denote the mean of the variable. High values of R^2 and low values of RMSE and MAE indicate the better performance of the ANN and MLR model.

After developing the final ANN model, a sensitivity test was applied for choosing the most influential input variables on the EOY% and TAY% as the outputs. For this, the dataset was run without any input variables (i.e., SY, NS, NDM, TSW, NU and NI), and the models' performance was assessed using R^2, RMSE and MAE. Neuro-Solutions software (version 5.0) was used for the ANN model developing, evaluating and sensitivity analysis.

3. Results and Discussions

3.1. Selection of Input Variables

Since the input variables have a significant effect on the weighted coefficients and the final architecture of the model, the selection of these variables is a very important step for the development of the model [35]. To this end, Pearson's correlation coefficient was used to consider the relationship between dependent variables (EOY% and TAY%) and the other characteristics (as the independent variables) (Figures 1 and 2). EOY% had the highest positive correlation with HI (R = 0.794), followed by SY/m^2 (R = 0.756), NU (R = 0.754), SY/P (R = 0.732), NDM (R = 0.706), NS/P (R = 0.699), LFI (R = 0.676) and NS/U (R = 0.467). A negative significant correlation coefficient was observed between EOY% and NDF50% (R = −0.842), NDF100% (R = −0.659), NI (R = −0.719), LLOI (R = −0.713), FPH (R = −0.661), LP (R = −0.616) and SP (R = −0.518) (Figure 1). Rahimmalek et al. [41] also reported a negative significant correlation between essential oil yield with plant height and flowering date of Iranian fennel accessions. Overall, the results of the correlation analysis showed that HI, SY/m^2, NU, SY/P and NS/P are the most important parameters to determine essential oil yield in fennel populations. As reported by Bahmani et al. [11,22] and Cosges [15], there is a significant correlation between the essential oil content of fennel and length of the peduncle, stem diameter, plant height, the weight of dry biomass, number of nodes, number of leaves, length of middle internodes, number of inflorescences and 1000-seed weight [11,15,22].

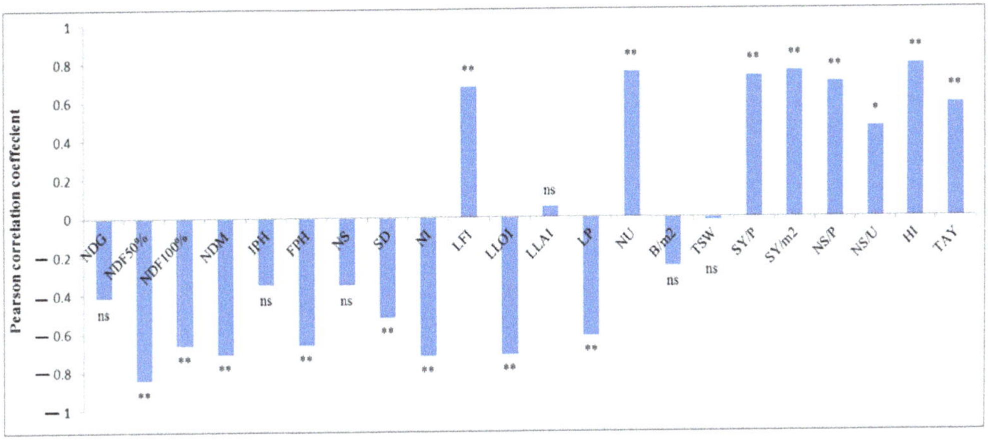

Figure 1. The Pearson correlation coefficient of input variables with EOY% of fennel populations in both ANN and MLR models. For detailed information on trait abbreviations, see Table 2. ns, * and **, non-significant, significant at the 0.05 and 0.01 probability level, respectively.

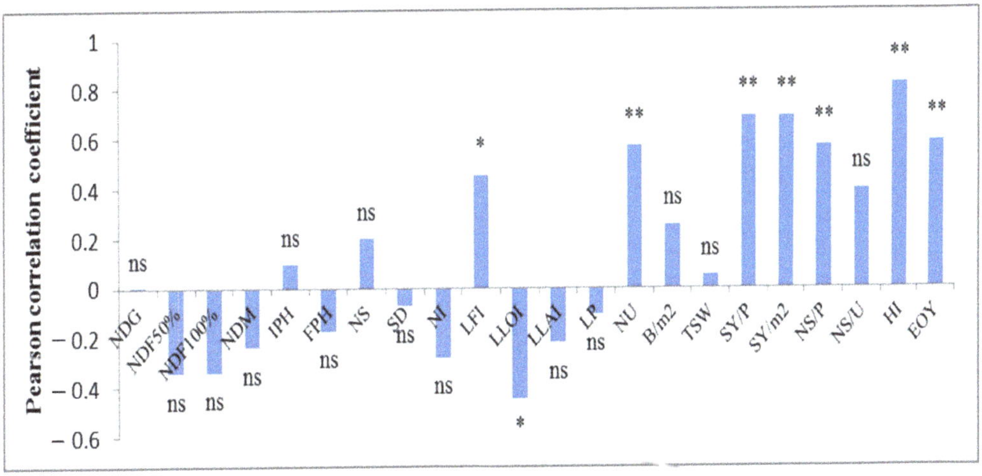

Figure 2. The Pearson correlation coefficient of input variables with TAY% of fennel populations in both ANN and MLR models. ns, * and **, non-significant, significant at the 0.05 and 0.01 probability level, respectively.

TAY% (as a second dependent variable) had the strongest positive correlation with HI (R = 0.823) followed by SY/P (R = 0.693), SY/m^2 (R = 0.693), EOY% (R = 0.590), NS/P (R = 0.573), NU (R = 0.572) and LFI (R = 0.456), as well as a negative significant correlation with LLOI (Figure 2). The correlation between various traits can be positively or negatively affected by other variables and these low coefficients can significantly reduce the capability of the correlation analysis to select the input variables [42,43]. However, there are parameters than other morphological and yield components that affect the oil yield and *trans*-anethole content of fennel. The correlation of climatic data (temperature) with oil yield and *trans*-anethole content of Iranian fennel accessions was assessed and a negative correlation between oil yield and T_{max} and a positive correlation between *trans*-anethole and Tmax (r = 0.459) were reported [41]. These results indicate the importance of environmental parameters in assessing the correlation analysis of fennel populations. Incorporating such data into the model can increase the decision-making power and accuracy of the predictive model. In addition to perform the correlation analysis, stepwise regression (SWR) analysis was employed in this study to optimize the number of input variables [43].

Based on the SWR (Tables 3 and 4), nine traits out of eleven (SY/m^2, NDF 50%, NS/P, NU, FPH, NS/U, NDM, TSW and NI) were entered into the models as the most suitable input variables (the dependent variables were EOY% and TAY%). The number of umbels is an important yield component characteristic that can affect grain yield and subsequently the essential oil yield of fennel populations. This characteristic affect both EOY% and TAY% of the evaluated fennel population of the present study. These results are consistent with the results of Sefidan et al. [44] and Kalleli et al. [45].

Table 3. Stepwise regression analysis for essential oil yield as dependent variable.

Step	Entered Variables in Model	Partial R^2	Model R^2
1	SY/m^2	0.1642	0.1642
2	SY/m^2, NDF 50%	0.1415	0.3057
3	SY/m^2, NDF 50%, NS/P	0.1276	0.4333
4	SY/m^2, NDF 50%, NS/P, NU	0.0781	0.5114
5	SY/m^2, NDF 50%, NS/P, NU, FPH	0.0742	0.5856

Adjusted R^2 = 0.5533.

Table 4. Stepwise regression analysis for trans anethole yield as the dependent variable.

Step	Entered Variables in Model	Partial R^2	Model R^2
1	SY/m$_2$	0.123	0.123
2	SY/m^2, NS/U	0.114	0.237
3	SY/m^2, NS/U, NDM	0.1056	0.3426
4	SY/m^2, NS/U, NDM, TSW	0.0561	0.3987
5	SY/m^2, NS/U, NDM, TSW, NU	0.052	0.4507
6	SY/m^2, NS/U, NDM, TSW, NU, NI	0.0441	0.4948

Adjusted R^2 = 0.4672.

Although HI and LFI were significantly correlated with EOY% and TAY%, according to the SWR analysis, these parameters could not be recognized as appropriate input variables (Tables 3 and 4). MLR results revealed the weakness of the correlation method, which can be due to the indirect positive and negative effects of other traits on the correlation between dependent traits (EOY% and TAY%) and other independent traits [43]. Bahmani et al. [11] also applied SWR to identify to most important characteristics affecting the essential oil content of Iranian fennel and reported low partial R^2 values for inserted variables in the model (partial R^2 = 0.32, 0.06, 0.03, 0.02 and 0.02 for number of leaves, length of peduncle, plant height and days to 50% flowering, respectively). The low estimated partial R^2 values for all independent variables indicate the insufficient efficiency of the linear regression model in interpreting the relationships between independent and dependent variables. Therefore, a non-linear model is needed to better interpret these relationships.

3.2. *Prediction of Dependent Variables Using MLP/ANN Model*

Several factors as numbers of the hidden layer (s) and their neurons (nods) and determination of the transfer function are important for the selection of input variables. They can be determined using trial and error [46]. There are four different transfer functions to running supervised neural networks, namely Sigmoid Axon, Linear Sigmoid Axon, Tangent Hyperbolic Axon and Linear Tangent Hyperbolic Axon, to select the proper transfer function (Table 5).

As presented in Table 6, the lowest values of MAE and RMSE and the highest R^2 values were obtained by the Sigmoid Axon function in both training and testing stages for the prediction of EOY%, as well as the TanhAxon transfer function to predict TAY%.

Table 5. Summary of the components of the neural networks used to predict essential oil and *trans*-anethole yield of fennel populations.

ANN Method	Number of Hidden Layers	Number of Neurons in Each Hidden Layer	Transfer Function	Learning Algorithm	Number of Epochs
Multi-layer perceptron (MLP)	1–5	1–20	Sigmoid Axon Linear Sigmoid Axon TanhAxon Liner TanhAxon	Levenberg–Marquardt Momentum Conjugate Gradient	50–2000

Table 6. The performance of the best artificial neural network models to predict outputs.

Output	Network Structure	Transfer Function	Learning Algorithm	Training			Testing			Cross Validation		
				R^{2} [a]	RMSE [b]	MAE [c]	R^{2} [a]	RMSE [b]	MAE [c]	R^{2} [a]	RMSE [b]	MAE [c]
Essential oil yield	11-9-7-1	Sigmoid Axon	Levenberg–Marquardt (LM)	0.953	0.522	0.375	0.929	0.544	0.385	0.904	0.552	0.389
Trans-anethole yield	11-10-1	TanhAxon	Momentum	0.794	0.246	0.334	0.777	0.264	0.352	0.764	0.258	0.359

[a]: Determination coefficient; [b]: root mean square error; [c]: mean absolute error.

The results of the linear transfer functions are not presented to maintain the clarity of the useful and applied results. In the testing and training phases, linear transfer functions reduced the efficiency of the models. These functions apply a simple linear conversion to the processed input variables and transfer it to the output stage, whereas nonlinear and TanhAxon functions produce outputs in the ranges 0 to 1 and −1 to 1, respectively [25]. Similar to the present study, Sigmoid Axon and TanhAxon functions have been applied in previous studies [24,25,32,35]. This is probably due to the high capability of these functions to justify nonlinear changes compared to the other functions. Various ANN models were implemented by selected transfer functions, with 1–5 hidden layers and 1–20 nodes per layer. As shown in Table 6, for the prediction of EOY%, the ANN model with two hidden layers provided the best performances in the training phases (R^2 = 0.953, RMSE = 0.522, and MAE = 0.375) and testing (R^2 = 0.929, RMSE = 0.544 and MAE = 0.385). Therefore, the results of Tables 5 and 6 reveal that the best EOY% (essential oil) predictive model consisted of an input layer with 11 input variables (NDF50%, NDF100%, NDM, FPH, NI, NU, SY/P, SY/m^2, NS/P, NS/U, and TSW) and two hidden layers with nine and seven neurons in each layer, i.e., the 11-9-7-1 structure (Figure 3). The TanhAxon transfer function, Momentum learning algorithm and one hidden layer (with 11-10-1 structure) were the best parameters in the ANN model to predict TAY% of fennel (Table 6). This topology had the minimum amounts of RMSE and MAE and the highest coefficient of determination (Table 6). Levenberg–Marquardt back-propagation and Logsig and Tansig transfer functions for hidden and output layers algorithm and the number of 10 neurons in the hidden layer have been reported as best parameters of an ANN for the modeling and optimization of anethole ultrasound-assisted extraction from fennel seeds [47]. One of the main objectives of ANN modeling studies is to achieve a simple model with the least number of hidden layers and neurons and the highest performance values [30,32]. Niazian et al. [25] reported an ANN model with a 4-4-1 structure, for the prediction of grain yield in ajowan (*Trachyspermum ammi* L.) belonging to the *Apiaceae* family [25]. These results will be useful to fit an excellent model structure of ANN in future research on the *Apiaceae* family.

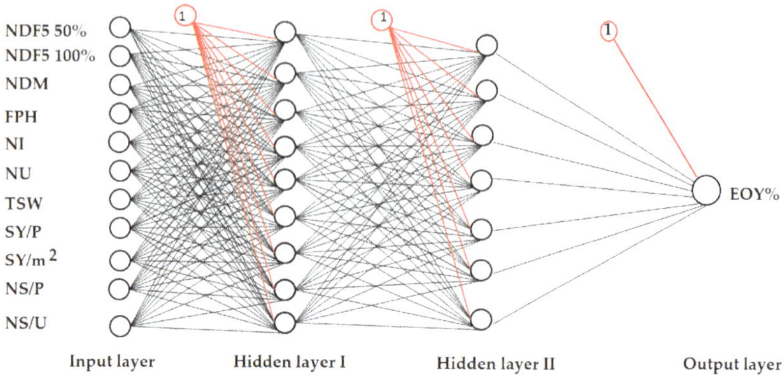

Figure 3. The topology of the ANN with 11 input parameters for prediction of essential oil yield.

An insufficient number of epochs can decrease the ANN performance and too many epochs can increase the risk of network overtraining and subsequent memorization [25]. To minimize the over-training and memorization, a pretest using two hidden layers and various numbers of epochs (50–2000) was conducted and the convergence point between training and validation was considered as the completion of training time to avoid over-training (Figure 4).

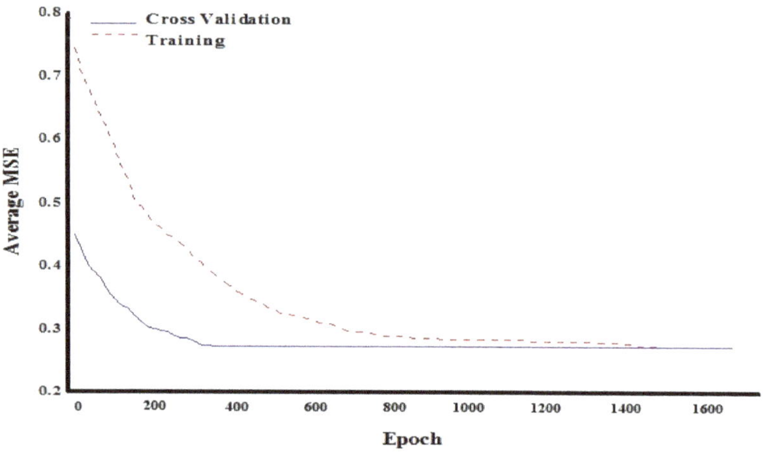

Figure 4. The convergence of the average MSE value during training and validation of the final 11-9-7-1 ANN structure to predict the essential oil yield of fennel populations.

The comparison of predicted and measured EOY% values is shown in Figures 5 and 6 for both training and testing datasets in the form of scatter plots. As shown in the scatter plots, the measured data and the ANN model had the same distribution. The EOY% values predicted by the ANN model tended to follow the corresponding actual ones quite closely.

A scatter plot was also applied to compare observed and predicted values of TAY% from the ANN model in both training and testing datasets. The ability of the ANN model to predict TAY% in training (R^2 = 0.794) and testing (R^2 = 0.777) stages are shown in Figures 7 and 8, respectively. According to the scatter plot, there was no significant difference between predicted data and measured data of TAY% in the ANN model in both training and testing datasets (Figures 7 and 8).

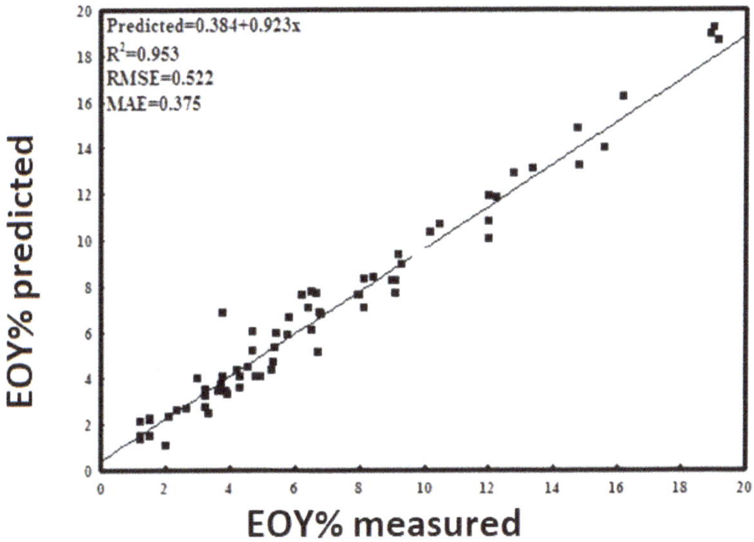

Figure 5. Scatter plot of measured and predicted essential oil yield of fennel populations in the training stage of ANN.

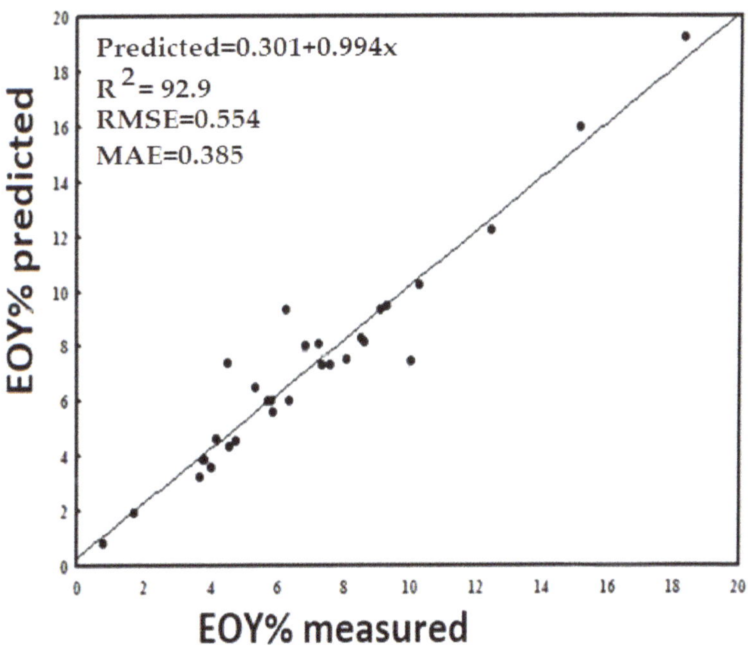

Figure 6. Scatter plot of measured and predicted essential oil yield of fennel populations in the testing stage of ANN.

3.3. Comparing MLR and ANN Models to Predict EOY% and TAY% of Fennel Populations

The MLR models, especially when there are linear relationships between the input and output variables are known as efficient modeling approaches [48]. In order to determine the strength of linear regression to predict EOY% and TAY%, two stepwise regression of

MLR models were determined and the following equations (Equations (7) and (8)) were computed to predict dependent variables.

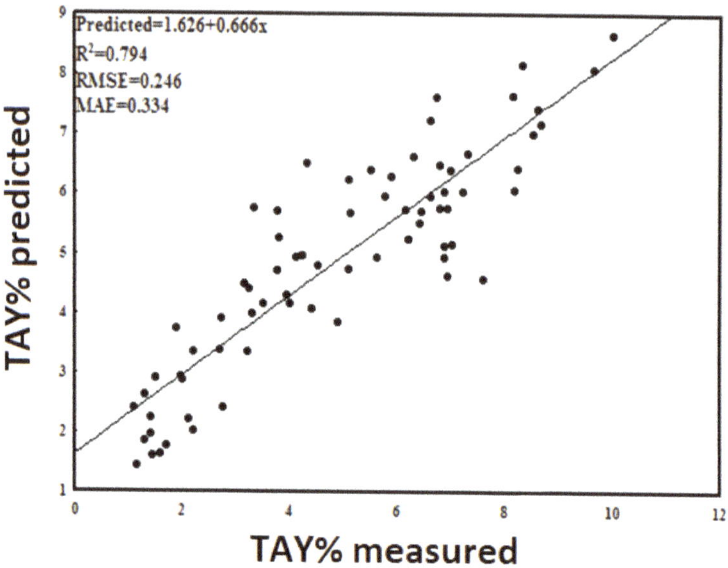

Figure 7. Scatter plot of measured and predicted *trans*-anethole yield of fennel populations in the training stage of the ANN.

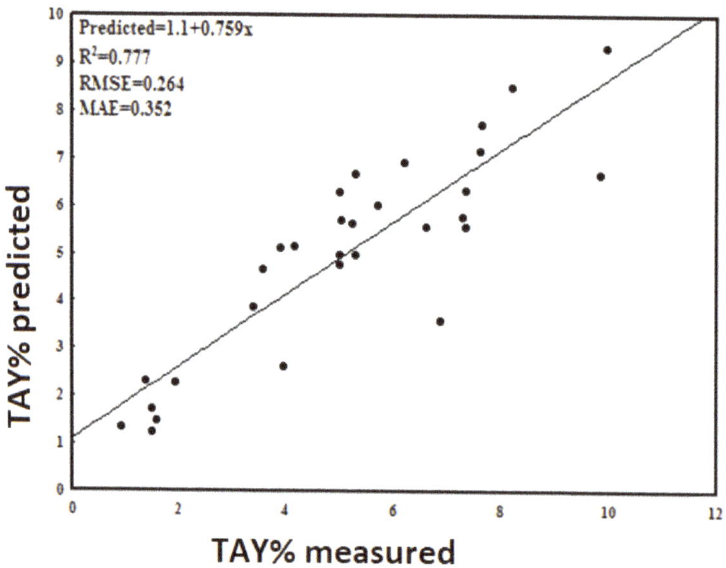

Figure 8. Scatter plot of measured and predicted *trans*-anethole yield of fennel populations in the testing stage of the ANN.

$$EOY\% = 0.037 + 0.216\ SY/m^2 + 0.113\ NDF\ 50\% + 0.126\ NS/P + 0.086\ NU + 0.079\ FPH \quad (7)$$

$$TAY\% = 0.915 + 0.196\ SY/m^2 + 0.108\ NS/U + 0.089\ NDM + 0.073\ TSW + 0.065\ NU + 0.044\ NI \quad (8)$$

According to Equation (7), the predicted value of EOY% is a linear transformation of seed yield per square meter, number of days to 50% flowering, number of seeds per plant, number of umbels and final plant height variables. On the other hand, seed yield per square meter, number of seeds per umbel, number of days to maturity, thousand seed weight, number of umbels and number of internodes were entered in the SWR model, when TAY% was considered as a dependent variable (Equation (8)). The depended variables are those that selected based on stepwise regression analysis (Tables 3 and 4). As shown in Tables 3 and 4, independent variables explained approximately 55% and 47% of variations in EOY% and TAY%, respectively. Therefore, the linear model was not strong enough to explain the variations of the dependent variables. The MLR constructions (Equations (7) and (8)) showed the importance and the effect of independent variables on dependent variables and these equations revealed that how the amounts of EOY% or TAY% in fennel can change by different amounts of independent variables. Bahmani et al. [11] used an MLR model to find the relationship between independent variables and grain yield of fennel and showed that 55.41%, 12.72%, 2.21% and 11.63% of total variance of grain yield was explained by weight of dry biomass, days to 50% flowering, number of inflorescent and days to 70% seed pasty, respectively [11]. Niazian et al. [25] studied the seed yield of ajowan using an MLR model and introduced shoot dry weight, number of umbellets in main inflorescence, number of branches and the biological yield as important independent variables [25].

The results of the ANN and MLR models' development based on performance evaluation indices including R^2, RMSE and MAE provide a set of the reasonable criteria for comparison between two modeling methods. Compared to the MLR model, the ANN models could predict EOY% and TAY% much better than the MLR model with 39.97% and 32.69% increases in R^2, reductions of 0.30 and 0.20 in RMSE and reductions of 0.25 and 0.12 in MAE, respectively. According to the obtained results, the ANN model had higher predictive power than the MLR model and was more efficient than MLR in predicting EOY% and TAY% traits in fennel populations. The different performance of the two models to predict EOY% and TAY% shows the importance of choosing the more suitable model. The superiority of the ANN modeling methods compared to the MLR methods has been reported in other previous studies [24,25,32,35]. The supremacy of ANN modeling seems to be due to the high capability of this model to capture the highly nonlinear and complex relationship between EOY% or TAY% and the relevant traits [23]. There is a considerable variation among different populations of fennel in terms of seed yield, yield components, essential oil content and essential oil composition [11,49]. This variation along with high genotype $\times$ environment interaction create a difficult situation to improve the fennel population for the desired traits in a short period using conventional statistical methods and direct selection [49]. However, using non-linear predicting methods, breeders are able to estimate the desired values of their desired traits in a faster and more confident way. Therefore, an advanced computational method can play a complementary role to conventional statistical methods previously employed to improve the fennel populations [11,49].

3.4. Sensitivity Analysis

A sensitivity analysis is a method of studying the behavior of a model and assessing the significance of each input variable on the values of the output variable of the model. Sensitivity analysis provides insight into the usefulness of individual variables. With the help of this kind of analysis, it is possible to judge which inputs for modeling EOY% or TAY% parameters should be considered as the most and least significant ones in the ANN model [36]. For this purpose, the sensitivity tests for ANN and MLR models were performed without a specific input variable, i.e., SY/m^2, NS/P, NDF50%, NS/U and NDM. The results of the sensitivity test for EOY% showed that the highest RMSE (0.608, 0.911) and MAE (0.439, 0.659) and the lowest R^2 (75.45, 39.12) were achieved without seed yield per square meter in both ANN and MLR models (Table 7). Number of days to 50% flowering and number of seeds per plant were the other most effective characteristics on the EOY% of fennel populations. As shown in Table 7, the ANN and MLR models for TAY% without the seed yield per square have the lowest R^2 (61.43 and 35.52) and highest RMSE (0.332 and 0.532) and MAE (0.487 and 0.471), respectively. As the results showed, SY/m^2 is the most influential factor to predict EOY% and TAY% in both models.

Bahmani et al. [11] investigated the direct and indirect effects of some morphological traits on the essential oil content of fennel populations by using path analysis. They reported that the number of leaves, days to 50% flowering and plant height had direct effects on essential oil content in the first year of the experiment as well as days to 50% flowering, stem diameter and the number of seeds per largest inflorescence in the second year of the experiment [11]. Abdipour et al. [24] used the sensitivity analysis in both ANN and MLR models to find the importance of each input variable on the oil content of sesame and reported capsule number per plant as the most important input variable that can significantly affect RMSE, MAE and R^2 of both ANN and MLR models [24]. In the other study, sensitivity tests were conducted in both MLR and ANN models and results showed that the highest RMSE and MAE and the lowest R^2 were achieved in the MLR and ANN models without biological yield [25].

Table 7. Sensitivity analysis and selecting three of the most influential inputs on the essential oil and *trans*-anethole yield of fennel populations.

Output	Method	ANN			MLR		
		$R^{2\,a}$ (%)	RMSE [b]	MAE [c]	$R^{2\,a}$ (%)	RMSE [b]	MAE [c]
Essential oil yield	The best ANN (with all input)	95.30	0.522	0.375	55.33	0.819	0.624
	ANN without SY/m^2	75.45	0.608	0.439	39.12	0.911	0.659
	ANN without NDF50%	84.70	0.585	0.421	42.18	0.747	0.571
	ANN without NS/P	85.98	0.578	0.416	44.25	0.812	0.583
trans-anethole yield	The best ANN (with all input)	79.41	0.246	0.334	46.72	0.448	0.452
	ANN without SY/m^2	61.43	0.332	0.487	35.52	0.532	0.471
	ANN without NS/U	66.40	0.316	0.459	37.14	0.431	0.384
	ANN without NDM	68.40	0.308	0.445	38.76	0.416	0.335

[a]: Determination coefficient; [b]: root mean square error; [c]: mean absolute error.

Since fennel is an indeterminate plant, having continuous growth of new leaves, flowers and seeds during the growing season, it can be said that indeterminate populations of fennel with a long period from flowering to maturity are the best in stable environmental conditions. This finding can be seen in the results of the sensitivity analysis (Table 7), indicating that the number of days to 50% flowering and number of days to maturity had significant effects on EOY% and TAY%, respectively. The findings of previous studies determined possible differences and similarities in essential oils and chemical composition of various plants at different phenological stages in the various medicinal plants [49–51]. The results showed that the EOY% and TAY% depend on different phenological stages. The effect of phenological stages on essential oil and its composition may be due to its effect on enzyme activity and the metabolism of essential oil production [52]. The contribution of each input trait to predict EOY% and TAY% are ranked from highest to lowest in both ANN and MLR models in Table 7.

4. Conclusions

Identifying high-heritable yield components and using efficient modeling methods can help breeders select the most advantageous traits in medicinal plant breeding programs. The results of the present study revealed that the ANN compared to the MLR was able to predict the EOY% and TAY% of fennel populations with more accuracy. The classical MLR model could not interpret the non-linear relationships between EOY% and TAY% and their corresponding independent variables. However, the ANN model showed more accuracy in interpreting complex relationships among EOY%, TAY% and other variables in the model, according to R^2, RMSE and MAE indicators. These results showed that the selected ANN model could surely replace MLR to predict EOY% and TAY% of fennel populations. Based on the sensitivity analysis, SY/m^2, NDF50%, and NS/P were the most important traits to predict EOY%, whereas SY/m^2, NS/U and NDM were the most important traits to predict the TAY% of fennel populations. The findings of the present study can provide important information to improve the EOY% of the other medicinal plants of the *Apiaceae* family. Plant breeders can also use the optimized artificial neural network models to model other complicated polygenic traits of medicinal plants, such as the content of various secondary metabolites that are more valuable for the food and pharmaceutics industries.

Author Contributions: Conceptualization, M.S.-N., G.N., M.Y.-H., S.A. and M.E.; methodology, M.S.-N., G.N., M.Y.-H., S.A. and M.A.; software, M.A.; validation, G.N., M.A. and S.K.; formal analysis, M.S.-N., G.N., M.Y.-H., S.A. and M.A.; investigation, M.S.-N., M.Y.-H., S.A. and M.A.; resources, M.S.-N., M.Y.-H., S.A. and M.E.; data curation, M.S.-N., M.Y.-H., S.A. and M.E.; writing—original draft preparation, M.S.-N., G.N., M.Y.-H. and S.A.; writing—review and editing, G.N., M.Y.-H., S.K. and M.N.; visualization, M.S.-N., M.Y.-H. and S.A.; supervision, M.S.-N., G.N., M.Y.-H., S.A. and M.E.; project administration, M.S.-N., M.Y.-H. and S.A.; funding acquisition, M.S.-N., M.Y.-H., S.A. and M.E. All authors have read and agreed to the published version of the manuscript.

Funding: This research received no external funding.

Institutional Review Board Statement: Not applicable.

Informed Consent Statement: Not applicable.

Data Availability Statement: The datasets generated during and/or analysed during the current study are available from the first corresponding author on reasonable request.

Acknowledgments: This paper was published as part of a research project (contract number 27.21) supported by the Research Affairs Office of the University of Tabriz.

Conflicts of Interest: The authors declare no conflict of interest.

Abbreviations:

ANN	Artificial neural network
EOY	Essential oil yield
FPH	Final plant height
HI	Harvest index
LFI	Length of the first internode
LLAI	Length of the last internode
LLOI	Length of the longest internode
LP	Length of the peduncle
MAE	Mean absolute error
MLR	Multilinear regression
NDF50%	Number of days to 50% flowering
NDF100%	Number of days to 100% flowering
NDG	Number of days to germination
NDM	Number of days to maturity
NI	Number of internodes
NS	Number of stems
NS/P	Number of seeds per plant
NS/U	Number of seeds per umbel
NU	Number of umbels
RMSE	Root mean square error
SD	Stem diameter
SWR	Stepwise regression
SY	Seed yield
SY/m^2	Seed yield per square meter
SY/P	Seed yield per plant
TAY	*Trans*-anethole yield
TSW	1000-seed weight

References

1. Nojadeh, M.S.; Pouresmaeil, M.; Younessi-Hamzekhanlu, M.; Venditti, A. Phytochemical profile of fennel essential oils and possible applications for natural antioxidant and controlling *Convolvulus arvensis* L. *Nat. Prod. Res.* **2020**, 1–5. [CrossRef]
2. Yaylayan, V.A. Flavor Technology: Recent Trends and Future Perspectives. *Can. Inst. Food Sci. Technol. J.* **1991**, *24*, 2–5. [CrossRef]
3. Guillén, M.D.; Manzanos, M.J. A study of several parts of the plant Foeniculum vulgate as a source of compounds with industrial interest. *Food Res. Int.* **1996**, *29*, 85–88. [CrossRef]
4. He, W.; Huang, B. A review of chemistry and bioactivities of a medicinal spice: Foeniculum vulgare. *J. Med. Plants Res.* **2011**, *5*, 3595–3600.
5. Younessi-Hamzekhanlu, M.; Abdipour, M.; Dejahang, A.; Sabzi-Nojadeh, M.; Amani, M. Herbals Used in Western Iran as Food and for Health Treatments. In *Biodiversity, Conservation and Sustainability in Asia*; Springer: Cham, Switzerland, 2021; pp. 547–599.
6. Younessi-Hamzekhanlu, M.; Ozturk, M.; Altay, V.; Nojadeh, M.S.; Alakbarli, F. Ethnopharmacological study of medicinal plants from khoy city of west Azerbaijan-Iran. *Indian J. Tradit. Knowl.* **2020**, *19*, 251–267.
7. Pouresmaeil, M.; Nojadeh, M.S.; Movafeghi, A.; Maggi, F. Exploring the bio-control efficacy of Artemisia fragrans essential oil on the perennial weed Convolvulus arvensis: Inhibitory effects on the photosynthetic machinery and induction of oxidative stress. *Ind. Crops Prod.* **2020**, *155*, 112785. [CrossRef]
8. Aghbashd, B.N.; Pouresmaeil, M.; Dehghan, G.; Nojadeh, M.S.; Mobaiyen, H.; Maggi, F. Chemical Composition, Antibacterial and Radical Scavenging Activity of Essential Oils from Satureja macrantha C.A.Mey. At different growth stages. *Foods* **2020**, *9*, 494. [CrossRef]
9. Aghbash, B.N.; Dehghan, G.; Movafeghi, A.; Talebpour, A.H.; Pouresmaeil, M.; Maggi, F.; Nojadeh, M.S. Chemical compositions and biological activity of essential oils from four populations of Satureja macrantha C.A.Mey. *J. Essent. Oil Res.* **2021**, *33*, 133–142. [CrossRef]
10. Younessi-Hamzekhanlu, M.; Sanjari, S.; Dejahang, A.; Karkaj, E.S.; Nojadeh, M.S.; Gönenç, T.M.; Ozturk, M. Evaluation of Essential Oil From Different Artemisia fragrans Willd. Populations: Chemical Composition, Antioxidant, and Antibacterial Activity. *J. Essent. Oil-Bear. Plants* **2020**, *23*, 1218–1236. [CrossRef]
11. Bahmani, K.; Darbandi, A.I.; Ramshini, H.A.; Moradi, N.; Akbari, A. Agro-morphological and phytochemical diversity of various Iranian fennel landraces. *Ind. Crops Prod.* **2015**, *77*, 282–294. [CrossRef]
12. Devesh, P.; Moitra, P.K.; Shukla, R.S. Correlation and path coefficient analysis for yield, yield components and quality traits in wheat. *Electron. J. Plant. Breed.* **2021**, *12*, 388–395. [CrossRef]
13. Togay, N.; Togay, Y.; Dogan, Y. Correlation and path coefficient analysis for yield and some yield components of wheat (Triticum aestivum l.). *Oxid. Commun.* **2017**, *40*, 946–951.
14. Piccaglia, R.; Marotti, M. Characterization of some Italian types of wild fennel (Foeniculum vulgare mill.). *J. Agric. Food Chem.* **2001**, *49*, 239–244. [CrossRef] [PubMed]
15. Coşge, B.; Ipek, A.; Gürbüz, B. Some phenotypic selection criteria to improve seed yield and essential oil percentage of sweet fennel (Foeniculum vulgare Mill. var. dulce). *Tarim. Bilim. Derg.* **2009**, *15*, 127–133. [CrossRef]
16. Coruzzi, G.M.; Burga, A.R.; Katari, M.S.; Gutiérrez, R.A. Systems Biology: Principles and Applications in Plant Research. In *Annual Plant Reviews Online*; Wiley-Blackwell: London, UK, 2018; pp. 3–40.
17. Gago, J.; Martínez-Núñez, L.; Landín, M.; Gallego, P.P. Artificial neural networks as an alternative to the traditional statistical methodology in plant research. *J. Plant. Physiol.* **2010**, *167*, 23–27. [CrossRef] [PubMed]
18. Prasad, V.S.S.; Gupta, S.D. Applications And Potentials Of Artificial Neural Networks In Plant Tissue Culture. In *Plan Tissue Culture Engineering*; Springer: Dordrecht, The Netherlands, 2007; pp. 47–67.
19. Gross, M.; Lewinsohn, E.; Tadmor, Y.; Bar, E.; Dudai, N.; Cohen, Y.; Friedman, J. The inheritance of volatile phenylpropenes in bitter fennel (Foeniculum vulgare Mill. var. vulgare, Apiaceae) chemotypes and their distribution within the plant. *Biochem. Syst. Ecol.* **2009**, *37*, 308–316. [CrossRef]
20. Kiruthika, S.; Narayanan, S.L.; Parameshwari, C.; Arunachalam, P.; Mini, M.L. Studies on trait association and path coefficient analysis of sesame (*Sesamum* sp.) for quantitative traits and oil quality parameters. *Electron. J. Plant. Breed.* **2020**, *11*, 18–24. [CrossRef]
21. Ibhrahim, E.; Khidir, O. Genotypic correlation and path coefficient analysis of yield and some yield components in sesame (*Sesamum indicum* L.). *Int. J. AgriScience* **2012**, *2*, 664–670.
22. Bahmani, K.; Izadi-Darbandi, A.; Noori, S.A.S.; Jafari, A.A.; Moradi, N. Determination of interrelationships among phenotypic traits of iranian fennel (foeniculum vulgare mill.) using correlation, stepwise regression and path analyses. *J. Essent. Oil-Bear. Plants* **2012**, *15*, 424–444. [CrossRef]
23. Khairunniza Bejo, S.; Mustaffha, S.; Khairunniza-Bejo, S.; Ishak, W.; Ismail, W. Application of Artificial Neural Network in Predicting Crop Yield: A Review. *J. Food Sci. Eng.* **2014**, *4*, 1–9.
24. Abdipour, M.; Ramazani, S.H.R.; Younessi-Hmazekhanlu, M.; Niazian, M. Modeling Oil Content of Sesame (*Sesamum indicum* L.) Using Artificial Neural Network and Multiple Linear Regression Approaches. *JAOCS J. Am. Oil Chem. Soc.* **2018**, *95*, 283–297. [CrossRef]
25. Niazian, M.; Sadat-Noori, S.A.; Abdipour, M. Modeling the seed yield of Ajowan (*Trachyspermum ammi* L.) using artificial neural network and multiple linear regression models. *Ind. Crops Prod.* **2018**, *117*, 224–234. [CrossRef]

26. Iquebal, M.A.; Ansari, M.S.; Sarika, S.; Dixit, S.P.; Verma, N.K.; Aggarwal, R.A.K.; Jayakumar, S.; Rai, A.; Kumar, D. Locus minimization in breed prediction using artificial neural network approach. *Anim. Genet.* **2014**, *45*, 898–902. [CrossRef]
27. Khoshnevisan, B.; Rafiee, S.; Mousazadeh, H. Application of multi-layer adaptive neuro-fuzzy inference system for estimation of greenhouse strawberry yield. *Meas. J. Int. Meas. Confed.* **2014**, *47*, 903–910. [CrossRef]
28. Silva, G.N.; Tomaz, R.S.; de Castro Sant'Anna, I.; Nascimento, M.; Bhering, L.L.; Cruz, C.D. Neural networks for predicting breeding values and genetic gains. *Sci. Agric.* **2014**, *71*, 494–498. [CrossRef]
29. Goel, P.; Bapat, S.; Vyas, R.; Tambe, A.; Tambe, S.S. Genetic programming based quantitative structure-retention relationships for the prediction of Kovats retention indices. *J. Chromatogr. A* **2015**, *1420*, 98–109. [CrossRef]
30. Zeng, W.; Xu, C.; Wu, J.; Huang, J. Sunflower seed yield estimation under the interaction of soil salinity and nitrogen application. *Field Crops Res.* **2016**, *198*, 1–15. [CrossRef]
31. Niazian, M.; Niedbała, G. Machine learning for plant breeding and biotechnology. *Agriculture* **2020**, *10*, 436. [CrossRef]
32. Mansouri, A.; Fadavi, A.; Mortazavian, S.M.M. An artificial intelligence approach for modeling volume and fresh weight of callus-A case study of cumin (*Cuminum cyminum* L.). *J. Theor. Biol.* **2016**, *397*, 199–205. [CrossRef]
33. Niazian, M.; Sadat-Noori, S.A.; Abdipour, M. Artificial neural network and multiple regression analysis models to predict essential oil content of ajowan (*Carum copticum* L.). *J. Appl. Res. Med. Aromat. Plants* **2018**, *9*, 124–131. [CrossRef]
34. Tanty, R.; Tanweer, S. Desmukh Application of Artificial Neural Network in Hydrology-A Review. *Int. J. Eng. Res.* **2015**, *4*, 184–188. [CrossRef]
35. Emamgholizadeh, S.; Parsaeian, M.; Baradaran, M. Seed yield prediction of sesame using artificial neural network. *Eur. J. Agron.* **2015**, *68*, 89–96. [CrossRef]
36. Naroui Rad, M.R.; Koohkan, S.; Fanaei, H.R.; Rad, M.R.P. Application of artificial neural networks to predict the final fruit weight and random forest to select important variables in native population of melon (Cucumis melo. Pahlavan). *Sci. Hortic.* **2015**, *181*, 108–112. [CrossRef]
37. Safa, M.; Samarasinghe, S.; Nejat, M. Prediction of wheat production using artificial neural networks and investigating indirect factors affecting it: Case study in canterbury province, New Zealand. *J. Agric. Sci. Technol.* **2015**, *17*, 791–803.
38. Hara, P.; Piekutowska, M.; Niedbała, G. Selection of independent variables for crop yield prediction using artificial neural network models with remote sensing data. *Land* **2021**, *10*, 609. [CrossRef]
39. Niedbała, G.; Piekutowska, M.; Weres, J.; Korzeniewicz, R.; Witaszek, K.; Adamski, M.; Pilarski, K.; Czechowska-Kosacka, A.; Krysztofiak-Kaniewska, A. Application of artificial neural networks for yield modeling of winter rapeseed based on combined quantitative and qualitative data. *Agronomy* **2019**, *9*, 781. [CrossRef]
40. White, B.W.; Rosenblatt, F. Principles of Neurodynamics: Perceptrons and the Theory of Brain Mechanisms. *Am. J. Psychol.* **1963**, *76*, 705. [CrossRef]
41. Rahimmalek, M.; Maghsoudi, H.; Sabzalian, M.R.; Pirbalouti, A.G. Variability of essential oil content and composition of different Iranian fennel (Foeniculum vulgare Mill.) accessions in relation to some morphological and climatic factors. *J. Agric. Sci. Technol.* **2014**, *16*, 1365–1374.
42. Li, C.C. The Concept of Path Coefficient and Its Impact on Population Genetics. *Biometrics* **1956**, *12*, 190. [CrossRef]
43. Samarasinghe, S. Neural Networks for Nonlinear Pattern Recognition. In *Neural Networks for Applied Sciences and Engineering*; Auerbach: New York, NY, USA, 2006; pp. 69–112.
44. Sefidan, A.Y.; Valizadeh, M.; Aharizad, S.; Sabzi, M. Path analysis of grain yield, some morphological traits and essential oil content in different fennel (*Foeniculum vulgare* Mill.) populations. *J. Biodivers. Environ. Sci.* **2014**, *4*, 10–15.
45. Kalleli, F.; Rebey, I.B.; Razgallah, N.; Tounsi, M.S.; M'Hamdi, M. Comparative Analysis of Morphological and Chemical Variability in Different Fennel (Foeniculum Vulgare Mill.) Cultivars. *Int. J. Agric. Environ. Biores.* **2019**, *4*, 200–217.
46. Baziar, M.H.; Ghorbani, A. Evaluation of lateral spreading using artificial neural networks. *Soil Dyn. Earthq. Eng.* **2005**, *25*, 1–9. [CrossRef]
47. Moradi, H.; Bahmanyar, H.; Azizpour, H.; Rezamandi, N.; Mirdehghan Ashkezari, S.M. Modeling and Optimization of Anethole Ultrasound-Assisted Extraction from Fennel Seeds using Artificial Neural Network. *J. Chem. Pet. Eng.* **2020**, *54*, 143–153.
48. Stangierski, J.; Weiss, D.; Kaczmarek, A. Multiple regression models and Artificial Neural Network (ANN) as prediction tools of changes in overall quality during the storage of spreadable processed Gouda cheese. *Eur. Food Res. Technol.* **2019**, *245*, 2539–2547. [CrossRef]
49. Moghaddam, M.; Miran, S.N.K.; Pirbalouti, A.G.; Mehdizadeh, L.; Ghaderi, Y. Variation in essential oil composition and antioxidant activity of cumin (*Cuminum cyminum* L.) fruits during stages of maturity. *Ind. Crops Prod.* **2015**, *70*, 163–169. [CrossRef]
50. Moghaddam, M.; Omidbiagi, R.; Sefidkon, F. Changes in content and chemical composition of tagetes minuta oil at various harvest times. *J. Essent. Oil Res.* **2007**, *19*, 18–20. [CrossRef]
51. Moghaddam, M.; Pirbalouti, A.G.; Mehdizadeh, L.; Pirmoradi, M.R. Changes in composition and essential oil yield of Ocimum ciliatum at different phenological stages. *Eur. Food Res. Technol.* **2015**, *240*, 199–204. [CrossRef]
52. Sellami, I.H.; Maamouri, E.; Chahed, T.; Wannes, W.A.; Kchouk, M.E.; Marzouk, B. Effect of growth stage on the content and composition of the essential oil and phenolic fraction of sweet marjoram (*Origanum majorana* L.). *Ind. Crops Prod.* **2009**, *30*, 395–402. [CrossRef]

agriculture

MDPI

Article

A Four Stage Image Processing Algorithm for Detecting and Counting of Bagworm, *Metisa plana* Walker (Lepidoptera: Psychidae)

Mohd Najib Ahmad [1,2], Abdul Rashid Mohamed Shariff [2,3,4,*], Ishak Aris [5] and Izhal Abdul Halin [5]

1 Agronomy & Geospatial Technology Unit, Biology & Sustainability Research Division, Malaysian Palm Oil Board, 6, Persiaran Institusi, Bandar Baru Bangi, Kajang 43000, Selangor, Malaysia; mnajib@mpob.gov.my
2 Department of Biological and Agricultural Engineering, Level 3, Faculty of Engineering, Universiti Putra Malaysia (UPM), Serdang 43400, Selangor, Malaysia
3 Smart Farming Technology Research Centre, Universiti Putra Malaysia, Serdang 43400, Selangor, Malaysia
4 Laboratory of Plantation System Technology and Mechanization (PSTM), Institute of Plantation Studies, Universiti Putra Malaysia, Serdang 43400, Selangor, Malaysia
5 Department of Electrical and Electronic Engineering, Level 4, Faculty of Engineering, Universiti Putra Malaysia (UPM), Serdang 43400, Selangor, Malaysia; ishak_ar@upm.edu.my (I.A.); izhal@upm.edu.my (I.A.H.)
* Correspondence: rashidpls@upm.edu.my; Tel.: +60-389466424

Citation: Ahmad, M.N.; Shariff, A.R.M.; Aris, I.; Abdul Halin, I. A Four Stage Image Processing Algorithm for Detecting and Counting of Bagworm, *Metisa plana* Walker (Lepidoptera: Psychidae). *Agriculture* **2021**, *11*, 1265. https://doi.org/10.3390/agriculture11121265

Academic Editors: Gniewko Niedbała and Sebastian Kujawa

Received: 29 September 2021
Accepted: 15 November 2021
Published: 14 December 2021

Abstract: The bagworm is a vicious leaf eating insect pest that threatens the oil palm plantations in Malaysia. The economic impact from defoliation of approximately 10 to 13% due to bagworm attack might cause about 33 to 40% yield loss over 2 years. Due to this, monitoring and detecting of bagworm populations in oil palm plantations is required as the preliminary steps to ensure proper planning of control actions in these areas. Hence, the development of an image processing algorithm for detection and counting of *Metisa plana* Walker, a species of Malaysia's local bagworm, using image segmentation has been researched and completed. The color and shape features from the segmented images for real time object detection showed an average detection accuracy of 40% and 34%, at 30 cm and 50 cm camera distance, respectively. After some improvements on training dataset and marking detected bagworm with bounding box, a deep learning algorithm with Faster Regional Convolutional Neural Network (Faster R-CNN) algorithm was applied leading to the percentage of the detection accuracy increased up to 100% at a camera distance of 30 cm in close conditions. The proposed solution is also designed to distinguish between the living and dead larvae of the bagworms using motion detection which resulted in approximately 73–100% accuracy at a camera distance of 30 cm in the close conditions. Through false color analysis, distinct differences in the pixel count based on the slope was observed for dead and live pupae at 630 nm and 940 nm, with the slopes recorded at 0.38 and 0.28, respectively. The higher pixel count and slope correlated with the dead pupae while the lower pixel count and slope, represented the living pupae.

Keywords: bagworms; image segmentation; color features; deep learning; faster R-CNN; false color

1. Introduction

In Malaysia, the palm oil sector is described as one of the key contributors to the national economy and currently, the palm oil industry has contributed a GNI of RM 79.9 billion in 2017 [1]. The palm oil industry spans the value chain from upstream plantations to downstream processing with the oil palm planted area in Malaysia at about 5.81 million hectares in 2017 [2].

The oil palm bagworm is a leaf eating insect that has caused 43% yield loss in two years period [3]. Based on bygone records of bagworm infestations (1986–2000), it has been confirmed that over 63,955 ha of oil palm planted in 69 estates in Peninsular Malaysia had been attacked by the *M. plana* and *P. pendula* bagworm species [4]. Integrated pest management of the bagworms has been implemented in the past, however, information on their population dynamics, incidences, biological characteristics and dispersion are still understudied.

Based on the literature, there are several examples of image processing that have been applied in agriculture. An exclusion of the non-leaf portion from the leaf's main area was applied by Mora et al. [5] to estimate the leaf area index (LAI) subsequently allowing more precise values of physiological information at the whole-plant level. An autonomous field robot was used for spraying and pollinating date clusters; it tracked its targets continuously whilst in motion. To realize the robot function, image processing algorithms were designed and implemented by Shapiro et al. [6] to be used in a spraying guidance system that was based on a proportional controller, evaluating the system dynamics and examining its behaviour under different parameters. Steward et al. [7] acquired sequential images of weed patches to switch nozzles on and off at correct locations in a plantation. The image processing algorithm, through image segmentation, was also used by Amatya et al. [8] to segment out branch and cherry regions from leaves and background for sweet cherry harvesting. They obtained 89.6% branch pixel classification exactness by applying RGB color structures on each pixel together with a Bayesian classifier for separating the branch pixels. Balch et al. [9] introduced fast color-based tracking and incorporated it with motion-based tracking to detect ants in a simultaneous video stream, and also presented new approaches for investigating the spatial movement of ant colonies.

Realizing the importance of precise data collection, efforts to develop a ground-based device using a deep learning image processing algorithm targeted to detect bagworms has been developed in this work. A Faster R-CNN was used to detect and count the bagworms on site. Robustness was evaluated by the ability to exclude moving objects which come in the form of flying insects in an oil palm plantation. As part of this study, an automatic detection system was the practical approach which used image processing techniques for the detection of the bagworms and classification algorithms to classify them according to the different properties of the images. The need for development of image processing algorithm for bagworm detecting and counting is crucial especially distinguishing the living and dead bagworms population. By meeting the target, planters can be assisted using an automated bagworm counter device to carry out census prior to control actions activity and can save the pesticides usage due to accurate timing of bagworms control.

2. Four Stage Methodology for the Bagworm Detection Algorithm

The proposed image processing algorithm comprises of four stages which involve image segmentation, shape extraction using morphological operators, object detection using deep learning with Faster R-CNN, image classification to distinguish between the stages of the bagworms, and finally, counting of the bagworms. Details of the designed procedure (Figure 1) followed by the experimental results of each stage is explained in the following subsections:

2.1. First Stage—Image Segmentation

The first stage involved the development of an image segmentation algorithm to localize/detect the region of interest (RoI) in the dataset based on color processing. The data obtained from the color processing was used to track the bagworms.

In stage 1100 RGB images from the site at Kg Teluk Bunot, Banting, Selangor, Malaysia was taken using a digital fixed-lens camera. The spatial resolution of each image was 4288×2845 pixels. The images were then resized to 300×199 pixels and filtered using a Gaussian smoothing technique with a 2D convolution operator to blur the image for noise reduction. These steps are comparable to using a mean filter, but with a different kernel that represents the shape of a Gaussian hump. The results were filtered again to sharpen, enhance the edges, and to allow detailed high-frequency image information to pass whilst lessening slow-changing data in the images. This was implemented by subtracting the blurry parts of the image [9]. Finally, the centre of focus where the bagworms resided was set as the target area [10].

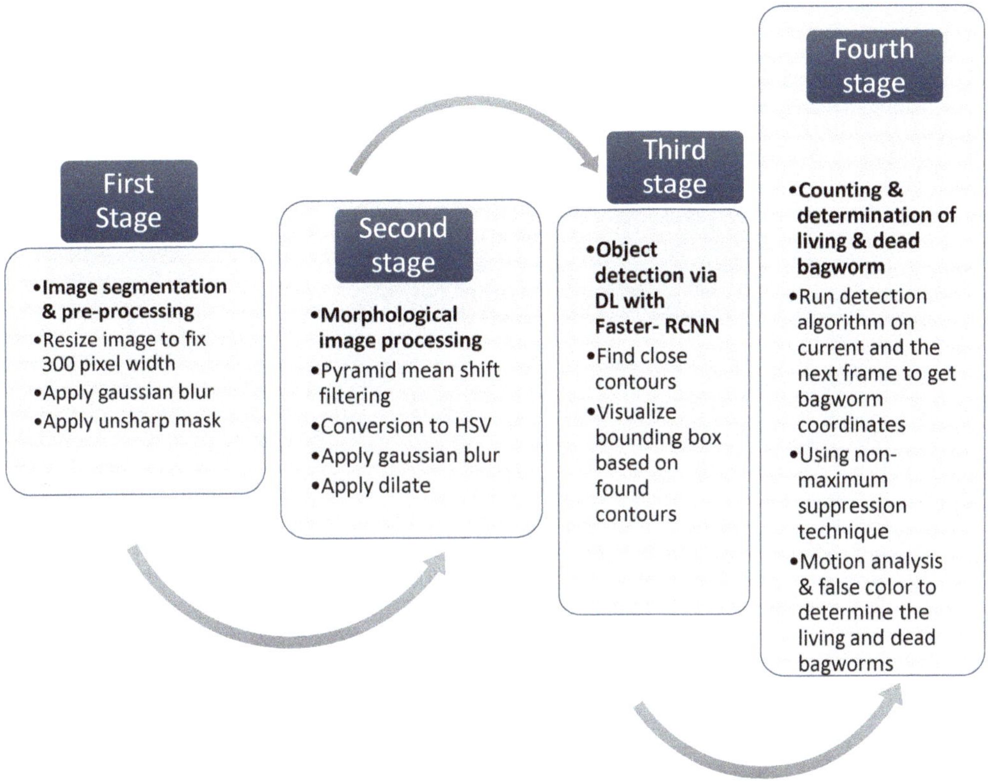

Figure 1. Big picture of the four stage methodology for bagworm detection.

2.2. Second Stage—Morphological Image Processing

The second stage used the morphological operator method, which focused on the extraction of shapes and patterns whilst removing non-targeted regions in the dataset.

In stage 2, the results were obtained using a Pyramid mean shift filter and Gaussian blur filter, which were both applied to the image in a Figure 2a, resulting in the images shown in Figure 2b,c.

(**a**) (**b**) (**c**)

Figure 2. Morphological image processing using: (**a**) Original RGB image, (**b**) Polarized image after Pyramid mean shift filter, and (**c**) Gaussian blur filter.

Next, a color space of the image (Figure 2b) was converted from the RGB to a hue, saturation, value (HSV) to search for a specific color that represented the bagworms (Figure 3).

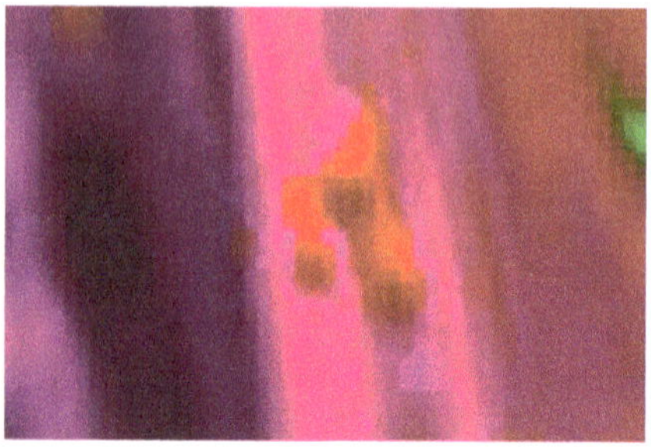

Figure 3. Color space conversion from RGB to HSV.

From the HSV color space, the Upper and Lower color range was determined using Color Picker software, where this range corresponded to the bagworm images. For the bagworms, the color ranges were as follows:

{Lower bagworms = [int(0.0 × 179), int(0.0 × 255), int(0.8 × 255)]}

{Upper bagworms = [int(0.2 × 179), int(1.0 × 255), int(1.0 × 255)]}

Subsequently, the threshold of the HSV image was found by subtracting the RoI with the HSV threshold image (Figure 4). Finally, the subtracted image was converted to grayscale and dilated to obtain the threshold image shown in Figure 5.

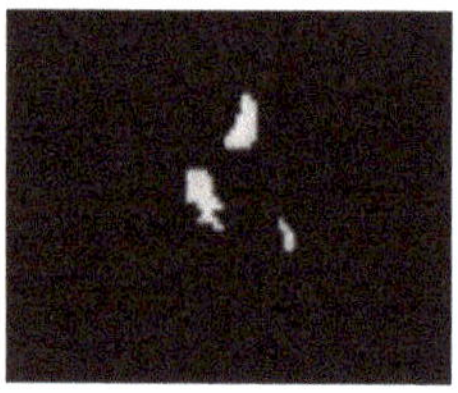

Figure 4. Targeted object with HSV.

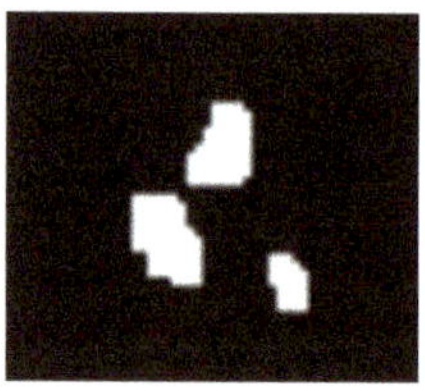

Figure 5. Threshold bagworms image.

2.3. Third Stage—Object Detection through Deep Learning and Classification

The third stage involved classification, which employed a supervised classifying algorithm. This algorithm was based on trained data, specific to size and shape recognition to distinguish between the stages of bagworms. Experiments were conducted on three groups of *M. plana* Walker, which were classified according to their stages of development as shown in Table 1. The first group contained the early larval stage, 1–3. The second group was made up of the late larval stage, 4–7. The third group was the pupal stage (positioned at the bottom part of the oil palm fronds).

In stage 3, an algorithm was developed to detect the bagworms and classify them into groups of stages. This was achieved by training the dataset from the images for object detection and recognition. A deep learning technique with Faster R-CNN coupled with a Region Proposal Network (RPN) [11] was used to predict the object bounds and exact scores at each position. The Faster R-CNN model with a RPN is a fully-convolutional network that simultaneously predicts object bounds and objectness scores at each position. RPNs are trained end-to-end to generate high quality region proposals, which are used by Faster R-CNN for detection. With a simple alternating optimization, RPN and Faster R-CNN can be trained to share convolutional features. It enables a unified and deep-learning-based object detection system to run at 5–17 fps. The learned RPN improves region proposal quality and object detection accuracy. With almost 8000 images plus videos, the dataset was trained to produce a trained detection model which was based on the Faster R-CNN algorithm and classified into sets of images with bagworms or without bagworms. There are two steps involved in preparing the system to detect the bagworm; training and testing. Training is the process to produce a trained detection model. It was based on Faster R-CNN algorithm. Each step of training reports the loss. It will start high and get lower and lower as training progresses. The model trains until the loss consistently drops below 0.05.

The progress of the training job is monitored using TensorBoard as shown in Figure 6. A total of 8000 samples were trained from 400 images with an average of 20 samples per image. Training images were manually marked with bounding boxes that indicated areas where bagworms exist. Once training is completed, the model is tested with another input image to verify the accuracy of bagworm detection.

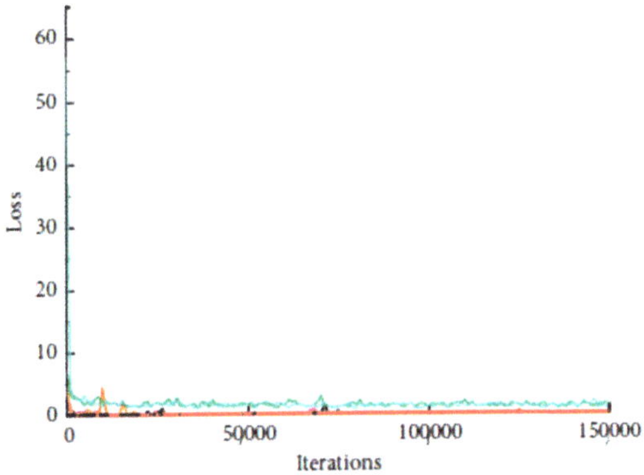

Figure 6. Loss of dataset during training progress.

TensorFlow had been opted for since it is an open source artificial intelligence library and it allows creation of large-scale neural networks with many layers. It has the fastest growing library with active support and a lot of pre-trained models including the Faster R-CNN. The deep learning (DL) with Faster R-CNN algorithm for detecting bagworms is illustrated in the flow chart as shown in Figure 7.

Table 1. Bagworm classification corresponding to real sizes subjected to the stages of development.

Group	Bagworm Stages	Real Sizes, mm
1	1–3	1.3–4.4
2	4–7	4.5–10.1
3	pupal	10.2–13.6

The classification of the larval stages was carried out by grouping the larval stages based on their real size. Purpose of grouping the bagworms according to their stages is to assist detection of the larvae and pupae based on their eating behavior, position on palm frond and movement characteristic. These groups are planned to be used as targeted objects in three mode of selection in automated detector device. Here, three groups have been defined and shown in Table 1. Based on their actual size, which ranged from 1–12 mm, the image dataset consisted of bagworm pixel intensity trained according to each stage. Then, the input RGB images were resized without distorting their aspect ratio. Since there were various shapes of the bounding box, a good size representation was achieved by measuring the diagonal distance of the box. This was to simplify the detection without investigating the orientation of the bagworm.

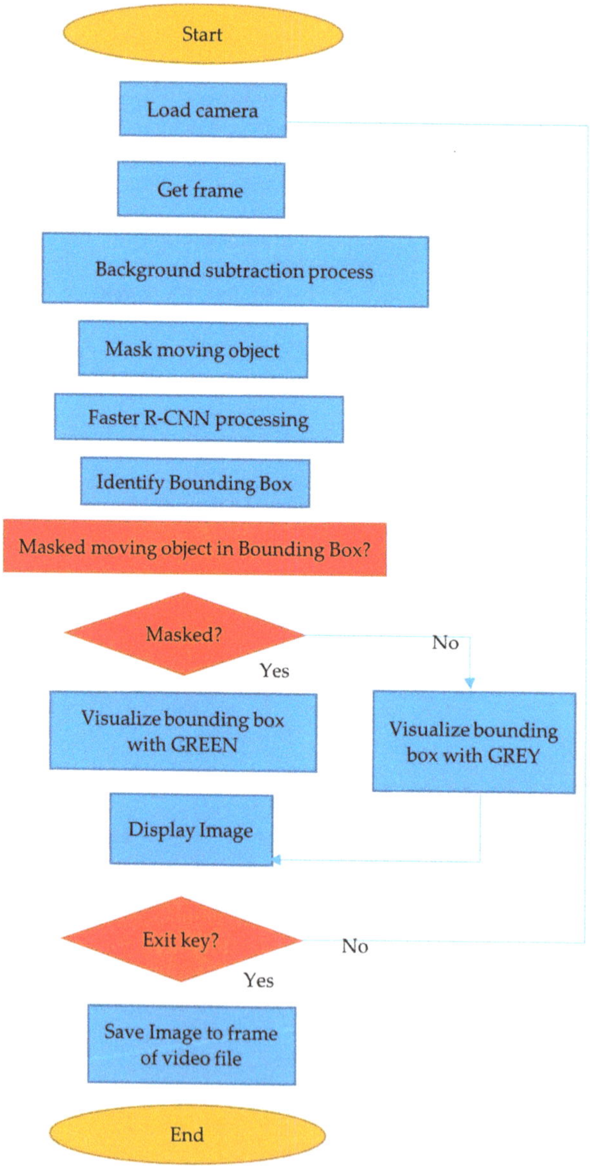

Figure 7. Process flow for the Faster R-CNN algorithm with moving object detection.

Figure 7 describes the process flow of the bagworm detection algorithm using the DL with Faster R-CNN. It detected any movement of the bagworm by substracting the moving foreground from static background by using BackgroundSubstractorMOG2. The image from camera was initially froze into sequence of frames and labeled as current and previous frame. The algorithm continued processing the next frame until video captured was stopped. Then, after identify and masking the bounding box for the moving object, the living bagworm was visualized in a green box, whilst the dead bagworm was identified in a grey box.

2.4. Fourth Stage—Counting and Determination of the Living and Dead Larvae and Pupae through Motion Tracking and False Color Analysis

At the fourth stage, the technique for distinguishing the living and dead larvae and pupae through the motion analysis algorithm and false color analysis is done. Then, the counting algorithm was used to count the bagworm populations corresponding to the specified groups.

All the images captured at the field consisted of bagworms and leaflets, and contained zero static backgrounds. A static background is a clean background without a subject to be detected. Therefore, the BackgroundSubtractorMOG2 technique was used to produce the static background. Images from the camera were frozen into a sequence of frames. For simplicity, two frames were labelled as the current frame and the previous frame. The algorithm kept on continuing to the next frame until the video capture was stopped. This condition generated an assumption, where the moving bagworm was identified as a living bagworm and the static bagworm was recognized as a dead bagworm [12].

The living and dead bagworms were determined through motion tracking and false color analysis. An image where the bagworms were detected contained the image of the bagworm size and mortality. Using the pixel arrangement as a Cartesian coordinate system, a bounding box was drawn over the detected regions in the image. Mortality was determined by analyzing the bounding box for moving objects by placing it on the masked images. If a box contained masked moving objects, then the bagworm was classified as alive (labelled in a green box) and a living bagworm counter was incremented. Dead bagworms were detected and accounted for when the bounding box analysis found no movement in the masked images and were represented in a grey box (Figure 7).

2.5. Motion Tracking for Determination of the Living and Dead Larvae

Motion analysis was used to detect dead and living larvae. It detected any movement of the larvae by subtracting the moving foreground from the static background. The following describes the motion analysis algorithm process.

i. Previous frame was processed to be the static background image using Background-SubtractorMOG2.
ii. Current frame was applied to the GaussianBlurr to filter the noise and become the foreground image.
iii. Masked the background image with the foreground image.
iv. Applied the cv2.countNonZero of the overlapping background image with the foreground image.
v. If there was a nonzero counter, it meant there was movement in the frame.

2.6. False Color for Determination of the Living and Dead Pupae

60 samples of dead and living pupae were randomly placed on a black ground canvas. Two light sources had been identified to be economical and practical, namely the 940 nm (IR) and 630 nm (red). These two wavelength points were selected based on the spectral reflectance properties of the living and dead pupae, which were significantly different between the 630 nm and the 940 nm. The data was achieved using a spectroradiometer in other experiments [13] to find the reflectance percentage at specific wavelengths for the pupae. The images for 630 nm and 940 nm were then captured in RGB format. The images were converted to grayscale before an average of all the pixel's values were picked within a pupa's boundary. The average pixel values were collected for all samples.

The steps for pixel counting are mentioned as follows:

1. Source captured in RGB
2. Location of each pupa was marked to calculate the slope value under the red vision (630 nm) as compared to the IR (940 nm) of the pupa at the same location.
3. Viewed image in grayscale using OpenCVimshow (img, imgfile, grayscale_option)
4. Zoomed each pupa until the pixel value was displayed.
5. Picked all the pixels from a pupa image.
6. Averaged the pixel values.

2.7. Counting

After all the bagworm detection procedures had been conducted, counting was carried out by averaging the total bagworms detected for 200 frames of an image. In this study, 200 frames were set during the snapshot.

2.8. Experimental Setup

The experiment was conducted under three different controlled conditions. The controlled conditions were set up as follows: (a) fully closed area, (b) half open area and (c) fully open area as shown in Figure 8. Meanwhile, the position of the camera from the targeted objects was set at 30 cm and 50 cm distance from the subjects to compensate for changes in lighting conditions and shadowing. Other factors that had to be accounted for were vibration and false capture of objects, such as other insect species that may have flown into the field of view of the camera setup.

(a) (b) (c)

Figure 8. Controlled conditions (**a**) fully closed area, (**b**) half open area, and (**c**) fully open area.

2.9. Data Analysis

Data on detection accuracy with different approaches were analyzed using a one-way ANOVA and the means were separated by the Least Significant Difference (LSD) test at $p < 0.05$. The Pearson product-moment correlation coefficient was calculated to evaluate detection performance of the live and dead larvae of the bagworms using deep learning algorithm and manual counting.

3. Results

3.1. Stage-1 and 2

In the color space processing method, which involved morphological image processing, the results showed that the accuracy of the detection was low; the averages were 40% and 34%, at the 30 cm and 50 cm camera distances, respectively (Table 2). This was due to the color similarity between the damaged leaflets and the bagworms (Figures 9 and 10).

Table 2. Color processing performance at different camera distances and conditions.

Camera Distance	Condition	Algorithm Detection	Human Detection	Detection, %
30 cm	Open	9	19	47
	Closed	10	19	53
	Half open	4	19	21
50 cm	Open	3	19	26
	Closed	7	19	36
	Half open	9	19	46

The different camera distances gave different coverage areas of the whole frond. The farther the distance (50 cm), the wider the area that could be covered on a single frond. However, it showed poor feature details of the bagworms (Figure 9). Meanwhile, the closer the distance (30 cm), the better the details were but the coverage area was limited. This condition might cause problems when the setup is applied in the field, where more camera snapshots will be needed to cover the whole frond. Based on site measurements, it was revealed that the 30 cm camera distance needed 9 snapshot sessions and the 50 cm camera distance required 6 snapshot sessions to cover the whole frond. By applying the color processing approach, the controlled conditions did not significantly affect the results (Figures 9 and 10). Although the camera distance was adjusted to 30 cm and 50 cm, the results had low detection accuracy as shown in Table 2.

The results for bagworms detection using a color processing technique at 30 cm and 50 cm camera distances is shown in Figures 9 and 10, respectively. This colorimetric approach gave a poor result on detecting bagworms, although, the input images had been controlled in terms of light source or surrounding light and snapshot distances from the subject. The HSV (Hue, Saturation, and Value) parameters had been set to their optimum values, with the lower limits set at [0,0,204] and the upper limits adjusted to [36,255,255], nevertheless, wrong and missed detections of the RoI were still observed. This was due to the damaged leaflets having a close intensity and color range with the bagworms [14] (Figure 11).

Based on Figure 11, the effect of the damaged palm leaflet which had the same color of the bagworms led to the difficulty to detect the RoI and attributed to the wrong detection of the bagworms.

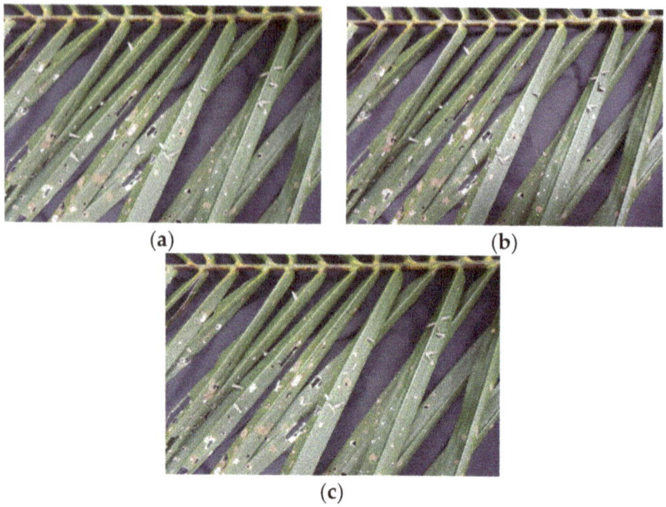

Figure 9. Object detection based on 30 cm camera distance with: (**a**) open, (**b**) closed and (**c**) half open areas.

Figure 10. Detection based on 50 cm camera distance with: (**a**) open, (**b**) closed and (**c**) half-open areas.

3.2. Stage-3

The results of the deep learning (DL) approach for detecting bagworms, captured at camera distances of 30 cm and 50 cm are shown in Figures 12 and 13. By adopting this technique, the detection accuracy improved tremendously as compared to the stage 1 and 2 methods. The green boxes represented the boundary boxes which contained the RoI of the bagworms. Figures 12b and 13b show better bagworm recognition as compared to the open and half open areas.

Figure 11. Wrong detection of targeted objects through the color processing technique.

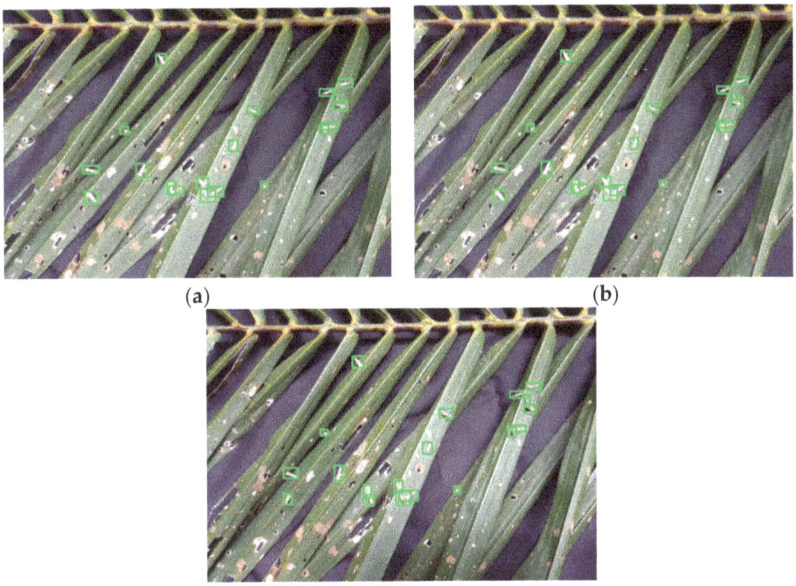

(**a**) (**b**)

(**c**)

Figure 12. Detection based on 30 cm camera distance through DL with controlled conditions: (**a**) open, (**b**) closed and (**c**) half-open areas.

The summary results of the DL approach with Faster R-CNN for stage 3 is illustrated as follows:

From Table 3, it was proven that the DL with Faster R-CNN gave better results. There was a significant difference in terms of detection accuracy between the 30 and 50 cm camera distances, where $p < 0.05$ for the closed condition, as compared to other conditions. It generated the highest detection accuracy, recorded at 100% and 90%, respectively (Figure 14). Whereas, there was a slightly low detection at 30 cm and 50 cm camera distances for open and half open conditions, resulting in 90% and 80% detection. The wrong detection was observed, 10% and 20%, due to insufficiently strong trained data.

Figure 13. Detection based on 50 cm camera distance through DL with: (**a**) open, (**b**) closed and (**c**) half-open areas.

Table 3. DL performance to detect bagworms at different camera distances and areas.

Camera Distance	Condition	Algorithm Detection	Human Detection	Detection, %
30 cm	Open	9	10	90 a
30 cm	Closed	10	10	100 b
30 cm	Half open	9	10	90 a
50 cm	Open	8	10	80 a
50 cm	Closed	9	10	90 b
50 cm	Half open	8	10	80 a

Note: Rows with different letters were significantly different ($p < 0.05$) after one-way ANOVA using the LSD test.

Based on Figure 14, the different image processing approaches between Stage 1&2 and Stage 3 gave different levels of detection accuracy and it was proven by the one-way ANOVA analysis with the Least Significant Difference (LSD) test at $p < 0.05$. By using the color processing technique, it was revealed that the percentage of the detection accuracy was low. The highest detection accuracy for the Stage 1&2 methods was recorded at the 55% detection accuracy at the 30 cm camera distance.

Meanwhile, by applying the Stage 3 method (DL), the percentage of detection accuracy increased up to 100% at the 30 cm camera distance. From both stages' techniques, it was revealed that the 30 cm camera distance resulted in better detection performance and showed more accurate feature details of the bagworms.

Figure 15 shows the results for the bagworm group detection based on their real sizes at the 30 cm camera distance. The RoI was detected based on the boundary boxes, which were labelled according to larval group, on Group 1 (G1) and group 2 (G2) that were tracked in real-time in the field and the results are shown in Table 4. The results of the algorithm detection correspond reasonably close with the results of human detection. The correlation coefficients of the detection rate for group 1 and group 2 were 0.914 and 0.891 at standard deviation of 0.581 and 0.162, respectively.

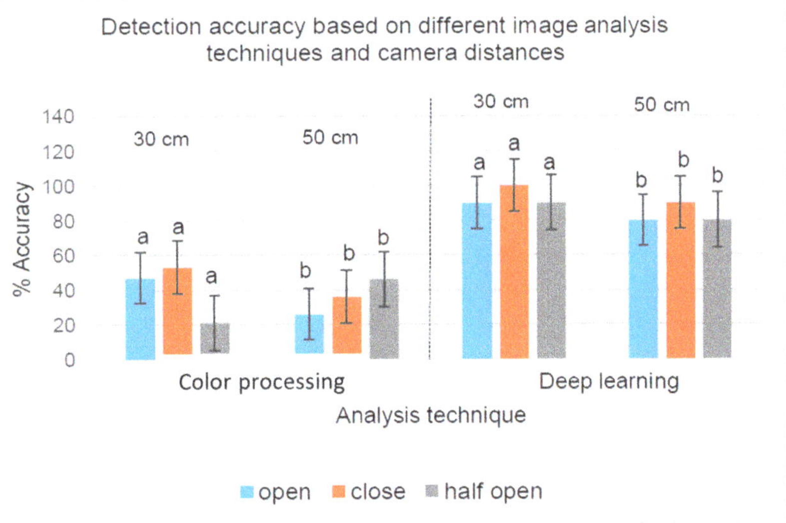

Figure 14. Detection accuracy based on different image analysis techniques according to the Stage 1&2 and Stage 3 methods at different camera distances. Note: Bars across a group with the different letters were significantly different ($p < 0.05$) after the LSD test.

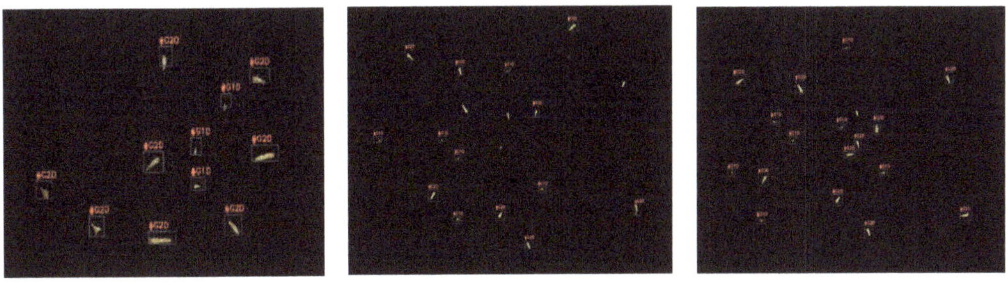

(**a**) (**b**) (**c**)

Figure 15. Classification of bagworm stages according to features extracted and marked with bounding box: (**a**) Test 1, (**b**) Test 2, and (**c**) Test 3.

Table 4. Bagworm group classification performance between algorithm and human.

Data Samples	Algorithm Detection		Human Detection		Wrong Detection of G1, %	Wrong Detection of G2, %
	Group 1	Group 2	Group 1	Group 2		
Figure 15a	3	8	3	8	0	0
Figure 15b	5	10	6	12	17	17
Figure 15c	6	8	6	12	0	33

Based on Table 4, the results of the algorithm detection were comparable with human detection for group 1 classification, and led to an almost 100% accurate detection. However, for Group 2 classification, 17–33% wrong detection was recorded due to the inconsistent distance of the image sensor to the bagworm and the moving orientation of the bagworms in Group 2.

3.3. Stage-4

Through motion tracking, the algorithm was able to distinguish between living and dead bagworms based on motion-tracking. Any motion detected in the bounding box was classified as a live bagworm. Bounding boxes created from previous detections were used to indirectly calculate the bagworm sizes and classified them into groups. The results of the detection of the living and dead bagworms are shown in Table 5.

Table 5. Algorithm performance on the determination of living and dead bagworms according to bagworm group.

Groups	Deep Learning Detection		Actual Number of Bagworms		Wrong Detection of Live Bagworms, %	Wrong Detection of Dead Bagworms, %
	Live	Dead	Live	Dead		
G1	3	1	3	1	0	0
G2	8	5	11	5	27	0

The results in Table 5 showed that the algorithm was able to distinguish between the living and dead larvae, 100% in Group 1. However, false detection amounting to 27% was observed when distinguishing Group 2. The reason for false detection is when bagworms overlapped each other. This false detection can also be a miss-detection because sometimes bagworms are still alive but they are parasitized.

A Pearson product-moment correlation coefficient was computed to assess the relationship between the detection of the live and dead G1 larvae using DL algorithm and manual counting. There was a positive correlation between the two variables (DL algorithm and manual counting), $r = 0.961$, $n = 30$, $p = 0.002$ (live larvae) and $r = 0.990$, $n = 30$, $p = 0.003$ (dead larvae). A scatterplot summarizes the results (Figure 16).

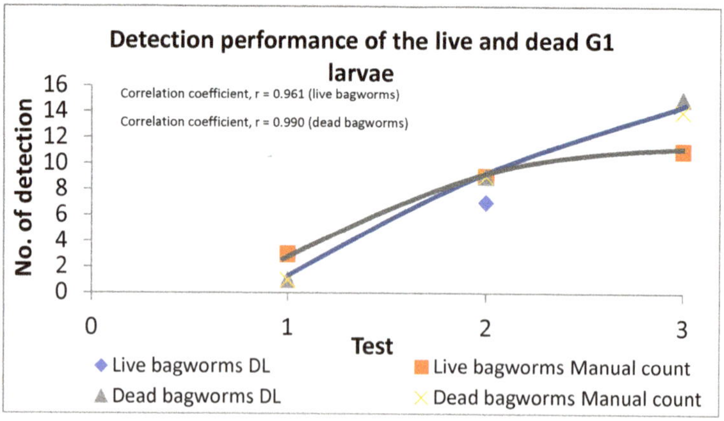

Figure 16. Detection performance of the live and dead G1 larvae by deep learning (DL) and manual counting.

From a scatterplot in Figure 17, there was a positive correlation between the two variables (DL algorithm and manual counting), $r = 0.961$, $n = 30$, $p = 0.004$ for live G2 larvae and $r = 0.982$, $n = 30$, $p = 0.002$ for dead G2 larvae. Overall, there was a strong or positive correlation between the detection of the live and dead G1 and G2 larvae via DL algorithm and manual counting. This indicated that DL algorithm showed a capability to be used in automatic detection system and able to produce high accuracy compared to manual counting.

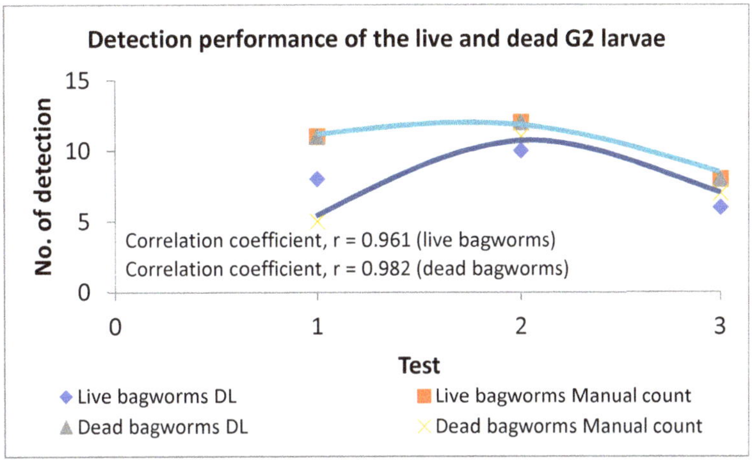

Figure 17. Detection performance of the live and dead G2 larvae by deep learning (DL) and manual counting.

The results of the living and dead pupae detection were achieved using our custom machine learning algorithm, which is presented in Table 6. A false color imaging method was used [15]. From the false color results, distinct differences in the pixel count based on the slope were observed for the dead (0.34) and live pupae (0.28) in the spectral band range of 630 nm (Red) and 940 nm (IR). This was possible due to the spectral reflectance properties between the dead and living specimens (0.38 and 0.26, respectively).

Table 6. Measured slope on reflectance of living and dead pupae specimens using 630 nm and 940 nm wavelengths.

Descriptive Statistics	Spectral Reflectance		False Color Imaging	
	Live	Dead	Live	Dead
Mean	0.26 ± 0.02	0.38 ± 0.02	0.28 ± 0.02	0.34 ± 0.02
Min	0.26 ± 0.02	0.38 ± 0.02	0.15 ± 0.01	0.30 ± 0.02
Max	0.27 ± 0.02	0.38 ± 0.02	0.46 ± 0.01	0.46 ± 0.01

4. Discussions

In color processing technique, it was revealed that percent detection of the bagworms was low, ranging from 30–40%. Although variables such as surrounding light effect and camera distance were controlled and minimized, the results seem to be similar due to difficulty to differentiate color between bagworm and damage palm leaflet. Within HSV colorspace, the bagworm color range was searched and Upper and Lower color range was determined, however, the damage leaflet has a close and same color range which is brownish color (normal practice observation) [14]. This nature condition contributed to low detection accuracy of the bagworm through color processing technique. In order to overcome this constraint, deep convolutional neural networks were practiced and applied. It has been broadly applied in various challenging and puzzling tasks namely segmentation [16], classification [17] and object detection [18]. In this study, by applying a deep CNN, the percentage of detection accuracy was increased, up to 100% (Figure 14). A closer camera distance resulted in a higher percentage of detection accuracy due to there being more details in the images. The result was almost the same as reported by Najib et al. [19], whereby a closer camera distance gave 100% detection accuracy and better detection performance. However, the imaging environment needs to be properly controlled to minimize the variables and increase the detection accuracy. Frankly, any movement such as shadow, noise and override from surrounding was the main challenge that needed to be considered and minimized [12]. A high resolution imagery, such as 1600 × 1200 pixels or more, is needed for strong detection. Thus, camera distance plays an important role to maximize details of the recognized object during snapshot [20]. By employing the closed conditions at the 30 cm and 50 cm camera distances, the highest percentages of detection accuracy were obtained, at 100 and 90%, respectively. Furthermore, the video processing or motion tracking method was applied at the 30 cm camera distance in the closed condition, with the aspect ratio of the video file being fixed to 16:9 to get an accurate result. Indeed, by classifying bagworms according to groups, the algorithm was able to successfully distinguish between the stages of the bagworms, which relied on a bounding box that possessed the orientation and stage of the bagworms. According to Gutierrez et al. [21], the best approach based on pests, *Bemisia tabaci* and *Trialeurodes vaporariorum* detection and identification accuracy is investigated using machine learning and deep learning models. By applying the ML algorithms, K-nearest neighbour (KNN) and Multilayer Perceptron (MLP), the best healthy and unhealthy pest detection average rate were obtained with an accuracy rate of 63.8% and 64.9%, respectively. Whilst, the DL algorithm was getting an accuracy rate of 85.5% for the pest detection, implying that the DL technique is a better solution than the ML technique. Ding and Taylor [22] has carried out a DL with a neural network model to detect, categorize and count the moths population and attained a good result. By working under idyllic experimental settings, it is predicted that the model can obtain better results, although, the sunlight and obstructions might affected the images quality. Then, Xie et al. [23] has integrated a sparse-coding method for encrypting insect images using a multi-kernel learning (MKL) approach to make an insect detection system, whereby obtained a mAP (mean average precision) result of 86% for 24 type of insects in the fields. Another approach was compared through a supervised classifier, which was generated and developed by Diago et al. [24] to portray the grapevine canopy, measured leaf area and yield by using RGB images, with referred to the *Mahalanobis* distance. The segmentation algorithm was successfully discriminate the leaves and grape cluster with 92% and 98% accuracy, respectively. The contrast between object, black grapes with leaves and surrounding determined the performance of segmentation. Another study done by Nuske et al. [25], successfully applying the image processing technique to estimate yield in the vineyard. Based on their results, different light conditions demonstrated an excellent color features and texture condition descriptors, aiming for grape bunch segmentation. With an average error between 3 to 11%, the total yield could be well projected. In addition, Xia et al. [26] reported that small insect pests detection should be carried out under controlled light conditions due to variable illumination effects. High resolution imagery such as 1600 × 1200 pixels or more is needed for strong detection. The visible imagery had a higher resolution and was able to display more detailed features, such as the target geometry and localization to discriminate the background. Furthermore, the Faster R-CNN which was created from SPP-Net, achieved bold-alteration performance in accuracy, speed and accomplish end-to-end networking as part of the main object detection work [27]. The convolutional layers create a set of anchors at diverse scales and aspect ratios, predicting the bounding box coordinates together with a probability score and determining the region whether it is an object or background. Through the feature maps, the anchors are developed by spatially sliding a 3 × 3 window. Then, these features are applied for object classification and bounding box regression layers, whereby it functions to determine or verify whether the region proposal is a targeted object or just a background. Meanwhile, the bounding box is forecasting the coordinates of the region. In our study, further analysis by pixel counting under greyscale images resulted in different slope values for both pupae, which successfully gave positive results, with an 89% detection accuracy.

5. Conclusions

We can conclude that a 4-stage image processing algorithm can be used to detect bagworms. Deep learning with Faster R-CNN is a feasible, practical, and reliable method for bagworm detection. An accuracy of 100% detection is possible as demonstrated by the accuracy achieved in stage 3 of the algorithm. A camera distance of 30 cm, in a controlled lighting environment resulted in the highest percentage of detection accuracy. The motion tracking and false color approaches have successfully distinguished between the living and dead larvae and pupae, respectively, with up to 100% and 89% accuracy.

Author Contributions: Conceptualization, M.N.A. and A.R.M.S.; methodology, M.N.A., I.A. and A.R.M.S.; software, M.N.A. and A.R.M.S.; validation, M.N.A., A.R.M.S., I.A. and I.A.H.; formal analysis, M.N.A.; investigation, M.N.A.; resources, M.N.A.; data curation, M.N.A.; writing—original draft preparation, M.N.A.; writing—review and editing, A.R.M.S. and I.A.H. All authors have read and agreed to the published version of the manuscript.

Funding: This research received Universiti Putra Malaysia Journal Publication Fund, project code 9001103.

Institutional Review Board Statement: This research received no external funding.

Informed Consent Statement: Not applicable.

Data Availability Statement: Not applicable.

Acknowledgments: The authors would like thank the Director-General of the Malaysian Palm Oil Board (MPOB) for permission to publish this article, the financial support, patenting of all data under the pending patent filing with PI No.: 2019006153 and protecting copyright of the algorithm under Voluntary Notification of Copyright No. CRLY 00022664. Thanks also to University Putra Malaysia (UPM) for all support rendered throughout this research. Our thanks to Sakina Bibi Mohamed Shariff for the professional editing services rendered.

Conflicts of Interest: No potential conflict of interest was reported by the authors.

References

1. National Transformation Programme Annual Report. Available online: https://www.pemandu.gov.my/assets/publications/annualreports/NTP_AR2017_ENG.pdf (accessed on 13 May 2018).
2. Kushairi, A.; Soh, K.L.; Azman, I.; Elina, H.; Meilina, O.A.; Zanal, B.M.N.I.; Razmah, G.; Shamala, S.; Ahmad Parveez, G.K. Oil palm economic performance in Malaysia and R&D progress in 2017. *J. Oil Palm. Res.* **2018**, *30*, 163–195.
3. Malaysian Palm Oil Board. *Standard Operating Procedures (SoP) Guidelines for Bagworm Control*; Malaysian Palm Oil Board (MPOB): Bangi, Malaysia, 2016; pp. 1–20, ISBN 978-967-961-218-9.
4. Ho, C.T.; Yusof, I.; Khoo, K.C. Infestations by the bagworms Metisa plana and Pteroma pendula for the period 1986–2000 in major oil palm estates managed by Golden Hope Plantation Berhad in Peninsular Malaysia. *J. Oil Palm Res.* **2011**, *23*, 1040–1050.
5. Mora, M.; Avila, F.; Carrasco-Benavides, M.; Maldonado, G.; Olguín-Cáceres, J.; Fuentes, S. Automated computation of leaf area index from fruit trees using improved image processing algorithms applied to canopy cover digital photography. *Comput. Electron. Agric.* **2016**, *123*, 195–202. [CrossRef]
6. Shapiro, A.; Korkidi, E.; Demri, A.; Ben-Shahar, O.; Riemer, R.; Edan, Y. Toward elevated agrobotics: Development of a scaled-down prototype for visually guided date palm tree sprayer. *J. Field Robot* **2009**, *26*, 572–590. [CrossRef]
7. Steward, B.L.; Tian, L.F.; Tang, L. Distance-based control system for machine vision-based selective spraying. *Trans. ASAE* **2002**, *45*, 1255–1262. [CrossRef]
8. Amatya, S.; Karkee, M.; Gongal, A.; Zhang, Q.; Whiting, M.D. Detection of cherry tree branches with full foliage in planar architecture for automated sweet-cherry harvesting. *Biosys. Eng.* **2015**, *146*, 3–15. [CrossRef]
9. Balch, T.; Khan, Z.; Veloso, M. Automatically tracking and analyzing the behaviour of live insect colonies. In Proceedings of the 5th International Conference on Autonomous Agents, Montreal, Canada, 28 May–1 June 2001; pp. 521–528.
10. Uvais, Q.; Chen, C.H. *Digital Image Processing: An Algorithmic Approach with MATLAB*; CRC Press, Taylor & Francis Group: Boca Raton, NJ, USA, 2010.
11. Ren, S.; He, K.; Girshick, R.; Sun, J. Faster R C-NN: Towards Real-Time Object Detection with Region Proposal Networks. In Proceedings of the Advances in Neural Information Processing System 25, 24th International Conference, ICONIP 2017, Guangzhou, China, 14–18 November 2017; pp. 91–99.
12. Yang, Y.; Peng, B.; Wang, J.A. System for Detection and Recognition of Pests in Stored-Grain Based on Video Analysis. In Proceedings of the Conference on Control Technology and Applications, Nanchang, China, 3–5 June 2010; pp. 119–124.
13. Najib, M.A.; Rashid, A.M.S.; Ishak, A.; Izhal, A.H.; Ramle, M. Identification and determination of the spectral reflectance properties of live and dead bagworms, Metisa plana Walker (Lepidoptera: Psychidae), using Vis/NIR spectroscopy. *J. Oil Palm Res.* **2020**, *33*, 425–435. [CrossRef]
14. Najib, M.A.; Rashid, A.M.S.; Ramle, M. Monitoring insect pest infestation via different spectroscopy techniques. *Appl. Spectrosc. Rev.* **2018**, *53*, 836–853.

15. Najib, M.A.; Rashid, A.M.S.; Ishak, A.; Izhal, A.H.; Ramle, M. A false colour analysis: An image processing approach to distinguish between dead and living pupae of the bagworms, Metisa plana Walker (Lepidoptera: Psychidae). *Trans. Sci. Technol.* **2019**, *6*, 210–215.
16. Badrinarayanan, V.; Kendall, A.; Cipolla, R. Segnet: A deep convolutional encoder-decoder architecture for image segmentation. *IEEE Trans. Pattern Anal. Mach. Intell.* **2017**, *37*, 2481–2495. [CrossRef] [PubMed]
17. He, K.; Zhang, X.; Ren, S.; Sun, J. Deep residual learning for image recognition. In Proceedings of the IEEE Conference on Computer Vision and Pattern Recognition, Las Vegas, NV, USA, 27–30 June 2016; pp. 770–778.
18. Dai, J.; Li, Y.; He, K.; Sun, J. R-FCN: Object detection via region based fully convolutional networks. *Adv. Neural Inf. Process. Syst.* **2016**, 379–387.
19. Najib, M.A.; Rashid, A.M.S.; Ishak, A.; Izhal, A.H.; Ramle, M. Development of an automated detection and counting system for the bagworms, Metisa plana Walker (Lepidoptera: Psychidae), census. In Proceedings of the 19th International Oil Palm Conference 2018: Nurturing People and Protecting the Planet, Cartagena, Colombia, 26–28 September 2018.
20. Simone, W.; Heinz, H.F.B.; Johannes, P.S.; Alexander, T.; Christian, S. Effects of infrared camera angle and distance on measurement and reproducibility of thermographically determined temperatures of the distolateral aspects of the forelimbs in horses. *JAVMA Sci. Rep.* **2013**, *242*, 388–395.
21. Gutierrez, A.; Ansuategi, A.; Susperregi, L.; Tubio, C.; Ivan Rankit, I.; Lenda, L. A Benchmarking of Learning Strategies for Pest Detection and Identification on Tomato Plants for Autonomous Scouting Robots Using Internal Databases. *J. Sens.* **2019**. [CrossRef]
22. Ding, W.; Taylor, G. Automatic moth detection from trap images for pest management. *Comput. Electron. Agric.* **2016**, *123*, 17–28. [CrossRef]
23. Xie, C.; Zhang, J.; Li, R.; Li, J.; Hong, P.; Xia, J.; Chen, P. Automatic classification for field crop insects via multiple-task sparse representation and multiple-kernel learning. *Comput. Electron. Agric.* **2015**, *119*, 123–132. [CrossRef]
24. Diago, M.; Correa, C.H.; Millan, B.; Barreiro, P.; Valero, C.; Tardaguila, J. Grapevine yield and leaf area estimation using supervised classification methodology on RGB images taken under field conditions. *Sensors* **2012**, *12*, 16988–17006. [CrossRef] [PubMed]
25. Nuske, S.; Wilshusen, K.; Achar, S.; Yoder, L.; Narasimhan, S.; Singh, S. Automated visual yield estimation in vineyards. *J. Field Robot.* **2014**, *31*, 837–860. [CrossRef]
26. Xia, C.; Lee, J.-M.; Li, Y. In situ detection of small-size insect pests sampled on traps using multifractal analysis. *Opt. Eng.* **2012**, *51*, 02700. [CrossRef]
27. Kai, H.; Muyi, S.; Xiaoguang, Z.; Guanhong, Z.; Hao, D.; Zhicai, L. A New Method in Wheel Hub Surface Defect Detection: Object Detection Algorithm Based on Deep Learning. In Proceedings of the International Conference on Advanced Mechatronic System, Xiamen, China, 6–9 December 2017; pp. 335–338.

agriculture

MDPI

Review

On Using Artificial Intelligence and the Internet of Things for Crop Disease Detection: A Contemporary Survey

Houda Orchi *, Mohamed Sadik and Mohammed Khaldoun

Networking Embedded Systems and Telecommunications (NEST) Research Group, Engineering Research Laboratory (LRI), Department of Electrical Engineering, National Higher School of Electricity and Mechanics (ENSEM), Hassan II University of Casablanca, Casablanca 8118, Morocco; m.sadik@ensem.ac.ma (M.S.); m.khaldoun@ensem.ac.ma (M.K.)

* Correspondence: houda.orchi@ensem.ac.ma

Abstract: The agricultural sector remains a key contributor to the Moroccan economy, representing about 15% of gross domestic product (GDP). Disease attacks are constant threats to agriculture and cause heavy losses in the country's economy. Therefore, early detection can mitigate the severity of diseases and protect crops. However, manual disease identification is both time-consuming and error prone, and requires a thorough knowledge of plant pathogens. Instead, automated methods save both time and effort. This paper presents a contemporary overview of research undertaken over the past decade in the field of disease identification of different crops using machine learning, deep learning, image processing techniques, the Internet of Things, and hyperspectral image analysis. Additionally, a comparative study of several techniques applied to crop disease detection was carried out. Furthermore, this paper discusses the different challenges to be overcome and possible solutions. Then, several suggestions to address these challenges are provided. Finally, this research provides a future perspective that promises to be a highly useful and valuable resource for researchers working in the field of crop disease detection.

Keywords: machine learning; deep learning; image processing; hyperspectral image analysis

Citation: Orchi, H.; Sadik, M.; Khaldoun, M. On Using Artificial Intelligence and the Internet of Things for Crop Disease Detection: A Contemporary Survey. *Agriculture* **2022**, *12*, 9. https://doi.org/10.3390/agriculture12010009

Academic Editors: Gniewko Niedbała and Sebastian Kujawa

Received: 21 November 2021
Accepted: 16 December 2021
Published: 22 December 2021

1. Introduction

Agriculture is the mainstay of many countries. Due to population growth, the demand for food is steadily increasing. To satisfy this pressing need, it is necessary to increase agricultural productivity and protect crops. Nevertheless, crops are highly prone to different diseases due to a large number of pathogens present in their environment. Some of these disease pathogens are virus organisms, whereas others are fungal or bacterial [1]. Crop diseases can reduce productivity by 10% to 95% [2], resulting in a significant decrease in the quantity and quality of agricultural production. Therefore, early identification of diseases is crucial to avoid huge losses and reduce the excessive use of pesticides, which can harm human health and the environment. In most cases, and especially in developing countries and small farms, farmers identify crop diseases with the naked eye based on visual symptoms. This is a tedious task that requires expertise in plant pathology and excessive treatment time [3]. Moreover, if the field is attacked by a rare disease, farmers seek expert advice to obtain an accurate and efficient diagnosis, which obviously generates additional treatment costs [4]. Thus, this method of visual observation is not practical and feasible for large farms and may even yield erroneous predictions due to biased decisions [5]. The restrictions of the traditional approach have motivated researchers to develop technological proposals for the early identification of crop diseases in an accurate, fast, and reliable manner, and in order to meet the increasing demands of consumers and alleviate the environmental impact of chemical inputs on the environment and health. In this regard, several methods [6–9] have been proposed to automate the process of disease detection. These methods for automatic recognition of crop diseases are divided into two groups, direct and indirect methods [10].

Direct methods comprise molecular [11] and serological techniques [12,13] that provide accurate and direct detection of the pathogens triggering the disease, although these techniques require a significant amount of time for the collection, processing, and analysis of the collected samples. By comparison, optical imaging techniques [14,15] are among the indirect methods that are able to identify diseases and predict the health of the crop through different parameters such as morphological change and transpiration rate. Fluorescence and hyperspectral imaging [16] are some of the most widely used indirect methods for early disease identification. Although hyperspectral images are a valuable source of data and contain more information than ordinary photos [17], hyperspectral devices are very expensive, bulky, and difficult to obtain for low-income farmers. Alternatively, other types of digital cameras are available at reasonable prices in electronics stores. As a result, most of the automatic identification processes considered so far are focused on visible domain images, which enables the use of very accurate and fast algorithms. Hence, this review focuses on various approaches based on image processing techniques and spectroscopy for automatic crop disease detection using numerous approaches and algorithms using deep and machine learning, fuzzy logic, and transfer learning.

The main objectives of this paper are, first, to identify the major issues that have not yet been properly explored in previous studies on the automation of the disease recognition process; and, second, to highlight future directions that may help circumvent these challenges. The upcoming sections are structured in the following order. Section 2 provides an insight into the current state of the art in disease recognition. Then, Section 3 is devoted to the comparative study of the various techniques used, identifying their advantages and drawbacks, followed by Section 4, in which the results are discussed and analyzed. In Section 5, the gaps in the existing literature are addressed. These shortcomings constitute possible avenues to explore in future research, which is addressed in Section 6. Eventually, the conclusion is drawn in Section 7.

2. Background

Manual identification of crop diseases is both fastidious and inaccurate, meaning it is only feasible in small farms [5]. In contrast, automatic disease detection is significantly more accurate and takes less time and labor [18]. As a result, numerous studies [19–22] have been conducted and are discussed in detail below. This section provides a review of different techniques applied in the identification of crop diseases, presents the taxonomy of various crop diseases, and describes the concept of image processing and machine learning. It also demonstrates the application of hyperspectral imagery, the Internet of Things, and deep and transfer learning in the field of disease recognition.

2.1. Taxonomy of Crop Diseases and Their Symptoms

The leaves of crops are highly prone to diseases, which are a natural phenomenon [23]. However, if corrective measures are not taken at the right time to stop the spread of the disease, it leads to a significant reduction in the quality and quantity of agricultural products [24]. Crops are affected by various pathogens [1] such as viruses, bacteria, fungi, and deficiencies. Thus, the pathogens responsible for the disease are classified into two categories [25]: autotrophs, which thrive on living tissue, or saprophytes, which dwell on dead tissue. The symptoms of the disease adversely affect the development and growth of crops and are easily visible. Leaf discoloration is the first symptom of disease in plants. In addition, the shape and texture of the leaves are highly useful in detecting various diseases. Therefore, different diseases, such as mildew, rust, and powdery mildew, can be detected by processing images of the leaves [26,27].

The following is a brief description of the three common types of plant diseases [28] that are illustrated in Figure 1 and described in Table 1:

- ***Virus diseases*** [1]: Among all plant diseases, those caused by infection are difficult to identify and diagnose; moreover, these symptoms are mistaken for signs of nutritional deficiency or injury, as there is no preconceived indicator that can be constantly monitored. Whiteflies, leafhoppers, aphids, and cucumber crawling insects are regular carriers of virus diseases.

- ***Fungal diseases*** [1]: Foliar diseases are caused by a fungus, such as downy mildew, anthracnose, and powdery mildew. It initially appears on old lower leaves, which have gray-green spots or are soaked in water. As the parasite matures, these spots darken and cause fungus to grow on them.
- ***Bacterial diseases*** [1]: Pathogens cause serious diseases in vegetables. They do not directly enter the vegetation, but rather through injuries or apertures in the crop. Crop injuries result from various pathogens, insects, and agricultural implements during tasks such as picking and pruning.

Figure 1. Different types of pathogens: viruses, fungi, and bacteria.

Table 1. Classification of some leaf diseases with their symptoms.

Plant Leaf	Diseases	Symptoms	Pathogen Category
Rice	- Brown spot/*Bipolaris oryzae* - Blast leaf/*Pyricularia oryzae Cavara*	- Whitish-gray center - An irregular dark brown	- Fungi
Cotton	- Faliar leaf/*Stemphylium solani* - Areolate mildew/*Cercospora* - Leaf spot/*Alternaria spot* - Bacterial blight/*Xanthomonas campestris*	- Spot of light-yellow color with dark brown margins - Tanned brown spot - Circular dark brown leaf spots to black - Halo yellowish green	- Fungi, bacterial, virus - Fungi - Fungi - Bacterial
Citrus	- Melanose/*Diaporthe citri* - Greasy spot/*Amycosphaerella africana* - Canker/*Xanthomonas citri* subsp	- The leaf becomes rough to the touch - Blister yellowish-brown - Includes flattened, swollen, cracked, round to irregular sunken	- Fungi - Fungi - Fungi, bacterial
Tomato	- Early blight/*Alternaria tomatophila* - Late blight/*Phytophthora infestans* - Powdery mildew/*Leveillula taurica* - Yellow curl/*tomato infectious chlorosis virus*	- Dark ring spot around it yellow - The dark spot is growing rapidly - Curly and yellowish leaf - Soaked in the water ringed by a yellow halo	- Fungi Fungi - Fungi - Virus
Maize	- Stalk rot/*Erwinia carotovora*	- Yellowing of dull green leaves and the lower parts of the stem	- Fungi
Wheat	- Rust/*Puccinia triticina Erikss.* - Powdery mildew/*Blumeria graminis* - Bacterial blight/*Pseudomonas syringae*	- Pale leaves spots - While gray or brown spot - Halo yellowish green	- Fungi - Fungi - Bacterial
Watermelon	- Anthracnose/*Colletotrichum obiculare* - Downey mildew/*Pseudoperonospora cubensis*	- Irregular yellow patches - Yellow to white spots	- Fungi

2.2. *Application of Machine Learning and Image Processing in Disease Identification*

Foliar images are an excellent and rich source of data on plant pathology and morphological behavior; thus, these data must be thoroughly extracted and analyzed. Image processing plays [28] a crucial role in the diagnosis and analysis of leaf diseases. The procedure adopted in this leaf disease identification process is illustrated in Figure 2, showing an insight into the different techniques employed by the authors to detect the disease by means of image processing and artificial intelligence.

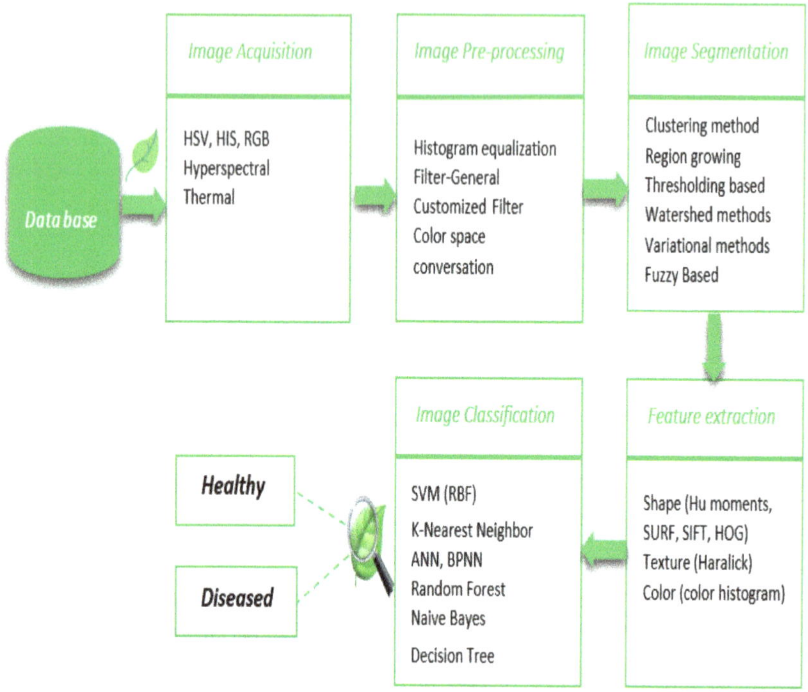

Figure 2. Different approaches for the identification of leaf diseases.

The primary step [29] in identifying diseases is the acquisition of images. In most cases, images can be fetched either from a digital camera or an imaging system. As raw images tend to contain noise, removing these impurities is required. As a result, the second step is known as image pre-processing, and involves the removal of unwanted distortions, in addition to contrast enhancement, to clarify and brighten the image features. For example, a Gaussian function that creates soft blur is commonly used to lessen the noise in the image. Subsequently, image segmentation [30] is the third step in which the image is segmented from its background, whereas the region of interest (ROI) is partitioned to emphasize the prominent features. The fourth step is feature extraction [31], which unveils the information and details of an image. As a side note, the leaf features usually include shape, texture, and color, which are used to diagnose the crop. Thus, these chosen features form an input feature vector which is then fed into the classifier. Using this vector, it is possible to discriminate one class of objects from another. The final step is classification [32]. Note that the choice of a suitable classifier depends on the specific problem. The classifier's aim is to recognize the images by sorting them into several predefined classes based on the resulting feature vector obtained in the fourth step. For this purpose, the classification task contains two phases, namely, training and testing. The training operation trains the classifier on a training dataset; thus, the greater the number of training sets, the better the accuracy obtained. It should be noted that the result, which is the crop's healthy state or diseased state associated with the species name, must be achieved as swiftly as possible.

2.3. Application of Deep and Transfer Learning in Disease Recognition

Over the past decade, deep learning [33–35] and transfer learning [19,36] applications in agriculture have gained widespread success and yielded highly promising outcomes due to their capability to reliably learn and discern visual features. Numerous intriguing studies [33,35,37–40] have been published on the employment of these promising approaches for identifying diseases. Of particular note, the use of transfer learning is a trend that is becoming increasingly popular and is widely used by researchers [41,42]. Furthermore, transfer learning is not a sole technique, but rather a set of fine-tuned techniques, which enables the development of highly accurate models on a more restrictive specialized dataset, such as those for plant diseases. Mohanty et al. [43] showed that the fine-tuning approach is far better than a CNN model that is trained from scratch. Another model is the Neural Network (NN), which is broadly employed and recommended to analyze hyperspectral data for the premature detection of diseases. Its basic mechanism was inspired by the human nervous system, and it possesses specific capabilities such as learning and generalization that aid in crop disease diagnosis. In contrast to other machine learning methods, it has a more accurate diagnostic capability because it is better able to combine training sets. Another similar comparative study was undertaken by Zhu et al. [44], wherein back-propagation neural networks (BPNNs) were tested with the support vector machine (SVM), random forest (RF), latent Dirichlet allocation (LDA), extreme learning machine (ELM), LS-SVM, and partial least squares discrimination analysis (PLS-DA) for pre-symptomatic detection and classification of tobacco mosaic virus (TMV) disease with the use of hyperspectral imaging. Similarly, Zhu et al. [45] studied the feasibility of hyperspectral imaging as a non-invasive technique for early detection of TMV disease with machine learning classifiers and the variable selection technique. The results revealed that the back-propagation neural network model (BPNN) achieved 95% accuracy, whereas the chemometric models achieved an accuracy of 80%. It is worth mentioning that it is possible to implement pattern identification methods such as the random forest and support vector machine by utilizing a new pattern recognition technique, named the Artificial Intelligent Nose. Cui et al. [46] provided a review of different invasive and non-invasive techniques, including their advantages and drawbacks, in which the authors noted that the smart nose is a non-invasive and fast method for plant disease diagnosis. In essence, neural networks ensure the highest quality and unaltered spectral information for hyperspectral data analysis. The most well-known study of the ANN spawned the concept of deep learning, which has recently become popular in farming applications. Deep learning has received increasing and widespread interest from many researchers, particularly since 2018, as shown in Table 2. Researchers have made remarkable progress in crop image classification; some of the most typical and representative models are the convolutional neural network (CNN), auto-encoder (AE) recurrent neural network (RNN), and restricted Boltzmann machine (RBM). Many fascinating studies have been published on deep learning for crop disease classification and detection. Among these works, that of Ma et al. [47] presented a deep convolutional neural network (DCNN) model able to detect over four types of cucumber disease. In a comparison with other traditional methods, such as the support vector machine, naive Bayes, and AlexNet, the DCNN was capable of identifying the different cucumber diseases with very high accuracy of up to 93.41%. Similarly, Tran et al. [48] offered a monitoring system for tomato growth and to maximize tomato yield. This system was able to classify nutritional deficiencies and diseases during growth. Thus, agricultural experts evaluate the symptoms based on the results to protect tomato crops. In a similar manner, to effectively monitor apple tree growth at each stage and estimate the yield, Tian et al. [49] deployed a dense YOLOV3 model that utilizes techniques for data augmentation to prevent overfitting. Their approach was found to be valid and applicable to apple orchards, although their study included wavy lights, interlaced fruit, and complex backgrounds.

Table 2. A brief summary of different research on transfer and deep learning since 2018 for identifying crop diseases.

Year	Authors	Model	Dataset				Accuracy
			Crop Name	Nb of Classes	Nb of Images	Name of Dataset	
2020	Singh et al. [37]	MobileNet, R-CNN	13 types	27	2598	PlantVillage	70.53%
2020	Al-bayati et al. [38]	DNN, SURF, GOA	Apple	6	2539	PlantVillage	98.28%
2019	Arsenovic et al. [39]	CNN-Multichannel	12 species of crops	42	79,265	PlantVillage	93.67%
2019	Costa et al. [40]	InceptionV3 and CNN using a Hierarchical Approach	Apple, Tomato, peach	16	24,000	PlantVillage	97.74%
2019	Geetharamani et al. [35]	9-layer deep CNN	14 species of crops	39	61,486	Leaf disease dataset	96.46%
2018	De Luna et al. [50]	CNN, Faster R-CNN	Tomato	4	4923	Own	91.67%
2018	Ferentinos et al. [33]	Overfeat, VGG16, AlexNet	25 species of crops	58	87,848	Open Dataset	99.53%

2.4. Hyperspectral Imaging Applied to Disease Recognition

The hyperspectral imagery method has been strongly developed during the past two decades [51], and used to identify abiotic and biotic stresses in cultivated plants [52]. Hyperspectral imaging is a technique combining spectroscopy and imagery, making it possible to simultaneously obtain the spatial and spectral information of an object. Disease infection causes changes in plant biochemical and biophysical properties, such as transpiration rate, tissue structure, water, and pigment content. These changes can then alter plant spectral properties, intercellular space, and water content [53]. However, the hyperspectral system is able to capture these spectral features. Zhu et al. [44,45] conducted a similar study to detect TSWV infection growth in tobacco, in which the authors reported that hyperspectral reflectance was gathered in the visible and near-infrared range to distinguish between healthy and infected TSWV tobacco leaves using statistical analysis methods. Primarily, the TSWV presence was identified at 14 DPI. Moreover, Zhu et al. [45] demonstrated that hyperspectral imaging is able to detect the tobacco mosaic virus (TMV) infection before showing any symptoms, while utilizing SPA for the selection of the effective EW wavelength and, most significantly, for identifying various diseases. Due to the huge number of spectral values that are highly correlated in the hyperspectral dataset, high dimensionality and multi-collinearity frequently appear in hyperspectral data [54,55]. Accordingly, the selection of EWs is crucial for hyperspectral analysis in order to lessen the computational complexity, increase the efficiency of using hyperspectral data, and reduce the computational complexity. Thus, to address this multi-collinearity issue, a variety of approaches and methods have been presented, such as the successive projection algorithm (SPA) [45,56], partial least squares regression (PLSR) models [54] and genetic algorithms (GAs).

2.5. Application of IoT in the Field of Leaf Disease Recognition

The Internet of Things (IoT) has improved agricultural capabilities. IoT applications can help farmers at any time during their farming activities and keep them updated with the latest crop and weather information to remotely monitor their fields. By means of IoT applications [57–59], farmers can make plans for the next season's harvest. Furthermore, they can detect crop diseases at an early stage to curb the spread of disease and save their yield. Agricultural IoT apps clearly play a major role in increasing agricultural production and decreasing crop losses due to diseases. In this context, a large amount of research has been conducted to identify diseases, as shown in Table 3. Truong et al. [60] devised an IoT-based system made up of various devices that is able to deliver real-time environmental information and send it to the cloud to be stored. These environmental data are processed and scrutinized to predict weather conditions by means of the SVM algorithm deployed in the cloud in order to detect crop fungal diseases. In addition, better results have been achieved when the Internet of Things and image processing have been combined and implemented in the area of disease recognition. Krishna et al. [58] implemented an IoT system featuring SMS alerts that enables automatic disease detection and pesticide spraying using the NodeMCU.

Table 3. Summary of the literature survey on Internet of Things systems.

Researchers	Detection Techniques and Algorithms	Parameter Evaluation
M. Mishra et al. [39] 2021	An IoT-based automated plant disease monitoring and detection system, using the median filter and a modified optimizer called the SCA-based RideNN Cycling Neural Network	The RideNN model based on SCA achieved accuracy of 91.56%
Devi et al. [37] 2019	IoT system using GLCM, RFC, and k-means clustering	The overall accuracy of disease detection and classification based on RFC-GLCM was almost 99.99%
Krishna et al. [38] 2019	IoT system using SVM and k-means clustering	An immediate SMS alert to the farmer
Chen et al. [36] 2019	RiceTalk platform using an AI model and IoT devices	Net prediction accuracy was 89.4%
Win et al. [41] 2018	IoT-based remote rice monitoring IoT system using deep learning and transfer learning	Real-time monitoring of environmental parameters
Truong et al. [40] 2017	IoT system using SVM	Real time analysis

3. Comparison of Various Crop Disease Detection Techniques

The primary objective of this section is to provide an overview of research carried out during the past decade for identifying crop diseases. Table 4 provides an outline of several methodologies adopted by researchers in the field of crop disease using machine learning, image processing, the Internet of Things, transfer learning, and deep learning techniques. It also indicates the limitations and gaps that need to be filled to help develop an automatic, efficient, accurate, and faster system in the future. Thus, in the conducted research, the authors conclude that deep learning provides accurate and highly promising results compared to other classification and detection methods. Additionally, the use of preprocessing techniques significantly improves segmentation accuracy. The k-means algorithm is the most widely and commonly used technique [29,58,61–63] for segmenting diseased leaves and classifying crop diseases. In practice, no generalizable algorithm is able to solve all issues, so choosing a suitable learning algorithm for a specific problem is a crucial step for the model efficiency. Note that the extracted texture features are the most relevant and most useful for representing the disease-affected regions in the images, which are then employed to train the support vector machine (SVM) and neural network (NN) classifier. It is further emphasized that these texture features are arithmetical parameters that are automatically calculated by means of the gray level co-occurrence matrix (GLCM) [64,65], as stated below:

1. *ASM:* The second angular momentum that stands for the total sum of squares in the GLCM.

$$Energy = \sum_{i,j=0}^{N-1} (P_{i,j})^2 \quad (1)$$

2. *Contrast:* Denotes the sum of the difference in local intensity, where $i \neq j$.

$$Contrast = \sum_{i,j=0}^{N-1} P_{i,j}(i-j)^2 \quad (2)$$

3. *Entropy:* The quantity of image information necessary for the compression.

$$Entropy = \sum_{i,j=0}^{N-1} -\ln(P_{i,j}) P_{i,j} \quad (3)$$

4. *Correlation:* Refers to the linear dependence of the adjacent pixels' gray levels.

$$Correlation = \sum_{i,j=0}^{N-1} P_{i,j}(i-\mu)(j-\mu)/\sigma^2 \quad (4)$$

where $P_{i,j}$ is the i, j component of the GLCM normalized symmetric matrix and N denotes the number of gray levels. σ^2 is the intensity variation of all pixels as given below:

$$\sigma^2 = \sum_{i,j=0}^{N-1} P_{i,j}(i-\mu)^2 \quad (5)$$

Moreover, the average of GLCM is μ given by:

$$\mu = \sum_{i,j=0}^{N-1} iP_{i,j} \quad (6)$$

5. *Homogeneity feature:* Represents the homogeneity of the voxel pairs of the gray level, and is equal to 1 for a diagonal GLCM.

$$Homogeneity = \sum_{i,j=0}^{N-1} P_{i,j}/1+(i-j)^2 \quad (7)$$

Ultimately, the models for identifying and classifying crop diseases were evaluated by means of various metrics, which were specific to the model used in each study, such as sensitivity, precision (*P*), recall *(R)*, quality measure *(QM)*, and F1-score. The statistical evaluation measures used to analyze the quantitative performance of crop disease detection models with deep and transfer learning can be calculated as follows:

$$Precision = \frac{TP}{(TP+FP)} \quad (8)$$

where *Precision (P)* is the fraction of true positives (*TP*) to the total amount of relevant results, that is, the sum of *TP* and false positives (*FP*). For multi-class classification problems, *P* is averaged across the classes.

$$Sensitivity = \frac{TP}{(TP+FN)} \quad (9)$$

Sensitivity/Recall (*R*) is the fraction of *TP* to the total amount of *TP* and false negatives (*FN*). For multi-class classification problems, R obtains the average of all classes.

$$Specificity = \frac{TN}{(TN+FP)} \quad (10)$$

Specificity is the proportion of true negative (*TN*) samples to all healthy samples (true negatives and false positives). This measure is utilized to evaluate the performance of a proposed model in forecasting true negatives.

$$Accuracy = \frac{TP+TN}{(TP+TN+FP+FN)} \quad (11)$$

Accuracy is the proportion of correctly classified samples to the total number of classified samples. This measure is employed to assess the overall performance of a suggested model.

$$F1\_score = \frac{2 \times (Sensitivity \times Precision)}{(Sensitivity + Precision)} \quad (12)$$

F1-score is the harmonic average of both precision and recall. For multi-class classification problems, *F*1 is averaged across all classes, where:

TP: represents the number of true positive image samples that are perfectly identified as infected.
FP: is the number of false-positive image samples that are incorrectly classified as infected.
TN: is the number of true-negative image samples that are correctly classified as healthy.
FN: is the number of false-negative image samples that are incorrectly identified as uninfected.

As revealed in the previous studies [66], it was found that accuracy was the most widely adopted metric; it was widely used in 72% of the articles reviewed, followed by the confusion matrix, then precision, recall, and the F1 measures. Some research has examined the root mean square error (RMSE), mean absolute error, R-squared, and mean squared error (MSE), among others. The authors note that it is difficult, if not impossible, to make a comparison across papers because different metrics were used for different tasks, different models, datasets, and parameters; in addition, different crops and diseases were analyzed under different conditions. Furthermore, it is very important to examine whether the researchers tested their implementation on an identical dataset (e.g., by splitting the dataset into training and validation sets), or whether they used different datasets to test their solution.

Table 4. Related work in the area of crop disease identification.

Title	Methodology	Advantages	Disadvantages
Tomato plant disease detection using transfer learning with C-GAN synthetic images (Abbas et al., 2021) [19]	This research paper provides a DL-based method for detecting tomato disease that uses the C-GAN to generate the synthetic images of tomato leaves for data augmentation purposes. Then, a pre-trained DenseNet121 model is fine-tuned on synthetic and real images to classify tomato leaves images into ten disease categories.	- The proposed method reached an accuracy of up to 99.51%, 98.65% and 97.11% in classifying tomato leaf images to the categories of 5, 7 and 10, respectively. It is shown that this method outperforms the current methodologies. - The C-GAN prevents overfitting and - enhances the network generalization	- The tomato disease detection was executed on the leaves, but other plant areas such as stems, and branches must be highly involved. - The study was only done on tomato plant disease.
MEAN-SSD: A novel real-time detector for apple leaf diseases using improved light-weight convolutional neural networks (Sun et al., 2021) [21]	- This paper proposes a lightweight CNN detection model suitable for mobile device deployment, namely MEAN-SSD, to detect apple leaf diseases in real time. Data annotation and augmentation techniques were used to generate 26,767 disease spot images for training by collecting 2230 original images with simple backgrounds from the laboratory and complex backgrounds images collected from the orchard. - The model is capable of automatically extracting the features of five common disease spots from apple leaves.	- The results showed that the MEAN-SSD model is able to detect apple diseases accurately by reaching 83.12 mAP and a speed of 12.53 FPS. - The MEAN block is used as a basic module to boost the detection speed and shrink the model's size.	- Disease detection was only devoted to apple leaf diseases and more specifically to 5 types of disease spots, such as Brown spot, grey spot, Mosaic, Alternaria blotch, and Rust.
Detection of oil palm leaf disease based on color histogram and supervised classifier (Hamdani et al., 2021)[20]	- A new method for detecting oil palm leaf disease is proposed in this paper to discriminate between two leaf classes: healthy and diseased. Then, feature extraction is carried out in the RGB (R, G, and B), LAB (a and b), HSI (H and S), and HSV (H and S) color spaces by splitting the histogram of the 8-bin color channel. This is further performed on the segmented leaf regions resulting from the k-means clustering. A total of 41 selected features are produced using PCA and subsequently fed into the ANN classifier.	- The classification results have shown that the proposed method performs satisfactorily, as evidenced by the high specificity, sensitivity, and accuracy values, which reach 100%, 99.3%, and 99.67% respectively. - The applied method produces a smaller number of features with discriminatory and more powerful characteristics.	- According to the classification outcomes, an error occurred due to a leaf being misclassified as a healthy one. - The study was only conducted on the oil palm leaf disease.
Detection of Rice Leaf Diseases Using Image Processing (Pothen et al., 2020) [22]	The proposed system identifies three diseases (bacterial leaf blight, leaf smut, and brown spot) that affect rice plant leaves using IP and ML techniques. This system, in turn, helps farmers save their crops at an early stage. As a first step in the process sequence, the images are collected and further pre-processed to ensure that the image features are upgraded and undesired distortions are eliminated, followed by segmenting the images through the Otsu thresholding algorithm. Using the segmented area, a range of features are extracted using the LBP and HOG. Then, these obtained features are classified using the SVM and reached 94.6%.	- SVM + HOG with polynomial kernel function can be used to detect other plant diseases. - The proposed work is relevant, offering better precision (of 94.6%) compared to other work.	- One downside of the LBP fundamental operator is its inability to capture certain prevailing features.
Image Processing Technologies for Automatic Detection of Plant Disease and Alerting System in Agricultural Farms (Mugithe et al., 2020) [7]	The authors developed a system able to detect leaf diseases and alert the farmer in case of the need to promptly act to circumvent the spread of the disease in the field. They employed IP techniques that entail six steps. Firstly, the leaf images were taken in real time via a webcam connected to RaspberryPi. Secondly, the images were pre-processed, segmented, and clustered using the k-means clustering algorithm, then features (e.g., perimeter and light intensity) were retrieved from the images. Lastly, these extracted features were evaluated to classify the leaf diseases. Once the disease is detected, the buzzer rings and an alert is generated so that the farmer can intervene promptly. Note that this system works in two ways, namely in the GUI and in real time.	- The results of the disease Alternaria Alternata achieved precision of 95.16313% in the graphical interface. - The use of a warning system when detecting a disease.	- The authors did not address any other disease and not even the reported results are obvious.

Table 4. *Cont.*

Title	Methodology	Advantages	Disadvantages
Plant Disease Detection Using Internet of Thing (IoT) (Usman et al., 2020) [67]	This paper presents the innovation of the IoT in agricultural infection and insect pest control. Data on insects and diseases are collected using a WSN, so an IoT-based control framework was proposed to obtain horticultural data from a farm with levels of trees and three frameworks. First, a computation framework was deployed to deem whether the plant is healthy or affected and second, an automated framework determined the disease closeness in the plants, and a mechanized framework was then set up to recognize the diseases through humidity, temperature, and shade sensors. Thus, through its sensors, the plants' progress is registered and then dissected using Arduino programming. Then, this collected information is transmitted to the cloud by WIFI to be processed and analyzed. Eventually, this information is compared to the whole data to determine if the studied plant is healthy or affected.	- The proposed IoT-based model is low-cost. - Low-income farmers can purchase it and take benefit of it for curtailing the disease spread.	- The authors restricted their model to only the three following parameters: temperature, moisture, and leaf shade. Another constraint is that the evaluated features for the considered parameters are not accurate. Moreover, a range of features was taken from these parameters that may fluctuate unexpectedly depending on the environmental conditions. - The leaves of diseased plants are not classified, so the disease types are still not able to be known.
Detection of Plant Leaf Disease using Digital Image Processing (Mojjada et al., 2020) [29]	This article focuses on the early identification of plant leaf diseases by image analysis. Thus, automated disease detection reduces the work for monitoring agricultural sites. The identification of diseases is carried out through various IP techniques and ML; in particular, a genetic optimization algorithm was used after image segmentation by k-means to obtain optimized results and they also exploited SVM for disease classification.	- The algorithm used was tested with an accuracy of 75% in five classes of infected leaf images identified for corn, tomato, bell pepper, peach, and grape.	- Very common methods were used in the paper.
Precision Method for Pest Detection in Plants using the Clustering Algorithm in Image Processing (Reddy et al., 2020) [68]	This work presents an accurate method for detecting pests in plants using the k-means clustering algorithm. Disease recognition involves steps such as image acquisition, image preprocessing, segmentation, and classification. First, the RGB leaf images were converted to HSV for partitioning, and then the median filter and boundary detection algorithm were both applied during the pre-processing step to suppress the clamor. Finally, the k-means clustering was employed to cluster the images.	- This paper provides an efficient and accurate framework for the detection of affected images. - The k-means clustering provides high accuracies compared to other methods and takes less time for the processing.	- Very typical and old techniques were considered in this paper.
Detection and Classification of Plant Diseases Using Image Processing and Multiclass Support Vector Machine (Khan et al., 2020) [69]	The authors described a framework for plant disease using ML and IP techniques. First, the suggested algorithm is applied to a 148-image dataset which contains 5 types of leaves diseases, namely, Alternaria, fire blight, Anthracnose, and Cercospora leaf spot, and the plant images were split up into two sets. A training set is composed of 73 images and a testing set is composed of 75 images. Then, image segmentation is performed to isolate the pathogenic parts of the leaf. Then, 13 texture features were extracted from the image, of which, nine features (standard deviation, variance, mean, entropy, smoothness, skewness, RMS root mean square, inverse difference, and kurtosis) are calculated using the assigned segment in RGB color space. The other four features (homogeneity, contrast, energy, and correlation) are determined from a grayscale image. Finally, the healthy and diseased leaves are classified based on the feature vector extracted using SVM.	- The obtaining results for plant disease detection showed that the proposed method yields a highly accurate rate of up to 92.8571%.	- The intervention of the operator is crucial to select the segment affected by the disease because this operation is not automatically executed in the system rather, it is performed by visual examination of the three segments.
Evolutionary feature optimization for plant leaf disease detection by deep neural networks (Al-bayati et al., 2020) [38]	The researchers used a DNN for apple leaf disease identification, namely black rot, apple scab, and cedar rust, by applying the GOA and Robust Accelerated Feature SURF, where GOA was employed for feature optimization and SURF was applied for feature extraction. Prior to the implementation of DNN, many steps were performed, such as the image improvement in the pre-processing process and the ROI segmentation. Then, the features were extracted using the SURF descriptor, followed by the optimization by the GOA algorithm, and, finally, the disease classification was carried out by the use of DNN.	- The experiments showed that the method based on DNN optimized by SURF provides a higher mean value of 98.28% in comparison with the other techniques; hence the accuracy of the model increases by 18.03%. Thus, the basic model has better transferability compared to the metric model.	- Only foliar diseases of apples were addressed in this paper.
Leaf Disease Detection using Image Processing (Karthikeyan et al., 2020) [65]	This research used IP techniques along with the SVM classifier to detect plant diseases. The identification of plant diseases requires the steps of transforming an RGB image to grayscale, then enhancing the image using the adaptive color histogram AHE, extracting 13 textural features using the GLCM and, finally, using SVM to classify the different types of diseases. Note that more than 500 images were taken for training and testing with intensity values ranging from 0 to 255.	- The system reveals the presence of disease in the leaves in a shorter time and at a lower cost than conventional systems.	- The accuracy rate is not given; also, the structure of the algorithm is complicated.
Plant disease detection using image processing techniques(Sawant et al., 2020) [70]	An IM and DL techniques-based approach was proposed for plant disease identification. First, pictures of healthy and unhealthy leaves are acquired and then stored in the database for preprocessing. Additionally, the images are pre-processed using different techniques such as histogram equalizer, Contrast Limited AHE, and image resizing. Then, once the RGB image is converted to CLAHE, it is resized to 70 × 70 for better resolution. Moreover, the leaf features are extracted and released to the CNN using the SoftMax function for plant disease classification. Note that the first layer has 1000 neurons.	- The CNN method enables accurate detection and classifies diverse plant diseases using IP techniques.	- Results were not provided in this paper.

Table 4. *Cont.*

Title	Methodology	Advantages	Disadvantages
Convolutional neural network for automatic identification of plant diseases with limited data (Afifi et al., 2020) [6]	Several approaches were developed in this study to identify plant diseases with little data. A DAML and triplet network approach was set up using three architectures of CNN (ResNet50, 34, and 18). Using a large dataset, the approaches were trained and then fitted from 5 to 50 images per disease for detecting new diseases.	- The model reached an accuracy of up to 99% when the change from the source domain to the targets was slight, but when the change was significant, the accuracy was up to 81%.	- The results showed a significant error rate for DAML methods of 22.2 per 50 shots and 42.6 per 5 shots compared to the other methods. - The basic model lags significantly behind the other methods.
Leaf disease detection using machine learning (Fulari et al., 2020) [71]	An efficient method for the identification of a healthy or infected leaf was presented using IP and ML techniques. The data were taken from the Kaggle website which contains more than 12,949 images. The method implies different steps such as image preprocessing, segmentation of the image, feature extraction (shape, color, and texture) with the use of GLCM, and classification with the use of SVM.	- The SVM offers a number of advantages over other classifiers, as it is efficient in high-dimensional spaces. - The SVM yielded an accuracy of up to 80% while the CNN provided an accuracy of up to 97.71%.	- The presented method, based on the CNN, provides good accuracy. Nevertheless, it is tedious and a lot of time is required to train the model.
Deep transfer learning models for tomato disease detection (Ouhami et al., 2020) [72]	The authors conducted a study to identify the most suitable DL model for identifying tomato diseases based on RGB leaf images, which were split into 6 different kinds of infections and parasitic attacks. Thus, two architectures of DL models, namely DensNet121 with 161 layers and VGG16, were used to perform the study.	- The results obtained were very promising with an accuracy of up to 95.65%, 94.93% and 90.5% respectively for the DensNet161, DensNet121, and VGG16 models, which shows that DensNet161 with 20 training periods surpassed the other two architectures. - DensNet models require significantly fewer parameters and calculations to achieve optimum performance.	- Adverse transfer can occur and may dramatically lower the model's accuracy.
A new segmentation method for plant disease diagnosis (Gurrala et al., 2019) [64]	IP and AI methods were used for the recognition of diseases such as scab, anthracnose, blight, and spots on plants. First, the RGB image was converted to grayscale, and the image was then segmented using the k-means and the modified CPDA algorithm. Thus, from the result obtained, a comparison was made between these two segmentation algorithms. The statistical parameters of the segmented image were calculated using the GLCM method; that is, the characteristics (entropy, mean, variance, type, RMS, contrast, correlation, energy, homogeneity, regularity, kurtosis, asymmetry, IDM moment of difference). Finally, the SVM classifier was trained with a dataset of about 100 images of leaves affected by the disease.	- The proposed modified CPDA detection algorithm yields more accurate results than the k-means clustering.	- Even if a disease is detected in a shorter time, the accuracy is limited.
IoT Enabled efficient Detection and Classification of Plant Diseases for Agricultural Applications (Devi et al., 2019) [61]	The authors proposed a simple and efficient IoT-based solution for the detection of bunch top and Sigatoka diseases in banana tree located on hills. First, 80 hill banana plant images are captured and then resized to 256 × 256 in the image preprocessing phase. Then, the preprocessed image is converted to a gray image. Thus, the histogram equalization technique is used to equalize the histogram of the resized gray image so that the intensities of the image are better distributed for better segmentation, which is performed by k-means clustering. From the segmented image, the GLCM features are extracted and uploaded to the cloud for further analysis. From these extracted characteristics, the hill banana diseases are classified using the RFC technique. Finally, the data is collected and analyzed by agricultural experts. In addition, this system allows remote monitoring of environmental parameters such as soil humidity and temperature to prevent diseases caused by climate change and pathogens as much as possible.	- The performance results showed an overall detection accuracy of 99.99% and demonstrate that RFC-GLCM-based leaf disease classification works best for the hill banana dataset. - Agricultural experts provide solutions to farmers in case of plant disease or massive changes in environmental parameters on the agricultural field.	- The accuracy of the system depends on the ambient conditions of the agricultural field, such as the angle of image capture and the lighting in the field.
A Preprocessing Approach for Accurate Identification of Plant Diseases in Leaves (Deepa et al., 2018) [73]	In this paper, the proposed method takes RGB images as the input and applies the preprocessing methods, such as image sharpness and median filters, to eliminate the noise from images, and for deblurring and edge detection. Then, these images were segmented using k-means clustering. It should be mentioned that they used the peak signal-to-noise ratio in order to measure the quality of the images.	- The results show that the method adopted allows better identification of plant leaf diseases.	- They studied only three diseases: Alternaria Alternata disease, Bacterial Blight, and Anthracnose disease. - They did not consider extracting characteristics or classifying diseases into different groups.
AI and IoT methods for plant disease detection in Myanmar (Win et al., 2018) [74]	The researchers developed two prototypes. The first is a mobile application that classifies diseases on rice plants. With this simple application, the farmer can easily identify the diseases or pests on rice plants, without using agronomists. The second is a system for monitoring temperature, atmospheric pressure, water level, and the sunlight level of rice fields. Using this system, the intensive work is reduced by remotely monitoring this environmental data from anywhere an Internet connection is available. For the development of the mobile app, they collected 6 kinds of rice pictures, of bacterial leaf blight, brown spot, rice blast, mice attack, insects, and healthy rice. In addition, they used the Arduino nano to turn off/on the solenoid valve, which was directly connected to the Raspberry Pi every 30 min, to ensure efficient power and a long system runtime. Diseases were classified using TL and DL models.	- They developed a simple Android app to monitor the temperature and battery level on a SensorTag. Thus, the farm sensor data reading can be easily viewed on smartphones or PCs.	- They installed 8 SensorTags in different areas. The distance between them and the Raspberry Pi was less than 50 m. To monitor a wide range of many rice fields, another type of communication system, rather than the Bluetooth technology of the SensorTags, must be considered. A problem was encountered with the SensorTag coil battery, Under normal conditions, the batteries last at least a year while keeping the SensorTag alive. However, firmware level changes were added to the SensorTag to advertise all the time so, the LEDs on the SensorTags blink and the batteries only last a few weeks. Hence, they needed to replace the button batteries many times.

Table 4. *Cont.*

Title	Methodology	Advantages	Disadvantages
Plant diseases recognition based on image processing technology (Sun et al., 2018) [75]	A multiple linear regression-based plant disease identification system is presented, along with a histogram-based segmentation method for accurate and automatic threshold calculation. The proposed system is based on IP techniques including four steps: preprocessing, image segmentation, feature extraction, and regression model. First, the spatial domain image denoising is used to filter the noise. Then, an improved histogram-based segmentation method is developed to distinguish lesions from normal foliage, which automatically determines the threshold and optimizes the segmentation process. Next, the regional growth method is used for multi-point selection to extract certain disease-affected areas. Then, lesion feature extraction is performed in terms of shape, texture, and color. Finally, a multiple linear regression model is implemented to determine the type of disease, and then the least-squares estimation algorithm is used to calculate the coefficients and confidence intervals to set up the disease recognition system.	- The obtained results proved that the proposed recognition system has high accuracy, reliability, and effective recognition ability of plant diseases. - The histogram segmentation method has great advantages, such as speed, efficiency, and accuracy.	- The error rate increases progressively as the disease state becomes more complex because, as the disease worsens, the characteristic parameters become more complex and so the results become unstable.
An IoT based smart solution for leaf disease detection (Thorat et al., 2017) [76]	This paper presents a solution based on a smart farming technique using WSN, a Raspberry PI module, and a camera to establish wireless communication. In addition, CV techniques were used, such as masking, segmentation, and feature extraction to identify leaf diseases.It should be noted that they used the Apache server to retrieve and send data.	- The proposed system allows remote monitoring of the farm. Thus, the recognition of different leaf diseases was carried out successfully.	- The power supply of the system is limited, so the whole process stops if the system fails, which is an inconvenience. Moreover, the images taken during the day can be affected by excessive sunlight or reflections, which means the leaf color cannot be identified by the camera or captured clearly at night.
Plant disease detection using hyperspectral imaging (Moghadam et al., 2017) [77]	The authors used hyperspectral imaging (SWIR and VNIR), ML techniques, and IP for detecting tomato wilt virus in capsicum plants. First, images were acquired from a hyperspectral imaging system consisting of two Headwall push-broom hyperspectral cameras, namely the SWIR hyperspectral camera that provides a spatial resolution of 384 pixels and a spectral range of 900 to 2500 with 168 spectral bands, and the VNIR hyperspectral camera, which provides a spatial resolution of 384 pixels. Then, these images were pre-processed using different pre-processing techniques such as the space-adaptive filtering approach for detection and grid removal. In addition, these images were segmented using an unsupervised k-means clustering algorithm. Further, discriminative feature extraction was performed using the full spectrum, VNIR, SWIR, and vegetation indices. Finally, these features were employed to train classifiers for discriminating leaves obtained from inoculated and healthy plants.They also used other techniques in the process of disease identification as follows:They used the KL divergence or relative entropy to estimate the distances between two distributions of the control group and of the inoculated group.Note that the high-pass filter used is a third-order Butterworth FIR filter, which is applied to flatten the power spectral density of the image in order to detect a known signal corrupted by additive white noise.	- The obtained results showed excellent discrimination based on the full spectrum. - The effectiveness of feature extraction techniques used for automatic disease classification in greenhouse experiments. - The cluster analysis was able to successfully classify the image spectra into two classes by using the significant difference in spectral profile between the vegetation and its surroundings.	- The researchers treated only one disease, TSWV. - They were not able to correlate the reduction in overall values of SWIR dissimilarity with a plant pathogenic biophysical interaction for DAI 7 and 10.
Deep Learning for Image-Based Cassava Disease Detection (Ramcharan et al., 2017) [41]	A new model for the identification of plant diseases is proposed based on TL to train a CNN using a dataset of 2756 images in order to identify two types of damage caused by pests and three diseases. This model was deployed on a mobile application.	- The CNN avoids the tedious and complex step of extracting features from images to train models on a mobile device. The results proved that the TL approach offers greater precision in cassava, and is also an affordable, fast, and easily deployable strategy for digital devices. The model accuracy was 96% for RMD and CMD, 95% for GMD, and 98% for brown spot and cassava.	- Due to several factors, such as the lighting in a complex environment, accurate identification of diseases is challenging.
Early detection and classification of tobacco leaves inoculated with tobacco mosaic virus based on hyperspectral imaging technique (Zhu et al., 2016) [44]	The authors proposed a procedure for the early detection of tobacco disease infected with the mosaic virus by different ML algorithms based on hyperspectral imaging techniques. Images of healthy leaves inoculated with TMV for a period of 7 days, i.e., after inoculation, were acquired by a hyperspectral imaging system every day with the VNIR wavelength region 380–1023 nm. In addition, the spectral reflectance of the predefined ROI was extracted from the hyperspectral images using the ENVI software. The different ML algorithms, namely, RF, SVM, BaBPNN, LS-SVM, PLS—DA, ELM, and LDA were used to quantitatively classify the stages of tobacco disease using EW that were selected using the SPA.	- BPNN and ELM models successfully detected healthy and diseased tobacco leaves (2 DPI, 4 DPI, 6 DPI); the detection rates were 98.33% and 96.67%, respectively. - The classification accuracy of the training set and the test set was 84.17% and 75% respectively.	- They used a single VNIR component from the electromagnetic spectrum. - Generation of average spectra from a single ROI rather than all pixels. - The performance of the SPA-PLS-DA model was relatively poor compared to other models and the accuracy was slightly lower 75%.
Plant disease detection using image processing (Khirade et al., 2015) [78]	This article discussed IP-based methods for the detection of plant diseases, in particular segmentation and feature extraction algorithms. Concerning the segmentation techniques, they exploited the k-mean clustering, boundary and spot detection algorithm, and the Otsu threshold algorithm. Regarding the feature extraction techniques, they studied various methods, such as the color co-occurrence method, and regarding the disease classification phase, they used various algorithms such as ANN, BBPN, and SVM to accurately classify various leaf diseases.	- A variety of IP and machine learning (ML) techniques are discussed.	- The authors did not suggest any system and no results were given.

Table 4. *Cont.*

Title	Methodology	Advantages	Disadvantages
Detection of unhealthy region of plant leaves and classification of plant leaf diseases using texture features (Arivazhaga et al., 2013) [79]	A software solution is proposed for the detection of unhealthy regions and the automatic classification of diseases using the extracted texture characteristics. Thus, the scheme of the process consists of four steps. First, the acquired RGB leaf images were converted to HSV format. Then, the green pixels were masked and removed, followed by a segmentation process. Further, the texture characteristics were computed using the Color-Co-Occurrence Matrix and, finally, the classification was first performed using the minimum distance criterion, which yielded a gain of 86.7%. Results were then improved by the SVM classifier.	- The proposed method was tested on ten species of plants: beans, mango, lemon, jackfruit, banana, sapota, potato, and tomato. The results gave an accuracy of 94.74% using the SVM classifier. Therefore, the proposed approach can classify leaf diseases with little computational effort.	- Various reasons lead to an erroneous classification, namely, the identifying vectors of the taken features have to be optimized, and the symptoms of the diseased plant leaves vary from the early to the late stage.
Early detection of diseases on leaves by image processing (Han et al., 2013)[80]	The authors worked on images of vine leaves affected by mildew at different stages that were acquired by photographic sensors. They were able to detect diseases that are visible and barely visible to the naked eye using techniques based on the image representation in color space, and hybrids including information on both color and texture. Thus, for visual plant diseases, they were oriented towards changing the color space, in particular, to facilitate detection. To ensure better visualization, they were interested in the bands of each image (the V band of the YUV color space). Regarding diseases barely visible to the eyes, as a solution, they used the image analysis method that combines color and texture information. Otherwise, the most relevant challenge was the disease detection at an early stage. As an optimal solution, they worked on thermal imaging, which is very effective in detecting water stress. Finally, they calculated the Mahalanobis distance for image segmentation.	- As an advantage of the texture analysis (conversion of the image into hybrid space), they discerned 3 classes of textures and, for each class, 20 learning patches, to choose from. - The use of a thermal imaging device allows the premature detection of leaf diseases.	- They only treated mildew disease. - The downside of using the color space change method is that when the mildew stains are at the early stage, this method does not work well; hence, the need to couple this type of color treatment with one based on texture analysis. - They did not calculate the area infected with mildew disease.
Classification of cotton leaf spot disease using image processing edge detection techniques (Revathi et al., 2012) [81]	This article describes how to identify the part affected by leaf diseases using IP techniques. First, to segment the image, the authors used the Canny and Sobel edge detection technique and finally, they proposed an HPCCDD to analyze the images and classify the different diseases.	- The percentage was indicated to reduce leaf diseases. - The validation was undertaken via MATLAB.	- The authors only discussed cotton crops.
Color image segmentation using K-Medoids Clustering(Yerpude et al., 2012) [82]	A color image segmentation method was proposed using k-medoids clustering; the idea is to find groups of objects by finding the medoids for each group. The main objective of this paper is the critical analysis of different disease segmentation techniques.	- The obtained result shows the effectiveness of the k-medoids algorithm on different types of images, such as grayscale images. Moreover, the proposed method is not sensitive to noise.	- The segmented images are highly reliant on the centroids. However, they did not consider finding the optimal number of segments to obtain more accurate results. - The k-means algorithm is very sensitive to outliers, as the data distribution can be significantly distorted if an object has an extremely large value.
Color transform-based approach for disease spot detection on plant leaf (Chaudhary et al., 2012) [83]	A comparison was made between the effect of YCbCr, HSV, and CIELAB color spaces in the disease spot detection process since disease spots are different in color but not in intensity level. First, the different images of wheat, soybeans, rice, corn, cotton, mustard, apple, magnolia, and cherry leaf were taken. Then, these RGB images were converted to YCbCr color space using the color transformation formula and further to CIELAB and HSV color space. Then, for smoothing and enhancing the image, a median filter was applied. Finally, the segmentation of the image was carried out using the Otsu method on the components of the color space: Cr for filtered YCbCr, component H for the filtered HSV space, and component A for the LAB filtered color space.	- The experimental results show that the noise generated by the camera flash, background, and vein can be effectively removed using the CIELAB color model.	- Unfortunately, due to an imbalance in pigment formation and micronutrient deficiency, the color of the vein is different from the leaf spots. Thus, in some cases, these disease spots cannot be detected with precision using the CIELAB method.
Fast and accurate detection and classification of plant diseases (Al-Hiary et al., 2011) [84]	The acquisition of the leaf RGB images is undertaken to perform color space transformation. Once the segmentation of these images is performed using k-means clustering, the value of the green pixels is masked using the threshold obtained through Otsu's method. In addition, the affected clusters were converted to the hue saturation value. For texture analysis, the SGDM matrix is used for the formation of each image. Finally, the disease recognition process is performed by the ANN classifier.	- The color occurrence method is an advantageous strategy that gives exact precision.	- The methodology can be improved to increase the recognition rate of the classification process and to automatically estimate the severity of the disease detected.
Early detection of Fusarium infection in wheat using hyper-spectral imaging (Bauriegel et al., 2011) [85]	In this paper, Fusarium head disease was detected by spectral analysis in wheat, barley, oat, and rye plants. The PCA distinguishes affected from healthy tissue in the wavelength ranges 927–931 nm, 682–733 nm, 560–675 nm, and 500–533 nm. The SAM method is used to classify the degrees of infection. Finally, the best time to identify ear blight is the stage between 71–85 on the BBCH scale. They analyzed 292 spectra and 80 spectra for diseased and healthy tissue, respectively, in time series experiments. Smoothing was performed for these spectra with the "proc expand" function. Then, the first derivative was calculated and 104 individual spectra were subjected to PCA to assess the relevant wavelengths for discriminating between healthy and blight-infected tissue.	- The robustness and efficiency of the proposed algorithm are proved by experimental results of a database of about 500 plant leaves. - The SAM image analysis method correctly classifies the degree of disease at 87%, and the visual assessment error is 10%. - The SAM method yields accurate classification results; it is not practical for an online application because the analysis of 512 spectral bands involves a significant amount of computation.	- It is impossible to distinguish between different degrees of infection using only spectral analysis, due to the lack of symptoms. - The disadvantage of using the SAM method is that it is time consuming. Indeed, it involves the configuration of the reference spectra for the classification and, at the same time, the analysis of all the spectral bands.

- ***The Difference between machine learning and deep learning***

The difference between machine and deep learning lies [86] first in the fact that machine learning algorithms deal with quantitative and structured data and, second, the operator is responsible for choosing the right algorithm to extract the features that will influence the prediction. Deep learning algorithms deal with unstructured data and the algorithm is trained to extract the influential elements in the prediction as shown in Figure S1 in the Supplementary Materials. It should be noted that deep learning algorithms, compared to ML algorithms, demand a large amount of data and high computational power.

4. Discussion

In this paper, the authors reviewed many research articles and identified 129 studies eligible for systematic review using the PRISMA statement as presented in Figure S2 in the Supplementary Materials, these studies involve methodologies in image processing, machine learning, and deep learning particularly focused on the identification and classification of plant diseases. The study showed that the techniques most used in the literature, in general, are the support vector machine [22,59,61,65] (SVM), random forest [87] (RF), artificial neural network [84] (ANN) and convolutional neural network (CNN) [35,39,50].

Additionally, many scientific contributions have focused on the prediction of major diseases affecting wheat, rice, and potatoes, such as powdery mildew [88,89], late blight [90,91], and blast [92,93]. The challenging aspect of this work is the evaluation and investigation of the computational efficiency of each study compared to other studies, because each paper applies different metrics to a variety of diseases in different crops. In addition, many techniques and pretreatment approaches are used to predict disease presence or severity. Accordingly, it is nearly impossible to generalize and compare different articles because it is paramount to follow the same experimental conditions. Thus, the present comparison of the different approaches used was strictly constrained, for example, by considering the types of crops on which the work was undertaken, in addition to the kinds of diseases considered during the work. Therefore, based on these constraints, from the results obtained in related works, it is observed that deep learning-based models have outperformed the classical approaches such as random forest, support vector machine, and k-nearest neighbors classifiers, knowing that the performance of these algorithms has been proven and validated using metrics such as accuracy, sensitivity, specificity, and F1-score etc.

Table 4 shows that several researchers applied spectral analysis using thermal and optical remote sensing images, in addition to multispectral and hyperspectral images. As shown by Duarte-Carvajalino et al. [94], multispectral images were found to be relevant for the early-stage detection of disease, whereas hyperspectral images, which are the most widely used in the existing literature, can predict disease even before symptoms are visible to the naked eye. Note that this difference is due to the spectral resolution used by the two technologies. However, compared to hyperspectral imaging, multispectral imaging offers less data complexity [95]. However, hyperspectral image analysis has various limitations. Several authors have highlighted the high dimensionality of the data as one of the difficulties encountered. As pointed out by Mahlein et al. [95], the high degree of interband correlation leads to information redundancy, generating convergence instability in multivariate prediction models. It is observed that the dataset used by most of the researchers is taken from PlantVillage, and, in the image preprocessing process, most researchers used the histogram equalization to improve the contrast, and the median, Gaussian filter, and Gabor filter for denoising and image enhancement. Furthermore, for image segmentation, researchers have focused on the hue using the k-means and fuzzy c-means algorithm to segment the images; this procedure enables extraction of the region of interest from the given image. Using this, plant features such as texture, shape, and color have often been extracted using the gray-level co-occurrence matrix (GLCM), local binary patterns (LBPs), and histogram of oriented gradients (HOG). This is the most prominent step in the classification process. The researchers used different classification algorithms based on machine learning, and deep and transfer learning, for the classification phase, such as the decision tree classifier (TC), random forest (RF), naive Bayes (NB), support vector machine (SVM), artificial neural network (ANN), probabilistic neural network

(PNN), back-propagation neural network (BPNN), convolutional neural network (CNN), and InceptionV3.

The SVM and NN are mainly used in disease classification. The main advantage of NNs is that they can tolerate noise and are built from available data. The SVM, in turn, offers outstanding classification performance because it nonlinearly maps the input feature vector into a high dimensional space where it can be easily separated. Nevertheless, SVMs are not suitable when the data is very noisy. Hence, when many redundant variables form the input vector, it is possible to the use principal component analysis (PCA) dimensionality reduction method, as used by Kadir et al. [96]. In addition, the convolutional neural network (CNN), faster R-CNN, Vgg16, and ResNet50 models have been used to fully automate the classification process. Moreover, a new approach employed by Turkoglu et al. [97] is the extreme learning machine (ELM), which offers faster learning and better performance and generalization with lower computational cost. The advantages and disadvantages of the classifiers used in the literature are summarized in Table 5.

Table 5. Comparison of various classifiers.

Classifier	Advantages	Drawbacks
Artificial Neural Network ANN	Faster and more accurate than KNN and MMC	Strict because the data can only belong to 1 class
Random Forest	Can classify a large data set with excellent accuracy	Constraints on storage and processing time
Multiclass-Support Vector Machine	Helps to classify the data in several classes	Not suitable when the data is noisy
Least-Square SVM	Fast and not complicated	Pruning techniques must be applied to be sparse
K-Nearest Neighbours KNN	No time spent on training	More time spent on testing and it is expensive to test each instance as well sensitive to noise and yields
Extreme learning machine ELM	Faster training and better generalization	Overfitting (occurs when a complex model has several parameters)
Naïve Bayes	Less training data is required. It works better than its counterparts when the assumption of an independent variable is true	Conditional independence may reduce accuracy
Penalized Discriminant Analysis PDA	Beneficial when the problem has a large number of noisy features	High calculation cost
Bag of Words	Uncomplicated, robust, efficient	It supposes that all words are independent of each other
CNN/Deep learning	It removes the need for a feature extraction step and classification time is shortened	A large amount of data is required for training and it is expensive to compute. They require better hardware such as Graphical Processing Unit (GPU).
Transfer Learning	This helps to apply CNN to problems with a small amount of training data	The pretrained model may not have classes with the desired labels all the time

In this regard, a study was carried out by Ngugi et al. [98] to compare the performance of 10 deep learning models using the PlantVillage dataset, namely AlexNet, ResNet-101, GoogleNet, DenseNet201, Vgg16, Inceptionv3, InceptionResNetv2, SqueezeNet, ShuffleNet, and MobileNets. We present the results obtained by the different architectures for all the performance measures in Table 6 below.

Table 6. The test set performance of 10 models considered in this comparative study.

Architecture	Recall	F1-score	Precision	Accuracy	Specificity
AlexNet	0.9843	0.9856	0.9871	0.9897	0.9997
InceptionV3	0.9906	0.9916	0.9926	0.9948	0.9999
GoogleNet	0.9874	0.9881	0.9891	0.9899	0.9997
SqueezeNet	0.9791	0.9787	0.9785	0.9837	0.9996
DenseNet201	0.9965	0.9961	0.9958	0.9973	0.9999
VGG16	0.9932	0.9930	0.9928	0.9951	0.9999
ResNet101	0.9936	0.9929	0.9924	0.9851	0.9999
ShuffleNet	0.9901	0.9897	0.9895	0.9929	0.9998
MobileNetv2	0.987	0.9862	0.9895	0.9905	0.9997
InceptionResNetv2	0.9887	0.9893	0.9901	0.9930	0.9998

According to Table 6, the DenseNet201 model is the most suitable because it requires less storage and has the advantage of having the best performance measures (accuracy = 0.9973, precision = 0.9958, recall = 0.9965, specificity = 0.9999, F1 score = 0.9961). However, it requires a longer learning time (82 h) compared to the InceptionV3 and ResNet-101 models; nonetheless, their accuracies are slightly lower than those of DenseNet201. Therefore, special care should be taken when choosing between these three architectures, as each model has certain advantages and limitations. By comparison, the small MobileNet, SqueezeNet, and ShuffleNet architectures are desirable in embedded and mobile applications where computing resources are limited, due to their short learning times and low storage requirements, while still achieving high accuracy.

In this regard, another comparative study of four machine learning algorithms—k-nearest neighbors, decision tree, naive Bayes, and logistic regression—was performed by Ahmed et al. [99] to detect three rice plant diseases where the images were taken from the same PlantVillage database.

As shown in Figure 3, the best accuracy (over 97% by applying it to the test dataset) was obtained by the decision tree algorithm after 10 cross-validations.

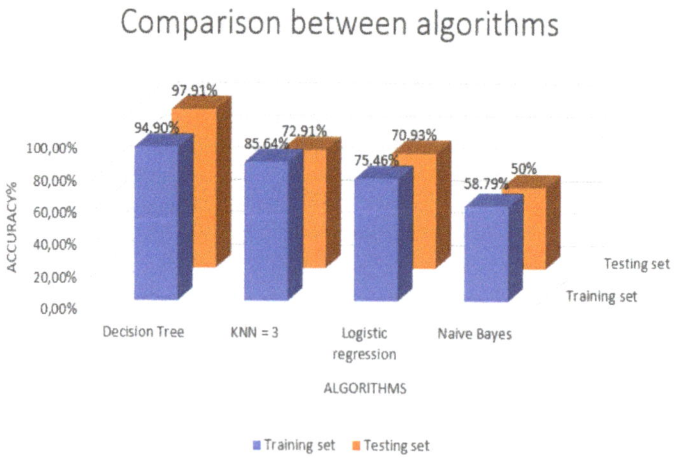

Figure 3. Comparison between machine learning algorithms.

To briefly summarize this section, it is inferred that multispectral and hyperspectral imagery represents a valuable source of useful information for developing autonomous non-invasive systems to predict abiotic and biotic stresses in plants. Additionally, the integration of multiple data sources will strengthen and increase the stability and generalization capabilities of the algorithms. Furthermore, from the results obtained in the literature, it appears that the automatic extraction of leaf features performed by deep learning-based models is more relevant and efficient than the process of extracting these features using traditional approaches such as the grey level co-occurrence matrix (GLCM), area-based techniques (ABTs), and scale invariant feature transform (SIFT). However, it is noted that there is a lack of validation of the models used in real-world scenarios. Therefore, proper validation is necessary for the studies to have an accurate and general impact.

5. Unresolved Challenges in the Crop Disease Detection Field

The above section presents a wealth of promising research undertaken in the past few years in the area of crop foliar disease recognition and detection using a range of techniques. In the existing literature to date, there are numerous unresolved challenges that remain to be address and overcome to derive robust and feasible crop disease detection systems that can operate accurately under various field conditions. The most prominent of these highlighted challenges are:

5.1. Insufficient Data

The major problem in the use of deep learning models for plant disease detection is the insufficiency of datasets in terms of both diversity and size [100] because these models have extremely large data requirements. In the majority of cases, the identification of plant diseases has been performed under ideal and controlled conditions [43], such as the presence of a single disease with a homogeneous background. In addition, environmental conditions are not considered; hence, the accuracy rate obtained will be higher than that actually obtained in a practical application. Additionally, image labeling is a very laborious and tedious task. Due to these factors, the production of a reliable, efficient, and comprehensive dataset is extremely challenging. At present, there are six ways to deal with the lack of a dataset: data augmentation techniques, data sharing, citizen science, transfer learning, synthetic data, and few-shot learning.

5.2. Imbalanced Data

The most commonly used datasets for crop disease detection are cleaned or their unbalanced nature is ignored to fully concentrate on training algorithms and avoid being distracted by other problems. However, in real-world settings, the distribution across classes is skewed and unbalanced [101], ranging from mildly biased to severely unbalanced. This poses a challenge for predictive modeling and may require specialized techniques, such as re-sampling techniques, because the machine learning algorithms typically employed for classification are built around the assumption of an equal number of examples for each class. As a result, some of the models have poor predictive performance, especially for the minority class which is more susceptible to misclassification than the majority class.

5.3. Vanishing Gradient Problem

Hochreiter's work [102] showed an issue called the "vanishing gradient problem" that arises during the training phase when employing back-propagation learning techniques with neural networks. Specifically, each weight of the neural network is updated based on the current weight and is proportionally related to the partial derivative of the error function. However, this updating of the weights may not take place in some cases due to an extremely small gradient that approaches zero. As a result, the gradient descent does not converge to the optimum and the neural network stops completely [103].

5.4. Exploding Gradient Problem

The opposite problem to the vanishing problem is the gradient explosion problem [104]. Specifically, the gradients become increasingly larger as the back-propagation algorithm advances. This will lead to extremely large updates of the network weights and causes the gradient descent to diverge, which means that the system becomes unstable [103]. Thus, the model will lose its ability to learn efficiently. In general, as we move up the network during back-propagation, the gradient grows exponentially by repeatedly multiplying the gradients. As a result, the weight values can become incredibly large and spill over to become a non-numerical value (NaN).

5.5. Overfitting and Underfitting Problem

Learning models have excessively high chances of overfitting and underfitting the data in the training stage due to the large number of parameters involved, which are correlated in complex ways. Such situations reduce the ability of the model to perform well on the tested data. Thus, it is considered that a learning algorithm is underfitting when it is unable to grasp the underlying trend in the data. Its occurrence simply means high bias, low variance, and that the model does not fit the data well enough. This usually occurs when fewer data are available to build an accurate model and also when a linear model is attempted to be built with non-linear data. Conversely, a model is said to be overfitted when it is trained with a large quantity of data; it then learns from noise and inaccurate data inputs in the dataset. Then, the model does not correctly categorize the data due to excessive detail and noise. Its occurrence simply means low bias and high variance. The overfitting occurs in nonlinear and nonparametric approaches, as these kinds of learning algorithms have greater leeway in setting up an unrealistic model. Ideally, both of these should not exist in models, but they are generally challenging to eliminate. This problem was noted by Ahmad et al. [105], whose model based on efficient convolutional neural networks tends to overfit during the training of the first epochs.

5.6. Image Acquisition: Conditions of Image Capture (Lighting, Spatial Location, Wind and Camera)

Ideally, images should be captured under similar conditions. However, in practice, this may only be feasible in the laboratory because it is extremely difficult to monitor the conditions of capture. Thus images may present unpredictable characteristics, making disease identification a daunting task. Moreover, the variable capture conditions have proven to be a challenging issue in measuring the severity of citrus leaf canker [106] and in identifying citrus diseases [107]. In light of this, several endeavors have been undertaken to develop methods of invariant illumination [108]. Nevertheless, their success to date is still relatively modest.

- ***Lighting Issue***

Crops grow in natural environments that fluctuate greatly. Thus, images are impacted by numerous factors, such as wind, illumination, and other climatic conditions. Consequently, lighting issues are inevitable, and completely eliminating the variations is almost impossible. Nevertheless, certain endeavors have been made to mitigate them, e.g., Pourreza et al. [109] developed a system to detect real-time citrus Huanglongbing disease using a narrow-band imaging and polarizing filter set. Specular lighting, the simultaneous presence of light and shadow, is the most difficult problem to deal with. However, the presence of specular lighting can be lessened by changing either the angle at which the image is taken or the leaf position, although this likely causes some degree of reflection. Furthermore, specular reflections and shadows were noted by Zhou et al. [110] as the main source of error in monitoring Cercospora leaf spot on sugar beets, which occurred because of the automatic captures that complicate the prevention of lighting problems.

- ***Camera***

The image resolution is one of the crucial factors that has a direct influence on the image features. A higher resolution enables the detection of small lesions and spores. Moreover, the device being used to capture the image also influences these features.

5.7. Image Preprocessing

During the preprocessing and storage of leaf images, more information is lost as the compression ratio is increased. This may not dramatically influence the analysis of large lesions, but may severely distort small symptoms. Therefore, compression should be kept to a minimum or even avoided, especially if the symptoms are tiny.

5.8. Image Segmentation and Symptom Discrimination

In general, symptoms do not have clear boundaries; they gradually disappear in normal tissue, making the distinction between healthy and diseased areas highly ambiguous. This clearly affects the accuracy of the threshold and extracted features. Although manual and visual representation cannot clearly determine the edges, any machine-based representation will be prone to many subjective issues. Notably, the issue of subjective delineation of affected regions was first addressed by Olmstead et al. [111] and later by Moya et al. [112], who emphasized that some sort of external reference needs to be established for proper validation of disease identification methods. However, without the use of a reference, Oberti et al. [113] observed for leaf powdery mildew that the number of false negatives or positives seen on the symptom discolored zones is too high. In summary, few solutions have been suggested for this problem because inconsistencies are intrinsic to the process. Furthermore, other difficulties are encountered when segmenting and locating regions of interest (ROIs):

- A leaf may overlap with another leaf or other parts of the plant, and they may even be tilted or covered with dew or dust.
- Images with complex backgrounds can render the segmentation of ROIs where symptoms appear challenging and intricate.

5.9. Feature Selection and Extraction

Although some plant species can be identified on the basis of leaf shape, other species have similar leaf shapes. Furthermore, symptoms do not necessarily arise in zones that are easily accessible; in practice, they can frequently be under the leaves or covered by other obstructions, or diseases can appear on the stems, fruits, or even flowers. Unfortunately, the latter problem has not attracted enough attention on the part of researchers. Furthermore, it is observed from the literature to date that researchers have mainly focused on the disease detection on the upper leaf surface. Nevertheless, Fuentes et al. [34] suggested using the faster network R-CNN for detecting a number of tomato plant diseases in several locations.

5.10. Disease Classification

In many of the cases listed below, the classifier used to identify plant diseases may not be able to distinguish between them; for example, if the symptoms presented by different diseases are visually very similar, as both Ahmad et al. [114] and Wiwartet al. [115] have stated. In addition, the difficulties stated below are highly relevant to measuring the disease severity:

- ***Differences in disease symptoms:*** According to the disease development stage, a specific disease can present very distinct characteristics in the symptoms' shape, color, and size, causing serious identification problems. It should be noted that many different diseases can occur at the same time, making it extremely complex to distinguish between combinations of symptoms and individual symptoms. This problem was noted by Camargo et al. [116] when handling symptoms produced by black streak disease on banana leaves, and Moya et al. [112] when evaluating powdery mildew severity on squash leaves.
- ***Diseases can occur simultaneously with many disorders, such as nutritional deficiencies, pests, and diseases:*** Typically, most techniques consider that there is only a single disease per image when, in reality, several other diseases can be present at the same time, in addition to other kinds of disorders, such as nutritional deficiencies and pests. These simultaneous symptoms can be either separate or physically combined, making

disease identification a significant challenge. In this regard, Bock et al. [106] observed the simultaneous presence of symptoms arising from different diseases and noted that this can lead to identification issues, and that more advances will be required to cope with this issue.

- ***The symptoms' similarity between different disorder types:*** Symptoms resulting from various disorders, such as diseases, phytotoxicity, presence of parasites, and nutritional deficiencies, can be visually similar. As a result, it can be extremely difficult to determine a symptom's source with certitude, particularly if only the visible spectrum is used in the identification process. This forces methods to rely on tiny differences to discriminate between the symptoms. Numerous researchers have stated that some disorders have close similarities, leading to major issues of discrimination. In this regard, Ahmad et al. [114] reported that symptoms resulting from Fusarium, Mosaic Potyvirus, Alternaria, and Phomopsis in soybean were very similar, and their classifier was unable to discern between them. This explains why the majority of studies conducted to date have chosen to tackle only diseases whose symptoms are quite dissimilar and, even then, their choices remain a significant challenge.

5.11. Other Challenges

Some other challenges facing automatic plant disease identification techniques cannot be categorized in the same way as those mentioned above. These challenges include reducing complexity, in addition to computational and memory demands [117], because low-cost computers and cameras have a very limited computational resource. At the same time, as image resolution is increasing, the computational resources are also growing. Another major concern is the lack of properly labeled [37] and sufficiently large datasets with high variability. This is notably the biggest hurdle when training recurrent neural network (RNN) models for plant disease detection, because collecting images in the field is not only a laborious task but also requires the guidance of agricultural experts for accurate annotation. Nevertheless, two free datasets exist [118]—PlantVillage and the Image Database of Plant Disease Symptoms dataset PDDB. Moreover, at present, no appropriate technology has been developed to automatically crop the leaf images around the affected area. A further issue is that hyperspectral data contain more than one hundred adjacent spectral bands and thus cannot be linearly trained [119]. Furthermore, these bands in different spectral regions are highly redundant [54,55] when extracting information to form an artificial neural network (ANN).

6. Future Work and Possible Solutions to Ongoing Limitations

In the previous section, gaps in the existing literature were highlighted to orient future research in this area. Thus, future work should first aim at acquiring diverse and large-size datasets to further promote research in this direction. Moreover, it is highly desirable to develop compact convolutional neural network CNN-based models that can achieve higher accuracy and promote the use of these technologies in the embedded platforms. Secondly, more emphasis should be placed in future research on the development of reliable methods and techniques to remove backgrounds and incorporate other forms of data, such as meteorological trends, disease occurrence history, and spatial location, to enhance the accuracy and reliability of disease identification systems. Additionally, disease recognition at different locations on plants and trees, such as the stems, blooms, and fruits, should receive greater attention from researchers due to its tremendous importance. One possible means to circumvent some of the limitations is to implement constraints to restrict variations in image capture conditions. However, even with very tight restrictions, many challenges will remain.

Some of the key challenges can be mitigated through the use of the most sophisticated approaches borrowed from the machine learning and computer vision fields. These include Markov random fields, mean shift, graph theory, and large margin nearest neighbor classification (LMNN), among other methods that have not yet been properly harnessed.

In this regard, the proposed solutions to remedy the challenges presented above can be summarized as follows.

6.1. Data Augmentation Techniques

If the aim is to avoid the overfitting problem and expand the size of the dataset without manually collecting new images, data augmentation techniques are a possible solution for any limited data problem [120,121]. Data augmentation incorporates a collection of methods that improve the attributes and size of training datasets. Thus, DL models can perform better when these techniques are exploited, such as rotation, canny edge detection, shear, image noise addition, shift, and flipping.

6.2. Tackling Overfitting Problem

Overfitting is one of the fundamental problems encountered when using learning models, and occurs due to the sensitivity to the scale of the cross-entropy loss and the continuous updating of the gradient. Three classes exist to avoid the overfitting problem. The first acts on both the model parameters and the model architecture. This includes the most familiar approaches, such as batch normalization [122], weight decay [123], and dropout [124]. Weight decay is the technique that is commonly used by default in all algorithms as a universal regularizer. The second class operates on model inputs such as data augmentation and corruption. One of the causes of the overfitting problem is the lack of training data; as a result, the learned distribution does not exactly reflect the real distribution. In contrast, the marginalized corrupted feature (MCF) exclusively improves the solution in data augmentation. MCF is a new approach to combat overfitting in supervised learning [125]. The main idea of the MCF is to allow the models to be regularized by training them on corrupted data copies, without raising the computational complexity. The final class works on the output of the model. A technique was recently proposed by Pereyra et al. [126] based on penalizing confident output distributions for model regularization. This method has demonstrated its high capacity to regularize CNN and RNN models. Hence, it will be judicious to explore these techniques in the field of crop disease detection.

6.3. Few-Shot Learning

In cases in which the dataset is extremely small, the techniques mentioned above may not be useful; that is, if there is a task in which the classification must be built with only one or two samples per class, and each sample is difficult to find. In such a case, innovative approaches are needed; one of these is few-shot learning (FSL) [127]. This is a relatively recent subfield of machine learning that needs more refinement and research. FSL allows the classification of new data when there are only a few training samples with supervised information. The approach of building an FSL classifier is suitable for solving the kind of problem related to rare plant pathologies, in which images are lacking for use in the training set. Typically, two major approaches are implicated in solving one-shot or few-shot machine learning issues, namely, the data-level approach and the parameter-level approach.

6.4. Transfer Learning

Recent research has revealed the extensive use of deep CNNs, which require a large quantity of data to perform effectively. The common challenge associated with the use of such models concerns the lack of training data. Specifically, collecting a large volume of data is an exhausting task, and no successful solution is available at this time. Therefore, in order to solve the fundamental dilemma of insufficient data, it is advisable to use TL models, which are highly effective in such cases [128]. In simple terms, transfer learning is the process by which the model trained for a specified task is reused as the starting point for training a new model. It attempts to transfer information from the original domain to the destination domain. This learning process is illustrated in Figure 4.

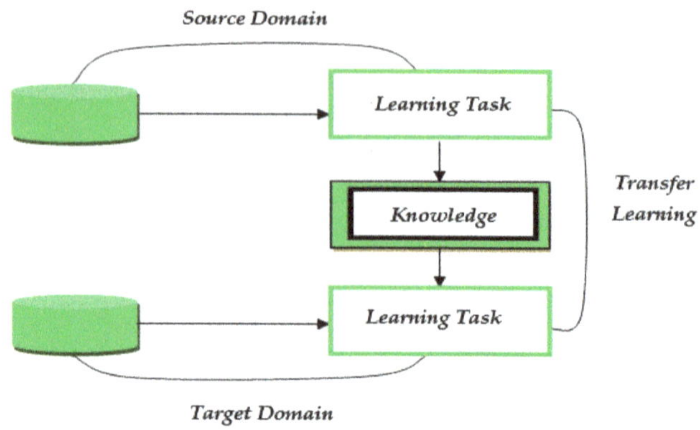

Figure 4. Transfer learning process.

Forthcoming research endeavors can be devoted to automatically estimate the detected disease severity, and expanded to attain the highest accuracy and speed via developing hybrid approaches such as genetic algorithms (GAs) and neural networks (NN) to increase the disease recognition rate, and combining particle swarm optimization (PSO) with other tools, such as gradient search techniques, to ensure a much higher speed. In addition, advanced and appropriate preprocessing techniques should be adopted to prevent noise interference in disease detection, in addition to partitioning the training and test data by employing more advanced techniques, such as stratified sampling, in order to create a well-balanced data partition, and avoid underfitting and overfitting. In addition, optimizing feature vectors should be contemplated to increase the disease recognition rate in these various stages, and recurrent neural network (RNN) models and the long-term memory function should be used to extract memory and temporal dimensions that can subsequently be harnessed for plant growth estimation. Finally, a web application can be designed with a range of features, such as displaying the identified diseases in the crops from leaf images taken by a smartphone camera. A discussion forum can also be developed for agronomists and farmers to talk about treatment and early preventive measures for the encountered diseases. Moreover, plant electrophysiology is a promising avenue for future research [129], i.e., the electrical signal response produced in plants can be used for real-time disease detection. This approach is based on the fact that plants perceive the environment, and this perception is translated by a generation of electrical signals that essentially represent changes in their underlying physiological processes. Under the influence of stress, the metabolic activities of plant tissues and cells are unstable, which is inevitably reflected in the plant's physiological electrical properties. As a result, the extraction of substantial characteristics from the generated electrical signals, such as impedance, varying capacitance, and conductivity, would be a highly interesting research direction for the classification of diseases in plants and crops.

7. Conclusions

Crop diseases are one of the main challenges in the farming sector. Thus, there is a need to identify crop diseases at the earliest stage to lessen disease severity and to curb disease propagation on farms. Accordingly, prominent and advanced research has been conducted in recent years on several kinds of disease identification techniques, as presented in this work. The main difference between other surveys and the present paper is the thorough technical analysis of the individual papers, and the approaches that have been applied to date. This provides a guideline and references to scientific communities. This paper also provides readers with insights into the automatic crop disease detection process and the key factors, namely the lack of sharp edges around the symptoms; fluctuating imaging conditions; variable symptoms presented by diseases; similar symptoms presented by different disorders; and the concomitant presence of symptoms arising from various disorders. These issues have a relevant impact on the effectiveness of both the image processing methods and the analytical tools that have been introduced to date. From this survey, it is concluded that image preprocessing directly impacts the segmentation process. Moreover, the k-means clustering algorithm was found to be the most suitable technique for segmenting disease-affected leaves. In addition, convolutional neural network (CNN) models were revealed to be extremely powerful and proficient in locating visual patterns in images. Notably, the use of computer vision and artificial intelligence in crop diagnostics in the agricultural sector is still recent, which implies that their numerous alternatives and opportunities remain to be explored, which may help mitigate the above-mentioned challenges. Additionally, with the increase in available computing power, previously demanding strategies can now be easily executed. Thus, based on this in-depth study of the existing literature on crop foliar disease automatic detection, in upcoming work the researchers intend to develop an efficient, accurate, low-cost, and swift system capable of identifying crop diseases from foliar images. In addition, this identifying system will be implemented in a mobile application, allowing an alert to be sent to the farmer once the disease is detected to enable him to intervene as soon as possible.

Supplementary Materials: The following are available online at https://www.mdpi.com/article/10.3390/agriculture12010009/s1, Figure S1: Difference between traditional machine learning and deep learning; Figure S2: PRISMA 2020 flow diagram for new systematic reviews which included searchers of databases and registers only.

Author Contributions: Writing—original draft preparation, H.O. and M.S.; writing—review and editing, M.S. and M.K. All authors have read and agreed to the published version of the manuscript.

Funding: This research received no external funding.

Institutional Review Board Statement: Not applicable.

Data Availability Statement: Not applicable.

Acknowledgments: We are grateful to the reviewers for their work and valuable comments that helped us improve the quality of this manuscript.

Conflicts of Interest: The authors declare no conflict of interest.

Abbreviations

The following abbreviations are used throughout this manuscript.

AI	Artificial Intelligence
ML	Machine Learning
DL	Deep Learning
TL	Transfer Learning
CV	Computer Vision
IoT	Internet of Things
IP	Image Processing
WSN	Wireless Sensor Network
NN	Neural Network
ANN	Artificial Neural Network
CNN	Convolutional Neural Network
DNN	Deep Neural Network
RNN	Recurrent Neural Network
PNN	Probabilistic Neural Network
BBPN	Back-Propagation Neural Network
GOA	Grasshopper Optimization Algorithm
DAML	Deep Adversarial Metric Learning
KNN	K-Nearest Neighbor
SVM	Support Vector Machine

M-SVM	Multiclass-Support Vector Machine
LS-SVM	Least Squares Supporting Vector Machine
RF	Random Forest
PDA	Penalized Discriminant Analysis
ELM	Extreme Learning machine
NB	Naïve Bayes
VGG	Visual Geometry Group
ResNet	Residual Neural Network
C-GAN	Conditional Generative Adversarial Network
GPU	Graphics Processing Unit
MEAN block	Mobile End Apple Net block
FSL	Few-Shot Learning
GOA	Grasshopper Optimization Algorithm
GA	Genetic Algorithms
PSO	Particle Swarm Optimization
PCA	Principal Component Analysis
LDA	Linear Discriminant Analysis
PLS—DA	Partial Least Squares Discrimination Analysis
SPA	Successive Projection Algorithm.
SAM	Spectral Angle Mapper
BBCH	Biologische Bundesanstalt, Bundessortenamt and CHemical industry
RBM	Restricted Boltzmann Machine
AE	Auto-Encoder
EW	Effective Wavelengths
GLCM	Gray Level Cooccurrence Matrix
CPDA	Color Processing Detection Algorithm.
GUI	Graphical User Interface
LBP	Local Binary Patterns
HOG	Histogram-Oriented Gradient
HSV	Hue Saturation Value
RoI	Region of Interest
HPCCDD	Homogeneous Pixel Counting technique for Cotton Disease Detection
CMD	Cassava Mosaic Disease
RMD	Red Mite Damage
GMD	Green Mite Damage
TMV	Tobacco Mosaic Virus
VNIR	Visible and Near-Infrared
SWIR	Short Wavelength Infrared
ENVI	Environment for Visualizing Images
KL	Kullback Leibler
ABT	Area-Based Techniques
SIFT	Scale Invariant Feature Transform
MFC	Marginalized Corrupted Features
LMNN	Large Margin Nearest Neighbor
PRISMA	Preferred reporting items for systematic reviews and meta-analyses

References

1. Lucas, G.B.; Campbell, C.L.; Lucas, L.T. Causes of plant diseases. In *Introduction to Plant Diseases*; Springer: Berlin/Heidelberg, Germany, 1992; pp. 9–14.
2. Shirahatti, J.; Patil, R.; Akulwar, P. A survey paper on plant disease identification using machine learning approach. In Proceedings of the 2018 3rd International Conference on Communication and Electronics Systems (ICCES), Coimbatore, India, 15–16 October 2018; pp. 1171–1174.
3. Liu, L.; Dong, Y.; Huang, W.; Du, X.; Ren, B.; Huang, L.; Zheng, Q.; Ma, H. A disease index for efficiently detecting wheat fusarium head blight using sentinel-2 multispectral imagery. *IEEE Access* **2020**, *8*, 52181–52191. [CrossRef]

4. Prasad, R.; Ranjan, K.R.; Sinha, A. AMRAPALIKA: An expert system for the diagnosis of pests, diseases, and disorders in Indian mango. *Knowl.-Based Syst.* **2006**, *19*, 9–21. [CrossRef]
5. Singh, V.; Misra, A.K. Detection of plant leaf diseases using image segmentation and soft computing techniques. *Inf. Process. Agric.* **2017**, *4*, 41–49. [CrossRef]
6. Afifi, A.; Alhumam, A.; Abdelwahab, A. Convolutional neural network for automatic identification of plant diseases with limited data. *Plants* **2021**, *10*, 28. [CrossRef] [PubMed]
7. Mugithe, P.K.; Mudunuri, R.V.; Rajasekar, B.; Karthikeyan, S. Image processing technique for automatic detection of plant diseases and alerting system in agricultural farms. In Proceedings of the 2020 International Conference on Communication and Signal Processing (ICCSP), Chennai, India, 28–30 July 2020; pp. 1603–1607.
8. Parikshith, H.; Rajath, S.N.; Kumar, S.P. Leaf disease detection using image processing and artificial intelligence—A survey. In Proceedings of the International Conference On Computational Vision and Bio Inspired Computing, Coimbatore, India, 25–26 September 2019; pp. 304–311.
9. Sethy, P.K.; Barpanda, N.K.; Rath, A.K.; Behera, S.K. Deep feature based rice leaf disease identification using support vector machine. *Comput. Electron. Agric.* **2020**, *175*, 105527. [CrossRef]
10. Mitra, D. Emerging plant diseases: Research status and challenges. *Emerg. Trends Plant Pathol.* **2021**, 1–17. [CrossRef]
11. Ichiki, T.U.; Shiba, T.; Matsukura, K.; Ueno, T.; Hirae, M.; Sasaya, T. Detection and diagnosis of rice-infecting viruses. *Front. Microbiol.* **2013**, *4*, 289.
12. Lacomme, C.; Holmes, R.; Evans, F. Molecular and serological methods for the diagnosis of viruses in potato tubers. In *Plant Pathology*; Springer: Berlin/Heidelberg, Germany, 2015; pp. 161–176.
13. Balodi, R.; Bisht, S.; Ghatak, A.; Rao, K. Plant disease diagnosis: Technological advancements and challenges. *Indian Phytopathol.* **2017**, *70*, 275–281. [CrossRef]
14. Bachika, N.A.; Hashima, N.; Wayayoka, A.; Mana, H.C.; Alia, M.M. Optical imaging techniques for rice diseases detection: A review. *J. Agric. Food Eng.* **2020**, 2. Available online: https://www.myjafe.com/wp-content/uploads/2020/04/MYJAFE2020-0001.pdf (accessed on 15 December 2021).
15. Galletti, P.A.; Carvalho, M.E.; Hirai, W.Y.; Brancaglioni, V.A.; Arthur, V.; da Silva, C.B. Integrating optical imaging tools for rapid and non-invasive characterization of seed quality: Tomato (*Solanum lycopersicum* L.) and carrot (*Daucus carota* L.) as study cases. *Front. Plant Sci.* **2020**, *11*, 577851. [CrossRef] [PubMed]
16. Bauriegel, E.; Herppich, W.B. Hyperspectral and chlorophyll fluorescence imaging for early detection of plant diseases, with special reference to Fusarium spec. infections on wheat. *Agriculture* **2014**, *4*, 32–57. [CrossRef]
17. Mishra, P.; Polder, G.; Vilfan, N. Close range spectral imaging for disease detection in plants using autonomous platforms: A review on recent studies. *Curr. Robot. Rep.* **2020**, *1*, 43–48. [CrossRef]
18. Neupane, K.; Baysal-Gurel, F. Automatic identification and monitoring of plant diseases using unmanned aerial vehicles: A review. *Remote Sens.* **2021**, *13*, 3841. [CrossRef]
19. Abbas, A.; Jain, S.; Gour, M.; Vankudothu, S. Tomato plant disease detection using transfer learning with C-GAN synthetic images. *Comput. Electron. Agric.* **2021**, *187*, 106279. [CrossRef]
20. Hamdani, H.; Septiarini, A.; Sunyoto, A.; Suyanto, S.; Utaminingrum, F. Detection of oil palm leaf disease based on color histogram and supervised classifier. *Optik* **2021**, *245*, 167753. [CrossRef]
21. Sun, H.; Xu, H.; Liu, B.; He, D.; He, J.; Zhang, H.; Geng, N. MEAN-SSD: A novel real-time detector for apple leaf diseases using improved light-weight convolutional neural networks. *Comput. Electron. Agric.* **2021**, *189*, 106379. [CrossRef]
22. Pothen, M.E.; Pai, M.L. Detection of rice leaf diseases using image processing. In Proceedings of the 2020 Fourth International Conference on Computing Methodologies and Communication (ICCMC), Erode, India, 11–13 March 2020; pp. 424–430.
23. Kelman, A.; Pelczar Michael, J.; Shurtleff Malcolm, C.; Pelczar Rita, M. Plant Disease. Available online: https://www.britannica.com/science/plant-disease (accessed on 15 December 2021).
24. Cerda, R.; Avelino, J.; Gary, C.; Tixier, P.; Lechevallier, E.; Allinne, C. Primary and secondary yield losses caused by pests and diseases: Assessment and modeling in coffee. *PLoS ONE* **2017**, *12*, e0169133. [CrossRef] [PubMed]
25. Kaur, M.; Bhatia, R. Leaf disease detection and classification: A comprehensive survey. In Proceedings of the International Conference on IoT Inclusive Life (ICIIL 2019), Chandigarh, India, 19–20 December 2019; pp. 291–304.
26. Singh, V.; Misra, A. Detection of unhealthy region of plant leaves using image processing and genetic algorithm. In Proceedings of the 2015 International Conference on Advances in Computer Engineering and Applications, Ghaziabad, India, 19–20 March 2015; pp. 1028–1032.
27. Karthika, J.; Santhose, M.; Sharan, T. Disease detection in cotton leaf spot using image processing. *J. Phys. Conf. Ser.* **2021**, *1916*, 012224. [CrossRef]
28. Devaraj, A.; Rathan, K.; Jaahnavi, S.; Indira, K. Identification of plant disease using image processing technique. In Proceedings of the 2019 International Conference on Communication and Signal Processing (ICCSP), Chennai, India, 4–6 April 2019; pp. 0749–0753.
29. Mojjada, R.K.; Kumar, K.K.; Yadav, A.; Prasad, B.S.V. Detection of plant leaf disease using digital image processing. *Mater. Today Proc.* **2020**. [CrossRef]

30. Iqbal, M.A.; Talukder, K.H. Detection of potato disease using image segmentation and machine learning. In Proceedings of the 2020 International Conference on Wireless Communications Signal Processing and Networking (WiSPNET), Chennai, India, 4–6 August 2020; pp. 43–47.
31. Srivastava, A.R.; Venkatesan, M. Tea leaf disease prediction using texture-based image processing. In *Emerging Research in Data Engineering Systems and Computer Communications*; Springer: Berlin/Heidelberg, Germany, 2020; pp. 17–25.
32. Whitmire, C.D.; Vance, J.M.; Rasheed, H.K.; Missaoui, A.; Rasheed, K.M.; Maier, F.W. Using machine learning and feature selection for alfalfa yield prediction. *AI* **2021**, *2*, 6. [CrossRef]
33. Ferentinos, K.P. Deep learning models for plant disease detection and diagnosis. *Comput. Electron. Agric.* **2018**, *145*, 311–318. [CrossRef]
34. Fuentes, A.; Yoon, S.; Kim, S.C.; Park, D.S. A robust deep-learning-based detector for real-time tomato plant diseases and pests recognition. *Sensors* **2017**, *17*, 2022. [CrossRef]
35. Geetharamani, G.; Pandian, A. Identification of plant leaf diseases using a nine-layer deep convolutional neural network. *Comput. Electr. Eng.* **2019**, *76*, 323–338.
36. Chen, J.; Chen, J.; Zhang, D.; Sun, Y.; Nanehkaran, Y.A. Using deep transfer learning for image-based plant disease identification. *Comput. Electron. Agric.* **2020**, *173*, 105393. [CrossRef]
37. Singh, D.; Jain, N.; Jain, P.; Kayal, P.; Kumawat, S.; Batra, N. PlantDoc: A dataset for visual plant disease detection. In Proceedings of the 7th ACM IKDD CoDS and 25th COMAD, Hyderabad, India, 5–7 January 2020; pp. 249–253.
38. Al-bayati, J.S.H.; Üstündağ, B.B. Evolutionary feature optimization for plant leaf disease detection by deep neural networks. *Int. J. Comput. Intell. Syst.* **2020**, *13*, 12–23. [CrossRef]
39. Arsenovic, M.; Karanovic, M.; Sladojevic, S.; Anderla, A.; Stefanovic, D. Solving current limitations of deep learning based approaches for plant disease detection. *Symmetry* **2019**, *11*, 939. [CrossRef]
40. Costa, J.; Silva, C.; Ribeiro, B. Hierarchical deep learning approach for plant disease detection. In Proceedings of the Iberian Conference on Pattern Recognition and Image Analysis, Madrid, Spain, 1–4 July 2019; pp. 383–393.
41. Ramcharan, A.; Baranowski, K.; McCloskey, P.; Ahmed, B.; Legg, J.; Hughes, D.P. Deep learning for image-based cassava disease detection. *Front. Plant Sci.* **2017**, *8*, 1852. [CrossRef] [PubMed]
42. Too, E.C.; Yujian, L.; Njuki, S.; Yingchun, L. A comparative study of fine-tuning deep learning models for plant disease identification. *Comput. Electron. Agric.* **2019**, *161*, 272–279. [CrossRef]
43. Mohanty, S.P.; Hughes, D.P.; Salathé, M. Using deep learning for image-based plant disease detection. *Front. Plant Sci.* **2016**, *7*, 1419. [CrossRef] [PubMed]
44. Zhu, H.; Cen, H.; Zhang, C.; He, Y. Early detection and classification of tobacco leaves inoculated with tobacco mosaic virus based on hyperspectral imaging technique. In Proceedings of the 2016 ASABE Annual International Meeting, Orlando, FL, USA, 17–20 July 2016; p. 1.
45. Zhu, H.; Chu, B.; Zhang, C.; Liu, F.; Jiang, L.; He, Y. Hyperspectral imaging for presymptomatic detection of tobacco disease with successive projections algorithm and machine-learning classifiers. *Sci. Rep.* **2017**, *7*, 4125. [CrossRef] [PubMed]
46. Cui, S.; Ling, P.; Zhu, H.; Keener, H.M. Plant pest detection using an artificial nose system: A review. *Sensors* **2018**, *18*, 378. [CrossRef] [PubMed]
47. Ma, J.; Du, K.; Zheng, F.; Zhang, L.; Gong, Z.; Sun, Z. A recognition method for cucumber diseases using leaf symptom images based on deep convolutional neural network. *Comput. Electron. Agric.* **2018**, *154*, 18–24. [CrossRef]
48. Tran, T.-T.; Choi, J.-W.; Le, T.-T.H.; Kim, J.-W. A comparative study of deep CNN in forecasting and classifying the macronutrient deficiencies on development of tomato plant. *Appl. Sci.* **2019**, *9*, 1601. [CrossRef]
49. Tian, Y.; Yang, G.; Wang, Z.; Wang, H.; Li, E.; Liang, Z. Apple detection during different growth stages in orchards using the improved YOLO-V3 model. *Comput. Electron. Agric.* **2019**, *157*, 417–426. [CrossRef]
50. De Luna, R.G.; Dadios, E.P.; Bandala, A.A. Automated image capturing system for deep learning-based tomato plant leaf disease detection and recognition. Proceedings of TENCON 2018-2018 IEEE Region 10 Conference, Jeju Island, Korea, 28—31 October 2018; pp. 1414–1419.
51. Bioucas-Dias, J.M.; Plaza, A.; Camps-Valls, G.; Scheunders, P.; Nasrabadi, N.; Chanussot, J. Hyperspectral remote sensing data analysis and future challenges. *IEEE Geosci. Remote Sens. Mag.* **2013**, *1*, 6–36. [CrossRef]
52. Mananze, S.; Pôças, I.; Cunha, M. Retrieval of maize leaf area index using hyperspectral and multispectral data. *Remote Sens.* **2018**, *10*, 1942. [CrossRef]
53. Rumpf, T.; Mahlein, A.-K.; Steiner, U.; Oerke, E.-C.; Dehne, H.-W.; Plümer, L. Early detection and classification of plant diseases with support vector machines based on hyperspectral reflectance. *Comput. Electron. Agric.* **2010**, *74*, 91–99. [CrossRef]
54. Ng, W.; Minasny, B.; Malone, B.P.; Sarathjith, M.; Das, B.S. Optimizing wavelength selection by using informative vectors for parsimonious infrared spectra modelling. *Comput. Electron. Agric.* **2019**, *158*, 201–210. [CrossRef]
55. Wei, C.; Huang, J.; Wang, X.; Blackburn, G.A.; Zhang, Y.; Wang, S.; Mansaray, L.R. Hyperspectral characterization of freezing injury and its biochemical impacts in oilseed rape leaves. *Remote Sens. Environ.* **2017**, *195*, 56–66. [CrossRef]
56. Xie, C.; Shao, Y.; Li, X.; He, Y. Detection of early blight and late blight diseases on tomato leaves using hyperspectral imaging. *Sci. Rep.* **2015**, *5*, 16564. [CrossRef]
57. Chen, W.-L.; Lin, Y.-B.; Ng, F.-L.; Liu, C.-Y.; Lin, Y.-W. RiceTalk: Rice blast detection using internet of things and artificial intelligence technologies. *IEEE Internet Things J.* **2019**, *7*, 1001–1010. [CrossRef]

58. Krishna, M.; Sulthana, S.F.; Sireesha, V.; Prasanna, Y.; Sucharitha, V. Plant disease detection and pesticide spraying using dip and IoT. *J. Emerg. Technol. Innov. Res.* **2019**, *6*, 54–58.
59. Mishra, M.; Choudhury, P.; Pati, B. Modified ride-NN optimizer for the IoT based plant disease detection. *J. Ambient Intell. Humaniz. Comput.* **2021**, *12*, 691–703. [CrossRef]
60. Truong, T.; Dinh, A.; Wahid, K. An IoT environmental data collection system for fungal detection in crop fields. In Proceedings of the 2017 IEEE 30th Canadian Conference on Electrical and Computer Engineering (CCECE), Windsor, ON, Canada, 30 April–3 May 2017; pp. 1–4.
61. Devi, R.D.; Nandhini, S.A.; Hemalatha, R.; Radha, S. IoT enabled efficient detection and classification of plant diseases for agricultural applications. In Proceedings of the 2019 International Conference on Wireless Communications, Signal Processing and Networking (WiSPNET), Chennai, India, 21–23 March 2019; pp. 447–451.
62. Kumar, S.; Prasad, K.; Srilekha, A.; Suman, T.; Rao, B.D.; Krishna, J.N.V. Leaf disease detection and classification based on machine learning. In Proceedings of the 2020 International Conference on Smart Technologies in Computing, Electrical and Electronics (ICSTCEE), Bengaluru, Karnataka, India, 10–11 July 2020; pp. 361–365.
63. Mallick, D.K.; Ray, R.; Dash, S.R. Detection and classification of crop diseases from its leaves using image processing. In *Smart Intelligent Computing and Applications*; Springer: Berlin/Heidelberg, Germany, 2020; pp. 215–228.
64. Gurrala, K.K.; Yemineni, L.; Rayana, K.S.R.; Vajja, L.K. A new segmentation method for plant disease diagnosis. In Proceedings of the 2019 2nd International Conference on Intelligent Communication and Computational Techniques (ICCT), Jaipur, India, 28–29 September 2019; pp. 137–141.
65. Karthikeyan, N.; Anjana, M.; Anusha, S.; Divya, R.; Vinod, A. Leaf disease detection using image processing. *Int. J. Innov. Sci. Eng. Technol.* **2020**, *7*. [CrossRef]
66. Kamilaris, A.; Prenafeta-Boldú, F.X. Deep learning in agriculture: A survey. *Comput. Electron. Agric.* **2018**, *147*, 70–90. [CrossRef]
67. Usman, A.; Bukht, T.F.N.; Ahmad, R.; Ahmad, J. Plant disease detection using internet of thing (IoT). *Plant Dis.* **2020**, *11*, 505–509.
68. Reddy, K.A.; Reddy, N.M.C.; Sujatha, S. Precision method for pest detection in plants using the clustering algorithm in image processing. In Proceedings of the 2020 International Conference on Communication and Signal Processing (ICCSP), Chennai, India, 28–30 July 2020; pp. 894–897.
69. Khan, M.A. Detection and classification of plant diseases using image processing and multiclass support vector machine. *Int. J. Comput. Trends Technol.* **2020**, *68*, 5–11. [CrossRef]
70. Sawant, C.; Shirgaonkar, M.; Khule, S.; Jadhav, P. Plant disease detection using image processing techniques. 2020. Available online: https://www.semanticscholar.org/paper/Plant-Disease-Detection-using-Image-Processing-Sawant-Shirgaonkar/d9a26b87f1879821fea0cd1944279cd51359d0c5 (accessed on 15 December 2021).
71. Fulari, U.N.; Shastri, R.K.; Fulari, A.N. Leaf disease detection using machine learning. *J. Seybold Rep.* **2020**, *1533*, 9211.
72. Ouhami, M.; Es-Saady, Y.; El Hajji, M.; Hafiane, A.; Canals, R.; El Yassa, M. Deep transfer learning models for tomato disease detection. In Proceedings of the International Conference on Image and Signal Processing, Marrakesh, Morocco, 4–6 June 2020; pp. 65–73.
73. Deepa. A pre processing approach for accurate identification of plant diseases in leaves. In Proceedings of the 2018 International Conference on Electrical, Electronics, Communication, Computer, and Optimization Techniques (ICEECCOT), Mysuru, India, 14–15 December 2018; pp. 249–252.
74. Win, T.T. AI and IoT Methods for Plant Disease Detection in Myanmar. Kobe Institute of Computing: Kobe, Tokyo, 2018.
75. Sun, G.; Jia, X.; Geng, T. Plant diseases recognition based on image processing technology. *J. Electr. Comput. Eng.* **2018**, *2018*, 6070129. [CrossRef]
76. Thorat, A.; Kumari, S.; Valakunde, N.D. An IoT based smart solution for leaf disease detection. In Proceedings of the 2017 International Conference on Big Data, IoT and Data Science (BID), Pune, India, 20–22 December 2017; pp. 193–198.
77. Moghadam, P.; Ward, D.; Goan, E.; Jayawardena, S.; Sikka, P.; Hernandez, E. Plant disease detection using hyperspectral imaging. In Proceedings of the 2017 International Conference on Digital Image Computing: Techniques and Applications (DICTA), Sydney, Australia, 29 November–1 December 2017; pp. 1–8.
78. Khirade, S.D.; Patil, A. Plant disease detection using image processing. In Proceedings of the 2015 International Conference on Computing Communication Control and Automation, Pune, India, 26–27 February 2015; pp. 768–771.
79. Arivazhagan, S.; Shebiah, R.N.; Ananthi, S.; Varthini, S.V. Detection of unhealthy region of plant leaves and classification of plant leaf diseases using texture features. *Agric. Eng. Int. CIGR J.* **2013**, *15*, 211–217.
80. Han, S.; Cointault, F. Détection précoce de maladies sur feuilles par traitement d'images. HAL Open Sci. Available online: https://hal.archives-ouvertes.fr/hal-00829402/document (accessed on 15 December 2021).
81. Revathi, P.; Latha, M.H. *Classification of Cotton Leaf Spot Diseases Using Image Processing Edge Detection Techniques*; IEEE: Piscataway, NJ, USA, 2012; pp. 169–173. [CrossRef]
82. Yerpude, A.; Dubey, S. Colour image segmentation using K-medoids clustering. *Int. J. Comput. Technol. Appl.* **2012**, *3*, 152–154.
83. Chaudhary, P.; Chaudhari, A.K.; Cheeran, A.; Godara, S. Color transform based approach for disease spot detection on plant leaf. *Int. J. Comput. Sci. Telecommun.* **2012**, *3*, 65–70.
84. Al-Hiary, H.; Bani-Ahmad, S.; Reyalat, M.; Braik, M.; Alrahamneh, Z. Fast and accurate detection and classification of plant diseases. *Int. J. Comput. Appl.* **2011**, *17*, 31–38. [CrossRef]

85. Bauriegel, E.; Giebel, A.; Geyer, M.; Schmidt, U.; Herppich, W. Early detection of Fusarium infection in wheat using hyper-spectral imaging. *Comput. Electron. Agric.* **2011**, *75*, 304–312. [CrossRef]
86. Janiesch, C.; Zschech, P.; Heinrich, K. Machine learning and deep learning. *Electron. Mark.* **2021**, 1–11. [CrossRef]
87. Panigrahi, K.P.; Das, H.; Sahoo, A.K.; Moharana, S.C. Maize leaf disease detection and classification using machine learning algorithms. In *Progress in Computing, Analytics and Networking*; Springer: Singapore, 2020; pp. 659–669.
88. Zhao, J.; Xu, C.; Xu, J.; Huang, L.; Zhang, D.; Liang, D. Forecasting the wheat powdery mildew (Blumeria graminis f. Sp. tritici) using a remote sensing-based decision-tree classification at a provincial scale. *Australas. Plant Pathol.* **2018**, *47*, 53–61. [CrossRef]
89. Zhang, J.; Pu, R.; Yuan, L.; Huang, W.; Nie, C.; Yang, G. Integrating remotely sensed and meteorological observations to forecast wheat powdery mildew at a regional scale. *IEEE J. Sel. Top. Appl. Earth Obs. Remote Sens.* **2014**, *7*, 4328–4339. [CrossRef]
90. Fenu, G.; Malloci, F.M. Artificial intelligence technique in crop disease forecasting: A case study on potato late blight prediction. In Proceedings of the International Conference on Intelligent Decision Technologies, Split, Croatia, 17–19 June 2020; pp. 79–89.
91. Fenu, G.; Malloci, F.M. An application of machine learning technique in forecasting crop disease. In Proceedings of the 2019 3rd International Conference on Big Data Research, Paris, France, 20–22 November 2019; pp. 76–82.
92. Hsieh, J.-Y.; Huang, W.; Yang, H.-T.; Lin, C.-C.; Fan, Y.-C.; Chen, H. *Building the Rice Blast Disease Prediction Model Based on Machine Learning and Neural Networks*; EasyChair: Manchester, UK, 2019.
93. Kim, Y.; Roh, J.-H.; Kim, H.Y. Early forecasting of rice blast disease using long short-term memory recurrent neural networks. *Sustainability* **2018**, *10*, 34. [CrossRef]
94. Duarte-Carvajalino, J.M.; Alzate, D.F.; Ramirez, A.A.; Santa-Sepulveda, J.D.; Fajardo-Rojas, A.E.; Soto-Suárez, M. Evaluating late blight severity in potato crops using unmanned aerial vehicles and machine learning algorithms. *Remote Sens.* **2018**, *10*, 1513. [CrossRef]
95. Mahlein, A.-K.; Kuska, M.T.; Behmann, J.; Polder, G.; Walter, A. Hyperspectral sensors and imaging technologies in phytopathology: State of the art. *Annu. Rev. Phytopathol.* **2018**, *56*, 535–558. [CrossRef]
96. Kadir, A.; Nugroho, L.E.; Susanto, A.; Santosa, P.I. Performance improvement of leaf identification system using principal component analysis. *Int. J. Adv. Sci. Technol.* **2012**, *44*, 113–124.
97. Turkoglu, M.; Hanbay, D. Recognition of plant leaves: An approach with hybrid features produced by dividing leaf images into two and four parts. *Appl. Math. Comput.* **2019**, *352*, 1–14. [CrossRef]
98. Ngugi, L.C.; Abelwahab, M.; Abo-Zahhad, M. Recent advances in image processing techniques for automated leaf pest and disease recognition–A review. *Inf. Process. Agric.* **2021**, *8*, 27–51. [CrossRef]
99. Ahmed, K.; Shahidi, T.R.; Alam, S.M.I.; Momen, S. Rice leaf disease detection using machine learning techniques. In Proceedings of the 2019 International Conference on Sustainable Technologies for Industry 4.0 (STI), Dhaka, Bangladesh, 24–25 December 2019; pp. 1–5.
100. Barbedo, J.G. Factors influencing the use of deep learning for plant disease recognition. *Biosyst. Eng.* **2018**, *172*, 84–91. [CrossRef]
101. Sambasivam, G.; Opiyo, G.D. A predictive machine learning application in agriculture: Cassava disease detection and classification with imbalanced dataset using convolutional neural networks. *Egypt. Inform. J.* **2021**, *22*, 27–34. [CrossRef]
102. Hochreiter, S. Untersuchungen zu dynamischen neuronalen Netzen. Master Thesis, Technische Universität München, München, Germany, 1991.
103. Or, B. The Exploding and Vanishing Gradients Problem in Time Series. Available online: https://towardsdatascience.com/the-exploding-and-vanishing-gradients-problem-in-time-series-6b87d558d22 (accessed on 15 December 2021).
104. Hasan, R.I.; Yusuf, S.M.; Alzubaidi, L. Review of the state of the art of deep learning for plant diseases: A broad analysis and discussion. *Plants* **2020**, *9*, 1302. [CrossRef]
105. Ahmad, M.; Abdullah, M.; Moon, H.; Han, D. Plant disease detection in imbalanced datasets using efficient convolutional neural networks with stepwise transfer learning. *IEEE Access* **2021**, *9*, 140565–140580. [CrossRef]
106. Bock, C.; Cook, A.; Parker, P.; Gottwald, T. Automated image analysis of the severity of foliar citrus canker symptoms. *Plant Dis.* **2009**, *93*, 660–665. [CrossRef] [PubMed]
107. Pydipati, R.; Burks, T.; Lee, W. Identification of citrus disease using color texture features and discriminant analysis. *Comput. Electron. Agric.* **2006**, *52*, 49–59. [CrossRef]
108. Guo, W.; Rage, U.K.; Ninomiya, S. Illumination invariant segmentation of vegetation for time series wheat images based on decision tree model. *Comput. Electron. Agric.* **2013**, *96*, 58–66. [CrossRef]
109. Pourreza, A.; Lee, W.S.; Raveh, E.; Ehsani, R.; Etxeberria, E. Citrus Huanglongbing detection using narrow-band imaging and polarized illumination. *Trans. ASABE* **2014**, *57*, 259–272.
110. Zhou, R.; Kaneko, S.I.; Tanaka, F.; Kayamori, M.; Shimizu, M. Image-based field monitoring of Cercospora leaf spot in sugar beet by robust template matching and pattern recognition. *Comput. Electron. Agric.* **2015**, *116*, 65–79. [CrossRef]
111. Olmstead, J.W.; Lang, G.A.; Grove, G.G. Assessment of severity of powdery mildew infection of sweet cherry leaves by digital image analysis. *HortScience* **2001**, *36*, 107–111. [CrossRef]
112. Moya, E.; Barrales, L.; Apablaza, G. Assessment of the disease severity of squash powdery mildew through visual analysis, digital image analysis and validation of these methodologies. *Crop. Prot.* **2005**, *24*, 785–789. [CrossRef]
113. Oberti, R.; Marchi, M.; Tirelli, P.; Calcante, A.; Iriti, M.; Borghese, A.N. Automatic detection of powdery mildew on grapevine leaves by image analysis: Optimal view-angle range to increase the sensitivity. *Comput. Electron. Agric.* **2014**, *104*, 1–8. [CrossRef]

114. Ahmad, I.S.; Reid, J.F.; Paulsen, M.R.; Sinclair, J.B. Color classifier for symptomatic soybean seeds using image processing. *Plant Dis.* **1999**, *83*, 320–327. [CrossRef] [PubMed]
115. Wiwart, M.; Fordoński, G.; Żuk-Gołaszewska, K.; Suchowilska, E. Early diagnostics of macronutrient deficiencies in three legume species by color image analysis. *Comput. Electron. Agric.* **2009**, *65*, 125–132. [CrossRef]
116. Camargo, A.; Smith, J. An image-processing based algorithm to automatically identify plant disease visual symptoms. *Biosyst. Eng.* **2009**, *102*, 9–21. [CrossRef]
117. Liu, J.; Wang, X. Plant diseases and pests detection based on deep learning: A review. *Plant Methods* **2021**, *17*, 1–18. [CrossRef]
118. Hughes, D.; Salathé, M. An open access repository of images on plant health to enable the development of mobile disease diagnostics. *arXiv* **2015**, arXiv:1511.08060.
119. Cucci, C.; Casini, A. Hyperspectral imaging for artworks investigation. In *Data Handling in Science and Technology*; Elsevier: Amsterdam, The Netherlands, 2020; Volume 32, pp. 583–604.
120. Alzubaidi, L.; Zhang, J.; Humaidi, A.J.; Al-Dujaili, A.; Duan, Y.; Al-Shamma, O.; Santamaría, J.; Fadhel, M.A.; Al-Amidie, M.; Farhan, L. Review of deep learning: Concepts, CNN architectures, challenges, applications, future directions. *J. Big Data* **2021**, *8*, 1–74.
121. Shorten, C.; Khoshgoftaar, T.M. A survey on image data augmentation for deep learning. *J. Big Data* **2019**, *6*, 1–48. [CrossRef]
122. Laurent, C.; Pereyra, G.; Brakel, P.; Zhang, Y.; Bengio, Y. Batch normalized recurrent neural networks. In Proceedings of the 2016 IEEE International Conference on Acoustics, Speech and Signal Processing (ICASSP), Shanghai, China, 20–25 March 2016; pp. 2657–2661.
123. Zhang, G.; Wang, C.; Xu, B.; Grosse, R. Three mechanisms of weight decay regularization. *arXiv* **2018**, arXiv:1810.12281.
124. Srivastava, N.; Hinton, G.; Krizhevsky, A.; Sutskever, I.; Salakhutdinov, R. Dropout: A simple way to prevent neural networks from overfitting. *J. Mach. Learn. Res.* **2014**, *15*, 1929–1958.
125. Maaten, L.; Chen, M.; Tyree, S.; Weinberger, K. Learning with marginalized corrupted features. In Proceedings of the International Conference on Machine Learning, Atlanta, GA, USA, 16–21 June 2013; pp. 410–418.
126. Pereyra, G.; Tucker, G.; Chorowski, J.; Kaiser, Ł.; Hinton, G. Regularizing neural networks by penalizing confident output distributions. *arXiv* **2017**, arXiv:1701.06548.
127. Kadam, S.; Vaidya, V. Review and analysis of zero, one and few shot learning approaches. In Proceedings of the International Conference on Intelligent Systems Design and Applications, Vellore, India, 6–8 December 2018; pp. 100–112.
128. Zhuang, F.; Qi, Z.; Duan, K.; Xi, D.; Zhu, Y.; Zhu, H.; Xiong, H.; He, Q. A comprehensive survey on transfer learning. *Proc. IEEE* **2020**, *109*, 43–76. [CrossRef]
129. Chatterjee, S.K.; Malik, O.; Gupta, S. Chemical sensing employing plant electrical signal response-classification of stimuli using curve fitting coefficients as features. *Biosensors* **2018**, *8*, 83. [CrossRef]

agriculture

MDPI

Article

Detection and Analysis of Sow Targets Based on Image Vision

Kaidong Lei [1], Chao Zong [1,*], Ting Yang [1], Shanshan Peng [1], Pengfei Zhu [1], Hao Wang [1,2] Guanghui Teng [1,*] and Xiaodong Du [3]

[1] College of Water Conservancy & Civil Engineering, China Agricultural University, Beijing 100083, China; leikaidong@cau.edu.cn (K.L.); sy20193091754@cau.edu.cn (T.Y.); pengss1007@126.com (S.P.); zpfmyl666@163.com (P.Z.); b20213090690@cau.edu.cn (H.W.)
[2] Chongqing Academy of Animal Sciences, Chongqing 402460, China
[3] New Hope Liuhe Co., Ltd., Beijing 100102, China; duxiaodong2@newhope.cn
* Correspondence: chaozong@cau.edu.cn (C.Z.); futong@cau.edu.cn (G.T.); Tel.: +86-010-62737583 (G.T.)

Abstract: In large-scale sow production, real-time detection and recognition of sows is a key step towards the application of precision livestock farming techniques. In the pig house, the overlap of railings, floors, and sows usually challenge the accuracy of sow target detection. In this paper, a non-contact machine vision method was used for sow targets perception in complex scenarios, and the number position of sows in the pen could be detected. Two multi-target sow detection and recognition models based on the deep learning algorithms of Mask-RCNN and UNet-Attention were developed, and the model parameters were tuned. A field experiment was carried out. The data-set obtained from the experiment was used for algorithm training and validation. It was found that the Mask-RCNN model showed a higher recognition rate than that of the UNet-Attention model, with a final recognition rate of 96.8% and complete object detection outlines. In the process of image segmentation, the area distribution of sows in the pens was analyzed. The position of the sow's head in the pen and the pixel area value of the sow segmentation were analyzed. The feeding, drinking, and lying behaviors of the sow have been identified on the basis of image recognition. The results showed that the average daily lying time, standing time, feeding and drinking time of sows were 12.67 h(MSE 1.08), 11.33 h(MSE 1.08), 3.25 h(MSE 0.27) and 0.391 h(MSE 0.10), respectively. The proposed method in this paper could solve the problem of target perception of sows in complex scenes and would be a powerful tool for the recognition of sows.

Keywords: computer vision; sow; image processing; behavior; precision livestock; animal welfare

Citation: Lei, K.; Zong, C.; Yang, T.; Peng, S.; Zhu, P.; Wang, H.; Teng, G.; Du, X. Detection and Analysis of Sow Targets Based on Image Vision. *Agriculture* **2022**, *12*, 73. https://doi.org/10.3390/agriculture12010073

Academic Editors: Gniewko Niedbała and Sebastian Kujawa

Received: 17 November 2021
Accepted: 1 January 2022
Published: 6 January 2022

1. Introduction

In China, African swine fever is ravaging farms across the country, the pig-raising industry is set to be upgraded with technological innovations [1]. At present, the management of livestock production mainly relies on experienced farmers [2], which is time-consuming and greatly dependent on individual judgments [3]. With the rapid expansion of the farms and the increased number of livestock, new technologies are urgently needed to improve production efficiency. The technologies of precision livestock farming have proved a success in many fields, and machine vision perception in pigs is one of the key applications [4]. In the production of sows, the sows' reproductive performance most important factor in the production efficiency of a pig farm [5–8]. The quick recovery of sows after giving birth can help to increase their follow-up breeding rate. In this process, the observation, identification, and tracking of sows are essential for better farm management and production performance [9]. Identifying sows through contact-free machine vision tracking is a first step in the development of smart pig farms [10–13]. With the development of machine vision technology and artificial intelligence (AI), the behavior of animals can even be perceived without human participation in daily management [14].

Object detection is a type of machine vision technique for detecting the region of interest (target object) in digital images, which has been of considerable interest in precision livestock farming [15–21]. In recent years, a series of object detection algorithms [22–24] based on deep learning have been developed with advantages of high precision, fast speed, and strong practicability in the field. Deep learning technology has been used for the detection of animal objects [14,25–28] and has also been used for the assessment of animal welfare [12,29–31]. The Convolutional Neural Network (CNN) based on deep learning has shown superior performance in pig image segmentation, behavior recognition, posture detection, and identification [29,32–35]. However, these applications are mostly in a simple scene and controlled condition with limited disturbances.

In a pig house, objects with different shapes, colors, and influencing factors, such as illumination, occlusion, and adhesion, will challenge the performance of target detection. Through literature, it is found that the YOLO method has been widely used in animal detections, but it also requires high hardware capacity which is very expensive [36]. In the follow-up study, the application of real-time sow detection will be performed in large-scale farms, therefore, the cost and running speed of the sow detection models needs to be comprehensively considered. The deep learning algorithms called Mask R-CNN and UNet-Attention have shown good detection capabilities on targets of various circumstances and have been successfully applied in many fields [37,38], with relatively low expense in hardware and high calculating speed. In the algorithm of UNet-Attention, an attention mechanism is added to improve the effectiveness of the model, which focuses on regional information conducive to object realization and suppresses irrelevant information. The UNet-Attention model has a simple algorithm statement, and even low-resolution information could be easily located and recognized. Its segmentation speed is fast, which can meet the purpose of real-time monitoring of sows. The algorithm based on the Mask Scoring R-CNN framework is supposed to solve image segmentation from complex backgrounds [39]. The Mask RCNN is designed for image instance segmentation, which has an advantage in instance segmentation from the complex background image [38]. Both algorithms do not require high hardware capacity and can rapidly recognize the target form images, which are suitable for real time and large-scale applications in the future.

Up until now, very few studies have been reported in the sow image segmentation using the UNet-Attention and Mask R-CNN algorithms. Therefore, it is necessary to analyze these two algorithms for the sow's object detection in complex scenes.

In this study, deep learning and image analysis technologies have been used to rapidly acquire sow information in complex scenes. The main contents are as follows: Development of sow target detection method based on deep learning algorithms; Assessment of the segmentation and recognition performances in sow detection between Unet-Attention and Mask-RCNN algorithms; Identity and analysis of sows' behaviors in the pen, based on the segmentation.

2. Materials and Methods

2.1. Animal and Housing

The experiment was conducted on a pig farm in Shandong province, China. The target objects were randomly selected Yorkshire pigs of 2nd–3rd Birth, weighing 245–260 kg. There were a number of 416,873 images were collected. These sows were raised in large pens and fed once a day at 10:00 a.m. and provided *ad libitum* water from the drinking trough. The sows can move freely in the pen. Due to free movement and contact of the sows, adhesion of sows in images often occurred. The room temperature was maintained at 29–30 °C. The size of the large pen was 3.15 m (L) × 2.06 m (W) × 2.44 m (H). The schematic diagram of the pig raising system is shown in Figure 1.

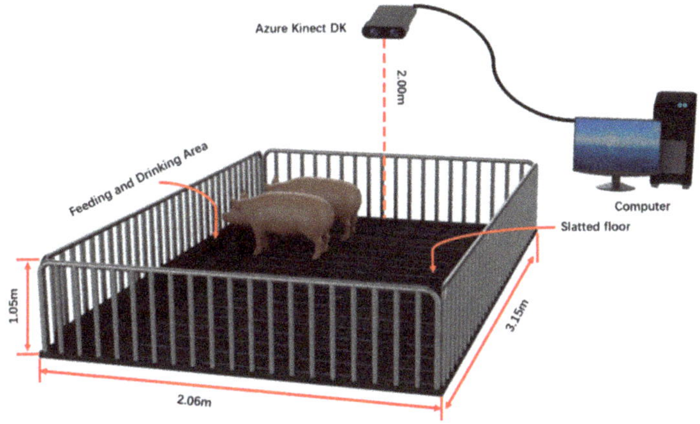

Figure 1. Data acquisition platform.

2.2. Data Acquisition Platform

The image acquisition equipment was Azure Kinect DK (Microsoft Corporation, U.S.). The camera continuously collected digital images and videos in .mkv format (16:9; 1920 ∗ 1080 pixels), with a sampling frequency of 60 Hz. As shown in Figure 1, the camera was connected to a computer through a USB 3.0 cable for storing data. The open source programming language of Python was used in the acquisition procedures.

The Kinect DK captured images of the sow as well as other things of feeder, drinking trough, railings, floor, etc., which formed a complex environment during the target detection process. The preprocessing procedures were as follow:

1. The collected video set in .mkv format was processed into .JPG format images through Python language, and then labeled with the Labelme software to form a .json format file.
2. In the training process, to increase the robustness of the Convolutional Neural Network, the input image of the sow was augmented. The data augmentation mainly used horizontal and vertical flipping, cropping, scale transformation, and rotation. The diversity of the data was also increased in this process.

2.3. Model Development

The program was run under the Pytorch framework in the Ubuntu16.04 system, and CUDA8.0 and cuDNN5.0 were used to accelerate training. The Central Processing Unit (CPU) of deep learning machines was Corei7-8700k (Intel), and the graphics processing unit (GPU) was GTX1080Ti (NVIDIA), and the memory was 32 GB. The data-set had a total number of 416,873 images. The image structure similarity algorithms were used for selection [40] to avoid a large number of similar images. A number of 15,094 images were sorted out. The data-set was divided into the training set and test set, according to the ratio of 9:1. The images were evenly divided into two data-set according to the complexity (stocking density and scale) during training and testing processes. Two optimized deep learning models Unet-Attention and Mask R-CNN were developed for analysis.

2.3.1. Development of the Sow Detection Model Based on Unet-Attention

The Unet-Attention was an algorithm based on the Unet network structure, which had been proposed by Ozan Oktay [37]. To better extract the sows from complex backgrounds in this study, the convolution kernel of stage2 to stage4 in the Attention module was replaced with deformable convolution (DCNV2). The network structure is shown in Figure 2 [37].

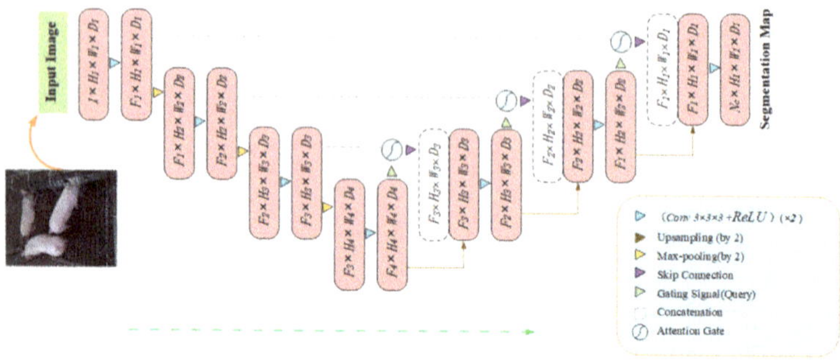

Figure 2. Block diagram of the Unet-Attention segmentation model.

Unet-Attention included two stages, with the first stage of down-sampling and the second stage of up-sampling. The down-sampling part was Visual Geometry Group Network 16 (VGG16), and the feature maps corresponding to the down-sampling layer were spliced in the up-sampling stage to achieve feature fusion. Before the down-sampling features were spliced with the up-sampling features, the attention block was used to re-adjust the down-sampling features. This module generated a gating signal to control the importance of features at different spatial locations. Specifically, the input features (x^l) were scaled with attention coefficients (α) computed in Attention Gate (AG), and the spatial regions were selected by analyzing both the activations and contextual information provided by the gating signal (g) collected from a wide scale. The grid resampling of attention coefficients was done using trilinear interpolation [37]. The Attention Block is shown in Figure 3:

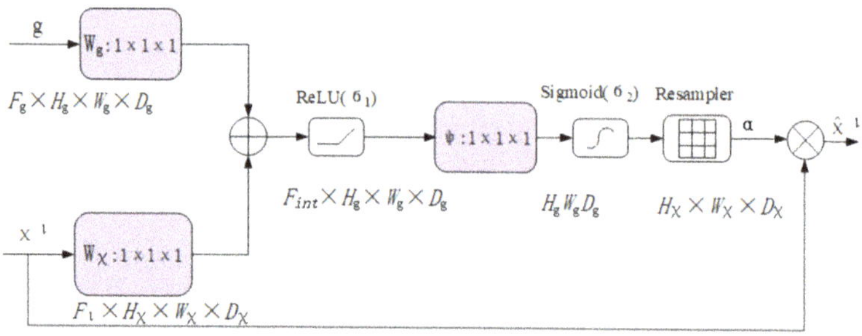

Figure 3. The proposed AG Attention Block.

In the application of Unet-Attention, the input image was progressively filtered and downsampled by factor 2 at each scale in the encoding part of the Unet network. The Ags filtered the features propagated through the skip connections. The feature selectivity in Ags was achieved by using contextual information (gating) extracted in wide scales (Figure 3).

The Equation (1) of Attention Block is as follows:

$$
\begin{aligned}
q_{att}^{l} &= \psi^{T}(\sigma_1(W_{\chi}^{T}\chi_i^{l} + W_g^{T}g_i + b_g)) + b_{\psi} \\
\alpha_i^{l} &= \sigma_2(q_{att}^{l}(\chi_i^{l}, g_i; \Theta_{att}))
\end{aligned}
\quad (1)
$$

where α_i is the attention coefficient, x^l is the input features of channels change, Ψ^T, W are operations of vectorization, g is the gating signal, l is convolution layer, Φ^l is the trainable kernel parameters, I is the pixel vector σ_2 corresponds to the sigmoid activation function, parameters Θ_{att} contains: linear transformations b_g, b_ψ are model bias terms.

In the process, g and x^l were multiplied by the weight matrix, which could be learned through backpropagation. The importance of each element was determined according to the goal of the algorithm. The term of attention was introduced to increase the weight matrix and to learn the importance of each element and the target.

The uNet-Attention algorithm used a deformable convolution kernel. Compared with the traditional convolution kernel whose size was fixed, the sampling method of the deformable convolution kernel was obtained through learning, mainly by adding offsets to the traditional convolution sampling points to obtain new sampling points and at the same time adding a modulation mechanism.

The deformable convolution can not only offset the input but also adjust the weight of each position input, as shown in Equation (2).

$$y(p) = \sum_{k=1}^{K} wk \cdot x(p + pk + \Delta pk) \cdot \Delta mk \tag{2}$$

where the convolution kernel has k sampling positions, w_k and p_k represent the weight of the k position and the preset offset respectively, Δmk is set to add the weight of each sampling point.

For the loss function of the uNet-Attention algorithm, the Focal Loss was adopted, For $y \geq 0$, p was the probability of the output of the model α, which balanced the contribution of positive and negative samples to the final loss, as shown in Equation (3).

$$\mathrm{L}(p) = \begin{cases} -\alpha(1-y')^{\gamma} log y', & y = 1 \\ -(1-\alpha)y'^{\gamma} log(1-y'), & y = 0 \end{cases} \tag{3}$$

Intuitively, the modulation factor reduced the loss contribution of the simple example and extended the low loss range of the samples [41].

2.3.2. Development of Sow Detection Model Based on Mask-RCNN

The Mask-RCNN is a model for instance segmentation, developed on the basis of Faster R-CNN, a region-based convolutional neural network. The Mask Branch of a small network of Fully Convolutional Networks for Semantic Segmentation (FCN) is supplemented in the Mask-RCNN model. During the process of Mask-RCNN, images were loaded and correspondingly preprocessed using pixel-level prediction to get the label map. The preprocessed images were then put into a pre-trained neural network (ResNeXt, etc.) to obtain the feature maps. A predetermined ROI (Region of interest) for each point in the feature map was set to get multiple candidate ROIs. Next, the candidate ROIs to the Region Proposal Network (RPN) were used for binary classification (foreground or background) and Bounding-Box (BB) regression and filtration. The ROI Align operation was performed on the remaining ROIs. The Mask-RCNN model used a per-pixel sigmoid. In the training step, for ROI in the Kth category, only the Kth mask contributes to L_{mask} of average binary cross-entropy loss. The overall process is shown in Figure 4 [38].

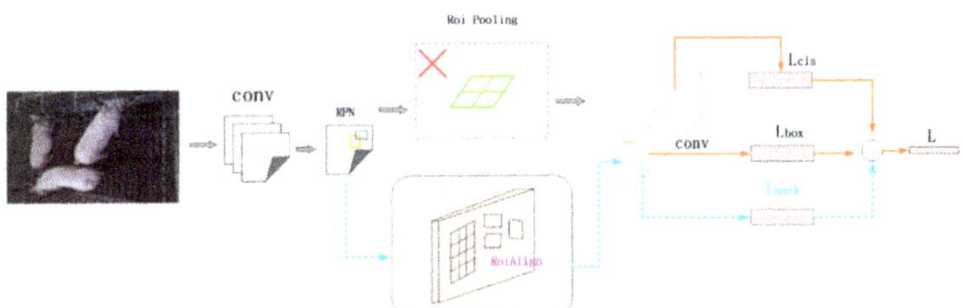

Figure 4. Mask-RCNN overall flow chart.

Backbone is a series of feature maps that convolutional layer was used to extract images. The backbone of Mask-RCNN was deep residual network 50 (ResNet-50), which provided a favorable train of thought for training deeper networks. ResNet used cross-layer connections to make training easier. The ResNet-50 backbone was used along with the improvement of training performance and speed. The FPN (feature pyramid networks) structure was composed of three parts: bottom-up, top-down, and horizontal connection. This structure could merge the features of each level to make it have both strong semantic and spatial information. The stage2 to stage4 in ResNet-50 convolution kernel was replaced with 5 identity blocks of deformable convolution DCN (Deep Cross Network) and ResNet-50.

Mask-RCNN was decomposed into three modules: Faster R-CNN, ROI Align, and FCN.

There was an ROI Align after 'head section', which was to enlarge the output dimension of ROI Align, to be more accurate when predicting mask. The ROI Align mainly canceled the quantization operation and used the method of bilinear interpolation to obtain the image value of pixels with floating-point coordinates, transforming the whole feature aggregation process into a continuous operation [38]. The process is illustrated in Figure 5.

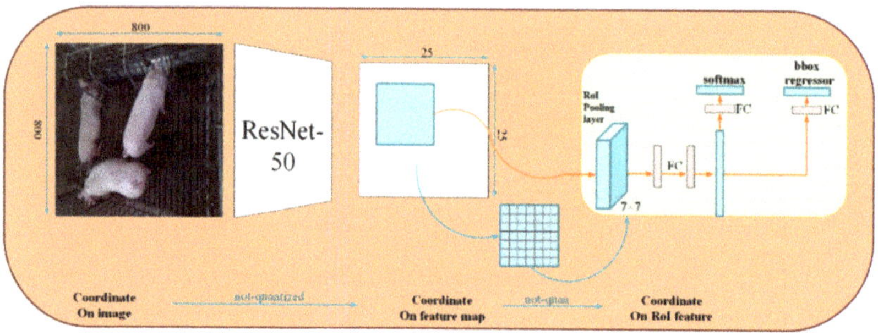

Figure 5. Characteristics of ROI Align.

There was no quantitative deviation for the ROI Align solution. The pixels in the original image and the pixels in the featured map were completely aligned, which could improve the detection accuracy of the model and promote the algorithm ability for instance segmentation.

For the loss function of the Mask-RCNN algorithm, on the basis of the Fast Region-based Convolutional Network method (Fast R-CNN), a third loss function for mask generation was added. The mask branch had an output of the $Km*m$ dimension for each ROI, where K represented two-class masks for encoding $m*m$ images, and each mask had K categories. The mask map had K channels, which represented the number of possible categories of the target. When calculating L_{mask}, only k-th map of these K maps were processed, and this k represented the object category of the ROI area located by another recognition branch. For an ROI belonging to the k-thcategory, L_{mask} only considered the k-th mask, and other mask inputs did not contribute to the loss function. Such a definition would allow a mask to be generated for each category, and there would be no inter-class competition. The multi-task loss function is as follow [38]:

$$L = L_{\text{cls}} + L_{\text{box}} + L_{\text{mask}} \tag{4}$$

2.4. Assessment of the Model

In this study, a dataset of images of sows in the pen was used to assess the proposed algorithms. The performance of the classification was measured by Equations (5) and (6). The obtained results from the detection models were assumed positive or negative.

$$\mathrm{P} = \frac{TP}{TP + FP} \tag{5}$$

$$\mathrm{R} = \frac{TP}{TP + FN} \tag{6}$$

where the number of correctly classified samples, that was the actually positive instances (samples) were classified as positive, as true positives (*TP*), and the number of incorrectly classified instances that were actually negative but classified as positive by the classifier was defined as false positives (*FP*). The number of erroneously classified instances that were actually positive but classified as negative by the classifier was defined as false negatives (*FN*), and the number of negative instances that were actually negative and classified as negative by the classifier was defined as true negatives (*TN*). P represented for precision, which was the accuracy rate. R was recall, which was the recall rate.

The models were also assessed using the indicator of Intersection-over-Union (IoU), which was a standard for measuring the accuracy of detecting corresponding objects in a specific data set. The IoU represented the overlap rate or degree of overlap between the generated candidate bound and the ground truth bound, in other words, the ratio of their intersection and union. *C* was Candidate bound and *G* was Ground truth bound. The expression is as follows:

$$\mathrm{IoU} = \frac{area(C) \cap area(G)}{area(C) \cup area(G)} \tag{7}$$

The Average Precision (AP) was used to evaluate the performance of the proposed models, which averaged the precision rates of all categories by combining IoU as the boundary. The AP measured the accuracy of the algorithm predictions and illustrated the percentage of algorithm positive predictions. The Average Recall (AR) referred to the maximum recall rate in a given number of the test results, and the average value was calculated on all IoUs and all categories. The IoU threshold value was defined as 0.50:0.75:0.95 to calculate the AP and AR and denoted as AP50/AR50, AP75/AR75 and AP95/AR95, respectively. To comprehensively analyze the model performance, different threshold values were used, and non-maximal suppression was applied during the calculations. The AP was also calculated based on area sizes in pixels with less than 32^2, between 32^2 and 96^2, and larger than 96^2, which corresponds to $\mathrm{AP}^{\mathrm{small}}$, $\mathrm{AP}^{\mathrm{medium}}$ and $\mathrm{AP}^{\mathrm{large}}$, respectively. The term of maxDets referring to the maximum detection number of targets in an image was used for defining the detection range.

2.5. Sow Behavior Recognition Based on Image Segmentation

Sows have a higher awareness of regional space in a group-raising state and could form a fixed area for behaviors of feeding and drinking, lying and moving around [42]. As shown in Figure 6, it is a schematic diagram of the position of the sows in the pen. The dark green part is the shared trough for the sows to feed and drink. The staff cleaned the water tank at 7:30 every morning and maintained a continuous water supply. At 9:50 in the morning, the water in the shared trough would be emptied, and the dry feed was then concentrated and fed at 10:00–10:30 in the morning. After 10:30, the staff cleaned the trough to prevent the left feed from becoming moldy. The drinking water was again switched on to maintain ad libitum water supply.

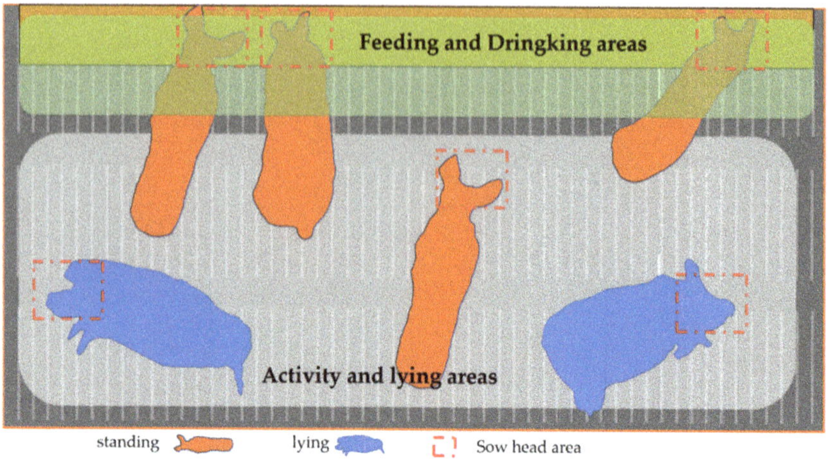

Figure 6. A schematic diagram of sows in the pen. Note: In order to ensure that each sow has a larger living space, three sows could be accommodated in the real pen at a time. In order to better illustrate the current situation, there are six sows drawn in this schematic diagram.

The green part in the image is the sow's feeding and drinking area. When the sow's head segmentation image was in the green area, the behavior of feeding or drinking behavior was recognized (the area in the red box in the figure). Due to the particularity of the sow feeding process, only the period of 10:00–10:30 in the morning was used for feeding, and if the head of the sow was detected in the feeding and drinking area during this period, it was judged as the sow's feeding behavior. On the contrary, in other periods, if the head of the sow was detected in this area, it would be recognized as the sow's drinking behavior.

In Figure 6, when the sow's head was detected in the gray area, and the sow is judged as standing and lying behaviors. Aiming at how to distinguish lying down and standing behaviors, this paper used the pixel value area method to identify the lying and standing behaviors of sows. The IMAQ tool in the LabVIEW software (National Instruments, United States) created the frame and calculated the image width (in pixels). The image type was set to Grayscale, and the IMAQ tool extracted a monochrome plane, and the image Dst Out function was a reference to the target image. When 'Image Dst' was connected, 'Image Dst Out' was the same as 'Image Dst'. Then, the ROI To Mask function in the IMAQ tool selected the region of interest, and finally, the pixel value was obtained. The block diagram in LabVIEW is as shown in Figure 7.

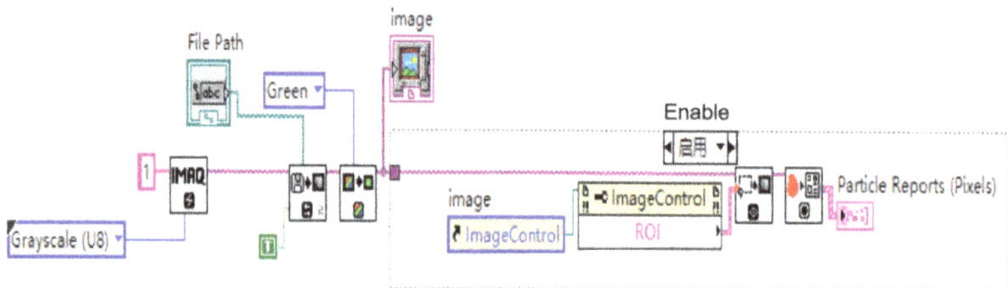

Figure 7. Block diagram of calculating pixel value in LabVIEW.

3. Results and Discussions

3.1. Performance of the Developed Models

In the complex pig house environment, the established deep learning target detection algorithms perfectly identified the target object of the sow as shown in Tables 1–3. The collected image data had been treated with preprocessing, model establishment, and model recognition. The AP and AR were used as the performance evaluation indices, as summarized in Tables 1 and 2. In the Tables, area referred to the number of pixels in the segmentation mask, and the maxDets was the maximum detection threshold of each image. It can be seen in Tables 1 and 2, the higher the value of IoU, the lower the accuracy and recall rates. Table 3 shows the average precisions of Mask-RCNN under different settings, where APs, APm, and APl stand for AP of small, medium and large objects, respectively. In the algorithm evaluation, AP_{50} was generally selected. The IoU threshold of the detector was greater than 0.5, and the accuracy of MaskRCNN reached 96.8%. At the same time, the model also defined different values, such as IoU and area for calculation. According to Tables 1 and 2, the final model operation results were not as good as the model with IoU greater than 0.5.

Table 1. The Average Precision (AP) rates of Mask-RCNN and UNet-Attention.

Model	AP	IoU	Area	maxDets
Mask-RCNN	0.772	0.50:0.95	all	100
	0.968	0.50	all	100
	0.948	0.75	all	100
	0.000	0.50:0.95	small	100
	0.083	0.50:0.95	medium	100
	0.792	0.50:0.95	large	100
	0.010	0.50:0.95	all	100
UNet-Attention	0.025	0.50	all	100
	0.006	0.75	all	100
	0.000	0.50:0.95	small	100
	0.000	0.50:0.95	medium	100
	0.046	0.50:0.95	large	100
	0.010	0.50:0.95	all	100

Table 2. The Average Recall (AR) rates of Mask-RCNN and UNet-Attention.

Model	AR	IoU	Area	maxDets
Mask-RCNN	0.291	0.50:0.95	all	1
	0.802	0.50:0.95	all	10
	0.802	0.50:0.95	all	100
	0.000	0.50:0.95	small	100
	0.135	0.50:0.95	medium	100
	0.823	0.50:0.95	large	100
UNet-Attention	0.004	0.50:0.95	all	1
	0.030	0.50:0.95	all	1
	0.167	0.50:0.95	all	5
	0.000	0.50:0.95	small	100
	0.000	0.50:0.95	medium	100
	0.176	0.50:0.95	large	100

Table 3. Mask -RCNN performance Indicators.

Task	AP	AP50	AP75	AP^{small}	AP^{medium}	AP^{large}
Bbox	0.6586	0.9668	0.8543	0.0000	0.1010	0.6751
Segm	0.7720	0.9682	0.9480	0.0000	0.0826	0.7925

A test set of sow images was used to verify the trained models. From Tables 1–3, it can be seen that the Mask-RCNN algorithm shows better object detection performance. The training and test images were filtered by the similarity algorithm [43] to get preprocessed image data for predictions. Some scholars used image processing to recognize the postures of animals [44,45], but the current segmentation performance was not good for animals in complex environments. Interference from floors, railings, brackets, feeders, drinking troughs affected the detection accuracy and could be misidentified as the sow target.

In this study, the Mask-RCNN algorithm accurately outlined the target object from the complex background. The sow segmentation outcome determined the performance of the algorithm. Figures 8 and 9 show the target object of perception using Mask-RCNN and UNet-Attention algorithms. The segmentation results were evaluated using the completeness of the outlines. It was found that the Mask-RCNN algorithm performed better than UNet-Attention in the segmentation.

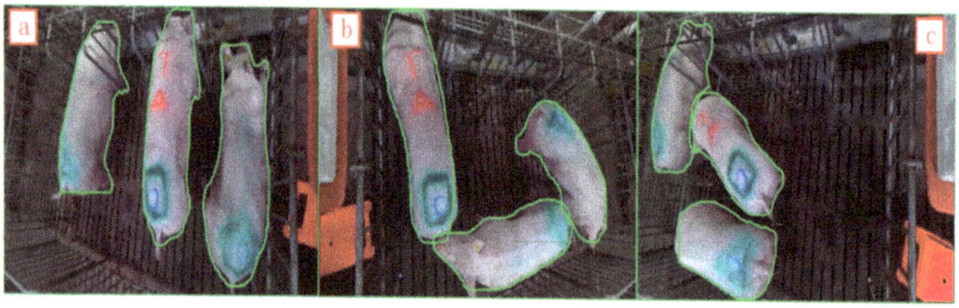

Figure 8. Target detection of Mask-RCNN: (**a**) without adhesion; (**b**,**c**) with adhesion. The green outline is the segmentation outcome.

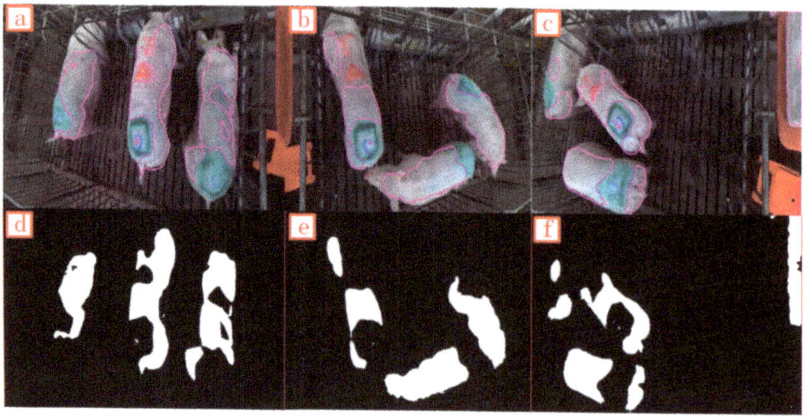

Figure 9. The target detection using UNet-Attention model, (**a**) without adhesion; (**b**,**c**) with adhesion. The pink outline is the segmentation outcome. The (**d**–**f**) are the processing binary diagrams of (**a**–**c**), respectively.

The targets in the image were usually overlapped, and the irrelevant objects could be mis-detected. In the algorithm process, non-target samples, such as railings, feeders and drinking troughs were often identified as the target. The UNet-Attention algorithm could also be affected by the patterns on the sows' backs. If there were overlaps or only part of the target in the image, the accuracy of detection and recognition could be poor. The model structure of the UNet-Attention algorithm could reduce its detection efficiency in the target of the sow. The optimization of the loss function and the network structure in the UNet-Attention algorithm was preferred, and certain layers could be flexibly customized according to the information of the target and surrounding environment [46,47]. The Mask-RCNN was based on He et al.'s (2018) work [38], which gained the detection results using Feature Pyramid Network (FPN) and a divided Network [48]. Mask RCNN had competitions among classes, and the network produced each class of mask. The addition of the ROI Align layer, the interpolation of the feature map, and the direct ROI Pooling quantization operation made the obtained mask have a slight offset from the actual object position. In this research, the Mask-RCNN algorithm of target segmentation performed well. Figure 10 shows the advantages of the Mask-RCNN algorithm in the detection: when the epoch was close to 30, the model had a better convergence effect. After starting training, the status of training was monitored by indicators, such as loss, loss-mask, loss-objectness, and others. Figure 10 shows that the loss has a downward trend after each complete iteration (a complete iteration means that all samples have been passed through). A proper learning rate could be ensured after each round of complete training, and the loss was reduced to a smaller extent after a period of time. The curves in Figure 10 also show some jumps up and down, which are related to the batch size set during stochastic gradient descent. When the batch size was very small, there would be a large degree of instability. If the batch size was set larger, it would be relatively stable.

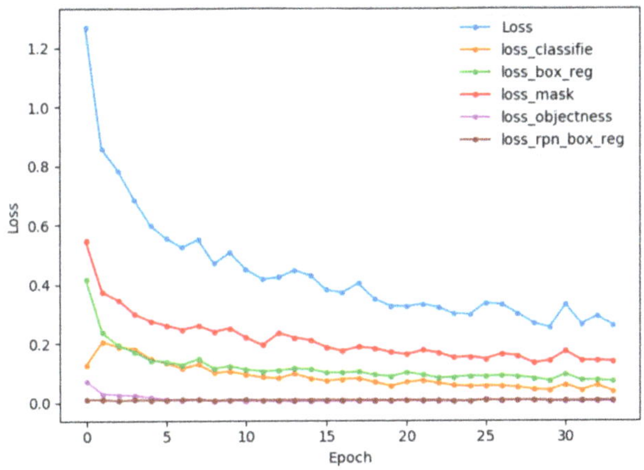

Figure 10. Performance index of the Mask-RCNN model algorithm. (Note: Loss is the total loss function of the model; loss-classifier and loss-box-reg are the rpn classification error; loss-mask is the segmentation error; loss-objectness and loss-rpn-box-reg are the model's detection error).

The UNet-Attention algorithm built a model on a fully connected convolutional layer, which only needed a small amount of training image data to get accurate segmentation. The computing speed of a computer would be slow to process the target with sizes and shapes in big differences. Moreover, the prediction result of the UNet-Attention model was not very satisfactory in treating a sow with patterns on its back.

In the literature on pig image segmentation, some authors have proposed an interactive image segmentation method based on improved graph cuts aiming at a specific target pig [49]. To segment pig images in a complex background, an image segmentation algorithm based on wavelet modulus maximum and edge growth has been proposed, but the algorithm processing process is relatively cumbersome [50]. The traditional detection algorithms of Haar + AdBoost and HOG + SVM only had the accuracy of 65.8 and 37.3, respectively [51]. The deep learning methods have achieved higher accuracy compared with traditional ones. A study proposed the use of infrared thermal imaging cameras to fuse infrared optical images to improve the effect of pig contour segmentation, with a success rate of 94% [52]. Zhang et al. (2019) proposed a real-time sow behavior detection algorithm based on deep learning (SBDA-DL), with an accuracy of 93.4% [51]. In this study, the Mask-RCNN algorithm had a high accuracy rate (96.8%) and could achieve sow contour segmentation well. The speed of Mask-RCNN was 720 ms, which could meet the requirement of real-time monitoring of sows.

3.2. Analysis of Poor Segmentation

Under the interference of railings, debris, and drinking troughs in the pig house, the segmentation effect of Mask-RCNN was good, as shown in Figure 8. It could realize image segmentation of sows in a complex pig pen, and the shape and position of sows in the pen could be further identified.

However, the segmentation effect of the UNet-Attention model was poor. As shown in Figure 11, the markings on the sow affected the segmentation performance (Figure 11a,c), and the railing, as well as the drinking trough in the pen (Figure 11b,c), would be incorrectly segmented as a sow target.

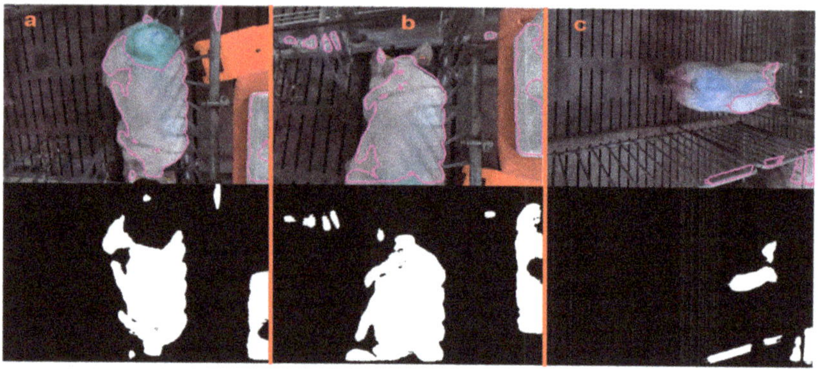

Figure 11. Performance effect of the UNet-Attention algorithm.

The UNet-Attention model is a classic network, which has a large number of applications in image segmentation tasks. However, the same network structure may show different performances in different scenarios. Increasing the network layers may improve the segmentation performance to specifically satisfy the segmentation of sows.

3.3. Image Recognition of Sow Behaviors and Positions

The activity behavior and location of the sow in the pen is an indicator to evaluate the animal's living conditions. In this study, the image segmentation algorithm was used to determine the position and contour of the sow's head, and then to determine the behavior of the sow. If the head of the sow was in the drinking and feeding area, it could be determined that the sow was drinking or feeding. The change in the pixel area of the contour of the sow could determine whether the sow was standing or lying down.

There were fixed areas for different behaviors of sows in group housing [42]. As shown in Figure 12, the sows are feeding and drinking at the pen trough, and the sows' heads were in the blue box, which represented the feeding and drinking area. When the sow's head was detected in the blue box, it could be judged as feeding or drinking behavior. As mentioned in the section on materials and methods, when the time was from 10:00–10:30 every morning, it was judged to be the feeding behavior of the sow. On the contrary, at other times of the day, it was judged as drinking behaviors.

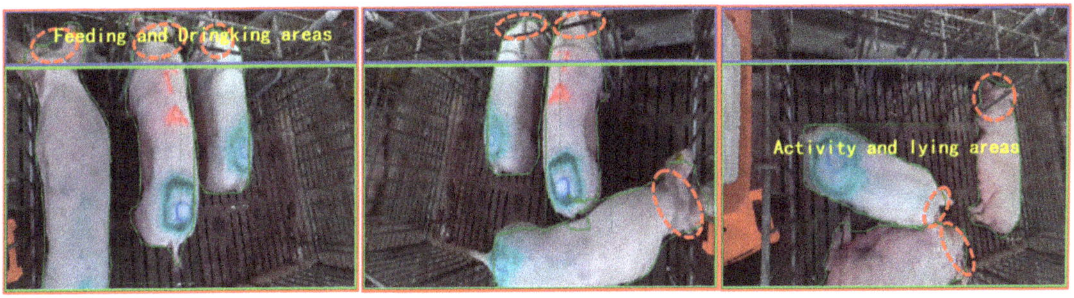

Figure 12. Image segmentation of sows with behaviors and different areas (The blue box: drinking and feeding areas; The green box: activity and lying areas).

The green box in Figure 12 is the lying and activity area of the sow pen. The lying and standing behaviors were distinguished based on the size of the image segmentation area. In addition, binary grayscale image of sows with different behaviors and positions (Figure 13). The pixel area value of the standing behavior was {2214–2641}, and the pixel value area range of lying behavior was {3025–3299}, therefore, for the data collected in this study, when the pixel value area of the segmented image of the sow was less than 3025, it was judged as a standing behavior. Meanwhile, a lying sow had a pixel value greater than 3025. However, the above data was based on a small amount of data, and there would also be the influence of individual body size differences, so this method was suitable for samples of sows of similar body size and weight.

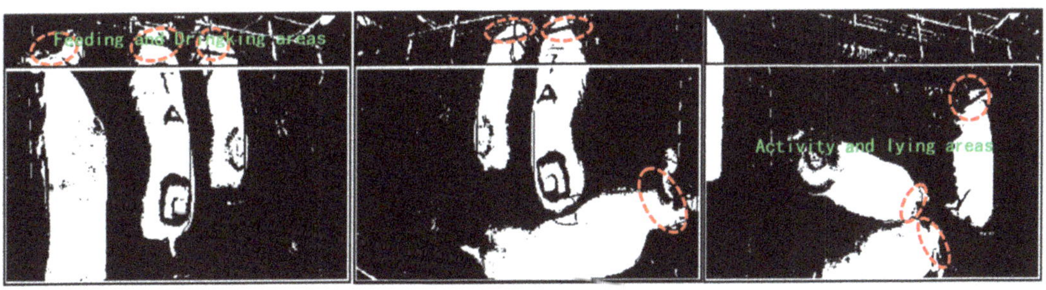

Figure 13. Binary grayscale image of sows with different behaviors and positions

Analyzing the active and resting behaviors of the sow after weaning could effectively get the physiological status of the sow. As shown in Figure 14, a sow is randomly selected, and within 24 h, the percentage of time spent on different behaviors has been analyzed. The circle inscribed in the blue rectangle, numbered 1 in the picture is the sow resting state in the lying area, and the time is more concentrated after the lights are turned off at night. Number 2 is that the sow stands in the active area, and the time is concentrated in the daytime. Numbers 3 and 4 are sows standing in the feeding and drinking area. Since the daily feeding time is a fixed 30 min, the sow's feeding time is from 10:00–10:30, numbered as 4. The rest time is the proportion of the sow's drinking behavior, numbered as 3. This analysis could effectively figure out the sow's daily activity and rest patterns within 24 h.

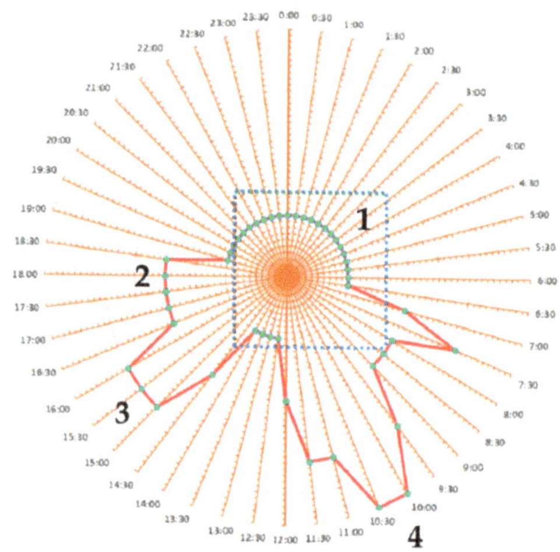

Figure 14. The proportion time of different sow behaviors in 24 h (1, lying down; 2, standing sows in the active area; 3, standing sows in the feeding and drinking area, and drinking water; 4 the sow being eats in the drinking and feeding area).

Table 4 shows the lying, standing, drinking and feeding time of 6 sows within 24 h. The average lying, standing, and drinking time of sows was 12.67 h, 11.33 h, and 3.25 h, respectively. Figure 15 shows the cumulative timetable of different behaviors of six sows in 24 h. The average time spent on lying down for sows in 24 h accounted for 52.8%, while the percentage of standing time was 47.2%. Due to the noise caused by the staff working in the house during the daytime, the sows spent much time standing. However, when the barn became quiet, sows were more likely to lie down and rest.

Table 4. Timetable for sows in standing, lying, feeding and drinking within 24 h.

	1	2	3	4	5	6	Average Value	Mean Square Error
Lying time/h	13.5	12	11	13	14	12.5	12.67	1.08
Standing time/h	10.5	12	13	11	10	11.5	11.33	1.08
Drinking time/h	3.5	3.0	3.0	3.5	3.5	3.0	3.25	0.27
Feeding time/h	0.5	0.4	0.5	0.4	0.25	0.3	0.391	0.10

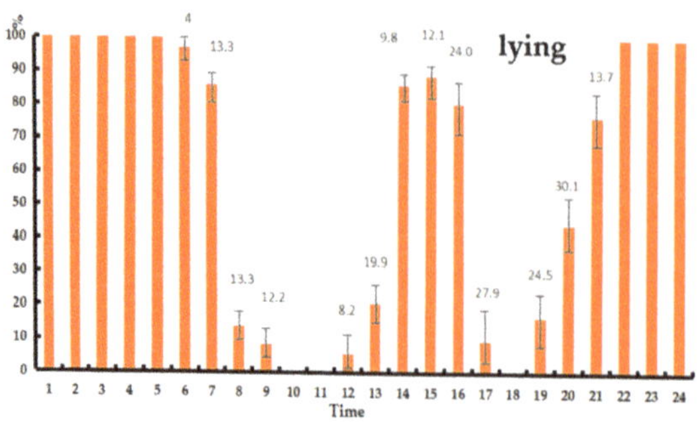

Figure 15. The distribution of time spent on lying, standing, feeding and drinking within 24 h.

Figures 15 and 16 show the proportion of time spent by sows in the lying behavior within 24 h. During the period from 21:00 to 7:00, there was no light in the pig house. The sows rested intensively, so the lying time accounted for a relatively high proportion, and the average hourly proportion of lying behavior accounts for 97.53%. The house was quiet from 14:00 to 16:00. The sows were mostly lying down and resting. The average hourly proportion of lying was 84.7% from 14:00 to 16:00. In Figure 17, sows mainly lied down and rested from 20:00 to 6: 00 the next morning, so the average drinking time was 0. From 7:00 to 11:00 and from 15:00 to 19:00, the proportion of drinking water increased, and the average hourly drinking behavior accounted for 43.73%.

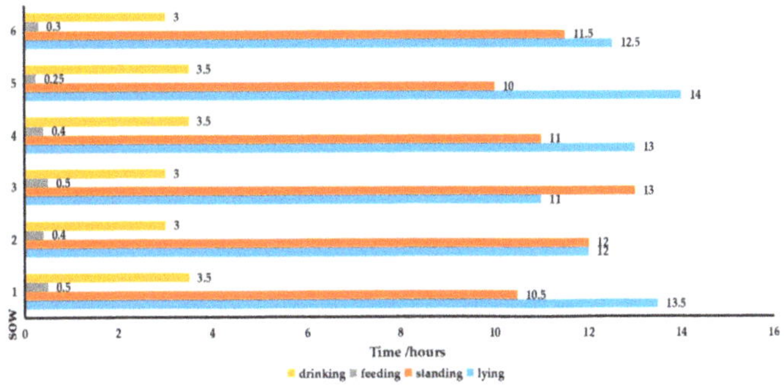

Figure 16. Percentage of lying time in 24 h.

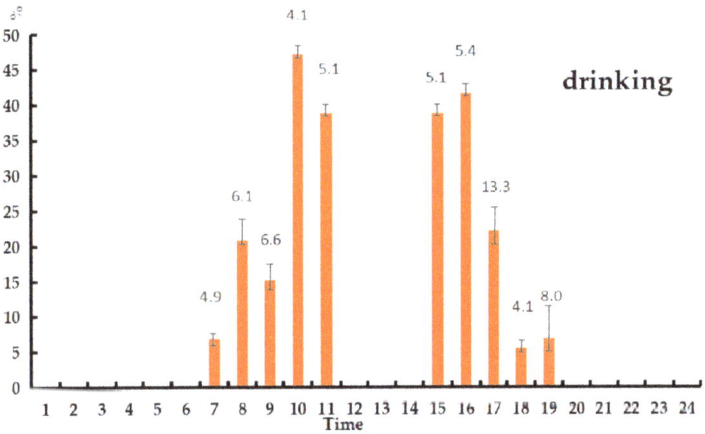

Figure 17. Percentage of drinking time in 24 h.

3.4. Application of the Sow Target Detection Model and Its Perspectives

With the development of livestock husbandry, precision livestock farming technology becomes more and more popular. In Europe, there are relevant policy requirements for the identification and traceability of farm animals, which mainly involved stress and welfare [53]. In particular, the technology has brought a lot of conveniences, greatly reduced the amount of labor and workload, improved production efficiency, and laid the foundation for large-scale operations. As mentioned in this article, sow behavior can be judged through image segmentation and further analyzing the sow's drinking, lying and standing time. At the same time, comprehensively judge the physiological state of the sow. In recent years, the Internet of Things (IoT) has also been widely applied in the precision farming process. Livestock production can be remotely monitored and controlled in real-time, which greatly improves production efficiency [54–57]. It is of great significance to apply the sow recognition algorithm to efficiently and accurately identify the physiological behaviors of the sow. The algorithm proposed in this paper is part of the follow-up precision sow perception system, which will be later deployed together with IoT technology for production evaluation and decision-making.

Meanwhile, the model should have a high accuracy rate, and portable controllers embedded in the algorithms can be optimized to adapt to various complex pig house environments. As shown in Figure 18, in the follow up study, a real-time monitoring system will be developed. The Mask-RCNN algorithm will be firstly used to segment the image of sows, and the sow's position and shape will be obtained after segmentation. After that, the sow's behavior recognition will be performed, and finally, precision control of the micro-environment and management in the barn will be conducted to better raise the sows.

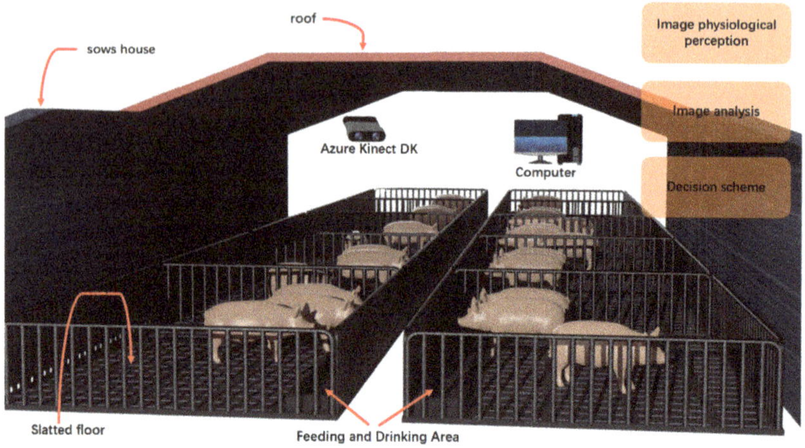

Figure 18. PLF (Precision Livestock Farming) of sow information perception diagram.

4. Conclusions

An approach for sow target detection based on deep learning for complex pig house environments has been proposed. A data acquisition system with two types of algorithms of UNet-Attention and Mask-RCNN was established. It has been found that Mask-RCNN had a better recognition rate, completeness, and faster running speed in analyzing the sow image dataset compared with UNet-Attention. The shape and position of sows in a pen can be detected through the segmentation, and the sow's behavior of eating, drinking and lying can also be identified. In the follow-up study, the network layer structure of the model will be optimized to achieve a better recognition effect. The Mask-RCNN algorithm will be further investigated for real-time monitoring of sows in large scale production.

Author Contributions: Data curation, K.L., C.Z.; Formal analysis, K.L., C.Z.; Investigation, K.L., C.Z.; Methodology, K.L., C.Z., X.D., S.P., P.Z., H.W. and T.Y.; Project administration, G.T. and C.Z.; Software, K.L., C.Z., T.Y.; Supervision, G.T. and C.Z.; Writing—original draft, K.L., C.Z.; Writing—review & editing, K.L., C.Z., X.D., G.T. and S.P. All authors have read and agreed to the published version of the manuscript.

Funding: This research was funded by the National Key Research and Development Program of China (grant number: 2016YFD0700204), Research on the Technology of Creating Comfortable Environment in Pig House (YF202103), and Chongqing Technology Innovation and Application Development Project (grant number: cstc2019jscx-gksbX0093).

Institutional Review Board Statement: All sows were managed by trained staff under standard guidelines, Chongqing, China. The study proposal was approved by The Laboratory Animal Ethical Committee of China Agricultural University (AW41211202-5-1).

Informed Consent Statement: Informed consent was obtained from all subjects involved in the study.

Data Availability Statement: The data presented in this study are available from the corresponding author on reasonable request.

Conflicts of Interest: The authors declare no conflict of interest.

References

1. Hoy, S. Precision pig farming. Innovative Technologies and decision models for pig farming. *Tierärztl. Prax. Ausg. Grosstiere/Nutztiere* **2009**, *37*, 184.
2. Peltoniemi, O.; Oliviero, C.; Yun, J.; Grahofer, A.; Björkman, S. Management practices to optimize the parturition process in the hyperprolific sow. *J. Anim. Sci.* **2020**, *98*, S96–S106. [CrossRef]
3. Kashiha, M.; Bahr, C.; Haredasht, S.A.; Ott, S.; Moons, C.P.; Niewold, T.A.; Ödberg, F.O.; Berckmans, D. The automatic monitoring of pigs water use by cameras. *Comput. Electron. Agric.* **2013**, *90*, 164–169. [CrossRef]
4. Lao, F.; Brown-Brandl, T.; Stinn, J.P.; Liu, K.; Teng, G.; Xin, H. Automatic recognition of lactating sow behaviors through depth image processing. *Comput. Electron. Agric.* **2016**, *125*, 56–62. [CrossRef]
5. Weng, R.-C. Variations in the body surface temperature of sows during the post weaning period and its relation to subsequent reproductive performance. *Asian-Australas. J. Anim. Sci.* **2020**, *33*, 1138–1147. [CrossRef]
6. Lopes, T.P.; Padilla, L.; Bolarin, A.; Rodriguez-Martinez, H.; Roca, J. Ovarian follicle growth during lactation determines the reproductive performance of weaned sows. *Animals* **2020**, *10*, 1012. [CrossRef] [PubMed]
7. Iida, R.; Pineiro, C.; Koketsu, Y. Removal of sows in Spanish breeding herds due to lameness: Incidence, related factors and reproductive performance of removed sows. *Prev. Veter.-Med.* **2020**, *179*, 105002. [CrossRef] [PubMed]
8. Hwang, J.; Yoe, H. Study of the Ubiquitous Hog Farm System Using Wireless Sensor Networks for Environmental Monitoring and Facilities Control. *Sensors* **2010**, *10*, 10752–10777. [CrossRef] [PubMed]
9. Thongkhuy, S.; Chuaychu, S.B.; Burarnrak, P.; Ruangjoy, P.; Juthamanee, P.; Nuntapaitoon, M.; Tummaruk, P. Effect of backfat thickness during late gestation on farrowing duration, piglet birth weight, colostrum yield, milk yield and reproductive performance of sows. *Livest. Sci.* **2020**, *234*, 103983. [CrossRef]
10. Chen, C.; Zhu, W.; Steibel, J.; Siegford, J.; Han, J.; Norton, T. Recognition of feeding behavior of pigs and determination of feeding time of each pig by a video-based deep learning method. *Comput. Electron. Agri.* **2020**, *176*, 105642. [CrossRef]
11. Ott, S.; Moons, C.P.H.; Kashiha, M.A.; Bahr, C.; Tuyttens, F.A.M.; Berckmans, D.; Niewold, T.A. Automated video analysis of pig activity at pen level highly correlates to human observations of behavioral activities. *Livest. Sci.* **2014**, *160*, 132–137. [CrossRef]
12. Riekert, M.; Klein, A.; Adrion, F.; Hoffmann, C.; Gallmann, E. Automatically detecting pig position and posture by 2D camera imaging and deep learning. *Comput. Electron. Agric.* **2020**, *174*, 105391. [CrossRef]
13. Marsot, M.; Mei, J.; Shan, X.; Ye, L.; Feng, P.; Yan, X.; Li, C.; Zhao, Y. An adaptive pig face recognition approach using Convolutional Neural Networks. *Comput. Electron. Agric.* **2020**, *173*, 105386. [CrossRef]
14. Zhuang, X.; Zhang, T. Detection of sick broilers by digital image processing and deep learning. *Biosyst. Eng.* **2019**, *179*, 106–116. [CrossRef]
15. Xiao, Y.; Tian, Z.; Yu, J.; Zhang, Y.; Liu, S.; Du, S.; Lan, X. A review of object detection based on deep learning. *Multimed. Tools Appl.* **2020**, *79*, 23729–23791. [CrossRef]
16. Hossain, S.; Lee, D.-J. Deep learning-based real-time multiple-object detection and tracking from aerial imagery via a flying robot with gpu-based embedded devices. *Sensors* **2019**, *19*, 3371. [CrossRef] [PubMed]
17. Tang, C.; Ling, Y.; Yang, X.; Jin, W.; Zheng, C. Multi-view object detection based on deep learning. *Appl. Sci.* **2018**, *8*, 1423. [CrossRef]
18. Algarni, A.D. Efficient object detection and classification of heat emitting objects from infrared images based on deep learning. *Multimed. Tools Appl.* **2020**, *79*, 13403–13426. [CrossRef]
19. Lu, S.; Wang, B.; Wang, H.; Chen, L.; Linjian, M.; Zhang, X. A real-time object detection algorithm for video. *Comput. Electr. Eng.* **2019**, *77*, 398–408. [CrossRef]
20. Aziz, L.; Salam, S.B.H.; Sheikh, U.U.; Ayub, S. Exploring deep learning-based architecture, strategies, applications and current trends in generic object detection: A comprehensive review. *IEEE Access* **2020**, *8*, 170461–170495. [CrossRef]
21. Bamne, B.; Shrivastava, N.; Parashar, L.; Singh, U. Transfer learning-based Object Detection by using Convolutional Neural Networks. In Proceedings of the 2020 International Conference on Electronics and Sustainable Communication Systems (ICESC), Coimbatore, India, 2–4 July 2020; pp. 328–332.
22. Yann, L.C.; Bengio, Y.; Hinton, G. Deep learning. *Nature* **2015**, *521*, 436–444.
23. Guo, Y.; Liu, Y.; Oerlemans, A.; Lao, S.; Wu, S.; Lew, M.S. Deep learning for visual understanding: A review. *Neurocomputing* **2016**, *187*, 27–48. [CrossRef]
24. Ren, S.; He, K.; Girshick, R.; Sun, J. Faster R-CNN: Towards real-time object detection with region proposal networks. *IEEE Trans. Pattern Anal. Mach. Intell.* **2017**, *39*, 1137–1149. [CrossRef]
25. Poursaberi, A.; Bahr, C.; Pluk, A.; Van Nuffel, A.; Berckmans, D. Real-time automatic lameness detection based on back posture extraction in dairy cattle: Shape analysis of cow with image processing techniques. *Comput. Electron. Agric.* **2010**, *74*, 110–119. [CrossRef]

26. Salau, J.; Haas, J.H.; Junge, W.; Thaller, G. Automated calculation of udder depth and rear leg angle in Holstein-Friesian cows using a multi-Kinect cow scanning system. *Biosyst. Eng.* **2017**, *160*, 154–169. [CrossRef]
27. Traffano-Schiffo, M.V.; Castro-Giraldez, M.; Colom, R.J.; Fito, P.J. Development of a spectrophotometric system to detect white striping physiopathy in whole chicken carcasses. *Sensors* **2017**, *17*, 1024. [CrossRef] [PubMed]
28. Girshick, R.; Donahue, J.; Darrell, T.; Malik, J. Region-based convolutional networks for accurate object detection and segmentation. *IEEE Trans. Pattern Anal. Mach. Intell.* **2015**, *38*, 142–158. [CrossRef]
29. Nasirahmadi, A.; Sturm, B.; Edwards, S.; Jeppsson, K.-H.; Olsson, A.-C.; Müller, S.; Hensel, O. Deep learning and machine vision approaches for posture detection of individual pigs. *Sensors* **2019**, *19*, 3738. [CrossRef] [PubMed]
30. Zhang, L.; Gray, H.; Ye, X.; Collins, L.; Allinson, N. Automatic individual pig detection and tracking in pig farms. *Sensors* **2019**, *19*, 1188. [CrossRef]
31. Kim, J.; Chung, Y.; Choi, Y.; Sa, J.; Kim, H.; Chung, Y.; Park, D.; Kim, H. Depth-based detection of standing-pigs in moving noise environments. *Sensors* **2017**, *17*, 2757. [CrossRef]
32. Han, S.; Zhang, J.; Zhu, M.; Wu, J.; Kong, F. Review of automatic detection of pig behaviors by using image analysis. *IOP Conf. Ser. Earth Environ. Sci.* **2017**, *69*, 012096. [CrossRef]
33. Gangsei, L.E.; Kongsro, J. Automatic segmentation of Computed Tomography (CT) images of domestic pig skeleton using a 3D expansion of Dijkstra's algorithm. *Comput. Electron. Agric.* **2016**, *121*, 191–194. [CrossRef]
34. Guo, Y.-Z.; Zhu, W.-X.; Jiao, P.-P.; Ma, C.-H.; Yang, J.-J. Multi-object extraction from topview group-housed pig images based on adaptive partitioning and multilevel thresholding segmentation. *Biosyst. Eng.* **2015**, *135*, 54–60. [CrossRef]
35. Hansen, M.F.; Smith, M.L.; Smith, L.N.; Salter, M.G.; Baxter, E.; Farish, M.; Grieve, B. Towards on-farm pig face recognition using convolutional neural networks. *Comput. Ind.* **2018**, *98*, 145–152. [CrossRef]
36. Hu, J.-Y.; Shi, C.-J.R.; Zhang, J.-S. Saliency-based YOLO for single target detection. *Knowl. Inf. Syst.* **2021**, *63*, 717–732. [CrossRef]
37. Oktay, O.; Schlemper, J.; Folgoc, L.L.; Lee, M.; Heinrich, M.; Misawa, K.; Rueckert, D. Attention u-net learning where to look for the pancreas. *arXiv* **2018**, arXiv:1804.03999.
38. He, K.; Gkioxari, G.; Dollár, P.; Girshick, R. Mask R-CNN. *arXiv* **2018**, arXiv:1703.06870v3.
39. Tu, S.; Liu, H.; Li, J.; Huang, J.; Li, B.; Pang, J.; Xue, Y. Instance Segmentation Based on Mask Scoring R-CNN for Group-housed Pigs. Presented at the 2020 International Conference on Computer Engineering and Application (ICCEA); Institute of Electrical and Electronics Engineers (IEEE), Piscataway Township, NJ, USA, 18–20 March 2020; pp. 458–462.
40. Lowe, D.G. Distinctive image features from scale-invariant keypoints. *Int. J. Comput. Vis.* **2004**, *60*, 91–110. [CrossRef]
41. Lin, T.Y.; Goyal, P.; Girshick, R.; He, K.; Dollár, P. Focal Loss for Dense Object Detection. *arXiv* **2018**, arXiv:1708.02002.
42. Simonsen, H.B. Behavior and distribution of fattening pigs in the multi-activity pen. *Appl. Anim. Behav. Sci.* **1990**, *27*, 311–324. [CrossRef]
43. Wang, Z.; Bovik, A.C.; Sheikh, H.R.; Simoncelli, E.P. Image quality assessment: From error visibility to structural similarity. *IEEE Trans. Image Process.* **2004**, *13*, 600–612. [CrossRef] [PubMed]
44. Porto, S.M.C.; Arcidiacono, C.; Anguzza, U.; Cascone, G. A computer vision-based system for the automatic detection of lying behavior of dairy cows in free-stall barns. *Biosyst. Eng.* **2013**, *115*, 184–194. [CrossRef]
45. Viazzi, S.; Bahr, C.; Van Hertem, T.; Schlageter-Tello, A.; Romanini, C.; Halachmi, I.; Lokhorst, C.; Berckmans, D. Comparison of a three-dimensional and two-dimensional camera system for automated measurement of back posture in dairy cows. *Comput. Electron. Agric.* **2014**, *100*, 139–147. [CrossRef]
46. Wang, X.; Xiao, T.; Jiang, Y.; Shao, S.; Sun, J.; Shen, C. Repulsion Loss: Detecting Pedestrians in a Crowd. *arXiv* **2018**, arXiv:1711.07752.
47. Zhang, S.; Wen, L.; Bian, X.; Lei, Z.; Li, S.Z. Occlusion-Aware R-CNN: Detecting Pedestrians in a Crowd. In Proceedings of the Transactions on Petri Nets and Other Models of Concurrency XV, Munich, Germany, 8–14 September 2018; pp. 657–674.
48. Lin, T.; Dollár, P.; Girshick, R.B.; He, K.; Hariharan, B.; Belongie, S. Feature Pyramid Networks for Object Detection. In Proceedings of the IEEE Conference on Computer Vision and Pattern Recognition, Honolulu, HI, USA, 21–26 July 2017.
49. Sun, L.; Li, Y.; Zou, Y. Pig image segmentation method based on improved Graph Cut algorithm. *Nongye Gongcheng Xuebao/Trans. Chin. Soc. Agric. Eng.* **2017**, *33*, 196–202.
50. Yang, L.; Zhu, W.X. Image segmentation of pig using wavelet modulus and edge growth. *Appl. Mech. Mater.* **2014**, *687–691*, 3695–3698. [CrossRef]
51. Zhang, Y.; Cai, J.; Xiao, D.; Li, Z.; Xiong, B. Real-time sow behavior detection based on deep learning. *Comput. Electron. Agric.* **2019**, *163*, 104884. [CrossRef]
52. Liu, B.; Zhu, W.; Ji, B.; Ma, C. Automatic registration of IR and optical pig images based on contour match of radial line feature points. *Trans. Chin. Soc. Agric. Eng.* **2013**, *29*, 153–160.
53. Carillo, F.; Abeni, F. An Estimate of the Effects from Precision Livestock Farming on a Productivity Index at Farm Level. Some Evidences from a Dairy Farms' Sample of Lombardy. *Animals* **2020**, *10*, 1781. [CrossRef]
54. Nasirahmadi, A.; Sturm, B.; Olsson, A.-C.; Jeppsson, K.-H.; Müller, S.; Edwards, S.; Hensel, O. Automatic scoring of lateral and sternal lying posture in grouped pigs using image processing and Support Vector Machine. *Comput. Electron. Agric.* **2019**, *156*, 475–481. [CrossRef]
55. Ammendrup, S.; Füssel, A.E. Legislative requirements for the identification and traceability of farm animals within the European Union. *Rev. Sci. Tech.* **2001**, *20*, 437–444. [CrossRef] [PubMed]

56. Zamora-Izquierdo, M.A.; Santa, J.; Martínez, J.A.; Martínez, V.; Skarmeta, A.F. Smart farming IoT platform based on edge and cloud computing. *Biosyst. Eng.* **2019**, *177*, 4–17. [CrossRef]
57. Banhazi, T.M.; Lehr, H.; Black, J.L.; Crabtree, H.; Schofield, P.; Tscharke, M.; Berckmans, D. Berckmans precision livestock farming: An international review of scientific and commercial aspects. *Int. J. Agric. Boil. Eng.* **2012**, *5*, 1–9.

agriculture

MDPI

Article

Multi-Temporal Data Fusion in MS and SAR Images Using the Dynamic Time Warping Method for Paddy Rice Classification

Tsu Chiang Lei [1], Shiuan Wan [2,*], You Cheng Wu [1], Hsin-Ping Wang [3] and Chia-Wen Hsieh [3]

1 Department of Urban Planning and Spatial Information, Feng Chia University, Taichung 40724, Taiwan; tclei@fcu.edu.tw (T.C.L.); steven159357@yahoo.com.tw (Y.C.W.)
2 Department of Information Technology, Ling Tung University, Taichung 40851, Taiwan
3 Construction and Disaster Prevention Research Center, Feng Chia University, Taichung 40724, Taiwan; t04111@mail.fcu.edu.tw (H.-P.W.), cwhsieh@mail.fcu.edu.tw (C.-W.H.)
* Correspondence: shiuan123@teamail.ltu.edu.tw

Abstract: This study employed a data fusion method to extract the high-similarity time series feature index of a dataset through the integration of MS (Multi-Spectrum) and SAR (Synthetic Aperture Radar) images. The farmlands are divided into small pieces that consider the different behaviors of farmers for their planting contents in Taiwan. Hence, the conventional image classification process cannot produce good outcomes. The crop phenological information will be a core factor to multi-period image data. Accordingly, the study intends to resolve the previous problem by using three different SPOT6 satellite images and nine Sentinel-1A synthetic aperture radar images, which were used to calculate features such as texture and indicator information, in 2019. Considering that a Dynamic Time Warping (DTW) index (i) can integrate different image data sources, (ii) can integrate data of different lengths, and (iii) can generate information with time characteristics, this type of index can resolve certain classification problems with long-term crop classification and monitoring. More specifically, this study used the time series data analysis of DTW to produce "multi-scale time series feature similarity indicators". We used three approaches (Support Vector Machine, Neural Network, and Decision Tree) to classify paddy patches into two groups: (a) the first group did not apply a DTW index, and (b) the second group extracted conflict predicted data from (a) to apply a DTW index. The outcomes from the second group performed better than the first group in regard to overall accuracy (OA) and kappa. Among those classifiers, the Neural Network approach had the largest improvement of OA and kappa from 89.51, 0.66 to 92.63, 0.74, respectively. The rest of the two classifiers also showed progress. The best performance of classification results was obtained from the Decision Tree of 94.71, 0.81. Observing the outcomes, the interference effects of the image were resolved successfully by various image problems using the spectral image and radar image for paddy rice classification. The overall accuracy and kappa showed improvement, and the maximum kappa was enhanced by about 8%. The classification performance was improved by considering the DTW index.

Keywords: remote sensing; time series; data fusion; feature extraction; dynamic time warping; crop phenological information

Citation: Lei, T.C.; Wan, S.; Wu, Y.C.; Wang, H.-P.; Hsieh, C.-W. Multi-Temporal Data Fusion in MS and SAR Images Using the Dynamic Time Warping Method for Paddy Rice Classification. *Agriculture* **2022**, *12*, 77. https://doi.org/10.3390/agriculture12010077

Academic Editors: Gniewko Niedbała and Sebastian Kujawa

Received: 12 December 2021
Accepted: 5 January 2022
Published: 7 January 2022

1. Introduction

Paddy rice takes up the largest crop area and has great significance for the global economy, society, and culture. Presently, farmland surveys in various countries are mainly manual surveys, which are time-consuming, labor-intensive, and extremely inefficient. It is hard to conduct a large-scale survey in a short time. However, with the advancement and development of satellite remote sensing detection technology in recent years, farmland monitoring methods have become well-accepted. The use of satellite image data as a monitoring tool along with the use of the machine learning approach has become a major solution for land cover measurements. This includes supervised learning and unsupervised learning in machine learning. This greatly reduces the manpower and material resources required for agricultural monitoring and management [1,2].

Many types of research have been dedicated to investigating satellite optical data to carry out the target GIS map for delineation of paddy field areas by image classification through the pixel-based method. This includes Maximum Likelihood, Neural Network, Decision Tree, Support Vector Machine, K-means, ISODAT, etc. Different classification methods can be applied by using by the material of Landsat TM and ETM+ series, SPOT series, MODIS, Sentinel-2 and 3, RADARSAT series, ERS-1 and ERS-2, ENVISAT/ASAR, IRS, AVHRR, Sentinel-1A, Aerial-Photo, UAV, etc. In addition, related research is focused on how to increase the accuracy of paddy rice fields. For instance, a series of Vegetation Indicators (VI) and Texture Indicators (TI) may become the proper material for interpretation results and classification accuracy, which can be reinforced. The VI indicators include the Ratio Vegetation Index (RVI), NDVI, Soil-adjusted Vegetation Index (SAVI), etc. The texture indicators include the Gray Level Co-Occurrence Matrix (GLCM), Fractal dimension, Semi-Variogram, etc. Hence, the benefits of statistical analysis and machine learning can be greatly improved. In addition to pixel-based classification, the Region-based Object Classification (ROC) is also well-accepted. There are two main steps for ROC: (1) image segmentation and (2) image classification. The method also renders an effective performance in classification. On the other hand, some scholars have imported Synthetic Aperture Radar (SAR) with multiple time-series data to detect the area of rice fields, which has attracted new attention in recent years. The SAR data are not affected by sunlight. Accordingly, there are many kinds of research that perform Image Fusion (IF) processing between MS and SAR as well [3,4]. Unfortunately, traditional image fusion methods seem to be unable to solve the large differences between the two images, and it is not easy to overcome some practical limitations. For example, the images after fusion are prone to produce unexpected noise [5,6]. The errors of data fusion by multi-period sequence images are often accumulated into classification progress. It generally results in unsatisfactory classification outcomes. That is, the image fusion can not obtain the crop phenological information in detail, which is important for image classification [7]. Rice patches may be affected by mixed crops of land-use on a single patch, different planting seasons, and different varieties, which are governed by different farmer behaviors. In addition, using a single image may result in the interference of cloud and fog effects, which can obstruct the classification results. Furthermore, it destroys the structure of the landscape for considering a single period of texture/vegetation indicator through image fusion.

Therefore, this research does not aim at using the IF methods but uses the concept of the DF method at the starting point. The DF method requires a set of integrated calculations, rather than a method of evaluating data through a single model [8,9]. More specifically, this is a hybrid model to employ different data sources or analysis methods. Related research shows that most of the past studies focused on the fusion of multiple methods [9]. In this study, the data fusion process extracts the variation of features based on various periods with the following considerations. First, the length of time-series changes of different data sources is different. Second, the different properties of data sources vary greatly. Whereas the quality of the original spectral bands must be effectively converted into rational indicators or textures, the different resource data have different resolutions and formats. Third, the method for changing characteristics of patches of different properties vary at different times. It has to effectively extract this relevant information for our research. Hence, DTW is applied in this study for multi-period images for data fusion [10]. It was successfully applied to rice area survey [11,12], landscape changes [10,13], forest type classification [14], farmland mapping [15–20], crop phenological period of factor analysis [21], crop yield estimation [22], etc. This research plans to extract the phenological information in the fragmented landscape from the image through the DTW method to achieve the purpose of rapid mapping, stability, and high accuracy when making rice farmland thematic maps.

Consequently, this study uses the Dynamic Time Warping (DTW) method to compare the similarity results of MS and SAR image datasets by vegetation index and texture index characteristics, respectively. The similarity index results can show the relations in the change of each different land-use of patches. That is, the DTW is examined by the DF method, which contains numerical responses (spectrum, indicators, and textures) to detect similar features in different time series. These relations can improve the classification results efficiently. Hence, three approaches (Support Vector Machine, SVM; Neural Network, NN; and Decision Tree, DT) are used to classify the paddy patches with two groups: (a) without applying the DTW index and (b) considering the DTW index. We adopt the most common classifiers such as SVM, NN, and DT. The goal is not to compare the performance of those three classifiers and determine which one is the best. The key point is to find a possible solution to rationally integrate them with consideration of the function of DTW employing two kinds of image data (optical and radar). The paper contains four steps: Step 1, the first stage of the accuracy of consistency classification; Step 2, the discussion of the accuracy of inconsistency classification; Step 3, examples of multi-scale features and description of integration results; Step 4, the overall accuracy of the hybrid classification.

2. Research Materials and Design

2.1. Research Materials

2.1.1. Research Site

The study was located at Yunlin County, which is in the Jianan Plain of western Taiwan. It is a major agricultural county for Taiwan. The annual grain output is noteworthy. It mainly produces rice, head, vegetables, peanuts, sweets, and other grains. The coordinates of longitude and latitude are 27′00″120 E, 48′00″23 N. The soil is rich in organic nitrogen, phosphorus, potassium, and other elements, which makes the land abundant in agricultural production. The soil is fertile, and the climate is suitable. Irrigation and rainfall are abundant. Therefore, this study selected Yunlin Xiluo, as in Figure 1. Figure 2a is a map of farmland patches in this area. The total area is about 5016.21 ha with 53,212 patches. Figure 2b is the ground truth data of the study area taken in the first half of 2019 by the Agriculture and Food Agency. Since the main axis of this research is to classify and interpret rice fields, the classification of ground truth data is only divided into paddy rice and non-paddy rice, as in Table 1.

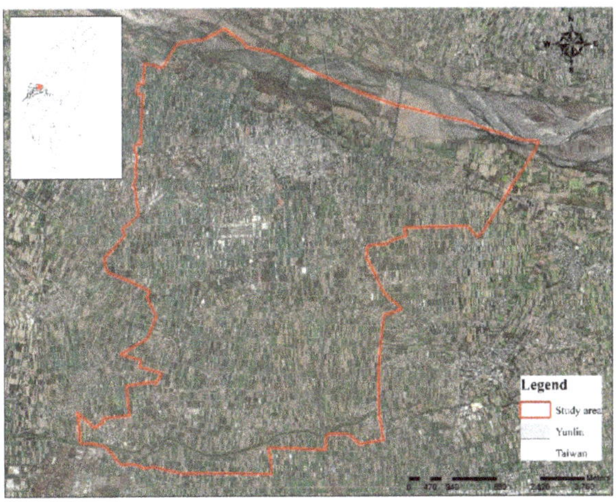

Figure 1. Study area.

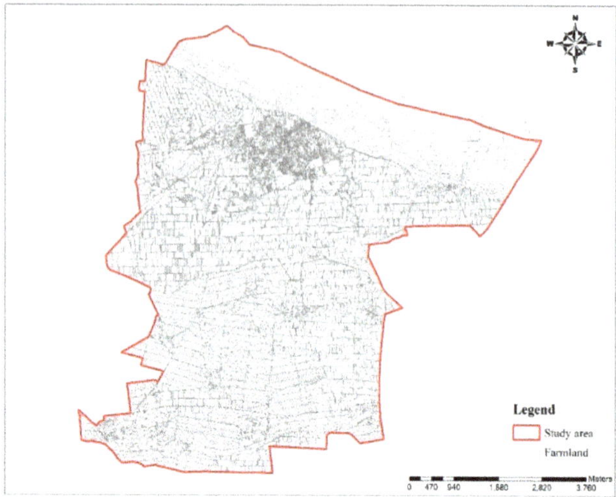

(**a**)

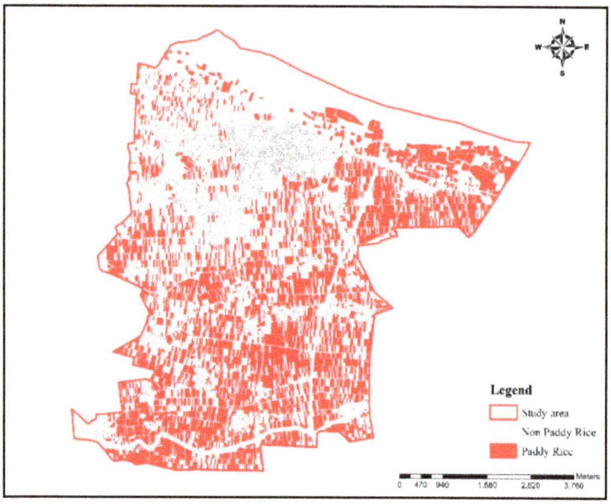

(**b**)

Figure 2. The patches are taken from the photo in 2019. (**a**) Farmland patches in study area. (**b**) The ground truth data from the Agriculture and Food Agency.

Table 1. The ground truth data of Xiluo in 2019.

Categories	Number of Patches	Area (ha)
Paddy Rice	7699	1634.96
Non-Paddy Rice	45,513	3379.83
Total	53,212	5014.80

2.1.2. Research Data

SPOT 6 Images

SPOT images basically have four bands, which are multi-spectral images of B, G, R, and IR, with a spatial resolution of 6 m. This study selected SPOT-6 images on 23 January, 1 March, and 9 April in 2019. The time differences between the three were 37 days and

39 days, respectively. Figure 3 below is the three SPOT6 optical satellite images selected for this study. The SPOT6 satellites are easy to attain in Taiwan, and they are already atmospherically and geometrically corrected. They are completed by the Central University Space Remote Sensing Center. Hence, this study decided to use them. However, in this study, we only used the three indicators of G, R, and IR to generate indicators of ancillary information. For detailed indicator descriptions, please see the content in Section 2.2.

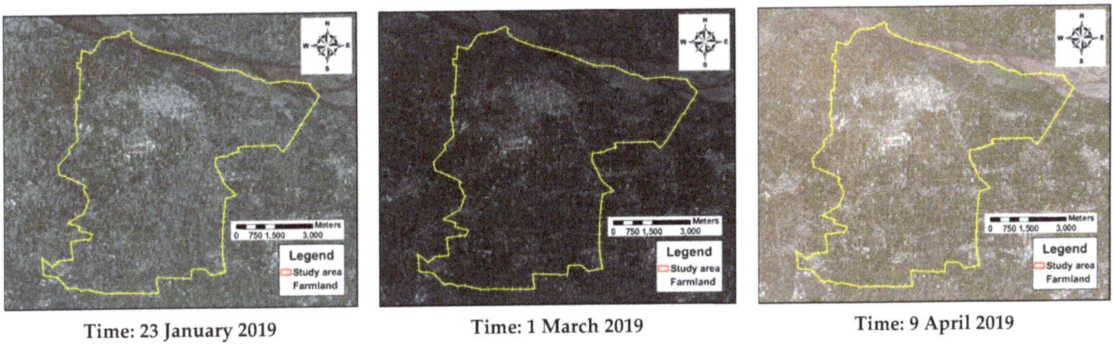

Figure 3. The selected SPOT6 satellite image.

Sentinel-1A Images

Sentinel-1A has a shooting period of 12 days, and Sentinel-1 is equipped on four sensors for different shooting purposes, namely Stripmap Mode (SM) and Interfero-metric Wide Swath Mode (IW), Extra Wide Swath Mode (EW), and Wave Mode (Wave Mode, WM). Its spatial resolution is 5 × 20 m (16 × 66 feet). To have a better understanding of the rice growth cycle time of this survey, the short time of Sentinel-1 radar image material in this study includes all radar images between 31 January and 7 May 2019. Since the radar images are 12 days old during the shooting cycle, nine radar images during this period are finally selected. All the data can be downloaded from the European Space Agency (ESA) website for free [23]. The downloaded images must be downloaded through the SNAP software. There are three pre-processing steps: radiometric correction, geometric correction, and image speckle noise removal. The selection of radar images was taken on 1/31, 2/12, 2/24, 3/8, 3/20, 4/1, 4/13, 4/25, 5/7. This study uses images in IW mode because this mode is the main operating mode for shooting on land. The shot type is IW and VV polarization and VH polarization images.

2.2. Research Design

In this study, 3 of SPOT6 optical satellites and 9 of Sentinel-1A synthetic aperture radar images were used (total 12 images) to identity features such as textures and indicators. The feature value information was extracted using patches as the smallest unit. In the meantime, the time-series features were constructed. Dynamic Time Warping was used to produce "multi-scale time series feature similarity indicators". DTW is dynamic programming [10,11]. The algorithm of this approach compares two sequences with different lengths. It effectively solves the deviation of time distortion in identification and calculates the Euclidean distance between the two sequences to determine the similarity of content information. The gap between the vector distances is rationally found. This indicator can convert an image in a time series of feature information on different scales and different sources into ancillary information.

Figure 4 outlines steps for research, and it has four parts: 1, image feature calculation and database construction; 2, classification algorithm; 3, comparison of classification results. The contents are as follows.

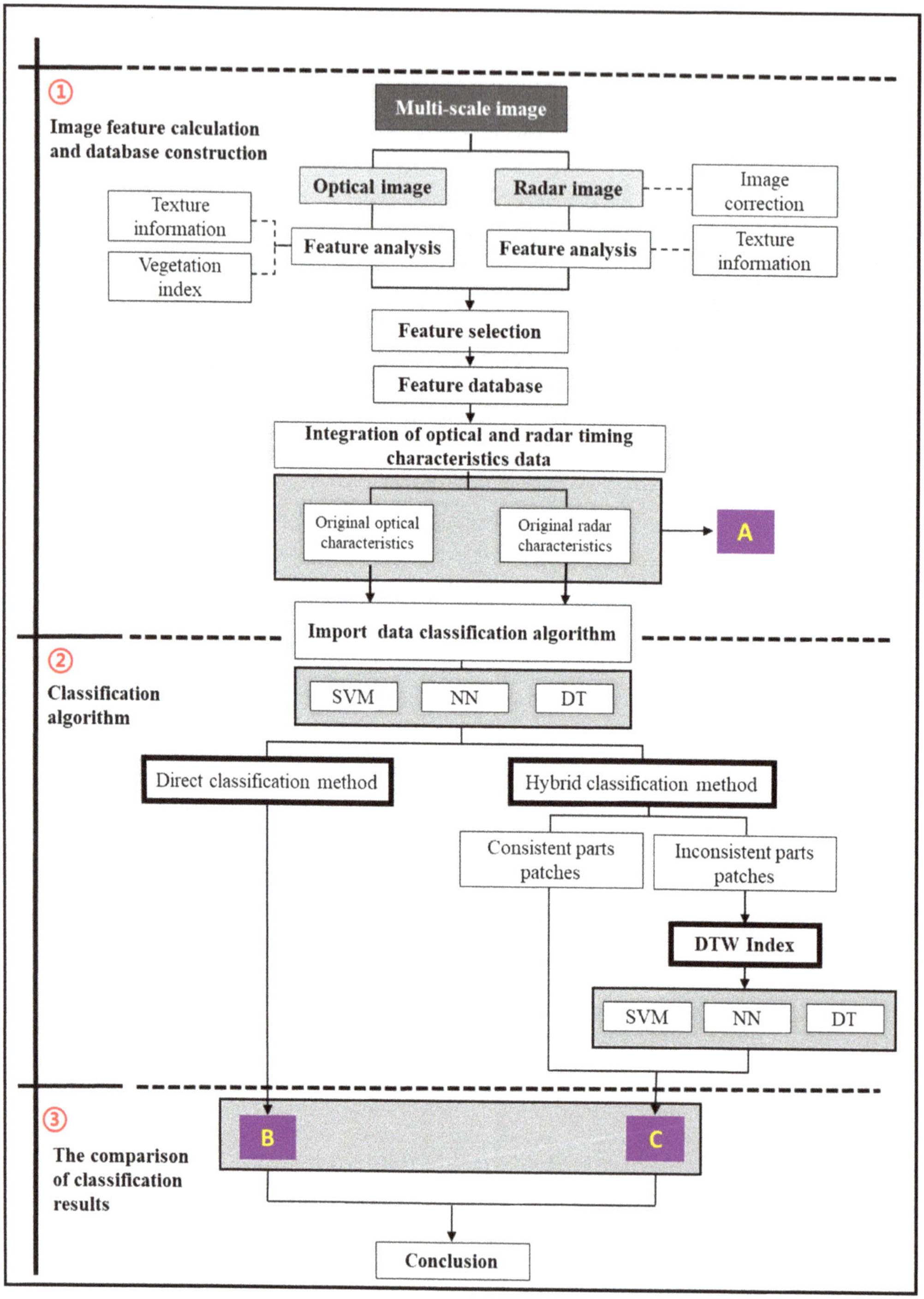

Figure 4. Steps for this study.

1. Image feature calculation and dataset construction: In the optical image part, in addition to the four basic bands of red light (Red), green light (Green), blue light (Blue), and near-infrared light (NIR), this research included Ratio Vegetation Index (RVI) and Normalized Difference Vegetation Index (NDVI) and four Co-occurrence matrix (Gray Level Co-Occurrence Matrix, GLCM) texture indexes, including Homogeneity, Contrast, Dissimilarity, and Entropy, making a total of 19 types of feature information. It is worth mentioning that GLCM and associated texture features are image analysis techniques. An image is composed of pixels, each with an intensity (a specific gray level) suitable to apply GLCM, as different combinations of gray levels often co-occur in an image or image section. Texture feature calculations use the contents of GLCM to measure the variation in intensity (image texture) at the pixel of interest; on the other hand, the radar part uses C-band synthetic aperture radar (SAR) of two polarized images, called VV and VH. In addition to VV and VH, the 4 aforementioned kinds of texture information were also adopted. There was a total of 10 types of feature information. The features are shown in Table 2. In addition, the variation of satellite images for different time features could become a key factor when observing paddy rice and non-paddy rice patterns. We assign this part of the data as Label A.
2. Classification algorithm: This study used Support Vector Machine (SVM), Neural Network (NN), and C5.0 Decision Tree (DT) machine learning algorithm models. We conducted the training and verification of the models. The model training and verification rate ranged from 70% (37,227 patches) to 30% (15,985 patches). This further illustrates that the design of this research is different from the traditional method. We applied the three classification methods with two direct classification methods and a hybrid classification method to perform better classification outcomes. Direct classification methods use three machine learning models to directly classify the image data information (optical, index, and texture) features (Label A data set). Generally speaking, paddy rice is a long-period crop that requires an indicator and involves the combination of optical, index, and texture features for recording the variation over a long period. Hence, this study adopted two stages to solve the problem. The first stage was to import the image data to three classifiers (SVM, DT, and NN). The second stage was to extract the confusion samples (or patches) from the first stage to employ a new factor (DTW), which considers the time variation factor to improve classification performance. Specifically, this new ancillary information (DTW) was used for the dynamic calculation of the two-time series, and Euclidean distance matrices were computed one by one. For example, optical image characteristic band B to band G is one group, band B to band R is another group, etc. All combinations had to be calculated. The total number of feature information groups reached 26 levels. It was necessary to add 351 combinations of optical feature similarity. The similarity index between the two-time series features was produced, and the dataset was consolidated by the combination of the time series in patch features. The 351 combinations contained Optical Image Feature Similarity (171 attributes), Radar Image Feature Similarity (28 attributes), and the two combinations of feature similarity indicator (optical image + radar image) groups (152 at-tributes). Therefore, in this study, all data combinations were generated to form a DTW index (Figure 5). We hope this process can resolve the confusion around classification patches (samples). In the meantime, the inconsistent patches from the classification model were further refined.
3. The comparison of classification: The training model accuracy had two parts. The first part was the result of the direct classification method, which was assigned as Label B. The second part was the result of the hybrid classification, which was assigned Label C. The comparison items were computed from an analysis of commission errors and omission errors. The overall accuracy and kappa values were also employed. We also compared the performance of Label B and Label C.

Table 2. Ancillary Information.

Vegetation Index	Formula
RVI (Ratio Vegetation Index)	$\frac{R}{NIR}$
NDVI (Normalized Difference Vegetation Index)	$\frac{NIR-R}{NIR+R}$
PVI (Perpendicular Vegetation Index)	$\frac{NIR-NIR_{soil}}{\sqrt{1+B^2}}$
SAVI (Soil-adjusted Vegetation Index)	$(1+L)\times\frac{NIR-R}{NIR+R+L}$
TSAVI (Transformed Soil-adjusted Vegetation Index)	$\frac{B(NIR-NIR_{Soil})}{R+B(NIR-A)+X(1+B^2)}$
CMFI (Cropping Management Factor Index)	$\frac{R}{NIR+R}$
GI (Greenness Index)	$\frac{NIR}{G}$
IPVI (Infrared Percentage Vegetation Index)	$\frac{NIR}{NIR+R}$
MSAVI (Modified Soil-adjusted Vegetation Index)	$\frac{\left(2NIR+1-\sqrt{(2NIR+1)^2-8(NIR-R)}\right)}{2}$
OSAVI (Optimization Soil adjusted Vegetation Index)	$\frac{NIR-R}{NIR+R+Y}$
GESAVI (Generalize Soil- adjusted Vegetation Index)	$\frac{NIR-NIR_{Soil}}{R+Z}$
HOM (Homogeneity)	$\text{Homogeneity}=\sum_{i=0}^{N}\sum_{j=0}^{N}\frac{1}{1+(i-j)^2}C_{ij}(d,\theta)$
CON (Contrast)	$\text{Contrast}=\sum_{i,j}\lvert i-j\rvert^2 p(i,j)$
DIS (Dissimilarity)	$\text{Dissimilarity}=\sum_{i=0}^{n}\sum_{j=0}^{n}C_{ij}\lvert i\ j\rvert$
ENT (Entropy)	$\text{Entropy}=\sum_{i=0}^{n}\sum_{j=0}^{n}C_{ij}\log C_{ij}$

Experienced Coefficient for $L = 0.5$; $X = 0.08$; $Y = 0.16$; $Z = 0.35$ considering multiple scattered conditions: $R_{Soil} = A + B \times R$; ($A = 0.011$, $B = 1.16$) [24]. NIR_{Soil} = BR − A; The A and B is the soil line parameters. BR = Blue light × B.

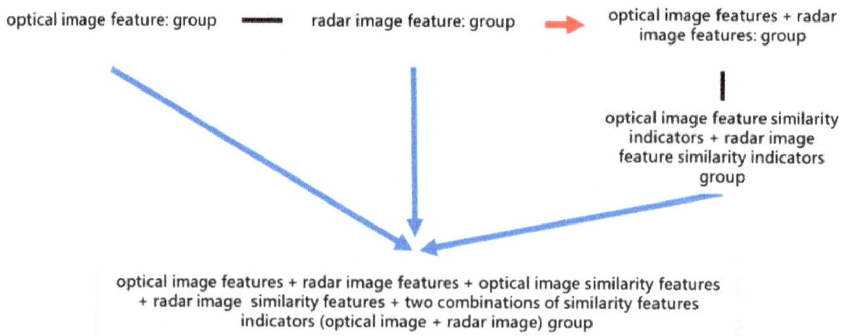

Figure 5. The DTW index of analyzed data sets for steps.

3. Research Methods

3.1. Support Vector Machine

The Support Vector Machine (SVM) is a popular machine learning tool that offers solutions for both classification and regression problems. Support vector machines (SVMs) are well-accepted supervised learning methods used for classification. The SVM classifier supports binary classification and multiclass classification, whereas the structured SVM trains a classifier for generally structured output labels. Moreover, there exist many hyperplanes that may be able to classify the data. One rational choice as the best hyperplane is to produce the largest separation, or margin, between the two classes. The optimal choice of the hyperplane is the distance from the selected sample to the nearest sample point on each side, which is maximized. The study considers the concept of improving statistical learning theory, generally applied as an effective classifier to solve many practical problems. The feature of these classifiers is to minimize the empirical classification error and maximize the geometric margin [25]. The support vector machine is requested to select an appropriate kernel function. The function of the kernel is to take data as input and convert it into the required form. This is because different types of data cannot be linearized in the original space. When separated, the data after nonlinear projection can be more separated in a higher-dimensional space, usually linear, polynomial, Radial Basis Function (RBF), and Sigmoid Function. We explain the classification methods of SVM in detail. The value of bias is set to 0. The core function is adopted as RBF. The c = 10 and gamma = 0.1 are used as the initial conditions for setting parameters. The stopping criterion of 0.001 was used as a standard to terminate the program and output the results.

3.2. Neural Network

The neural networks are information processing networks inspired by the way biological neural systems process data. Neural Networks (NN) were first proposed in the early 1940s as an attempt to simulate human brain cognitive learning processes [26]. They are programmed with a primary function, which is to develop models of problems based on trial and error or learning procedures. In the last decades, Back Propagation Neural Network has been widely applied in many fields. The relations among massive data and a certain phenomenon are obtained through a learning system (instead of calculation). In the past, scientists and researchers experienced that the inputs of attributes (included ancillary information) for remote sensing images are usually used to apply to image classification. If a paddy area spatial dataset was well developed to perform the input variables and output categories rationally, it may be appropriate to apply Back Propagation Neural Network as a learning machine [27]. Basically, the neural network consists of many nodes to connect input neurons and output neurons to three sorts of layers: input layer, hidden layer, and output layer. The study adopted Multi-Layer Perception. Our MLP consists of at least three layers of nodes: an input layer, a hidden layer with 13 neurons, and an output layer of classification. The optical input has 19 neurons, and the radar input has 8 neurons. The active function used the sigmoid function. The output used 800 epochs or 0.02% difference as a criterion to obtain the classification outcomes.

3.3. Decision Tree

A decision tree is a tree structure containing internal and external nodes connected by branches. A decision tree is a data-driven predictive model where it is mapped from the observation of samples about an item to conclusions about its target value. It is usually used as a tool for scientists and engineers to generate "rules" [28]. The internal node is a decision-making point to investigate a decision function to determine which child node to visit next. On the other hand, an external node is also known as a leaf or terminal node, which has no child nodes and, with respect to a label, characterizes the given data that lead to being visited. In general, a decision tree is employed as follows. It presents a datum (a vector composed of several attributes) to the root node of the decision tree. It may depend on the result of a decision function used by an internal node, and the tree will branch to one of the children of the node. This process will repeat until a terminal node is approached and a label or value is then assigned to the given data. The height of DT is limited to 17 layers, which uses the Exhaustive Algorithm to display all the possibilities of the condition the samples should fit in.

3.4. DTW Methods

DTW is one of the algorithms for computing the similarity on two temporal sequences. Hence, DTW can be successfully applied to temporal sequences of video, audio, and graphics data. That is, any data that can be turned into a linear sequence can be analyzed by DTW as well. A well-known application has been automatic speech recognition for different speaking speeds. It can also be used in partial shape matching applications. Reviewing the DTW past research, Petitjean et al. (2012) used the DTW algorithm for SPOT time-series satellite images to classify the land-use coverage. This research incorporated the K-means and DTW into image measurement to obtain the classification. The results show that the similarity of multi-period images is matched by using DTW, which performs better in classification outcomes of multi-period images than that of a single image use [10].

Similarity measurements between the two sequences are named as "warping path". In this path the two signals may be aligned at the same time. The signal with an original set of points X (original), Y (original) is converted to X (warped), Y (warped). Related technique sequences of varying speed may be averaged using this technique. While two different time series data are matched with each other, it can be seen directly through a line chart or other visualized graphs whether there is a strong similarity between the two. It is possible to objectively quantify the degree of similarity between the two images.

To calculate the DTW similarity of two-time series, one can establish an m × n Euclidean distance matrix. Then, a cost matrix or cumulative matrix M_c based on the distance matrix is generated. The cumulative matrix is M_c (i, j) defined as follows:

$$M_C(i,j) = \min\begin{cases} M_C(i-1,j-1) \\ M_C(i-1,j) \\ M_C(i,j-1) \end{cases} + M(i,j) \tag{1}$$

In Equation (1), $M_C(i,j)$ represents a matrix from a point (i, j) of the route. The accumulated value of the minimal value in (1) has three terms. The optimal value can be found by considering $M_C(i,j)$ with two periods of distance in DTW by computing from $(1, 1)$ to (i, j).

Since the DTW can analyze two sets of timing information with different scales and timing lengths, it produces the most intuitive numerical value to show the degree of similarity of the timing fluctuations between the two sets of information [10–22]. This study presumes that different sources of information have their contributions for classification. We utilized the "multi-scale time series feature similarity" indicators from the concept of data fusion, especially considering the ancillary information of radar data, to compare the optical image data to produce the similarities. According to Equation (1), this study uses Python to write a "multi-scale time series feature similarity indicators" program that can process multiple time series feature information in batches. The program merges all the characteristics into it as well. The features are computed as a feature similarity index, and then each of the rest of the features step-by-step will be imported. All the aforementioned data are applied to the classification outcomes from three approaches (SVM, NN, and DT) by inconsistent classified results. The best way to resolve the inconsistent classified results is to use new ancillary information (such as DTW).

3.5. Accuracy Verification

To test the accuracy of the final automated classification model, this study uses the Confusion Matrix and kappa value in image interpretation and classification accuracy of the final results of this research. Four different regions are randomly selected in this study area as verification regions to check the final results of the developed new classification model.

3.6. Model Software

In this study, we used IBM SPSS Modeler 18.1 to carry out the analysis. The software is user-friendly with graphical interference to display how the outcome is obtained.

4. Results and Discussion

4.1. Examples of Optical and Radar Timing Characteristics Data

The results of this research are shown in Label A in Figure 4. Conventionally, vegetation indices and texture information can successfully classify paddy rice through image classification. However, this study made further progress. Taking a closer look at the radar image, Figure 6 shows the texture characteristic curve of Sentinel-1A (VH) and (VV) polarization images, respectively. Overall, entropy and homogeneity display dramatic differences in the time series analysis for paddy rice and non-paddy rice. In addition, rice and non-rice also show differences among these indicators. By carefully examining the texture analysis of Figure 6a,c in the changes of rice growth, it can be found that rice transplanting (transplanting rice seedlings) happens after 31 January. While the rice grows, the rice leaves gradually cover the surface soil. Then the leaves continuously have edges broken, and the bright spots and flat areas increase at the same time. The changes in the entire texture information are inconsistent, and the texture value decreases. After paddy heading from 24 February to 1 April as the ears of rice grow, the degree of texture disorder (see the entropy indicator) gradually increases, while the homogeneity decreases slightly and the homogeneity area produces a smaller value. From 1 April to 25 April as the rice leaves grow to cover the soil reflection, the texture tends to be consistent. The entropy decreases while the homogeneity increases. On 27 May when the rice ears are mature and exposed, the paddy harvest period begins. The rice ears of these highly reflective objects reduce the uniformity, so the texture value increases sharply [29]. According to the analysis of the aforementioned waveband information, we notice certain temporal characteristics of the trend information t. In the past, we tended to ignore these changes in the time axis. In Figure 6, the x axis is the observation time, and y axis is the normalization value of each type of texture information. Previously, there was a lack of an ideal tool to effectively integrate the information. If we employ the subsequent classification algorithms to increase the effective image information, we can certainly provide some help for the classification of rice fields.

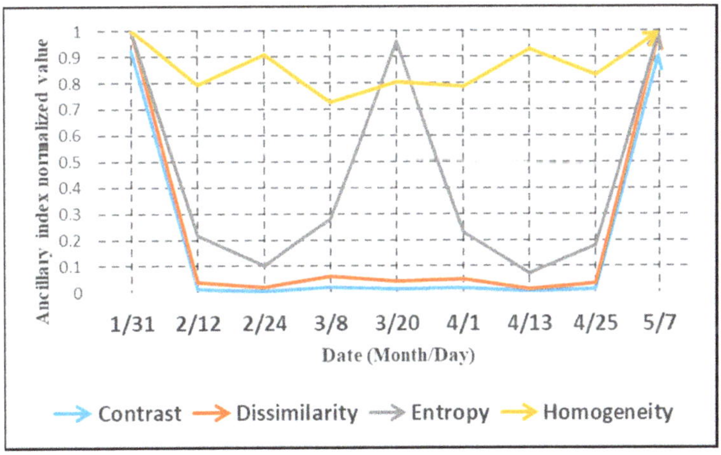

(**a**)

Figure 6. *Cont.*

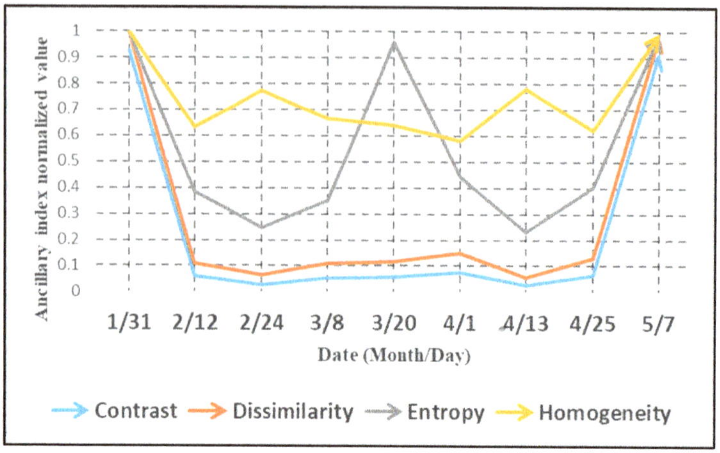

(**b**)

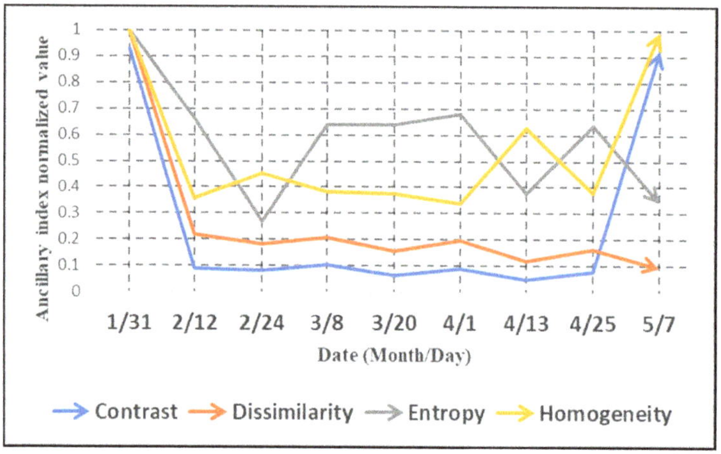

(**c**)

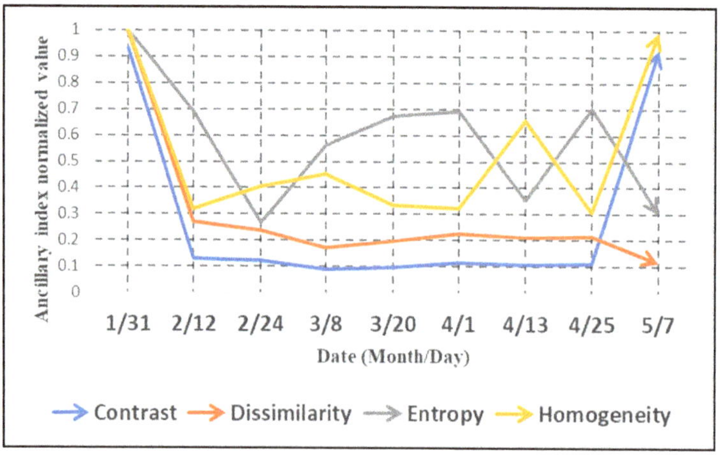

(**d**)

Figure 6. Sentinel-1A VH and VV variation, (**a**) paddy rice (VH), (**b**) non-paddy rice (VH), (**c**) paddy rice (VV), and (**d**) non-paddy rice (VV).

4.2. Comparison on Direct Classification Method and Hybrid Classification Method

Since the range of study area is too large, we chose a small area to display the classification performance result. However, the confusion matrix is generated by the entire study area. The range is presented as the red frame in Figure 7.

Figure 7. Display area with detailed information.

4.2.1. Direct Classification Method

The results of this research are shown in Label B in Figure 4. Table 3 shows the analysis results of the direct classification methods of this study. From top to bottom, the methods are SVM, NN, and DT. Among the three classification methods, DT achieved the best accuracy and kappa value of 93.26% and 0.76, respectively. The worst result was NN; overall accuracy and kappa value were, respectively, 89.51% and 0.66. It seems that the overall median value is the SVM method. Figure 8 shows the results displayed by the three algorithms. The blue frame in Figure 8 indicates that the calculation result is inconsistent compared to the ground truth data. In terms of direct classification, no matter which algorithm we used, the commission errors of rice still accounted for a certain number of cases.

However, we decided to examine how to integrate the optical information and radar information by considering multiple data resources with multiple algorithms. See Table 3 for further information. For this algorithm, the commission error of rice is quite serious. Although a large number of texture images in the analysis process are used, it seems that there are still many commission errors in the classification problem. Even if we use radar information at the same time, this does not seem to enhance the performance. On the other hand, this shows that the current analysis results may tend to be over-trained in the non-rice part. There are many reasons for the result of rice misclassification. This reason is a common phenomenon in the problem of rice classification because the rice samples are grown on the ground in different time scenarios. To solve the problem of commission errors, we employ a hybrid classification method in the next step.

Table 3. Direct classification.

SVM		Ground Truth			Producer's Accuracy
		Paddy Rice	**Non-Paddy Rice**	**Sum of Columns**	
Direct Classification	Paddy Rice	7173	3868	11,041	0.65
	Non-Paddy Rice	526	41,645	42,171	0.99
	Sum of Rows	7699	45,513	53,212	
	User's Accuracy	0.93	0.92		
	Accuracy	91.74%			
	kappa	0.72			
NN		**Ground Truth**			**Producer's Accuracy**
		Paddy Rice	**Non-Paddy Rice**	**Sum of Columns**	
Direct Classification	Paddy Rice	7075	4958	12,033	0.59
	Non-Paddy Rice	624	40,555	41,179	0.99
	Sum of Rows	7699	45,513	53,212	
	User's Accuracy	0.92	0.89		
	Accuracy	89.51%			
	kappa	0.66			
DT		**Ground Truth**			**Producer's Accuracy**
		Paddy Rice	**Non-Paddy Rice**	**Sum of Columns**	
Direct Classification	Paddy Rice	7250	3139	10,389	0.70
	Non-Paddy Rice	449	42,374	42,823	0.99
	Sum of Rows	7699	45,513	53,212	
	User's Accuracy	0.94	0.93		
	Accuracy	93.26%			
	kappa	0.76			

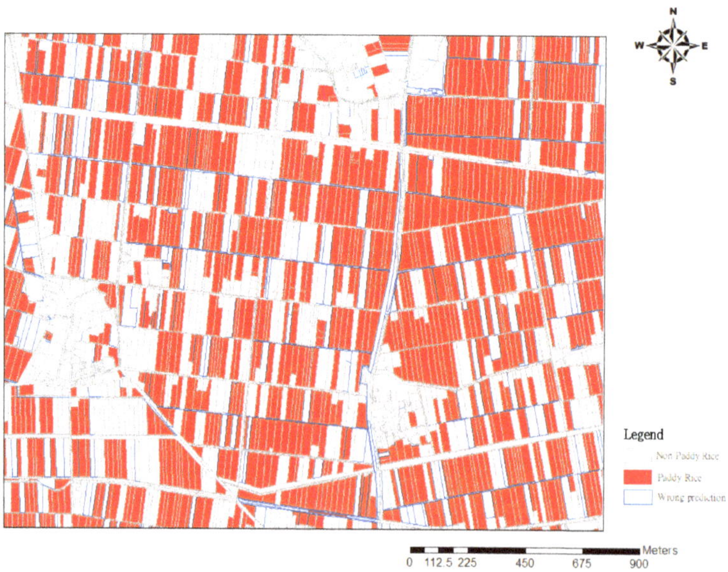

(a)

Figure 8. *Cont.*

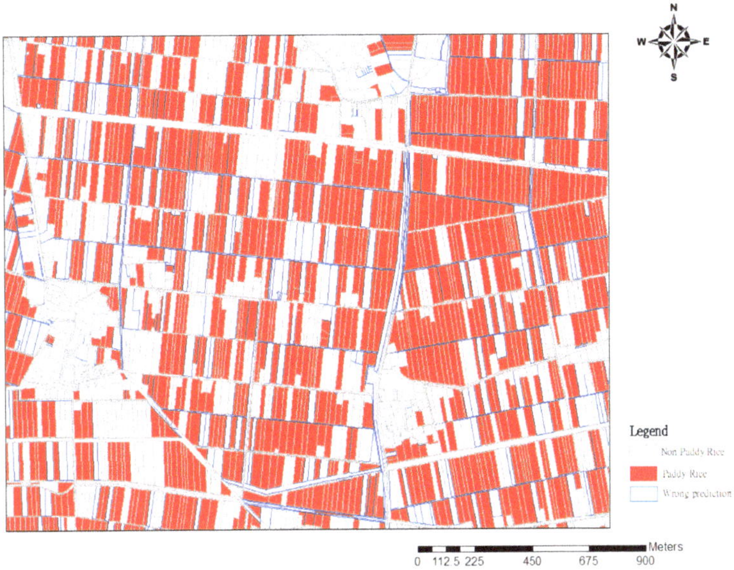

(b)

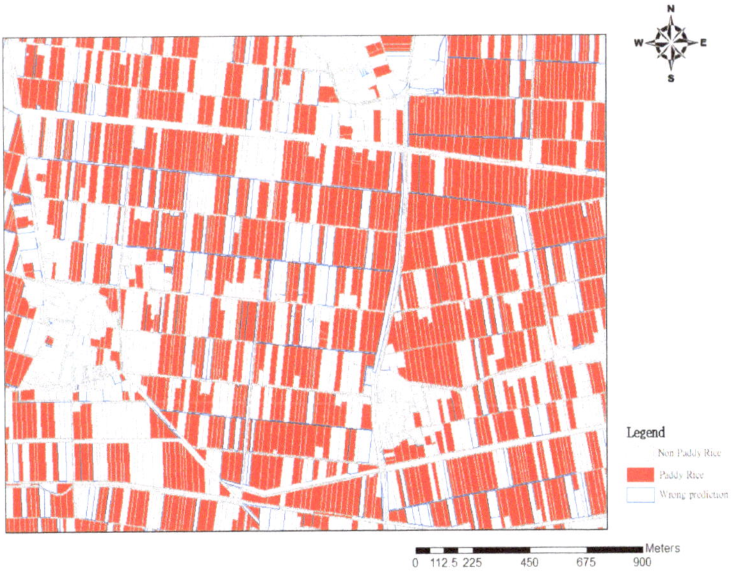

(c)

Figure 8. Comparison of direct classification and ground truth data, (**a**) SVM, (**b**) NN, and (**c**) DT.

4.2.2. Hybrid Classification

The results are shown in Label C in Figure 4. The hybrid classification method is divided into two parts. In the first, the patches show consistency after the first stage of classification, regardless if they show rice or non-rice (49,084 patches). The number of patches with inconsistent parts is 4128, so we execute re-classification. In the meantime, the DTW index information of "multi-scale time series feature similarity indicators" is employed. The DTW is calculated based on the values for the entire image, and the extraction of the inconsistent patches is carried out individually. Then, they are newly plugged into the dataset and re-classified. Therefore, we explain the results of the three parts as follows: 1, the accuracy of consistency classification; 2, inconsistency classification patches for discussion; 3, DTW indicator results; 4, hybrid classification method integration accuracy results.

Step 1. The first stage of accuracy of consistency classification.

This study uses a two-stage classification. Showing the results of the first stage of the multi-calculation classification method, Table 4 presents the patches that were consistent in the classification among the three algorithms. The meaning of this analysis is that current input feature variables for classification reach the best limitation of classification. In other words, considering a classification model under the condition of maximizing the accuracy of image data, the maximum classification approach of the machine learning model may be found.

Table 4. Consistency of classification outcome.

		Ground Truth			
		Paddy Rice	**Non-Paddy Rice**	**Sum of Columns**	**Producer's Accuracy**
Consistency of Classification	Paddy Rice	6904	2208	9112	0.76
	Non-Paddy Rice	292	39,680	39,972	0.99
	Sum of Rows	7196	41,888	49,084	
	User's Accuracy	0.96	0.95		
	Accuracy	94.91%			
	kappa	0.82			

Step 2. Discuss accuracy of consistency classification.

Table 5 presents the patches that were inconsistent in the classification among the three algorithms. From top to bottom, they are SVM, NN, and DT. Among these three classification methods, DT had the best accuracy and kappa value of 73.64% and 0.26, respectively. The worst result was NN. Its overall accuracy and kappa value were, respectively, 25.34% and −0.14. Table 5 shows that the patches had inconsistent classifications under different classification approaches. Both rice and non-rice samples had extreme commission errors and omission errors. Furthermore, there are many reasons for the resulting commission errors of rice. The complications of image quality and planting methods (time difference, mixed planting) are the major reasons. It is very difficult to resolve them by using an existing classifier unless time-history data are employed. Hence, in this study, we decided to incorporate DTW and time feature variables to provide three algorithms for the second stage classification. According to the results when employing DTW in the classification process, the proposed approach enhanced classification to the maximum classification level.

Table 5. Inconsistency of classification outcome.

SVM		Ground Truth			
		Paddy Rice	**Non-Paddy Rice**	**Sum of Columns**	**Producer's Accuracy**
Inconsistency of Classification	Paddy Rice	269	1660	1929	0.14
	Non-Paddy Rice	234	1965	2199	0.89
	Sum of Rows	503	3625	4128	
	User's Accuracy	0.53	0.54		
	Accuracy	54.12%			
	kappa	0.03			

Table 5. *Cont.*

NN		Ground Truth			Producer's Accuracy
		Paddy Rice	**Non-Paddy Rice**	**Sum of Columns**	
Inconsistency of Classification	Paddy Rice	171	2750	2921	0.06
	Non-Paddy Rice	332	875	1207	0.72
	Sum of Rows	503	3625	4128	
	User's Accuracy	0.34	0.24		
	Accuracy	25.34%			
	kappa	−0.14			

DT		Ground Truth			Producer's Accuracy
		Paddy Rice	**Non-Paddy Rice**	**Sum of Columns**	
Inconsistency of Classification	Paddy Rice	346	931	1277	0.27
	Non-Paddy Rice	157	2694	2851	0.94
	Sum of Rows	503	3625	4128	
	User's Accuracy	0.69	0.74		
	Accuracy	73.64%			
	kappa	0.26			

Step 3. Examples of multi-scale time series feature similarity indicators and description of integration results.

Table 6 presents a dataset that converts the time series dynamic relationship of the information. That is, each patch is generated by an index for the calculation of "multiscale time series feature similarity indicators" in this research. Because the amount of data is too large, we extracted a part of the data to present the research results.

Table 6. The relations of patches and multi-scale of features.

		The number of multi-scale of feature on similarity index												
		1	2	3	4	5	6	7	8	9	10	……	349	350	351
Patch number	1	0.1027	0.7318	0.2544	0.0869	0.0868	0.0870	0.0872	0.0874	0.0874	0.0874		0.0124	0.0044	0.0088
	2	0.1319	0.5630	0.1165	0.1041	0.1039	0.1043	0.1046	0.1049	0.1049	0.1049		0.0104	0.0016	0.0078
	3	0.0448	0.4065	0.0985	0.2306	0.2305	0.2310	0.2312	0.2316	0.2315	0.2315		0.0028	0.0017	0.0029
	4	0.0901	0.2802	0.0601	0.2640	0.2639	0.2644	0.2646	0.2651	0.2649	0.2649		0.0084	0.0015	0.0064
	5	0.0892	0.4852	0.1634	0.1190	0.1188	0.1193	0.1194	0.1196	0.1196	0.1196	⋮	0.0020	0.0035	0.0030
	6	0.2003	0.6422	0.2344	0.0276	0.0276	0.0278	0.0280	0.0284	0.0283	0.0283		0.0011	0.2824	0.2861
	7	0.0974	0.4458	0.1200	0.1045	0.1044	0.1047	0.1049	0.1053	0.1052	0.1052		0.0020	0.3954	0.4006
	8	0.0784	0.4373	0.0988	0.1061	0.1060	0.1064	0.1066	0.1069	0.1068	0.1068		0.0022	0.1980	0.2008
	9	0.0598	0.4317	0.1187	0.1217	0.1216	0.1219	0.1221	0.1224	0.1223	0.1223		0.0037	0.3297	0.3345
	10	0.0870	0.4811	0.1436	0.1090	0.1089	0.1093	0.1095	0.1098	0.1098	0.1098		0.0031	0.4233	0.4297
	⋮	……													
	53210	0.0545	0.2967	0.0845	0.1275	0.1274	0.1278	0.1279	0.1282	0.1282	0.1282		0.0148	0.0157	0.0062
	53211	0.0920	0.4646	0.1588	0.0749	0.0748	0.0752	0.0753	0.0756	0.0755	0.0755		0.0113	0.0116	0.0042
	53212	0.1268	0.4366	0.1357	0.0902	0.0900	0.0905	0.0906	0.0909	0.0909	0.0909		0.0046	0.0052	0.0037

In Table 6, the y axis is the number of patches in the demonstration area, including number of 53,212. The x axis is the number of combinations of multi-scale time series features, i.e., 351 features, and each grid value in Table 6 is the time series between the two-time series features of different patches. The higher the similarity value is, the more similarity between them. For instance, taking patch ID = 1 as an example, one to four groups of feature groups of similar value indexes are sorted by size as follows: feature dataset 2 (0.7318) > dataset 3 (0.2544) > dataset 1 (0.1027) > dataset 4 (0.0869). The information of the feature number is derived from dataset one to four as (1) Dataset1: SPOT6 red light band vs. SPOT6 green light; (2) Dataset2: SPOT6 red light band vs. SPOT6 blue light; (3) Dataset3: SPOT6 red light band vs. SPOT6 near-infrared light; (4) Dataset4: SPOT6 near-infrared light vs. crop management factor index (CMFI).

The above example shows the calculation of this indicator. The similarity between time series features of different scales can be converted into actual values, and the dataset's multiple features can be integrated into a worksheet, which greatly increases the analysis among different time domains and various data sources. It is worth mentioning that our analysis at this stage analyzes the entire image. The above analyses are easy to govern numerically because we can trace the IDs for their corresponding locations.

Step 4. The overall accuracy of the hybrid classification.

Table 7 shows the analysis results of the hybrid classification method of this study. The results are presented for SVM, NN, and DT. From the results, when we compare the three classification methods, DT had the best classification result. The accuracy and kappa values were 94.71% and 0.81, respectively. NN showed the worst result again, with overall accuracy and kappa values of 92.63% and 0.74, respectively. The NN approach still greatly improved the classification accuracy when applying DTW. The other two classifiers (SVM and DT) also had increased performance in classification for DTW. The final classification results of DT and SVM are largely the same. This also shows that the new ancillary information of DTW can sustainably improve the classification results. Figure 9 shows the results of the three algorithms. By zooming in on the selected area, it can be found that the original direct classification method achieved more significant improvement in commission errors than in omission errors. Figure 9 shows that there are many yellow frames indicating correction of NN, which also means that the prediction accuracy of NN in the hybrid classification method is improved when compared to the direct classification method. In other words, the result shows that the DTW indicator can provide better classification performance. To sum up, we decided to employ the DTW index in the classification process. Our results show how the DTW index resolved the confusing parts of the image. As usual, if a pixel is classified as the same pattern by different classifiers, very few errors are produced [9]. However, if a pixel is not classified as the same category for a different classifier, deployment of a new indicator (DTW) can be expected to update the erroneous pattern.

Table 7. Outcomes for hybrid classification.

SVM		**Ground Truth**			**Producer's Accuracy**
		Paddy Rice	**Non-Paddy Rice**	**Sum of Columns**	
Hybrid Classification	Paddy Rice	7360	2626	9986	0.74
	Non-Paddy Rice	339	42,887	43,226	0.99
	Sum of Rows	7699	45,513	53,212	
	User's Accuracy	0.96	0.94		
	Accuracy	94.43%			
	kappa	0.80			

Table 7. *Cont.*

NN		Ground Truth			Producer's Accuracy
		Paddy Rice	**Non-Paddy Rice**	**Sum of Columns**	
Hybrid Classification	Paddy Rice	7184	3407	10,591	0.68
	Non-Paddy Rice	515	42,106	42,621	0.99
	Sum of Rows	7699	45,513	53,212	
	User's Accuracy	0.93	0.93		
	Accuracy	92.63%			
	kappa	0.74			
DT		**Ground Truth**			**Producer's Accuracy**
		Paddy Rice	**Non-Paddy Rice**	**Sum of Columns**	
Hybrid Classification	Paddy Rice	7397	2512	9909	0.75
	Non-Paddy Rice	302	43,001	43,303	0.99
	Sum of Rows	7699	45,513	53,212	
	User's Accuracy	0.96	0.94		
	Accuracy	94.71%			
	kappa	0.81			

Usually, there are two ways to express classification accuracy, the first is the overall accuracy (OA), and the second is the kappa value. OA represents the proportion of the number of correctly classified samples to the total number of samples, but such indicators are easily affected by the omission error and commission error rate. Thus, the kappa value must be considered. The kappa is a better reference than OA to observe commission errors and omission errors. For instance, kappa results for SVM (0.72 vs. 0.80), NN (0.66 vs. 0.74), and DT (0.76 vs. 0.81) are adopted in Tables 3 and 7 which proved that the three classifier models are satisfactory in terms of applying DTW.

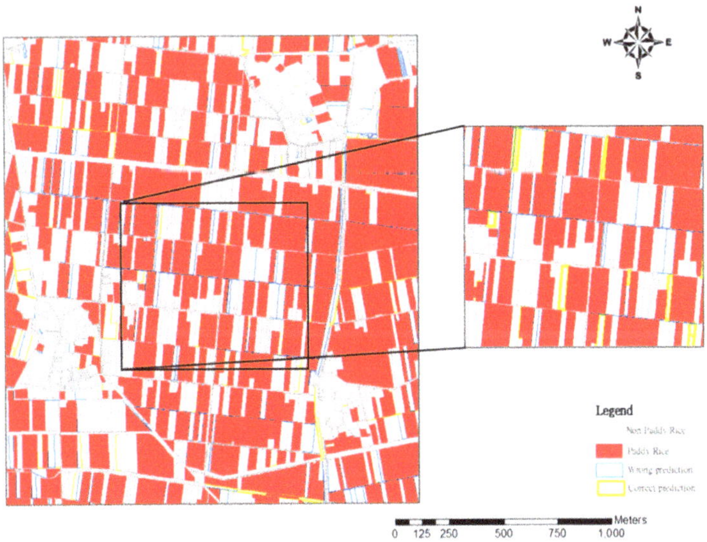

(a)

Figure 9. *Cont.*

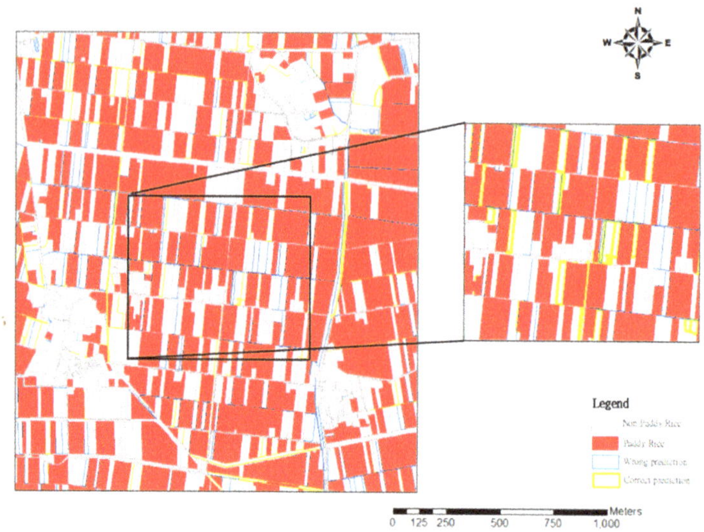

(**b**)

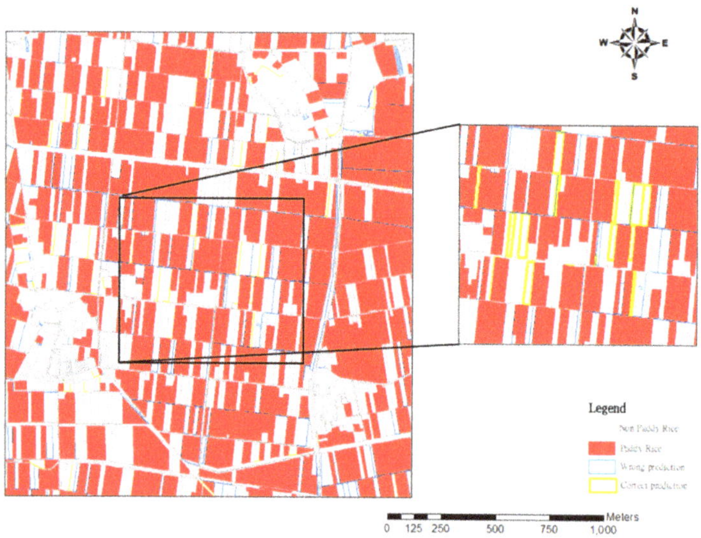

(**c**)

Figure 9. Comparison on hybrid classification and ground truth data, (**a**) SVM, (**b**) NN, and (**c**) DT.

Overall, DTW is based on the dynamic programming method for effectively reducing search and comparison time. The multi-scale time series feature similarity indicators developed in this research have the ability to transform multi-dimensional data into two-dimensional information. The reason for applying DTW is because the status of crops is a long-term characteristic. This research shows that time features are helpful for long-term characteristic image classification. In particular, it can be used in small farmland areas and fragile landscapes. Through the integration of DTW data, it can overcome the limitations of the large difference between optical images and radar images. In addition, the different spatial resolutions of the two types of images are integrated. Moreover, the various limitations of different atmospheric conditions in the shooting process of the two images are resolved. This indicator has extremely high potential in the detection of crop phenology.

5. Summary and Conclusions

This study developed a multi-scale time series feature similarity index through the Dynamic Time Wrapping (DTW) theory to integrate multi-source scale time-series image information. The training/test dataset was analyzed through a verification process proving that the original feature information was added to the time series similarity index of the multi-scale time series feature data. The conclusions of this paper are as follows:

1. This study used SPOT6 optical images and Sentinel-1A radar images as the materials of research, which differs from the mainstream use of image fusion in the interpretation in past studies. The massive time series features in the datasets are integrated into a simple index to present the data dimensions in a single dataset. This approach provides new possibilities for subsequent analysis of information considering different scales of data.
2. The homogeneity and entropy in radar images provides some new information in time series analysis, which greatly helps the classification of paddy rice. It is found that the behavior of time variations can distinguish paddy rice and non-paddy rice easily.
3. This study uses the "direct classification method" and "hybrid classification method" for comparison. The characteristic information of optical satellite images and radar images is applied to directly perform classification methods for their behaviors. The results show that the overall accuracy results of the direct classification method are 91.7% (kappa value 0.72, SVM), 89.5% (kappa value 0.66, NN), and 93.26% (kappa value 0.76, DT). In the second stage of classification, the patches were classified optically with DTW feature information using three approaches, and neutral patches were added in the first stage, producing the overall accuracy results of 94.43% (kappa value is 0.80, SVM), 92.63% (kappa value is 0.74, NN), and 94.71% (kappa value is 0.81, DT). This also proves the DTW is robust.
4. This result renders a feasible way to integrate radar feature information with optical feature information, especially in multi-period data. The optical images in different periods are difficult to obtain due to the influence of weather conditions. Radar images can be obtained regularly since cloud and fog interference can be avoided. A possible solution has been designed to overcome their disadvantages, which could lead to better classification performance. Considering those various restrictions, it is especially suitable for small farmland areas and fragile landscapes.

Author Contributions: T.C.L. was responsible for the plan and design of this study. He analyzed the data and discussion. S.W. helped with writing the manuscript and discussion of results. Y.C.W. wrote the computer program. H.-P.W. and C.-W.H. used the program to plot the thematic map and generate tables. All authors have read and agreed to the published version of the manuscript.

Funding: This research was funded by the Ministry of Science and Technology (MOST) 103-2119-M-035 -002 –.

Acknowledgments: The authors would like to thank the MOST for providing image data and related information. The authors are also very grateful to Z. H. Zhu, Department of Geography, National Taiwan University, for his advice and suggestions for this project.

Conflicts of Interest: The authors declare no conflict of interest.

References

1. Kumar, S.; Mishra, S.; Khanna, P. Precision sugarcane monitoring using SVM classifier. *Procedia Comput. Sci.* **2017**, *122*, 881–887. [CrossRef]
2. Wan, S.; Wang, Y.P. The comparison of density-based clustering approach among different machine learning models on paddy rice image classification of multispectral and hyperspectral image data. *Agriculture* **2020**, *10*, 465. [CrossRef]
3. De Bernardis, C.; Vicente-Guijalba, F.; Martinez-Marin, T.; Lopez-Sanchez, J.M. Contribution to real-time estimation of crop phenological states in a dynamical framework based on NDVI time series: Data fusion with SAR and temperature. *IEEE J. Sel. Top. Appl. Earth Obs. Remote Sens.* **2016**, *9*, 3512–3523. [CrossRef]
4. Onojeghuo, A.O.; Blackburn, G.A.; Wang, Q.; Atkinson, P.M.; Kindred, D.; Miao, Y. Mapping paddy rice fields by applying machine learning algorithms to multi-temporal Sentinel-1A and Landsat data. *Int. J. Remote Sens.* **2018**, *39*, 1042–1067. [CrossRef]
5. Betbeder, J.; Laslier, M.; Corpetti, T.; Pottier, E.; Corgne, S.; Hubert-Moy, L. Multi-temporal optical and radar da-ta fusion for crop monitoring: Application to an intensive agricul-tural area in Brittany (France). In Proceedings of the 2014 IEEE Geoscience and Remote Sensing Symposium, Quebec City, QC, Canada, 13–18 July 2014; pp. 1493–1496.
6. Esteban, J.; Starr, A.; Willetts, R.; Hannah, P.; Bryanston-Cross, P. A Review of data fusion models and architectures: Towards engineering guidelines. *Neural Comput.* **2005**, *14*, 273–281. [CrossRef]
7. Zhou, G.; Liu, X.; Liu, M. Assimilating remote sensing phenological information into the WOFOST model for rice growth simulation. *Remote Sens.* **2019**, *11*, 268. [CrossRef]
8. Joshi, N.; Baumann, M.; Ehammer, A.; Fensholt, R.; Grogan, K.; Hostert, P.; Waske, B. A review of the application of optical and radar remote sensing data fusion to land use mapping and monitoring. *Remote Sens.* **2016**, *8*, 70. [CrossRef]
9. Lei, T.C.; Wan, S.; Wu, S.C.; Wang, H.P. A new approach of ensemble learning technique to resolve the uncertainties of paddy area through image classification. *Remote Sens.* **2020**, *12*, 3666. [CrossRef]
10. Petitjean, F.; Ketterlin, A.; Gancearski, P. A global averaging method for dynamic time warping, with applications to clustering. *Pattern Recognit.* **2011**, *44*, 678–693. [CrossRef]
11. Wang, M.; Wang, J.; Chen, L. Mapping paddy rice using weakly supervised Long Short-term Memory Network with time Series sentinel optical and SAR Images. *Agriculture* **2020**, *10*, 483. [CrossRef]
12. Gella, G.W.; Bijker, W.; Belgiu, M. Mapping crop types in complex farming areas using SAR imagery with dynamic time warping. *ISPRS J. Photogramm. Remote Sens.* **2021**, *175*, 171–183. [CrossRef]
13. Viana, C.M.; Girão, I.; Rocha, J. Long-term satellite image time-series for land use/land cover change detection using refined open source data in a rural region. *Remote Sens.* **2019**, *11*, 1104. [CrossRef]
14. Cheng, K.; Wang, J. Forest-type classification using time-weighted Dynamic Time Warping analysis in mountain areas: A case study in southern China. *Forests* **2019**, *10*, 1040. [CrossRef]
15. Guan, X.; Huang, C.; Liu, G.; Meng, X.; Liu, Q. Mapping rice cropping systems in Vietnam using an NDVI-based time-series similarity measurement based on DTW distance. *Remote Sens.* **2016**, *8*, 19. [CrossRef]
16. Moola, W.S.; Bijker, W.; Belgiu, M.; Li, M. Vegetable mapping using fuzzy classification of Dynamic Time Warping distances from time series of Sentinel-1A images. *Int. J. Appl. Earth Obs. Geoinf.* **2021**, *102*, 102405. [CrossRef]
17. Guan, X.D.; Liu, G.H.; Huang, C.; Meng, X.L.; Liu, Q.S.; Wu, C.; Ablat, X.; Chen, Z.R.; Wang, Q. An Open-boundary locally weighted Dynamic Time Warping method for cropland mapping. *ISPRS Int. J. Geo-Inf.* **2018**, *7*, 75. [CrossRef]
18. Manabe, V.D.; Melo, M.R.; Rocha, J.V. Framework for mapping integrated crop-livestock systems in Mato Grosso, Brazil. *Remote Sens.* **2018**, *10*, 1322. [CrossRef]
19. Csillik, O.; Belgiu, M.; Asner, G.P.; Kelly, M. Object-Based time-constrained Dynamic Time Warping classification of crops using Sentinel-2. *Remote Sens.* **2019**, *11*, 1257. [CrossRef]
20. Dong, Q.; Chen, X.; Chen, J.; Zhang, C.S.; Liu, L.; Cao, X.; Zang, Y.Z.; Zhu, X.F.; Cui, X.H. Mapping winter wheat in North China using Sentinel 2A/B data: A method based onvphenology-Time Weighted Dynamic Time Warping. *Remote Sens.* **2020**, *12*, 1274. [CrossRef]
21. Zhao, F.; Yang, G.; Yang, X.; Cen, H.; Zhu, Y.; Han, S.; Yang, H.; He, Y.; Zhao, C. Determination of key phenological phases of winter wheat based on the time-weighted Dynamic Time Warping algorithm and MODIS time-series data. *Remote Sens.* **2021**, *13*, 1836. [CrossRef]
22. Zhao, F.; Yang, G.J.; Yang, H.; Zhu, Y.H.; Meng, Y.; Han, S.Y.; Bu, X.L. Short and medium-term prediction of winter wheat NDVI based on the DTW–LSTM combination method and MODIS time series data. *Remote Sens.* **2021**, *13*, 4660. [CrossRef]
23. European Space Agency—ESA. Available online: https://step.esa.int/main/toolboxes/snap/ (accessed on 1 May 2018).
24. Huete, A.R. A soil-adjusted vegetation index (SAVI). *Remote Sens. Environ.* **1988**, *25*, 295–309. [CrossRef]
25. Wan, S.; Yeh, M.L.; Ma, H.L. An innovative intelligent system with integrated CNN and SVM: Considering various crops through hyperspectral image data. *ISPRS Int. J of Geo-Inform.* **2021**, *10*, 242. [CrossRef]
26. Werbos, P.J. Beyond Regression: New Tools for Prediction and Analysis in the Behavioral Sciences. Ph.D. Thesis, Harvard University, Cambridge, MA, USA, 1974.
27. Kumar, A.; Kim, J.; Lyndon, D.; Fulham, M.; Feng, D. An ensemble of fine-tuned convolutional Neural Networks for medical image classification. *IEEE J. Biomed. Health Inform.* **2017**, *21*, 31–40. [CrossRef] [PubMed]

28. Bazzi, H.; Baghdadi, N.; Hajj, E.M.; Zribi, M.; Minh, D.H.T.; Ndikumana, E.; Courault, D.; Belhouchette, H. Mapping paddy rice using Sentinel-1 SAR time series in camargue, France. *Remote Sens.* **2019**, *11*, 887. [CrossRef]
29. Yang, K.; Gong, Y.; Fang, S.; Duan, B.; Yuan, N.; Peng, Y.; Zhu, R. Combining Spectral and Texture Features of UAV Images for the Remote Estimation of Rice LAI throughout the Entire Growing Season. *Remote Sens.* **2021**, *13*, 3001. [CrossRef]

agriculture

Review

Evaluating the Effectiveness and Efficiency of Climate Information Communication in the African Agricultural Sector: A Systematic Analysis of Climate Services

Chidiebere Ofoegbu [1,*] and Mark New [2]

[1] Department of Forest Ecology and Management, Swedish University of Agricultural Sciences (SLU), 90183 Umeå, Sweden
[2] African Climate and Development Initiative, University of Cape Town, Rondebosch 7700, South Africa; Mark.New@uct.ac.za
* Correspondence: chidiebere.ofoegbu@slu.se; Tel.: +46-720227610

Abstract: The use of climate services (CS) for the provisioning of climate information for informed decision-making on adaptation action has gained momentum. However, a comprehensive review of the literature to evaluate the lessons and experiences of CS implementation in the African agriculture sector is still lacking. Here, we present a systematic review (mapping) of 50 pieces of literature documenting lessons and experiences of CS adoption in the agriculture sector of 20 African countries. The qualitative analysis of the reviewed literature revealed: (1) CS implementation overwhelmingly relied on a participatory process through workshops and participatory scenario planning meetings to connect users with actors along the CS value chain of forecast production, translation, integration, and application. Additionally, innovations such as mobile phones and internet service are increasingly being integrated with CS to strengthen the relationship between CS providers and users. They are, however, mostly at the trial stage and tend to have a varying impact depending on available facilities and infrastructure in the community. (2) Although there is a growing recognition of the need for the integration of indigenous and scientific knowledge systems in the production of climate information, such integration is currently not happening. Rather, indigenous knowledge holders are engaged in a participatory process for insight on modalities of making scientific climate information locally relevant and acceptable. Given the aforementioned findings, we recommend further research on modalities for facilitating indigenous knowledge mainstreaming in climate information production, and investigation of options for using innovations (e.g., mobile) to enhance the interactions between CS users and CS providers. Such research will play a great role in scaling up the adoption of CS in the African agricultural sector.

Keywords: Africa; knowledge systems; climate change; agriculture; adaptation

Citation: Ofoegbu, C.; New, M. Evaluating the Effectiveness and Efficiency of Climate Information Communication in the African Agricultural Sector: A Systematic Analysis of Climate Services. *Agriculture* **2022**, *12*, 160. https://doi.org/10.3390/agriculture12020160

Academic Editors: Gniewko Niedbała and Sebastian Kujawa

Received: 22 November 2021
Accepted: 14 January 2022
Published: 24 January 2022

1. Introduction

Agricultural development and sustainability in Africa are linked to the discourse around poverty and wellbeing [1,2]. Moreover, the agriculture sector in most developing countries is highly vulnerable to climate change due to compounding factors, including lack of capital, poor infrastructure, dependence on rain-fed agriculture, insecure land rights, and degradation of natural resources [3,4]. Consequently, several initiatives aimed at enhancing stakeholders' access to tailored and contextual climate information for adapting farming practices to climate and socioeconomic risks are being promoted [5–7]. In this regard, Climate Services (CS) has become a popular initiative. (In this study, we adopt the American Meteorological Society's definition of CS, which defines CS as scientifically-based information and products that enhance users' knowledge and understanding about the impacts of climate on their decisions and actions.) "Climate services, involve the timely production, translation, and delivery of useful climate data, information and knowledge for societal decision-making and climate-smart policy and planning" [8].

Several initiatives for scaling up the implementation of CS in the African agriculture sector have been adopted. Some of these initiatives include: the African Center of Meteorological Applications for Development (in 1995), the Climate Services Partnership (in 2011), the Global Framework of Climate Services (in 2012), and the Climate Services for Resilient Development Partnership (in 2017). These initiatives have been used in several approaches to facilitate the production and dissemination of climate information to stakeholders in the agricultural sector. The Global Framework of Climate Services, for example, implemented several projects in many African countries that aimed to facilitate timely delivery of contextual climate information to stakeholders through a collaborative participatory process [5,9–11]. Similarly, the African Centre of Meteorological Application for Development initiative implemented several projects aimed at producing forecasts of an appropriate timeline that are most suitable to decision-making in the agricultural sector [12–14].

Despite the growing number of initiatives promoting the adoption of CS in the African agricultural sector, there have been no regional level syntheses (Africa scale) of issues driving the adoption of CS and how CS has impacted adaptation actions in the African agricultural sector. Nevertheless, there have been several country-level studies on the contributions of CS to climate risk management in the agriculture sector [15,16]. Ref [17] investigated users' needs for CS. Ref [18] investigated the role of participatory processes in enhancing CS implementation in the agriculture sector. Ref [16] investigated the importance of CS for food security in East Africa. Ref [19] analyzed the process of forecast production and translation into relevant information for the agriculture sector. Ref [20] investigated the value of forecasts to farming activities, with the view of identifying content and timescale forecasts which are more appropriate for adapting farming operations to climate and socioeconomic risks.

Although these studies have provided important insight into the knowledge and experience of CS implementation in the African agriculture sector, they are, however, mostly at a national to sub-national scale. Thereby, they run the risk of missing important information for robust/generalizable interpretation of CS' impact on the African agriculture sector [21], which may lead to a biased interpretation of CS contributions to the African agriculture sector.

There is, therefore, a dearth of systematic review on the lessons and experiences of CS implementation in the African agriculture sector [22,23]. In order to fill this gap, we apply a systematic mapping review, drawing from lessons and experience in the literature on CS implementation across Africa. We ask: (1) how do the CS models characterize the process of climate information production and dissemination? (2) What types of information are provided by the CS? (3) How does CS facilitate indigenous knowledge systems' integration in the provisioning of these interventions? (4) How does the adoption of CS promote two-way learning (bottom-up and top-down) about climate-smart agricultural practices?

The rest of this paper is organized as follows: Section 2 describes the methods used for data extraction and analysis. Section 3 presents the results by (i) giving insights into how existing CS models shapes the process of climate information production and dissemination; (ii) elaborating on the characteristics of the CS and the types of information they disseminate; (iii) exploring how CS facilitates knowledge systems' integration; and (iv) providing insights on how the CS facilitates two-way learning about climate-smart agricultural practices. Section 4 concludes with a summary of the main findings and reflections on areas of future research.

2. Methods

2.1. Systematic Mapping Review (SMR) Data Selection Process

This paper adopts an SMR approach, wherein a thematic content analysis is used to analyze extracted data from all the reviewed literature. The execution of the SMR approach entailed six steps: definition of research questions, literature search, literature screening for papers, data extraction, coding, and mapping (Figure 1). A systematic literature search was conducted using the following databases: AGRIS, CAB Abstracts, ISI Web of Science, Scopus, Emerald, Open Access Theses and Dissertations (OATD), and Directory of Open Access Journals (DOAJ). We limited our selections to literature published between 2000 and 2019. We additionally searched for published reports, policy briefs, and working papers using the following databases and organization websites: Food and Agriculture Organisation of the UN (FAO), International Fund for Agricultural Development (IFAD), Centre for International Forestry Research (CIFOR), World Agroforestry Centre (ICRAF), Food and Agriculture Organization of the United Nations, International Food Policy Research Institute (IFPRI), International Institute for Environment and Development, International Centre for Tropical Agriculture (CIAT), Bioversity International, International Centre for Agricultural Research in the Dry Areas (ICARDA), International Crops Research Institute for the Semi-Arid Tropics (ICRISAT), International Institute of Tropical Agriculture (IITA), International Livestock Research Institute (ILRI), and World Agroforestry Centre.

The focus was on literature addressing the application of climate services in the agricultural sector in Africa. The study considered only studies published in English, which we acknowledge as a limitation. We used the following search terms [(Farm* OR Agr*) AND (Climate* OR Global change OR Resilience) AND (climate services OR climate info* OR knowledge systems OR knowledge network OR knowledge flow)] to search for literature. The literature search, including searches for published journal papers, reports, book chapters, conference papers, working papers, and policy briefs, resulted in 6713 papers.

The papers collated through the literature search were assessed for inclusion through a multi-tiered process: firstly based on the title, then by abstract, and finally by full-text review (Figure 1). The papers that were screened and included for full-text review must fulfill the following criteria: (1) Published between 2000 and 2019, (2) subject area must be related to CS use and management in the agricultural sector in Africa, (3) studies that addressed different models of CS used in the African agricultural sector and the knowledge systems that inform the functioning of the CS models were adjudged as meeting the relevant intervention criteria, and (4) studies that evaluated the impact and outcomes of CS on the sustainability of the African agricultural sector were judged as meeting the relevant outcomes criteria.

The screening for full-text review resulted in 50 published studies (Figure 1) which were subjected to the full-text review and used in the literature mapping analysis. Interestingly, the screening exercise resulted in the identification of 359 studies on CS use and management in the agricultural sectors of developing countries other than Africa. These 359 studies were not used in the literature mapping analysis but were reviewed to gain comparative insight into CS use and management in the agricultural sectors of other developing countries.

2.2. Review Analysis

The selected 50 pieces of literature (Table 1) focusing on Africa were analyzed using qualitative (thematic content) analysis with the view of generating all possible responses to the study's four research questions. The information extracted from the studies comprised households'/farmers' socioeconomic profile, conceptual framework, data collection methods, the process and type of information content produced and disseminated via the CS, and the dissemination pathways. The thematic content analysis was performed in Microsoft Excel to guide the thematic grouping of collected data and substantiate the interpretation of results. Repeated crosschecking during the data extraction and coding process served to reduce the risk of error.

Thematic content analysis was conducted by coding the extracted data into themes and categorizing the codes into broader themes. The coding was applied at three levels [24]: initial/open coding, focused coding, and thematic coding. The study's research questions guided the initial/open coding of the extracted data until no further new codes emerged (thematic saturation) [25]. Abductive reasoning was used to allow themes to emerge from the data to provide reasonable answers to all four research questions. In an abductive reasoning process, logical inferences are made by finding the simplest and most likely explanation to an observation or set of observations [26].

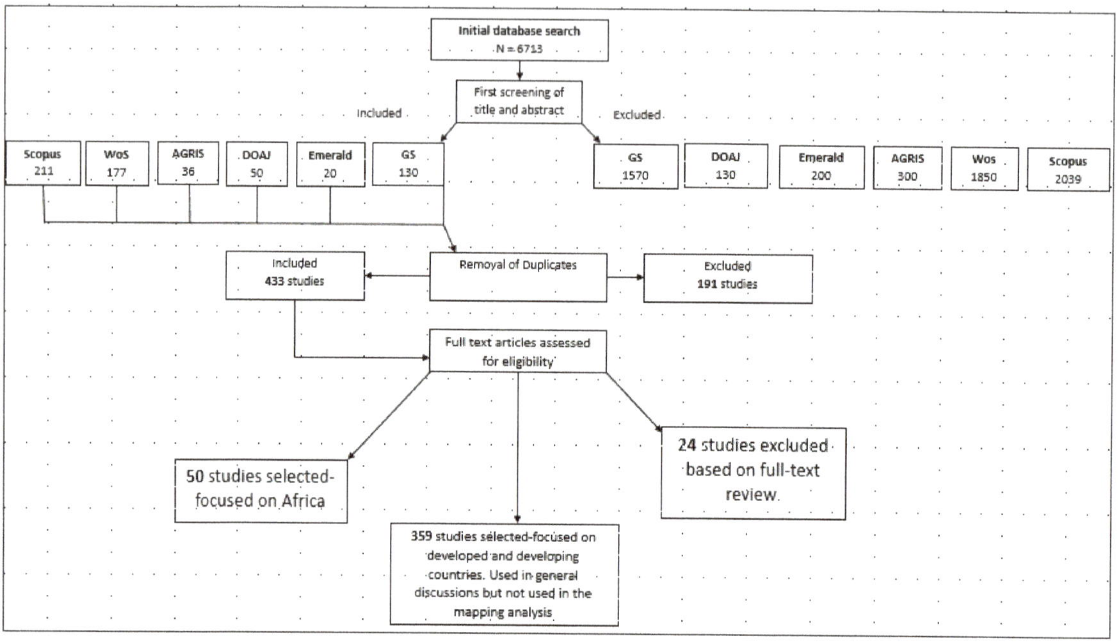

Figure 1. Summary diagram of the literature screening process.

2.3. Overview of Extracted Data

The unit of analysis (Table 1) in the selected studies is mostly sectoral targeting of the entire agricultural sector. Some of the studies are, however, more specific, targeting rural households, farmers, and/or pastoralists. The analysis method adopted in the selected studies varied. Most of the selected literature applied a mixed method, combining household surveys with focus group discussions (14). Some of the studies were based on the review and evaluation of implemented CS projects across Africa (9). The third most common methodological approach used in the selected studies is a model-based approach to improving the tailoring of climate forecasts to the farmers' contexts, and understanding the value of climate forecasts to crop and livestock productivity (4). Some of the studies are based on a traditional literature review to understand the conceptualization and application of CS (6). Some of the studies used a workshop approach to tease out factors shaping CS application and effectiveness (see Figure 2).

The distribution of the study site across regions in Africa varies, with most of the studies located in the Western Africa region (26%). This is followed by the Southern African region (23%). Twenty percent of the studies are located in the East Africa region. Although we did not record any study focusing specifically on the Central and Northern Africa region, 20% of the selected studies focused on Africa in general while the geographical region of 11% of the selected remains unspecified.

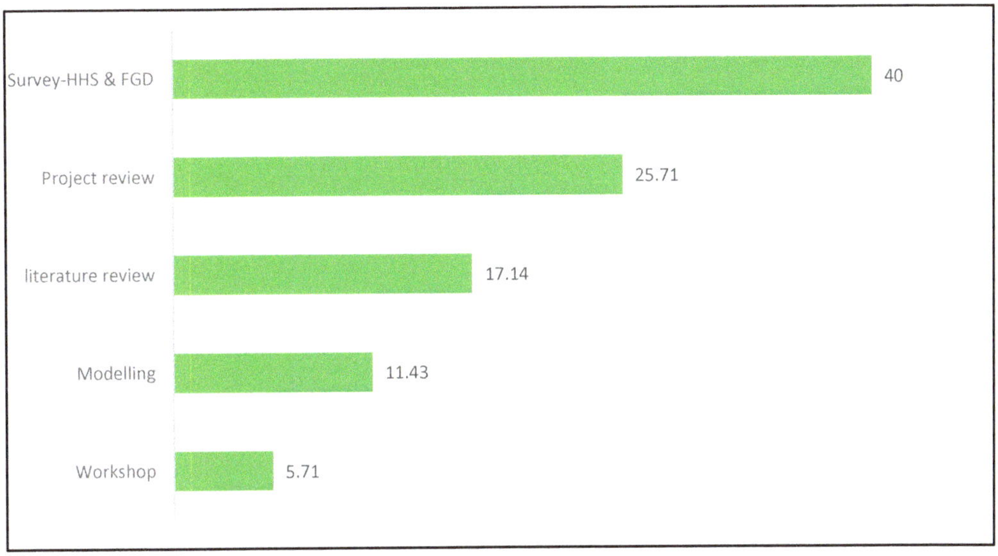

Figure 2. Methodological approach usage in CS project analysis (%).

Table 1. List of the 50 kinds of literature analyzed for this study.

SN	Publication Type	Unit of Analysis	Study Scale	Study Country
1	Research article	Farmers	National	Senegal
2	Research article	Agriculture sector	National	Kenya
3	Research article	Agriculture sector	Regional–Southern Africa	Swaziland, Zambia
4	Research article	Farmers	National	Burkina Faso
5	Research article	Agricultural institutions	National	Burkina Faso
6	Research article	Households	National	Kenya
7	Research article	Sectoral–Agriculture	National	Kenya
8	Research article	Sectoral–Agriculture	Regional–West Africa	Mauritania, Niger, Ivory Coast, and Ghana
9	Research article	Sectoral–Agriculture	Regional–East Africa	Kenya and Ethiopia
10	Research article	Farmers	Regional	Ghana, Uganda and Cameroon
11	Research article	Farmers	Regional–East Africa	Kenya, Ethiopia, IGAD countries
12	Research article	Households	National	South Africa
13	Research article	Farmers	National	Zimbabwe
14	Research article	Farmers	National	Lesotho
15	Research article	Households	National	Lesotho
16	Research article	Households	Sub-national	Mali
17	Research article	Sectoral–Agriculture	Global, but have Kenya case studies	Global
18	Research article	Sectoral–Agriculture	Regional–Africa	NA
19	Research article	Sectoral–Agriculture	National	Malawi
20	Research article	Sectoral–Agriculture	Regional–Africa	NA

Table 1. *Cont.*

SN	Publication Type	Unit of Analysis	Study Scale	Study Country
21	Research article	Farmers	National	Zimbabwe
22	Research article	Sectoral–Agriculture	Regional–Africa	Africa
23	Research article	Sectoral–Agriculture	global	Global
24	Research article	Pastoralists	Regional	Kenya, Ethiopia
25	Research article	Households	National	Uganda
26	Research article	Pastoralists	National	Burkina Faso
27	Research article	Pastoralists	National	Burkina Faso
28	Research article	Sectoral–Agriculture	West Africa	West Africa countries
29	Research article	Households	National	Burkina Faso
30	Research article	Farmers	National	Senegal
31	Research article	Sectoral–Agriculture	National	Uganda
32	Research article	Sectoral–Agriculture	National	Ghana
33	Research article	Pastoralists	National	Senegal
34	Research article	Households	National	Zambia
35	Research article	Sectoral-Agriculture, water	Global	Africa
36	Research article	Households	National	Uganda
37	Research article	Farmers	National	Mali
38	Technical report	Sectoral–Agriculture	Regional–Africa	Africa
39	Book chapter	Households	National	Uganda
40	Book chapter	Households	National	Uganda
41	Review paper	Sectoral–Agriculture	Regional–Sub-Saharan Africa	Africa
42	Review paper	Sectoral–Agriculture	Global	Africa-South Africa, Zimbabwe
43	Review paper	Sectoral–Agriculture	Global	Africa
44	Review paper	Sectoral–Agriculture	Global	Sudan, Kenya
45	Review paper	Sectoral–Agriculture	Global	unspecified
46	Review paper	Sectoral–Agriculture	Regional	West Africa countries
47	Review paper	Sectoral–Agriculture	Global	NA
48	Conference proceedings	Sectoral–Agriculture	Sub-national	Kenya
49	Working paper	Households	Regional	Africa
50	Working paper	Farmers and pastoralists	National	Tanzania

3. Results

3.1. Characterizing the Process of Climate Information Production

The organizational collaboration process underpinning the process of climate information production and dissemination in all the reviewed case studies followed a similar value chain approach. The value chain has three distinct phases: forecast production, forecast translation and integration into agriculture relevant climate information, and climate information communication and adoption in decision-making [5,9,11,27–29]. The nature of the relationship among the actors along the three phases of the value chain is the key factor that shapes the nature and function of the CS model in the African agriculture sector. Figure 3 presents the identified key challenges associated with each of the three phases of the CS value chain.

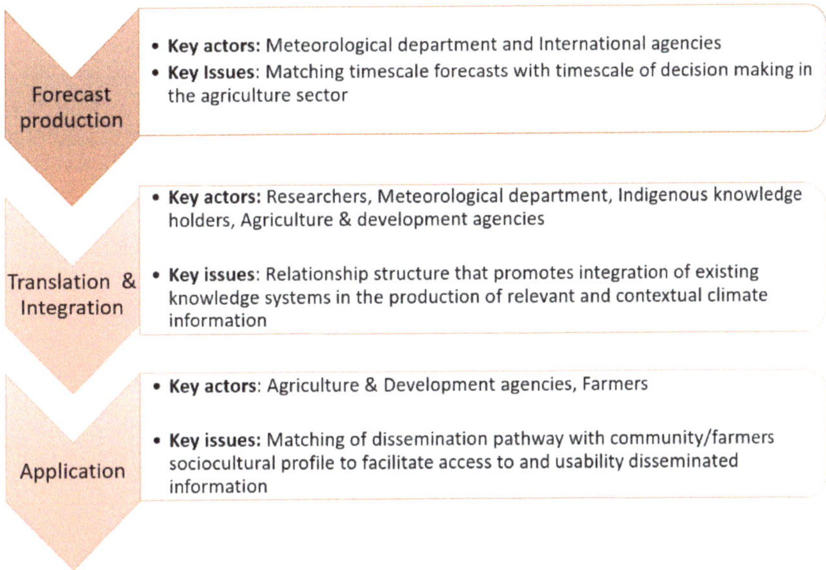

Figure 3. CS value chain characterizing phases of forecast production and translation climate information.

Forecast Production: Most activity at the forecast production stage focuses on the production of forecasts that are in accord with the timescale at which decisions are made in the agriculture sector [30–32]. The assumption is that such harmony can increase the relevance and adoption of forecasts in the agriculture sector. This is further discussed in Section 3.2. The key actors are the national meteorological departments and regional/international forecast-producing organizations.

Translation and Integration: The translation and integration stage focuses on the application of a transdisciplinary approach through the encouragement of a collaborative process that facilitates interaction among diverse actors of the varying disciplinary fields to promote the translation and integration of forecasts into information on climate risk warning and risk response strategy relevant for the agriculture sector [13,15,33–35]. The key issue includes the establishment of an appropriate modality for the integration of scientific and indigenous knowledge systems in produced climate information to facilitate its sociocultural relevance and acceptance. This is elaborated in Sections 3.2 and 3.3

Application: The application stage focuses on a two-way interaction with farmers to understand their socio-cultural characteristics to implement a dissemination pathway that will facilitate access and uptake of disseminated information [36–40]. The key issue identified in this phase includes the development of a comprehensive socioeconomic profile of the community to implement an appropriate dissemination pathway that facilitates rural people's access to disseminated information while also providing them with a platform for feedback on their experience and expectation for revision and reformation of disseminated information. This issue is elaborated in Sections 3.3 and 3.4

3.2. Types of Interventions and Climate Information Provided through CS

The types of interventions provided in the African agriculture sector through CS were thematically grouped into 11 categories (Figure 4). Most of the interventions are within the thematic group of analysis of adoption pathways (19%). The type of interventions within this group mostly focussed on analyzing the adoption pathways for effective CS uptake in the agriculture sector [9,40,41]. To optimize CS adoption by relevant actors' across scales in the African agriculture sector, project implementers employ several strategies including socioeconomic characterization of households to identify efficient and effective information dissemination pathways [9,11,42,43], while some projects used peer-to-peer and social learning to promote and facilitate CS use awareness among potential users [42,44]. Some projects chose an economic pathway by analyzing households willingness to pay for CS to identify the cost-effective strategy for promoting CS' adoption in the agriculture sector [45,46].

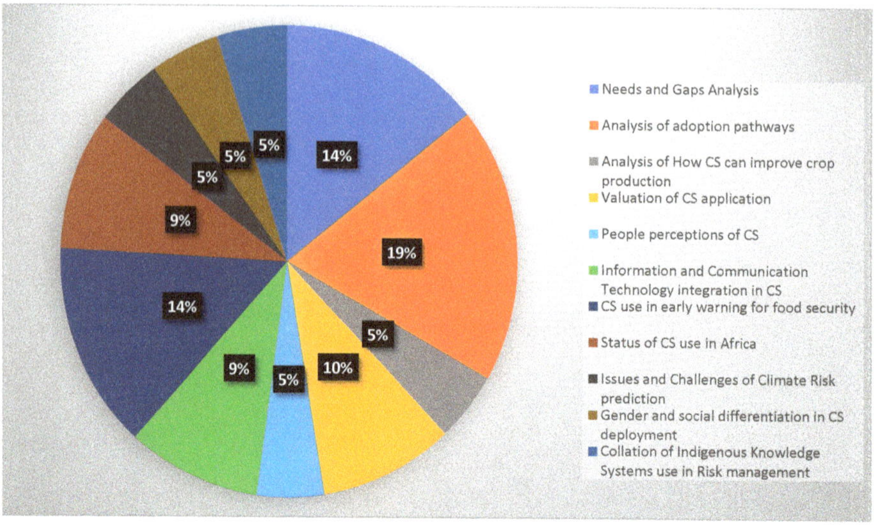

Figure 4. Thematic groupings of CS according to issues of focus.

The second most common thematic groups of interventions are needs and gaps analysis (14%) and early warnings for food security (14%). In the context of needs and gaps analysis, CS is used to specifically target the local contextual needs of farmers/users concerning their information needs for livelihood system adaptation to climate change. In the context of early warning, CS is used for timely and relevant information communication to users to facilitate their early preparation for various risks including drought, wildfire, erratic rainfall, etc., to help ensure food security at household and community levels. The third most common thematic group of intervention is the valuation of CS' application in the agriculture sector (10%). Interventions within this thematic group are mostly focused on the evaluation of CS' contribution to the economic viability of agricultural practices and CS' contribution to agriculture productivity, e.g., crop yield. The fourth most common thematic group of interventions are Information and Communication Technology (ICT) integration in CS (9%) and status of CS use in Africa (9%). Interventions focusing on ICT integration mostly focused on the investigation of options for using ICT to enhance farmers' access to relevant climate information in a cost-effective and timely manner. These types of studies are relatively recent but are growing exponentially in number. A positive trend was observed between the year of project implementation and the type of intervention the projects provide. This is especially true for project interventions focused on ICT integration in CS applications. Although this type of intervention is currently the least common, nevertheless, most of the projects providing this type of intervention are recent, with most occurring between 2011 to 2019. This is a strong indication that this type of intervention is steadily growing and may become the dominant type of intervention in the future. Other thematic groups of interventions that are sparingly provided through CS include: analysis of how CS can improve crop production (5%), people's perceptions of CS (5%), issues and challenges of climate risk prediction (5%), gender and social differentiation in CS deployment (5%), and collation of indigenous knowledge systems use in risk management (5%).

3.3. Types of Climate Information Provided through CS

There are three main types of climate information commonly provided through CS. These are forecasts, agrometeorological services, and early warnings.

Forecasts use in farming operations: Several CS projects in the African agriculture sector have applied forecasts of varying timescale in providing risk warning and risk response advisory services to farmers [36,47,48]. The literature synthesis and mapping process identified five timescale forecasts as the most widely used forecasts in the production of climate risk warnings and risk response advisory services: (1) Weather forecasts (daily to weekly), (2) Seasonal forecasts (on a timescale of 1–6 months), (3) Short-term forecasts (1–5 years), (4) Intra-decadal/Medium-term forecasts (5–10 years), and (5) Decadal forecasts. The most commonly used among these forecasts are short-term, seasonal, and weather forecasts [11,28,33]. Intra-decadal and decadal forecasts are sparingly used in agricultural risk management, even though they may be more useful for making a strategic decision and anticipatory adaptation plans [26,27,38].

Table 2 presents a summary of how these forecasts are used in risk warning and risk response advisory services for informing stakeholders in the agriculture sector. We also present additional information on how the forecasts are used in decision making and the identified gaps based on users' information needs and the type of information communicated to them [49].

Table 2. Observed forecasts and their use in the agriculture sector.

Forecast	Typical Content	Application in Decision Making	Gaps	Reference
Weather forecasts (daily to weekly)	They normally contain detailed likelihood of the occurrence of climate events, e.g., rainfall possibility	Decision making on daily farming operations: Timing of fertilizer and chemical applications, timing of fungicide applications.	None identified	[10,27,50]
Seasonal forecasts (on a timescale of 1–6 months)	Seasonal rainfall onset and cessation, the rainfall amounts, rainfall duration, rainfall distribution, and anticipated extreme weather events such as drought, flood, fire risk, strong wind/wind gusts, hail, frost, among others.	Used in making tactical decisions on the scheduling of: When to plow the fields, when to sow, when to add fertilizers, when to irrigate, when to provide pesticides, when to harvest, when to sell, and choice of seed variety for planting.	Desired but not widely available information include optimal sowing date, evapotranspiration, insolation, soil water availability (to inform the scheduling of irrigation).	[10,13,31,36,45,48]
Short-term forecasts (1–5 years)		Mostly used in the livestock sector for preparedness messages and education on: fodder availability, water resource availability, potential disease occurrence zone.	Desired but not yet widely available information includes: Forecasts of parasite and animal diseases	[13,31,36,41,45,46,48]
Intra-decadal/ Medium-term forecasts (5–10 years)	Sectoral decision making			[10,13,46]
Decadal forecasts			No record of the use of decadal to medium-term projections. Although it acknowledged that such could inform future agricultural research investments, irrigation and water resource management planning, and training needs for agricultural extension staff.	[46,51]

The generation of timescale forecasts that are relevant to the timeline at which decisions are made in the agriculture sector is increasingly relying on the nature of the partnership and collaboration among the transdisciplinary actors (climatologists, meteorologists, and agriculturalists) operating in the agriculture and climate information space. Although this partnership has played a key role in advancing the uptake of CS in the Africa agriculture sector, there are, however, gaps reported in the selected studies concerning differences between the type of information desired by users and the type of information they receive via the CS.

Most of the reported gaps are associated with seasonal forecast usage. This includes several additional important pieces of information to optimize the resilience of agricultural operations to climate change's impact. An example is a desire for information on crop water requirements and evapotranspiration rate, which users believe will enhance the efficiency of the use of irrigation systems as a climate change response strategy [4,50]. There were also identified gaps in the literature that are associated with the use of short-term forecasts. Many studies reported the absence of forecasts on the precise occurrence of parasite and/or livestock disease as a result of climate change. Users believe that such information will enable them to anticipate and adjust their management strategy to manage climate change's impact on their livestock [5,52]. There are, however, differences in the extent to which seasonal and short-term forecasts are used across African countries. For example, a study in Malawi reported that there is no availability of models to predict the different periods when the rains can set in [34,35], whereas this type of limitation is not an issue in many other African countries [9]. In general, we did not record any gap associated with the use of intra-decadal and decadal forecasts in the agriculture sector; this is largely because such forecasts are currently seldom used in the African agriculture sector.

Agrometeorological services: Agrometeorological services are the second most common type of climate information provided by the CS [36,53]. Included in this category is information provided to manage the impact of both climate change and climate variability. This includes advisory information on the scheduling of planting operations, weeding, fertilizer applications, etc. CS is also, in some cases, used to provide information on climate-smart agriculture practices (CSA). The type of CSA information communicated includes conservation farming practices like ridging, minimum tillage, soil conservation practices, etc. [52,54]. The use of CS to communicate agrometeorological services and CSA to farmers is acknowledged as a valuable innovation to assist decision-making and develop farmers' specific adaptive capacities [36]. Table 3 presents a summary of how agrometeorological services are used in farming operations and the associated benefits. The benefits associated with agrometeorological services integration in farming operation decision-making can be summarized by an increase in crop productivity and a decrease in cropping costs in terms of inputs and working time [36,37].

Table 3. Commonly used agrometeorological services.

Farming Operation	Agro-Meteorological Services	CSA	Benefits	References
Land preparation	Advice on weather and seasonal forecasts, and crop calendar	Land conservation practices		[36,55]
Weeding	Soil moisture and weather forecasts	Land conservation practices to reduce weed infestation		[52,54]
Sowing	Forecasts on onset and offset of rain seasons with sowing calendar		Avoid loss due to crop failure to germinate or establish because of dry spells	[44,52,56]
Crop variety choice	Insight from forecasts on rain distribution, average annual rainfall, and seasonal forecasts in combination with crop calendar is used to advise farmers on crop type and variety to sow			[45,52]

Early warning interventions: The third type of climate information provided through CS are early warnings. Early warning intervention provisioning is commonly used for drought, flood, and wildfire risk warnings [27,40,44,52,54]. Early warnings are rarely solely disseminated to users; rather, they are provided in combination with agrometeorological services [33,36,37,45]. The early and timely delivery of early warnings is increasingly being facilitated through the integration of ICT CS dissemination strategy.

3.4. Extent of Scientific and Indigenous Knowledge Systems Integration in CS

The review of studies to understand the extent to which scientific and indigenous knowledge systems are integrated into CS revealed that knowledge system integration is not yet an issue of significant emphasis in CS adoption in the African agriculture sector. Of the reviewed studies, 72% applied only scientific knowledge systems and did not in any way integrate indigenous knowledge system CS applications [9–11,41]. However, 17% of the reviewed studies collated indigenous practices of climate risk prediction and risk response strategy [8,13,56]. Although the documented indigenous knowledge system is not included in the content of information disseminated via the CS, they are nevertheless, on rare cases, used to fine-tune the statistical forecast of risks and risk response strategy [55,57].

Nonetheless, indigenous knowledge systems are not entirely neglected in the current model of CS deployment. Eleven percent of the reviewed literature reported the inclusion of indigenous knowledge holders in the process of a forecast's translation into relevant climate information for actors in the agriculture sector. The study deduces that the inclusion of indigenous knowledge holders in the process of CS deployment is mainly for two purposes. The first is to promote the acceptance of CS by rural farmers, because rural farmers in Africa overwhelmingly rely on indigenous knowledge systems for their operations [27,41,55]. The second purpose is for the fine-tuning of statistical forecasts to suit the local context of climate risk warning and risk response strategy.

Resistance to information adoption often occurs when new knowledge interplays negatively with old knowledge [30,57,58]. This assertion can be attributed to the challenge of meteorological forecasts' acceptance, especially by rural farmers in many African countries where CS is sometimes rejected in favor of the old way of farming because the new information tends to interfere with the traditional way of farming [55]. This type of resistance is very common among the elderly, who tend to favor the traditional way of farming that is rooted in their indigenous knowledge system [59]. However, there is a growing trend towards the co-production of forecasts, whereby indigenous knowledge holders collaborate with researchers and meteorologists to generate plausible forecasts for their locality [8,27,49]. The approach currently tends to focus on using a participatory process for consensus on plausible risk scenarios for the local community as a way of securing the people's trust and confidence in the disseminated information [27,35]. As a result, most CS' lack information on the indigenous system of risk prediction and risk response.

The study, therefore, infers that an actionable point of entry for indigenous knowledge system integration into CS would be to integrate scientific risk response strategy with local sociocultural farming coping practices. Integration must occur across all three phases of the CS value chain. This means much needs to be done to encourage integration at forecasts production and forecast translation phase.

3.5. CS' Role in Facilitating Two-Way Learning for Robust Adaptation Action

To analyze how the adoption of CS has facilitated two-way learning (bottom-up and top-down) about climate change mitigation and adaptation in the agriculture sector, we analyzed the selected literature for information on methods used to facilitate a feedback relationship among actors in the value chain of CS. This is because the production and dissemination of contextual climate information for actors in the agriculture sector relies mainly on the structure and feedback loop of the network of relations that exists among the actors [7,16].

The analysis of the selected literature indicates that the relationships among the actors in the CS value chain operate mainly on a participatory collaborative process. This collaborative process is primarily through workshops and participatory scenario planning meetings. The participatory collaborative process is used for the production of relevant climate information, development of appropriate channels for information dissemination, and promotion of local ownership in climate information production and dissemination [9,49,60]. This ultimately influences learning and revisiting to ensure the relevancy, suitability, and usability of information disseminated via the CS [35,61,62]. Table 4 provides a summary of recorded evidence of how the participatory process approach in CS deployment facilitates two-way learning (bottom-up and top-down).

Table 4. Evidence of participatory process influence in CS application.

Case	Key Impact	Reference
The participatory process is targeted at facilitating the relationship between CS providers and local farmers to enable CS providers to understand the user's socio-cultural context to provide contextual information	User's context	[11,41]
The participatory process was used to spur farmers group interest in CS which resulted in them taking ownership and initiative of the process of CS dissemination and application in farming practices in their locality	Ownership and taking the initiative	[6,11,28,49]
The participatory process was used to improve local people understanding of and trust meteorological weather and climate forecasts	Trust and confidence in meteorological forecasts	[49]
There are several recorded case studies where participatory processes have successfully been used to improve the rate of CS application in farming practices by local farmers	CS usage	[10,49]
There are several recorded case studies where participatory processes have successfully been used to provide inclusive training to users to enhance their capacity to understand and apply disseminated information via the CS	Capacitation of users	[10,27]

Nevertheless, the cost and difficulty of gathering all relevant stakeholders in a workshop are limiting the effectiveness of this approach. This is evident in the reported gaps (Table 2) in the information disseminated through the CS [32,40]. CS providers, therefore, need to be proactive in interacting with the farmers regarding their needs for climate information and in determining a more suitable feedback mechanism for maintaining the relevancy of CS [63,64]. To this end, several methods for reaching smallholder farmers have been attempted by various agencies, but a scalable solution has yet to be found [16,31]. The internet and mobile phone (SMS) are the two prominent new and innovative methods being used to facilitate collaborations among the actors. They are, however, still in infancy and need a lot of research to improve their efficiency. The use of the internet, for example, has been constrained by lack of facilities and, in some cases, by unwillingness on the part of the local people to pay the internet fee for accessing CS [44,65]. SMS, on the other hand, has also been constrained by poor signal/reception in many regions and, in most cases, the feedback communication between CS providers and users via SMS has been reported to be inefficient and inadequate [66]. There is, therefore, a need for further investigation for insight on appropriate modalities for facilitating impactful and sustainable reciprocated relationships among the actors along the CS value chain via the use of SMS and the internet, particularly within the context of African rural communities.

4. Conclusions

The integration of CS in climate information in the decision-making process in the agriculture sector is a valuable innovation that can enhance the resilience of the sector to the impacts of climate change. In the reviewed literature on selected African countries' agricultural sectors, considerable progress has been made in the integration of CS in the decision-making process. The outcomes of the analysis of the study's four research questions are summarized as follows:

The role of CS in climate information integration in decision-making in the agricultural sector is explained through the three distinct phases in the value-chain: forecast production, forecast translation and integration, and climate information communication and adoption in decision-making (Section 3.1). The key findings from the study show that the nature of collaboration among the actors operating in each phase is a key determinant of the efficiency of the CS model and the usability of the communicated climate information.

Forecasts, agrometeorological services, and early warnings are the three main types of climate information communicated through CS. The five timescale forecasts widely used in CS for communication of information on climate risk warning and risk response advisory services are: (1) Weather forecasts (daily to weekly), (2) Seasonal forecasts (on a timescale of 1–6 months), (3) Short-term forecasts (1–5 years), (4) Intra-decadal/Medium-term forecasts (5–10 years), and (5) Decadal forecasts. Agrometeorological services communicated via the CS are comprised mostly of advisory information on the scheduling of planting operations, weeding, fertilizer applications, and climate-smart agriculture practices. The third type of climate information provided through CS is early warnings. The early warning system uses an integrated communication system to help farmers and decision-makers prepare for climate risks. Common climate risks event managed by early warning interventions includes drought, flood, and wildfires.

Reflecting on insights from the review of the various timescale forecasts used in CS deployment, two contemporary realities are observed to drive the high demand for shorter timescale forecasts in the African agriculture sector:

1. Most decision-making in the agriculture sector, such as the scheduling of planting or harvest operations that could benefit from integrated and targeted climate forecasts, is made at a range of temporal and spatial scales that are matching with a shorter-term forecast timeline. Not surprisingly, several authors [11,15,60,66] reported that farmers are more interested in weather and seasonal forecasts. This is in contrast to CS' demand in the forest sector, where the majority of foresters are interested in longer timescale forecasts, due mostly to the fact that product harvesting is usually on a long timescale range of 8–30 years.
2. Another factor is the socio-economic profile of the farmers. The majority of farmers in the agriculture sector in the African countries are subsistence farmers, with poor and limited capacity for long-term planning. Hence, the timescale of their management decisions is often based on a short timescale. Consequently, the farmer mostly demands forecasts of shorter timescale horizons (weather and seasonal). This demand/supply factor plays a crucial role in tilting CS providers to mostly focus on providing forecasts for short timescales. Nevertheless, there is a gradual increase in the demand for forecasts of a longer timeline horizon.

The dominant use of forecasts of a shorter timescale in CS is also an indication of a dearth of effort towards long-term anticipatory adaptation actions in most African countries' agriculture sectors. This may pose an unforeseen challenge to the sustainability of the agriculture sector. Without a considerable effort to understand what the future scenario and outlook will be for the African agriculture sector in the face of a changing climate, it may be that the government is unknowingly setting up the sector for a massive failure. There is a need for enhanced capacity towards being able to anticipate, predict the future scenario of the agriculture sector in the face of climate change across the scale from national to local so that appropriate anticipatory adaptation action can be devised and implemented.

The review of the extent to which CS facilitates indigenous knowledge systems' integration in climate information communication revealed that indigenous knowledge system integration is not yet an issue of significant emphasis in CS adoption in the African agriculture sector. Nonetheless, indigenous knowledge systems are not entirely neglected in the current model of CS deployment. Little of the reviewed literature reported the inclusion of indigenous knowledge holders in the process of forecasts translation into relevant climate information for actors in the agriculture sector.

CS facilitates two-way learning (bottom-up and top-down) about climate-smart agricultural practices through the adoption of a participatory process in the generation and communication of climate information. This collaborative process is primarily through workshops and participatory scenario planning meetings. This ultimately influences learning and revisiting to ensure the relevancy, suitability, and usability of information disseminated via the CS.

Though CS usage has improved the communication and use of climate information in the African agricultural sector, the study findings have shed light on research gaps and opportunities that should be explored to maximize the benefit of CS application in the African agriculture sector. These are summarized as follow:

1. The most commonly used forecasts in the African agricultural sector are short-term, seasonal, and weather forecasts. Intra-decadal and decadal forecasts are sparingly used in agricultural risk management. This is a worrying trend because intra-decadal and decadal forecasts are useful for making strategic decisions and anticipatory adaptation plans. In order, therefore, to shift away from reactionary adaptation actions, the study recommends a further investigation on appropriate modalities for facilitating the integration of intra-decadal and decadal forecasts in climate information communication and usage in the African agricultural sector.
2. The poor integration of indigenous knowledge systems in CS adoption is a concern that warrants an increased emphasis on knowledge systems integration in CS deployment. This can be adopted as a strategy for facilitating CS acceptance, especially at the local scale. To this end, the study recommends further research on modalities of aligning scientific climate risk response strategies with farmers' sociocultural farming/coping practices.
3. Participatory process (e.g., workshop and participatory scenario planning) is increasingly used in CS adoption for facilitating forecast translation into relevant climate information and promoting two-way learning on climate risk and risk response strategy. To this end, several methods for reaching smallholder farmers (e.g., mobile phone) have been attempted by various agencies, but a scalable solution has yet to be found. Mobile phone usage in rural Africa is faced with challenges that impact its efficiency. The study, therefore, recommends a further investigation on appropriate modalities for facilitating impactful and sustainable reciprocated relationships among the actors along the CS value chain, with a specific focus on modalities for enhancing the efficiency of mobile phone usage in climate information communication in rural Africa.
4. The particular time-scale at which management decisions are made and the available forecast timescale greatly influences the integration of climate information in management action in the agriculture sector.

Author Contributions: M.N. developed the idea and received funding for the project. Both authors further developed the idea, permofrmed the analysis and wrote the manuscript. All authors have read and agreed to the published version of the manuscript.

Funding: This research was funded by the International Development Research Centre, Canada (IDRC) and Dept. for International Development, United Kingdom (DFID) through the ASSAR project (Adaptation at Scale in the Semi-Arid Region), [ACDI-ASSAR-1034].

Institutional Review Board Statement: This study was approved by the Ethics Research Committee of the faculty of Science, University of Cape Town, South Africa.

Informed Consent Statement: Not applicable.

Data Availability Statement: The data that support the findings of this study are available from the corresponding author upon reasonable request.

Acknowledgments: This research was funded by the Adaptation at Scales in Semi-Arid Regions (ASSAR) project. ASSAR project received funding from the International Development Research Centre, Canada (IDRC) and Dept. for International Development, United Kingdom (DFID). We would also like to thank all those who provided feedback on the scope of the review question, including the peer reviewers of a draft version of this document.

Conflicts of Interest: The authors declare no conflict of interest.

References

1. Milder, J.C.; Hart, A.K.; Dobie, P.; Minai, J.; Zaleski, C. Integrated landscape initiatives for african agriculture, development, and conservation: A region-wide assessment. *World Dev.* **2013**, *54*, 68–80. [CrossRef]
2. Roncoli, C.; Jost, C.; Kirshen, P.; Sanon, M.; Ingram, K.T.; Woodin, M.; Somé, L.; Ouattara, F.; Sanfo, B.J.; Sia, C.; et al. From accessing to assessing forecasts: An end-to-end study of participatory climate forecast dissem-ination in Burkina Faso (West Africa). *Clim. Chang.* **2009**, *92*, 433–460. [CrossRef]
3. Cadger, K.; Quaicoo, A.K.; Dawoe, E.; Isaac, M.E. Development interventions and agriculture adaptation: A social network analysis of farmer knowledge transfer in ghana. *Agriculture* **2016**, *6*, 32. [CrossRef]
4. Cooper, P.J.M.; Dimes, J.; Rao, K.P.C.; Shapiro, B.; Shiferaw, B.; Twomlow, S. Coping better with current climatic variability in the rain-fed farming systems of sub-Saharan Africa: An essential first step in adapting to future climate change? *Agric. Ecosyst. Environ.* **2008**, *126*, 24–35. [CrossRef]
5. Dayamba, D.S.; Ky-Dembele, C.; Bayala, J.; Dorward, P.; Clarkson, G.; Sanogo, D.; Mamadou, L.D.; Traoré, I.; Diakité, A.; Nenkam, A.; et al. Assessment of the use of Participatory Integrated Climate Services for Agriculture (PICSA) approach by farmers to manage climate risk in Mali and Senegal. *Clim. Serv.* **2018**, *12*, 27–35. [CrossRef]
6. Sivakumar, M.V.K.; Motha, R.P. Managing Weather and Climate Risks in Agriculture Summary and Recommendations. In *Managing Weather and Climate Risks in Agriculture*; Sivakumar, M.V.K., Motha, R.P., Eds.; Springer: Berlin, Heidelberg, 2007.
7. Vaughan, C.; Dessai, S. Climate services for society: Origins, institutional arrangements, and design elements for an evaluation framework. *WIREs Clim. Chang.* **2014**, *5*, 587–603. [CrossRef]
8. Machingura, F.; Nyamwanza, A.; Hulme, D.; Stewart, E. Climate information services, integrated knowledge systems and the 2030 Agenda for Sustainable Development. *Sustain. Earth* **2018**, *1*, 1. [CrossRef]
9. Tall, A.; Coulibaly, J.Y.; Diop, M. Do climate services make a difference? A review of evaluation methodologies and practices to assess the value of climate information services for farmers: Implications for Africa. *Clim. Serv.* **2018**, *11*, 1–12. [CrossRef]
10. Dougill, V.K.; Dixon, A.J.; Stringer, J.L.; Cull, T.L.C. Identifying climate services needs for national planning: Insights from Malawi. *Clim. Policy* **2017**, *17*, 189–202. [CrossRef]
11. Tall, A.; Jay, A.; Hansen, J. *Scaling Up Climate Services for Farmers in Africa and South Asia Workshop Report*; CCAFS Working Paper no. 40; CGIAR Research Program on Climate Change, Agriculture and Food Security (CCAFS): Copenhagen, Denmark, 2013; Available online: https://cgspace.cgiar.org/bitstream/handle/10568/27833/WP40.pdf?sequence=1&isAllowed=y (accessed on 5 December 2018).
12. Ogallo, L. *Integrating Indigenous Knowledge in Climate Risk Management to Support Community-Based Adaptation*; Final Technical Report; The International Development Research Centre (IDRC): Ottawa, ON, Canada, 2010.
13. Ongoma, V.; Shilenje, Z.W. The effectiveness of agrometeorological information in the realization of Kenya's Vision 2030; lessons learnt from China. *Ital. J. Agrometeorol.* **2016**, *1*, 67–72.
14. Washington, R.; Harrison, M.; Conway, D. *African Climate Report—A Report Commissioned by the UK Government to Review African Climate Science, Policy, and Options for Action*; Department for Environment, Food and Rural Affairs: London, UK, 2004; p. 45.
15. Thornton, P.K.; Kristjanson, P.; Förch, W.; Barahona, C.; Cramer, L.; Pradhan, S. Is agricultural adaptation to global change in lower-income countries on track to meet the future food production challenge? *Glob. Environ. Chang.* **2018**, *52*, 37–48. [CrossRef]
16. Kadi, M.; Njau, L.N.; Mwikya, J.; Kamga, A. *The State of Climate Information Services for Agriculture and Food Security in East African Countries*; Working Paper No. 5; Climate Change, Agriculture and Food Security (CCAFS): Copenhagen, Denmark, 2011; Available online: www.ccafs.cgiar.org (accessed on 9 December 2018).
17. Bernard, M. Understanding user needs for climate services in agriculture. *WMO Bull.* **2011**, *60*, 67–72.
18. Dorward, P.; Clarkson, G.; Stern, R. *Participatory Integrated Climate Services for Agriculture (PICSA): Field Manual*; Walker Institute, University of Reading: Reading, UK, 2015; ISBN 9780704915633.
19. Street, R.; Buontempo, C.C.; Mysiak, J.; Karali, E.; Pulquério, M.; Murray, V.; Swart, R. How could climate services support disaster risk reduction in the 21st century. *Int. J. Disaster. Risk Reduct.* **2010**, *34*, 28–33. [CrossRef]
20. Wanders, N.; Wood, E.F. Assessing seasonal climate forecasts over africa to support decision-making. *World Sci. Ser. Asia-Pac. Weather Clim.* **2017**, *10*, 1–15. [CrossRef]
21. Kettle, N.P.; Dow, K.; Tuler, S.; Webler, T.; Whitehead, J.; Miller, K. Integrating scientific and local knowledge to inform risk-based management approaches for climate adaptation. *Clim. Risk Manag.* **2014**, *4–5*, 17–31. [CrossRef]
22. Qasim, M. Sustainability and wellbeing: A scientometric and bibliometric review of the literature. *J. Econ. Surv.* **2017**, *31*, 1035–1061. [CrossRef]

23. Venkatasubramanian, K.; Tall, A.; Hansen, J.; Aggarwal, P.K. *Assessment of India's Integrated Agrometeorological Advisory Service Program from a Farmer Perspective*; CCAFS Working Paper no. 54; CGIAR Research Program on Climate Change, Agriculture and Food Security (CCAFS): Copenhagen, Denmark, 2014. Available online: https://assets.publishing.service.gov.uk/media/57a089baed915d622c000385/CCAFS_WP_54.pdf (accessed on 7 January 2019).
24. Strauss, A.; Corbin, J. *Basics of Qualitative Research: Grounded Theory Procedures and Techniques/Anselm Strauss, Juliet Corbin*; Sage Publications: Newbury Park, CA, USA, 1990.
25. Charmez, K. Grounded theory methods in social justice research. In *The SAGE Handbook of Qualitative Research*, 4th ed.; Denzin, N., Lincoln, Y., Eds.; Sage: Thousand Oaks, CA, USA, 2011; pp. 359–380.
26. Haig, B.D. Précis of 'an abductive theory of scientific method'. *J. Clin. Psychol.* **2008**, *64*, 1019–1022. [CrossRef]
27. Ofoegbu, C.; New, M.G.; Staline, K. The effect of inter-organizational collaboration networks on climate knowledge flows and communication to pastoralists in Kenya. *Sustainability* **2018**, *10*, 4180. [CrossRef]
28. Coulibaly, Y.J.; Kundhlande, G.; Amosi, N.; Tall, A.; Kaur, H.; Hansen, J. *What Climate Services Do Farmers and Pas-Toralists Need in Tanzania? Baseline Study for the GFCS Adaptation Program in Africa*; CCAFS Working Paper no. 110; CGIAR Research Program on Climate Change, Agriculture and Food Security (CCAFS): Copenhagen, Denmark, 2015; Volume 112.
29. Kirui, C.; Oseni, S.V.; Bebe, O.B. Characterizing Access to Climate Information and Services by the Vulnerable Groups in Semi-arid Kenya. In Proceedings of the Third RUFORUM Biennial Meeting, Entebbe, Uganda, 24–28 September 2012; pp. 2077–2082.
30. Nyamwanza, A.M.; New, M.; Fujisawa, M.; Johnston, P.; Hajat, A. Contributions of decadal climate information in agriculture and food systems in east and southern Africa. *Clim. Chang.* **2017**, *143*, 115–128. [CrossRef]
31. Hansen, J.W.; Mason, S.J.; Sun, L.; Tall, A. Review of seasonal climate forecasting for agriculture in Sub-Saharan Africa. *Exp. Agric.* **2011**, *47*, 205–240. [CrossRef]
32. Ziervogel, G.; Zermoglio, F. Climate change scenarios and the development of adaptation strategies in Africa: Challenges and opportunities. *Clim. Res.* **2009**, *40*, 133–146. [CrossRef]
33. Vermeulen, S.J.; Aggarwala, P.K.; Ainslie, A.; Angelone, C.; Campbell, B.M.; Challinor, A.J.; Hansen, J.W.; Ingram, J.S.I.; Jarvis, A.; Kristjanson, P.; et al. Options for support to agriculture and food se-curity under climate change. *Environ. Sci. Policy* **2012**, *15*, 136–144. [CrossRef]
34. Weichselgartner, J.; Kasperson, R. Barriers in the science-policy-practice interface: Toward a knowledge-action-system in global environmental change research. *Glob. Environ. Chang.* **2010**, *20*, 266–277. [CrossRef]
35. Waiswa, M.; Mulamba, P.; Isabirye, P. Climate information for food security: Responding to user's climate information needs. In *Climate Prediction and Agriculture: Advances and Challenge*; Sivakumar, M.V.K., Hansen, J., Eds.; Springer: Berlin/Heidelberg, Germany, 2007; pp. 225–248.
36. Tarchiani, V.; Camacho, J.; Coulibaly, H.; Rossi, F.; Stefanski, R. Agrometeorological services for smallholder farmers in West Africa. *Adv. Sci. Res.* **2018**, *15*, 15–20. [CrossRef]
37. Roudier, P.; Muller, B.; D'Aquino, P.; Roncoli, C.; Soumaré, M.; Batté, L.; Sultan, B. The role of climate forecasts in smallholder agriculture: Lessons from participatory research in two communities in Senegal. *Clim. Risk Manag.* **2014**, *2*, 42–55. [CrossRef]
38. Ziervogel, G.; Johnston, P.; Matthew, M.; Mukheibir, P. Using climate information for supporting climate change adaptation in water resource management in South Africa. *Clim. Chang.* **2009**, *103*, 537–554. [CrossRef]
39. Risbey, J.; Kandlikar, M.; Dowlatabadi, H. Scale, context, and decision-making in agricultural adaptation to climate variability and change. *Mitig. Adapt. Strateg. Glob. Chang.* **1994**, *4*, 137–165. [CrossRef]
40. Singh, C.; Daron, J.; Bazaz, A.; Ziervogel, G.; Spear, D.; Krishnaswamy, J.; Zaroug, M.; Kituyi, E. The utility of weather and climate information for adaptation decision-making: Current uses and future prospects in Africa and India. *Clim. Dev.* **2018**, *10*, 389–405. [CrossRef]
41. Ouédraogo, M.; Barry, S.; Zougmoré, R.B.; Partey, S.T.; Somé, L.; Baki, G. Farmers' willingness to pay for climate information services: Evidence from cowpea and sesame producers in Northern Burkina Faso. *Sustainability* **2018**, *10*, 611. [CrossRef]
42. Cornell, S.; Berkhout, F.; Tuinstra, W.; Tàbara, J.D.; Jäger, J.; Chabay, I.; de Wit, B.; Langlais, R.; Mills, D.; Moll, P.; et al. Opening up knowledge systems for better responses to global environmental change. *Environ. Sci. Policy* **2013**, *28*, 60–70. [CrossRef]
43. Stone, R.C.; Meinke, H. Weather, climate, and farmers: An overview. *Meteor. Appl.* **2006**, 3 (Suppl. 1), 7–20. [CrossRef]
44. Oladele, O. Agricultural extension and rural advisory services: Proactiveness or reactiveness on climate change for food security in Africa. *Life Sci. J.* **2013**, *10*, 593–597.
45. Ouedraogo, I.; Diouf, N.S.; Ouédraogo, M.; Ndiaye, O.; Zougmoré, R.B. Closing the gap between climate information producers and users: Assessment of needs and uptake in Senegal. *Climate* **2018**, *6*, 13. [CrossRef]
46. Zongo, B.; Diarra, A.; Barbier, B.; Zorom, M.; Yacouba, H.; Dogot, T. Farmers' perception and willingness to pay for climate information in Burkina Faso. *J. Agric. Sci.* **2016**, *8*, 175–187. [CrossRef]
47. Brasseur, G.P.; Gallardo, L. Climate services: Lessons learned and future prospects. *Earth's Future* **2016**, *4*, 79–89. [CrossRef]
48. Nesheim, I.; Barkved, L.J.; Bharti, N. What is the role of agro-met information services in farmer decision-making? Uptake and decision-making context among farmers within three case study villages in Maharashtra, India. *Agriculture* **2017**, *7*, 70. [CrossRef]
49. Kniveton, D.; Visman, E.; Tall, A.; Diop, M.; Ewbank, R.; Njoroge, E.; Pearson, L. Dealing with uncertainty: Inte-grating local and scientific knowledge of the climate and weather. *Disasters* **2014**, *39*, S35–S53. [CrossRef]
50. Jones, L.; Harvey, B.; Godfrey-Wood, R. *The Changing Role of NGOs in Supporting Climate Services*; BRACED Resilience Intel, 4; London Overseas Development Institute: London, UK, 2016.

51. Nyamwanza, A.M.; New, M. Anticipatory adaptation and the role of decadal climate information in rural African livelihood systems: Lessons from the Mid-Zambezi Valley, Zimbabwe. *Int. J. Clim. Change Strateg. Manag.* **2015**, *8*, 236–252. [CrossRef]
52. Cramer, L. Evaluation of Climate Services Interventions in the GFCS Adaptation Programme for Africa. Part II: Bene-ficiary Assessment Final Evaluation Summary Report. Statistics for Sustainable Development and Cramer-Njihia Consultants for the World Food Programme and CGIAR Research Program on Climate Change, Agriculture and Food Security (CCAFS). 2017. Available online: https://gfcs.wmo.int/sites/default/files/projects/Global%20Framework%20for%20Climate%20Services%20Adaptation%20Programme%20in%20Africa%20%28GFCS%20APA%29%2C%20Phase%20I%2C%20-%20Building%20Resilience%20in%20Disaster%20Risk%20Management%2C%20Food%20Security%20and%20Health/Final-report_1-Nov-2017-min.pdf (accessed on 6 January 2019).
53. Nadine, A.; Fanny, H.; Ousmane, N.; Armando, M.; Carlos, M.; Guy, F. Climate smart services: Case studies in Senegal, Burkina, and Colombia. In *Climate Smart Agriculture-Global Science Conference*; CIRAD: Montpeiller, France, 2015; p. 113. Available online: http://csa2015.cirad.fr/var/csa2015/storage/fckeditor/file/L3%20Towards%20Climate-smart%20Solutions(1).pdf (accessed on 18 December 2018).
54. Mwaniki, F.; Gichuki, C.; Mwangi, M.; Mburia, P.; Wandago, B.O. Addressing challenges in communicating adaptation practices to smallholder farmers in Kenya through a radio intervention. *J. Agric. Environ. Int. Dev.* **2017**, *111*, 279–322. [CrossRef]
55. Kruk, M.C.; Parker, B.; Marra, J.J.; Werner, K.; Heim, R. Engaging with users of climate information and the copro-duction of knowledge. *Weather Clim. Soc.* **2017**, *9*, 839–849. [CrossRef]
56. Aker, J.C. Dial "A" for agriculture: A review of information and communication technologies for agricultural extension in developing countries. *Agric. Econ.* **2011**, *42*, 631–647. [CrossRef]
57. Kanno, H.; Sakurai, T.; Shinjo, H.; Miyazaki, H.; Ishimoto, Y.; Saeki, T.; Umetsu, C.; Sokotela, S.; Chiboola, M. Indigenous climate information and modern meteorological records in Sinazongwe district, Southern Province, Zambia. *Jpn. Agric. Res. Q.* **2013**, *47*, 191–201. [CrossRef]
58. Makondo, C.C.; Thomas, D.S.G. Climate change adaptation: Linking indigenous knowledge with western science for effective adaptation. *Environ. Sci. Policy* **2018**, *88*, 83–91. [CrossRef]
59. Buizer, J.; Jacobs, K.; Cash, D. Making short-term climate forecasts useful: Linking science and action. *Proc. Natl. Acad. Sci. USA* **2016**, *113*, 4597–4602. [CrossRef] [PubMed]
60. Best, A.; Holmes, B. Systems thinking, knowledge and action: Towards better models and methods. *Evid. Policy J. Res. Debate Pract.* **2010**, *6*, 145–159. [CrossRef]
61. Naaba, F.Z.; Abubakari, Z.; Ahmed, A. The role of climate services in agricultural productivity in Ghana: The per-spectives of farmers and institutions. *Clim. Serv.* **2019**, *13*, 24–32. [CrossRef]
62. Kalafatis, S.E.; Lemos, M.C.; Lo, Y.-J.; Frank, K.A. Increasing information usability for climate adaptation: The role of knowledge networks and communities of practice. *Glob. Environ. Chang.* **2015**, *32*, 30–39. [CrossRef]
63. Never, B. *Regional Power Shifts and Climate Knowledge Systems: South Africa as a Climate Power?* Working Paper No. 125; German Institute of Global and Area Studies: Hamburg, Germany, 2010; p. 311.
64. Sivakumar, M. Climate prediction and agriculture: Current status and future challenges. *Clim. Res.* **2006**, *33*, 3–17. [CrossRef]
65. Rasmussen, L.V.; Mertz, O.; Rasmussen, K.; Nieto, H. Improving how meteorological information is used by pastoralists through adequate communication tools. *J. Arid. Environ.* **2015**, *121*, 52–58. [CrossRef]
66. Mittal, S.; Hariharan, V.K. Mobile-based climate services impact on farmers risk management ability in India. *Clim. Risk Manag.* **2018**, *22*, 42–51. [CrossRef]

agriculture

MDPI

Article

Modelling of Mechanical Properties of Fresh and Stored Fruit of Large Cranberry Using Multiple Linear Regression and Machine Learning

Józef Gorzelany [1], Justyna Belcar [1], Piotr Kuźniar [1], Gniewko Niedbała [2] and Katarzyna Pentoś [3,*]

[1] Department of Food and Agriculture Production Engineering, University of Rzeszow, 4 Zelwerowicza Street, 35-601 Rzeszów, Poland; gorzelan@ur.edu.pl (J.G.); jbelcar@ur.edu.pl (J.B.); pkuzniar@ur.edu.pl (P.K.)

[2] Department of Biosystems Engineering, Faculty of Environmental and Mechanical Engineering, Poznań University of Life Sciences, Wojska Polskiego 50, 60-627 Poznań, Poland; gniewko.niedbala@up.poznan.pl

[3] Institute of Agricultural Engineering, Wrocław University of Environmental and Life Sciences, 37b Chełmonskiego Street, 51-630 Wrocław, Poland

* Correspondence: katarzyna.pentos@upwr.edu.pl; Tel.: +48-71-320-5970

Abstract: The study investigated the selected mechanical properties of fresh and stored large cranberries. The analyses focused on changes in the energy requirement up to the breaking point and aimed to identify the apparent elasticity index of the fruit of the investigated large cranberry fruit varieties relating to harvest time, water content, as well as storage duration and conditions. After 25 days in storage, the fruit of the investigated varieties were found with a decrease in mean acidity, from 1.56 g·100 g^{-1} to 1.42 g·100 g^{-1}, and mean water content, from 89.71% to 87.95%. The findings showed a decrease in breaking energy; there was also a change in the apparent modulus of elasticity, its mean value in the fresh fruit was 0.431 ± 0.07 MPa, and after 25 days of storage it decreased to 0.271 ± 0.08 MPa. The relationships between the cranberry varieties, storage temperature, duration of storage, x, y, and z dimensions of the fruits, and their selected mechanical parameters were modeled with the use of multiple linear regression, artificial neural networks, and support vector machines. Machine learning techniques outperformed multiple linear regression.

Keywords: large cranberry; mechanical properties; cranberry compression; water content; mathematical modelling; machine learning

Citation: Gorzelany, J.; Belcar, J.; Kuźniar, P.; Niedbała, G.; Pentoś, K. Modelling of Mechanical Properties of Fresh and Stored Fruit of Large Cranberry Using Multiple Linear Regression and Machine Learning. *Agriculture* **2022**, *12*, 200. https://doi.org/10.3390/agriculture12020200

Academic Editor: Andrea Colantoni

Received: 28 December 2021
Accepted: 28 January 2022
Published: 31 January 2022

1. Introduction

The large cranberry (*Vaccinium macrocarpon* Aiton) is mainly cultivated as an industrial crop in North America (Canada and north-eastern regions of the USA) since peat soils and wetlands, commonly occurring there, constitute the optimum substrate for cranberry cultivation; the crop is also grown in Europe, mainly in Latvia and Belarus, and, in recent years, in south-eastern Poland [1]. The highest harvest in 2019 was recorded in the USA and amounted to 359,110 tons from a cultivation area of 15,580 ha. In Canada 172,440 tons were harvested from an area of 6393 ha, while in Belarus 235 tons of cranberries from an area of 101 ha [2].

Depending on the variety, the fruit of the large cranberry is spherical with a common diameter of 20 mm and skin color ranging from pink to dark purple [3]. Cranberry fruit and juice are highly valued in medicine because of their antimicrobial and antimycotic properties, producing beneficial effects, for instance, in the treatment of kidney diseases and inflammation of the urinary tract [4,5]. The high contents of polyphenols, including anthocyanins, flavonoids, stilbenes, phenolic acids, and proanthocyanidins also confirm the fact that cranberry fruits offer health benefits [6,7].

The harvesting of cranberries in large plantations is carried out using highly efficient mechanical shakers. "Wet harvesting" technology (where a given section of the plantation is flooded with water) makes it possible to increase the effectiveness of the harvest and to reduce damage to fruit collected mechanically, compared to fruit collected using the "dry harvest" method [8–10]. Mechanical defects occurring during harvest, transportation to a processing plant, and handling activities such as cleaning, rinsing, and other post-harvest operations adversely affect the quality of the raw material, eliminating a part of or an entire batch designated for commerce or food processing. This is linked to the fact that the fruit is crushed or bruised leading to a decrease in water content, as well as changes in the texture and firmness of the berries [11]. A decrease in water content in soft fruit after storage was reported in prior literature [12,13].

There is increased usage of this raw material and higher demands regarding quality requirements defined by food and pharmaceutical industries for cranberries, which predominantly are harvested mechanically in commercial plantations. For this reason, it is necessary to investigate the mechanical properties, which will make it possible to determine the timing of harvest and storage conditions. A study investigating the mechanical properties of cranberry skin and flesh was conducted by Gorzelany et al. [14]. Skin and flesh puncture testing was carried out with a cylindrical stamp, with a diameter of 2 mm, which was pressed into the fruit. The recorded measurements included puncture strength and energy and absolute lengthwise deformation.

Given the spherical shape of cranberries of the selected varieties, their mechanical properties may be determined using methodology applicable to spherical fruit, such as high-bush blueberries, redcurrants, tomatoes, onions, and Brussels sprouts [15,16]. The analysis of force-displacement relation showed the value of acceptable force and relative displacement as a reaction to preset load.

In a uniaxial compression test, applied with quasi-static loading, cranberries were compressed between two flat horizontal planes until breaking point [17]. To interpret the results, it is necessary to apply the appropriate statistical methods which make it possible to accurately assess the effect of the selected factors on the relevant mechanical parameters of the fruit of large cranberries as well as the chemical characteristics of the raw material.

Mathematical modelling is a very common method used in food technology and agriculture [18–22]. Accurate models allow the prediction of the physicochemical properties of food and optimize storage conditions. There are two approaches to the mathematical modelling of the mechanical properties of food. The first is based on experimental data (empirical), and the second is based on the physical nature of the phenomenon (theoretical). Since empirical modelling is more precise and easier to develop [23], this approach is becoming increasingly popular. Various methods and techniques are employed for empirical modelling in the food industry, including artificial intelligence techniques such as artificial neural networks (ANNs) and support vector machines (SVMs). Many applications of empirical modelling in the food industry were reported in prior literature. The ANN technique was useful for the determining changes in the water content, protein, and gluten in stored wheat [24], for accurate and rapid prediction of the moisture and fat content of tofu [25], for the development of a crispness prediction model of crunchy food [26], or the estimation of sugar concentration in food products [27]. Chauchard et al. [28] proposed the sensor for acidity prediction in grapes based on NIR spectroscopy and Least-Squared Support Vector Machine regression. The same technique was employed to predict the mechanical properties of prawns [29]. The Support Vector Machine was reported as a regression technique for the development of an accurate model of the soluble solid content of apples [30].

The aim of this research was to investigate the effect of the duration of storage on the fruits' selected physical (content of water and total acidity) and mechanical properties, including deformation, breaking energy, and apparent elasticity index. The results obtained can be considered a useful tool when the harvesting, transportation, and processing of cranberry fruits is developed and optimized. The acquired results were assessed using statistical analysis and mathematical modelling. The comparison of models developed with the use of multiple linear regression and machine learning was the additional aim of the research.

2. Materials and Methods

2.1. Characteristics of the Research Material

The research material consisted of three cranberry fruit varieties: Ben Lear, Pilgrim, and Stevens, obtained from a plantation located in Radomyśl nad Sanem (50°40′52″ N 21°56′41″ E; Stalowa Wola District, Podkarpackie Region). The cranberry plantation was established in 2014 on plots with a modified substrate, the top layer of which was fine-grained washed sand with a thickness of 0.4 m. The meteorological conditions in the year of research are characterized in Table 1 [31].

Table 1. Weather conditions during cranberries vegetation in the year 2018.

Weather Parameters	Months								Period
	III	IV	V	VI	VII	VIII	IX	X	III-X
Air temperature (°C)	0.6	14.5	17.5	19.3	20.6	20.8	16.2	10.8	15.0
Rainfalls (mm)	32.4	16.5	41.3	40.2	134.9	74.4	36.1	39.5	415.3
Air relative humidity (%)	73.4	60.6	63.0	66.6	73.5	74.4	75.3	78.0	70.6

Fruits were collected in the maturation stage (fruit being ready for harvest). The date of fruit harvest depended on the variety: 29 September 2018 for the Stevens variety and 10 October 2018 for the Pilgrim and Ben Lear varieties. Fruits of each cranberry variety were divided into three batches: two were placed in cold storage at temperatures of 4 °C and 10 °C, and one sample was stored at a temperature of 20 °C.

2.2. Measurement of the Chemical Properties

The water content of the fresh cranberries and the material kept in cold storage (randomly selected plant samples) was determined using the dryer method (105 °C)—PN-90/A-75101-03 [32], whereas total acidity of the material was determined in accordance with PN-90/A-75101-04 [33].

2.3. Measurement of the Mechanical Properties

The selected mechanical parameters of cranberry fruit (randomly selected samples) were determined in the uniaxial compression test between two horizontal plates using a Zwick/Roell 2010 testing machine (ZwickRoell GmbH & Co. KG, Ulm, Germany). The following parameters were defined for the measurement process: initial stress applied to the sample was 0.1 N, and the speed of the loading panel during the test was 0.5 mm·s^{-1}. Values of maximum breaking force F (N) and deformation λ [mm] were recorded after each series of measurements. Characteristics of the force—deformation were determined based on the strength test. A summary value of the unit energy inputs (work) was considered in calculating the values of the apparent elasticity index Ec as a measure of the effective value of mechanical resistance of the investigated material.

$$E_c = \frac{\int_0^{\lambda} F_{(\lambda)} \cdot d\lambda}{0.26 \cdot x \cdot y \cdot \lambda} \quad (1)$$

where E_c is an apparent modulus of elasticity, F is a maximum breaking force (N), x and y are the dimensions of the ellipsoid in the direction perpendicular to the acting load (mm), and λ is a deformation in the direction of the loading applied (mm).

2.4. Method of Building Linear Models Using Multiple Linear Regression

Multiple linear regression (MLR) is the most commonly used linear regression. It is often used as a predictive tool, and it helps to explain the relationship between multiple independent variables ($X_1, X_2 \ldots X_k$) and the tested dependent variable (Y). The model's coefficient of determination, R^2, explains the percentage variation in the dependent variable explained by the model; in other words, it is a measure of model fit.

The computational problem of multiple regression is to fit a straight line to a set of points. The most frequently used method for its implementation is the method of least squares. The method enables adjusting parameters of the regression equation so that the sum of squares of distances of measurement points from the determined line is as small as possible.

The equation of the regression line is in the form:

$$Y = \beta_0 + \beta_1 X_1 + \beta_2 X_2 + \cdots + \beta_k X_k + \varepsilon \quad (2)$$

where Y is the dependent (explained) variable, $[X_1, X_2, \ldots, X_k]$ are the independent (explanatory) variables, $[\beta_0, \beta_1, \beta_2, \ldots, \beta_k]$ are the parameters of the equation, and ε is the random component (model residual).

In this study, three regression models were built. The models were named Rλ, RW, and RE$_c$, according to the labels of the independent characteristics that were analyzed (λ—deformation, W—breaking energy, and E_c—modulus of elasticity). The whole set of available measurement data consisted of 244 records and included six independent traits: variety, storage temperature, duration of storage, and the x, y, and z dimensions were used for the analyses.

2.5. Artificial Neural Networks

Artificial neural networks (ANNs) are a group of tools that are very useful for regression, classification, clustering, and other tasks. The most important advantage of ANNs is the fact that they are trained based on a data set (e.g., experimental data set) and, therefore, don't have to be programmed. This means that no prior knowledge about modelled phenomenon is necessary. In this research, a multilayer perceptron (MLP) was used for nonlinear regression. MLP consists of layers: an input layer, one or more hidden layers, and an output layer. The hidden layers and the output layer are composed of very simple units called artificial neurons. The input layer is composed of nodes that transfer input signals into the structure of ANN. In MLP, signals are forwarded only from the input layer through the hidden layers to the output layer where the output signals of ANN are produced (with no feedback loops). For this research, MLP with one hidden layer was used. The error backpropagation algorithm was employed for MLP training to adjust the connection weights in the network starting from their initial random values. The training process minimizes the error between the target output vector and output signals calculated by the ANN. Some parameters need to be adjusted in the MLP development process, namely the number of neurons in the hidden layers and transfer functions of neurons. During this research, these parameters were adjusted using a trial-and-error method. For each regression model, 5000 ANNs were trained with the use of Statistica v. 13 software. The number of neurons in the hidden layer was changed from 10 to 40. Different activation functions were used, namely sigmoid, hyperbolic tangent, and exponential. The experimental data set of 244 vectors were first normalized and then divided randomly into training, test, and validation sets at a 70:15:15 ratio.

2.6. Sensitivity Analysis

Sensitivity analysis in neural networks is a method that provides information about the relative importance of independent input variables in the model. In this research, the sensitivity analysis implemented in a Statistica v. 13 environment was used to calculate the influence of the input parameters on the output parameter of the ANN model. This method consists of two steps. First, the values of each input variable are replaced by its mean value, calculated based on the training data set. Then an error ratio is calculated. The error ratio is a quotient of the network error with a certain input changed by its mean value, and the network error with the input with the original value is calculated. Based on the error ratios of all input parameters, a percentage influence of independent input variables on the output of an ANN model can be determined. A similar sensitivity analysis method was used by Hadzima-Nyarko et al. [34] to model and analyze the structural damage after an earthquake.

2.7. Support Vector Machines

The support vector machine (SVM) was first proposed by Vapnik [35]. This technique is generally used for classification or nonlinear regression. There are many papers describing details of the underlying concept and the theoretical background of SVM [36,37]. Two types of SVM models can be used for the regression process: ε-type and υ-type support vector regression method. For this research, the ε-SVM regression model was used. The regression function is defined as:

$$y = f(x) = w^T \varphi(x) + b \tag{3}$$

where w is the weights vector and b is the bias, x is an input feature vector, and y is target vector. The objective function (4) is minimized.

$$\frac{1}{2} w^T w + C \sum_{i=1}^{N} \xi_i + C \sum_{i=1}^{N} \xi_i^* \tag{4}$$

subject to the constraints

$$\begin{array}{c} w^T \varphi(x_i) + b - y_i \le \varepsilon + \xi_i \\ y_i - w^T \varphi(x_i) - b \le \varepsilon + \xi_i^* \\ \xi_i,\ \xi_i^* \ge 0,\ i = 1,\ \ldots,\ N \end{array} \tag{5}$$

where C is the capacity constant, ε is the size of the ε-insensitive tube which can be interpreted as the accuracy demanded for the approximation, $\phi(x)$ is the kernel function, and ξ_i, ξ_i^* are slack variables. The kernel function is crucial for the performance of the SVM model. Kernel functions used for SVM regression models are e.g., polynomial, sigmoid, and radial basis function (RBF). RBF kernel was reported as the most appropriate for nonlinear regression [38]. Gaussian radial basis function (6) was used as a kernel function in this study.

$$K(x_i, x) = \exp(-\gamma ||x - x_i||_2) \tag{6}$$

When the ε-SVM regression model with RBF kernel function is used, the three parameters should be adjusted: C, ε, and γ. Proper tuning of these parameters can greatly improve the generalization capacity of the model. The correct value of γ parameter in RBF kernel can avoid under-fitting and over-fitting phenomena in prediction [39]. The ε influences the bias significantly, and its optimal value depends on the type of noise present in the dataset [40,41]. The C parameter affects the number of support vectors, and the proper value of C can minimize the over-fitting problem [42]. In this research, C, ε, and γ were adjusted by the trial-and-error method. The dataset was randomly divided into training and validation sets in a ratio of 3:1. The ten-fold cross-validation method was used. All experiments were performed in Statistica v. 13 software.

2.8. Criteria of Accuracy Assessment of Models

The accuracy of the models developed in this research was evaluated based on two criteria, namely coefficient of correlation (R) and root mean squared error (RMSE) which are calculated as follows:

$$R = \frac{\sum(Y_{meas} - \overline{Y}_{meas})\left(Y_{pred} - \overline{Y}_{pred}\right)}{\sqrt{\sum(Y_{meas} - \overline{Y}_{meas})^2 \sum\left(Y_{pred} - \overline{Y}_{pred}\right)^2}} \tag{7}$$

$$RMSE = \sqrt{\frac{1}{n}\sum_{i=1}^{n}\left(Y_{pred} - Y_{meas}\right)^2} \quad (8)$$

where: Y_{pred} is the absolute predicted value, $\overline{Y}_{pred}$ is the average predicted value, Y_{meas} is the absolute measured (experimental) value, and $\overline{Y}_{meas}$ is the average of measured values.

The better a model is, the closer to 1 the *R*-value is and the closer to 0 the RMSE value is.

3. Results

3.1. Water Content

The mean water content in the fresh fruits of the relevant large cranberry varieties is detailed in Table 2. The results differed slightly and ranged between 89.19 and 90.05%. After 14 days in storage, the water content of the berries decreased on average by 1% compared to fresh fruit. After 25 days of storage the lowest water content was found in the Pilgrim variety (87.66%) and the highest value of the parameter was identified in the Stevens variety (88.12%).

Table 2. The mean water content [%] in fruit of the selected cranberry varieties, relative to duration of storage.

Cranberry Variety	**Duration of Storage**		
	0 Days	**14 Days**	**25 Days**
Ben Lear	89.19 ± 0.76	88.51 ± 0.82	88.06 ± 0.66
Pilgrim	90.05 ± 0.90	88.92 ± 0.79	87.66 ± 0.77
Stevens	89.88 ± 0.67	89.54 ± 0.70	88.12 ± 0.65
Mean	89.71 ± 0.78	88.99 ± 0.77	87.95 ± 0.69

Statistical data are expressed as means ± SD.

3.2. Acidity of Cranberries

The mean acidity of the fresh fruit representing the selected cranberry varieties is presented in detail in Table 3. The mean acidity was in the range of 1.50–1.60 g·100 g^{-1}. After 14 days in storage, the acidity of the berries decreased on average by 0.08 g·100 g^{-1} compared to fresh fruit. After 25 days of storage the lowest acidity was found in the Stevens variety (1.30 g·100 g^{-1}) and the highest value of the parameter was identified in the Pilgrim variety (1.50 g·100 g^{-1}). Compared to fresh fruit, the most significant decrease in acidity following 25 days of storage was observed in the case of the Stevens variety.

Table 3. The mean total acidity (g·100 g^{-1}) of fruit of the selected cranberry varieties, relative to duration of storage.

Cranberry Variety	**Duration of Storage**		
	0 Days	**14 Days**	**25 Days**
Ben Lear	1.60 ± 0.13	1.57 ± 0.11	1.47 ± 0.12
Pilgrim	1.57 ± 0.09	1.44 ± 0.11	1.50 ± 0.10
Stevens	1.50 ± 0.14	1.42 ± 0.10	1.30 ± 0.08
Mean	1.56 ± 0.12	1.48 ± 0.11	1.42 ± 0.10

Statistical data are expressed as means ± SD.

3.3. Mechanical Properties of Cranberry Fruit

Based on the test results (Table 4), it was observed that irrespective of the variety and storage conditions (4 °C, 10 °C, 20 °C), there was a relationship between the selected mechanical parameters of cranberry fruit and duration of storage. The lowest values of the mechanical parameters were identified in the fruit kept in storage for 25 days. Analysis of the specific mechanical parameters showed a significant decrease in their values, which were as follows:

- deformation at the breaking point (λ) 7.83 ± 1.47 mm, after 25 days in storage 4.20 ± 1.54 mm;
- breaking energy (W) 230.74 ± 56.96 mJ, after 25 days in storage 94.96 ± 43.25 mJ;
- apparent elasticity module (E_c) 0.4306 ± 0.07 MPa, after 25 days in storage 0.2715 ± 0.08 MPa.

Table 4. Mean values of the selected mechanical parameters of selected cranberry varieties in a process of uniaxial compression, relative to duration of storage.

Variety	Duration of Storage	λ (mm)	W (mJ)	E_c (MPa)
Ben Lear	0 days	$8.51^{c} \pm 2.48$	$249.02^{c} \pm 92.87$	$0.45^{b} \pm 0.11$
Ben Lear	14 days	$6.67^{b} \pm 1.12$	$154.28^{b} \pm 49.48$	$0.29^{a} \pm 0.07$
Ben Lear	25 days	$4.55^{a} \pm 1.24$	$103.15^{a} \pm 36.94$	$0.25^{a} \pm 0.06$
	Mean	6.58 ± 1.61	168.82 ± 59.76	$0.33^{a} \pm 0.08$
Pilgrim	0 days	$6.49^{b} \pm 0.85$	$199.50^{b} \pm 37.98$	$0.37^{b} \pm 0.05$
Pilgrim	14 days	$4.21^{a} \pm 1.02$	$102.94^{a} \pm 42.77$	$0.32^{a} \pm 0.05$
Pilgrim	25 days	$3.78^{a} \pm 1.75$	$84.06^{a} \pm 33.32$	$0.30^{a} \pm 0.11$
	Mean	4.83 ± 1.21	128.83 ± 38.02	0.33 ± 0.07
Stevens	0 days	$8.50^{b} \pm 1.07$	$243.69^{b} \pm 40.03$	$0.46^{c} \pm 0.06$
Stevens	14 days	$7.50^{b} \pm 0.90$	$234.51^{b} \pm 32.17$	$0.36^{b} \pm 0.04$
Stevens	25 days	$4.26^{a} \pm 1.63$	$97.66^{a} \pm 59.50$	$0.25^{a} \pm 0.08$
	Mean	6.75 ± 1.2	191.95 ± 43.9	0.36 ± 0.06
All groups	0 days	$7.83^{c} \pm 1.47$	$230.74^{c} \pm 56.96$	$0.43^{c} \pm 0.07$
	14 days	$6.13^{b} \pm 1.01$	$163.91^{b} \pm 41.47$	$0.32^{b} \pm 0.05$
	25 days	$4.20^{a} \pm 1.54$	$94.96^{a} \pm 43.25$	$0.27^{a} \pm 0.08$
	Mean	6.05 ± 1.34	163.20 ± 47.27	0.34 ± 0.07

Statistical data are expressed as means ± SD. Means in a column followed by different letters show significant differences (α = 0.05) according to the LSD test.

3.4. The Results of Multiple Linear Regression

The developed Rλ, RW, and RE_c regression models were based on six independent variables (variety, storage temperature, duration of storage, x, y and z dimensions). Detailed results of the multiple regression analysis for the presented independent variables and the dependent variables are presented in Table 5.

Table 5. Regression coefficients, standard errors, and probability levels for the developed regression models.

Factor	Rλ: R = 0.578 RMSE = 2.934 Free Term = 17.272				RW: R = 0.579 RMSE = 56.027 Free Term = 508.691				RE_c: R = 0.475 RMSE = 0.138 Free Term = 0.810			
	b	Standard Error b	*p*	Significance	b	Standard Error b	*p*	Significance	b	Standard Error b	*p*	Significance
Variety (Vr)	0.299	0.139	0.033	+	13.223	4.857	0.007	+	0.007	0.012	0.551	-
storage temperature (ST)	0.013	0.021	0.522	-	−0.13	0.721	0.857	-	−0.001	0.002	0.627	-
duration of storage (DS)	0.002	0.023	0.932	-	−1.809	0.805	0.026	+	0.001	0.002	0.681	-
y dimension	−0.131	0.044	0.003	+	−7.273	1.531	0	+	−0.01	0.004	0.009	+
x dimension	−0.586	0.072	0	+	−16.374	2.514	0	+	0	0.006	0.97	-
z dimension	0.041	0.038	0.278	-	3.736	1.325	0.005	+	−0.013	0.003	0	+

Determination of the level of statistical significance: - non-significant. + significant for α = 0.05.

The variable for which statistical significance was not confirmed at the α = 0.05 level in all models was storage temperature. In the Rλ model, the statistically significant traits were variety, y and x. Whereas in the RW model, the statistically significant traits were: variety, storage time, and the y, x, and z dimensions. The situation was different in the RE_c model, where only two variables were statistically significant, namely the y and z dimensions.

Based on the results from Table 5, considering only statistically significant traits, multiple regression equations were constructed for each model, which took the form:

$$R\lambda = 17.272 + 0.299 \cdot Vr - 0.131 \cdot y - 0.586 \cdot z \quad (9)$$

$$RW = 508.691 + 13.223 \cdot Vr - 1.809 \cdot ST - 7.273 \cdot y - 16.374 \cdot x + 3.736 \cdot z \quad (10)$$

$$RE_c = 0.810 - 0.01 \cdot y - 0.013 \cdot z \quad (11)$$

3.5. *Artificial Neural Networks*

For each output parameter (λ, W and E_c) a separate neural model was developed (NNλ, NNW, NNE_c). In Table 6, the structure and quality metrics of the best neural models are presented. Model structure means the number of neurons in each layer of MLP: input—hidden—output. The number of nodes in the input layer equals the number of input parameters of model, which is six. The number of neurons in the hidden layer was selected by trial-and-error method. In the output layer there is one neuron calculating the value of the output value of neural model.

Table 6. Structures and error metrics of best neural models.

Model	Model Structure	Train		Validation	
		RMSE	*R*	RMSE	*R*
NNλ	6-48-1	1.392	0.685	1.176	0.677
NNW	6-10-1	41.048	0.805	53.281	0.752
NNE_c	6-47-1	0.065	0.890	0.067	0.878

For deformation, the best architecture of the model is 48 neurons in the hidden layer. This model is of rather low accuracy with an R-value of 0.69 for the train data set and 0.68 for the validation data set. Better performance was obtained in the case of NNW model. The best architecture of ANN is network containing 10 neurons in the hidden layer with R value of 0.80 for train data set and 0.74 for the validation data set. For the NNE_c the best model was achieved for architecture with 47 neurons in the hidden layer. The accuracy of this model can be stated as satisfactory with a relatively high R-value (0.89 for the train data set and 0.88 for the validation data set).

3.6. *Sensitivity Analysis*

The best MLP models described in Table 6 were used for sensitivity analysis. For better readability, the results are presented as a percentage influence of certain input variables on the output parameter.

As presented in Figure 1, in the case of deformation, a cranberry variety influences this parameter the most (23.04%). The geometrical dimensions, duration of storage, and storage temperature similarly affect deformation (from 14.21% to 16.97%). The parameter influencing breaking energy the most was duration of storage (45.64%). A significantly lower impact was observed for the variety and x dimensions. The influence of the y and z dimensions and storage temperature was minimal. The modulus of elasticity was affected the most by storage temperature (41.45%). Lower influence was noticed for the duration of storage (19.25%) and variety (14.96%). The impact of geometrical dimensions on E_c was very low.

3.7. *Support Vector Machines*

The three separate SVM models (SVMλ, SVMW and $SVME_c$) were developed with the same six input parameters (variety, storage temperature, duration of storage, and the x, y, and z dimensions) and different output parameters (deformation λ, breaking energy *W*, or modulus of elasticity E_c). Model parameters (*C*, ε and γ) were adjusted by trial-and-error approach. For all three models these parameters were as follows: *C* = 10, ε = 0.03, and γ = 0.24. Error metrics of models of the best accuracy are detailed in Table 7.

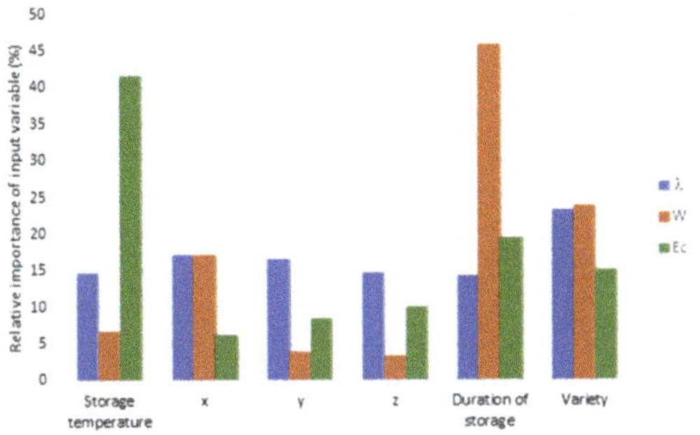

Figure 1. The relative importance of input variables of MLP model on deformation λ, breaking energy W, and modulus of elasticity E_c.

Table 7. Error metrics of best SVM models.

Model	Train		Validation	
	RMSE	*R*	RMSE	*R*
SVMλ	1.581	0.595	1.451	0.705
SVMW	51.905	0.626	53.869	0.758
SVME_c	0.113	0.702	0.122	0.665

The best accuracy was obtained for the SVMW model (R = 0.76 for validation data set). A lower correlation between experimental results and model prediction was observed for the SVMλ model (0.71). The SVME_c model was low accuracy with R-value of 0.67.

In this research, the three methods of modelling were used to develop models of relationships between large cranberry variety, storage temperature, duration of storage, x, y, z dimensions of fruits, and mechanical parameters, namely deformation, breaking energy, and modulus of elasticity. The regression method produced models of very low accuracy (R = 0.578 for deformation, R = 0.579 for breaking energy, and R = 0.475 for modulus of elasticity). The better models were developed with the use of artificial intelligence techniques. The best model for deformation and breaking energy was produced with the use of the SVM method. The error metrics for the validation data set calculated for these models are significantly better than for regression models (R = 0.705, RMSE = 1.451, and R = 0.758, RMSE = 53.869, respectively). Neural networks produced slightly worse models for these mechanical parameters. In the case of the modulus of elasticity, the ANN model was found to be the most accurate model (R = 0.878, RMSE = 0.067 for validation data set). In Figures 2–4, the performance of the best models of deformation λ, breaking energy W, and modulus of elasticity Ec for the validation data set is presented.

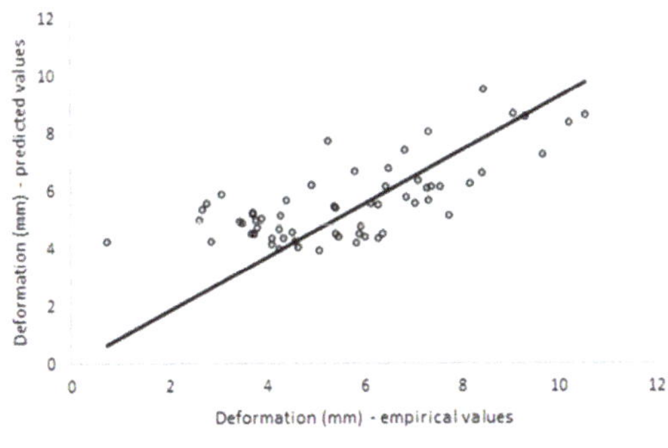

Figure 2. Predicted values versus measured values of deformation for validation data set (SVMλ model).

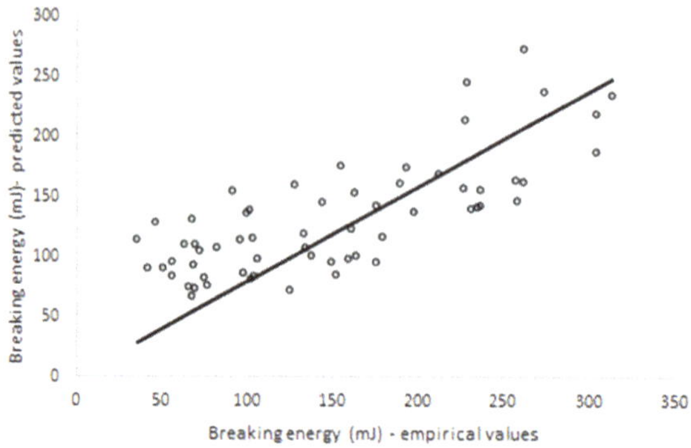

Figure 3. Predicted values versus measured values of breaking energy for validation data set (SVMW model).

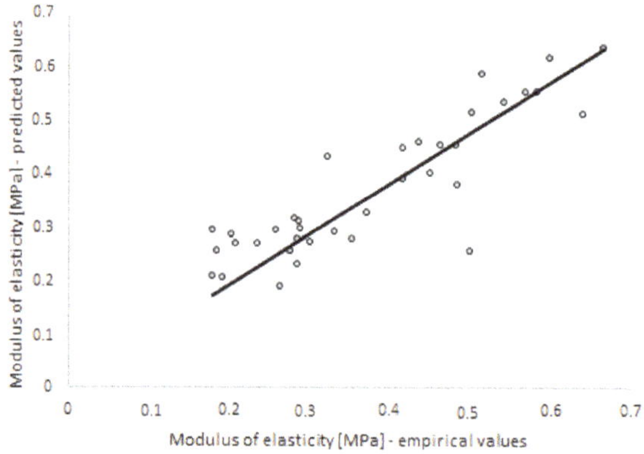

Figure 4. Predicted values versus measured values of modulus of elasticity for validation data set (NNE_c model).

4. Discussion

Compared to fresh fruit, the most significant decrease in water content after 25 days of storage was observed in the Pilgrim variety. The fresh cranberries examined in a study by Oszmiański et al. (2017) [42] were found with varied water content, ranging from 87.21% in the Pilgrim variety and 87.52% in the Stevens variety to 89.94% in the Ben Lear variety. Similar water content in the fresh cranberries of the investigated varieties (87.0–87.5%) was reported by Oszmiański et al. (2018) [43]. Paniagua et al. [12] found a decrease in water content in blueberry fruit by 1.34% after three weeks of storage. Ruse et al. [13] reported a decrease in moisture content by 2% in cranberries stored in closed PP boxes in air ambiance for six months.

Compared to this study, a study by Teleszko [44] reported slightly higher acidity in the fresh cranberries of the Ben Lear variety amounting to 2.18 $g \cdot 100\ g^{-1}$, whereas Oszmiański et al. [42] found that total acidity in the fresh cranberries ranged from 1.95 $g \cdot 100\ g^{-1}$ in Pilgrim and 2.25 $g \cdot 100\ g^{-1}$ in the Stevens variety to 2.29 $g \cdot 100\ g^{-1}$ in the Ben Lear variety. In another study, Oszmiański et al. [43] reported slightly higher acidity in the investigated cranberry varieties, with values in the range of 2.1–2.4 $g \cdot 100\ g^{-1}$.

Modelling and the development of regression models are very common and important in many scientific fields. Model accuracy is crucial in real-life applications, and experimental data is used for model development. Therefore, besides traditional techniques, artificial intelligence algorithms are often used for modeling. In this research, ANN and SVM provided much better results for the estimation of the mechanical parameters of large cranberry when compared to the MLR technique. Similar results were reported in state of art literature, where these techniques were used to develop regression models for various relationships. Karsavran and Erdik [45] developed sea-level prediction models and revealed that the ANN and SVM models outperformed MLR. The best performance resulted from ANN model with a coefficient of correlation $R = 0.76$. The same techniques were used by Mohammed et al. [46] to estimate time and cost indexes to predict the site overhead cost. They reported that the ANN and SVM techniques produced more accurate models than the MLR technique. A slightly higher accuracy of the ANN model ($R = 0.99$) when compared to SVM ($R = 0.97$) was reported by Afradi and Ebrahimabadi [47] who used AI methods to predict the penetration rate of tunnel boring machine. Sabzi-Nojadeh et al. [48] compared the accuracy of ANN and MLR models used to predict the oil yield and trans-anethole yield of fennel populations; ANN performed better ($R = 0.96$ and $R = 0.88$) than MLR ($R = 0.74$ and $R = 0.68$).

5. Conclusions

Knowledge of the mechanical parameters of fruits is crucial to optimize the storage process. Especially in the case of delicate fruit such as large cranberry. Therefore, the mechanical parameters of cranberry fruits in relation to storage conditions were investigated, and mathematical models of relationships under study were developed. The results of this study revealed that the water content in the fresh fruit of the relevant large cranberry varieties ranged between 89.19 and 90.05%. After 25 days of storage, the lowest water content was found in the Pilgrim variety (87.66%), and the highest value of this parameter was identified in the Stevens variety (88.12%). The mean acidity of fresh fruit representing the selected cranberry varieties was in the range of 1.50–1.60 $g \cdot 100\ g^{-1}$. After 25 days of storage, the lowest acidity was found in the Stevens variety (1.30 $g \cdot 100\ g^{-1}$), and the highest value of this parameter was identified in the Pilgrim variety (1.50 $g \cdot 100\ g^{-1}$). It was observed that irrespective of the variety and storage conditions (4 °C, 10 °C, 20 °C), there was a relationship between the selected mechanical parameters of cranberry fruit and the duration of storage. The lowest values of the mechanical parameters were identified in fruit kept in storage for 25 days. The ANN and SVM prediction models of relationships under study outperformed MLR models. The accuracy of the ANN and SVM models was comparable. In the case of deformation and breaking energy, the best performance was observed for the SVM model ($R = 0.705$ and $R = 0.758$, respectively). ANN produced the best model for modulus of elasticity ($R = 0.878$).

Author Contributions: Conceptualization, K.P. and J.G.; data curation, J.G., J.B. and P.K.; formal analysis, K.P., G.N. and J.G.; investigation, K.P., J.G., J.B., P.K. and G.N.; methodology, K.P., J.G. and G.N.; writing—original draft, K.P., J.G. and G.N.; writing—review and editing, J.G. and G.N.; visualization, K.P. All authors have read and agreed to the published version of the manuscript.

Funding: Wrocław University of Environmental and Life Sciences.

Institutional Review Board Statement: Not applicable.

Informed Consent Statement: Not applicable.

Data Availability Statement: Data are available by contacting the authors.

Conflicts of Interest: The authors declare no conflict of interest.

References

1. Krzewińska, D.; Smolarz, K. Wpływ nawożenia azotem na wzrost i plonowanie żurawiny wielkoowocowej (Vaccinium macrocarpon AIT). *Zesz. Nauk. Inst. Sadow. Kwiaciarstwa* **2008**, *16*, 135–144.
2. FAOSTAT. Available online: https://www.fao.org/faostat/en/#data/QV (accessed on 1 December 2021).
3. Crescent Bloom. Hierarchical Position of the Genus Oxycoccus. In *The Compleat Botanica*; Crescent Bloom: Sebastopol, CA, USA, 2009.
4. Allison, D.G.; Cronin, M.A.; Hawker, J.; Freeman, S. Influence of cranberry juice on attachment of *Escherichia coli* to glass. *J. Basic Microb.* **2000**, *40*, 3–6. [CrossRef]

5. Howell, A.B.; Vorsa, N.; Marderosian, A.D.; Foo, L.Y. Inhibition of the adherence of P-fimbriated *Escherichia coli* to uroepithelial-cell surfaces by proanthocyanidin extracts from cranberries. *New Engl. J. Med.* **1998**, *339*, 1085–1086. [CrossRef]
6. Oszmianski, J.; Kolniak-Ostek, J.; Lachowicz, S.; Gorzelany, J.; Matlok, N. Effect of dried powder preparation process on polyphenolic content and antioxidant capacity of cranberry (*Vaccinium macrocarpon* L.). *Ind. Crop. Prod.* **2015**, *77*, 658–665. [CrossRef]
7. Bonarska-Kujawa, D.; Cyboran, S.; Kleszczyńska, H.; Oszmiański, J. Aktywność przeciwutleniająca ekstraktów polifenolowych z owoców czerwonej porzeczki i żurawiny w odniesieniu do błony erytrocytów. *Żywność Nauka Technol. Jakość* **2015**, *3*, 148–159.
8. Averill, A.; Caruso, F.; DeMoranville, C.; Jeranyama, P.; LaFleur, J.; McKenzie, K.; Rinta, L.; Sandler, H.; Wick, B. *Cranberry Production Guide*; Paper 8; University of Massachusetts: Amherst, MA, USA, 2008.
9. Girard, K.K.; Sinha, N.K. Cranberry, blueberry, currant, and gooseberry. In *Handbook of Fruits and Fruit Processing*; John Wiley & Sons: Hoboken, NJ, USA, 2006; pp. 399–417.
10. Sandler, H.A.; DeMoranville, C.J. *Cranberry Production: A Guide for Massachusetts*; CP-08; University of Massachusetts: Amherst, MA, USA, 2008.
11. Rabcewicz, J. Perspectives of mechanical harvesting of stone and berry fruit. *Informator Sadowniczy*, 5 May 2013. (In Polish)
12. Paniagua, A.C.; East, A.R.; Hindmarsh, J.P.; Heyes, J.A. Moisture loss is the major cause of firmness change during postharvest storage of blueberry. *Postharvest Biol. Technol.* **2013**, *79*, 13–19. [CrossRef]
13. Ruse, K.; Rakcejeva, T.; Dukalska, L. Changes in Physically-Chemical Parameters of Latvian Cranberries During Storage. *Proc. Latv. Univ. Agric.* **2013**, *30*, 11–19. [CrossRef]
14. Gorzelany, J.; Kotlicka, M.; Migut, D.; Witek, G.; Matłok, N.; Lachowicz, S. Ocena właściwości chemicznych i mechanicznych świeżych owoców wybranych odmian żurawiny wielkoowocowej. In *Rolnictwo XXI Wieku—Problemy i Wyzwania*; Idea Knowledge Future: Wroclaw, Poland, 2018; pp. 111–122.
15. Bohdziewicz, J.; Czachor, G. The impact of load on deformation progress for ball-shaped vegetables. *Agric. Eng.* **2010**, *1*, 85–91.
16. Bohdziewicz, J.; Czachor, G. The Rheological Properties of Redcurrant and Highbush Blueberry Berries. *Agric. Eng.* **2016**, *2*, 15–22. [CrossRef]
17. Migut, D.; Gorzelany, J.; Matłok, N.; Kotlicka, M.; Kuźniar, P. Ocena wybranych właściwości mechanicznych owoców żurawiny wielkoowocowej. In *Rolnictwo XXI Wieku—Problemy i Wyzwania*; Idea Knowledge Future: Wroclaw, Poland, 2016; pp. 200–211.
18. Piekutowska, M.; Niedbala, G.; Piskier, T.; Lenartowicz, T.; Pilarski, K.; Wojciechowski, T.; Pilarska, A.A.; Czechowska-Kosacka, A. The Application of Multiple Linear Regression and Artificial Neural Network Models for Yield Prediction of Very Early Potato Cultivars before Harvest. *Agronomy* **2021**, *11*, 885. [CrossRef]
19. Hara, P.; Piekutowska, M.; Niedbala, G. Selection of Independent Variables for Crop Yield Prediction Using Artificial Neural Network Models with Remote Sensing Data. *Land* **2021**, *10*, 609. [CrossRef]
20. Niedbala, G.; Piekutowska, M.; Weres, J.; Korzeniewicz, R.; Witaszek, K.; Adamski, M.; Pilarski, K.; Czechowska-Kosacka, A.; Krysztofiak-Kaniewska, A. Application of Artificial Neural Networks for Yield Modeling of Winter Rapeseed Based on Combined Quantitative and Qualitative Data. *Agronomy* **2019**, *9*, 781. [CrossRef]
21. Wojciechowski, T.; Niedbała, G.; Czechlowski, M.; Nawrocka, J.R.; Piechnik, L.; Niemann, J. Rapeseed seeds quality classification with usage of VIS-NIR fiber optic probe and artificial neural networks. In Proceedings of the 2016 International Conference on Optoelectronics and Image Processing (ICOIP), Warsaw, Poland, 10–12 June 2016; pp. 44–48. [CrossRef]
22. Kujawa, S.; Dach, J.; Kozłowski, R.J.; Przybył, K.; Niedbała, G.; Mueller, W.; Tomczak, R.J.; Zaborowicz, M.; Koszela, K. Maturity classification for sewage sludge composted with rapeseed straw using neural image analysis. In Proceedings of the Eighth International Conference on Digital Image Processing (ICDIP 2016), Chengu, China, 20–22 May 2016. [CrossRef]
23. Moradi, M.; Balanian, H.; Taherian, A.; Khaneghah, A.M. Physical and mechanical properties of three varieties of cucumber: A mathematical modeling. *J. Food Process Eng.* **2020**, *43*, e13323. [CrossRef]
24. Szwedziak, K.; Polanczyk, E.; Grzywacz, Z.; Niedbala, G.; Wojtkiewicz, W. Neural Modeling of the Distribution of Protein, Water and Gluten in Wheat Grains during Storage. *Sustainability* **2020**, *12*, 5050. [CrossRef]
25. Xia, A.L.; Zhang, Y.; Zhao, L.Z.; Qin, P. Simultaneous, Rapid and Nondestructive Determination of Moisture, Fat Content and Storage Time in Leisure Dried Tofu Using LF-NMR. *Anal. Sci.* **2021**, *37*, 301–307. [CrossRef] [PubMed]
26. Chen, L.; Ding, J.F. Analysis on Food Crispness Based on Time and Frequency Domain Features of Acoustic Signal. *Trait. Signal* **2021**, *38*, 231–238. [CrossRef]
27. Gonzalez-Viveros, N.; Gomez-Gil, P.; Castro-Ramos, J.; Cerecedo-Nunez, H.H. On the estimation of sugars concentrations using Raman spectroscopy and artificial neural networks. *Food Chem.* **2021**, *352*, 129375. [CrossRef]
28. Chauchard, F.; Cogdill, R.; Roussel, S.; Roger, J.M.; Bellon-Maurel, V. Application of LS-SVM to non-linear phenomena in NIR spectroscopy: Development of a robust and portable sensor for acidity prediction in grapes. *Chemom. Intell Lab.* **2004**, *71*, 141–150. [CrossRef]
29. Dai, Q.; Cheng, J.H.; Sun, D.W.; Zeng, X.A. Potential of hyperspectral imaging for non-invasive determination of mechanical properties of prawn (*Metapenaeus ensis*). *J. Food Eng.* **2014**, *136*, 64–72. [CrossRef]
30. Guo, W.C.; Shang, L.; Zhu, X.H.; Nelson, S.O. Nondestructive Detection of Soluble Solids Content of Apples from Dielectric Spectra with ANN and Chemometric Methods. *Food Bioprocess Technol.* **2015**, *8*, 1126–1138. [CrossRef]
31. Meteomodel. Available online: https://meteomodel.pl/dane/srednie-miesieczne/?imgwid=350210585&par=tm&max_empty=2 (accessed on 5 December 2021).

32. *PN-90/A-75101-03*; Fruit and Vegetable Products—Preparation of Samples and Testing Methods—Determination of Dry Matter Content By Gravimetric Method. Polish Committee for Standardization: Warsaw, Poland, 1990. Available online: http://www.ydylstandards.org.cn/static/down/pdf/PN%20A75101-03-1990_3750.pdf(accessed on 22 January 2022).
33. *PN-90/A-75101-04*; Fruit and Vegetable Products—Preparation of Samples and Testing Methods—Determination of Total Acidity. Polish Committee for Standardization: Warsaw, Poland, 1990. Available online: http://www.ydylstandards.org.cn/static/down/pdf/PN%20A75101-04-1990_5000.pdf(accessed on 22 January 2022).
34. Hadzima-Nyarko, M.; Nyarko, E.K.; Moric, D. A neural network based modelling and sensitivity analysis of damage ratio coefficient. *Expert Syst. Appl.* **2011**, *38*, 13405–13413. [CrossRef]
35. Vapnik, V. *Nature of Statistical Learning Theory*; Springer: New York, NY, USA, 1995.
36. Yu, P.S.; Chen, S.T.; Chang, I.F. Support vector regression for real-time flood stage forecasting. *J. Hydrol.* **2006**, *328*, 704–716. [CrossRef]
37. Noori, R.; Yeh, H.D.; Abbasi, M.; Kachoosangi, F.T.; Moazami, S. Uncertainty analysis of support vector machine for online prediction of five-day biochemical oxygen demand. *J. Hydrol.* **2015**, *527*, 833–843. [CrossRef]
38. Dibike, Y.B.; Velickov, S.; Solomatine, D.; Abbott, M.B. Model induction with support vector machines: Introduction and applications. *J. Comput. Civil Eng.* **2001**, *15*, 208–216. [CrossRef]
39. Han, D.; Chan, L.; Zhu, N. Flood forecasting using support vector machines. *J. Hydroinform.* **2007**, *9*, 267–276. [CrossRef]
40. Zeng, J.Y.; Tan, Z.H.; Matsunaga, T.; Shirai, T. Generalization of Parameter Selection of SVM and LS-SVM for Regression. *Mach. Learn. Knowl. Extr.* **2019**, *1*, 43. [CrossRef]
41. Akbarzadeh, A.; Naseh, M.R.V.; NodeFarahani, M. Carbon Monoxide Prediction in the Atmosphere of Tehran Using Developed Support Vector Machine. *Pollution* **2020**, *6*, 43–57. [CrossRef]
42. Oszmianski, J.; Kolniak-Ostek, J.; Lachowicz, S.; Gorzelany, J.; Matlok, N. Phytochemical Compounds and Antioxidant Activity in Different Cultivars of Cranberry (*Vaccinium Macrocarpon* L). *J. Food Sci.* **2017**, *82*, 2569–2575. [CrossRef]
43. Oszmianski, J.; Lachowicz, S.; Gorzelany, J.; Matlok, N. The effect of different maturity stages on phytochemical composition and antioxidant capacity of cranberry cultivars. *Eur. Food Res. Technol.* **2018**, *244*, 705–719. [CrossRef]
44. Teleszko, M. Żurawina wielkoowocowa- możliwości wykorzystania do produkcji biożywności. *Żywność Nauka Technol. Jakość* **2011**, *6*, 132–144.
45. Karsavran, Y.; Erdik, T. Artificial Intelligence Based Prediction of Seawater Level: A Case Study for Bosphorus Strait. *Int. J. Math. Eng. Manag. Sci.* **2021**, *6*, 1242–1254. [CrossRef]
46. Mohammed, S.J.; Abdel-Khalek, H.A.; Hafez, S.M. Predicting Performance Measurement of Residential Buildings Using Machine Intelligence Techniques (MLR, ANN and SVM). *Iran. J. Sci. Technol.-Trans. Civ. Eng* **2021**. [CrossRef]
47. Afradi, A.; Ebrahimabadi, A. Comparison of artificial neural networks (ANN), support vector machine (SVM) and gene expression programming (GEP) approaches for predicting TBM penetration rate. *SN Appl. Sci.* **2020**, *2*, 2004. [CrossRef]
48. Sabzi-Nojadeh, M.; Niedbała, G.; Younessi-Hamzekhanlu, M.; Aharizad, S.; Esmaeilpour, M.; Abdipour, M.; Kujawa, S.; Niazian, M. Modeling the Essential Oil and Trans-Anethole Yield of Fennel (*Foeniculum vulgare* Mill. var. vulgare) by Application Artificial Neural Network and Multiple Linear Regression Methods. *Agriculture* **2021**, *11*, 1191. [CrossRef]

agriculture

Article

LA-DeepLab V3+: A Novel Counting Network for Pigs

Chengqi Liu [1], Jie Su [1], Longhe Wang [2], Shuhan Lu [3] and Lin Li [1,2,*]

1 Department of Computer Science and Technology, College of Information and Electrical Engineering, China Agricultural University, Beijing 100083, China; cieelcq@cau.edu.cn (C.L.); sujiework@cau.edu.cn (J.S.)
2 Office of Model Animals, National Research Facility for Phenotypic and Genotypic Analysis of Model Animals, China Agricultural University, Beijing 100083, China; phil.wang@cau.edu.cn
3 Department of Information, School of Information, University of Michigan, Ann Arbor, MI 48109, USA; shuhanlu@umich.edu
* Correspondence: lilinlsl@cau.edu.cn

Abstract: Accurate identification and intelligent counting of pig herds can effectively improve the level of fine management of pig farms. A semantic segmentation and counting network was proposed in this study to improve the segmentation accuracy and counting efficiency of pigs in complex image segmentation. In this study, we built our own datasets of pigs under different scenarios, and set three levels of number detection difficulty—namely, lightweight, middleweight, and heavyweight. First, an image segmentation model of a small sample of pigs was established based on the DeepLab V3+ deep learning method to reduce the training cost and obtain initial features. Second, a lightweight attention mechanism was introduced, and attention modules based on rows and columns can accelerate the efficiency of feature calculation and reduce the problem of excessive parameters and feature redundancy caused by network depth. Third, a recursive cascade method was used to optimize the fusion of high- and low-frequency features for mining potential semantic information. Finally, the improved model was integrated to build a graphical platform for the accurate counting of pigs. Compared with FCNNs, U-Net, SegNet, and DenseNet methods, the DeepLab V3+ experimental results show that the values of the comprehensive evaluation indices P, R, AP, F_1-score, and MIoU of LA-DeepLab V3+ (single tag) are higher than those of other semantic segmentation models, at 86.04%, 75.06%, 78.67%, 0.8, and 76.31%, respectively. The P, AP, and MIoU values of LA-DeepLab V3+ (multiple tags) are also higher than those of other models, at 88.36%, 76.75%, and 74.62%, respectively. The segmentation accuracy of pig images with simple backgrounds reaches 99%. The pressure test of the counting network can calculate the number of pigs with a maximum of 50, which meets the requirements of free-range breeding in standard piggeries. The model has strong generalization ability in pig herd detection under different scenarios, which can serve as a reference for intelligent pig farm management and animal life research.

Keywords: complex background; pigs; DeepLab V3+; attention mechanism; count

Citation: Liu, C.; Su, J.; Wang, L.; Lu, S.; Li, L. LA-DeepLab V3+: A Novel Counting Network for Pigs. *Agriculture* **2022**, *12*, 284. https://doi.org/10.3390/agriculture12020284

Academic Editors: Gniewko Niedbała and Sebastian Kujawa

Received: 19 January 2022
Accepted: 15 February 2022
Published: 17 February 2022

1. Introduction

Group free-range breeding will be the mainstream breeding method of pig farms in the future, and the increase in the number of pigs will lead to an increase in manual inspection [1]. The achievement of automatic pig identification, trajectory tracking, and quantity statistics by using computer vision technology has become a current research hotspot [2]. In this field, foreground segmentation of pig herd images and separation of adhesive individual images are the basis for achieving automatic inventory of pig numbers [3]. Owing to the complexity of pig images, such as light changes, crowding, stacking, and occlusion, the existing semantic segmentation technology still faces problems, such as missing segmentation and mis-segmentation, which result in inaccurate counting [4,5]. Therefore, enhancing the characterization ability of high- and low-frequency detail information of images, along with improving the utilization rate of individual characteristic information of pigs, are the key research directions to improve the semantic segmentation accuracy of pig images [6].

Fully convolutional neural networks (FCNNs) [7] have promoted the rapid development of semantic segmentation algorithms. Deformable convolution [8] enhances the adaptability of the model to scale transformation by adding direction vectors to each parameter of the convolution kernel and adaptively adjusting the scale and receptive field [9]. U-Net [10] introduces different scale features in the coding layer through the jump connection structure to recover the lost information and achieve accurate positioning of pixels. SegNet [11], based on a codec structure, saves the pooled index in the coding stage, accurately recovers image size and spatial information, and effectively retains the integrity of high-frequency details. Compared with U-Net, SegNet cannot capture multi-scale information effectively, while U-Net has many learning parameters, so its training is relatively slow. PSPNet [12], based on a multiscale feature aggregation structure, uses a pyramid pooling module to capture feature information of different regions for alleviating the problem of multiple scale changes. DeepLab series models [13–15] combine DCNNs and dense CRFs to achieve better detection accuracy. Google introduced deep detachable convolution in atrous spatial pyramid pooling (ASPP) [13] and a decoder [16] to reduce the computational complexity of the network, in order to achieve a better DeepLab V3+ network [14]. However, the internal parallel structure of ASPP in DeepLab V3+ models makes the branch information independent, and lacks spatial correlation. The decoding phase only fuses one of the multistage shallow features on the backbone network, resulting in partial loss of effective information, segmentation discontinuity, and rough segmentation boundaries. In feature fusion, high-level feature output is directly fused with shallow features in the backbone network, thus ignoring the noise problem introduced into the semantic feature graph due to the misalignment of high–low features, and damaging the semantic segmentation accuracy [17].

The attention mechanism module has been widely applied in image classification, target detection, and tracking tasks [18–20]; it has recurrent models of visual attention [21] and residual attention networks for image classification [22], all of which use the attention mechanism to generate high-level feature maps to guide the forward propagation of the network. Squeeze-and-excitation networks (SENet) [23] compress the feature graph channels into a single point to obtain the category properties between the channels. Finally, the gate mechanism fuses the channel relationship into the original feature graph to obtain the final feature graph. EncNet [24] and DFNs [25] use the channel attention mechanism to obtain global context information of the image and construct dependencies between categories. Subakan et al. [26] first proposed the self-attention mechanism and acquired the global dependence of input information, which was eventually applied to the field of machine translation. In addition, self-attention generative adversarial networks [27] introduce the self-attention mechanism module to provide a better image generator for generating better images. DANet [28] adopts self-attention and channel attention mechanisms to establish long-term context-dependent relationships in the spatial and channel dimensions, respectively. At present, few attention mechanisms are applied to complex images of pigs—in particular, the extraction of rich detail information and small target information in complex images of pigs needs to be improved.

On the basis of the abovementioned existing methods, this study proposes a semantic segmentation network with a light attention mechanism. The main contributions of this work are as follows:

(1) The current DeepLab V3+ semantic segmentation framework has too many network layers and slow speed. Thus, by embedding the attention module based on rows and columns into the backbone network, we can achieve the lightweight and fast network computing efficiency that the traditional semantic segmentation algorithm and other attention modules do not have;

(2) In view of the problem of detail information loss in semantic segmentation algorithms, a recursive cascade mechanism is introduced to supplement the detail information

of the unit input feature graph to the output feature graph. This approach better integrates the high-level semantic information into the low-level high-resolution feature graph, improving the segmentation accuracy;

(3) This study integrates deep learning models and attention mechanisms, and it preliminarily achieves the application of complex image inventory of pigs.

2. Materials and Methods

2.1. Self-Built Datasets

The basic dataset was collected from large-scale breeding farms in Nanyang city, Henan Province. The collection period was from June to December 2019. The collection device was a Hikvision Smart Ball Camera (DS-2DE4320IW-DEDS-2DE4320IW-D, made in Hangzhou, China), which had 3 million pixels and 20 × optical zoom; its infrared radiation at night could reach 100 m. As shown in Figure 1, the pigs were 80-day-weaned Yorkshire piglets. The basic dataset was characterized by single-pig and multi-pig (5–7) scenarios. The collection environment included different weather conditions—such as cloudy day, sunny day, and rainy day—to test the robustness of the algorithm.

(**a**) (**b**) (**c**)

Figure 1. Video capture scene: (**a**) piggery during the day; (**b**) another piggery at night; (**c**) camera position.

The extended dataset included different species and orders of magnitude in all weather, captured via the Internet. The self-built dataset consisted of 5000 images with a normalized resolution of 512 × 256 pixels. In this study, 4500 typical images were selected for annotation, among which 3500 were selected as training set images and 1000 as verification set images. The remaining 500 unlabeled images were used as the test set. As shown in Figure 2, the extended dataset was characterized by a typical complex environment, including human beings, weeds, trees, and light, as well as adhesion and occlusion of the images themselves. In addition, different orders of magnitude of pig groups were set to verify the segmentation and counting ability of the model. This approach increased the generalization and robustness of the segmentation model.

2.2. Experimental Design

In this work, the software environment was the Windows 10 operating system. Python was the programming language, and TensorFlow was the open-source framework for deep learning. The hardware environment was an NVIDIA RTX3060 16 G graphics card, Intel(R) Core(TM) I7-11800H CPU, and 16 GiB DDR.

To better evaluate the model correctly, this study adopted the evaluation indices commonly used in semantic segmentation: precision (P, %), recall (R, %), average precision (AP, %), F_1 score (F_1), mean intersection over union (MIoU, %), and algorithm running efficiency in frames per second (fps). Calculation of each evaluation index is shown in Equations (3)–(5):

$$P = \frac{TP}{TP + FP} \times 100\% \tag{1}$$

$$R = \frac{TP}{TP + FN} \times 100\% \tag{2}$$

$$\mathrm{AP} = \int_0^1 \mathrm{P(R)}d\mathrm{R} \times 100\% \tag{3}$$

$$\mathrm{F}_1 - \mathrm{score} = 2\frac{\mathrm{P \cdot R}}{\mathrm{P + R}} \times 100\% \tag{4}$$

$$\mathrm{MIoU} = \frac{1}{k+1}\sum_{i=0}^{k}\frac{\mathrm{TP}}{\mathrm{TP + FP + FN}} \tag{5}$$

where TP represents the positive sample for which the model prediction is positive, FP represents the positive sample for which the model prediction is negative, FN represents the positive sample for which the model prediction is negative, TN represents the negative sample for which the model prediction is negative, and AP is the integration of precision in recall. The model performance is better when the AP value is higher. F_1 score is the harmonic average of precision and recall, and its value range is (0,1). MIoU is the most direct evaluation index in image segmentation; it is the average union ratio of two sets of real value and predicted value, and $k + 1$ is the number of categories (including empty classes).

Figure 2. Pig images in different environments: (**a**) overlap of pigs; (**b**) object occlusion; (**c**) adhesion of pigs; (**d**) light; (**e**) nature; (**f**) other interfering factors; (**g**) lightweight herd (1–10 pigs); (**h**) middleweight herd (10–20 pigs); (**i**) heavyweight herd (over 20 pigs).

2.3. *Improved Light Attention DeepLab V3+ Method*

2.3.1. Original DeepLab V3+ Model Analysis

The encoding module of the DeepLab V3+ network extracted high-level semantic feature maps of images through ResNet101 and connected to ASPP modules with multiple cavity convolution dilation rates. After multiscale sampling of the high-level feature graph and combination in the channel dimension, the multiscale low-dimensional feature matrix was obtained using a convolution kernel with a size of 1 × 1. The decoding module sampled the feature graph four times and fused it with the low-level feature image in the middle of ResNet101. After the bilinear insertion sampling, the segmentation graph was output. The network structure of DeepLab V3+ is shown in Figure 3. When the DeepLab V3+ network was used in the field of pig segmentation, problems such as rough contour segmentation and complex background segmentation errors could be found in this network.

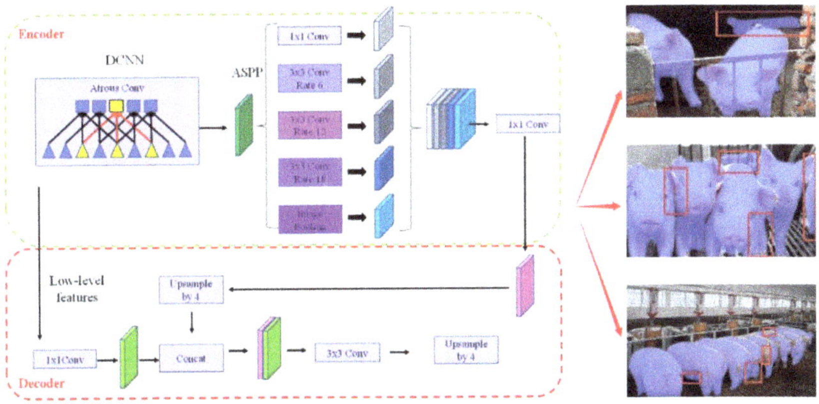

Figure 3. The DeepLab v3+ network structure.

2.3.2. Lightweight Attention Mechanism

The feature map of the coding module acquired the high-level semantic information of the image, while ResNet101 sampled the low-level details of the image. This not only increased the amount of feature computation, but also lost many key features, leading to the noise problem in the fusion of high- and low-frequency features. Domestic and foreign scholars have proposed various attention mechanisms—such as nonlocal attention mechanisms [29], dual attention mechanisms [30], and cross-attention mechanisms [31]—to improve the performance of segmentation models. All of the abovementioned attention mechanisms improved the segmentation model to varying degrees, but greatly increased the required computational resources at the same time. Therefore, this study proposed an attention module and a recursive cascade mechanism based on rows and columns. On the one hand, this method could aggregate global information more effectively and increase the network's receptive field. On the other hand, the attention module had very little effect on video memory and computation due to the lightweight module design method.

Given that the feature graph had a total of W $\times$ H pixels (where W and H were the width and height of the feature graph, respectively), the size of the relational matrix between the pixels was WH $\times$ WH. The size of this matrix was very large, and the attention mechanism was usually placed in the depths of the network; otherwise, too many pixels would make the relational matrix too large, which would significantly increase the computation required by the GPU, and even lead to incapability of the limited video memory to store the matrix. Therefore, in this study, the attention mechanism was introduced into the dimensional reduction feature graph, and the feature parameters were rescreened. Only the relationships between rows or the relationships between columns were calculated to decompose the original relationship matrix of WH $\times$ WH size into two small matrices—namely, the WW and HH size matrices. After decomposition, the space occupied by the two small matrices and the amount of computation required by the large matrix were geometrically reduced (WW + HH $\leq$ WH $\times$ WH). Figures 4 and 5 show the row- and column-based attention modules, respectively.

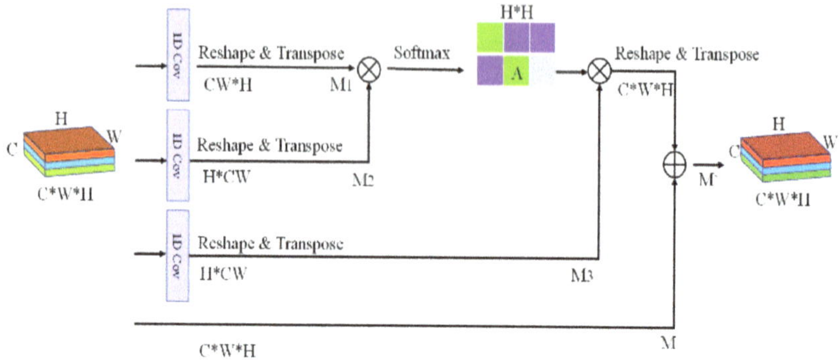

Figure 4. Row-based attention module.

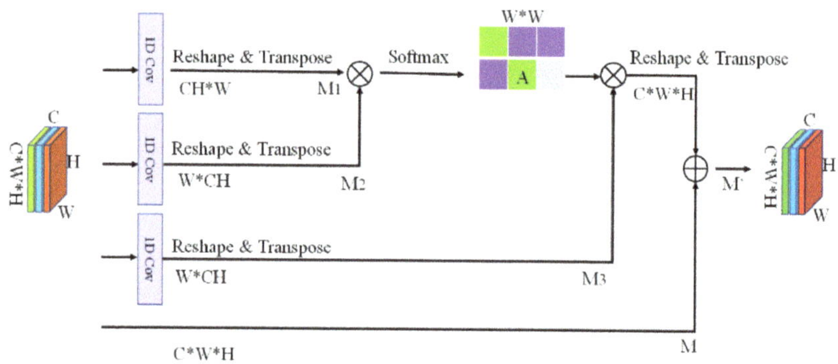

Figure 5. Column-based attention module.

We supposed the input of the module was a feature graph $M \in R^{C \times H \times W}$, where C, H, and W are the number of channels, height, and width of the feature graph, respectively. First, the input feature graph was transformed by the feature tensor. M was reshaped and transposed to obtain $M_1 \in R^{CW \times H}$. M_2 and M_3 were then obtained by M in the same way. Next, matrix multiplication and Softmax operation were performed on M_2 and M_1 to obtain the relational matrix A. The process can be described by the following formula:

$$A_{ij} = \frac{\exp(M_2[i,:]M_1[:,j])}{\sum_{k=1}^{D}\exp(M_2[i,:]M_1[:,k])} \tag{6}$$

where A_{ij} computes the relationship between the ith row and the jth row. Each row in attention map A refers to the relationship between this pixel feature and all of the other pixel features, which can be used to aggregate new features. M_2 $[i,:]$ refers to the ith row of the matrix M_2; M_1 $[:, j]$ refers to the jth column of matrix M_1. After obtaining the relational matrix A, we can use A to perform feature aggregation operations on the original feature graph. This supposes that RT( ) is a function of shape remodeling and transposition. The new aggregated feature graph was obtained by the following formula:

$$M' = \alpha \mathrm{RT}(AM_3) + \mathrm{M} \tag{7}$$

where α is a scale parameter used to adjust the weight of polymerization features. The two formulae mentioned above could be used to obtain the line-based attention module. The column-based attention module is similar to the row-based attention module.

This study cascaded the column-based module with the row-based module. First, the input feature graph M was fed into the row-based attention module to obtain the output feature graph M'. Second, M' was fed into the column-based attention module as an input to obtain the final aggregation feature E. Through recursion, each pixel feature in the E feature graph was the weighted sum of all other pixel features. In this study, the features were effectively aggregated and the latent semantic information was fully mined. At the same time, the time and space consumption were much smaller than those of the traditional attention module. The recursive cascade mechanism structure is shown in Figure 6.

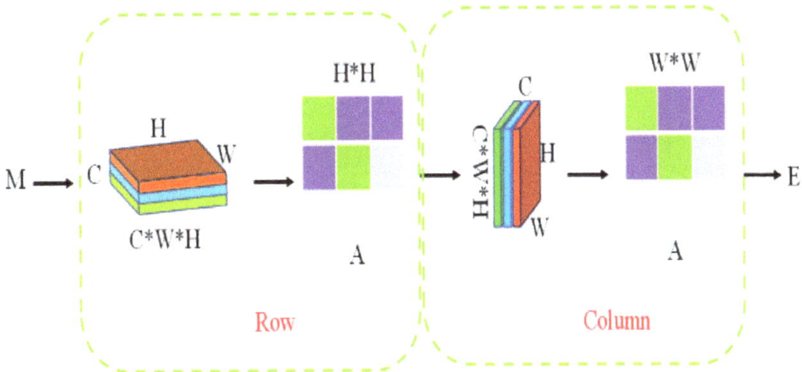

Figure 6. Recursive mechanism.

2.3.3. Improved Network Model

On the basis of the traditional DeepLab V3+ model, this study mainly improved the feature fusion part deep in the model network. After the initial high-dimensional feature images were extracted through the ASPP module and the dimensionality was reduced, the initial feature images were fused by a recursive cascade mechanism. First, the row-based attention mechanism was used to extract semantic information. Then, the column-based attention mechanism was cascaded to deepen the feature information. Theoretically, the computation of the model could be reduced exponentially, and the high-dimensional feature information of the image could be further optimized to provide support for the subsequent high- and low-frequency feature fusion. The improved model structure is shown in Figure 7.

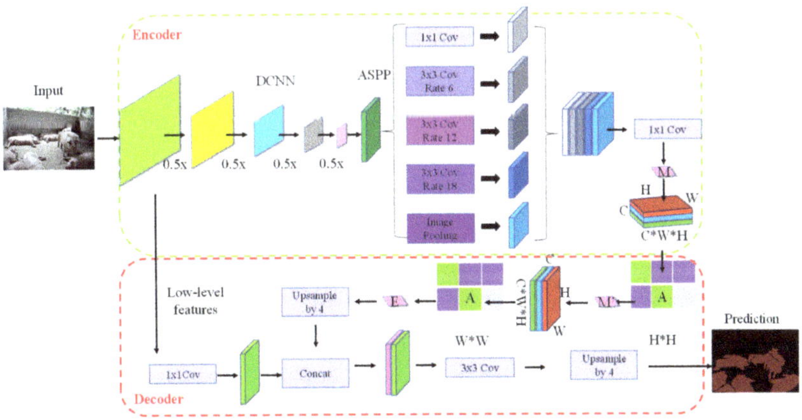

Figure 7. Light attention DeepLab V3+ network structure.

3. Results

3.1. Model Training Experiment

In this study, five representative semantic segmentation models including FCNNs, SegNet, U-Net, DenseNet, and DeepLab V3+ were reproduced. The initial learning rate was set to 0.01, and the regularization coefficient was set to 0.001. With the increase in epoch times, the learning rate followed the principle of exponential decay, and decreased to 0.05 times the original. The comparison of the loss value of model training is shown in Figure 8. After approximately 1000 iterations, the models converged rapidly and the loss function curve was still declining. After 2000 iterations of training, the model was stable. The loss value of this method was 0.002, which meant that the training effect of the model was the best. The comparative experiment of operation efficiency of the proposed method is shown in Figure 9. Images with different resolutions were introduced into models of different batch sizes for training. The results showed that the segmentation speed was faster when the resolution was smaller, and the fastest was up to 441 fps. However, smaller resolution did not mean higher segmentation accuracy, because smaller scale images lost more information. Therefore, we finally chose a suitable size of 512 × 256 pixels for the model training set.

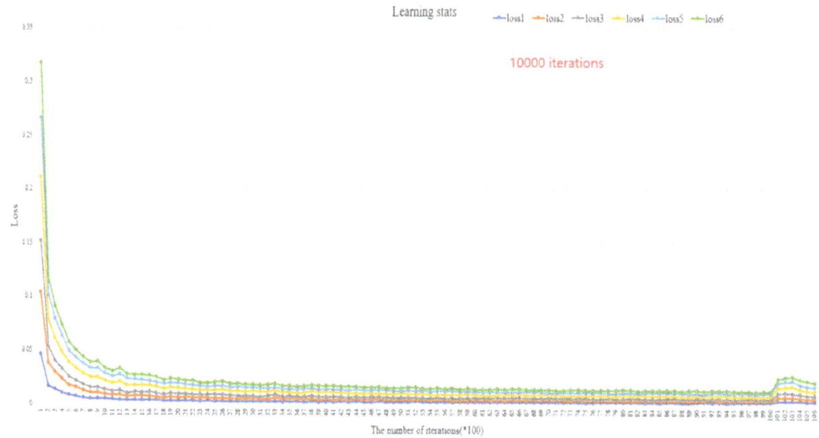

Figure 8. Comparison of training loss of different models: Here, loss1 represents our method in this study; loss2 represents the traditional DeepLab v3+ result; loss3 represents the DenseNet result; loss4 represents the SegNet result; loss5 represents the U-Net result; loss6 represents the FCNNs result.

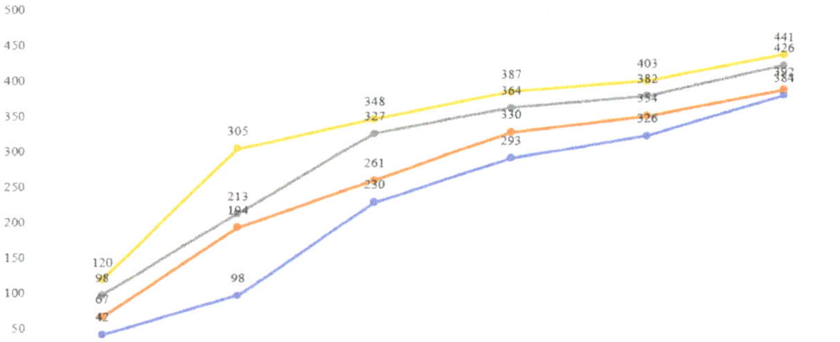

Figure 9. Processing speed of our method.

3.2. *Segmentation Experiment*

3.2.1. Qualitative Comparative Analysis

The proposed method was further compared with the existing segmentation methods of FCNNs, U-Net, SegNet, DenseNet, and traditional DeepLab V3+. Image segmentation results were presented in lightweight, middleweight, and heavyweight difficulty. The visualization results are shown in Figure 10, illustrating the advantages of the new model more intuitively.

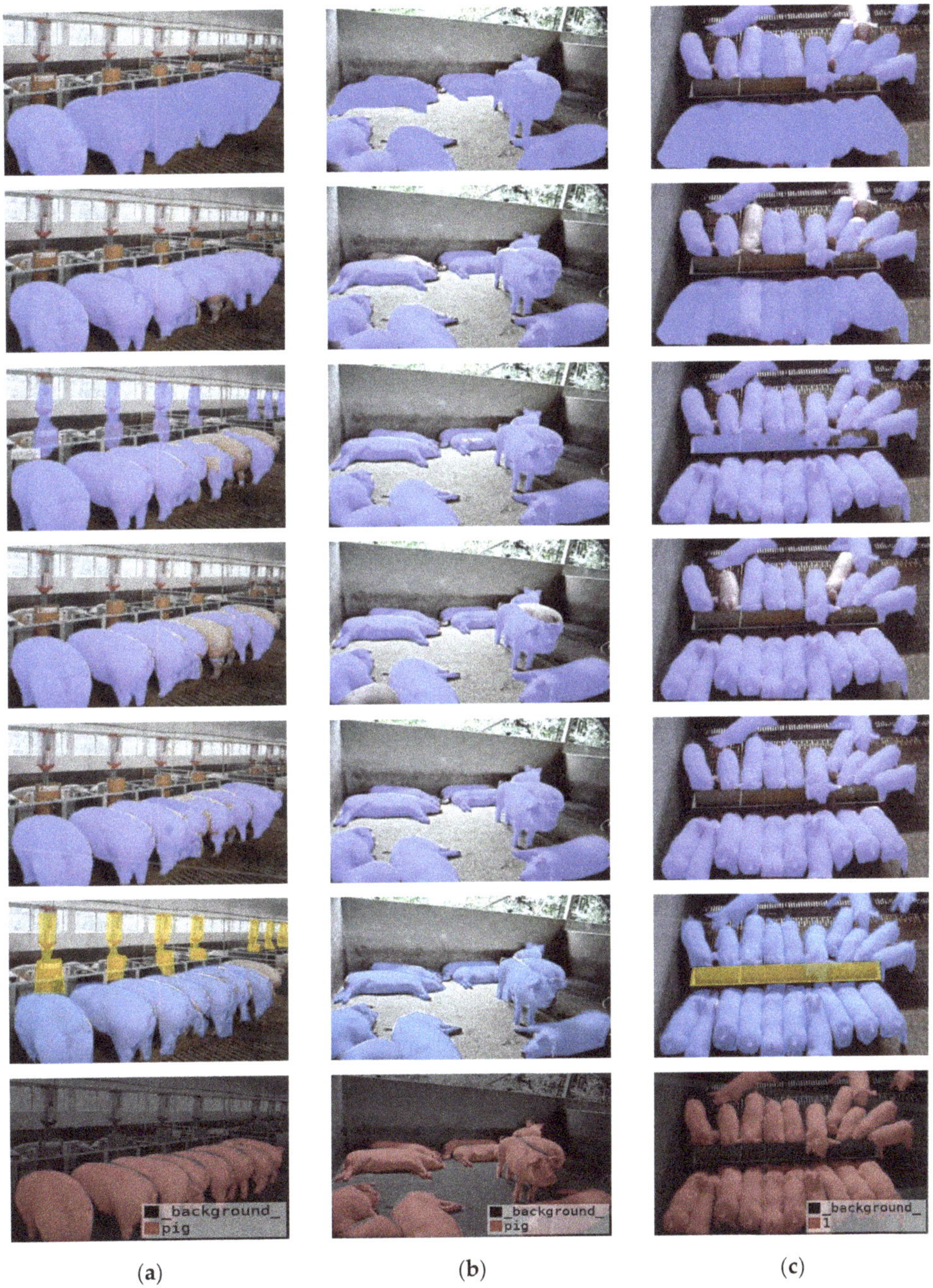

Figure 10. Comparison results of different segmentation models: (**a**) lightweight image; (**b**) middleweight image; (**c**) heavyweight image.

The pictures in the first line are the segmentation results of FCNNs. As observed, the algorithm had an effect on the overall segmentation of pigs; however, it could not achieve fine-grained individual differentiation, resulting in serious overlap.

The pictures in the second line are the segmentation results of U-Net. Here, the phenomenon of overlapping segmentation decreased significantly; however, the processing of details was still insufficient, and the processing of lightweight images was imperfect. With the increase in difficulty, the phenomena of missing segmentation and overlapping segmentation appeared simultaneously, and the problem of adhesion had not been properly solved.

The pictures in the third line are the segmentation results of SegNet. The algorithm solved the overlapping problem; however, an oversegmentation phenomenon was observed in the segmentation results of lightweight and heavyweight images, and the feeder was mistakenly separated. In addition, the processing of middleweight images was improved compared with that of the former models.

The pictures in the fourth line are the segmentation results of DenseNet. The problem of oversegmentation and overlapping had been solved, but the problem of missing segmentation existed to different degrees.

The pictures in the fifth line are the segmentation results of the traditional DeepLab V3+, which had a good segmentation effect on the whole. However, the segmentation of the detailed parts of the pig itself—such as legs, back, and outline—was rough. In particular, the typical overlapping occlusion phenomenon was observed in the lightweight image, and the segmentation accuracy needed to be improved.

In comparison, the pictures in the sixth line are the segmentation results of the improved LA-DeepLab V3+ model, which had well inherited the complex background segmentation capability of the former. After the label sample was expanded, this study added separate feeders, drinkers, people, and other external labels, which not only retained the image details better, but also made the multi-category prediction results more accurate and comprehensive.

The pictures in the seventh line are the truth value of manual segmentation. Therefore, the proposed improved model properly dealt with the abovementioned shortcomings by accurately representing the details of the image; it also solved the problem of missing segmentation and segmentation discontinuity.

The segmentation results of different complex backgrounds in this study are shown in Figure 11. The method in this study extended several other typical labels—including feeders, drinkers, people, sky, and trees—on the basis of the pig label. The purpose was to solve the problem of misidentification and classification, deepen the understanding of specific scenes, enhance the ability of the model to further distinguish different complex background factors, and provide support for deeper semantic segmentation, such as behavior. The results showed that the proposed algorithm could deal well with different scenes, complex lighting, occlusion, and overlap problems, and it had a certain degree of generalization ability. The model maintained its segmentation accuracy under the complex background, and provided effective support for individual behavior recognition and pig farm counting in the future.

3.2.2. Quantitative Comparative Analysis

Segmentation evaluation indices of different models were given in this study to quantify the performance of the models. Table 1 shows that the P, R, AP, F_1 score, and MIoU values of LA-DeepLab V3+ (single tag) in this study were the highest, at 86.04%, 75.06%, 78.67%, 0.8, and 76.31%, respectively. Among them, AP and MIoU showed obvious difference and significance. Compared with the FCNNs, U-Net, SegNet, DenseNet, and DeepLab V3+ methods, AP improved by 27.24%, 22.63%, 13.28%, 6.03%, and 2.79%, respectively. MIoU improved by 14.10%, 13.67%, 7.79%, 2.33%, and 5.80% over the FCNNs, U-Net, SegNet, DenseNet, and traditional DeepLab V3+ methods, respectively. In addition, the P, AP, and MIoU values of LA-DeepLab V3+ (multiple tags) in this study were the second highest, at 88.36%, 76.75%, and 74.62%, respectively. The traditional DeepLab V3+ method maintained its advantages in R and F_1 score of 74.75% and 0.79, respectively; however, the differences were insignificant. Overall, the proposed algorithm was absolutely optimal. The difficulty of the pig dataset with complex background set in this study meant that achieving an accuracy of more than 90% with other simple scenes was impossible, because the sample size of the simple background in our dataset was itself very low. We specially tested the image segmentation of pig herds under a simple background in order to further verify the accuracy and effectiveness of the algorithm in this study. The accuracy remained above 99%, which could better realize the accurate inventory of pig herds.

(**a**) Feeder scene (**b**) Drinker scene (**c**) People scene (**d**) Tree scene (**e**) Sky scene

Figure 11. Segmentation results of different scenes.

Table 1. Segmentation results of the different methods.

Methods	P	R	AP	F_1-Score	MIoU
FCNNs	71.60%	68.46%	51.43%	0.69	62.21%
U-Net	74.46%	68.66%	56.04%	0.71	62.64%
SegNet	78.04%	62.66%	65.39%	0.69	68.52%
DenseNet	76.14%	68.94%	72.70%	0.72	73.98%
DeepLab v3+	84.10%	**74.75%**	75.88%	**0.79**	70.51%
LA-DeepLab v3+ (single tag)	**86.04%**	**75.06%**	**78.67%**	**0.80**	**76.31%**
LA-DeepLab v3+ (multiple tags)	**88.36%**	70.03%	**76.75%**	0.78	**74.62%**

3.2.3. Generalization Experiment

The improved model was tested on the public dataset PASCAL VOC 2012 (VOCdevkit) [32] after extended training. Not all images in VOC2012 were suitable for segmentation in this model. We selected some images including people, sky, and trees to verify the robustness of this model against complex backgrounds.

In Figure 12, the first line is the original image, the second line is the real label corresponding to the original image, and the third line is the segmentation result of our method. The results showed that the self-built dataset in this study mainly focused on the segmentation and counting of pigs. Thus, the segmentation effect of background factors was rather rough—in particular, the adhesion of people and bicycles, the fine segmentation of trees' internal cavities, and the processing of contours were imperfect. However, rough segmentation of complex backgrounds could be realized on the whole, which was helpful for the model to further learn the complex background of pigs. In conclusion, the proposed LA-DeepLab V3+ model still achieved good segmentation performance on the PASCAL VOC 2012 datasets, further verifying the generalization of the proposed model.

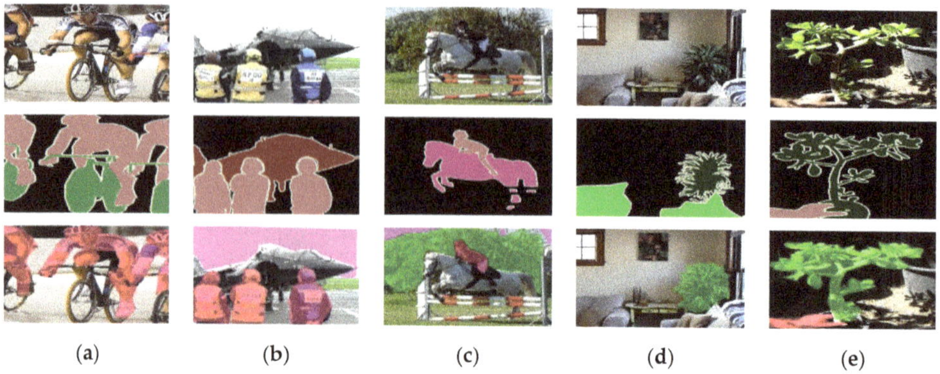

Figure 12. Segmentation results of different scenes on PASCAL VOC 2012: (**a**) people scene; (**b**) sky and people scene; (**c**) sky, people, and tree scene; (**d**) tree scene; (**e**) tree and people scene.

3.3. Model Deployment and Visual Counting Applications

The purpose of semantic segmentation was to obtain the whole life cycle law of pig quantity, behavior, category, and trajectory tracking. In this study, the H5 program was used to build a graphical user interface for the application of pig counting. As shown in Figure 13, the model selected in the figure was the DeepLab V3+ model, which introduced a lightweight attention mechanism.

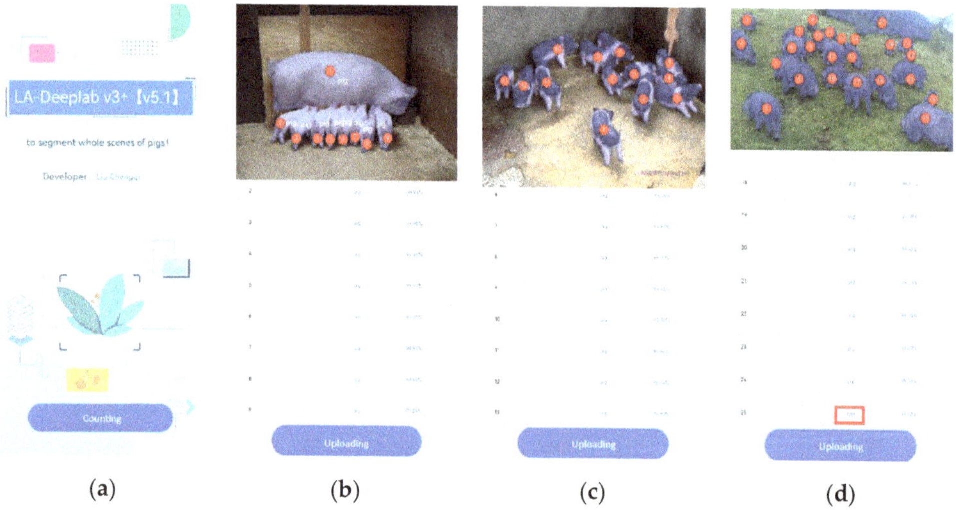

Figure 13. Counting application of pigs in complex environments based on the H5 program. (**a**) The home page; (**b**) The lightweight image (1–10 pigs); (**c**) The middle-weight images (10–20 pigs); (**d**) The heavyweight images (more than 20 pigs)

In the figure, panel (a) shows the home page. A photo was uploaded through shooting or an album, and then it was submitted to the program for segmentation and counting.

Panel (b) shows the count result of the lightweight image (1–10 pigs). If we could accurately detect the number of piglets in the actual breeding process, then early warning of the death and injury caused by extrusion in the production process could be provided. The accuracy of this model for lightweight image segmentation was more than 99.8%.

Panel (c) shows the counting result of middleweight images (10–20 pigs). The accuracy of the small pig group with black and white color could reach 99.9%, and 13 pigs could be accurately counted.

Panel (d) shows the counting result of heavyweight images (more than 20 pigs). The proposed model could effectively segment the wild black pigs with complex background and multiscale targets, and the segmentation accuracy was approximately 97%. In addition to the segmentation of all 24 objects, other kinds of tags could be effectively distinguished.

At present, the average recognition accuracy of the pig and human labels is 97.65% and 95.86%, respectively. For other labels—such as trees, sky, drinkers, and feeders—the recognition was relatively low. The number of model labels had some significance; thus, the overall evaluation index of the model was lowered. The application results showed that the proposed model had a good application effect on fine-grained segmentation tasks.

4. Discussion

4.1. Analysis of Each Model

The comparison results showed that each model had unique advantages. FCNNs used a deconvolution process to restore image resolution and optimize segmentation results; however, the downsampling operation of this method weakened the feature extraction ability of the model, resulting in poor segmentation ability of details of pig images with complex backgrounds; therefore, its segmentation performance index was poor. U-Net and SegNet are U-shaped codec structures. The segmentation model based on dilated convolution could enlarge the local receptive field of the original convolution kernel; however, the proportion of some pig targets in the overall image was small; thus, these two segmentation methods were still imperfect in the performance method. DenseNet greatly reduced the number of parameters, which not only consumed memory but also led to insufficient extraction of low-frequency features; as a result, missing segmentation occurred. On the basis of the traditional DeepLab V3+, the proposed method further integrated high- and low-frequency features to bridge the semantic gap between different feature graphs; thus, better accuracy and stronger feature expression were achieved. In addition, the lightweight attention mechanism not only retained the advantage of attention, but also avoided excessive consumption. Next, the algorithm could be further optimized based on the extraction ability of image depth features and the complexity of the network.

4.2. Analysis of Improved Segmentation Methods

As shown in Figure 14, very few cases of mis-segmentation and missing segmentation occurred in the test set when using the proposed method in this study.

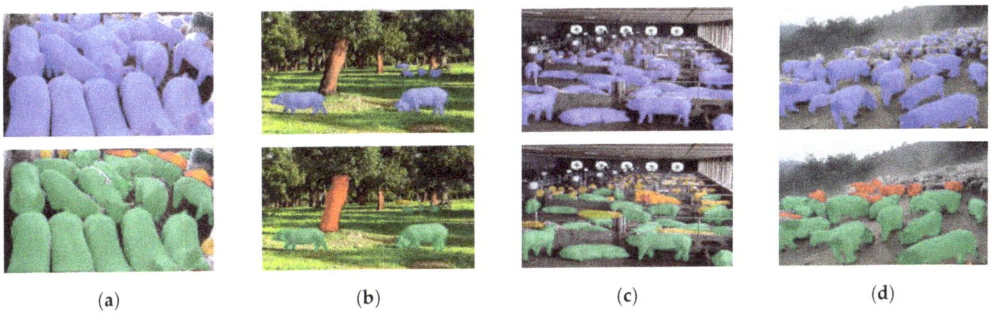

Figure 14. Results of false and missing segmentation. (**a**) missing segmentation indoors; (**b**) mis-segmentation outdoors; (**c**) mis-segmentation indoors; (**d**) missing segmentation outdoors

Among them, the main reason for (a), (c), and (d) was that the image of the pigs was too difficult. The label comparison in the first line showed that ensuring the accuracy of the label was difficult. In addition, the target was very dense and the image depth information was low. As a result, identification in the case of serious occlusion was difficult. As shown in (b), mis-segmentation of tree trunks and missing segmentation of distant small targets occurred due to the influence of the real environment in the field.

More sample data and label quantity could be added to improve its segmentation ability.

4.3. Pressure Test of the Counting Application System

This study conducted pressure test analysis to further test the robustness and generalization of the counting application. Representative images with different scenes were input into the counting system to detect their robustness against occlusion, overlap, adhesion, illumination, and multiscale targets. Figure 15 shows an example of an error on the current application platform.

(a) (b) (c) (d)

Figure 15. Error examples in counting applications: (**a**) missing segmentation; (**b**) mis-segmentation; (**c**) missing count; (**d**) pressure test.

In the figure, panel (a) represents missing segmentation after the small-scale target was blocked in the image perspective. As shown in panel (b), mis-segmentation occurred due to overlapping and adhesion problems. Panel (c) shows the count error caused by the leg target being too small after severe occlusion. Panel (d) shows the cases of missing segmentation and mis-segmentation in the field environment containing all of the abovementioned problems. This shows that the counting accuracy gradually decreased with the increase in the number of pigs.

The pressure test results showed that the application could calculate the number of pigs up to a maximum of 50, and that the counting accuracy of pigs with less than 30 was high. This could meet the requirements of standard free-range piggery, but the counting of large-scale free-range piggery requires further study. The errors of the model were mainly caused by insufficient feature extraction of small-scale targets in the foreground, or after occlusion.

However, the application can meet the requirements of pig population identification and counting in most common scenarios. The optimization can be further improved by setting a minimum target scale threshold, supplementing large-scale sample data, and optimizing the model network.

5. Conclusions

In this study, we proposed a novel semantic segmentation method with a lightweight attention mechanism. By fusing high- and low-frequency features and reducing redundant parameters, the DeepLab V3+ semantic segmentation method was optimized, and a pig counting system was built.

First, this study constructed pig datasets for different scenarios, including field, indoor, day and night, white pigs, black pigs, humans, trees, and other images of the real environment. According to the number of pigs, three kinds of quantity detection difficulty were set: lightweight (1–10), middleweight (10–20), and heavyweight (more than 20).

Second, a lightweight attention mechanism was introduced based on the DeepLab V3+ deep learning method to improve the segmentation accuracy of complex images of pigs. The attention module based on rows and columns improved the efficiency of feature calculation. Recursive cascade mode was adopted to optimize the fusion of high- and low-frequency features, and potential semantic information was mined in order to reduce time and space consumption. In this study, the values of the segmentation evaluation indices P, R, AP, F_1 score, and MIoU of LA-DeepLab V3+ (single tag) were the highest, and the P, AP, and MIoU values of LA-DeepLab V3+ (multiple tags) were the second highest. Quantitative and qualitative experiments showed that the segmentation effect of the model was improved significantly.

Finally, the improved model was integrated to enhance the efficiency of pig counting, and a graphical counting platform was built to achieve accurate pig counting. The counting network could calculate the number of pigs up to a maximum of 50, and the counting accuracy of pigs with less than 30 was higher, meeting the requirements of free-range counting in standard piggery.

The optimization can be further improved by setting a minimum target detection threshold, supplementing large-scale sample data, and optimizing the model network.

Author Contributions: Conceptualization, C.L. and L.L.; methodology, C.L.; software, C.L.; validation, J.S.; resources, C.L. and J.S.; writing—original draft preparation, C.L.; writing—review and editing, L.W.; visualization, S.L.; supervision, L.L.; funding acquisition, L.L. All authors have read and agreed to the published version of the manuscript.

Funding: This research was funded by the National Major Science and Technology Project (Innovation 2030) of China (No.2021ZD0113701), the National Key Research and Development Program of China (No.2021YFD1300101), and the National Research Facility for Phenotypic and Genotypic Analysis of Model Animals (Beijing) (No.2016-000052-73-01-001202).

Institutional Review Board Statement: Not applicable.

Informed Consent Statement: Not applicable.

Data Availability Statement: Not applicable.

Acknowledgments: We thank all of the funders and all of the reviewers.

Conflicts of Interest: The authors declare no conflict of interest.

References

1. Yang, A.; Huang, H.; Zheng, C.; Zhu, X.; Yang, X.; Chen, P. High-accuracy image segmentation for lactating sows using a fully convolutional network. *Biosyst. Eng.* **2018**, *176*, 36–47. [CrossRef]
2. Zhang, L.; Gray, H.; Ye, X.; Collins, L.; Allinson, N. Automatic individual pig detection and tracking in pig farms. *Sensors* **2019**, *19*, 1188. [CrossRef] [PubMed]
3. Yang, W.; Yan, Y.; Chen, S. Multi-scale generative adversation network based pedestrian reidentification method for occlusion. *J. Softw.* **2020**, *31*, 1943–1958.
4. Zhao, Y.; Rao, Y.; Dong, S. A review of deep learning target detection methods. *J. Image Graph.* **2020**, *25*, 629–654.
5. Yan, H.; Lu, H.; Ye, M. Segmentation of pulmonary nodules by sobel operator and mask R-CNN. *J. Chin. Comput. Syst.* **2020**, *41*, 161–165.
6. Tian, Q.; Meng, Y. Image semantic segmentation based on convolutional neural network. *J. Chin. Comput. Syst.* **2020**, *41*, 1302–1313.
7. Long, J.; Shelhamer, E.; Darrell, T. Fully convolutional networks for semantic segmentation. *IEEE Trans. Pattern Anal. Mach. Intell.* **2015**, *39*, 640–651. [CrossRef]
8. Yu, F.; Koltun, V.; Funkhouser, T. Dilated residual networks. In Proceedings of the IEEE Conference on Computer Vision and Pattern Recognition, Honolulu, HI, USA, 21–26 July 2017; pp. 472–480.
9. Dai, J.; Qi, H.; Xiong, Y.; Li, Y.; Zhang, G.; Hu, H.; Wei, Y. Deformable convolutional networks. In Proceedings of the IEEE Conference on Computer Vision and Pattern Recognition, Honolulu, HI, USA, 21–26 July 2017; pp. 764–773. [CrossRef]

10. Ronneberger, O.; Fischer, P.; Brox, T. U-Net: Convolutional Networks for Biomedical Image Segmentation. In *International Conference on Medical Image Computing and Computer-Assisted Intervention*; Springer International Publishing: Cham, Switzerland, 2015; pp. 234–241. [CrossRef]
11. Badrinarayanan, V.; Kendall, A.; Cipolla, R. Segnet: A deep convolutional encoder-decoder architecture for image segmentation. *IEEE Trans. Pattern Anal. Mach. Intell.* **2017**, *39*, 2481–2495. [CrossRef] [PubMed]
12. Zhao, H.; Shi, J.; Qi, X.; Wang, X.; Jia, J. Pyramid scene parsing network. In Proceedings of the IEEE Conference on Computer Vision and Pattern Recognition, Honolulu, HI, USA, 21–26 July 2017; pp. 2881–2890. [CrossRef]
13. Chen, L.C.; Papandreou, G.; Kokkinos, I.; Murphy, K.; Yuille, A.L. Deeplab: Semantic image segmentation with deep convolutional nets, atrous convolution, and fully connected crfs. *IEEE Trans. Pattern Anal. Mach. Intell.* **2018**, *40*, 834–848. [CrossRef] [PubMed]
14. Chen, L.C.; Papandreou, G.; Schroff, F.; Adam, H. Rethinking atrous convolution for semantic image segmentation. *arXiv* **2017**, arXiv:1706.05587.
15. Chen, L.C.; Zhu, Y.; Papandreou, G.; Schroff, F.; Adam, H. *Encoder-Decoder with Atrous Separable Convolution for Semantic Image Segmentation*; Springer: Cham, Switzerland, 2018. [CrossRef]
16. Chollet, F. Xception: Deep Learning with Depthwise Separable Convolutions. In Proceedings of the IEEE Conference on Computer Vision and Pattern Recognition, Honolulu, HI, USA, 21–26 July 2017; pp. 1800–1807. [CrossRef]
17. Zhang, R.; Li, J.T. A survey on algorithm research of scene parsing based on deep learn. *J. Comput. Res. Dev.* **2020**, *57*, 859–875.
18. Peng, X.; Yin, Z.; Yang, Z. Deeplab_v3_plus-net for Image Semantic Segmentation with Channel Compression. In Proceedings of the IEEE 20th International Conference on Communication Technology, Nanning, China, 28–31 October 2020.
19. Guo, L.; Zhang, T.S.; Sun, W.Z. Image Semantic Description Algorithm with Integrated Spatial Attention Mechanism. *Laser Optoelectron. Prog.* **2021**, *58*, 1210030.
20. Lou, T.; Yang, H.; Hu, Z. Grape cluster detection and segmentation based on deep convolutional network. *J. Shanxi Agric. Univ. Nat. Sci. Ed.* **2020**, *40*, 109–119.
21. Mnih, V.; Heess, N.; Graves, A. Recurrent Models of Visual Attention. *Adv. Neural Inf. Processing Syst.* **2014**, *3*, 27.
22. Wang, F.; Jiang, M.; Chen, Q. Residual Attention Network for Image Classification. In Proceedings of the IEEE Conference on Computer Vision and Pattern Recognition, Honolulu, HI, USA, 21–26 July 2017; pp. 3156–3164.
23. Jie, H.; Li, S.; Gang, S. Squeeze-and-Excitation Networks. In Proceedings of the IEEE/CVF Conference on Computer Vision and Pattern Recognition, Salt Lake City, UT, USA, 18–23 June 2018; pp. 7132–7141. [CrossRef]
24. Zhang, H.; Dana, K.; Shi, J.; Zhang, Z.; Wang, X.; Tyagi, A. Context Encoding for Semantic Segmentation. In Proceedings of the IEEE/CVF Conference on Computer Vision and Pattern Recognition, Honolulu, HI, USA, 21–26 July 2017; pp. 7151–7160. [CrossRef]
25. Yu, C.; Wang, J.; Peng, C. Learning a Discriminative Feature Network for Semantic Segmentation. In Proceedings of the IEEE/CVF Conference on Computer Vision and Pattern Recognition, Honolulu, HI, USA, 21–26 July 2017; pp. 1857–1866.
26. Subakan, C.; Ravanelli, M.; Cornell, S. Attention Is All You Need in Speech Separation. In Proceedings of the IEEE International Conference on Acoustics, Speech and Signal Processing, Toronto, ON, Canada, 6–11 June 2021.
27. Zhang, H.; Goodfellow, I.; Metaxas, D. Self-Attention Generative Adversarial Networks. In Proceedings of the International Conference on Machine Learning, Vancouver, BC, Canada, 13 December 2019; pp. 7354–7363.
28. Fu, J.; Liu, J.; Tian, H.; Li, Y.; Bao, Y.; Fang, Z. Dual Attention Network for Scene Segmentation. In Proceedings of the IEEE/CVF Conference on Computer Vision and Pattern Recognition, Long Beach, CA, USA, 15–20 June 2019; pp. 3146–3154.
29. Wang, X.; Girshick, R.; Gupta, A. Non-local Neural Networks. In Proceedings of the IEEE Conference on Computer Vision and Pattern Recognition, Salt Lake City, UT, USA, 18–23 June 2018; pp. 7794–7803.
30. Chen, J.; Lu, Y.; Yu, Q. TransUNet: Transformers Make Strong Encoders for Medical Image Segmentation. *arXiv* **2021**, arXiv:2102.04306.
31. Huang, Z.; Wang, X.; Huang, L.; Huang, C.; Wei, Y.; Liu, W. CCNet: Criss-Cross Attention for Semantic Segmentation. In Proceedings of the International Conference on Computer Vision, Seoul, Korea, 27–28 October 2019; pp. 603–612.
32. Everingham, M.; Eslami, S.; Gool, L.V.; Williams, C.; Winn, J.; Zisserman, A. The pascal visual object classes challenge: A retrospective. *Int. J. Comput. Vis.* **2015**, *111*, 98–136. [CrossRef]

agriculture

MDPI

Review

The Role of FAIR Data towards Sustainable Agricultural Performance: A Systematic Literature Review

Basharat Ali [1,2,*] **and Peter Dahlhaus** [1,2]

1 Centre for eResearch and Digital Innovation (CeRDI), Greenhill Enterprise Centre Ballarat Technology Park, Federation University, Mount Helen, VIC 3350, Australia; p.dahlhaus@federation.edu.au
2 Food Agility Cooperative Research Centre, Ultimo, NSW 2007, Australia
* Correspondence: basharat.ali2@federation.edu.au

Abstract: Feeding a growing global population requires improving agricultural production in the face of multidimensional challenges; and digital agriculture is increasingly seen as a strategy for better decision making. Agriculture and agricultural supply chains are increasingly reliant on data, including its access and provision from the farm to the consumer. Far-reaching data provision inevitably needs the adoption of FAIR (Findable, Accessible, Interoperable, and Reusable) that offer data originators and depository custodians with a set of guidelines to safeguard a progressive data availability and reusability. Through a systematic literature review it is apparent that although FAIR data principles can play a key role in achieving sustainable agricultural operational and business performance, there are few published studies on how they have been adopted and used. The investigation examines: (1) how FAIR data assimilate with the sustainability framework; and (2) whether the use of FAIR data by the agriculture industry, has an impact on agricultural performance. The work identifies a social science research gap and suggests a method to guide agriculture practitioners in identifying the specific barriers in making their data FAIR. By troubleshooting the barriers, the value propositions of adopting FAIR data in agriculture can be better understood and addressed.

Keywords: FAIR data; findable; accessible; interoperable; reusable; sustainability

Citation: Ali, B.; Dahlhaus, P. The Role of FAIR Data towards Sustainable Agricultural Performance: A Systematic Literature Review. *Agriculture* **2022**, *12*, 309. https://doi.org/10.3390/agriculture12020309

Academic Editors: Gniewko Niedbała and Sebastian Kujawa

Received: 15 January 2022
Accepted: 17 February 2022
Published: 21 February 2022

1. Introduction

The global need for agricultural production has been increasing [1], and most food production remains soil-based. With nearly all arable land under cultivation [2], agriculture is projected to encounter several challenges including: sustaining maximum production, limited natural resources, endangered environments and ecosystems, soil degradation and erosion [3,4]. It is quantitatively confirmed that farming strategies, methods and decision making are key factors in the future of sustainable and enhanced agricultural production [5,6]. Operationally, digital agriculture and data/information mutually rely on each other [3,7,8]. The volume of digital data in agricultural landscapes has grown exponentially, much of it collected by sensors (both remote sensing and the Internet of Things). Agricultural knowledge building, appropriate management responses [9] and farm management decisions [10] highly depend on the data collected through the use of digital technologies, and ubiquitous internet technologies provide access to all these data, delivered on demand via high-speed broadband to mobile tablet devices [11]. Precision agriculture technology employs data [12] to perform operations such as economising crop inputs, optimising machinery performance and appropriate location finding [13]. The digital agricultural revolution has led to a plethora of websites and mobile applications (Apps) that are now available to assist the farmer, agronomist, agribusiness investor, landscape manager and researcher in decision making. However, the applications and tools are only as good as the data they use, and because of the disparity of data collection, formats, and storage only a fraction of the required data are utilised. Current common limitations in system models for decision support are: (1) data scarcity (quantity, resolution, and quality) and (2) inadequate knowledge systems to effectively communicate the results to the end-user. These limitations are greater obstacles to the use of the tools than gaps in theory or technology [14,15]. Seamless automated data collection (from both public and private sources), data interoperability and the federation of multidisciplinary data (plant, animal, soil, land, climate, weather, machinery, farm business, economics, marketing, trade, etc.) are required, preferably utilising open cloud-based systems for data storage and open standards for data exchange. Combining these data in new technologies, such as those deploying data mining, machine learning, artificial intelligence algorithms and digital twins, will ultimately provide the holistic viewpoint needed for sustainable agricultural production [3,7,16].

Improving agricultural knowledge, appropriate management responses [9] and farm management decisions [10] requires the stakeholders to increasingly depend on data collected through the use of digital technologies, such as the internet, sensors and mobile computing at a more localised scale [11]. Precision agricultural technologies massively employ data [12] to perform operations such as economised crop inputs based on high resolution location finding [3,16–18], hence requiring a robust database management system with far-reaching data provision [19]. Since financial benefits highly depend upon the suitability of the chosen farming techniques and technology, access to data, data literacy and/or technical support are important needs in the adoption of innovative practices in the agricultural sector [20–22]. FAIR data principles, first published by Wilkinson [23], provide data originators and data depository custodians with a set of guidelines to ensure data availability and reusability. While recognising an ever-growing need for automation of wide-ranging data encounter, recovery, integration and analysis, the FAIR data principles make this goal possible, by combining multidisciplinary, cross-disciplinary data, with disparate data formats, from different sources, as well as by emphasizing that each of the principles should be equally valid to both humans and machines [23–25].

FAIR (findable, accessible, interoperable, reusable) data principles add more value to data by enhancing data utility, especially for legacy data sets. Data *findability* focuses on maintaining its worldwide uniqueness through persistent identifiers that are machine-readable and index-able to assist individuals and artificially intelligent systems. It is an exclusive, persistent way to refer to the data using standard digital object identifiers (DOI), uniform resource identifiers (URI) or uniform resource locators (URL). Accessibility ensures that the data and other digitised information are available subject to specified conditions of access. It has three major components, i.e., access protocol, access permission and metadata permanency. Accessibility confirms that the data can be accessed by humans and devices using standard internet protocols, provided that the access controls allow that. The controls may require managing data licensing in a convenient way. *Interoperability* makes sure that the data and other digitised information is unambiguously understood machine-to-machine. In accordance with FAIR principles, data and metadata should be conveyed using syntactic and semantic data structures covering the raw and highly processed data. Ontologies, communicated through a resource description framework (RDF) or other open-source frameworks, can carry data integration across the board. Through interoperability, heterogeneous data distributed across disparate databases and devices can be brought together in standardised and harmonised formats. *Reusability* segregates conventional data management from FAIR data stewardship, which requires a multi-layered approach that addresses the demand of data to be reusable. Fully described contextual and descriptive machine-readable metadata is required to allow new consumers to reuse data for new needs and applications, decades after those data were collected. It also requires enhanced provenance metadata for tracing changes to names, editions, and parameters of analysis. Principally, if data are not comprehensively and unambiguously described in a machine-readable form, they have little benefit, regardless of whether they are published with open access [23–28]. The adoption of FAIR data principles, a thorough review of literature reveals, is in its infancy in agricultural systems despite recognition of its value and development of guidelines.

This study conducts a systematic literature review aimed at: (1) understanding the adoption and use of FAIR data principles in the agriculture industry; (2) demonstrating by what means the sub-indicator/characteristics of FAIR data complement sustainable agricultural performance indicators, and the way the sustainability indicators assimilates with the sustainable agricultural performance indicators; and (3) whether FAIR data have impacted agricultural industry performance. Furthermore, this study uniquely describes how the adoption of FAIR data contribute to sustainable agricultural (operational and business) performance, by summarising the literature with comprehensive methodological approaches that demonstrate FAIR data implementation processes. The research identifies a social science research gap and suggests a method to guide agriculture practitioners in identifying the specific barriers to making their data FAIR, and to duly analyse the barriers to better understand and address the value propositions of adopting FAIR data in agriculture.

2. Methodology

This systematic literature review follows more rigorous and transparent methodology, following guidelines by Van der Knaap et al. [29], Moher et al. [30] and Koutsos et al. [31].

2.1. Scoping

This study attempts to answer two related research questions: To what extent have FAIR data principles been adopted by the agricultural industry? and what role do FAIR data principles play in agricultural performance?

The research uses the Web of Science digital database, initially created by the Institute for Scientific Information (ISI) but now maintained by Clarivate [32], to search for more authentic and comprehensive scientific literature. Considering the multidisciplinary nature of FAIR data in agriculture, the search includes all the relevant Web of Science categories and pertinent publications/journals. The broad literature search shows a clearer research gap in this area, with no previous studies found that address the above research questions.

2.2. Planning

To find the qualified publications from the Web of Science, a far-reaching search approach was carried out by using the most appropriate terms or keywords (given below), combining with the Boolean operators (AND and OR). The search query we used is as follows:

> ("FAIR data" OR "FAIR data principles" OR "FAIR data guidelines" OR " FAIR principles" OR "FAIR guidelines" OR "findability" OR "accessibility" OR "interoperability" OR "reusability" OR "findable" OR "accessible" OR "interoperable" OR "reusable" OR "datasets" OR "data sources") AND ("metadata standards" OR "metadata schema" OR "metadata schenes" OR "big date" OR "data management" OR "database")

The search was limited by document type (article, review), years (2016–present) and language (English). Following eligibility criteria, an additional search was conducted through Google Scholar [33], applying the 'snowballing' (or 'backward and forward snowball') technique [34]. The search was broadened because: (1) FAIR data are a relatively new subject in the agriculture disciplines; and (2) agriculture is a multidimensional and multidisciplinary subject that encompasses sociology, data science, economic, environmental sciences, etc. The suitability criterium of the additional literature was assessed as being peer-reviewed (approved by two independent reviewers), following Van derWindt et al. [35] guidelines.

The selection criteria of the selected studies through the Web of Science were based on: (1) the role of FAIR data towards agricultural performance; (2) the implementation of FAIR data principles: i.e., applied research; (3) a comprehensive methodological approach, tools and/or rules, etc. To exclude studies not related to agricultural research (exclusion criteria), the query results were refined by choosing most appropriate Web of Science categories: (1) Agriculture Multidisciplinary; (2) Agriculture Dairy Animal Science; (3) Soil Science; (4) Plant Sciences; (5) Green Sustainable Science Technology; and (6) Computer Science Interdisciplinary Applications.

This review includes all the studies (with document type article, review) from the last six years (2016–present) since the FAIR data principles were first published in 2016. The search was confined to English language papers published by the top five publishing houses, i.e., Elsevier, Springer Nature, MDPI, Wiley and Taylor & Francis, since that covered most high-quality peer-reviewed papers.

At the initial stage, the total number of studies obtained was 1042. To further confine the research focus to agriculture, the selection excluded numerous journals having entirely different scope areas, for instance medical sciences, astronomy, industrial engineering, computation and mathematics, etc. bringing the total number of studies to be assessed to 469. There were five studies included through the 'Google Scholar' search. More details are provided in the PRISMA (Preferred Reporting Items for Systematic reviews and Meta-Analyses) statement (Figure 1).

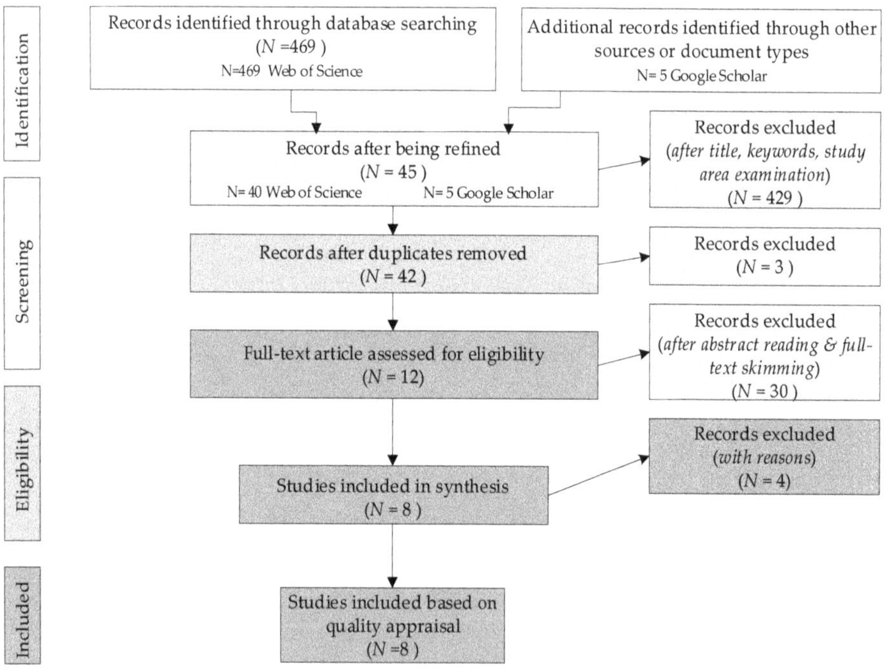

Figure 1. Systematic review based on PRISMA flowchart [30].

2.3. *Identification/Search*

Details of the studies found through the 'advanced search option' of the Web of Science were exported into a reference manager (Endnote) and spreadsheet (Excel) for further assessments. A Google Scholar search was again used to ensure that the maximum number of relevant papers were captured, given the multidisciplinary and multidimensional nature of the research objective. The snowballing technique was helpful for additional searching as there was a limit on searching imposed in the Web of Science database. All studies obtained from the search queries were fully checked based on their titles, abstracts, keywords, year of publication and the research area and so on, to confirm their eligibility for inclusion.

2.4. Screening

All the eligible literature found though the Web of Science and Google Scholar searches were saved (with title, keywords, abstracts) in two separate files. Duplicates (3 studies) were deleted. That resulted in 42 studies (full text) downloaded for a thorough assessment.

2.5. Eligibility/Assessment

The inclusion criteria comprise studies that: (1) cover the role of FAIR data in relation to agricultural performance; (2) take a FAIR data approach to agricultural data; and (3) include a comprehensive methodological segment on the practical implementation of FAIR data principles. As shown in Figure 1, out of 469 studies, 45 records remained after being refined. After deletion of duplicate studies, abstract reading and full-text skimming, there were 12 studies to be considered for full-text assessment.

Since the primary process of this systematic review has been to find studies that clearly describe the role of FAIR data in relation to agricultural performance, as well as evidence (i.e., a methodology approach) of the practical implementation of FAIR data principles, the selected papers were assigned a strength of evidence based on a grading system, described in Table 1:

Table 1. Strength of evidence based on a grading system.

Grade	Criteria
Substantiated	include a clear role of FAIR data in relation to agricultural performance include a comprehensive methodological approach that demonstrates FAIR data implementation processes scientific, evidence based, empirical, quantitative and/or case study
Partially substantiated	include a clear role of FAIR data in relation to agricultural performance include a comprehensive methodological approach that demonstrates FAIR data implementation processes scientific, evidence based, empirical, quantitative and/or case study
Unsubstantiated	studies discussing the role of FAIR data in other contexts and do not qualify for the eligibility criteria

Finally, eight studies met the final appraisal criteria. The eight finalists were examined and assessed for their strength of evidence. Of the final eight studies, two (N = 2 or 25%) were graded as partially substantiated with the strength of evidence II, and six (N = 6 or 75%) were graded as substantiated with the strength of evidence I (Table 2). Average citation rate of the included studies is calculated as 7.

Table 2. Assessment of the selected studies based on their strength of evidence.

#	Citation [a]	Research Areas [b]	Methodology	Scientific	Empirical	Case Study	Descriptive	Evidence [c]	Strength of Evidence [d]	Cited by [e]
1	Wijk et al. [36]	Sci. and Tech; Other topics	*	*	*	*		+++	I	03
2	Harrison et al. [37]	Agri.; Genetics and Heredity	*	*	*	*		+++	I	14
3	Dorich et al. [38]	Sci. and Tech; Env. Sci. & Ecology	*	*		*	*	+++	I	05
4	Giuliani et al. [39]	Remote Sensing	*	*	*	*		+++	I	19
5	Specka et al. [40]	Computer Science; Geology	*	*	*	*		+++	I	02
6	Arnaud et al. [41]	Computer Science	*	*		*	*	+++	I	04
7	Hackett et al. [42]	Plant Sciences		*		*	*	++	II	01
8	Singh et al. [43]	Plant Sciences		*	*	*		++	II	11

* Shows study descriptions (with respect to methodology, types) [a] Authors; [b] Web of Science research areas of the included studies; [c] Substantiated (+++); Partially substantiated (++); Unsubstantiated (+); [d] Strength; [e] Citations.

In this literature review, the included studies were shortlisted based on the inclusion criteria that were confined to the studies: (1) classified under the six Web of Science categories: Agriculture Multidisciplinary, Agriculture Dairy Animal Science, Soil Science, Plant Sciences, Green Sustainable Science Technology and Computer Science Interdisciplinary Applications; (2) listed as article and/or review; (3) written in the English language; (4) comprise applied implications of FAIR data, its implementation and comprehensive methodological approach; (5) published with the top-ranked publishers, i.e., Elsevier, Springer Nature, MDPI, Wiley and Taylor & Francis, in all the journals nominated for the research area; and (6) published within last six (2016–present) years. There were three studies selected through Google Scholar, using the snowballing search technique, based on the reference lists of the studies obtained from Web of Science search. Notably, during our literature review process, we found three studies, i.e., Capalbo et al. [15], Weersink et al. [11] and Wolfert et al. [44], describing the importance of data/big data in the agricultural performance without relating to FAIR data principles. Our search also found four research papers, i.e., Koers et al. [45], Robinson et al. [46], Roitsch et al. [47], Ingram et al. [48] and Bahlo et al. [49], recommending the adoption of FAIR data principles to enhance agricultural performance, without explaining how or why. These papers do not meet the selections criteria; hence, they were excluded during screening and eligibility processes.

It is acknowledged that farming is a multifaceted business with no 'best bet' or 'one method fits all' solution due to the multiplicity of unforeseen factors. All the above criteria may introduce bias in this systematic review that may influence or contrast with the findings.

2.6. Presentation/Interpretation

According to the Web of Science research areas, two studies were categorised under 'Plant Sciences', one listed under 'Remote Sensing', and five studies relates to Agriculture, Science and Technology, Genetics & Heredity, Environmental Science, Ecology, and Geology categories Moreover, the included eight studies have been published to different prominent journals (Table 3).

Table 3. Web of science categories and journals included.

ID	Web of Science Categories	Journal
1	Multidisciplinary Sciences	Scientific Data
2	Agri., Dairy & Animal Science; Genetics & Heredity	Animal Genetics
3	Green & Sustainable Science & Technology; Env. Sci.	Current Opinion in Env. Sustainability
4	Remote Sensing	Int'l Journal of Applied Earth Observation & Geoinformation
5	Computer Sci., Interdisciplinary Applications; Geosciences, Multidisciplinary	Computers & Geosciences
6	Computer Sci., Artificial Intelligence; Computer Sci., Information Systems; Computer Sci., Interdisciplinary Applications	Patterns
7	Plant Sciences	Applications in Plant Sciences
8	Plant Sciences	Trends in Plant Science

Although the trend is increasing (Figure 2), the minimal number of studies demonstrates very slow adoption of FAIR data principles reported in agricultural research publications, which may reflect on the slow uptake by the agricultural researchers and practitioners.

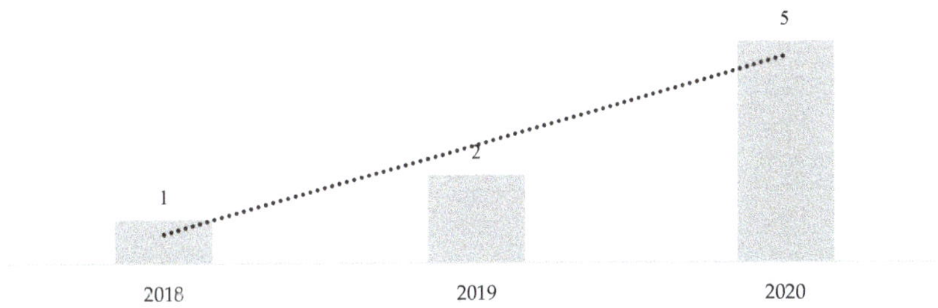

Figure 2. Total number of the studies included and growing trendline.

3. Findings

Since the publication of the FAIR data guidelines [23], there are only eight published studies found through this systematic literature review approach of their adoption in agricultural data. Though these studies confirm the significant role that FAIR data can have on agricultural operational and business performance, the actual implementation of FAIR data principles in the agricultural industry appears to be minimal. Figure 3 demonstrates indicators, sub-indicators of FAIR data, their impacts on sustainable agriculture performance and their assimilation with the sustainability framework [50,51]. The detailed analysis of the final selected papers with respect to their objectives, roles towards agricultural performance, scope and the sources or tools employed, is listed in Table 4.

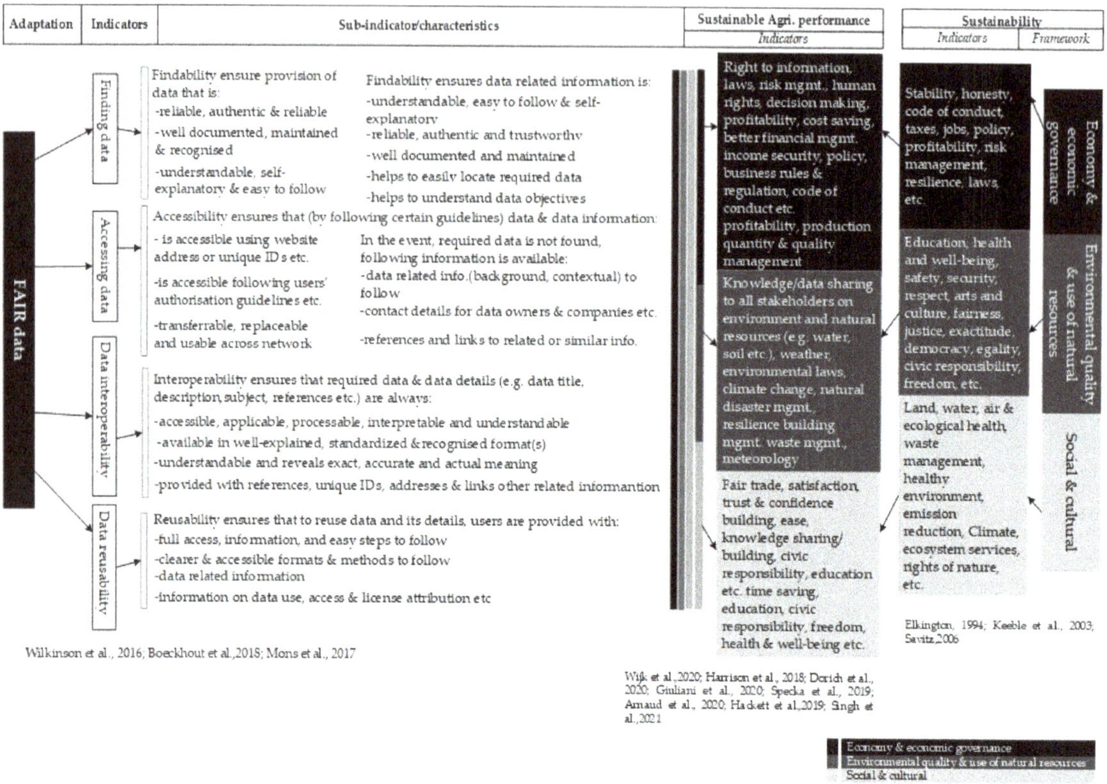

Figure 3. Indicators, sub-indicators, their impacts on sustainable agri. performance, & assimilation with sustainability framework.

Table 4. Authors, objectives, and roles of FAIR data towards agricultural performance, scopes and sources or tools in the included studies.

Author, Objective & Scope	FAIR Data Role Towards Agricultural Performance
1. Wijk et al. [36] *Objective:* To list a well-coherent and interoperable dataset (following FAIR data principles) to help standardization of agricultural household surveys approach by collecting information on 758 variables, to better quantify more than 40 different indicators on farm and household characteristics, welfare, productivity and economic performance. *Scope:* - In line with the rural household multiple indicator survey (RHoMIS). - A total of 21 countries in Central America, sub-Saharan Africa and Asia.	Rural Household Multiple Indicator Survey (RHoMIS) aims to: ▪ promptly characterise a sequence of key indicators across the range of agricultural products and off farm activities, together with marketplace integration, nutrition, food security, poverty and greenhouse gas (GHG) emissions. ▪ measure on- and off-farm paths to food security, various diets and variations in poverty for rural smallholder farm households. ▪ attain sustainable development goals that require more improved sustainable food production and development of rural economies. ▪ better understand the relationships between farming practices, livelihood and the influences on farm performance and household welfare that help with developing targeted investment to advance in agronomic development. Trustworthy indicators at farm-household level of both agri. performance and household wellbeing help to better understand and model these relationships, and to review, revise and update the strategy and execution of intermediations involved officials through an extensive array of changing geographies and socio-economic features. ▪ manage propagation of survey tools and indicators heading to illogical, incoherent and un-interoperable datasets.

Table 4. *Cont.*

Author, Objective & Scope	FAIR Data Role Towards Agricultural Performance
2. Harrison et al. [37] *Objective:* To develop high-quality and rich-supporting metadata (in line with FAIR data guidelines) to describe the project's animals, specimens, cell cultures and experimental assays. *Scope:* - Functional annotation of animal genomes (FAANG) with an initial focus on farmed and companion animals. - UK, USA.	Functional annotation of animal genomes (FAANG) metadata helps in: ▪ creating sample and experiment metadata standards to enhance data recording. ▪ standardising global vocabularies and or expressions by using ontologies. ▪ utilising the wide-ranging livestock datasets produced outside of the project by employing less rigorous legacy guidelines. ▪ developing authentication software to help the community in fulfilling the metadata standards and to input their data to the public archives, hence actively support the community. ▪ providing a community data portal that classifies all sample and experimental datasets by using a single-focused user interface. ▪ successfully dealing with the challenges faced by the consumers in creating infrastructure that mutually and efficiently coordinate genome-to-phenotype research activities. ▪ maximising the usefulness and inter-comparison of assay data. ▪ creating a powerful genome-to-phenotype resource and supports on-going developments in animal data standards as a whole, to support the community.
3. Dorich et al. [38] *Objective:* To create the Global N_2O Database (following FAIR data principles) to serve as a repository for N_2O datasets to be publicly available data and for analytical advances. *Scope:* - Global N_2O Database. - Nitrous oxide emissions. - Global.	The Global N_2O Database deals with farming-oriented (nearly 20% of the total global) GHG emissions and is likely to improve evaluations level by improving annual N_2O estimates. The Global Nitrous oxide (N_2O) Database aims to: ▪ improve Nitrous oxide (N_2O) emission estimates that are primarily obtained from agriculture and is an intoxicating greenhouse gas (GHG) that is roughly 300 times more intense than CO_2 and is the most hazardous ozone-draining material. ▪ improve N_2O assessments, detection of hotspots and alleviation priority zones, and better interpretation of climate change feedbacks. ▪ help with providing an opportunity for methods comparisons by collecting exact data from all practices. ▪ provide value-added computation of annual emissions within a monitoring, reporting and verification process (MRV) scheme for enhanced policy making, for instance greenhouse gas exchanges and to reduce N_2O emissions. ▪ help to moderate and better prepare for climate change by developing and verifying practical abatement strategies in the land use sector, as agriculture has a cause and effect' correlation with the climate change.

Table 4. *Cont.*

Author, Objective & Scope	FAIR Data Role Towards Agricultural Performance
4. Giuliani et al. [39] *Objective:* To develop an innovative, scalable and flexible framework to monitor land degradation at various scales by using various components of the Global Earth observation system of systems (GEOSS) platform to leverage EO resources. *Scope:* - Monitoring land degradation at various scales system. - Land degradation. - Global.	Monitoring of land degradation at various (national, regional, global) scales system, in accordance with the UN SDG 15.3.1 framework, is a successful milestone that effectively embed science into the decision-making process. This system enables users to use EO-based resources more effectively and efficiently. It further aims to: ▪ reduce climate change and biodiversity losses, as well as ensuring food security and sufficient provision of ecosystem services, at the same time. ▪ produce multidisciplinary reliable knowledge on quantifiable objectives, at different scales, in order to proficiently support applied policymaking to ensure balanced functioning of the ecosystem. ▪ achieve sustainable development goals (SDG) 15.3.1., by following the data-information-knowledge pattern using the Trends.Earth model [52] and several data sources to produce the indicator. ▪ provide more flexible and scalable version of Trends.Earth to enhance decision-making processes and to scope of our planet, for instance natural resources, etc. ▪ strengthen respective regional capacities to effectively assess and map the degraded lands as per the UN sustainable development goals (SDGs). ▪ institute 'data analytics' podia that can potentially help nations to discover, access and use the necessary datasets to evaluate land degradation.
5. Specka et al. [40] *Objective:* To frame new model, BonaRes metadata schema (following FAIR principles), by integrating the INSPIRE and DataCite metadata schemas. *Scope:* - Model based on two schemas, i.e., INSPIRE and DataCite metadata. - The BonaRes metadata schema for geospatial soil-agricultural research. - Global.	In compliance with the INSPIRE and DataCite metadata schemes and FAIR data principles, a modern research data management, BonaRes metadata: ▪ supports cross-portal metadata interoperability with other INSPIRE-compliant spatial data infrastructures (SDIs). ▪ increases the visibility and findability of researchers' investigation, as the data can be assigned a digital object identifier (DOI), which is essential for data publications and data citations. ▪ allows targeted dataset queries and to better the discovery and reusability of research data.
6. Arnaud et al. [41] *Objective:* To annotate multidisciplinary research data with the appropriate ontologies to stimulate the ontology content to fill the gap rather than developing completely new ontologies. *Scope:* - Consultative group on international agricultural research (CGIAR) methodology. - Multidisciplinary ontologies—Focuses on ontologies in the fields of agronomy, crop, environment, plant, and socio-economic. - Global.	Annotation of and integrative, multifaceted, versatile, associative research data with the most suitable ontologies aims to comply with the FAIR data principles, and to strengthen the findability of data for further reuse, hence adding to the return on investment (RoI) for information collection and storage. It further aims to: ▪ provide most appropriate ontologies with more applicable and reliable data control, including data stewardship, ownership and robust policy. ▪ enhance technological structure (essential softwares, connectivity and servers) that plays key role in arranging and classifying the tangible data structures, i.e., ontologies, taxonomies and structured vocabularies to finally form a supportable data management system. ▪ exclusively increase the influence of agronomic research and development (R&D) by following FAIR data guidelines to build community trust level.

Table 4. *Cont.*

Author, Objective & Scope	FAIR Data Role Towards Agricultural Performance
7. Hackett et al. [42] *Objective:* To present biodiversity data workflow, i.e., data collection and curation, from multiple sources while abiding by FAIR data principles, to enable researchers, managers, and policymakers to address issues of global and future concern. *Scope:* - Biodiversity associated with globally vulnerable prairie fen wetlands. Plant diversity research and species-focused studies concerning the biology, ecology and behaviour of the federally endangered Poweshiek skipperling. - Plant communities and the federally endangered Poweshiek skipperling (Oarisma poweshiek). - Michigan prairie Fens, USA.	Global biodiversity information facility (GBIF) data sets: ▪ help to deal with the heterogeneity of biodiversity data issues, faced by data collections units, research and management communities, by implementing FAIR data principles for large-scale research. ▪ provides new opportunities to better understand vibrant natural systems and help establishing more applied data resource management. ▪ successfully address specific queries in the fields of phylogenomics, biogeography climatology, ecology and evolution, etc.
8. Singh et al. [43] *Objective:* To develop a single database for annotated plant stress images that supports FAIR principles (of accessibility and reusability) *to* propose an overarching strategy for utilizing ML techniques that methodically enables the application of plant stress phenotyping at multiple scales across different types of stresses, program goals, and environments. *Scope:* - Plant stress phenotyping 'plant stress severity' to encompass both biotic and abiotic stresses. - Plant stress phenotyping, Maize plants. - Iowa, USA.	Plant stress evaluations measure the visible signs and/or indications of stress and its progress on different plant units (e.g., leaf, stem, or roots) at the leaf, canopy, plot and field levels. A comprehensive database for annotated plant stress images, embedded with FAIR data principles, aims to: ▪ choose stress-resistant varieties and to develop better stress-management schemes. ▪ standardise visual evaluations and to utilise imaging techniques to better enhance the precision and trustworthiness of stress assessment in contrast with single-handed visual measurement. ▪ enhance machine learning (ML) approaches combined with image-based phenotyping to get up-to-date insights from highly organised, annotated (supported with explanations and/or comments), and high-dimensional (provided with the staggeringly higher number of dimensions) datasets across wide-ranging stresses and crops. ▪ concurrently build up the pace, precision, trustworthiness and scalability of stress phenotyping, and agility for highly varying program objectives; whereas innovative ML algorithms offer extended plant stress phenotyping techniques to deal with these challenges. ▪ advance mechanisation and accuracy of plant stress gravity evaluations that improve the proportion of genetic gain within crops, providing comprehensive management approaches. Notably, stress gravity evaluations in plants are important for appraising management strategies, plant breeding choice approaches and checking novel varieties for their capacity to alleviate crop damages. Additionally, efficiently quantify plant traits under different environmental circumstances with a stipulated precision and accuracy at various scales from organs to canopies.

Sustainable and enhanced agricultural production highly rely on the farming strategies, methods and decision making made by the stakeholders [5,6,53]. Accomplishment of FAIR data principles ensures data findability, accessibility, interoperability and reusability that extensively contributes better decision making towards both the operations and business management. Figure 3 demonstrates on how the sub-indicator/characteristics of FAIR data lead to sustainable agricultural performance indicators, and the way the sustainability indicators assimilate with the sustainable agricultural performance indicators, ultimately complementing the sustainability triple bottom line framework.

4. Discussion

The analysis of papers found through this systematic literature review clearly demonstrate that the provision of FAIR data can help improve agriculture operational and business performances. The multidimensional challenges in agriculture demand better decision making, which in turn relies on better access to data for wiser decision making. In this regard, a big responsibility lies on the shoulders of the stakeholders, i.e., farmers, farm managers, agronomists, service providers, researchers, etc. to make their data FAIR.

However, the systematic literature review also found only a few documented cases of practical examples that relate FAIR data to agricultural performance. While there may be many reasons for the low number of published case studies of adoption of FAIR data in agriculture, this review strongly indicates a need for social science research to explore what those reasons may be. Clearly, the stakeholders' value propositions for the adoption of FAIR data in agriculture need to be better understood.

In accordance with Rogers' [54], diffusion of innovation theory simply having knowledge of a new idea is not enough. By and large, informative individuals go through a persuasion stage to build either positive or negative attitudes towards the new idea, in accordance with innovation decision processes. Every novel idea carries a certain level of hesitation as the individual consumers of various personality types weigh up their appetite for risk and capacity for adopting change and so on. It is keenly realised that researchers should incorporate and comprehend the farmers' social perspectives, seeing that the comprehension of attitude and social capacity are the key indicators towards viable agriculture and central to better explore the community attitudes and behaviour [55,56].

Firstly, agriculture practitioners' attitude towards data and data sources in addition to their capability to employ the information are critical factors for the useful utilisation of those data [57,58]. Data are key in decision making [27], whereas the practitioners' attitude towards data authenticity, data source, its genuineness and applicability are critical factors towards its usefulness [57,58]. Secondly, practitioners and stakeholders play their respective roles towards the betterment of agriculture. If agricultural data—as a key input in decision-making—are not findable, and accessible, they will be of no use even though they are well researched. Stakeholders' data comprehension and knowledge exchange at all levels are important factors towards agricultural improvement [59]. If the data are not comprehendible to the end-users, they will not be used in accordance with the notion of the behavioural economics [60]. Thirdly, the available agriculture data are found disparate, disorganised and disarrayed [61]. The required datasets are either not available [62,63] or the consumers are provided with restricted and/or limited access [19].

Theory of reasoned action (TRA) uncovers that an individual's objective and decision making depends upon the level of information or data they have [64]. Stakeholders' knowledge building, including that of FAIR data, is essential. In this way, those with limited or no technical knowledge are asked to educate themselves with the emerging technology to better adapt with the modern advancements. Adoption of digital agriculture in general, and the adoption of FAIR data principles in particular, are facing several challenges. Politically, in Australia for instance, digital agriculture needs more proactive steps towards policy making, cooperation, administration and cross industry collaboration. Socially, digital literacy among all the key stakeholders can play a vital role to better comprehend value proposition of digital agriculture, and to bridge the trust gap between consumers, technology providers and data custodians. While device connectivity and the required data input are key prerequisites to make the technology work, the lack of mobile connectivity and internet telecommunications infrastructure, and limited or no access to the required data, curtail practitioners to use the technology at its best [65]. Economically, the Australian agriculture sector estimates an increase by 42% in its technology-oriented capital by 2030, whereas the farming sector needs to adopt the latest farming technologies and techniques to meet future production demands. Digital agriculture can add a gross value of AUD 20.3 billion [63], while better agricultural planning and appropriate farming techniques have a potential to further supplement AUD 100 billion to the Australian economy by 2030 [66].

The provision of FAIR data is a well-documented constraint in modern agriculture [19,67,68], but there is little to no research on why FAIR data are not made available. If FAIR data are widely regarded as important, why is it not widely adopted? Is stakeholders' knowledge and perception towards FAIR data a barrier? To comprehend these questions, a FAIR data process flowchart has been devised as a guiding template to direct the social research. Using plain language, this flowchart attempts to step an agricultural practitioner through the requirements for FAIR data to ascertain where they perceive barriers and difficulties in adopting FAIR data processes.

The flowchart (Figure 4) intends to test the extents to which each of the required FAIR data components (findability, accessibility, interoperability, reusability) present barriers to adoption, whether these barriers are equally distributed across the four components and whether there are specific steps within each component that present higher barriers than others. By troubleshooting the specific barriers of each requirement, the value propositions of adopting FAIR data in agriculture can be better understood and addressed.

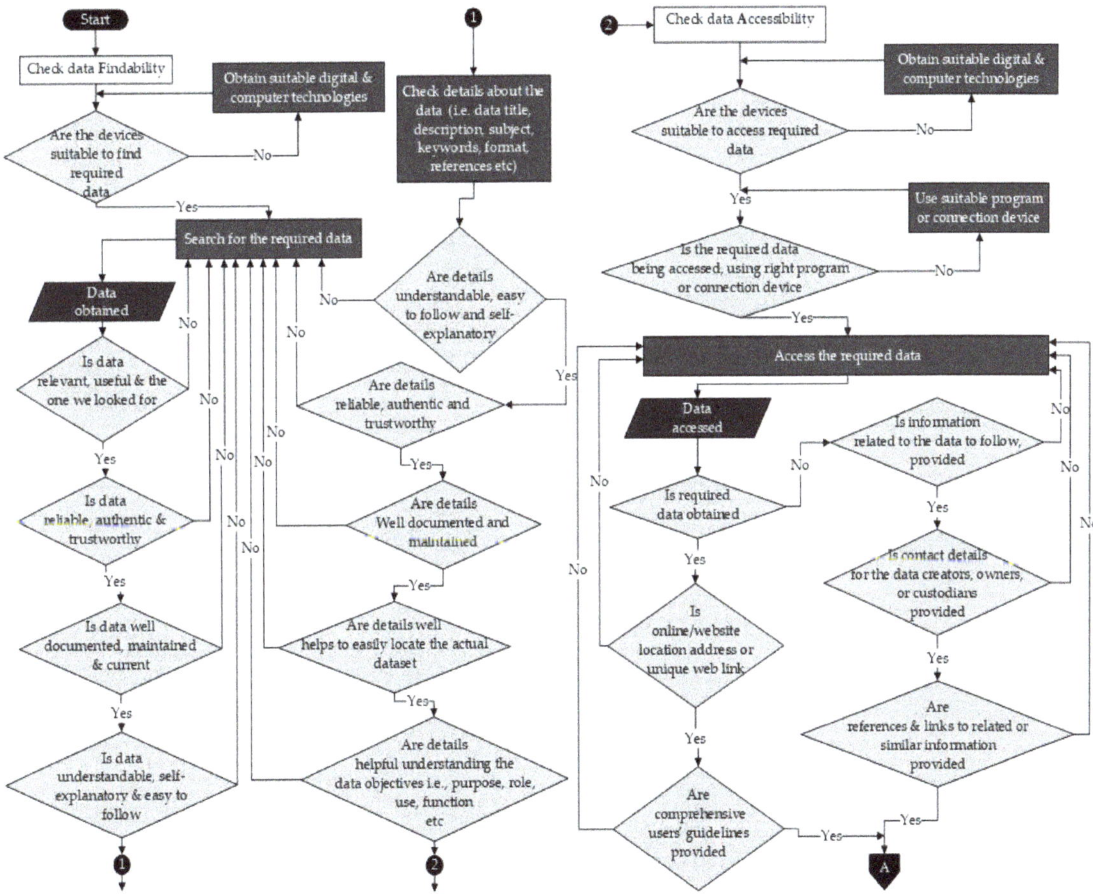

Figure 4. *Cont.*

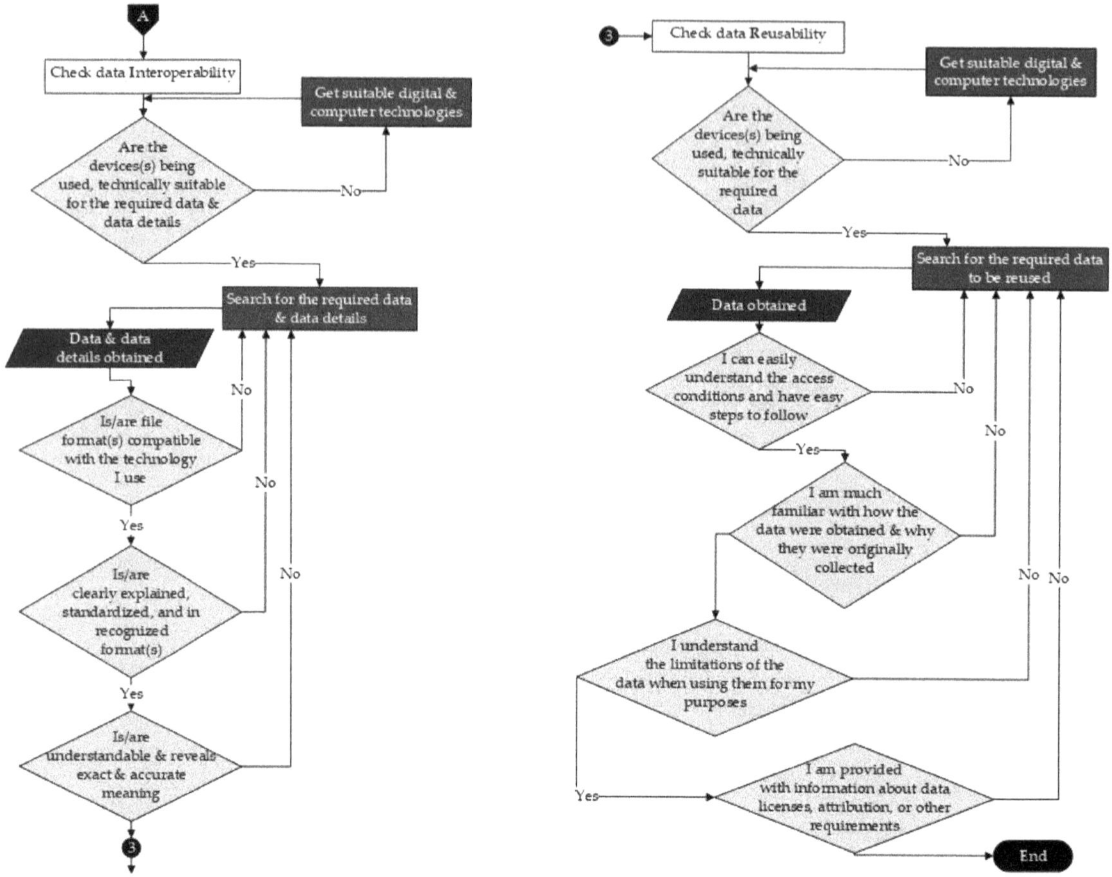

Figure 4. FAIR data process flowchart diagram in plain language.

5. Conclusions

The advent of digital agriculture has provided agricultural practitioners with access to a plethora of internet resources, sensors and applications for decision support in their operations and businesses. The FAIR data principles were designed to address the challenge of how to harness the increasing volume of disparate, but relevant data to improve decision making and enhance agricultural performance. This systematic literature review reinforces this hypothesis but finds very few published examples of how adopting the principles are related to agricultural performance, indicating that the value proposition is yet to be realised. To explore this tardiness will require social science research to find the specific barriers in making data findable, accessible, interoperable and reusable, so that benefits of digital agricultural can be more broadly gained.

Author Contributions: This project is supervised by P.D.; B.A. developed research framework, methodology and strategy; original draft preparation was prepared by B.A.; P.D. reviewed and edited the draft. All authors have read and agreed to the published version of the manuscript.

Funding: This research was funded by the Centre for eResearch and Digital Innovation (CeRDI), Federation University and Food Agility Cooperative Research Centre, Australia, under Grant Project CERDI-19-136.

Institutional Review Board Statement: Not applicable.

Informed Consent Statement: Not applicable.

Acknowledgments: We thank the Centre for eResearch and Digital Innovation (CeRDI), Federation University, and Food Agility Cooperative Research Centre, Australia. We are grateful to the whole CeRDI research team for their support in completing this project.

Conflicts of Interest: The authors declare no conflict of interest.

References

1. FAO, Food and Agriculture Organization of the United Nations. *The State of Food Security and Nutrition in the World 2018*; Food and Agriculture Organization of the United Nations: Rome, Italy, 2018.
2. Baligar, V.; Fageria, N.; He, Z. Nutrient use efficiency in plants. *Commun. Soil Sci. Plant Anal.* **2001**, *32*, 921–950. [CrossRef]
3. Capmourteres, V.; Adams, J.; Berg, A.; Fraser, E.; Swanton, C.; Anand, M. Precision conservation meets precision agriculture: A case study from southern Ontario. *Agric. Syst.* **2018**, *167*, 176–185. [CrossRef]
4. Rijswijk, K.; Klerkx, L.; Turner, J.A. Digitalisation in the New Zealand Agricultural Knowledge and Innovation System: Initial understandings and emerging organisational responses to digital agriculture. *NJAS-Wagening. J. Life Sci.* **2019**, *90–91*, 100313. [CrossRef]
5. Kusunose, Y.; Mahmood, R. Imperfect forecasts and decision making in agriculture. *Agric. Syst.* **2016**, *146*, 103–110. [CrossRef]
6. Acosta, M.; van Wessel, M.; van Bommel, S.; Ampaire, E.L.; Twyman, J.; Jassogne, L.; Feindt, P.H. What does it Mean to Make a 'Joint' Decision? Unpacking Intra-household Decision Making in Agriculture: Implications for Policy and Practice. *J. Dev. Stud.* **2020**, *56*, 1210–1229. [CrossRef]
7. Shepherd, M.; Turner, J.A.; Small, B.; Wheeler, D. Priorities for science to overcome hurdles thwarting the full promise of the 'digital agriculture' revolution. *J. Sci. Food Agric.* **2020**, *100*, 5083–5092. [CrossRef]
8. Birner, R.; Daum, T.; Pray, C. Who drives the digital revolution in agriculture? A review of supply-side trends, players and challenges. *Appl. Econ. Perspect. Policy* **2021**, *43*, 1260–1285. [CrossRef]
9. Whitelock, V. Business analytics and firm performance: Role of structured financial statement data. *J. Bus. Anal.* **2018**, *1*, 81–92. [CrossRef]
10. Paustian, M.; Theuvsen, L. Adoption of precision agriculture technologies by German crop farmers. *Precis. Agric.* **2017**, *18*, 701–716. [CrossRef]
11. Weersink, A.; Fraser, E.; Pannell, D.; Duncan, E.; Rotz, S. Opportunities and challenges for Big Data in agricultural and environmental analysis. *Annu. Rev. Resour. Econ.* **2018**, *10*, 19–37. [CrossRef]
12. Kitchen, N.; Sudduth, K.; Myers, D.; Massey, R.; Sadler, E.; Lerch, R.; Hummel, J.; Palm, H. Development of a conservation-oriented precision agriculture system: Crop production assessment and plan implementation. *J. Soil Water Conserv.* **2005**, *60*, 421–430.
13. Fountas, S.; Blackmore, S.; Ess, D.; Hawkins, S.; Blumhoff, G.; Lowenberg-Deboer, J.; Sorensen, C. Farmer experience with precision agriculture in Denmark and the US Eastern Corn Belt. *Precis. Agric.* **2005**, *6*, 121–141. [CrossRef]
14. Antle, J.M.; Basso, B.; Conant, R.T.; Godfray, H.C.J.; Jones, J.W.; Herrero, M.; Howitt, R.E.; Keating, B.A.; Munoz-Carpena, R.; Rosenzweig, C. Towards a new generation of agricultural system data, models and knowledge products: Design and improvement. *Agric. Syst.* **2017**, *155*, 255–268. [CrossRef]
15. Capalbo, S.M.; Antle, J.M.; Seavert, C. Next generation data systems and knowledge products to support agricultural producers and science-based policy decision making. *Agric. Syst.* **2017**, *155*, 191–199. [CrossRef]
16. Yost, M.; Kitchen, N.; Sudduth, K.; Sadler, E.; Drummond, S.; Volkmann, M. Long-term impact of a precision agriculture system on grain crop production. *Precis. Agric.* **2017**, *18*, 823–842. [CrossRef]
17. Delgado, J.A.; Khosla, R.; Mueller, T. Recent advances in precision (target) conservation. *J. Soil Water Conserv.* **2011**, *66*, 167A–170A. [CrossRef]
18. Berry, J.K.; Delgado, J.; Khosla, R.; Pierce, F. Precision conservation for environmental sustainability. *J. Soil Water Conserv.* **2003**, *58*, 332–339.
19. Barry, S.; Darnell, R.; Grundy, M.; Moore, A.; Robertson, M.; Brown, J.; Gaire, R.; George, A. *Precision to Decision–Current and Future State of Agricultural Data for Digital Agriculture in Australia*; CSIRO: Canberra, Australia, 2018.
20. Bronson, K.; Knezevic, I. Big Data in food and agriculture. *Big Data Soc.* **2016**, *3*, 2053951716648174. [CrossRef]
21. Leite, A.E.; De Castro, R.; Jabbour, C.J.C.; Batalha, M.O.; Govindan, K. Agricultural production and sustainable development in a Brazilian region (Southwest, São Paulo State): Motivations and barriers to adopting sustainable and ecologically friendly practices. *Int. J. Sustain. Dev. World Ecol.* **2014**, *21*, 422–429. [CrossRef]
22. Delgado, J.A.; Bausch, W. Potential use of precision conservation techniques to reduce nitrate leaching in irrigated crops. *J. Soil Water Conserv.* **2005**, *60*, 379–387.
23. Wilkinson, M.D.; Dumontier, M.; Aalbersberg, I.J.; Appleton, G.; Axton, M.; Baak, A.; Blomberg, N.; Boiten, J.-W.; da Silva Santos, L.B.; Bourne, P.E. The FAIR Guiding Principles for scientific data management and stewardship. *Sci. Data* **2016**, *3*, 1–9. [CrossRef] [PubMed]
24. Boeckhout, M.; Zielhuis, G.A.; Bredenoord, A.L. The FAIR guiding principles for data stewardship: Fair enough? *Eur. J. Hum. Genet.* **2018**, *26*, 931–936. [CrossRef] [PubMed]

25. Wilkinson, M.D.; Verborgh, R.; da Silva Santos, L.O.B.; Clark, T.; Swertz, M.A.; Kelpin, F.D.; Gray, A.J.; Schultes, E.A.; van Mulligen, E.M.; Ciccarese, P. Interoperability and FAIRness through a novel combination of Web technologies. *PeerJ Comput. Sci.* **2017**, *3*, e110. [CrossRef]
26. Foundation, G.F. GO FAIR Foundation. Available online: https://www.gofairfoundation.org/ (accessed on 1 November 2021).
27. Mons, B.; Neylon, C.; Velterop, J.; Dumontier, M.; da Silva Santos, L.O.B.; Wilkinson, M.D. Cloudy, increasingly FAIR; revisiting the FAIR Data guiding principles for the European Open Science Cloud. *Inf. Serv.* **2017**, *37*, 49–56. [CrossRef]
28. Commons, A.R.D. FAIR Self Assessment Tool. Available online: https://ardc.edu.au/resources/working-with-data/fair-data/fair-self-assessment-tool/ (accessed on 20 August 2021).
29. Van der Knaap, L.M.; Leeuw, F.L.; Bogaerts, S.; Nijssen, L.T. Combining Campbell standards and the realist evaluation approach: The best of two worlds? *Am. J. Eval.* **2008**, *29*, 48–57. [CrossRef]
30. Moher, D.; Liberati, A.; Tetzlaff, J.; Altman, D.G.; Grp, P. Preferred Reporting Items for Systematic Reviews and Meta-Analyses: The PRISMA Statement. *PLoS Med.* **2009**, *6*, e1000097. [CrossRef]
31. Koutsos, T.M.; Menexes, G.C.; Dordas, C.A. An efficient framework for conducting systematic literature reviews in agricultural sciences. *Sci. Total Environ.* **2019**, *682*, 106–117. [CrossRef]
32. Clarivate. Available online: https://clarivate.com/webofsciencegroup/solutions/web-of-science/ (accessed on 16 February 2022).
33. Google Scholar. Available online: https://scholar.google.com.au/ (accessed on 16 February 2022).
34. Wohlin, C. Guidelines for Snowballing in Systematic Literature Studies and a Replication in Software Engineering. In Proceedings of the 18th International Conference on Evaluation and Assessment in Software Engineering, London, UK, 13–14 May 2014; pp. 1–10.
35. van der Windt, D.; Thomas, E.; Pope, D.P.; de Winter, A.F.; Macfarlane, G.J.; Bouter, L.M.; Silman, A.J. Occupational risk factors for shoulder pain: A systematic review. *Occup. Environ. Med.* **2000**, *57*, 433–442. [CrossRef]
36. Wijk, v.M.; Hammond, J.; Gorman, L.; Adams, S.; Ayantunde, A.; Baines, D.; Bolliger, A.; Bosire, C.; Carpena, P.; Chesterman, S. The Rural Household Multiple Indicator Survey, data from 13,310 farm households in 21 countries. *Sci. Data* **2020**, *7*, 1–9.
37. Harrison, P.W.; Fan, J.; Richardson, D.; Clarke, L.; Zerbino, D.; Cochrane, G.; Archibald, A.L.; Schmidt, C.J.; Flicek, P. FAANG, establishing metadata standards, validation and best practices for the farmed and companion animal community. *Anim. Genet.* **2018**, *49*, 520–526. [CrossRef]
38. Dorich, C.D.; Conant, R.T.; Albanito, F.; Butterbach-Bahl, K.; Grace, P.; Scheer, C.; Snow, V.O.; Vogeler, I.; van der Weerden, T.J. Improving N_2O emission estimates with the global N_2O database. *Curr. Opin. Environ. Sustain.* **2020**, *47*, 13–20. [CrossRef]
39. Giuliani, G.; Mazzetti, P.; Santoro, M.; Nativi, S.; Van Bemmelen, J.; Colangeli, G.; Lehmann, A. Knowledge generation using satellite earth observations to support sustainable development goals (SDG): A use case on Land degradation. *Int. J. Appl. Earth Obs. Geoinf.* **2020**, *88*, 102068. [CrossRef]
40. Specka, X.; Gärtner, P.; Hoffmann, C.; Svoboda, N.; Stecker, M.; Einspanier, U.; Senkler, K.; Zoarder, M.A.M.; Heinrich, U. The BonaRes metadata schema for geospatial soil-agricultural research data—Merging INSPIRE and DataCite metadata schemes. *Comput. Geosci.* **2019**, *132*, 33–41. [CrossRef]
41. Arnaud, E.; Laporte, M.-A.; Kim, S.; Aubert, C.; Leonelli, S.; Miro, B.; Cooper, L.; Jaiswal, P.; Kruseman, G.; Shrestha, R. The ontologies community of practice: A CGIAR initiative for big data in agrifood systems. *Patterns* **2020**, *1*, 100105. [CrossRef]
42. Hackett, R.A.; Belitz, M.W.; Gilbert, E.E.; Monfils, A.K. A data management workflow of biodiversity data from the field to data users. *Appl. Plant Sci.* **2019**, *7*, e11310. [CrossRef]
43. Singh, A.; Jones, S.; Ganapathysubramanian, B.; Sarkar, S.; Mueller, D.; Sandhu, K.; Nagasubramanian, K. Challenges and Opportunities in Machine-Augmented Plant Stress Phenotyping. *Trends Plant Sci.* **2021**, *26*, 56–69. [CrossRef]
44. Wolfert, S.; Ge, L.; Verdouw, C.; Bogaardt, M.-J. Big data in smart farming–a review. *Agric. Syst.* **2017**, *153*, 69–80. [CrossRef]
45. Koers, H.; Bangert, D.; Hermans, E.; van Horik, R.; de Jong, M.; Mokrane, M. Recommendations for Services in a FAIR Data Ecosystem. *Patterns* **2020**, *1*, 100058. [CrossRef]
46. Robinson, N.J.; Dahlhaus, P.G.; Wong, M.; MacLeod, A.; Jones, D.; Nicholson, C. Testing the public–private soil data and information sharing model for sustainable soil management outcomes. *Soil Use Manag.* **2019**, *35*, 94–104. [CrossRef]
47. Roitsch, T.; Cabrera-Bosquet, L.; Fournier, A.; Ghamkhar, K.; Jiménez-Berni, J.; Pinto, F.; Ober, E.S. Review: New sensors and data-driven approaches—A path to next generation phenomics. *Plant Sci.* **2019**, *282*, 2–10. [CrossRef]
48. Ingram, J.; Maye, D.; Bailye, C.; Barnes, A.; Bear, C.; Bell, M.; Cutress, D.; Davies, L.; de Boon, A.; Dinnie, L. What are the priority research questions for digital agriculture? *Land Use Policy* **2022**, *114*, 105962. [CrossRef]
49. Bahlo, C.; Dahlhaus, P.; Thompson, H.; Trotter, M. The role of interoperable data standards in precision livestock farming in extensive livestock systems: A review. *Comput. Electron. Agric.* **2019**, *156*, 459–466. [CrossRef]
50. Elkington, J. Towards the sustainable corporation: Win-win-win business strategies for sustainable development. *Calif. Manag. Rev.* **1994**, *36*, 90–100. [CrossRef]
51. Keeble, J.J.; Topiol, S.; Berkeley, S. Using indicators to measure sustainability performance at a corporate and project level. *J. Bus. Ethics* **2003**, *44*, 149–158. [CrossRef]
52. Trends.Earth. Available online: https://trends.earth/docs/en/ (accessed on 16 February 2022).
53. Tilman, D.; Balzer, C.; Hill, J.; Befort, B.L. Global food demand and the sustainable intensification of agriculture. *Proc. Natl. Acad. Sci. USA* **2011**, *108*, 20260–20264. [CrossRef]

54. Rogers, E.M. *Diffusion of Innovations*, 5th ed.; The Free Press: New York, NY, USA, 2003.
55. CCMA. *North Central Victoria Regional Sustainable Agriculture Strategy*; Central Catchment Management Authority: Huntly, Australia, 2016.
56. Tomer, M. How do we identify opportunities to apply new knowledge and improve conservation effectiveness? *J. Soil Water Conserv.* **2010**, *65*, 261–265. [CrossRef]
57. Milovanović, S. The role and potential of information technology in agricultural improvement. *Econ. Agric.* **2014**, *61*, 471–485. [CrossRef]
58. Zhang, Y.; Wang, L.; Duan, Y. Agricultural information dissemination using ICTs: A review and analysis of information dissemination models in China. *Inf. Process. Agric.* **2016**, *3*, 17–29. [CrossRef]
59. Webb, N.P.; Marshall, N.A.; Stringer, L.C.; Reed, M.S.; Chappell, A.; Herrick, J.E. Land degradation and climate change: Building climate resilience in agriculture. *Front. Ecol. Environ.* **2017**, *15*, 450–459. [CrossRef]
60. Teitelbaum, J.C.; Zeiler, K. *Research Handbook on Behavioral Law and Economics*; Edward Elgar Publishing: Northampton, MA, USA, 2018.
61. Leonard, E.; Rainbow, R.; Laurie, A.; Lamb, D.; Llewellyn, R.; Perrett, E.; Sanderson, J.; Skinner, A.; Stollery, T.; Wiseman, L. *Accelerating Precision Agriculture to Decision Agriculture: Enabling Digital Agriculture in Australia*; Cotton Research and Development Corporation: Narrabri, Australia, 2017.
62. Keogh, M.; Henry, M. *The Implications of Digital Agriculture and Big Data for Australian Agriculture*; Australian Farm Institute: Sydney, Australia, 2016; pp. 1–51. ISBN 978-1-921808-38-8.
63. Nolet, S. *Seeds of Success: Advancing Digital Agriculture from Point Solutions to Platforms*; United States Studies Centre at the University of Sydney: Sydney, Australia, 2018; pp. 20–25.
64. Ajzen, I.; Fishbein, M. *Understanding Attitudes and Predicting Social Behavior*; Prentice-Hall: Englewood Cliffs, NJ, USA, 1980.
65. Wiseman, L.; Sanderson, J. *P2D Project: Accelerating Precision Agriculture to Decision Agriculture*; Griffith University: Gold coast, Australia, 2017; ISBN 978-0-6482462-1-3.
66. NFF. *Budget Roadmap Charts Course for $100 Billion in Farm Production by 2030*, 2018.
67. Allemang, D.; Teegarden, B. A global data ecosystem for agriculture and food. *FResearch* **2017**, *6*. [CrossRef]
68. CRDC, Cotton Research & Development Corporation. *A Big Data Reference Architecture for Digital Agriculture in Australia*; Cotton Research & Development Corporation: Narrabri, Australia, 2017; ISBN 978-0-6481983-0-7.

agriculture

MDPI

Article

Application of Artificial Neural Network Sensitivity Analysis to Identify Key Determinants of Harvesting Date and Yield of Soybean (*Glycine max* [L.] Merrill) Cultivar Augusta

Gniewko Niedbała [1,*], Danuta Kurasiak-Popowska [2], Magdalena Piekutowska [3], Tomasz Wojciechowski [1], Michał Kwiatek [2] and Jerzy Nawracała [2]

[1] Department of Biosystems Engineering, Faculty of Environmental and Mechanical Engineering, Poznań University of Life Sciences, Wojska Polskiego 50, 60-627 Poznań, Poland; tomasz.wojciechowski@up.poznan.pl

[2] Department of Genetics and Plant Breeding, Faculty of Agronomy, Horticulture and Bioengineering, Poznań University of Life Sciences, Dojazd 11, 60-632 Poznań, Poland; danuta.kurasiak-popowska@up.poznan.pl (D.K.-P.); michal.kwiatek@up.poznan.pl (M.K.); jerzy.nawracala@up.poznan.pl (J.N.)

[3] Department of Geoecology and Geoinformation, Institute of Biology and Earth Sciences, Pomeranian University in Słupsk, Partyzantów 27, 76-200 Słupsk, Poland; magdalena.piekutowska@apsl.edu.pl

* Correspondence: gniewko.niedbala@up.poznan.pl

Abstract: Genotype and weather conditions play crucial roles in determining the volume and stability of a soybean yield. The aim of this study was to identify the key meteorological factors affecting the harvest date (model M_HARV) and yield of the soybean variety Augusta (model M_YIELD) using a neural network sensitivity analysis. The dates of the start of flowering and maturity, the yield data, the average daily temperatures and precipitation were collected, and the Selyaninov hydrothermal coefficients were calculated during a fifteen-year study (2005–2020 growing seasons). During the experiment, highly variable weather conditions occurred, strongly modifying the course of phenological phases in soybean and the achieved seed yield of Augusta cultivar. The harvesting of mature soybean seeds took place between 131 and 156 days after sowing, while the harvested yield ranged from 0.6 t·ha^{-1} to 2.6 t·ha^{-1}. The sensitivity analysis of the MLP neural network made it possible to identify the factors which had the greatest impact on the tested dependent variables among all the analyzed factors. It was revealed that the variables assigned ranks 1 and 2 in the sensitivity analysis of the neural network forming the M_HARV model were total rainfall in the first decade of June and the first decade of August. The variables with the highest impact on the Augusta soybean seed yield (model M_YIELD) were the mean daily air temperature in the second decade of May and the Seljaninov coefficient values calculated for the sowing–flowering date period.

Keywords: soybean; yield; sensitivity analysis; vegetation period; weather conditions; artificial neural network

Citation: Niedbała, G.; Kurasiak-Popowska, D.; Piekutowska, M.; Wojciechowski, T.; Kwiatek, M.; Nawracała, J. Application of Artificial Neural Network Sensitivity Analysis to Identify Key Determinants of Harvesting Date and Yield of Soybean (*Glycine max* [L.] Merrill) Cultivar Augusta. *Agriculture* **2022**, *12*, 754. https://doi.org/10.3390/agriculture12060754

Academic Editor: William A. Payne

Received: 11 April 2022
Accepted: 24 May 2022
Published: 25 May 2022

1. Introduction

Soybean (*Glycine max* [L.] Merrill) is the most important legume crop worldwide with a forecast of production at 353.8 million tonnes [1]. It is also the main source of valuable plant protein and the second source of oil, and the global demand for soybean has been constantly growing. Poland is highly dependent on soybean meal imports, a current volume of around 2.5 million tons. Independence from protein imports can be ensured by an increase of the acreage of soybean cultivation. Over the last ten years, the cultivation area has increased from <1000 to 25,552 hectares (in 2021) [2,3], but soybean is still considered to be a new crop for Polish farmers. One of the reasons for such a small acreage is the location of Poland, which is over 49 degrees latitude, north of the world's main soybean cultivation regions. There are several major factors which limit soybean's fitness for its purposes, and they are as follows: long daytime duration, low temperature at the time of germination and flowering, and requirements for rainfall [4]. The lack of suitable cultivars adapted to climatic conditions is the main problem for soybean cultivation in Poland. Primarily, early maturing cultivars ("000") [5] have been promoted in Poland. However, the key issue for the higher latitude adaptation is the proper combination of allelic variants at the *E1*, *E2*, *E3*, and *E4* loci [6]. The varieties having all four recessive alleles are insensitive in terms of photoperiod; for example Nawiko and Augusta, which was bred at the Department of Genetics and Plant Breeding, Poznań University of Life Sciences. Further, the very high variability and diversity of weather conditions observed in individual years, which is a characteristic of Poland's transitional climate, can be considered an additional obstacle for soybean adaptation, causing significant fluctuations in the dates of the flowering initiation and maturity [7]. Unfavorable growing conditions—including cold stress—cause a reduction in soybean yield and its nutritional value [8,9].

There are many different factors that affect soybean adaptiveness around the world. In Central and South Germany, a positive correlation between seed yield with solar radiation (r = 0.32) and precipitation (r = 0.33) was found to be significant, but the same factor was negatively correlated with Crop Heat Units (CHU) (r = −0.42). Varieties from maturity group MG 00 were less correlated with the tested environmental factors than varieties from maturity group MG 000 [10]. In the far east, the yield-limiting, environmental factor is temperature, but for the Krasnodar region, the yield was positively related to the hydrothermal coefficient; a lack of moisture becomes a significant disadvantage for soybean in this region [11]. Also, in Argentina, the moisture availability during the period from flowering to pod formation is critical for productivity [12]. Precipitation is considered a major factor in the formation of soybean yield components in most regression models [13,14], and water deficiency is reported to be one of the most important environmental factors, reducing crop (including soybean) productivity more than any other factor [15,16]. Both too-high and too-low temperatures can reduce the yield of soybean. Cold stress at the flowering stage negatively affects the elements of the plant habits and seed yield of soybean, which results in a high level of yield decrease shown in late cultivars, while a smaller and similar yield decrease was observed in early and medium–early cultivars [17]. Both elevated temperature and water stresses post-flowering significantly affected plant growth and yield parameters negatively. The combined effects of the two factors were more severe than the individual stresses [18].

Moreover, global warming has been a new factor that has increased the incidence of extreme weather events in recent years. The effect of temperature rise may vary. Annual global mean temperatures varied from 15.0 to 15.3 °C and are likely to exert a positive impact on the average yield [19]. Tacarindua et al. [20,21] reported that temperature rise during the growing season from 26 to 30 °C affected the reduction of dry mass production, harvest index, seed number, pod number, and single-seed size, and thereby seed yield. Predicting models demonstrated a nonsignificant decrease in the global average soybean yield of 3.1% per °C increase with large uncertainties [22].

Thus, genotype and weather conditions have a significant impact on the amount and stability of soybean yield, which depends on many other cultivation factors [23,24]. Therefore, breeding new soybean cultivars for such conditions—as well as selecting European cultivars for cultivation—is much more difficult and requires long-term experiments. Analyses of the influence of weather factors on the phenological data and yield should be carried out on the same genotype. It is reported that Augusta is the only variety that has been cultivated in Poland for a long period of time (since 2002). For this reason, the results from 16 years of cultivation of this variety were used to determine the impact of meteorological conditions on the harvest date and soybean yield.

In this pilot study, the neural modeling method was used. Artificial neural networks (ANN) are a tool designed to implement various types of problems, including the performance of prognostic and deterministic analyses [25–31]. The reason for the great interest in neural networks is the fact that they are called "universal function estimators", and they are capable of solving problems of a linear and non-linear nature. Often, the simultaneous use of multiple linear regression (MLR) and artificial neural networks can be found in the literature. Unfortunately, linear methods are characterized by much lower analysis results than ANNs [32–34]. It should be noted that artificial neural networks operate on a "black box" principle; that is, they do not provide complete information regarding the method of obtaining specific answers or detailed relations between the input and output variables [35]. To be able to extract as many clues and messages as possible from a trained network, several techniques were used, including neural network sensitivity analysis. This analysis is used to determine how "sensitive" the model is to changes in model parameter values and to changes in the model structure. The so-called "sensitivity of the network" is determined, among others, by the error ratio. A high network sensitivity to a given parameter suggests that the system's performance may change drastically with a small change in that parameter. Conversely, a low sensitivity suggests a small change in performance [35,36]. In this way, it is easy to identify variables of high importance in influencing the variability of the output factors—i.e., the main problems set by the model developers, which are then solved by the network.

The aim of this study was to identify the key meteorological factors affecting the harvest date and yield of soybean using a neural network sensitivity analysis based on two deterministic models.

2. Materials and Methods

2.1. Plant Material

Polish soybean cultivar Augusta, one of the earliest soybean cultivars in Europe, was used as the plant material in this study. It was developed at the Department of Genetics and Plant Breeding of the Poznań University of Life Sciences (PULS) and registered in Poland in 2002. Augusta was selected from two crosses: (1) in the first step, the cross between Fiskeby V and line PI 194,643 was made and the line 104 was obtained; (2) in the second step, the line 104 was crossed with line 11, belonging to *G. soja* (Siebold & Zucc.) syn. *G. ussuriensis* (Regel & Maack) wild species. Line 11 of *G. soja* is growing in a natural environment in the far east latitude region of Russia, similar to Poland, and is a day-long tolerant genotype. Thus, Augusta has two sources of photoperiod insensitivity and chilling tolerance.

2.2. Field Test

The field experiment was conducted at the Agricultural Research Station Dłoń, Poznań University of Life Sciences, Poland (51°41′37″ N, 17°04′06″ E) during the 2005–2020 growing seasons. The plot soils are classified as Haplic Luvisols (LVh, WRB Soil Classification—FAO) [37] and the previous crop for the experiment was wheat. The Augusta seeds were sown from 20 to 28 of April at the density of 60 seeds per 1 m^2. Just after sowing, a pre-emergence herbicide that contained linuron (0.1 $g \cdot m^{-2}$) and S—metolachlor (0.14 $g \cdot m^{-2}$) was applied. The fertilizer was used according to the conventional farming practices in this area (N 30 $kg \cdot ha^{-1}$, P 80 $kg \cdot ha^{-1}$, K 120 $kg \cdot ha^{-1}$). The dates of the beginning of flowering and maturity were recorded due to the BBCH scale. The yield results were collected from the fields measuring from 0.5 to 5.0 hectares, on which the seeds of Augusta were multiplied.

The average daily temperatures and precipitation, measured according to the WMO guidelines for 2005–2020, were obtained from a Vantage Vue 6357 UE 9 meteorological station (Davis Instruments, United States) located approximately 400 m from the experimental field. Atmospheric conditions and information on vegetation length and yield from 2005 to 2020 are shown in Figures A1 and A2.

2.3. Division of Experimental Data into Sets Used in the Analyses

All experimental data collected in the database were divided into two sets. This division resulted from the assumptions made about the use of deterministic models in indicating the independent variables with the greatest influence on the evaluated feature, i.e., harvest date (model M_HARV) and yield (model M_YIELD). The procedure for the experimental data intended for the development and specific verification of each of the deterministic models is presented below (Figure 1).

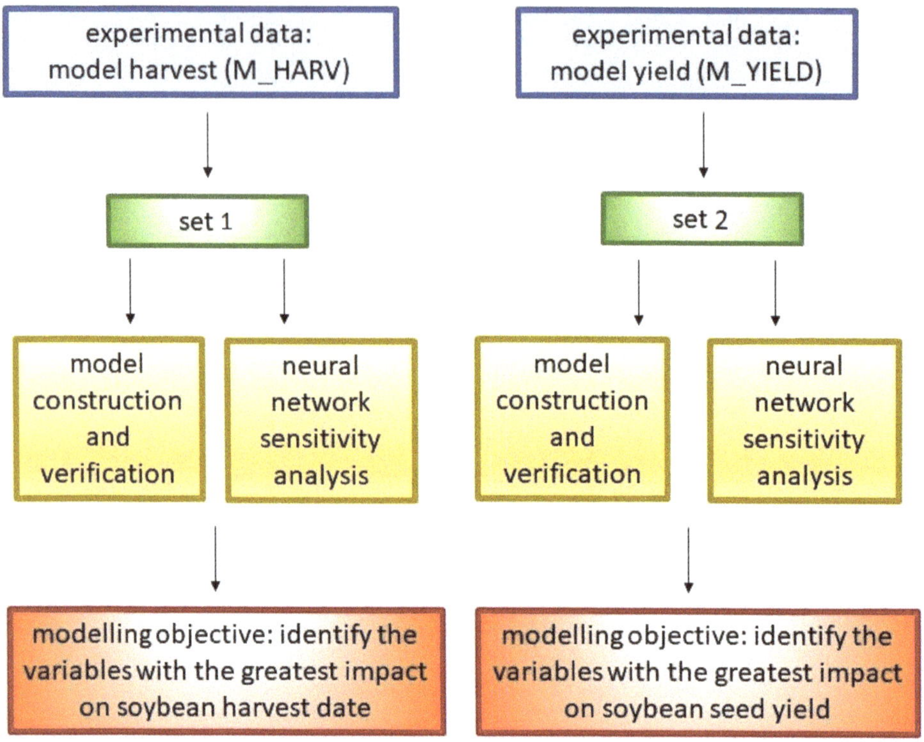

Figure 1. The division of experimental data into sets according to the assumptions made about the construction and application of deterministic models.

2.4. Methodology for Predictive Model Development

Primary meteorological data were used to develop deterministic neural models (M_HARV, M_YIELD). These included: mean air temperature and precipitation totals for each decade, starting from the first decade of April to the third decade of September (M_HARV, M_YIELD) in the current agronomic season. Some of the proposed independent variables required additional calculations. For example, these included the values of Selyaninov hydrothermal coefficients (HTC) (Equation (1)), calculated for different time intervals depending on the deterministic assumptions of the selected models, growing degree days (GDD) > 6 °C and the total precipitation for selected vegetation periods.

$$HTC = (P \cdot 10)/\Sigma t \quad (1)$$

where:

P—total monthly rainfall (mm),
Σt—sum of monthly average daily air temperatures > 6 °C.

The duration of the soybean growing season (M_HARV, M_YIELD) was also determined. The independent variables in the developed models were the date of harvest (M_HARV) and yield (M_YIELD). The date of harvest was presented as a number of days since the beginning of the year, while the yield was $t \cdot ha^{-1}$. A detailed list of independent and dependent variables taken into account in the development of each model, along with the range of their values, is presented in Table 1.

Table 1. The neural models' data structure.

Symbol	Unit of Measure	Variable Name	Model M_HARV	Model M_YIELD	The Scope of Data
T-IV-1	°C	Average air temperature in the 1st decade of April	+	+	1.8–12
T-IV-2	°C	Average air temperature in the 2nd decade of April	+	+	5.9–15.9
T-IV-3	°C	Average air temperature in the 3rd decade of April	+	+	6.9–17.4
T-V-1	°C	Average air temperature in the 1st decade of May	+	+	9.4–18
T-V-2	°C	Average air temperature in the 2nd decade of May	+	+	10.7–17.9
T-V-3	°C	Average air temperature in the 3rd decade of May	+	+	11.1–22
T-VI-1	°C	Average air temperature in the 1st decade of June	+	+	14.1–22.5
T-VI-2	°C	Average air temperature in the 2nd decade of June	+	+	15.8–24.6
T-VI-3	°C	Average air temperature in the 3rd decade of June	+	+	15.5–24.1
T-VII-1	°C	Average air temperature in the 1st decade of July	+	+	16.4–25.2
T-VII-2	°C	Average air temperature in the 2nd decade of July	+	+	17.9–25.6
T-VII-3	°C	Average air temperature in the 3rd decade of July	+	+	16.3–26.9
T-VIII-1	°C	Average air temperature in the 1st decade of August	+	+	17.2–26.4
T-VIII-2	°C	Average air temperature in the 2nd decade of August	+	+	17.8–24.5
T-VIII-3	°C	Average air temperature in the 3rd decade of August	+	+	15.8–22.3
T-IX-1	°C	Average air temperature in the 1st decade of September	+	+	12.9–20.7
T-IX-2	°C	Average air temperature in the 2nd decade of September	+	+	11.5–18.8
T-IX-3	°C	Average air temperature in the 3rd decade of September	+	+	10.3–16.4
O-IV-1	mm	Total precipitation in the 1st decade of April	+	+	0–28.5
O-IV-2	mm	Total precipitation in the 2nd decade of April	+	+	0–32.5
O-IV-3	mm	Total precipitation in the 3rd decade of April	+	+	0–22.4
O-V-1	mm	Total precipitation in the 1st decade of May	+	+	3–40.2
O-V-2	mm	Total precipitation in the 2nd decade of May	+	+	0–71
O-V-3	mm	Total precipitation in the 3rd decade of May	+	+	0.4–62
O-VI-1	mm	Total precipitation in the 1st decade of June	+	+	0–61.8
O-VI-2	mm	Total precipitation in the 2nd decade of June	+	+	0.4–67.7

Table 1. *Cont.*

Symbol	Unit of Measure	Variable Name	Model M_HARV	Model M_YIELD	The Scope of Data
O-VI-3	mm	Total precipitation in the 3rd decade of June	+	+	0–69
O-VII-1	mm	Total precipitation in the 1st decade of July	+	+	0–94
O-VII-2	mm	Total precipitation in the 2nd decade of July	+	+	4–109
O-VII-3	mm	Total precipitation in the 3rd decade of July	+	+	1.2–76
O-VIII-1	mm	Total precipitation in the 1st decade of August	+	+	0–189.5
O-VIII-2	mm	Total precipitation in the 2nd decade of August	+	+	1.5–53
O-VIII-3	mm	Total precipitation in the 3rd decade of August	+	+	2.3–74.5
O-IX-1	mm	Total precipitation in the 1st decade of September	+	+	0–41.4
O-IX-2	mm	Total precipitation in the 2nd decade of September	+	+	0–72
O-IX-3	mm	Total precipitation in the 3rd decade of September	+	+	0–52.5
STE_SK	°C	Growing Degree-Days (GDD) in the sowing-flowering period	+	+	873.67–1146.53
STE_SZ	°C	Growing Degree-Days (GDD) in sowing-harvest period	+	+	2309.87–2818.43
SO_SK	mm	Total precipitation in the sowing-flowering period	+	+	37.5–236.9
SO_SZ	mm	Total precipitation in the sowing-harvest period	+	+	180–584.2
S_SK	-	HTC in the sowing-flowering period	+	+	0.37–2.59
S_SZ	-	HTC in the sowing-harvest period	+	+	0.66–2.1
S	Day	Sowing date	+	+	108–116
K	Day	Flowering date	+	+	160–174
Z	Day	Harvest date	-	+	240–271
W	Day	Length of vegetation	+	+	128–156

"+"—the variable exists in the model, "-"—the variable does not exist in the model.

The next step in performing the appropriate analyses was the selection of appropriate neural network architectures that make up the M_HARV and M_YIELD models. By using Statistica v.7.1. [38], it was possible to test the Automatic Network Designer, a tool that automatically evaluates a large number of different network architectures of varying complexity, selecting a set of those that best suit a given problem. In the first stage of work with the Automatic Network Designer, several types of neural networks were selected in order to test them in terms of the quality of implementation of deterministic problems. The tool allows for verifying 5 types of networks, i.e., multilayer perceptron (MLP) (three-layer and four-layer), radial basis function network (RBF), probabilistic neural network (PNN), and generalized regression neural network (GRNN). According to the literature data, the most popular type of network selected for the implementation of prognostic and deterministic issues is MLP with two hidden layers [39,40]. After verification of the preliminary results obtained during the pilot analyses, 3 types of networks were selected for further, more detailed testing: RBF and MLP (three-layer or four-layer). The linear transfer function and two activation functions—linear and logistic—were chosen for the MLP network. In the next step, the complexity of each type of network was determined.

For the RBF network, the minimum number of neurons was assumed to be 5, while the maximum number was 80. For the MLP (three-layer) network, a minimum of 3 neurons and a maximum of 25 were assumed in the second layer. For the MLP (four-layer) network, the third layer contained a minimum of 3 neurons and a maximum of 25 neurons. After establishing the above assumptions, an analysis was carried out for 10,000 networks. This number of tested networks is most common in other studies. The interpretation of the values that characterize the learning quality and the error values for the networks developed allowed for the selection of the final network type, for which the analyses continued. The final analysis was of an MLP network with two hidden layers. The selection of the best final networks forming deterministic models was based on the most favourable values of parameters relating to their quality, i.e., standard deviation, mean value from error modules, the quotient of standard deviations, and correlation coefficient. With results which are ambiguous or difficult to evaluate, networks with high correlation coefficients and a low value of mean absolute error were sought. Finally, two MLP networks were selected with the following ratios: MLP 45:45-21-21-1:1 (M_HARV) and MLP 46:46-21-21-1:1 (M_YIELD). The structure of the two selected MLP networks, including the independent and dependent variables, is shown in Figure 2.

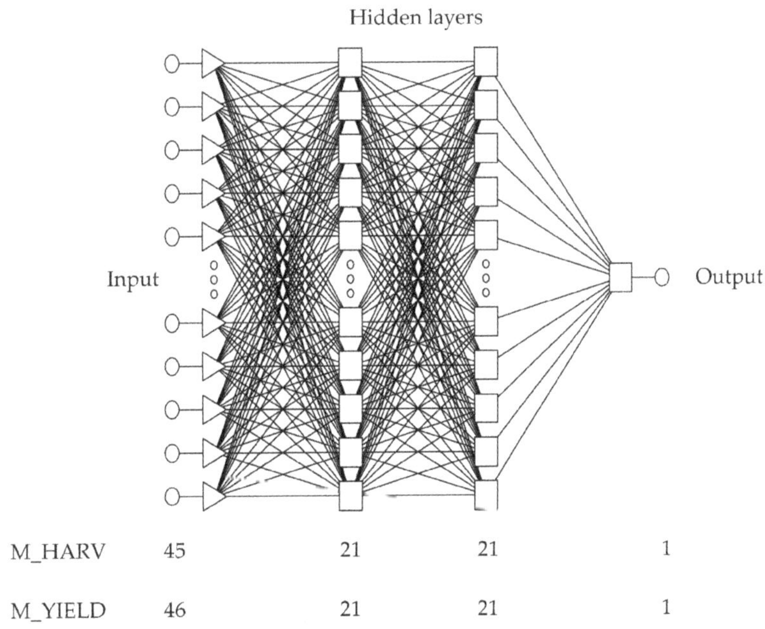

Figure 2. The network structure for models Harvest (M_HARV) and Yield (M_YIELD).

The model Harvest (M_HARV) contained 45 neurons (nods) in the input, 21 in the first hidden layer, 21 in the second hidden layer, and 1 in the output.

The Yield model (M_YIELD) contained 46 neurons in the input, 21 in the first hidden layer, 21 in the second hidden layer, and 1 in the output.

To train and validate the selected MLP networks that formed the M_HARV and M_YIELD models, sets 1 and 2 were randomly divided into two sets: a training set (70% of cases) and a validation set (30% of cases). The data collected in the training set enabled the calculation of the gradient, weight, and value of any loads on the network. The role of the validation set was to control the training error of the network during the training procedure. If the validation set's error increased for several consecutive epochs, the training process was halted. The most important task of this set was to prevent the overfitting of the neural network. Two error backpropagation methods were chosen to train the network forming

M_HARV and M_YIELD models. As can be seen in Table 2, individual networks forming M_HARV and M_YIELD models were taught with different conditions ending the network training process. The best results were achieved at different epochs.

Table 2. A number of epochs and training methods for neural networks.

Predictive Model	M_HARV	M_YIELD
Neural network architecture	MLP 45:45-21-21-1:1	MLP 46:46-21-21-1:1
	The training epochs	
Back-propagation method	2 *	30 *

* means the best result in the indicated training epoch

2.5. Neural Network Sensitivity Analysis

The analysis of the sensitivity of the neural network makes it possible to identify the factors with the greatest impact on the tested dependent variables among all the analyzed factors. After removing a specific explanatory variable (independent feature) from the model, its influence on the value of the total error of the neural network is observed. This allows for the significance (validity) of the tested factors to be determined. To accomplish the above task, the error quotient and rank are used. The error quotient expresses the ratio of the error to the total error of all independent variables. As its value increases, the importance of a given variable increases. If for any of the independent variables the quotient drops below 1, such a variable should be removed from the model to improve its quality. A rank that acts as a place in the ranking list indicates the characteristics according to decreasing error. The rank value of 1 proved to be the most important influence on the explanation of the variability of the dependent variable.

3. Results

During the experiment, highly variable weather conditions were observed, which strongly altered the course of phenological phases in soybean and the achieved seed yield of Augusta cultivar. The soybean flowering phase was observed between 11 and 25 June in the interval from 52 to 62 days from the sowing of seeds. Harvesting of mature soybean seeds took place from 30–31 August (in 2012, 2016, and 2017) to 30 September in 2020, i.e., from 131 to 156 days after sowing in 2012 and 2020, respectively. Soybean yields obtained during the study years ranged from 0.6 $t{\cdot}ha^{-1}$ and 2.6 $t{\cdot}ha^{-1}$ in 2015 and 2020, respectively. Soybean yields were generally low mainly due to very variable weather conditions (Figure A1). During the sixteen study years, harvested yield ranged from 0.6 $t{\cdot}ha^{-1}$ during extreme drought in 2015 to 2.6 $t{\cdot}ha^{-1}$ in 2020. During the eight years, harvested yield was below 2 tons, from 1.2 to 1.8 $t{\cdot}ha^{-1}$. In only six years of the study, yields ranged from 2.0–2.2 $t{\cdot}ha^{-1}$ (Figure A2).

The growing degree days (GDD) from soybean sowing to flowering ranged from 873–893 in 2017 and 2009 to over 1100 in 2007 and 2012. It was during this period in the six years of the study (2008, 2015, 2011, 2006, and 2017) that very low rainfall was recorded to be less than 100 mm; and in 2020, the sum of rainfall in the period from sowing to flowering was 236 mm. Thus, dry or very dry years (only 37.5 mm of precipitation in 2008) were observed, associated with the occurrence of droughts in the first growing season, and extremely rainy years causing the flooding of the experimental fields in 2020.

GDD over the entire soybean growing season ranged from 2309 in 2017 to 2818 in 2018. During the entire soybean growing season in 2008, 2015, and 2019, a total of less than 200 mm of precipitation was recorded, while 584 mm of precipitation fell in the rainy year of 2020. A calculated HTC indicates catastrophic drought; 0.3–0.5-drought; 0.5–1.0 humidity below balance; 1–2 sufficient amount of water; 2–4 excess of water. The lowest value of HTC in the period from sowing to flowering was recorded in 2008 (0.369), while the highest value was 2.59 in 2020. Humidity below balance occurred until flowering in 2006, 2011, 2015, and 2019. Analyzing the results of HTC in the whole growing period of soybean, no drought was observed, while a level of humidity below balance was recorded in 2008, 2015, 2018, and 2019, as well as an excess of water in 2020.

Sensitivity analysis for the M_HARV model identified the factor "total precipitation in the first decade of June" as the factor that most influenced the timing of soybean harvest (rank 1). The second important factor of the M_HARV model (rank 2) was the total rainfall in the first decade of August. The third important variable was the value of the Seljaninov coefficient, calculated for the period from sowing to harvest (Table 3).

Table 3. A sensitivity analysis of the neural networks.

Variable	Model			
	M_HARV		M_YIELD	
	Quotient	Rank	Quotient	Rank
T-IV-1	1.019	38	0.991	35
T-IV-2	0.977	44	1.024	26
T-IV-3	1.060	33	0.990	36
T-V-1	1.065	31	1.006	30
T-V-2	1.325	6	1.225	1
T-V-3	1.281	12	1.021	27
T-VI-1	1.091	26	1.036	22
T-VI-2	1.013	39	1.006	31
T-VI-3	1.052	34	0.979	39
T-VII-1	1.266	13	1.134	5
T-VII-2	1.320	7	1.097	10
T-VII-3	1.201	16	0.962	45
T-VIII-1	0.930	45	1.013	28
T-VIII-2	1.111	22	1.101	7
T-VIII-3	1.065	30	0.971	42
T-IX-1	1.172	18	1.034	23
T-IX-2	1.008	42	1.134	4
T-IX-3	1.075	29	1.053	18
O-IV-1	1.185	17	1.100	9
O-IV-2	1.008	40	0.972	41
O-IV-3	1.286	10	0.914	46
O-V-1	1.338	5	1.127	6
O-V-2	1.338	4	1.042	21
O-V-3	1.028	37	1.067	15
O-VI-1	1.707	1	0.966	44
O-VI-2	1.298	8	1.087	11
O-VI-3	1.147	19	1.101	8
O-VII-1	1.064	32	1.032	24
O-VII-2	1.087	27	0.976	40
O-VII-3	1.209	15	0.969	43
O-VIII-1	1.486	2	1.056	17
O-VIII-2	1.129	20	1.076	13
O-VIII-3	1.034	35	1.069	14
O-IX-1	1.008	41	1.002	33
O-IX-2	1.029	36	1.004	32
O-IX-3	1.096	24	1.027	25
STE_SK	1.127	21	0.984	38
STE_SZ	1.083	28	1.049	19
SO_SK	1.286	9	0.989	37
SO_SZ	1.099	23	1.012	29
S_SK	1.282	11	1.163	2
S_SZ	1.345	3	1.153	3
S	1.007	43	1.080	12
K	1.216	14	1.064	16
Z	-	-	0.995	34
W	1.094	25	1.042	20

The factor with the greatest influence on soybean seed yield (M_YIELD model) was the mean air temperature in the second decade of May. This variable was given a rank of 1 in the sensitivity analysis of the neural network. The factor that received rank 2 was the HTC values calculated for the period from sowing to flowering. The variable with rank 3 was also the HTC value but was determined for a different time interval, i.e., from sowing to harvesting (Table 3).

Comparison of Models M_HARV and M_YIELD Quality Characteristics

The best neural networks that allowed for the identification of factors with the greatest influence on harvest date (M_HARV) and soybean seed yield (M_YIELD model) were selected based on a detailed analysis of the quality parameters of the generated networks. Detailed results of the analyses are presented in Table 4.

Table 4. The quality and structure of the neural models produced.

Quality Parameter	M_HARV	M_YIELD
Neural network structure	45:45-21-21-1:1	46:46-21-21-1:1
Learning error [-]	0.1875	0.1273
Validation error [-]	0.0259	0.0062
Mean [day], [t·ha^{-1}]	250.5625	1.825
Standard deviation [day], [t·ha^{-1}]	8.3289	0.4322
Average error [day], [t·ha^{-1}]	1.2849	0.0152
Deviation error [day], [t·ha^{-1}]	3.6788	0.2865
Mean Absolute error [day], [t·ha^{-1}]	2.8846	0.2034
Quotient deviations [-]	0.4416	0.6629
Correlation [-]	0.8976	0.7503

The results of the presented neural models were characterized in each of the considered cases by the best values of quality measures of the generated neural networks. When it was difficult to indicate the best network, the values of two quality parameters were considered: the correlation coefficient (r) and the mean absolute error. The principle followed was that the value of the correlation coefficient should be the highest with a simultaneous low value of the mean absolute error. In both analyzed cases, the values of the correlation coefficient were very high, i.e., for the model M_HARV: 0.898 and for the model M_YIELD: 0.75. The value of the mean absolute error was the highest for the model M_YIELD and it was 0.203 t·ha^{-1}. Another important parameter in assessing the quality of the generated neural networks was the error quotient, defined as the quotient of the standard deviation of the prediction errors and the standard deviation of the output variable. For the model to be useful for forecasting purposes, the value of this parameter should not exceed 0.7. In the three analyzed cases, this assumption was fulfilled.

The response plots are a visual representation of the results of the sensitivity analysis of the neural networks. It shows the relationship between variables of rank 1 and 2 from the sensitivity analysis and the dependent variable. On the x and y axes of the three-dimensional plot are placed the values of the selected independent variables, and on the z-axis are the values taken by the dependent variable. The response plots for MLP 45:45-21-21-1:1 and MLP 46:46-21-21-1:1 networks are shown in Figures 3 and 4, respectively.

Figure 3 shows the response surface for the MLP network 45:45-21-21-1:1, where the explanatory variable is harvest date and the explanatory variables are precipitation in the first decade of June and precipitation in the first decade of August. The graph shows that the persistence of low average precipitation in the second decade of April and the first decade of April delays the date of the soybean harvest.

Figure 4 shows the response surface for the MLP 46:46-21-21-1:1 network forming the M_YIELD model, which shows the relationship between soybean seed yield levels and the average air temperature in the second decade of May, as well as the HTC values calculated for the sowing–flowering period. The highest soybean seed yield can be expected when the average air temperature in the second decade of May is about 10–12 °C, and the value of the HTC for the sowing–flowering period is relatively low, ranging from 0.4 to 0.8. It can be concluded that the first factor in question determines soybean yield to a greater extent than the second independent variable.

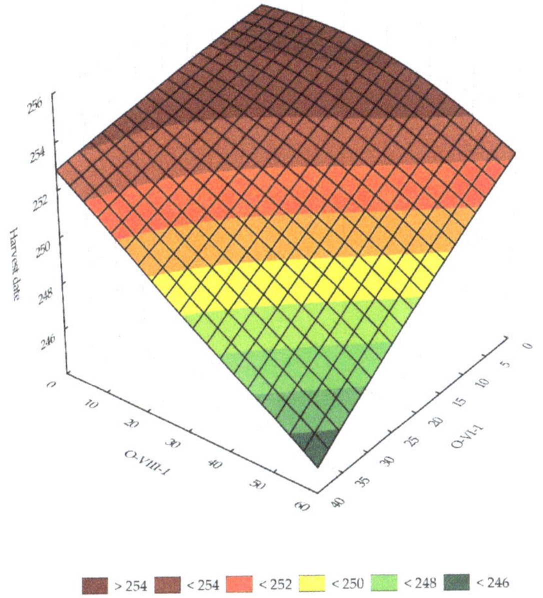

Figure 3. The response surface for the harvest date and two variables, O-VII-1 and O-VI-1.

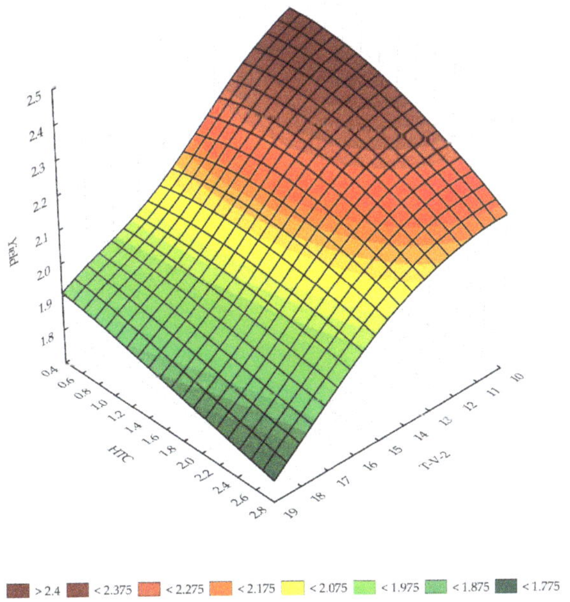

Figure 4. The response surface for a yield and two variables, T-V-2 and HTC.

The results presented in the previous stages were supplemented with additional analyses and visualisations of the relations between the observed and predicted values of the harvest date and soybean yield. The results of the analyses are presented in Figures 5 and 6.

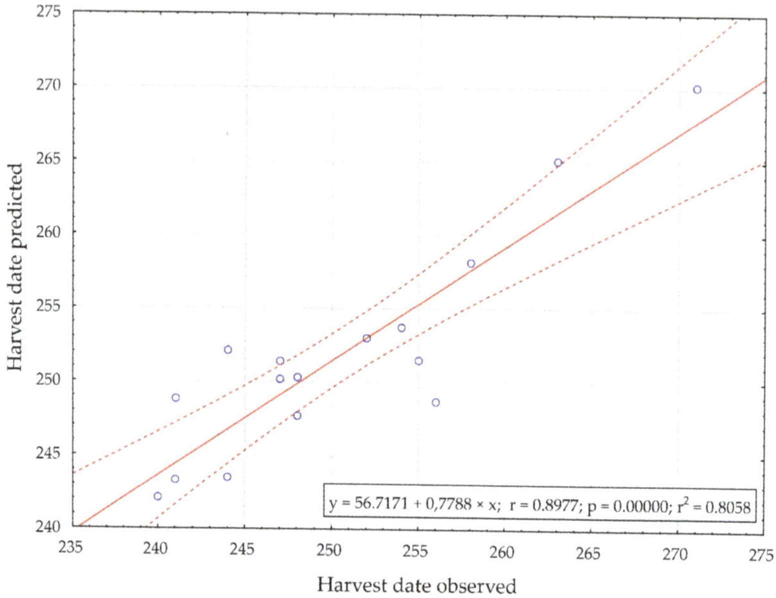

Figure 5. The scatter plot between the observed and predicted values of the soybean harvest date in model M_HARV.

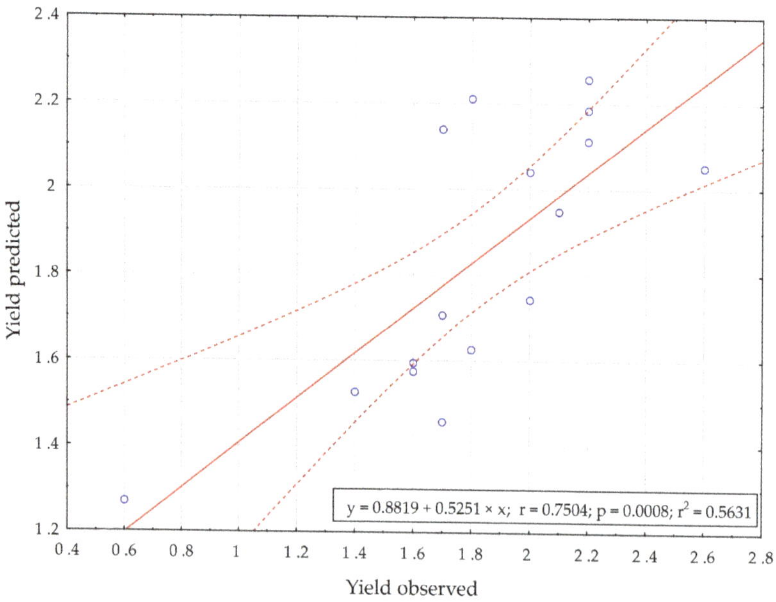

Figure 6. The scatter plot between the observed and predicted values of the soybean seed yield in model M_YIELD.

4. Discussion

An important aspect of the use of neural networks in implementing deterministic problems is the choosing of an appropriate network topology and training method. A complex phase of testing different neural network topologies allowed us to indicate the most suitable type of network for the problems presented in this paper. Finally, the MLP network with two hidden layers was chosen. Training a neural network allows for the combining of certain behaviors of the model based on many experiences. The user enforces specific responses to given input signals from the network. The network remembers questions and answers based on selected patterns of behavior so that when a new "question" is asked, it gives an answer that is most similar to the original one. In the presented results, all neural networks were taught using the method of backward error propagation. This method allows for the creation of neural networks with very favorable quality parameters [25,35].

The selected neural networks forming the M_HARV and M_YIELD models were characterized by standard values of their quality metrics. The quality parameters of the M_YIELD model were less accurate than M_HARV. However, it turns out that the values of correlation coefficient (0.75), mean absolute error (0.2), and deviation quotient (0.663) obtained for the M_YIELD model fall within the generally accepted criteria related to the application of this type of tool in agricultural practice [41].

One of the most difficult steps in developing deterministic neural models is choosing the right independent variables to form the model. These variables should have a real influence on the development of the explained variable. The correct identification of explanatory variables requires excellent knowledge of the research object. Admittedly, the significance of selected variables can be verified by additional analyses and calculations [42], but it is practical experience that is the most valuable way to correctly match independent variables to explain the complexity and variability of a specific phenomenon. In our study, meteorological and phenological data were used with 45 and 46 selected variables for the M_HARV and M_YIELD models, respectively. We conclude that such a detailed approach to explaining the influence of weather conditions on the phenology and yield of soybean in Wielkopolska allowed for the precise identification of variables that have the greatest influence on harvest timing and seed yield.

Sensitivity analysis of neural networks was used to fully implement the issues presented in this paper. It is a method that allows for distinguishing important variables in the model from those that contribute little to the outcome of the network [43]. This method is widely used in typing the most important variables in issues related to the phenology and yield of crop species [44,45]. The result of the analysis is the value of the error quotient, based on which a rank (ranking place) is assigned to a particular trait. It is assumed that traits with an error quotient below one are not considered when interpreting the importance of variables.

The sensitivity analysis performed for the two described neural networks indicated different independent variables that determined to the greatest extent the variability of the next explained variables: harvest date and soybean seed yield. For the M_HARV model, these were precipitation amounts in the first decade of June and August. It should be noted, however, that the factors with an error quotient higher than 1.3 included the Seljaninov coefficient in the sowing–harvesting range, precipitation in the first and second decade of May, and temperature in the second decade of May and the second decade of July. These values testify to the high importance of the mentioned variables in the work to determine their influence on the optimal harvest date of soybean cv. Augusta. In the case of the M_YIELD model, all of the highly important factors responsible for yield were characterized by an error quotient above 1.1.

Both rainfall deficiency and significant excess strongly modify plant development and yield, and the developmental stage most sensitive to drought stress varies according to the cultivar used [46]. Drought occurring from late flowering to the beginning of pod filling results in a reduction in the number of seeds per pod, and drought occurring late in pod filling results in a reduction in seed size [47]. The results of our study are consistent with these findings.

The most important trait assessed by growers and later by farmers is yield potential, which depends on many factors, including climatic conditions during the growing season [48]. In the second M_YIELD model, the average air temperature in the second decade of May, the Seljaninov coefficient values calculated for the period from sowing to flowering, and then the values of this coefficient calculated for the whole growing season were found to be the most important variables. Multi-criteria analysis of the results generated by deterministic models should be carried out for each model separately. The values of the error quotients assigned to explanatory variables of rank 1 or 2 cannot be compared between the models M_HARV and M_YIELD. Still, these values must be analyzed for each model independently. In the M_HARV model, the significance of 43 out of 45 tested independent variables considered in the construction of the model was confirmed. For two variables, i.e., T-IV-2 and T-VIII-1, error quotient values below one were calculated. These results testify to a very good fit of the variables to explain the variability of the modeled phenomenon. Besides, in further analyses, it would be advisable to exclude the participation of variables T-IV-2 and T-VIII-1 in developing new deterministic or predictive models. In turn, for the M_YIELD model, the neural network sensitivity analysis confirmed the significance of 34 out of 46 tested independent variables. In the next stage of work with improving these models, the contribution of these variables can be eliminated. A detailed interpretation of the results of the sensitivity analysis of neural networks allows us to conclude that the harvest date is a factor more dependent on meteorological conditions than the yield of soybean of the Augusta cultivar.

For many years, the suitability of soybean genotypes for cultivation in Poland depended, among other things, on the tolerance of lower temperatures during the flowering and the harvest dates. The Augusta variety was bred in Poland, and during its breeding special attention was paid to these factors. The Fiskeby V cultivar from Sweden was used in crossbreeding, which is characterized by photoneutrality and resistance to cold stress. Despite this, as our research showed, it was the temperature in the initial period of plant growth that had the most significant effect on the yields obtained.

5. Conclusions

Presented deterministic models—M_HARV and M_YIELD—allowed us to use artificial neural networks for the preliminary identification of major factors affecting the harvest date and yield of soybean cultivar Augusta.

The sensitivity analysis of the neural network makes it possible to initially select the factors with the greatest influence on the explained variable while maintaining the adopted level of significance.

Total precipitation in the first decade of June and the first decade of August were the variables assigned ranks 1 and 2 in the sensitivity analysis of the neural network forming the M_HARV model. On the other hand, the variables with the highest impact on the Augusta soybean seed yield (model M_YIELD) were mean daily air temperature in the second decade of May and Seljaninov coefficient values calculated for the sowing–flowering date period.

Further research on the improvement of deterministic models in soybean cultivation should be carried out on multiple levels. It is worth exploring other analytical methods that optimize important production factors (controllable) which have a significant impact on soybean seed yield in terms of quantity and quality.

Author Contributions: Conceptualization, G.N. and D.K.-P.; methodology, G.N., D.K.-P. and M.P.; software, G.N., M.P. and T.W.; validation, G.N., D.K.-P., M.P., T.W., M.K. and J.N.; formal analysis, G.N.; investigation, D.K.-P., M.K. and J.N.; resources, G.N., D.K.-P., M.P., T.W., M.K. and J.N.; data curation, G.N. and D.K.-P.; writing—original draft preparation, G.N., D.K.-P., M.P., T.W. and J.N.; writing—review and editing, G.N., D.K.-P., M.P., T.W., M.K. and J.N.; visualization, G.N.; supervision, G.N., D.K.-P. and J.N.; project administration, G.N. All authors have read and agreed to the published version of the manuscript.

Funding: This research received no external funding.

Institutional Review Board Statement: Not applicable.

Informed Consent Statement: Not applicable.

Data Availability Statement: Not applicable.

Conflicts of Interest: The authors declare no conflict of interest.

Appendix A

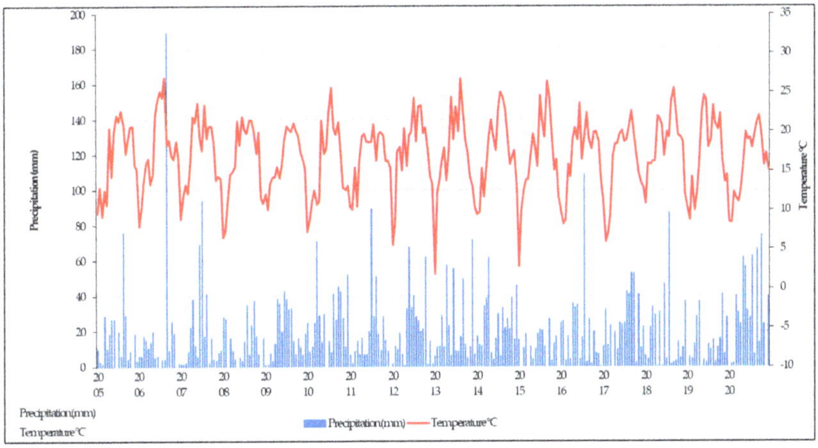

Figure A1. Rainfall and average air temperatures between January 2005 and December 2020, ARS Dłoń, Poland.

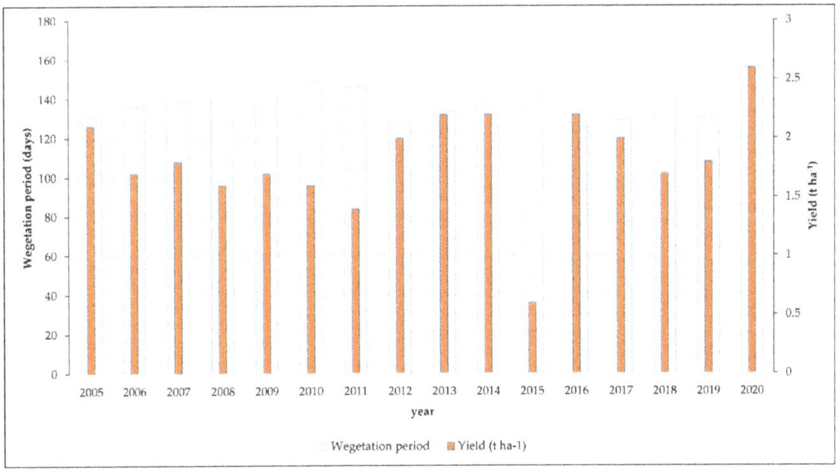

Figure A2. The length of vegetation and the yield of soybean variety Augusta in the period 2005–2020, ARS Dłoń, Poland.

References

1. World Agricultural Production. Available online: https://apps.fas.usda.gov/psdonline/circulars/production.pdf (accessed on 5 April 2022).
2. Powierzchnie Upraw W Gminach. Available online: https://rejestrupraw.arimr.gov.pl/ (accessed on 5 April 2022).
3. Niwińska, B.; Witaszek, K.; Niedbała, G.; Pilarski, K. Seeds of n-GM Soybean Varieties Cultivated in Poland and Their Processing Products as High-Protein Feeds in Cattle Nutrition. *Agriculture* **2020**, *10*, 174. [CrossRef]
4. Gawęda, D.; Nowak, A.; Haliniarz, M.; Woźniak, A. Yield and Economic Effectiveness of Soybean Grown Under Different Cropping Systems. *Int. J. Plant Prod.* **2020**, *14*, 475–485. [CrossRef]
5. Boerma, H.R.; Specht, J.E. *Soybeans: Improvement, Production, and Uses*, 3rd ed.; American Society of Agronomy, Crop Science Society of America, Soil Science Society of America: Madison, WI, USA, 2004.
6. Miladinović, J.; Ćeran, M.; Đorđević, V.; Balešević-Tubić, S.; Petrović, K.; Đukić, V.; Miladinović, D. Allelic Variation and Distribution of the Major Maturity Genes in Different Soybean Collections. *Front. Plant Sci.* **2018**, *9*, 1286. [CrossRef]
7. Kružel, J.; Ziernicka-Wojtaszek, A.; Borek, Ł.; Ostrowski, K. The changes in the duration of the meteorological vegetation period in Poland in the years 1971–2000 and 1981–2010. *Inż. Ekol.* **2015**, *44*, 47–52. [CrossRef]
8. Michałek, S.; Borowski, E. Yielding, oil, fatty acids and protein content in the seeds of polish soybean cultivars under drought conditions. *Acta Agrophysica* **2006**, *8*, 459–471.
9. Kołodziej, J.; Pisulewska, E. Effect of climatic factors on seed yield, fat yield and fat content in seeds of two soybean cultivars. *Oilseed Crop.* **2000**, *XXI*, 759–776.
10. Sobko, O.; Stahl, A.; Hahn, V.; Zikeli, S.; Claupein, W.; Gruber, S. Environmental Effects on Soybean (*Glycine Max* (L.) Merr) Production in Central and South Germany. *Agronomy* **2020**, *10*, 1847. [CrossRef]
11. Novikova, L.Y.; Bulakh, P.P.; Nekrasov, A.Y.; Seferova, I.V. Soybean Response to Weather and Climate Conditions in the Krasnodar and Primorye Territories of Russia over the Past Decades. *Agronomy* **2020**, *10*, 1278. [CrossRef]
12. Penalba, O.C.; Bettolli, M.L.; Vargas, W.M. The impact of climate variability on soybean yields in Argentina. Multivariate regression. *Meteorol. Appl.* **2007**, *14*, 3–14. [CrossRef]
13. Choi, D.-H.; Ban, H.-Y.; Seo, B.-S.; Lee, K.-J.; Lee, B.-W. Phenology and Seed Yield Performance of Determinate Soybean Cultivars Grown at Elevated Temperatures in a Temperate Region. *PLoS ONE* **2016**, *11*, e0165977. [CrossRef]
14. Gao, X.-B.; Guo, C.; Li, F.-M.; Li, M.; He, J. High Soybean Yield and Drought Adaptation Being Associated with Canopy Architecture, Water Uptake, and Root Traits. *Agronomy* **2020**, *10*, 608. [CrossRef]
15. Lambers, H.; Chapin, F.S.; Pons, T.L. The Plant's Energy Balance. In *Plant Physiological Ecology*; Springer: New York, NY, USA, 2008; pp. 225–236.
16. Miladinov, Z.; Maksimovic, I.; Tubic, S.B.; Miladinovic, J.; Djordevic, V.; Vasiljevic, M.; Radic, V. The Impact of Water Deficit on The Soybean (*Glycine max* L.) Reproductive Stage of Development. *Legum. Res.-AN Int. J.* **2020**, *43*, 693–697. [CrossRef]
17. Staniak, M.; Czopek, K.; Stępień-Warda, A.; Kocira, A.; Przybyś, M. Cold Stress during Flowering Alters Plant Structure, Yield and Seed Quality of Different Soybean Genotypes. *Agronomy* **2021**, *11*, 2059. [CrossRef]
18. Ogunkanmi, L.; MacCarthy, D.S.; Adiku, S.G.K. Impact of Extreme Temperature and Soil Water Stress on the Growth and Yield of Soybean (*Glycine max* (L.) Merrill). *Agriculture* **2021**, *12*, 43. [CrossRef]
19. Cheng-Zhi, C.; Cong-Jian, L.; Dan, X.; Xiao-Shan, Z.; Jin, Z. Global warming and world soybean yields. *J. Agrometeorol.* **2021**, *23*, 367–374. [CrossRef]
20. Tacarindua, C.R.P.; Shiraiwa, T.; Homma, K.; Kumagai, E.; Sameshima, R. The response of soybean seed growth characteristics to increased temperature under near-field conditions in a temperature gradient chamber. *F. Crop. Res.* **2012**, *131*, 26–31. [CrossRef]
21. Tacarindua, C.R.P.; Shiraiwa, T.; Homma, K.; Kumagai, E.; Sameshima, R. The effects of increased temperature on crop growth and yield of soybean grown in a temperature gradient chamber. *F. Crop. Res.* **2013**, *154*, 74–81. [CrossRef]
22. Zhao, C.; Liu, B.; Piao, S.; Wang, X.; Lobell, D.B.; Huang, Y.; Huang, M.; Yao, Y.; Bassu, S.; Ciais, P.; et al. Temperature increase reduces global yields of major crops in four independent estimates. *Proc. Natl. Acad. Sci. USA* **2017**, *114*, 9326–9331. [CrossRef] [PubMed]
23. Nawracała, J. *Analiza Genetyczno-Hodowlana Mieszańców i Linii Soi Otrzymanych z Krzyżowania Międzygatun-Kowego Glycine Max x Glycine Soja*; Rozprawa naukowa w serii Rozprawy Naukowe Uniwersytetu Przyrodniczego w Poznaniu nr 394; Wydawnictwo Uniwersytetu Przyrodniczego w Poznaniu: Poznań, Poland, 2008.
24. Mandić, V.; Đorđević, S.; Đorđević, N.; Bijelić, Z.; Krnjaja, V.; Petričević, M.; Brankov, M. Genotype and Sowing Time Effects on Soybean Yield and Quality. *Agriculture* **2020**, *10*, 502. [CrossRef]
25. Piekutowska, M.; Niedbała, G.; Piskier, T.; Lenartowicz, T.; Pilarski, K.; Wojciechowski, T.; Pilarska, A.A.; Czechowska-Kosacka, A. The Application of Multiple Linear Regression and Artificial Neural Network Models for Yield Prediction of Very Early Potato Cultivars before Harvest. *Agronomy* **2021**, *11*, 885. [CrossRef]
26. Hara, P.; Piekutowska, M.; Niedbała, G. Selection of Independent Variables for Crop Yield Prediction Using Artificial Neural Network Models with Remote Sensing Data. *Land* **2021**, *10*, 609. [CrossRef]
27. Pentoś, K. The methods of extracting the contribution of variables in artificial neural network models—Comparison of inherent instability. *Comput. Electron. Agric.* **2016**, *127*, 141–146. [CrossRef]
28. Niazian, M.; Niedbała, G. Machine Learning for Plant Breeding and Biotechnology. *Agriculture* **2020**, *10*, 436. [CrossRef]

29. Kujawa, S.; Dach, J.; Kozłowski, R.J.; Przybył, K.; Niedbała, G.; Mueller, W.; Tomczak, R.J.; Zaborowicz, M.; Koszela, K. Maturity classification for sewage sludge composted with rapeseed straw using neural image analysis. In Proceedings of the SPIE—The International Society for Optical Engineering, Chengu, China, 20–22 May 2016; Volume 10033, p. 100332H.
30. Wojciechowski, T.; Niedbala, G.; Czechlowski, M.; Nawrocka, J.R.; Piechnik, L.; Niemann, J. Rapeseed seeds quality classification with usage of VIS-NIR fiber optic probe and artificial neural networks. In Proceedings of the 2016 International Conference on Optoelectronics and Image Processing, ICOIP 2016, Warsaw, Poland, 10–12 June 2016.
31. Niedbała, G.; Piekutowska, M.; Rudowicz-Nawrocka, J.; Adamski, M.; Wojciechowski, T.; Herkowiak, M.; Szparaga, A.; Czechowska-Kosacka, A. Application of artificial neural networks to analyze the emergence of soybean seeds after applying herbal treatments. *J. Res. Appl. Agric. Eng.* **2018**, *63*, 145–149.
32. Majkovič, D.; O'Kiely, P.; Kramberger, B.; Vračko, M.; Turk, J.; Pažek, K.; Rozman, Č. Comparison of using regression modeling and an artificial neural network for herbage dry matter yield forecasting. *J. Chemom.* **2016**, *30*, 203–209. [CrossRef]
33. Gorzelany, J.; Belcar, J.; Kuźniar, P.; Niedbała, G.; Pentoś, K. Modelling of Mechanical Properties of Fresh and Stored Fruit of Large Cranberry Using Multiple Linear Regression and Machine Learning. *Agriculture* **2022**, *12*, 200. [CrossRef]
34. Sabzi-Nojadeh, M.; Niedbała, G.; Younessi-Hamzekhanlu, M.; Aharizad, S.; Esmaeilpour, M.; Abdipour, M.; Kujawa, S.; Niazian, M. Modeling the Essential Oil and Trans-Anethole Yield of Fennel (*Foeniculum vulgare* Mill. var. *vulgare*) by Application Artificial Neural Network and Multiple Linear Regression Methods. *Agriculture* **2021**, *11*, 1191. [CrossRef]
35. Lu, M.; AbouRizk, S.M.; Hermann, U.H. Sensitivity Analysis of Neural Networks in Spool Fabrication Productivity Studies. *J. Comput. Civ. Eng.* **2001**, *15*, 299–308. [CrossRef]
36. Nourani, V.; Sayyah Fard, M. Sensitivity analysis of the artificial neural network outputs in simulation of the evaporation process at different climatologic regimes. *Adv. Eng. Softw.* **2012**, *47*, 127–146. [CrossRef]
37. IUSS Working Group WRB. *World Reference Base for Soil Resources 2014, Update 2015 International Soil Classification System for Naming Soils and Creating Legends for Soil Maps. World Soil Resources Reports No. 106*; FAO: Rome, Italy, 2015; ISBN 978-92-5-108369-7.
38. TIBCO Statistica® Automated Neural Networks. Available online: https://community.tibco.com/wiki/tibco-statistica-automated-neural-networks (accessed on 10 January 2022).
39. Bhojani, S.H.; Bhatt, N. Wheat crop yield prediction using new activation functions in neural network. *Neural Comput. Appl.* **2020**, *32*, 13941–13951. [CrossRef]
40. Niedbała, G.; Piekutowska, M.; Weres, J.; Korzeniewicz, R.; Witaszek, K.; Adamski, M.; Pilarski, K.; Czechowska-Kosacka, A.; Krysztofiak-Kaniewska, A. Application of Artificial Neural Networks for Yield Modeling of Winter Rapeseed Based on Combined Quantitative and Qualitative Data. *Agronomy* **2019**, *9*, 781. [CrossRef]
41. Schober, P.; Boer, C.; Schwarte, L.A. Correlation Coefficients. *Anesth. Analg.* **2018**, *126*, 1763–1768. [CrossRef] [PubMed]
42. Mas, D.M.L.; Ahlfeld, D.P. Comparing artificial neural networks and regression models for predicting faecal coliform concentrations. *Hydrol. Sci. J.* **2007**, *52*, 713–731. [CrossRef]
43. Hadzima-Nyarko, M.; Nyarko, E.K.; Morić, D. A neural network based modelling and sensitivity analysis of damage ratio coefficient. *Expert Syst. Appl.* **2011**, *38*, 13405–13413. [CrossRef]
44. Farjam, A.; Omid, M.; Akram, A.; Fazel Niari, Z. A neural network based modeling and sensitivity analysis of energy inputs forpredicting seed and grain corn yields. *J. Agric. Sci. Technol.* **2014**, *16*, 767–778.
45. Niedbała, G.; Kozłowski, R.J. Application of Artificial Neural Networks for Multi-Criteria Yield Prediction of Winter Wheat. *J. Agric. Sci. Technol.* **2019**, *21*, 51–61.
46. Ku, Y.-S.; Au-Yeung, W.-K.; Yung, Y.-L.; Li, M.-W.; Wen, C.-Q.; Liu, X.; Lam, H.-M. Drought Stress and Tolerance in Soybean. In *A Comprehensive Survey of International Soybean Research—Genetics, Physiology, Agronomy and Nitrogen Relationships*; InTech: Palm Beach, FL, USA, 2013; Available online: https://books.google.co.jp/books?hl=zh-CN&lr=&id=87WiDwAAQBAJ&oi=fnd&pg=PA209&dq=Drought+Stress+and+Tolerance+in+Soybean&ots=fzzy8-mhTX&sig=Eqnob_LJ7Xh4MsPOQCWxpjwH6ng#v=onepage&q=Drought%20Stress%20and%20Tolerance%20in%20Soybean&f=false (accessed on 10 January 2022).
47. Oya, T.; Lima Nepomuceno, A.; Neumaier, N.; Bouças Farias, J.R.; Tobita, S.; Ito, O. Drought Tolerance Characteristics of Brazilian Soybean Cultivars—Evaluation and characterization of drought tolerance of various Brazilian soybean cultivars in the field. *Plant Prod. Sci.* **2004**, *7*, 129–137. [CrossRef]
48. Kucharik, C.J.; Serbin, S.P. Impacts of recent climate change on Wisconsin corn and soybean yield trends. *Environ. Res. Lett.* **2008**, *3*, 34003. [CrossRef]

agriculture

MDPI

Article

Performance Evaluation of Communication Systems Used for Internet of Things in Agriculture

Geovanny Yascaribay [1], Mónica Huerta [1,*], Miguel Silva [2] and Roger Clotet [3]

1 Grupo De Investigación en Telecomunicaciones y Telemática (GITEL), Universidad Politécnica Salesiana, Cuenca EC10105, Ecuador; gyascaribay@est.ups.edu.ec
2 Department of Communications and Networking, Universidad de Cuenca, Cuenca EC010107, Ecuador; miguel.silva@ucuenca.edu.ec
3 GRID Research Group, Universidad Internacional de Valencia, 46002 Valencia, Spain; roger.clotet@campusviu.es
* Correspondence: mhuerta@ups.edu.ec

Abstract: The rapid development of Internet of Things (IoT) technology has provided ample opportunity for the implementation of intelligent agricultural production. Such technology can be used to connect various types of agricultural devices, which can collect and send data to servers for analysis. These tools can help farmers optimize the production of their crops. However, one of the main problems that arises in agricultural areas is a lack of connectivity or poor connection quality. For these reasons, in this paper, we present a method that can be used for the performance evaluation of communication systems used in IoT for agriculture, considering metrics such as the packet delivery ratio, energy consumption, and packet collisions. To achieve this aim, we carry out an analysis of the main Low-Power Wide-Area Networks (LPWAN) protocols and their applicability, from which we conclude that those most suited to this context are Long Range (LoRa) and Long Range Wide Area Network (LoRaWAN). After that, we analyze various simulation tools and select Omnet++ together with the Framework for LoRa (FLoRa) library as the best option. In the first stage of the simulations, the performances of LoRa and LoRaWAN are evaluated by comparing the average propagation under ideal conditions against moderate propagation losses, emulating a rural environment in the coastal region of Ecuador. In the second phase, metrics such as the package delivery ratio and energy consumption are evaluated by simulating communication between an increasing number of nodes and one or two gateways. The results show that using two gateways with the Adaptive Data Rate technique can actively increase the delivery ratio of the network while consuming the same amount of energy per node. Finally, a comparison is made between the results of the simulation scenario considered in this project and those of other research works, allowing for the validation of our analytical and simulation results.

Keywords: Internet of Things (IoT); LPWAN; LoRaWAN; Omnet++; FLoRa; agriculture; rural applications

Citation: Yascaribay, G.; Huerta, M.; Silva, M.; Clotet, R. Performance Evaluation of Communication Systems Used for Internet of Things in Agriculture. *Agriculture* **2022**, *12*, 786. https://doi.org/10.3390/agriculture12060786

Academic Editors: Gniewko Niedbała, Massimo Cecchini

Received: 8 February 2022
Accepted: 26 May 2022
Published: 30 May 2022

1. Introduction

The Food and Agriculture Organization of the United Nations (FAO) has estimated that by the year 2050 the global population will reach 9.7 billion people. It is foreseen that for the decade 2050–2060 only one-third of the population will live in rural areas, whereas the remaining 66% will move to ever-growing cities, thus decreasing the size of the agricultural workforce. Furthermore, there are many regions worldwide that experience limited rainfall, meaning that between 80% and 90% of the available water is used by farmers, which in turn leads to shortages of primary sources of this resource (e.g., rivers and water reserves) [1].

The analysis of scarce water resources and the decline in agricultural labor, combined with soil infertility among other factors, has led researchers to consider how productivity could be improved in this sector to meet the growing demand for food [2]. Different studies focused on optimizing production in the agricultural sector have shown that the implementation of different systems based largely on the automation of processes is required [3–7]. Other investigations have focused on improving production systems according to the demand for quality foods and optimizing the use of water and other resources necessary in this field [8–11]. Advances in technological and communication solutions allow for the monitoring of field conditions in agricultural sectors through the use of various systems, including sensors that acquire agricultural data. These acquisition systems allow for the local transmission of information, potentially through the implementation of Internet of Things (IoT) technology based on Internet connectivity [12–14]. This interconnection entails a degree of intelligence, which allows for controlling and processing variables. The number of IoT applications has increased exponentially in recent years, playing an important role in the agricultural sector under the concepts of Precision Agriculture (PA) and Agriculture 4.0 [15,16]. One of the highlights of IoT devices in PA and Agriculture 4.0 is the low power consumption of devices, the low sampling rate, data transmission/reception, and advances in communication systems [17–19].

In the scientific literature, few studies have focused on the use of communication protocols such as RFID/WiFi/Bluetooth for monitoring variables related to the agricultural sector. This is due to the short range and low coverage of these technologies [20]. Traditional long-range communication systems, such as satellite communications [21], Wimax [22] and LTE/4G [23], are most commonly used in this sector. Solutions based on cellular phone communication could provide a greater range of coverage at the cost of a higher energy consumption by the end devices. On the other hand, Low-Density Parity-Check (LDPC) codes, Polar codes, and Ultra-Reliable and Low-Latency Communications (URLLC) are of particular importance in improving the transmission reliability of wireless networks; hence, they have been included in the 5G New Radio Standard and in 6G standard currently under development [24–27]. The requirements of IoT applications have led to the emergence of Low-Power Wide-Area Networks (LPWANs) [28].

LPWANs are perfect for devices that need to send small amounts of information over long distances. These distances can be up to 15 km in rural areas and 1–5 km in urban areas, with the added advantage of low energy consumption (i.e., the associated batteries can last for 10 years). These communication networks are seen as highly suitable for IoT applications [29]. Among the most popular LPWAN networks are Weightless, Ingenu RPMA, Symphony link Sigfox, Long Range (LoRa)/Long Range Wide Area Network (LoRaWAN), and Narrowband Internet of Things (NBIoT) [30–32].

In recent years, Sigfox and LoRa have been positioning themselves in the world of LPWANs. Sigfox was developed in France in 2010 and is currently operating in different countries such as the United States, Ecuador, Mexico, Colombia, South Africa, and Australia, among others. It is characterized as using an Ultra-Narrow Band (UNB), occupying little space within the frequency range and employing low data rates.

As for the frequency band, it uses the Industrial, Scientific, and Medical (ISM) frequency bands [33]. Sigfox uses Differential Binary Phase Shift Keying (DBPSK) modulation for ascending messages and Gaussian Frequency Shift Keying (GFSK) modulation for descending messages. Its main features are low power consumption [34], low cost of devices, and capacity for bidirectional communication [35,36].

LoRa/LoRaWAN are also networks that play big roles within LPWAN. LoRa was developed by the startup Cycle in 2010 and was later acquired by Semtech (USA), while LoRaWAN became a network specification proposal by the LoRa Alliance in 2015. It offers a MAC layer based on LoRa modulation [37]. LoRa operates in ISM bands, uses spread spectrum technology, is suitable for transmitting small amounts of data over long distances, and consumes little energy [38]. LoRaWAN defines three classes of devices for bidirectional communication through LoRa: Class A, Class B, and Class C. The choice of class depends on the application [39].

NB-IoT is a LPWAN which is currently one of the three main LPWANs. It is a narrowband network that can coexist with LTE or GSM networks on licensed frequencies. It is standardized by the Third Generation Partnership Project (3GPP), and its specifications were published in the 3GPP Release 13 in June 2016 [40]. NB-IoT operates on the same frequencies as LTE and uses QPSK and BPSK modulation, as well as the LTE architecture, but features some optimizations to meet the requirements of massive IoT users. NB-IoT devices consume additional energy due to their synchronous communication and Quality of Service (QoS) handling. This system also provides low latency connectivity for IoT applications [41].

Latency is considered one of the parameters that allows for measuring the performance of a communications network; however, there are other variables, such as throughput, latency, speed, propagation delay, network capacity, range coverage, device lifetime, duration of useful life, service quality, and cost, that serve as indicators of the quality of links for data transmission. If a network does not have good performance, it can cause delays, loss of information, and information transfer limitations [42].

Implementing this system of networks for innovation in the agricultural sector allows for greater profitability in production while at the same time reducing fertilizer use and environmental impacts. Farmers can readily obtain information and statistics on crop growth, and smartphones may be used to remotely control crops, equipment, and decision making.

The remainder of this document is structured as follows: An overview of LPWAN networks is presented in Section 2, including a comparison of the three main technologies used, followed by the analysis and results in Section 3. Finally, our conclusions and direction for future research are provided in Section 4.

2. Methodology

In this section, we analyze the main LPWAN technologies for IoT applications. Additionally, the different features that define LoRaWAN and its use in the agricultural sector are studied. Furthermore, the most important simulators used with this technology are studied, as well as the testing scenarios in this work.

2.1. Internet of Things

The IoT, which is defined as the connection of all types of objects (i.e., elements that send information) to the Internet, has been gaining in importance and participation in many fields [43]. IoT provides connectivity for thousands of devices such as sensors and actuators, among others, and allows for connectivity to a network, allowing the objects to exchange data. The information provided by the devices is stored in the cloud [44]. One of its application fields that has gained traction in recent years is agriculture. The use of such technology has allowed traditional agriculture to be transformed into intelligent agriculture based on detection, measurement, and response, which allows farmers to obtain a higher productivity for their crops and better quality products [45]. Some implementations of IoT in agriculture are: real-time monitoring of farms, meteorological prediction modelling, customized fertilizer profiles based on soil chemistry, and water conservation. For this purpose, farmers can implement different types of sensors to monitor pH levels, temperature levels, and humidity levels, among others. These collected data are stored in the cloud for later use and decision making [46].

With the continuous growth of devices with Internet connectivity, IoT applications require long-range communication technologies, large-scale connectivity, low energy consumption, and low cost. LPWAN technologies have the required parameters to cover these needs [47].

2.2. LPWAN Analysis: Technologies

The IoT boom in recent years has led to an increase in the number of connected devices, giving rise to new technologies capable of interconnecting large numbers of devices. Current wireless communication technologies such as Bluetooth and ZigBee have a short range, while cellular communication technologies (e.g., 3G, 4G, and 5G) have a medium range but are limited by high power consumption; however, this limitation has been mitigated with the emergence of LPWAN long-range technologies that have lower energy consumption [48]. Figure 1 shows the relationship between coverage and data transmission rate for the most important wireless networks.

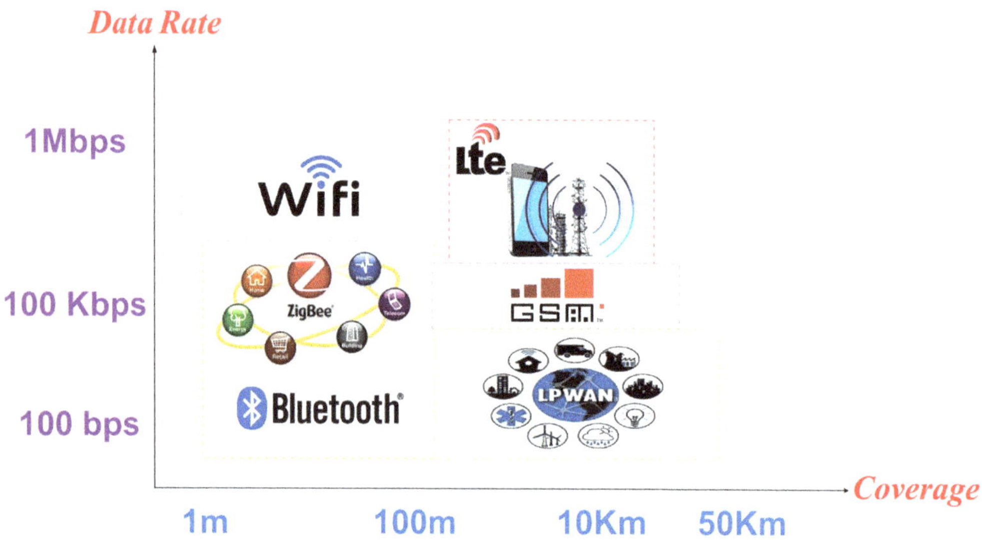

Figure 1. Relationship between coverage and data rate for various wireless communication technologies.

LPWAN technologies are wireless networks that allow for the transmission of small amounts of data over long distances with low power consumption on their final devices. The technologies also provide connectivity for a large number of devices, making them ideal candidates for IoT applications [49].

Several types of LPWAN technologies are currently available, some using licensed frequencies, and others operating in unlicensed spectra. The most popular technologies in this category include LoRa [50,51] and NB-IoT [52], each with their own unique technical features. Table 1 compares the key technical features of LPWAN technologies.

To choose an appropriate LPWAN technology for IoT applications, several factors, including range, coverage, device lifetime, latency, scalability, payload duration, implementation complexity, quality of service, and cost, must be considered [53]. Consideration of these parameters allowed us to analyze which of the considered technologies was most suitable for this project.

From the analysis provided in Table 1, we can conclude that LoRa has advantages in terms of cost, battery life, and implementation; furthermore, it is commercially available in several countries. For these reasons, we focus on the use of this technology, which is adapted to the conditions of rural areas, making it particularly suited to agricultural applications. Implementation of LoRa will allow for the connection of a large number of devices over a long range with low power consumption, favoring the monitoring of climate variables, water consumption in irrigation systems, and crop fertilization. These parameters are expected to allow for improvements in the quality of products.

Table 1. Comparative chart of the three main Low-Power Wide-Area Networks (LPWANs) technologies.

Types of LPWAN Technology	LoRaWAN	SIGFOX	NB-IoT
Coverage	2–5 km urban zone 10–20 km rural zone	3–10 km urban zone 20–40 km rural zone	1 km urban zone 10 km rural zone
Standard	LoRa Alliance	Sigfox	3GPP release 13
Licensed spectrum	No	No	Yes
Frequency	ISM Bands 433 MHz Asia, 868 MHz Europe, 915 MHz N. America	ISM Bands 433 MHz Asia, 868 MHz Europe, 915 MHz N. America	Cell Band LTE
Modulation	Chirp Spread Spectrum (CSS)	DBPSK/GFSK	QPSK/BPSK
Data speed	250 bps–50 kbps	100 bps	200 Kbps
Bandwidth	125–250 KHz	100 Hz	200 KHz
Topology	star	star	LTE network
Capacity Connected device	50 K per cell	50 K per cell	100 K per cell
Bidirectional communication	yes/Half duplex	Limited/Half duplex	yes/Half Duplex
Protocol	asynchronous	asynchronous	synchronous
Message per day	unlimited	140 uplink 4 downlink	unlimited
Maximum payload length	243 bytes	12 bytes uplink 8 bytes downlink	1600 bytes
Security	Yes (AES 128b)	No	Yes (LTE)
Geolocation QoS	TDoA No	RSSI No	OTDoA Yes
Energy consumption	low	low	high
Latency	Low with class C	high	low
Interference immunity	high	high	low
Installation cost per base station	>EUR 1000	>EUR 4000	>EUR 15,000
Final device cost	EUR 3–5	<EUR 2	>EUR 20

2.3. LoRa/LoRaWAN

LoRa is a physical layer technology developed by the Semtech Corporation [54]. LoRaWAN is the access layer developed by the LoRa Alliance, which employs LoRa technology for communication and device management [55].

LoRa allows for the transmission and reception of point-to-point information. Its low-power and long-range communication, patented by Semtech, is based on spectrum widening using the Chirp Spread Spectrum (CSS) technique, together with Forward Error Correction (FEC), making LoRa a robust technology against interference. The application of CSS is based on modulated pulses of linear frequency bandwidth whose frequency increases or decreases depending on the encoded information [56]. LoRa operates in several frequency ranges without license in the ISM bands according to the region [57]. In Table 2 a comparison of frequencies is given for areas in which it has higher penetration; Ecuador's frequency range is also included in the table.

Table 2. Comparative LoRa frequency chart for Europe, North America, and Ecuador.

Parameters	Europe	North America	Ecuador
Frequency	863–870 MHz	902–928 MHz	902–928 MHz
Channel plan	EU863–870	US902–928	AU915–928
Duty cycle	<1%	No limit	No limit
Channel uplink	125/250 KHz	125/500 KHz	125/500 KHz
Channel downlink	125 KHz	500 KHz	500 KHz
Channels	10	64 + 8 + 8	64 + 8 + 8
SF	7–12	7–10	7–12

Table 2 provides parameters based on the ETSI standard, such as the frequency, number of channels, duty cycle, and Spreading Factor (SF), analyzed for Europe, North America, and Ecuador. The specifications define that there are 64 channels with a 125 kHz bandwidth and 8 uplink channels with 500 kHz bandwidth for a total of 72 uplink channels; however, the 8 channels with 500 kHz bandwidth overlap with the 64 channels. A Duty cycle is the fraction of time during which the device is occupied. SF allows for improved network efficiency and capacity while also admitting 7 to 12 bits per symbol; by varying this parameter, one can achieve long-range communications at the expense of decreased data speed. Alternatively, decreasing the SF will result in increased data speed at the expense of shorter distances. Furthermore, the speed varies depending on the area. In Table 3, a comparison of SF values and data speeds according to the location is provided. For Europe (EU), the speed ranges from 250 bps to 50 kbps; for North America (NA), it ranges from 980 bps to 21.9 kbps; for Ecuador (EC), it ranges from 250 bps to 21.9 kbps [58].

Table 3. Comparative chart of Long Range (LoRa) Spreading Factor (SF) parameters according to the analyzed areas.

	SF	BW	Bitrate (EU)	Bitrate (NA)	Bitrate (EC)
LoRa	SF12	125 kHz	250 bps	-	250 bps
LoRa	SF11	125 kHz	440 bps	-	440 bps
LoRa	SF10	125 kHz	980 bps	980 bps	980 bps
LoRa	SF9	125 kHz	1.7 Kbps	1.7 Kbps	1.7 Kbps
LoRa	SF8	125 kHz	3.1 Kbps	3.1 Kbps	3.1 Kbps
LoRa	SF7	125 kHz	5.4 Kbps	5.4 Kbps	5.4 Kbps
LoRa	SF7	250 kHz	11 Kbps	-	-
LoRa	SF7	500 kHz	-	21.9 Kbps	21.9 Kbps
FSK	-	-	50 Kbps	-	-

Table 3 indicates that the bitrate obtained under a given SF may vary with different bandwidth (BW); for example, SF7 allows us to obtain higher data speeds when using a higher BW. This analysis was carried out to indicate the maximum values that can be obtained. In summary, when the SF is varied with the bandwidth, the maximum and minimum values for the data speed can be defined.

LoRa utilizes a dynamic adjustment method called Adaptive Data Rate (ADR), which allows the final device to dynamically adjust the transmission power parameters and data rate according to the distance between the end device and the gateway. This helps to optimize energy consumption in end devices.

The most important parameters of the LoRa physical layer are the bandwidth, the scattering coefficient, the frequency, and the Coding Rate (CR). The modes supported for the CR are 4/5, 4/6, 4/7, and 4/8. When increasing the CR, the transmission time will be longer as the packet size is larger [59].

2.3.1. LoRaWAN Architecture

The LoRaWAN architecture uses a star network topology, which allows for ease of implementation and management as there is no need for routing elements.

The devices that make up a LoRaWAN network are: end devices (e.g., sensors and actuators), gateways, network servers, and application servers [60]. Its operation starts when the end devices send encrypted messages to the gateways, which then forward information to the network server using TCP/IP protocols. In this way, the network server receives and processes information from the end devices, discarding messages that were received more than once from different gateways. Additionally, the network server is responsible for security between the end device and the application server. The application server decrypts messages and makes information available to the user [61]. In Figure 2, we show the architecture of a LoRaWAN network. Additionally, LoRaWAN allows for two-way communication through LoRa.

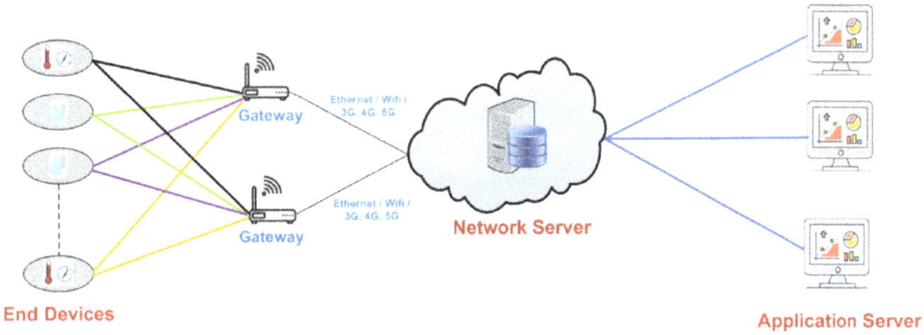

Figure 2. Long Range Wide Area Network (LoRaWAN) architecture.

2.3.2. LoRaWAN Device Classes

In LoRaWAN, the study of energy consumption in end devices is very important. There are three types of device classes used to expand the battery life: Class A, Class B, and Class C.

- **Class A.** Class A devices are bidirectional end devices with greater energy efficiency; most of the time, they are in sleep mode. The transmission for the uplink is followed by two descending link windows within a short period of time. They are used for applications which do not require the continuous receiving of data, and by default, all devices come pre-defined as Class A. Figure 3 shows the transmission type of Class A devices.
- **Class B.** Class B devices are two-way devices with programmed reception slots that open additional receiving link windows at programmed times, where the time is synchronized with beacons transmitted by the gateway. These devices have additional power consumption. Figure 4 shows the transmission type of Class B devices.
- **Class C.** Class C devices are two-way devices with a maximum reception slot, which keep their reception windows open continuously, only closing them when transmitting. The energy consumption of these devices is excessive, and it is recommended that they only be used in places where energy is not limited. These are employed for applications requiring low latency. Figure 5 shows the transmission type of Class C devices.

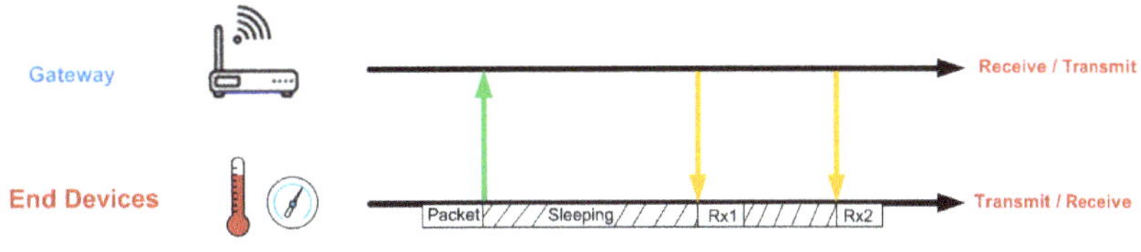

Figure 3. Class A transmission.

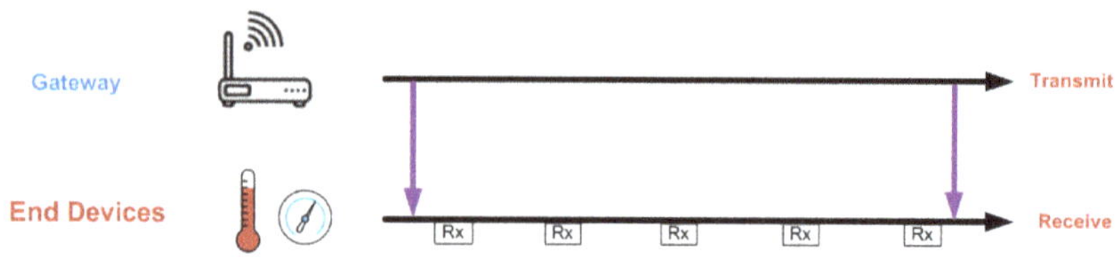

Figure 4. Class B transmission.

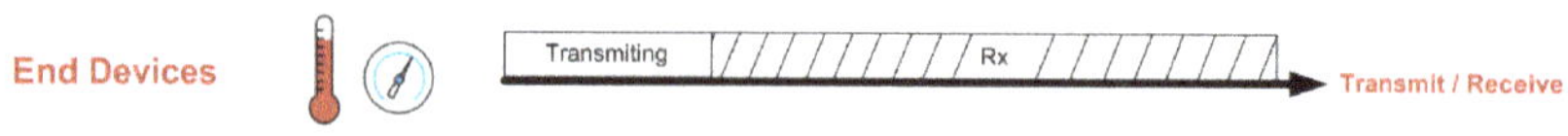

Figure 5. Class C transmission.

2.3.3. Security

Security is one of the main features in wireless communications, which is why LoRaWAN is one of the few networks in IoT that employs Advanced Encryption Standard (AES) encryption for applications that require the secure transmission of data. The security protocol needs to meet the criteria of LoRAWAN, such as low energy consumption, installation, and implementation. LoRaWAN devices employ two types of session keys in the network:

- **NetworK Session Key (NwKSKey)**, consisting of an AES-128 bit encryption key that is unique to the network server and is shared between the final device and the network server.
- **Application Session Key (AppSKey)**, implementing end-to-end encryption between the final device and the application server, is an AES-128 bit encryption key that is unique to the application server [62].

The security types are depicted in Figure 6.

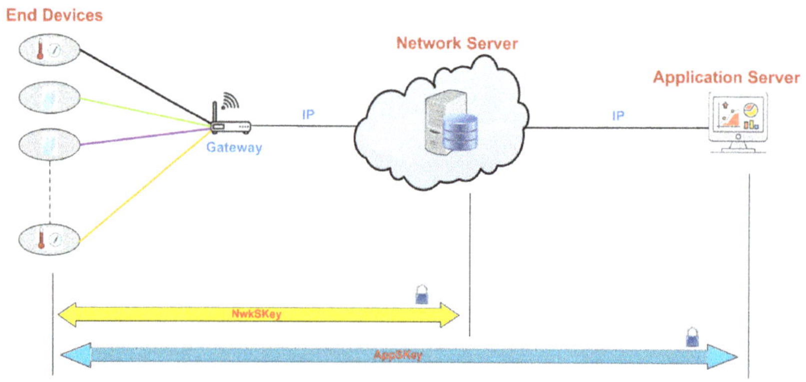

Figure 6. LoRaWAN Security.

To be operational and part of the LoRaWAN network, the end device must be activated and authenticated, which can be accomplished in the following ways:

- **Air Activation.** In this type of activation, the final device exchanges MAC messages with the server (e.g., request and acceptance). In the process, an address (DevAddr) and security key are assigned. This is performed each time the final device loses connection.
- **Activation By Personalization.** In this type of activation, configuration is manual. When starting up, the device connects directly to the network. This application is not commonly used [44].

2.4. LoRaWAN Simulators

There are simulation tools for communication networks that allow users to specify specific scenarios they want to simulate for a given technology, such as WiFi, Ethernet, LTE, and so on. These simulations allow for the determination of the virtual behavior of the network, which is very close to the real conditions of evaluation scenarios. Unlike implementation scenarios, this field of research allows us to obtain reliable information at low cost.

There are several types of simulators, depending on the method used:

- **Continuous Simulation.** These simulators show results produced at all points during the simulation, not in intervals.
- **Discrete-Event Simulation.** These simulators model the operation of a system as a sequence of discrete events at different points in time.

It is important to know the simulators that are available for LoRAWAN, even though it is a technology that emerged in 2015. Work related to this technology started from 2018, in the physical layer of LoRa as a first stage [63]. In recent years, work has been ongoing to update simulators to include MAC support, energy consumption, and other LoRAWAN parameters [64]. The most commonly used simulators are detailed in the following:

- **NS3** is a discrete-event simulator that uses open-source code and is based on the C++ programming language coupled with Python. This simulator allows for the evaluation of LoRaWAN media access capabilities in comparison to a common ALOHA scheme [65]. By default, NS3 lacks a graphical interface. Another attribute of the simulator is that it allows for the generation of PCAP files (a file format used for capturing packets). It is available for Windows and Unix platforms [55,66].
- **Omnet++** is an open-source, free simulation software based on C++ programming for discrete events, which additionally uses a specific high-level language named NEtwork Description (NED). It has specific functionalities, such as the simulation of sensor networks, ad-hoc wireless networks, optical networks, and Internet protocols. OMNET++ has a graphical interface for modelling topologies and analyzing results. Another consideration is its modular architecture; the simulation kernel can easily be integrated with other applications. It is compatible with Windows and Unix. This simulator has been used in both academic and industrial environments. The components of OMNET++ are: kernel library (C++), NED language, eclipse-based IDE simulator, command line interface for running simulations (Cmdenv), time execution GUI for interactive simulations (Qtenv), utilities (file creation tools MAKE), documentation, and simulation examples [67,68].
- There are also frameworks based on **INET** which extend it in specific directions, such as the case of LoRa with FLoRa, which allows for point-to-point simulations. Its functionality creates a LoRaWAN network with nodes for LoRa, gateways, and the LoRa server.
- **LoRaSim** is a SimPy-based discrete event simulator employed for collision simulations and scalability analysis. Its functionality is limited to simulation between the final devices and gateways [55,69].

Table 4 compares the three simulators in which their main features for implementation in LoRa can be seen.

Table 4 indicates that the most notable simulators are NS3 and OMNET++; their use will depend on the required applications. In our case, we decided to use OMNET++ for our LoRa applications as it contains a high-level language allowing for analysis of the network. Furthermore, it contains the FLoRa library. Being an open-source simulator, changes or modifications to the network can be made by adding modules and extra functionalities. In our case, we evaluate our network by varying the number of nodes (or end devices), the number of gateways, and distances.

Table 4. Comparison of simulators. Adapted from [64] and completed with new data by authors

	NS-3	OMNET++	LoRaSim
Discrete event simulator	Yes	Yes	Yes
Open source simulator	Yes	Yes	No
Language	C++/Python	C++/NED	Python/SimPy
Graphic interface	No	Yes	Yes
Operating system	Windows/Unix/macOs	Windows/Linux/macOs	Linux
Application	investigative/academic	investigative/academic	investigative
LPWAN	NB-IoT/LoRa	LoRa	LoRa
Framework	LoraPhy/Loramac	LoRa	loraDir.py/loraDirMulBs.py/ directionalLoraIntf.py
ADR	Yes	Yes	No
Energy Consumption	Yes	Yes	Yes
Bidirectional communication	Yes	Yes	No
Medium spread	Yes	Yes	No

2.5. *Use of LoRaWAN Technology in the Agricultural Sector*

The use of LoRaWan for IoT applications in agriculture is very helpful, especially for water irrigation optimization in fields. This situation arises from limitations of access to this resource or due to droughts. With the implementation of IoT in agriculture, the efficient use of water and quality food production can be guaranteed. This technology allows farmers to cover large areas of farmland due to its long-range transmission ability; moreover, its star topology allows for easy implementation. The cost of implementation is low compared to the other technologies already analyzed. Another important factor is the low power consumption, allowing batteries to be used for years, making it ideal for such applications.

Below is a brief synopsis of some use cases in the agricultural sector:

- In a vineyard for wine production, where air temperature and humidity monitoring is conducted for the optimal growing of grapes, the preferred range can be indicated to ensure quality. Other measurement methods were not optimal or were even harmful to production. This allows the growers to determine the best growing season and improve their production [70].
- Organic fertilizer production is based on vermicomposting, a system that transforms organic matter through the combined action of earthworms and micro-organisms, yielding a natural fertilizer with physical, chemical, and biological properties that benefit soil crops. It is monitored for variables such as temperature and humidity, which must remain within specific ranges to ensure the survival and reproduction of earthworms [71].
- Another use-case is temperature and humidity monitoring in a horse stable. Some tests were carried out, in which there were fluctuations in temperature that caused more messages to be sent. It was also considered that the fluctuations depended on the location of the sensors being near entrances and exits [39].

2.6. *Evaluation Scenario*

The analysis and evaluation in this work is based on an agricultural scenario applied in rural areas. The coastal region of Ecuador, which is made up of 7 provinces, was chosen. Manabí is the province with the highest extent of surface area, with 18,893 km^2 [72]. The technology employed for this scenario was LoRaWAN as it allows for wide coverage, scalability, low cost, and easy implementation. A coverage area of 5 $\times$ 5 km was considered, with a homogeneous distribution of nodes across the simulation area as well as a standard transmission packet size of 10 bytes. The assigned frequency band for Ecuador is

902–928 MHz. This frequency does not have a duty cycle restriction, which allowed it to be considered in this study. In Figure 7, the evaluation scenario is shown.

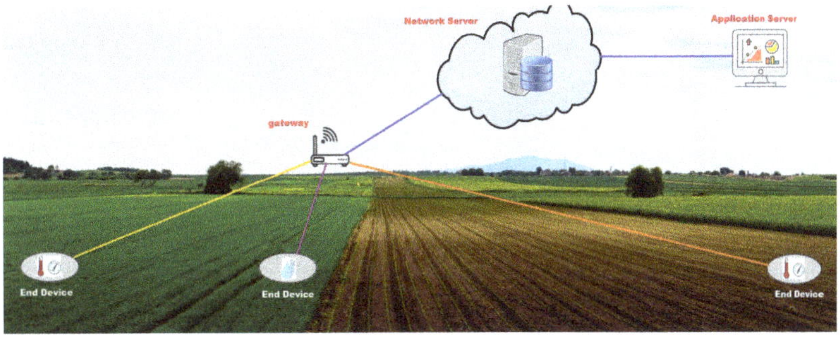

Figure 7. Evaluation scenario.

Some of the particular features of the chosen agricultural environment are:

- Large areas of land with few obstacles or interferences from other networks as is often the case in urban environments;
- High density of devices, which should be placed a few meters apart to achieve better resolution of data from climate monitors on crops;
- Low heterogeneity of the data collected; this means that only a few climatic parameters, such as humidity, temperature, and so on, need to be acquired. For this reason, similar devices are needed for the implementation of the network.

For each measurement presented, the simulation window time was 1 day, and we ran the simulation five times. The results shown in this paper are the average of the five trials. We performed a simulation to measure the scalability of the LoRa/LoRaWAN network; so sensor data were not included in the simulation. Therefore, the size of each transmitted data packet was constant (i.e., 10 bytes long).

2.7. Simulation Environment

To evaluate the proposed scenario, a simulator called Omnet++ was used. This simulator includes a library of open code called FLoRa, developed by researchers at Aalto University in Finland for their studies on LoRa and LoRaWAN [73]. This simulation framework allows for the deployment of a large number of nodes and gateways, configuring the maximum distance in the coordinates X and Y, as well as the object coordinates; propagation loss coefficients in the wireless communication medium; and activating or deactivating the ADR technology in both nodes and the network server (see Section 2.3).

Figure 8 shows the implementation of the LoRa network architecture design, where the link connected to the *GwRouter* comes from the gateways located in the simulation area. This router is connected through any network link to the network server (*networkServer* in Figure 8) which, in this case, is represented by *InternetCloud* and *nsRouter*. As mentioned before, we could change the transmission media parameters—in our case, the propagation losses (*LoRaMedium*)—as well as other network configurations, such as the transmission latency, IPv4 routing, and so on, which we kept constant.

For simulations, the tool provides metrics that can be collected to analyze the behavior of the network that has been simulated. Some of the metrics considered in this study are:

- **Packet Delivery Ratio.** This is the relationship between the supply of packets sent by all nodes and the successfully received packets by the network server.
- **Energy consumption.** This is the total energy consumption of all nodes, expressed in Joules (J).
- **Packet collisions.** This is the total number of packet collisions in all gateways present on the network.

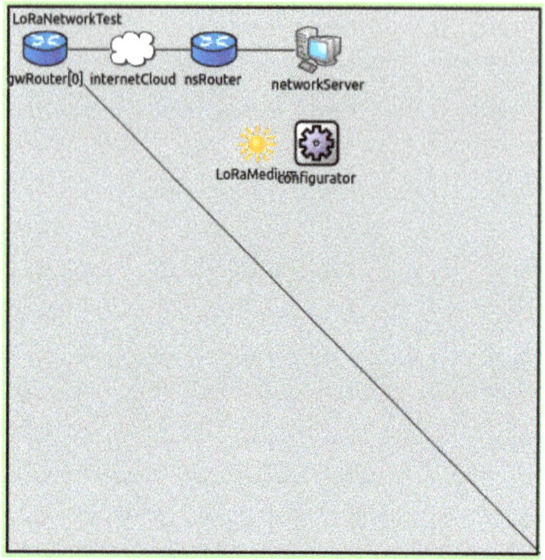

Figure 8. Configuration of the network architecture for LoRa in the FLoRa simulation environment.

One of the elements analyzed in this project is the scalability of the network, which was analyzed using the simulator. In the simulation environment, between 100 and 3000 LoRa devices were deployed. The simulation of scenarios was carried out by varying the following input parameters:

- **Number of nodes.** A simulation was started with 100 nodes (Figure 9), and the number was increased by 100 nodes for each subsequent simulation. Considering computational power issues, we were able to achieve up to 3000 nodes in most cases, a scenario in which each simulation took more than 2 h.
- **Number of gateways.** Two scenarios were considered—with 1 and 2 gateways—to observe the network behavior and determine which situation was better for each alternative.
- **ADR.** The Adaptive Data Rate can be activated or disabled in the simulation configurations. Both options were checked to observe the impact of this mechanism.
- **Propagation losses.** We considered configurations that allow for the simulation of ideal propagation conditions (i.e., with almost no losses), as well as moderate propagation conditions, which better represent wireless transmission in rural environments.

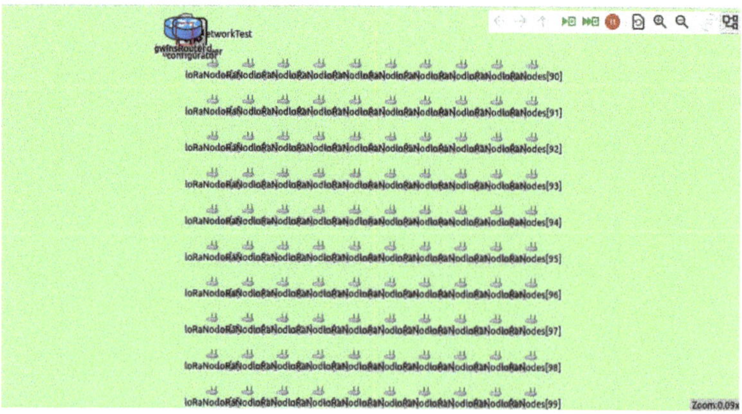

Figure 9. FLoRa simulation environment with 100 nodes placed in the simulation area.

3. Results

The following section details the results obtained from the methodology proposed for assessing the performance of communication systems used in IoT, such as propagation losses, energy consumption, and collisions.

First, we detail the impact that the wireless propagation medium has on the packet delivery factor, contrasting ideal conditions (without losses) with medium conditions (with moderate losses) simulating the considered rural scenario. Next, simulations are performed with two gateways within the network to compare their metrics with those obtained with a single gateway. In the latter case, a rural propagation medium is considered to simulate as realistic an environment as possible. It should be noted that in all cases the network with ADR technology enabled is compared to that without (i.e., where transmission power parameters and the scattering coefficient *SF* are kept constant).

3.1. Impact of Wireless Medium Propagation with a Single Gateway

Simulating the impact of the propagation medium with a single gateway is one of the first parameters to consider. The simulation tool includes several scenarios, including an ideal wireless medium with no propagation losses, another with moderate losses, and a third with very high losses due to obstruction by buildings and trees, as well as interference from other networks using the same frequency band. The moderate loss scenario models an environment with few obstacles and negligible interference. This represents the rural/agricultural setting of the given LPWAN application. The positive impact of ADR commands on packet delivery ratio was also measured. When ADR technology is active, the network server receives information from the gateway about the Signal-to-Noise Ratio (SNR) to adjust the transmission power or SF. In this way, when a node has difficulty reaching the network server, adjustments are made such that the communication improves; conversely, when a node has excellent transmission parameters—either because it is close to the gateway or there is low interference—power consumption is reduced, thereby conserving energy. Thus, ADR optimizes energy consumption while guaranteeing communication between nodes.

3.1.1. Simulating Wireless Medium without Losses in Propagation: Ideal Environment

Figures 10 and 11 show the results under ideal and rural conditions for the propagation medium, respectively. In both graphs, it can be seen that communication was maintained with high performance when ADR technology was active (*ADR On*), while it quickly declined when ADR was not active (*No ADR*), that is, when the nodes kept the power parameters and scattering factor settings with which they were originally configured.

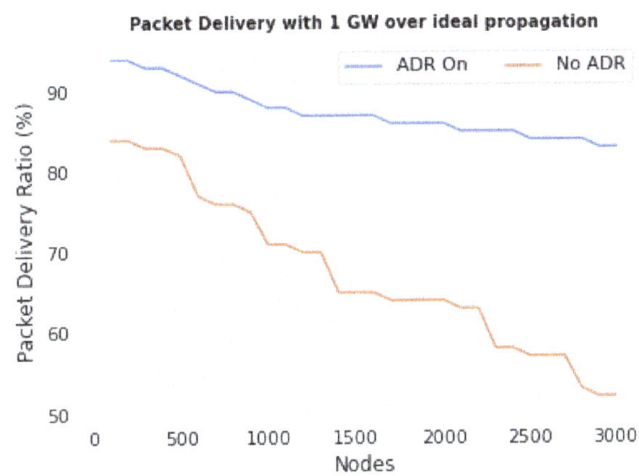

Figure 10. Packet delivery ratio to the network server with a single gateway under ideal propagation conditions.

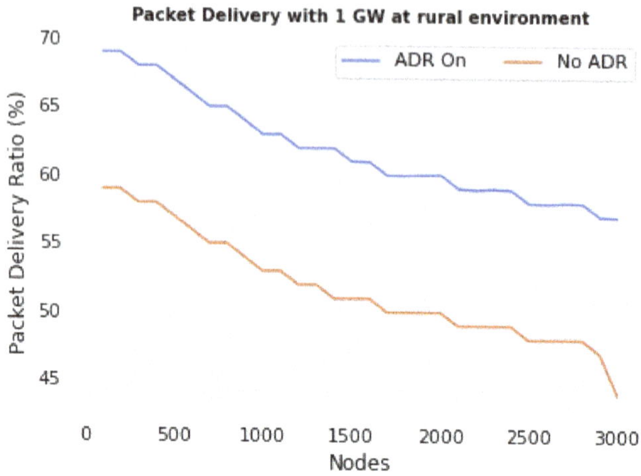

Figure 11. Packet delivery ratio to the network server with a single gateway under propagation conditions simulating a rural environment.

To analyze the results of the simulation with a gateway, we will place the point at which there are 100 nodes and the ADR in state *No ADR*. It can be observed that the delivery ratio is 85%, and for the ADR in *On* it increases to 93%, an improvement of 9%.

For the case of 3000 nodes, the delivery of packets with the ADR in *No ADR* is 51%. When the ADR is in *On* for 3000 nodes, the packet delivery factor is 84%, an improvement of 64%.

3.1.2. Simulating Wireless Medium Moderate Propagation Losses: Rural Environment

In Figure 11, the packet delivery ratio results for the rural environment can be observed. To analyze these results, we located the point at which there were 100 nodes with the ADR in state *No ADR*. From this graph, it can be seen that the packet delivery ratio was 58%, while with the ADR in *On* mode, it increased to 68%, comprising an improvement of 17%. For 3000 nodes, parcel delivery with the ADR in state *No ADR* was 44% while, when the ADR was in *On* mode for 3000 nodes, the package delivery factor increased to 57%, representing a 30% improvement.

From these results, it can be seen that as the number of nodes increased—both for the ideal simulation environment and for rural settings—the delivery ratio also decreased, having a negative impact on the transmission quality. It was observed that ADR technology improved communication, although it struggled to maintain packet transmission performance as the number of nodes increased. This is consistent with the simulated obstacles and interference in the propagation medium.

3.2. *Packet Delivery Ratio Simulation with Two Gateways*

To achieve redundancy against incidents, one solution would be to place two gateways in the simulation area. In this case, if one of the gateways fails, communication between the nodes and the network server can be maintained. In LoRaWAN, when there are two gateways, the nodes transmit their packets in broadcast mode and are received indiscriminately by either gateway, with the network server removing duplicate packets. The latter also decides through which gateway a packet should be sent to a node based on the gateway with the least congestion and best SNR. In this simulation, we added two gateways and compared the results for various factors, such as the packet delivery ratio, when using a single gateway versus when using multiple gateways.

The advantage of having two gateways is that even if one gateway fails, the other can receive data from and send data to the network server. However, as shown in Figure 12, this is only true if ADR technology is activated *On*. The graphs show that the packet delivery ratio was the same with one or two gateways when ADR was not active (*No ADR*, green and red lines). In this case, the nodes maintained their communication parameters. This situation is also a result of the simulation area. However, it can be observed that the packet delivery ratio was further improved when ADR technology was active and both gateways were operational.

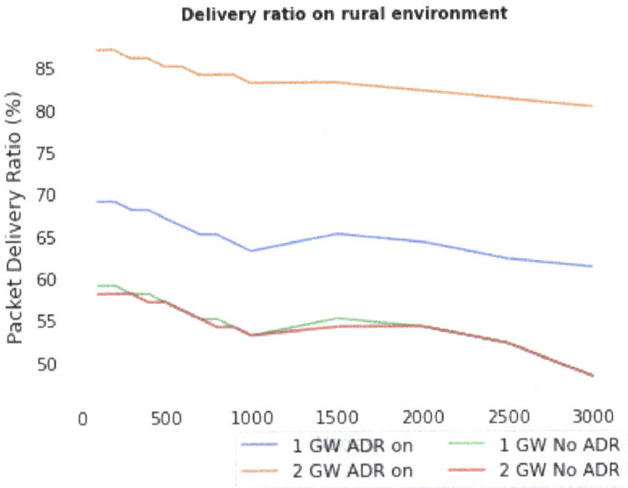

Figure 12. Packet delivery ratio measured with one and two gateways, with and without Adaptive Data Rate (ADR).

From the graph, it can be seen that the delivery ratio with two gateways, ADR off, and 100 nodes was 58%, while with two gateways, ADR on, and 100 nodes, it was 87%, comprising a 50% improvement. For the case of two gateways, ADR off, and 3000 nodes, the delivery ratio was 45%, while with two gateways, ADR on, and 3000 nodes, it was 80%, comprising a 77% improvement.

This means that commands sent from ADR when there is only one gateway may not reach certain nodes; therefore, communication for them will not improve. With two gateways, however, the improvement in the rate of successful packet deliveries, as well as scalability and consistency when the number of nodes increases, can be seen; only a 5% drop in successful packet delivery rate occurred.

ADR technology seeks to increase energy consumption performance by improving communications. While it is true that the communication performance improves when it is activated, the simulation results show that the variation in energy consumption was minimal. In Figure 13, it can be clearly seen that using two gateways or ADR technology had a minimal impact on energy consumption per node. What can be observed instead, is that the average energy consumption per node increased linearly with the number of nodes in the network. This plainly means that with a larger number of nodes transmitting in the network the interference on each channel increases dramatically, causing each node to have to re-send packets and to listen to transmissions not directed at the particular node. This increases the time that the nodes are active, therefore increasing energy consumption.

Having two gateways can in some ways be counterproductive. As shown in Figure 14, collisions remain low when only one gateway is present, regardless of whether ADR technology is active or not. When a new gateway is added, packet collisions at each gateway naturally increase as the packets sent by each node are duplicated. This situation worsens as the number of nodes increases. As communication is more effective and a greater number of packets arrive at gateways when ADR technology is active, collisions within each gateway become much higher and increase linearly as the number of nodes present in the network increases. Although the server in Figure 12 received packets successfully, having high numbers of collisions within the network affects the scalability of the gateways. This situation could be improved if there were some mechanism by which each node in the network knows which gateway to transmit packets to the server through, thereby making the other gateways ignore those packets.

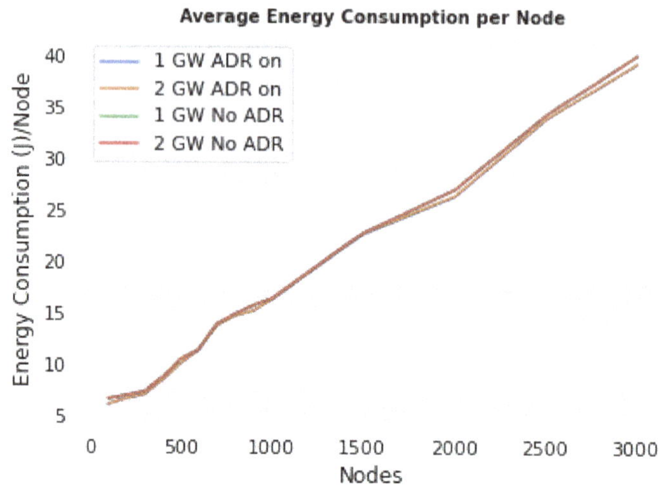

Figure 13. Average energy consumption per node with one and two gateways deployed, with and without Adaptive Data Rate (ADR).

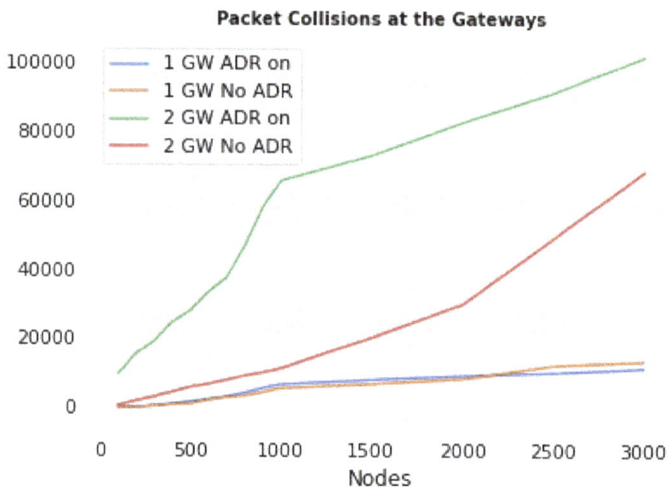

Figure 14. Packet collisions measured in all gateways of each simulation.

It is necessary to emphasize that for all the results collected in these simulations we considered the use of a single channel in the LoRa communication band as the FLoRa tool does not have a feature or option to divide communications among the available channels. Thus, if we take into account that the channel plan for Ecuador contains 64 channels for node–gateway communication, network scalability could be increased considerably, being able to count 100 nodes per channel and, thus, reducing energy consumption while having high-quality communications.

3.3. Validation of the Proposed Simulation Environment

To validate the realism of the simulation results, we conducted a review of other studies on the capabilities of LoRaWAN protocols obtained both analytically and through simulation. Although the objectives of the presented articles were not the same as those of this study, the results serve to reinforce the reliability of the simulated results that have been collected.

The effect that ADR technology has on the packet delivery ratio has also been studied in [74]. This study analyzed how the Packet Delivery Ratio (PDR) decays with respect to an increase in distance, over a range between 50 and 300 m, with 1000 nodes deployed. In our study, the packet delivery rate had low values because the nodes were distributed within a wider space of 5000 m; but it was validated by the results of the aforementioned study, where the PDR was lower than 70% with 2000 nodes deployed and ADR activated.

The impact of packet collisions was analyzed in [75], where a technology called *Adaptive Data Payload* (or *ADP*) was proposed as an alternative to ADR, for which they conducted a comparison within a simulation environment using algorithms developed in the same study. The percentage of packet collisions on uplink packets at the gateway remained low when ADR technology was active, resembling the results obtained in our simulation.

In the article [76], a comparison of delivery ratio results was conducted, both analytically and through simulation, with respect to the distance and node number. There, the maximum distance presented was 3000 m, for which a delivery ratio of around 65% was achieved both theoretically and through simulation with 3000 nodes deployed.

The consumption of energy has also been reviewed, with respect to several scenarios, in [77]. Consumption was shown to grow linearly with respect to the number of nodes deployed—a trend very similar to that observed in the present study. Additionally, the *PDR* was also analyzed, indicating that it remains consistent as the number of nodes increases. The simulator used in the aforementioned document was LoRaSim, and a study of the scalability of LoRa/LoRaWAN networks was also conducted.

The cited studies provide evidence that the simulation environment proposed for the coastal region comes close to reality regarding the behaviour of LoRa networks; therefore, the validity of our results can be concluded. However, deployment under actual conditions and with appropriate instrumentation is always the most reliable way to characterize the parameters and ranges of a network in an agricultural setting.

4. Conclusions

In this work, we evaluated the performance of communication systems used for Internet of Things applications in the agricultural sector. LPWAN technology was chosen as the communication system paradigm for simulation as it allows for the deployment of wide-area networks with low power specifically for IoT. Within LPWAN, LoRaWAN was chosen as it met the requirements of the project in terms of scalability, energy consumption, frequency, data speed, and cost, among other aspects. The OMNET++ simulator was used, making use of the FLoRa library, which allowed for the modeling of the behavior of LoRaWAN networks. After analyzing other simulators, this was found to be the one that best suited the conditions of the evaluation scenario. For our evaluation, we proposed two propagation scenarios: ideal conditions and moderate losses. The number of nodes varied from 100 to 3000, and the scenarios were analyzed with respect to one or two gateways.

Based on the results of the simulations performed for evaluation of the LoRaWAN network in the proposed scenarios, we reached the following conclusions:

- The results show that in an ideal wireless medium and with a single gateway, 100 nodes, and without ADR (*No ADR*), the packet delivery was 85%. With the activation of ADR (*ON*), this number increased to 93%. In the case of 3000 nodes, delivery with *No ADR* was 51% and that with ADR *ON* was 84%. Thus, with ADR, a better delivery of packets can be achieved.
- For a medium wireless network with moderate losses and a single gateway, the results indicated that with 100 nodes and *No ADR*, delivery of packets was 58%, while that

with ADR *ON* increased to 68%. For 3000 nodes, the delivery of packets with *No ADR* reached 47%, while with ADR *ON*, we obtained 62%. As was observed in both scenarios, as we increase the number of nodes, the packet delivery decreases, having a negative impact on transmission quality.

- The results with two gateways indicated that with 100 nodes and *No ADR*, the delivery of packets was 58%; meanwhile, with the activation of ADR *ON*, this increased to 87%. In the case of 3000 nodes, the delivery of packets with *No ADR* reached 45%, while that with ADR *ON* was 80%. Thus, it was observed that the improvement in packet delivery was especially pronounced as the number of nodes increased.
- The data obtained from the simulations indicated that energy consumption did not significantly change when ADR technology was turned on with two gateways in the network. However, the average node consumption increased linearly with the number of nodes. Additionally, it was found that with two gateways, the communication quality improves and the coverage radius is extended, albeit at a higher installation cost.
- The number of collisions increased dramatically when the number of nodes in the network increased and there were two gateways present. This is due to the fact that all gateways receive packets transmitted by nodes using the broadcast mode then forward these packets to the network server, which is responsible for eliminating duplicates. This can be counterproductive if two gateways are present when there are few nodes in the network. Another situation that could improve this situation is if each node knows which of the gateways in the network it should communicate with—a mechanism that does not exist in the protocol, which should be studied further.
- In Ecuador, the frequency band used offers 64 channels for communication uplink. Taking into account that the scenarios presented here were only simulated with the use of one channel, the results could be improved if a high number of nodes were spread among the available channels. LoRaWAN technology has high scalability and allows for a high density of nodes in a wide terrain area; these are highly desirable characteristics for applications in rural areas such as precision agriculture, which can contribute to improving the transmission of data acquired from climatic variables, visibility state of crops, efficiency in water usage, fertilizer levels, and agricultural product production capacity.
- Farmland environments have few obstacles to data connectivity, with the SNR received by gateways directly correlating to the distance from the node, thus influencing the spreading factor and transmission power. This scenario does not apply in urban environments, where a node may be close to a gateway but have low SNR due to nearby obstructions such as buildings.

A comparison was made with the results of other studies, such as those considering the ADR, particularly in evaluating the effect that it has on the packet delivery ratio. The results of our study showed low values as the nodes in our scenario were distributed across 5 km. This demonstrates that as the distance between nodes increases, the packet delivery ratio decreases. Another case studied packet collisions, where the ADR was used; the results of the previous study were similar to those obtained in our analysis. Additionally, energy consumption has also been analyzed in other studies, showing that it grows linearly with respect to the number of deployed nodes; this trend was also observed in our case.

Further Studies

In future work, we expect to simulate the LoRaWAN network behavior with the presence of several channels, as well as testing frequency change mechanisms when unfavorable communication conditions are present—situations that the simulator used in this study cannot handle, despite being one of the most complete simulators available.

The simulation of downlink packets was beyond the scope of this study; this is a situation that must be considered to simulate communication with devices that, in addition to collecting data with several sensors, have actuators that allow for the regulation of water and fertilizer supply in agriculture.

The LoRaWAN protocol is characterized by having devices that send packets to all nodes simultaneously, which leads to an increase in packet collisions as the number of nodes increases, directly impacting the protocol's scalability. Therefore, future studies should focus on implementing mechanisms that reduce this weakness while preserving low energy consumption and high communication quality levels.

Finally, a real-life scalability study in a rural setting would be of utmost importance as it could reveal characteristics that simulators do not allow for or do not model adequately. A deployment with fewer than 50 nodes could be carried out at a reasonable cost, while a study involving 5000 to 10,000 nodes would require significant financial investment—not counting the deployment land rights and instrumentation needed to carry out measurements.

Author Contributions: Conceptualization, G.Y., M.H. and R.C.; data curation, G.Y., M.H. and M.S.; formal analysis, G.Y., M.H. and R.C.; funding acquisition, M.H.; investigation, G.Y. and M.H.; methodology, G.Y. and M.H.; project administration, M.H.; software, M.S.; supervision, M.H.; validation, G.Y., M.H., M.S. and R.C.; visualization, M.H. and R.C.; writing—original draft, G.Y., M.H. and M.S.; writing—review and editing, M.H. and R.C. All authors have read and agreed to the published version of the manuscript.

Funding: This research was funded by projects CYTED 788 [REDTPI4.0-320RT0006], CYTED [EFIMOVI-720RT0014], PLAGRI project by Telecommunications and Telematics Research Group (GITEL) from Universidad Politécnica Salesiana, Cuenca, Ecuador.

Conflicts of Interest: The authors declare no conflict of interest.

References

1. Organización de las Naciones Unidas para Alimentación y la Agricultura. *El Futuro de la Alimentación y la Agricultura*; Tendencias y desafíos; Food and Agriculture Organization: Rome, Italy, 2017.
2. Feng, X.; Yan, F.; Liu, X. Study of wireless communication technologies on Internet of Things for precision agriculture. *Wirel. Pers. Commun.* **2019**, *108*, 1785–1802. [CrossRef]
3. Giri, A.; Dutta, S.; Neogy, S. Enabling agricultural automation to optimize utilization of water, fertilizer and insecticides by implementing Internet of Things (IoT). In Proceedings of the 2016 International Conference on Information Technology (InCITe)-The Next Generation IT Summit on the Theme-Internet of Things: Connect your Worlds, Noida, India, 6–7 October 2016; pp. 125–131.
4. Benke, K.; Tomkins, B. Future food-production systems: vertical farming and controlled-environment agriculture. *Sustain. Sci. Pract. Policy* **2017**, *13*, 13–26. [CrossRef]
5. Lavanya, R.; Chakkaravarthy, G.V.; Alli, P. An Integrated GIS and Knowledge-Based Automated Decision Support System for Precision Agriculture Using IoT. In *IoT and WSN Applications for Modern Agricultural Advancements: Emerging Research and Opportunities*; IGI Global: Hershey, PA, USA, 2020; pp. 86–98.
6. Erazo, M.; Rivas, D.; Pérez, M.; Galarza, O.; Bautista, V.; Huerta, M.; Rojo, J.L. Design and implementation of a wireless sensor network for rose greenhouses monitoring. In Proceedings of the 2015 6th International Conference on Automation, Robotics and Applications (ICARA), Queenstown, New Zealand, 17–19 February 2015; pp. 256–261.
7. Guillermo, J.C.; García-Cedeño, A.; Rivas-Lalaleo, D.; Huerta, M.; Clotet, R. Iot architecture based on wireless sensor network applied to agricultural monitoring: A case of study of cacao crops in ecuador. In Advances in Intelligent Systems and Computing, Proceedings of the 2nd International Conference of ICT for Adapting Agriculture to Climate Change (AACC'18), Cali, Colombia, 21–23 November 2018; Springer: Cham, Switzerland, 2018; pp. 42–57.
8. Tzounis, A.; Katsoulas, N.; Bartzanas, T.; Kittas, C. Internet of Things in agriculture, recent advances and future challenges. *Biosyst. Eng.* **2017**, *164*, 31–48. [CrossRef]
9. Bell, J.M.; Schwartz, R.C.; McInnes, K.J.; Howell, T.A.; Morgan, C.L. Effects of irrigation level and timing on profile soil water use by grain sorghum. *Agric. Water Manag.* **2020**, *232*, 106030. [CrossRef]
10. Kirby, R.M.; Bartram, J.; Carr, R. Water in food production and processing: quantity and quality concerns. *Food Control* **2003**, *14*, 283–299. [CrossRef]
11. Wallace, J. Increasing agricultural water use efficiency to meet future food production. *Agric. Ecosyst. Environ.* **2000**, *82*, 105–119. [CrossRef]
12. Pérez, D.; Risc, R. Implementación de Lora y Lorawan como escenario futuro de la industrias 4.0 en el sector agroindustrial peruano. *Rev. Campus* **2020**, *25*. Available online: https://www.aulavirtualusmp.pe/ojs/index.php/rc/article/view/1829 (accessed on 25 May 2022).
13. Muangprathub, J.; Boonnam, N.; Kajornkasirat, S.; Lekbangpong, N.; Wanichsombat, A.; Nillaor, P. IoT and agriculture data analysis for smart farm. *Comput. Electron. Agric.* **2019**, *156*, 467–474. [CrossRef]

14. Sreekantha, D.; Kavya, A. Agricultural crop monitoring using IOT-a study. In Proceedings of the 2017 11th International Conference on Intelligent Systems and Control (ISCO), Coimbatore, India, 5–6 January 2017; pp. 134–139.
15. Sarker, V.K.; Queralta, J.P.; Gia, T.N.; Tenhunen, H.; Westerlund, T. A survey on LoRa for IoT: Integrating edge computing. In Proceedings of the 2019 Fourth International Conference on Fog and Mobile Edge Computing (FMEC), Rome, Italy, 10–13 June 2019; pp. 295–300.
16. Liu, Y.; Ma, X.; Shu, L.; Hancke, G.P.; Abu-Mahfouz, A.M. From Industry 4.0 to Agriculture 4.0: Current Status, Enabling Technologies, and Research Challenges. *IEEE Trans. Ind. Informatics* **2020**, *17*, 4322–4334. [CrossRef]
17. Samie, F.; Tsoutsouras, V.; Bauer, L.; Xydis, S.; Soudris, D.; Henkel, J. Computation offloading and resource allocation for low-power IoT edge devices. In Proceedings of the 2016 IEEE 3rd World Forum on Internet of Things (WF-IoT), Reston, VA, USA, 12–14 December 2016; pp. 7–12.
18. Huerta, M.; Garcia, A.; Guillermo, J.C.; Martinez, R.C. Wireless Sensor Networks Applied to Precision Agriculture: A worldwide literature review with emphasis on Latin America. *IEEE Geosci. Remote Sens. Mag.* **2021**, *9*, 209–222. [CrossRef]
19. Erazo-Rodas, M.; Sandoval-Moreno, M.; Muñoz-Romero, S.; Huerta, M.; Rivas-Lalaleo, D.; Naranjo, C.; Rojo-Álvarez, J.L. Multiparametric monitoring in equatorian tomato greenhouses (I): Wireless sensor network benchmarking. *Sensors* **2018**, *18*, 2555. [CrossRef] [PubMed]
20. Chi, T.; Chen, M. A frequency hopping method for spatial RFID/WiFi/Bluetooth scheduling in agricultural IoT. *Wirel. Netw.* **2019**, *25*, 805–817. [CrossRef]
21. Stombaugh, T. Satellite-based Positioning Systems for Precision Agriculture. In *Precision Agriculture Basics*; Wiley Online Library: Hoboken, NJ, USA, 2018; pp. 25–35.
22. Rani, M.U.; Suganya, C.; Kamalesh, S.; Sumithra, A. An integration of wireless sensor network through Wi-max for agriculture monitoring In Proceedings of the 2014 International Conference on Computer Communication and Informatics, Coimbatore, India, 3–5 January 2014; pp. 1–5.
23. Beritelli, F.; Capizzi, G.; Sciuto, G.L.; Napoli, C.; Scaglione, F. Rainfall estimation based on the intensity of the received signal in a LTE/4G mobile terminal by using a probabilistic neural network. *IEEE Access* **2018**, *6*, 30865–30873. [CrossRef]
24. Ji, H.; Park, S.; Yeo, J.; Kim, Y.; Lee J.; Shim, B. Ultra-Reliable and Low-Latency Communications in 5G Downlink: Physical Layer Aspects. *IEEE Wirel. Commun.* **2018**, *25*, 124–130. [CrossRef]
25. Dai, L.; Fang, Y.; Yang, Z. ; Chen, P.;Li, Y. Protograph LDPC-Coded BICM-ID With Irregular CSK Mapping in Visible Light Communication Systems. *IEEE Trans. Veh. Technol.* **2021**, *70*, 11033–11038. [CrossRef]
26. Shao, S.; Hailes, P.; Wang, T.; Wu, J.; Maunder, R.; Bashir, M.; Hanzo, L. Survey of Turbo, LDPC, and Polar Decoder ASIC Implementations. *IEEE Commun. Surv. Tutor.* **2019**, *21*, 2309–2333. [CrossRef]
27. Fang, Y.; Bu, Y.; Chen, P.; Lau, F. C. M.; Otaibi,S. A. Irregular-Mapped Protograph LDPC-Coded Modulation: A Bandwidth-Efficient Solution for 6G-Enabled Mobile Networks. *IEEE Trans. Intell. Transp. Syst.* **2021**. [CrossRef]
28. Mekki, K.; Bajic, E.; Chaxel, F.; Meyer, F. Overview of cellular LPWAN technologies for IoT deployment: Sigfox, LoRaWAN, and NB-IoT. In Proceedings of the 2018 Ieee International Conference on Pervasive Computing and Communications Workshops (Percom Workshops), Athens, Greece, 19–23 March 2018; pp. 197–202.
29. Liya, M.L.; Arjun, D. A Survey of LPWAN Technology in Agricultural Field. In Proceedings of the 2020 Fourth International Conference on I-SMAC (IoT in Social, Mobile, Analytics and Cloud) (I-SMAC), Palladam, India, 7–9 October 2020; pp. 313–317. [CrossRef]
30. Mermer, G.B.; Zeydan, E. A comparison of LP-WAN technologies: An overview from a mobile operators' perspective. In Proceedings of the 2017 25th Signal Processing and Communications Applications Conference (SIU), Antalya, Turkey, 15–18 May 2017; pp. 1–4.
31. Vejlgaard, B.; Lauridsen, M.; Nguyen, H.; Kovács, I.Z.; Mogensen, P.; Sorensen, M. Coverage and capacity analysis of sigfox, lora, gprs, and nb-iot. In Proceedings of the 2017 IEEE 85th Vehicular Technology Conference (VTC Spring), Sydney, NSW, Australia, 4–7 June 2017; pp. 1–5.
32. García-Cedeño, A.; Guillermo, J.C.; Barzallo, B.; Punín, C.; Soto, A.; Rivas, D.; Clotet, R.; Huerta, M. PLATANO: Intelligent technological support platform for azuay province farmers in Ecuador. In Proceedings of the 2019 IEEE International Conference on Engineering Veracruz (ICEV), Boca del Rio, Mexico, 14–17 October 2019; Volume 1, pp. 1–7.
33. Pérez Hernández, R. Desarrollo de prototipo de sensor IoT usando la red SigFox. In *Trabajo Fin de Grado en Ingeniería de las Tecnologías de Telecomunicación (pp. 171)*; Universidad de Sevilla. Departamento de Ingeniería Telemática: Seville, Spain, 2015.
34. Gomez, C.; Veras, J.C.; Vidal, R.; Casals, L.; Paradells, J. A Sigfox energy consumption model. *Sensors* **2019**, *19*, 681. [CrossRef]
35. Aernouts, M.; Berkvens, R.; Van Vlaenderen, K.; Weyn, M. Sigfox and LoRaWAN datasets for fingerprint localization in large urban and rural areas. *Data* **2018**, *3*, 13. [CrossRef]
36. Fourtet, C.; Ponsard, B. An introduction to Sigfox radio system. In *LPWAN Technologies for IoT and M2M Applications*; Elsevier: Amsterdam, The Netherlands, 2020; pp. 103–118.
37. Pérez García, R. Avaluació de LoRa/LoRaWAN per a Escenaris de Smart City. Bachelor's Thesis, Universitat Politècnica de Catalunya, Barcelona, Spain, 2017.
38. Ji, M.; Yoon, J.; Choo, J.; Jang, M.; Smith, A. LoRa-based Visual Monitoring Scheme for Agriculture IoT. In Proceedings of the 2019 IEEE Sensors Applications Symposium (SAS), Sophia Antipolis, France, 11–13 March 2019; pp. 1–6.

39. Grunwald, A.; Schaarschmidt, M.; Westerkamp, C. LoRaWAN in a rural context: Use cases and opportunities for agricultural businesses. In Proceedings of the Mobile Communication-Technologies and Applications; 24. ITG-Symposium, Osnabrueck, Germany, 15–16 May 2019; pp. 1–6.
40. Wang, Y.P.E.; Lin, X.; Adhikary, A.; Grovlen, A.; Sui, Y.; Blankenship, Y.; Bergman, J.; Razaghi, H.S. A primer on 3GPP narrowband Internet of Things. *IEEE Commun. Mag.* **2017**, *55*, 117–123. [CrossRef]
41. Ayoub, W.; Samhat, A.E.; Nouvel, F.; Mroue, M.; Prévotet, J.C. Internet of mobile things: Overview of lorawan, dash7, and nb-iot in lpwans standards and supported mobility. *IEEE Commun. Surv. Tutor.* **2018**, *21*, 1561–1581. [CrossRef]
42. Chaudhari, B.; Borkar, S. Design considerations and network architectures for low-power wide-area networks. In *LPWAN Technologies for IoT and M2M Applications*; Elsevier: Amsterdam, The Netherlands, 2020; pp. 15–35.
43. Rubio-Aparicio, J.; Cerdan-Cartagena, F.; Suardiaz-Muro, J.; Ybarra-Moreno, J. Design and implementation of a mixed IoT LPWAN network architecture. *Sensors* **2019**, *19*, 675. [CrossRef]
44. Devalal, S.; Karthikeyan, A. LoRa technology-an overview. In Proceedings of the 2018 Second International Conference on Electronics, Communication and Aerospace Technology (ICECA), Coimbatore, India, 29–31 March 2018; pp. 284–290.
45. Pallavi, S.; Mallapur, J.D.; Bendigeri, K.Y. Remote sensing and controlling of greenhouse agriculture parameters based on IoT. In Proceedings of the 2017 International Conference on Big Data, IoT and Data Science (BID), Pune, India, 20–22 December 2017; pp. 44–48.
46. Kodali, R.K.; Yerroju, S.; Sahu, S. Smart farm monitoring using LoRa enabled IoT. In Proceedings of the 2018 Second International Conference on Green Computing and Internet of Things (ICGCIoT), Bangalore, India, 16–18 August 2018; pp. 391–394.
47. Sari, E.K.; Wirara, A.; Harwahyu, R.; Sari, R.F. Lora characteristics analysis for IoT application using NS3 simulator. In Proceedings of the 2019 IEEE R10 Humanitarian Technology Conference (R10-HTC)(47129), Depok, West Java, Indonesia, 12–14 November 2019; pp. 205–210.
48. Chen, M.; Miao, Y.; Jian, X.; Wang, X.; Humar, I. Cognitive-LPWAN: Towards intelligent wireless services in hybrid low power wide area networks. *IEEE Trans. Green Commun. Netw.* **2018**, *3*, 409–417. [CrossRef]
49. Sanchez-Gomez, J.; Gallego-Madrid, J.; Sanchez-Iborra, R.; Santa, J.; Skarmeta, A.F. Impact of schc compression and fragmentation in lpwan: A case study with lorawan. *Sensors* **2020**, *20*, 280. [CrossRef] [PubMed]
50. Hattarge, S.; Kekre, A.; Kothari, A. LoRaWAN based GPS tracking of city-buses for smart public transport system. In Proceedings of the 2018 First International Conference on Secure Cyber Computing and Communication (ICSCCC), Jalandhar, India, 15–17 December 2018; pp. 265–269.
51. Lavric, A.; Petrariu, A.I.; Popa, V. Long range sigfox communication protocol scalability analysis under large-scale, high-density conditions. *IEEE Access* **2019**, *7*, 35816–35825. [CrossRef]
52. Duangsuwan, S.; Takarn, A.; Nujankaew, R.; Jamjareegulgarn, P. A study of air pollution smart sensors lpwan via nb-iot for thailand smart cities 4.0. In Proceedings of the 2018 10th International Conference on Knowledge and Smart Technology (KST), Chiang Mai, Thailand, 31 January–3 February 2018; pp. 206–209.
53. Casals, L.; Mir, B.; Vidal, R.; Gomez, C. Modeling the energy performance of LoRaWAN. *Sensors* **2017**, *17*, 2364. [CrossRef] [PubMed]
54. Yousuf, A.M.; Rochester, E.M.; Ousat, B.; Ghaderi, M. Throughput, coverage and scalability of LoRa LPWAN for internet of things. In Proceedings of the 2018 IEEE/ACM 26th International Symposium on Quality of Service (IWQoS), Banff, AB, Canada, 4–6 June 2018; pp. 1–10.
55. Hernández Caballero, S. Estudio en Detalle de LoRaWAN. Comparación con otras Tecnologías LPWAN Considerando Diferentes Patrones de tráfico. 2020. Available online: http://hdl.handle.net/10609/106369 (accessed on 25 May 2022).
56. Angelov, K.; Manchev, N.; Kogias, P.; Sadinov, S. Design and development of a platform for test applications in LoRa/LoRaWAN. Papers of International Conference in Telecommunications, Informatics, Energy and Management. 2019; pp. 27–31. Available online: http://www.jestr.org/downloads/SpecialIssue2020/fulltext3SE.pdf (accessed on 25 May 2022).
57. Zorbas, D. Design Considerations for Time-Slotted LoRa (WAN). 2020. Available online: https://cora.ucc.ie/handle/10468/9723 (accessed on 25 May 2022).
58. Polonelli, T.; Brunelli, D.; Marzocchi, A.; Benini, L. Slotted aloha on lorawan-design, analysis, and deployment. *Sensors* **2019**, *19*, 838. [CrossRef]
59. Moya Quimbita, M.A. Evaluación de Pasarela LoRa/LoRaWAN en Entornos Urbanos. 2018. Available online: http://hdl.handle.net/10251/109791 (accessed on 25 May 2022).
60. Fraga-Lamas, P.; Celaya-Echarri, M.; Azpilicueta, L.; Lopez-Iturri, P.; Falcone, F.; Fernández-Caramés, T.M. Design and empirical validation of a lorawan IoT smart irrigation system. *Multidiscip. Digit. Publ. Inst. Proc.* **2020**, *42*, 62.
61. Petrariu, A.I.; Lavric, A.; Coca, E. LoRaWAN Gateway: Design, Implementation and Testing in Real Environment. In Proceedings of the 2019 IEEE 25th International Symposium for Design and Technology in Electronic Packaging (SIITME), Cluj-Napoca, Romania, 23–26 October 2019; pp. 49–53.
62. Rizzi, M.; Ferrari, P.; Flammini, A.; Sisinni, E. Evaluation of the IoT LoRaWAN solution for distributed measurement applications. *IEEE Trans. Instrum. Meas.* **2017**, *66*, 3340–3349. [CrossRef]
63. Haxhibeqiri, J.; De Poorter, E.; Moerman, I.; Hoebeke, J. A survey of LoRaWAN for IoT: From technology to application. *Sensors* **2018**, *18*, 3995. [CrossRef]

64. Marais, J.M.; Abu-Mahfouz, A.M.; Hancke, G.P. A review of LoRaWAN simulators: Design requirements and limitations. In Proceedings of the 2019 International Multidisciplinary Information Technology and Engineering Conference (IMITEC), Vanderbijlpark, South Africa, 21–22 November 2019; pp. 1–6.
65. Capuzzo, M.; Magrin, D.; Zanella, A. Confirmed traffic in LoRaWAN: Pitfalls and countermeasures. In Proceedings of the 2018 17th Annual Mediterranean Ad Hoc Networking Workshop (Med-Hoc-Net), Capri, Italy, 20–22 June 2018; pp. 1–7.
66. NS3. Available online: https://www.nsnam.org (accessed on 25 May 2022).
67. Omnet++. Available online: https://omnetpp.org (accessed on 25 May 2022).
68. Ruz Nieto, A. Simulación realista de Comunicaciones IoT en Entornos Urbanos. 2021. Available online: https://repositorio.upct.es/handle/10317/9646 (accessed on 25 May 2022)).
69. LoRaSim. Available online: https://www.lancaster.ac.uk/scc/sites/lora/lorasim.html (accessed on 25 May 2022).
70. Davcev, D.; Mitreski, K.; Trajkovic, S.; Nikolovski, V.; Koteli, N. IoT agriculture system based on LoRaWAN. In Proceedings of the 2018 14th IEEE International Workshop on Factory Communication Systems (WFCS), Imperia, Italy, 13–15 June 2018; pp. 1–4.
71. Girón, N.A.G.; LópezHurtado, M.A. Selección de tecnologías LPWAN para la implementación de un sistema IoT aplicado a la lombricultura. *Memorias* **2020**. Available online: https://hemeroteca.unad.edu.co/index.php/memorias/article/view/4170 (accessed on 25 May 2022)).
72. Zambrano Aguayo, M.D.; Pérez Ruano, M.; Rodríguez Villafuerte, X. Brucelosis Bovina en la Provincia Manabí, Ecuador: Estudio de los Factores de Riesgo. *Rev. Investig. Vet. Del Perú* **2016**, *27*, 607–617. [CrossRef]
73. Slabicki, M.; Premsankar, G.; Di Francesco, M. Adaptive configuration of LoRa networks for dense IoT deployments. In Proceedings of the NOMS 2018-2018 IEEE/IFIP Network Operations and Management Symposium, Taipei, Taiwan, 23–27 April 2018; pp. 1–9.
74. Zhou, Q.; Xing, J.; Hou, L.; Xu, R.; Zheng, K. A novel rate and channel control scheme based on data extraction rate for LoRa networks. In Proceedings of the 2019 IEEE Wireless Communications and Networking Conference (WCNC), Marrakesh, Morocco, 15–18 April 2019; pp. 1–6.
75. Sisinni, E.; Bellagente, P.; Depari, A.; Ferrari, P.; Flammini, A.; Marella, S.; Pasetti, M.; Rinaldi, S.; Cagiano, A. A new LoRaWAN adaptive strategy for smart metering applications. In Proceedings of the 2020 IEEE International Workshop on Metrology for Industry 4.0 & IoT, Roma, Italy, 3–5 June 2020; pp. 690–695.
76. Gusev, O.; Turlikov, A.; Kuzmichev, S.; Stepanov, N. Data delivery efficient spreading factor allocation in dense LoRaWAN deployments. In Proceedings of the 2019 XVI International Symposium "Problems of Redundancy in Information and Control Systems" (REDUNDANCY), Moscow, Russia, 21–25 October 2019; pp. 199–204.
77. Bor, M.; Roedig, U.; Voigt, T.; Alonso, J. Do LoRa Low-Power Wide-Area Networks Scale? In Proceedings of the 19th ACM International Conference on Modeling, Analysis and Simulation of Wireless and Mobile Systems, Malta, 13–17 November 2016; pp. 59–67.

agriculture

Article

Lightweight Detection Algorithm of Kiwifruit Based on Improved YOLOX-S

Junchi Zhou, Wenwu Hu *, Airu Zou, Shike Zhai, Tianyu Liu, Wenhan Yang and Ping Jiang

College of Mechanical and Electrical Engineering, Hunan Agricultural University, Changsha 410128, China; jcz9858@stu.hunau.edu.cn (J.Z.); zar@stu.hunau.edu.cn (A.Z.); zsk0371@stu.hunau.edu.cn (S.Z.); liutianyu@hunau.edu.cn (T.L.); yangwenhan@stu.hunau.edu.cn (W.Y.); 1233032@hunau.edu.cn (P.J.)
* Correspondence: 1233087@hunau.edu.cn

Abstract: Considering the high requirements of current kiwifruit picking recognition systems for mobile devices, including the small number of available features for image targets and small-scale aggregation, an enhanced YOLOX-S target detection algorithm for kiwifruit picking robots is proposed in this study. This involved designing a new multi-scale feature integration structure in which, with the aim of providing a small and lightweight model, the feature maps used for detecting large targets in the YOLOX model are eliminated, the feature map of small targets is sampled through the nearest neighbor values, the superficial features are spliced with the final features, the gradient of the SiLU activation function is perturbed, and the loss function at the output is optimized. The experimental results show that, compared with the original YOLOX-S, the enhanced model improved the detection average precision (AP) of kiwifruit images by 6.52%, reduced the number of model parameters by 44.8%, and improved the model detection speed by 63.9%. Hence, with its outstanding effectiveness and relatively light weight, the proposed model is capable of effectively providing data support for the 3D positioning and automated picking of kiwifruit. It may also successfully provide solutions in similar fields related to small target detection.

Keywords: YOLOX; small target scale; loss function; feature integration; fruit picking

Citation: Zhou, J.; Hu, W.; Zou, A.; Zhai, S.; Liu, T.; Yang, W.; Jiang, P. Lightweight Detection Algorithm of Kiwifruit Based on Improved YOLOX-S. *Agriculture* **2022**, *12*, 993. https://doi.org/10.3390/agriculture12070993

Academic Editors: Gniewko Niedbała and Sebastian Kujawa

Received: 15 June 2022
Accepted: 7 July 2022
Published: 9 July 2022

1. Introduction

Agriculture is the source of human clothing, food, housing, and transportation; an important foundation for people's lives; the backbone that supports the national economy; and the guarantee for the country's stable development. At present, the application of artificial intelligence in agriculture mainly includes intelligent farm systems with management and decision-making capabilities based on the background of agricultural big data [1], motion obstacle target detection and path recognition [2], crop growth and pest detection [3], weed recognition [4], fruit and vegetable quality detection [5], and automatic picking based on agricultural robots and other related fields.

Fruit-picking robots can automate picking work, effectively resolving issues related to labor shortages, high costs, and low efficiency in the manual picking process [6,7]. Determination of the critical criteria of picking robots involves studying the visual system, while the efficiency and stability of such robots predominantly depend on the speed and accuracy of fruit recognition, along with the accuracy and adaptability in complex environments [8,9]. Therefore, research on visual systems that possess the capability to accurately identify the fruit on trees in complex environments is of substantial value and practical significance for achieving automatic picking and yield estimation.

Numerous scholars across the world have conducted extensive research on target object recognition technology [10–13]. In the field of fruit crop detection in natural environments, feature extraction and recognition have predominately targeted tomato [14,15], apple [16–18], cucumber [19,20], strawberry [21], sugarcane [22], pineapple [23], and various other fruits. Among the various fruits, the planting area and yield of kiwifruit in particular have continued to increase over time. With its high yield and rich nutritional value, kiwifruit has been widely planted and become popular among consumers. Methods for detection and recognition are predominantly segregated into traditional machine vision methods and deep learning methods. As an example of such machine vision, Cui Yongjie et al. [24] utilized the L*a*b* color space a* channel for kiwifruit image segmentation, and adopted the elliptical Hough transform to fit the contour of a single fruit for segmentation and recognition. In addition, Fu et al. [25] have proposed the use of 1.1R-G color characteristics for nighttime kiwifruit image segmentation, and combined the minimum circumscribed rectangle method and elliptical Hough transform to identify each fruit. However, both methods presented unsatisfactory results for fruit segmentation and unfavorable results for multi-fruit cluster recognition compared to traditional algorithms, such as SIFT [26], HOG [27], and texture extraction algorithms [28–30]. Kiwifruit images in the field environment possess vastly diverse features, complex backgrounds, and substantial differences in morphological features. Traditional machine vision methods are mainly constructed based on experience and are influenced by samples and human subjectivity; hence, they are unable to effectively meet the demands of applications in complex field environments.

Deep learning target detection algorithms have experienced significant leaps in performance and accuracy, and various model networks have substantially enhanced their ability to resist scale changes and translation. Song Zhenzhen et al. [31] have constructed a fast VGG16 model to achieve the detection of kiwifruit in live images by integrating a region proposal network (RPN) and a fast R-CNN network, while Fu Longsheng et al. [32] have proposed a network-based LeNet convolutional neural network deep learning model for multi-cluster kiwifruit images with general applicability to the recognition of multi-cluster kiwifruits. Although, as research on deep learning-based target detection methods has focused more on the construction of a deeper networks for the purpose of enhancing detection accuracy, the associated network models have generally suffered from an overly large number of parameters. This has led to slow detection speeds, meaning that the algorithms can only be run on high-performance graphics processors and generally have high equipment requirements. Concurrently, according to analysis of the growth characteristics of kiwifruit, most targets in kiwifruit detection tasks have predominantly been on small targets (both absolute and relative scales are relatively small).

Therefore, in the interest of reducing the number of model parameters and enhancing the model detection speed, the YOLOX-S network, which possesses excellent multi-scale detection performance and takes into account both the detection speed and accuracy as its basis, was selected for this research. This work aims to improve on the original network model, in order to maintain the target detection accuracy while compressing the model, thus effectively achieving the detection of small kiwifruit targets.

2. Experimental Platform and Materials

2.1. Vision Platform System

In this paper, we primarily focus on the object detection task in the image processing field. The image recognition module was a Jetson Nano embedded development board, as presented in Figure 1. The improved model algorithm, which was trained in advance, was embedded in the board, and wireless communications, remote monitoring, and remote control were achieved through the 4G network module. The communication system is mainly divided into the picking-machine end, cloud server end, and client end, ensuring the transmission and storage of information. Remote wireless control of the picking robot can also be achieved. In addition, the left and right imagers of the depth camera capture video or image data, which are sent to a depth imaging processor. This processor correlates points in the left image with those in the right image, and calculates the depth value of each pixel in the image by shifting the points in the left image to match with the right image. Finally, it returns the result to the terminal in order to command the manipulator to act accordingly.

Figure 1. Image-recognition embedded module.

2.2. Hardware Platform

A test platform was independently developed by our team, which can be applied as a fruit picking and transferring platform in hilly and mountainous areas (Figure 2). The platform has a pure electric drive and a CAN interface for chassis speed regulation, steering, and attitude feedback. It is capable of stable driving and meets the hardware requirements of the platform positioning test for the chassis in hilly and mountainous areas.

Figure 2. Electric fruit picking platform used for experimental trials.

2.3. Experimental Configuration and Environment

The used graphics card was an NVIDIA GeForce GTX 3060, and the CPU was an AMD Ryzen 7 5800H with 16 GB memory. The experimental configuration was Windows 10, Python 3.8, PyTorch 1.8.1, and CUDA 10.1. The parameter settings are presented in Table 1.

Table 1. Training parameter settings.

Parameter	Value
Momentum	0.937
Weight_decay	0.0005
Batch_size	45
Learning_rate	0.0001
Epochs	500

2.4. Experimental Sample Dataset

The experimental data in this paper were collected from the Internet and from on-site filming. A total of 1500 images were collected. The photos taken on the spot are all taken from the orchard. Each picture contained a significantly large number of kiwifruit target fruits, and the total number of targets was 41,687. The targets in each image were labeled with fine granularity, in order to facilitate subsequent enhancements in the detection of small targets.

3. Principles and Methods

YOLOX [33] is a brand new high performance real-time target detection network, recently launched by Beijing Megvii Technology. It adopts cutting-edge technologies such as the anchor-free mechanism, decoupled heads, multi-positives, advanced label assignment strategy, and strong data augmentation. Hence, it has faster speed, higher recognition accuracy, smaller weight files, and can be easily mounted on mobile devices with lower configurations, thereby offering high research value. The structure of the YOLOX-S network selected for this paper is depicted in Figure 3.

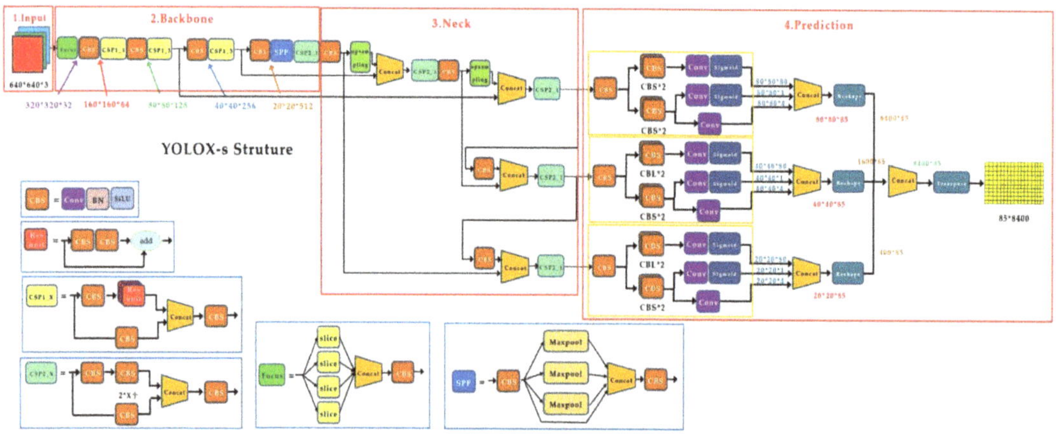

Figure 3. Structure of the YOLOX-S network model.

CSPDarknet is the backbone feature extraction network of the YOLOX algorithm, which is predominantly composed of three modules: Focus, CSPNet, and a spatial pyramid pooling network. The model first slices an input image for the operation. By sampling the complete image at equal intervals, multiple sampled images of appropriate size can be obtained. Subsequently, these images are combined in the channel dimension and the information in the image is transferred to the channel space, resulting in a down-sampled image with no information loss. The CSPNet module contains the backbone feature extraction and residual structure, which can effectively extract image features and significantly reduces the computational effort while maintaining high accuracy. The SPP network convolutes the output of the last CSPNet once, then utilizes three different scales of maximum pooled kernels to integrate the features of the feature image under different receptive fields. FPN + PAN is a circular pyramid structure composed of convolution, sampling, and feature fusion operations, which repeatedly extracts the input image features, performs feature fusion at different scales, and finally outputs the three feature maps at different scales to the decoupled head for accurate prediction.

3.1. Pre-Processing of the Data Set

YOLOX utilizes mosaic and mix-up data augmentation methods to substantially enrich the detection dataset; in particular, random scaling is conducted to supplement the many small targets and make the network more robust. Mosaic augmentation involves performing a series of operations, such as flipping, scaling, and color shifting, on multiple different pictures, followed by cropping and splicing to recombine them into a new image. Hence, the generated images often contain more targets. Therefore, this kind of augmentation technique can significantly enrich the background and alleviate the imbalance of positive and negative samples in the detection process, to a certain extent. Mix-up augmentation refers to the fusion of two pictures, to some degree, in which the labels of the samples are also weighted. The prediction results are weighted using the weighted labels in order to calculate the loss; subsequently, the backpropagation update parameters can be enhanced. The effect is shown in Figure 4.

(a) (b)

Figure 4. Data enhancement effects: (**a**) mosaic data enhancement and (**b**) mix-up data enhancement.

3.2. Improved YOLOX-S Network

3.2.1. Perturbing the Activation Function Gradient

The predominant function of the activation function is to provide non-linearity in the network structure. Considering that the difference between the gradient propagation effects of the SiLU and Mish loss functions utilized in the YOLOX model is slight, gradient perturbation was considered based on the SiLU activation function. As presented in Figure 5, the SiLU→SiLU-1 gradient change led to a smoother curve, while the SiLU→SiLU + 1 gradient change became steeper. Given that the Mish activation function worked relatively well in YOLOv4, it was considered to increase or decrease the gradient change based on SiLU. Introducing a gradient increase can enhance the generalization ability of the model more robustly, and as such, we found that the SiLU + 1 activation function enhanced the generalization ability of the model to a certain extent.

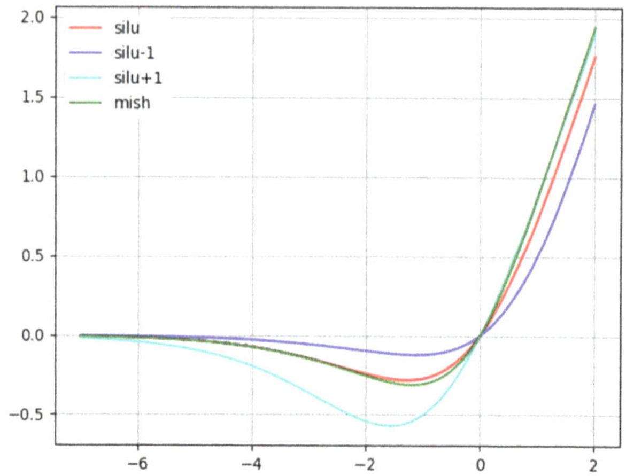

Figure 5. Variation between the perturbed gradient functions.

The dynamic positive and negative sample allocation algorithm utilized by YOLOX, SimOTA, is fast and effective. When determining the candidate areas for positive samples, the center point of a grid (20 × 20, 40 × 40, 80 × 80) was selected as the circle inside the ground truth (GT), with r being the radius centered on the center point of the GT. In Figure 6, the green box denotes the GT. It can be observed that there may be mismatches when using a small feature map. Subsequently, it can be observed that GTs are more likely to match smaller GTs in larger feature maps, but small feature maps can match a significantly small number of GTs.

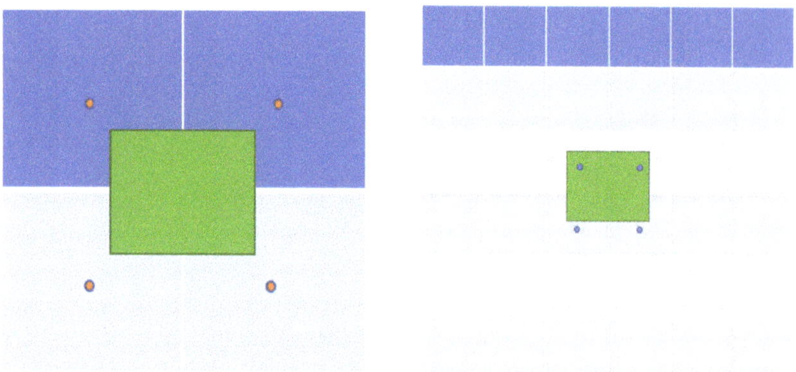

Figure 6. Feature matching candidate sample situation.

3.2.2. Nearest Neighbor Interpolation Up-Sampling of 80 × 80 Feature Map

Through in-depth research on the allocation strategy of positive and negative samples in the YOLOX model, the YOLOX model was found to reduce the number of predicted samples of the feature map in the confidence loss calculation, where almost all of the reduced samples were negative. Hence, the problem of imbalance in quantity caused by too many negative samples was alleviated, thereby suggesting that most targets in the kiwifruit detection task are small targets (i.e., both the absolute scale and relative scale are relatively small). Therefore, with the goal of reducing the number of model parameters and improving the model detection speed, the feature maps (20 × 20, 40 × 40) used for detecting large targets in the YOLOX model were eliminated. Subsequently, only the 80 × 80 feature map was retained, and a larger feature map size was introduced on this basis to match the

GT more effectively. When acquiring the final output from the 80 × 80 feature map, nearest-neighbor interpolation was utilized for up-sampling. This allows the model to provide more predictions and better match GTs, thereby extensively reducing the complexity of the model and the number of parameters. Figure 7 demonstrates the structure of the network before and after the improvement.

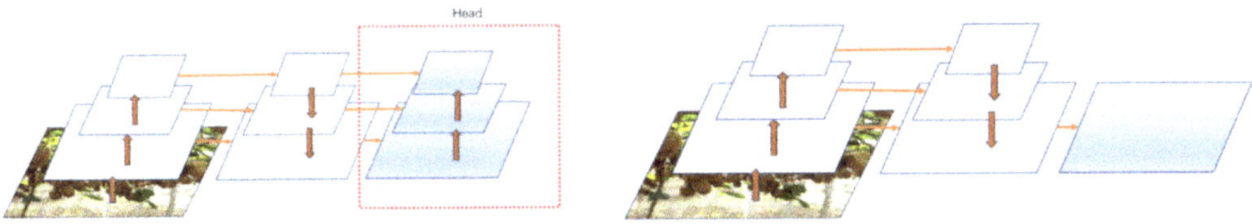

(**a**) (**b**)

Figure 7. (**a**) Original YOLOX feature fusion structure and (**b**) improved structure, in which only the 80 × 80 feature map structure is preserved.

3.2.3. Transfer of Shallow Features

The performance when using a single output feature map may be unstable under specific conditions. Considering that the low-level feature semantic information is relatively small but the target position is accurate, the final feature map and the feature map in the shallow network were concatenated, in order to better integrate the semantic and representation information to a certain extent, such that the accuracy of the regression box could be significantly enhanced (see Figure 8).

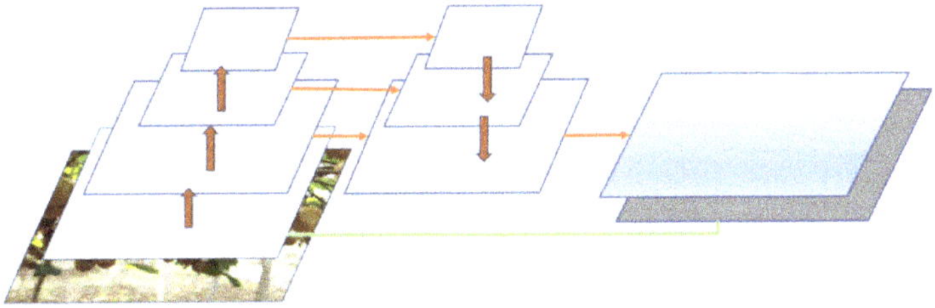

Figure 8. Structure of the final feature map.

3.2.4. Enhancing the Loss Function

Equations (1)–(5) are the loss functions of the YOLOX-S algorithm. The bounding box loss functions GIOU_loss and IOU_loss for predicting Reg have certain limitations, resulting in an inability to effectively optimize the overlap between the detection box and the real box when one is included in the other. Subsequently, for the confidence degree and category loss, the original algorithm adopts a binary cross-entropy loss function, which is not conducive to the classification of positive and negative samples.

$$Loss = GIOU\_Loss + Loss_{conf} + Loss_{class}, \tag{1}$$

$$GIOU\_Loss = 1 - GIOU = 1 - (IOU - \frac{|Q|}{C}), \tag{2}$$

$$\begin{aligned} Loss_{conf} = & \sum_{i=0}^{S^2}\sum_{j=0}^{B} I_{ij}^{obj}[\overset{\wedge j}{C_i}\log(C_i^j) + (1-\overset{\wedge j}{C_i})\log(1-\overset{\wedge j}{C_i})] - \\ & \lambda_{noobj}\sum_{i=0}^{S^2}\sum_{j=0}^{B} I_{ij}^{noobj}[\overset{\wedge j}{C_i}\log(C_i^j) + (1-\overset{\wedge j}{C_i})\log(1-\overset{\wedge j}{C_i})] \end{aligned}, \tag{3}$$

$$Loss_{class} = \sum_{i=0}^{S^2} I_{ij}^{obj} \sum_{c \in classes} [\overset{\wedge j}{P_i} \log(P_i^j) + (1 - \overset{\wedge j}{P_i}) \log(1 - \overset{\wedge j}{P_i})]. \tag{4}$$

In Equation (2), C represents the minimum circumscribed rectangle of the detection frame and the priori frame, and Q represents the difference between the minimum circumscribed rectangle and the concatenation of the two frames.

In Equations (3) and (4), I_{ij}^{obj} and I_{ij}^{noobj} indicate whether the target falls into detection frame j of grid i, and λ_{noobj} represents the loss weight of the localization error. Subsequently, C_j^i and P_j^i refer to the training values, and $\overset{\wedge j}{C_i}$ and $\overset{\wedge j}{P_i}$ refer to the prediction values.

Therefore, we adopted CIOU_loss as the Reg bounding box loss function and increased the aspect-ratio restriction mechanism, compared with the previous one, such that the prediction box was more in line with the real box, as demonstrated in Equation (5). Equation (6) was used to measure the consistency of the aspect ratio, and the confidence degree and category loss function utilized the PolyLoss function based on the Taylor expansion approximation of the focal loss [34]. Thus, it not only took into account the superior binary classification performance of the focal function, but also achieved enhancement of the accuracy and performance on this basis. The convergence speed was also effectively accelerated.

$$CIOU\_Loss = 1 - (IOU - \frac{\rho^2(b, b^{gt})}{c^2} - \alpha v), \tag{5}$$

$$\begin{gathered} v = \frac{4}{\pi^2}(\arctan\frac{w^{gt}}{h^{gt}} - \arctan\frac{w^p}{h^p})^2 \\ \alpha = \frac{v}{(1-IOU)+V} \end{gathered} \tag{6}$$

where $\rho()$ is the Euclidean distance between the center points of the two boxes, c is the diagonal length of the smallest circumscribed rectangle of the two, α is the weight coefficient, and v is the aspect ratio distance between the two frames.

$$L_{Ploy-1} = -\log(P_t) + \varepsilon_1(1 - P_t), \tag{7}$$

where P_t represents the probability of target label prediction.

4. Results and Discussion

4.1. Evaluation of Model Performance

In order to evaluate the effectiveness of the proposed method for kiwifruit detection in different aspects, the mean average accuracy, the number of model parameters utilized, and the *FPS*, along with the detection time per sheet, were selected as evaluation metrics. *mAP* refers to a comprehensive consideration of precision and recall, which is used to evaluate the effectiveness of the model, while *FPS* refers to the number of frames per second, which can be utilized to measure the real-time performance of the model. Finally, the number of model parameters reflects the lightness of the model.

$$\begin{gathered} recall = \frac{TP}{TP + FN} \\ precision = \frac{TP}{TP + FP}, \\ mAP = \frac{\sum_{i=0}^{N-1} \int_0^1 P(R)dR}{N} \end{gathered} \tag{8}$$

where *TP* represents the number of correctly identified images, *FP* represents the number of misidentified images, and *FN* represents the number of missed images. When there is only one category, *mAP* is equal to the *AP*.

4.2. Analysis and Comparison of the Enhanced Model

In order to efficiently verify the effectiveness of the proposed model, a comparative experiment was conducted on the enhanced YOLOX-S network using the same training parameters. Table 2 provides the detailed scores of each evaluation index before and after the enhancement. Figure 9 presents the *AP* diagram of the model before and after enhancement.

Table 2. The comparison of model scores before and after enhancement.

Model	*AP*	Param	Time/ms	*FPS*
IMPROVED	82.62	5,483,590	15.6	101
ORI	76.10	9,937,682	43.2	88

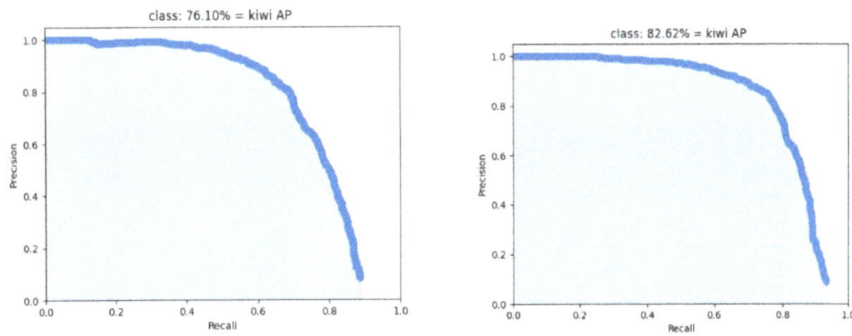

Figure 9. The YOLOX-S training curve.

From the perspective of model lightness, the improved model parameters were reduced by 44.8% and the model detection speed was increased by 63.9%, verifying the feature expression ability of the model. Feature map up-sampling and nearest-neighbor interpolation reduced the computational complexity by omitting unnecessary computations, thus achieving the effect of making the network lightweight.

So as to more intuitively depict the improvement in various aspects for the considered models, we created a performance comparison diagram with respect to the model improvement strategies, as shown in Figure 10.

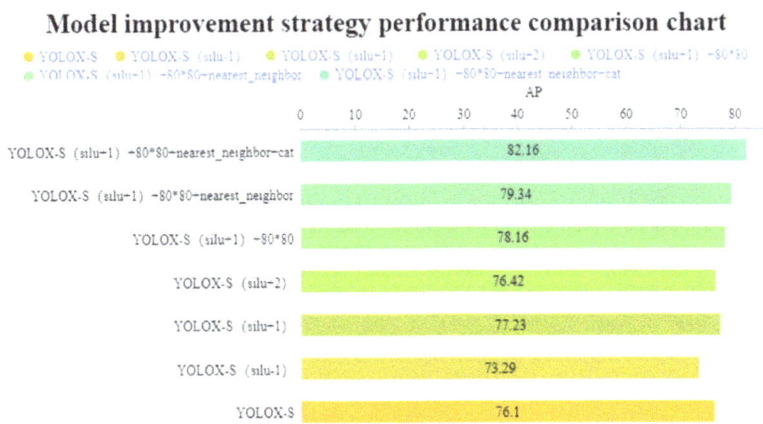

Figure 10. Model improvement strategy performance comparison chart.

As shown by Figure 10, in terms of model effectiveness and accuracy, the expressiveness growth of the model is mainly divided into three stages. The first stage is that the perturbation of the activation function enhances the generalization ability of the model by selecting the SILU+1 function, which increases the *AP* value by 1.13%. The improvement in the second stage is due to the cancellation of the feature map of the redundant large target in this detection task, so that the network detection is all concentrated on the small target, which reduces the calculation of negative samples and the misjudgment of positive samples, so that the *AP* value continues to increase by 2.01%. The improvement in the last stage comes from the design of the new network fusion structure. By splicing the final output feature map and shallow-level features, the semantic information of the two is combined, and the loss function is improved in the prediction segment. Compared with the original model, the enhanced model significantly improved the *AP* value on the kiwifruit images by 6.52%, which is a substantial increase.

Figure 11 presents the before and after images for comparison. By comparing the groups of images, it can be seen that in the (Figure 11a) group of experiments, the fruit could be effectively identified by the enhanced model, even when there were tree trunks, branches, and leaves in the way.

(a)
(b)
(c)

Figure 11. (**a**) Enhancement of fruit recognition under tree trunk and leaf occlusion. (**b**) Enhancement of low-density fruit missed recognition. (**c**) Enhancement of non-target fruit misrecognition.

However, for the (Figure 11b) group of experiments, the original algorithm was not able to recognize the low-density fruits effectively. Additionally, for the (Figure 11c) group of experiments, the original algorithm misidentified the tree trunk as a fruit, but the enhanced model corrected it, and was able to accurately identify more fruit.

In light of the above, the enhanced algorithm significantly improved the ability to detect small-scale target fruit and reduced the false recognition and misrecognition rates. In addition, we compared several state-of-the-art algorithms and conducted training tests under the same conditions. The proposed enhanced model provided improved results in all aspects. The performance comparison is given in Table 3. In addition, Table 4 compares our findings with those of various scholars around the world, and details the advantages and disadvantages of their techniques.

Table 3. Comparison of mainstream models.

Model	mAP@0.5/%	*FPS*
Ours	82.62	101
YOLOv5s	74.12	83
YOLOv3	69.46	68
YOLOv2	67.83	63
Fast R-CNN	80.15	52

Table 4. Comparison and analysis of advantages and disadvantages of methods.

Reference	Description	Advantages	Disadvantages
[24]	Utilized the L*a*b* color space a* channel for kiwifruit image segmentation	Accurate segmentation of a single fruit	susceptible to external changes
[25]	Combines least circumscribed rectangle method and elliptic Hough transform	Accurate segmentation of a single fruit	Not ideal for fruit cluster identification
[35]	Improved K-means algorithm	Multi-target detection possible	Easily disturbed by shape and texture
[36]	Built a color classifier	Low hardware requirements	slow detection
[37]	Construction of the positional relationship between fruit and calyx in linear clusters	high speed	Does not meet the multi-robot collaborative operation
[31]	Merge Region Proposal Network (RPN) and Fast R-CNN Network	high speed	Poor adaptability in complex situations
[32]	Convolutional Neural Network Based on LeNet	Multiple fruit clusters can be recognized	High equipment requirements and large amount of parameters
Ours	Cancel the large object detection layer Concatenate shallow features with final features	high speed few parameters Suitable for embedded mobile devices	AP can be further improved

It can be seen from the table that the algorithm proposed in this paper can solve the problem of poor recognition of multiple fruit clusters compared with the traditional image processing method used in past research [24,25,35–37]. Compared with studies based on deep learning methods [31,32], the improved accuracy of our algorithm alleviates the problems of the network model being too large and the equipment requirements being too high. The recognition in the case of fruit occlusion and misjudgment is improved, the recognition accuracy and speed of the fruit are further improved, and the parameter amount of the model is reduced. It can effectively complete the detection task of kiwifruit in agricultural production, and has a positive impact on future picking of kiwifruit.

5. Conclusions

The research ultimately proposed an enhanced YOLOX-S target detection algorithm for kiwifruit picking robots. In order to effectively improve the detection of small-scale targets, the YOLOX-S algorithm was enhanced through fine-grained annotation of the target frame of the data set, as well as mosaic and mix-up data augmentation methods. Through up-sampling of the nearest-neighbor value in the small target feature map and the splicing of superficial features with the final features, in addition to the optimization of the loss function, the number of parameters of the enhanced YOLOX-S were significantly reduced while the *AP* values was increased. We demonstrated that the proposed enhancement method is applicable to actual fruit-picking environments, and is beneficial for embedment in mobile devices.

This research predominantly focused on the detection of kiwifruit. Simultaneously, the critical key to effective picking is to locate the target and return its three-dimensional coordinate points.

In further research, we intend to focus on:

(a) At present, the *AP* of the model has not reached the ideal state. Next, the data set will be enriched to further improve the performance and accuracy of the model.
(b) We will use pre-processing of the depth image data and color image data by utilizing the camera's external and internal parameters, triangulation principles, and the conversion of pixel coordinates to 3D spatial coordinates to carry out fruit localization.
(c) The proposed algorithm effectively met the basic requirements for fruit picking using a large-end actuator. However, due to the large number of kiwifruit that need to be picked, in order to further enhance the efficiency of the manipulator, it is necessary to further research the picking sequence allocation for kiwifruit.
(d) We will analyze the correlation between data, identify a variety of other types of fruit through transfer learning, and design a multi-classification general picking model for orchards.

Author Contributions: Conceptualization, J.Z. and W.H.; methodology, P.J.; software, J.Z. and T.L.; validation, J.Z., A.Z. and S.Z.; formal analysis, J.Z.; investigation, J.Z.; resources, W.H. and P.J.; data curation, J.Z. and W.Y.; writing—original draft preparation, J.Z.; writing—review and editing, J.Z., P.J. and W.H.; visualization, J.Z.; supervision, P.J., T.L. and W.H.; project administration, P.J. and W.H.; funding acquisition, P.J. All authors have read and agreed to the published version of the manuscript.

Funding: This research was funded by the Excellent Youth Project of the Hunan Education Department (No. 20B292), the research and application of key technologies for remote active monitoring of wild animals under national key protection in forests, the Hunan Agricultural Machinery Equipment and Technology Innovation R&D Project (Xiang Cai Nong Zhi [2020] No.107), and the Hunan Province Science and Technology Achievement Transformation and Industrialization Plan Project (2020GK4075).

Institutional Review Board Statement: Not applicable.

Informed Consent Statement: Not applicable.

Data Availability Statement: Not applicable.

Conflicts of Interest: The authors declare no conflict of interest.

References

1. Junaid, M.; Shaikh, A.; Hassan, M.U.; Alghamdi, A.; Rajab, K.; Al Reshan, M.S.; Alkinani, M. Smart Agriculture Cloud Using AI Based Techniques. *Energies* **2021**, *14*, 5129. [CrossRef]
2. Liu, C.; Feng, Q.; Tang, Z.; Wang, X.; Geng, J.; Xu, L. Motion Planning of the Citrus-Picking Manipulator Based on the TO-RRT Algorithm. *Agriculture* **2022**, *12*, 581. [CrossRef]

3. Kong, J.; Wang, H.; Yang, C.; Jin, X.; Zuo, M.; Zhang, X. A Spatial Feature-Enhanced Attention Neural Network with High-Order Pooling Representation for Application in Pest and Disease Recognition. *Agriculture* **2022**, *12*, 500. [CrossRef]
4. Jiang, H.; Zhang, C.; Qiao, Y.; Zhang, Z.; Zhang, W.; Song, C. CNN feature based graph convolutional network for weed and crop recognition in smart farming. *Comput. Electron. Agric.* **2020**, *174*, 105450. [CrossRef]
5. Mesa, A.R.; Chiang, J.Y. Multi-Input Deep Learning Model with RGB and Hyperspectral Imaging for Banana Grading. *Agriculture* **2021**, *11*, 687. [CrossRef]
6. Jia, W.; Tian, Y.; Luo, R.; Zhang, Z.; Lian, J.; Zheng, Y. Detection and segmentation of overlapped fruits based on optimized mask R-CNN application in apple harvesting robot. *Comput. Electron. Agric.* **2020**, *172*, 105380. [CrossRef]
7. Fu, L.; Feng, Y.; Wu, J.; Liu, Z.; Gao, F.; Majeed, Y.; Al-Mallahi, A.; Zhang, Q.; Li, R.; Cui, Y. Fast and accurate detection of kiwifruit in orchard using improved YOLOv3-tiny model. *Precis. Agric.* **2020**, *13*, 754–776. [CrossRef]
8. Jia, W.; Zhang, Y.; Lian, J.; Zheng, Y.; Zhao, D.; Li, C. Apple harvesting robot under information technology: A review. *Int. J. Adv. Robot. Syst.* **2020**, *17*, 1729881420925310. [CrossRef]
9. Kang, H.; Chen, C. Fruit detection and segmentation for apple harvesting using visual sensor in orchards. *Sensors* **2019**, *19*, 4599. [CrossRef] [PubMed]
10. Yang, L.; Luo, J.; Song, X.; Li, M.; Wen, P.; Xiong, Z. Robust Vehicle Speed Measurement Based on Feature Information Fusion for Vehicle Multi-Characteristic Detection. *Entropy* **2021**, *23*, 910. [CrossRef]
11. Zhou, J.; Jiang, P.; Zou, A.; Chen, X.; Hu, W. Ship Target Detection Algorithm Based on Improved YOLOv5. *J. Mar. Sci. Eng.* **2021**, *9*, 908. [CrossRef]
12. Liu, T.; Ma, Y.J.; Yang, W.; Li, W.; Wang, R.; Jiang, P. Spatial-temporal interaction learning based two-stream network for action recognition. *Inform. Sci.* **2022**, *606*, 864–876. [CrossRef]
13. Liu, C.; Su, J.; Wang, L.; Lu, S.; Li, L. LA-DeepLab V3+: A Novel Counting Network for Pigs. *Agriculture* **2022**, *12*, 284. [CrossRef]
14. Arefi, A.; Motlagh, A.; Mollazade, K.; Teimourlou, R. Recognition and Localization of Ripen Tomato Basedon Machine Vision. *Aust. J. Crop. Sci.* **2011**, *5*, 1144–1149.
15. Xiang, R.; Ying, Y.; Jiang, H.; Rao, X.; Peng, Y. Recognition of Overlapping Tomatoes Based on Edge Curvature Analysis. *Trans. Chin. Soc. Agric. Mach.* **2012**, *43*, 157–162.
16. Si, Y.; Liu, G.; Gao, R. Segmentation Algorithm for Green Apples Recognition Based on K-means Algorithm. In Proceedings of the 3rd Asian Conference on Precision Agriculture, Beijing, China, 14–17 October 2009; pp. 100–105.
17. Zulkifley, M.A.; Moubark, A.M.; Saputro, A.H.; Abdani, S.R. Automated Apple Recognition System Using Semantic Segmentation Networks with Group and Shuffle Operators. *Agriculture* **2022**, *12*, 756. [CrossRef]
18. Jing, W.; Leqi, W.; Yanling, H.; Yun, Z.; Ruyan, Z. On Combining DeepSnake and Global Saliency for Detection of Orchard Apples. *Appl. Sci.* **2021**, *11*, 6269. [CrossRef]
19. Henten, E.; Tuijl, B.; Hoogakker, G.J.; Weerd, M.; Hemming, J.; Kornet, J.G.; Bontsema, J. An autonomous robot for de-leafing cucumber plants in a high-wire cultivation system. *Biosyst. Eng.* **2006**, *94*, 317–323. [CrossRef]
20. Liu, C.; Zhao, C.; Wu, H.; Han, X.; Li, S. ADDLight: An Energy-Saving Adder Neural Network for Cucumber Disease Classification. *Agriculture* **2022**, *12*, 452. [CrossRef]
21. Xie, Z.; Zhang, T.; Zhao, J. Ripened Strawberry Recognition Based on Hough Transform. *Trans. Chin. Soc. Agric. Mach.* **2007**, *38*, 106–109.
22. Lu, S.; Wen, Y.; Ge, W.; Peng, H. Recogntition and Features Extraction of Suagrcane Nodes Based on Machine Vision. *Trans. Chin. Soc. Agric. Mach.* **2010**, *41*, 190–194.
23. Li, B.; Wang, M.; Li, L. In-field pineapple recognition based on monocular vision. *Trans. Chin. Soc. Agric. Eng.* **2010**, *26*, 345–349.
24. Cui, Y.; Su, S.; Lyu, Z.; Li, P.; Ding, X. A Method for Separation of Kiwifruit Adjacent Fruits Based on Hough Transformation. *J. Agric. Mech. Res.* **2012**, *34*, 166–169.
25. Fu, L.; Wang, B.; Cui, Y.; Gejima, Y.; Taiichi, K. Kiwifruit recognition at nighttime using artificial lighting based on machine vision. *Int. J. Agric. Biol. Eng.* **2015**, *8*, 52–59.
26. Lowe, D.G. Object Recognition from Local Scale-Invariant Features. In Proceedings of the Seventh IEEE International Conference on Computer Vision (ICCV), Kerkyra, Greece, 20–27 September 1999; pp. 1150–1157.
27. Dalal, N.; Tniggs, B. Histograms of Oriented Gradients for Human Detection. In Proceedings of the 2005 IEEE Computer Society Conference on Computer Vision and Pattern Recognition (CVPR'05), San Diego, CA, USA, 20–25 July 2005; pp. 886–893.
28. Lin, C.; Xu, G.L.; Cao, Y.J.; Liang, C.H.; Li, Y. Improved contour detection model with spatial summation properties based on nonclassical receptive field. *J. Electron. Imaging* **2016**, *25*, 043018. [CrossRef]
29. Suh, H.K.; Hofstee, J.W.; De Ijsselmui, N.J.; Henten, E.V. Sugar beet and volunteer potato classification using Bag-of-Visual Words model, Scale-Invariant Feature Transform, or Speeded Up Robust Feature descriptors and crop row information. *Biosyst. Eng.* **2018**, *166*, 210–226. [CrossRef]
30. Mukherjee, P.; Lall, B. Saliency and KAZE features assisted object segmentation. *Image Vis. Comput.* **2017**, *65*, 82–97. [CrossRef]
31. Song, Z.; Fu, L.; Wu, J.; Liu, Z.; Li, R.; Cui, Y. Kiwifruit detection in field images using Faster R-CNN with VGG16. *IFAC-PapersOnLine* **2019**, *52*, 76–81. [CrossRef]
32. Fu, L.; Feng, Y.; Elkamil, T.; Liu, Z.; Li, R.; Cui, Y. Image recognition method of multi-cluster kiwifruit in field based on convolutional neural networks. *Trans. Chin. Soc. Agric. Eng.* **2018**, *34*, 205–211.
33. Ge, Z.; Liu, S.; Wang, F.; Li, Z.; Sun, J. YOLOX: Exceeding Yolo Series in 2021. *arXiv* **2021**, arXiv:2107.08430.

34. Leng, H.; Tan, M.; Liu, C.; Cubuk, E.; Shi, X.; Cheng, S.; Anguelov, D. PolyLoss: A Polynomial Expansion Perspective of Classification Loss Functions. In Proceedings of the Tenth International Conference on Learning Representations (ICLR), Virtual, 25–29 April 2022.
35. Chen, L. The Multi-Objectrecognition Method of Cluster Kiwifruits Based on Machinevision. Master's Thesis, Northwest A&F University, Yangling, China, 2018.
36. Cui, Y.; Su, S.; Wang, X.; Tian, Y.; Li, P.; Zhang, F. Recognition and Feature Extraction of Kiwifruit in Natural Environment Based on Machine Vision. *Trans. Chin. Soc. Agric. Mach.* **2013**, *44*, 247–252.
37. Fu, L.; Tola, E.; Al-Mallahi, A.; Li, R.; Cui, Y. A novel image proo cessing algorithm to separate linearly clustered kiwifruits. *Biosyst. Eng.* **2019**, *183*, 184–195. [CrossRef]

agriculture

MDPI

Article

Tea Category Identification Using Wavelet Signal Reconstruction of Hyperspectral Imagery and Machine Learning

Qiang Cui [1], Baohua Yang [1,*], Biyun Liu [1], Yunlong Li [1] and Jingming Ning [2,*]

[1] School of Information and Computer, Anhui Agricultural University, Hefei 230036, China; cuiqiang@stu.ahau.edu.cn (Q.C.); liubiyun@stu.ahau.edu.cn (B.L.); liyunlong@stu.ahau.edu.cn (Y.L.)

[2] State Key Laboratory of Tea Plant Biology and Utilization, Anhui Agricultural University, 130 Changjiang West Road, Hefei 230036, China

* Correspondence: ybh@ahau.edu.cn (B.Y.); ningjm@ahau.edu.cn (J.N.)

Abstract: Accurately distinguishing the types of tea is of great significance to the pricing, production, and processing of tea. The similarity of the internal spectral characteristics and appearance characteristics of different types of tea greatly limits further research on tea identification. However, wavelet transform can simultaneously extract time domain and frequency domain features, which is a powerful tool in the field of image signal processing. To address this gap, a method for tea recognition based on a lightweight convolutional neural network and support vector machine (L-CNN-SVM) was proposed, aiming to realize tea recognition using wavelet feature figures generated by wavelet time-frequency signal decomposition and reconstruction. Firstly, the redundant discrete wavelet transform was used to decompose the wavelet components of the hyperspectral images of the three teas (black tea, green tea, and yellow tea), which were used to construct the datasets. Secondly, improve the lightweight CNN model to generate a tea recognition model. Finally, compare and evaluate the recognition results of different models. The results demonstrated that the results of tea recognition based on the L-CNN-SVM method outperformed MobileNet v2+RF, MobileNet v2+KNN, MobileNet v2+AdaBoost, AlexNet, and MobileNet v2. For the recognition results of the three teas using reconstruction of wavelet components LL + HL + LH, the overall accuracy rate reached 98.7%, which was 4.7%, 3.4%, 1.4%, and 2.0% higher than that of LH + HL + HH, LL + HH + HH, LL + LL + HH, and LL + LL + LL. This research can provide new inspiration and technical support for grade and quality assessment of cross-category tea.

Keywords: redundant discrete wavelet transform; tea; convolutional neural network; classification

Citation: Cui, Q.; Yang, B.; Liu, B.; Li, Y.; Ning, J. Tea Category Identification Using Wavelet Signal Reconstruction of Hyperspectral Imagery and Machine Learning. *Agriculture* **2022**, *12*, 1085. https://doi.org/10.3390/agriculture12081085

Academic Editors: Gniewko Niedbała and Sebastian Kujawa

Received: 16 June 2022
Accepted: 18 July 2022
Published: 23 July 2022

1. Introduction

Tea is a drink widely loved by consumers because of its unique flavor and health function [1]. In particular, unfermented green tea, lightly fermented yellow tea, and fully fermented black tea have received extensive attention and research. Among them, green tea contains polyphenols to help prevent cancer [2], and yellow tea and black tea have strong antioxidant activity [3–5]. Different categories of tea have different characteristics and quality standards. Even if the same type of tea leaves, it is difficult to distinguish the type of tea due to different processing techniques and geographical indications of origin. Therefore, fast and accurate tea classification has always been an active research hotspot, which is of great significance to the control of tea fermentation time, production process, and processing links.

The traditional methods of tea classification were generally based on physical and chemical indicators, and sensory evaluation. Among them, methods based on physical and chemical indicators, such as gas chromatography-mass spectrometry [6], were not easy popularized and applied because of time-consuming operations and destructive experiments. Some methods were costly and complicated, such as near-infrared spectroscopy detection [7], fluorescence spectroscopy detection [8], electronic tongue, and electronic nose [9–11], which might limit the universality of tea classification methods. Of course, the sensory evaluation method is one of the common methods in the tea field [12]. Although it could make up for the limitations of physical and chemical testing methods, it was susceptible to the influence of the subjective factors of review experts, which could lead to inaccurate tea classification [13,14]. Therefore, it is necessary to develop a method to classify different categories of tea objectively and reliably.

In fact, classification based on the appearance characteristics of tea was a relatively simple and effective method. Especially the classification of tea based on machine vision has been widely used [15]. However, the resolution of the tea sample image and the shooting environment may affect the classification accuracy [16]. With the rapid development of hyperspectral image processing technology, its application in food analysis is becoming more and more attractive [17,18]. Studies have shown that analysis based on hyperspectral images has achieved better results not only in tea internal quality, such as tea polyphenols, catechins, and tea amino acids [19–21] but also in tea classification, such as the classification of green tea [22–24], the classification of tieguanyin tea [25], oolong tea varieties [26]. However, most of the studies mentioned above were based on common features, such as texture features and spectral reflectance, which severely limited further research on tea classification.

Multi-resolution analysis based on wavelet transform has been extensively studied in various fields, such as image enhancement, image decomposition and reconstruction, signal-to-noise separation, and signal filtering based on wavelet transform have been applied [27]. In addition, wavelet transform has good analysis ability in both the time domain and frequency domain [28]. Wu et al. proposed entropies from the wavelet coefficient to successfully classify green, black, and oolong [29]. Bakhshipour et al. extracted wavelet features to classify black and green teas [30]. Borah et al. extracted wavelet textures to classify black teas of different levels [31]. Regardless of the form of the wavelet feature, the research mentioned above provides all the retrieved features of tea biochemistry or composition for tea classification. Obviously, they neither tried to extract more advanced wavelet semantic features to classify tea more effectively nor did they try to better understand the contribution of wavelet coefficient components to tea classification.

In fact, the essence of wavelet transform is to solve the signal energy at different decomposition scales to form feature vectors for identification. In particular, the wavelet coefficient components of the image obtained by the discrete wavelet transform method respectively represent the approximate features and detailed features in different directions. As we all know, some methods use wavelet coefficients to reconstruct images, and other methods use wavelet coefficients to achieve signal denoising. At present, there is no universally accepted consensus on which wavelet coefficient components perform well. Nevertheless, the research mentioned above was based on the manual wavelet feature extraction method, which lacked stability and robustness. Therefore, the research on tea classification still faces many technical issues.

With the wide application of deep learning in different fields, the research ideas of tea classification have been greatly expanded. To more fully extract the spectral-spatial joint features in hyperspectral images, the researchers applied deep learning techniques to the task of tea classification [32]. Deep neural networks are composed of many network layers and have powerful feature extraction capabilities from low-level to high-level, which could solve the problem of insufficient and unstable features extraction by traditional methods [33]. In particular, the combination of deep learning and wavelet transform has been widely concerned and applied. Some research focused on image stitching based on convolutional neural network (CNN) and wavelet transform methods [34]. For example, El-Latif et al. respectively carried out research on the strategies of high-frequency sub-band splicing and low-frequency sub-band splicing based on wavelet decomposition. In contrast, some studies applied deep learning methods to process wavelet components to achieve image reconstruction. For example, Qi et al. used CNN to restore image sub-bands corresponding to different wavelet coefficient components including low frequency (LL) and high frequency (LH, HL, and HH) [35]. Wang et al. used CNN to generate corresponding weight maps for low-frequency and high-frequency wavelet coefficient components [36]. Although the convolutional neural network and wavelet transform exhibited the analysis potential of different frequency sub-band for the wavelet components obtained by wavelet decomposition of the image, they have not been applied to the combination of wavelet sub-bands of different frequencies to achieve accurate classification. In addition, studies have shown that discrete wavelet transform decomposing time-frequency domain signals from images could improve classification accuracy [37]. However, it remains unclear how the optimal combination of multiple wavelet components affects the classification performance of different teas.

Therefore, a method of tea classification was proposed with the deep semantic features from wavelet component combination based on the lightweight CNN model, aiming to further extract features from hyperspectral images. The purpose of this research is to (1) decompose tea hyperspectral images through redundant wavelet transform to obtain different wavelet coefficients in different time and frequency domains, thereby improving the feature expression ability, (2) reconstruct the image with the optimal combination of wavelet components, and an improved lightweight convolutional neural network model was proposed, aiming to achieve tea recognition accuracy, and (3) compare and evaluate the classification results of different kinds of tea to verify the effectiveness of the method proposed in this study.

2. Materials and Methods

2.1. Sample Collection and Data Set Construction

2.1.1. Collection of Tea Samples

The tea samples used in the experiment were three categories of tea purchased from large supermarkets and online, including yellow tea, green tea, and black tea. Among them, yellow tea includes Mogan Huangya (MGHY, produced in Huzhou, China), Mengding Huangya (MDHY, produced in Ya'an, China), Huoshan Huangya (HSHY, produced in Lu'an, China), Pingyang Huangtang (PYHT, produced in Wenzhou, China), Junshan Yinzhen (JSYZ, produced in Yueyang, China). Green tea includes Maofeng (MF, produced in Huangshan, China) and Liuan Guapian (LAGP, produced in Lu'an, China). In order to obtain more samples, it is necessary to collect the same category of tea from different manufacturers. For example, although the geographical indication origin of Qimen black tea is Qimen County, Huangshan City. We still collect black tea from different company, including Anchi Tea limited company, Chizhou, China (ACBT), Xiaolukou Tea limited company (XLBT), Gaoxiang Black Tea Factory, Huangshan, China (GXBT), Qihong Tea limited company, Huangshan, China (QMQH), and Qimen Tea limited company, Huangshan, China (HSBT). Moreover, Maofeng (green tea) was collected from different production companies, including Guangming Tea limited company, Huangshan, China (GMMF), Beijing Zhangyiyuan Jingtailong Tea limited company, Huangshan, China (ZYYMF), Ziwei Tea limited company, Huangshan, China (ZWMF), Yijiangyuan Tea limited company, Huangshan, China (YJYMF). In total, 15 kinds of tea samples collected were produced in four provinces (Anhui Province, Zhejiang Province, Hunan Province, Sichuan Province) in China (Table 1).

Table 1. Geographical sources of tea.

Tea Category	Tea Variety	Abbreviations	Number	Geographical Origins
Black tea	QMBT from ACBT	ACBT	30	Anhui
	QMBT from XLBT	XLBT	30	Anhui
	QMBT from GXBT	GXBT	30	Anhui
	QMBT from QMQH	QMQH	30	Anhui
	QMBT from HSBT	HSBT	30	Anhui
Green tea	Maofeng from ZYYMF	ZYYMF	30	Anhui
	Maofeng from ZWMF	ZWMF	30	Anhui
	Maofeng from YJYMF	YJYMF	30	Anhui
	Maofeng from GMMF	GMMF	30	Anhui
	Liuan Guapian	LAGP	30	Anhui
Yellow tea	Junshan Yinzhen	JSYZ	30	Hunan
	Huoshan Huangya	HSHY	30	Anhui
	Mengding Huangya	MDHY	30	Sichuan
	Mogan Huangya	MGHY	30	Zhejiang
	Pingyang Huangtang	PYHT	30	Zhejiang

2.1.2. Acquisition of Hyperspectral Images of Tea

The near-infrared hyperspectral imaging (NIR-HSI) system was used to acquire hyperspectral images of tea samples. The system mainly consists of an image spectrograph (Imspector V17E, Spectral Imaging Ltd., Oulu, Finland), two 150W fiber optic halogen lamps (Model 3900, Illumination Technologies Inc., New York, NY, USA), a camera obscura, and mobile platform. The hyperspectral image obtained by this system has a total of 616 wavelength bands ranging from 908 to 1735 nm. The structure diagram of the hyperspectral imaging system is shown in Figure 1.

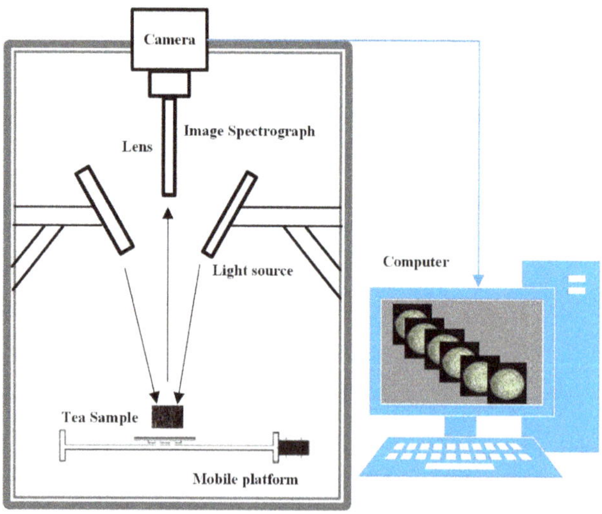

Figure 1. The structure diagram of hyperspectral imaging system.

The preparations before collecting hyperspectral images are as follows: turn on the light source for preheating 30 min before the experiment, set the distance between the camera and the tea sample to 38.4 cm, set the sample moving speed on the conveyor belt to 1 cm/s, and set the exposure time to 20 ms. Set the frame rate to 13 Hz. The tea samples were evenly spread in a Petri dish with a diameter of 9 cm × 1 cm. The Petri dish was pre-built with black rubber with approximately zero reflectivity, so as not to affect the experimental data. Then, open the operation interface of the hyperspectral image acquisition software to collect hyperspectral images of the tea samples. Furthermore, the hyperspectral images were white-boarded and dark-current corrected. Finally, the Environment for Visualizing Images (ENVI 5.1, ITT visual information solutions, Boulder, CO, USA) was used to analyze the spectral features of tea.

2.2. Method

2.2.1. Decomposition and Reconstruction of Image Signal by Wavelet

Wavelet transform is a signal time-frequency analysis method, and a common method is wavelet multi-resolution analysis. When a two-dimensional grayscale image is processed, a one-dimensional discrete wavelet transform is used to operate the image, two directions are selected to pass through a filter bank, and the data is reduced by downsampling [38]. That is, the different characteristics of images on various scales are described from the perspective of space.

Redundant wavelet transform (RWT) is a kind of wavelet transform in which the decomposition results of signals or images on adjacent scales have redundancy. That is, the high- and low-frequency information of the signal or image is separated, and finally, it is decomposed into approximate signals and wavelet surfaces on different frequency channels. Moreover, the length of the approximation signal and the detail signal after the signal transformation are the same as the original signal length [39].

The redundant discrete wavelet transform is represented by a filter bank. The output coefficient obtained after each level of decomposition is twice the input coefficient. The decomposition formulas of redundant discrete wavelet transform are as Formulas (1) and (2), and the mathematical expressions for reconstruction are as Formula (3):

$$c_{j+1}[k] = c_j[k] * h[-k] \tag{1}$$

$$d_{j+1}[k] = c_j[k] * g[-k] \tag{2}$$

$$c_{j+1}[k] = \frac{c_j[k] * h[k] + d_j[k] * g[k]}{2} \tag{3}$$

Among them, $h[-k]$ and $g[-k]$ represent low-pass and high-pass decomposition filters, respectively, c_j and d_j represent the coefficients of the low-band and high-band output of the jth level, $*$ means convolution, the low-pass and high-pass synthesis filters are $h[k]$ and $g[k]$, respectively. A is the original signal, A^{τ} is the reconstructed signal.

An image can be viewed as a two-dimensional signal. Applying wavelet theory to image processing is to use multi-resolution decomposition to decompose the image into sub-images of different spaces and frequencies, and then encode the coefficients of the sub-images. At the same time, wavelet transform can better solve the contradiction between time and frequency resolution, so wavelet transform is very beneficial to the decomposition and reconstruction of image signals.

2.2.2. Classification Model Based on Improved Lightweight CNN

MobileNet is a lightweight deep neural network based on a depthwise separable convolution design, which has performed very well in the classification. Among them, MobileNetV2 introduces an inverted residual and linear bottleneck structure, which makes the number of network parameters and lower computing costs. The parameters of the network structure are shown in Table 2.

The MobileNetV2 network includes ordinary convolution (Conv), inverse residual structure deep separation convolution (Bottleneck), and average pooling (Avgpool). To enhance the applicability of the network in tea classification and improve the accuracy of target classification. Based on MobileNetV2 in this study, the following improvements have been made (as shown in Figure 2 and Table 3). To further reduce computing resources and save memory space when training the network. Remove the network layer after the 9th layer, and reduce the number of channels of the convolutional layer from 1280 to 128. The three-dimensional feature map was converted into one-dimensional through the Flatten layer. To better adapt to the problem of tea hyperspectral image classification, SoftMax was replaced with SVM classifier to improve the generalization ability of the model.

Table 2. Detailed parameters of each layer of CNN (Origin).

Input	Operator	Channel	N	Stride	Out
224 × 224 × 3	conv2d	32	1	2	112 × 112 × 32
112 × 112 × 32	bottleneck	16	1	1	112 × 112 × 16
112 × 112 × 16	bottleneck	24	2	2	56 × 56 × 24
56 × 56 × 24	bottleneck	32	3	2	28 × 28 × 32
28 × 28 × 32	bottleneck	64	4	2	14 × 14 × 64
14 × 14 × 64	bottleneck	96	3	1	14 × 14 × 96
14 × 14 × 96	bottleneck	160	3	2	7 × 7 × 160
7 × 7 × 160	bottleneck	320	1	1	7 × 7 × 320
7 × 7 × 320	conv2d	1280	1	1	7 × 7 × 1280
7 × 7 × 1280	avgpool	-	1	-	1 × 1 × 1280
1 × 1 × 1280	conv2d	3	1	1	1 × 1 × 3
1 × 1 × 3	softmax	3	1	-	3

N: represents the number of repetitions of the Operator.

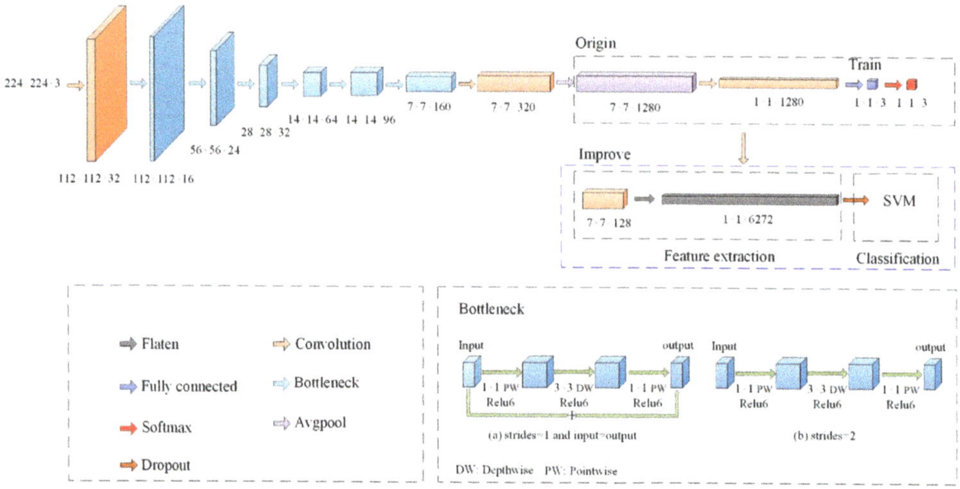

Figure 2. The optimized lightweight CNN.

Table 3. Detailed parameters of improved lightweight CNN.

Input	Operator	Channel	N	Stride	Out
224 × 224 × 3	conv2d	32	1	2	112 × 112 × 32
112 × 112 × 32	bottleneck	16	1	1	112 × 112 × 16
112 × 112 × 16	bottleneck	24	2	2	56 × 56 × 24
56 × 56 × 24	bottleneck	32	3	2	28 × 28 × 32
28 × 28 × 32	bottleneck	64	4	2	14 × 14 × 64
14 × 14 × 64	bottleneck	96	3	1	14 × 14 × 96
14 × 14 × 96	bottleneck	160	3	2	7 × 7 × 160
7 × 7 × 160	bottleneck	320	1	1	7 × 7 × 320
7 × 7 × 320	conv2d	128	1	1	7 × 7 × 128
7 × 7 × 128	flatten	6272	1	-	1 × 1 × 6272
1 × 1 × 6272	SVM	3	1	-	3

N: represents the number of repetitions of the Operator.

2.2.3. Tea Classification Model Based on Optimized L-CNN

The technical route of this research was shown in Figure 3. Firstly, the hyperspectral images of the tea samples were acquired, the wavelet components corresponding to the hyperspectral image were extracted by redundant wavelet transform, and the combination of different wavelet components was used as the input of the deep convolutional neural network. Secondly, the deep convolutional neural network model undergoes transfer learning, model training, and parameter optimization to generate an optimized classification model. Finally, the classification model was tested on the selected tea samples. The specific steps were as follows: The transfer learning process of this research was mainly realized by using the improved CNN model pre-trained on the large-scale ImageNet dataset as the source domain. The trained network parameters were used as the initial parameters for the training of the tea classification model based on hyperspectral images, and the self-built data sets (the combination of wavelet coefficients extracted from the tea hyperspectral images) were used to fine-tune the parameters of the L-CNN model to improve the automatic tea identification ability.

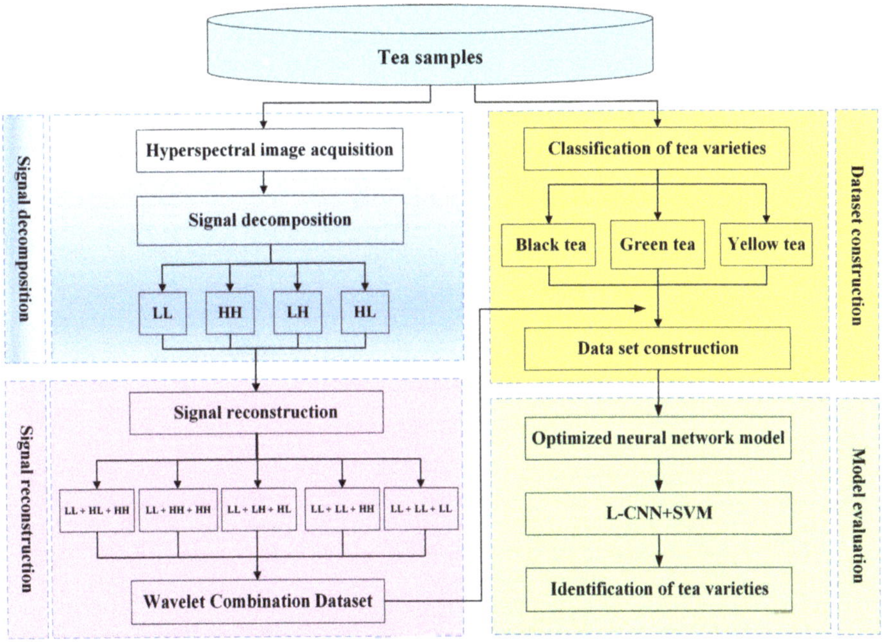

Figure 3. The technology roadmap for this study.

The experimental environment of this study is shown in Table 4. The parameters of the L-CNN were set as follows: the learning rate is 0.0001, the epoch is 50, and the batch size is 10.

Table 4. Parameters of the experimental environment.

Settings	Parameters
CPU	Intel (R) Core (TM) i7-8700 CPU @ 3.20G Hz
GPU	NVIDIA GeForce GTX 1070 Ti
RAM	16.0 GB
Operating system	Win 10_64 bit
MATLAB version	MATLAB R2019a
Lab environment	Deep Learning Toolbox

2.2.4. Evaluation Indicator of Classification Model

To accurately evaluate the tea classification model, precision, recall, and accuracy are used as indicators for model performance evaluation. Precision measures the classification accuracy of positive samples, recall represents the proportion of correctly classified positive samples to the total positive samples, and accuracy measures the proportion of all samples that are accurately classified.

$$\text{precision} = \frac{\text{TP}}{\text{TP} + \text{FP}} \tag{4}$$

$$\text{recall} = \frac{\text{TP}}{\text{TP} + \text{FN}} \tag{5}$$

$$\text{accuracy} = \frac{\text{TP} + \text{TN}}{\text{TP} + \text{TN} + \text{FP} + \text{FN}} \tag{6}$$

In the classification task, when it is determined that the category of a certain type of tea is positive, the other categories are negative.

TP means that the predicted category is a positive category and the model judges it as a positive category; that is, the positive category is judged correctly;

FP means that the predicted category is a positive category and the model judges it to be a negative category; that is, the judgment of the positive category is wrong;

FN means that when the predicted category is a negative category and the model judges it as a negative category; that is, the negative category is judged correctly;

TN means that when the predicted category is a negative category and the model judges it as a positive category, the negative category is judged wrong.

In addition, the ROC (receiver operating characteristic) curve is used to compare different classification results. The horizontal axis of the ROC curve represents the false-positive ratio (FPR), and the vertical axis represents the true-positive rate (TPR). The closer the ROC curve is to the upper left corner, the higher the accuracy of the prediction. TPR and FPR are defined as shown in Equations (7) and (8).

$$\text{TPR} = \frac{\text{TP}}{\text{TP} + \text{FN}} \tag{7}$$

$$\text{FPR} = \frac{\text{FP}}{\text{TN} + \text{FP}} \tag{8}$$

In addition, the ROC curve is divided into two parts according to the position of the curve. The area under the curve and the horizontal axis part is called AUC (area under the roc curve), and the value is between [0, 1]. The closer the AUC is to 1, the better the classification effect of the model.

To more fully demonstrate the classification effect, a confusion matrix is used to indicate the classification visualization, which describes the relationship between the true category attributes of the sample data and the classification results in the form of a matrix. Suppose that for the classification task of type m pattern, V is the test sample set, the number of samples is n the number of categories is m, and $C = \{C_1, C_2, \ldots C_L\}$ is the classifier set. L classifiers are used to test in set V, respectively, to obtain the confusion of each classifier matrix (confusion matrix) $\text{CM}^k (k = 1, 2, \ldots L)$, the confusion matrix of the k classifier C_k is expressed as follows:

$$\text{CM}^k = \begin{bmatrix} n_{1,1}^k & n_{1,2}^k & \cdots & n_{1,m}^k \\ n_{2,1}^k & n_{2,2}^k & \cdots & n_{2,m}^k \\ \cdots & \cdots & \cdots & \cdots \\ n_{m,1}^k & n_{m,2}^k & \cdots & n_{m,m}^k \end{bmatrix} \tag{9}$$

Among them, the element in the i-th row and the j-th column represents the number of the i-th class recognized as the j-th class by the classifier C_k in the sample. If $i = j$, it means that the classifier could correctly identify the number of samples, so diagonal elements represent the number of correctly classified by classifier C_k, and off-diagonal elements represent the number of errors classified by classifier C_k. In the confusion matrix, each column represents the predicted category, and each row represents the true attribution category (sample label) of the tea. The total number of columns is equal to the total number of rows, which is the total number of label categories of the data sample. The larger the diagonal value of the confusion matrix, the greater the probability that the model is correctly classified, and the better the model effect.

3. Results

3.1. Hyperspectral Images and Spectral Reflectance of Different Types of Tea

One hundred and fifty grams of each sample was sealed, stored, and sent to the hyperspectral laboratory for hyperspectral imaging collection in time. The hyperspectral images of all samples were acquired using a hyperspectral instrument, which accessories were reported in the previous research of our team [28]. The self-built data set using the collected hyperspectral images from 15 teas (as shown in Figure 4) contained 450 hyperspectral images of three categories of tea, including 150 images of black tea, 150 images of green tea, and 150 images of yellow tea. The data sets were divided into training set, validation set, and test set according to the ratio of 7:3:5.

Figure 4. Hyperspectral images of three tea categories including black tea, green tea, and yellow tea.

Different parts of the tea samples, such as stems, leaves, and buds, contain significant differences in the content of substances, so the spectral data between each pixel point is quite different. To address this issue, 50 regions of interest (ROI) of 20 × 20 pixels were randomly selected in the sample area. A pixel contains a piece of spectral information, and the average spectrum of all pixels in the ROI was calculated as the spectral reflectance of a sample. The original spectrum collection of tea samples is shown in Figure 5.

In Figure 6, the wavelength range is 944–1688 nm, all tea samples had similar trends in the whole spectral region, most of the reflectance is from 0.4–0.7, but the size of absorbance is different. In particular, in the three absorption bands located at 1200, 1380, and 1450 nm, the absorption peak at 1200 nm is attributed to the C-H second overtone; the peak bands near 1400 nm are attributed to the O-H vibration. It can be seen that black tea, green tea, and yellow tea have similar spectral features. It is difficult to distinguish different types of tea based on spectral information alone. To effectively identify different kinds of tea, further analysis was carried out according to the different spatial information of tea.

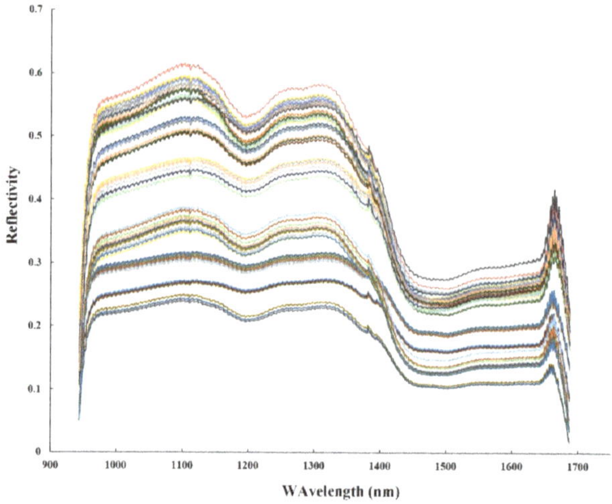

Figure 5. Hyperspectral images of three tea categories including black tea, green tea, and yellow tea.

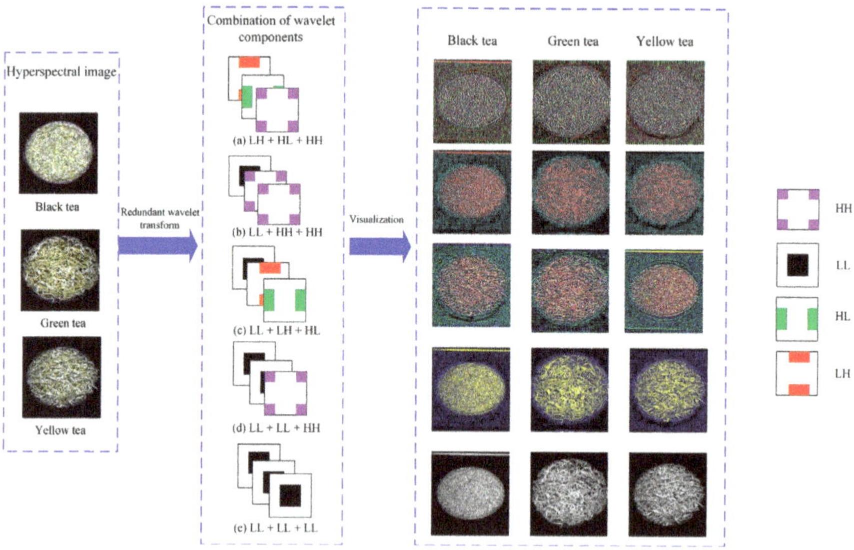

Figure 6. Combinations of different wavelet components.

3.2. Multi-Component Combination of Wavelet Decomposition of Hyperspectral Image of Tea

To extract the wavelet coefficient combination information of the hyperspectral image of tea. First, from the full-band hyperspectral image, the wavelet function db2 was used to perform the redundant discrete wavelet transform (RDWT) in the horizontal direction to obtain the high-frequency component and the low-frequency component, then the two components in the vertical direction were respectively performed RDWT. The above process realized the two-dimensional discrete wavelet transform of the hyperspectral image. Therefore, the approximate component LL, the horizontal component HL, the vertical component LH, and the diagonal component HH are obtained from the hyperspectral image of the tea based on the wavelet transform. Three of the four components are selected to form a three-channel data as the input of the L-CNN. According to the results of multiple tests, the five combinations performed well, as shown in Figure 6. Figure 6a shows the combination of the three high-frequency components LH, HL, and HH. Figure 6b shows the combination of the low-frequency component LL and the high-frequency components HH and HH. Figure 6c shows a combination of low-frequency component LL, high-frequency component LH, and HL. Figure 6d shows the combination of two low-frequency components LL and one high-frequency component HH. Figure 6e shows the combination of three low-frequency components LL. From the visualization map corresponding to the five wavelet combinations of the tea hyperspectral image in Figure 6, it could be seen that the features displayed by the different combinations were still significantly different.

3.3. Tea Classification Results Based on the L-CNN-SVM Model

Different combinations based on wavelet components were used as the input of the three channels of the CNN model, and the classification accuracy is shown in Table 5. It could be seen from Table 5 that the classification accuracy was 0.940–1.000 for black tea, 0.923–0.980 for green tea, and 0.918–0.980 for yellow tea. The classification effect based on the wavelet component was better than that of the original hyperspectral image. Among the five combinations, the classification result based on LL + HL + LH was the best, with an accuracy of 0.987, and only 0.013% of tea samples were misidentified. The possible reason was that the three wavelet components represent wavelet information from different angles, which expanded the expressive ability of tea features. In addition, the contribution of the HH component to tea classification was relatively low.

Table 5. The accuracy of tea classification based on the combination of different wavelet components.

Input Data	Kappa Coefficient	Overall	Black Tea	Green Tea	Yellow Tea
Original	0.90	0.933	0.940	0.958	0.904
LH + HL + HH	0.91	0.940	0.980	0.923	0.918
LL + HH + HH	0.93	0.953	0.980	0.941	0.939
LL + LL + HH	0.96	0.973	0.980	0.962	0.980
LL + HL + LH	0.98	0.987	1.000	0.980	0.980
LL + LL + LL	0.95	0.967	1.000	0.980	0.925

In addition, the kappa coefficients [40] were obtained by using reconstructed figures based on different wavelet components. It could be seen from Table 5 that the Kappa coefficient based on the LL + HL + LH component was the highest, reaching 0.98, which was 8.2% higher than that of the original hyperspectral image, 7.1% than that of LH + HL + HH, 5.1% than that of LL + HH + HH, 2% than that of LL + LL + HH, and 3.1% than that of LL + LL + LL.

To better display the results of tea classification, the confusion matrix was used to analyze the classification results. In the confusion matrix, the y-axis represented the real tea category, and the x-axis represented the result of the model classification. A total of 150 images were used as tests, including 50 each for black tea, green tea, and yellow tea. As shown in Figure 7, the number of misclassifications of black tea, green tea, and yellow tea using wavelet components did not exceed 5. Black tea, green tea, and yellow tea were misclassified as 3, 4, and 3 for the original hyperspectral image, 2, 3, and 5 for LH + HL + HH, 1, 2, and 4 for LL + HH + HH, 2, 0, and 0 for LL + LL + HH, 0, 1, and 1 for LL + LH + HL, 2, 2, and 1 for LL + LL + LL. It was easy to see that the tea classification results of the CNN model based on the combined information of the wavelet components performed well. Although there were some errors in the identification of individual figures, most of the tea samples were correctly identified.

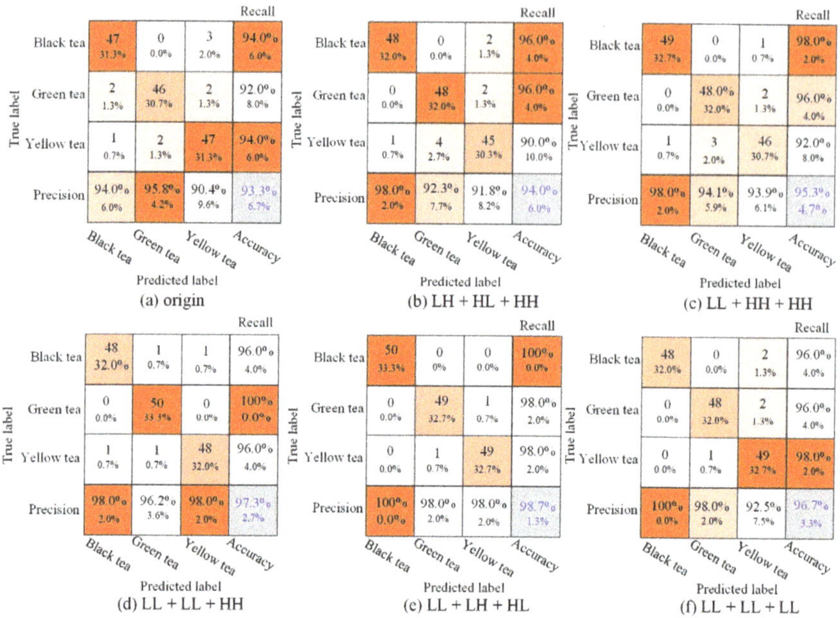

Figure 7. Confusion matrix of tea classification based on different components.

4. Discussion

4.1. Compare Tea Classification Based on Different Wavelet Component Combinations

Wavelet transform has obvious advantages in local time-frequency analysis. It could decompose the hyperspectral image of tea leaves with non-stationary signals into high-frequency components and low-frequency components. Although Fourier transform or contourlet transform is also a common method in resolution analysis [41], However, Fourier transform lacks time domain processing and has a very bad effect on non-stationary information. In addition, the contour of the image is transformed with more redundancy, which makes it difficult to achieve perfect image reconstruction.

Two-dimensional wavelet transform could be used to characterize the target object in the image [42]. Li et al. used the texture features obtained by wavelet transform to classify the famous green tea in China [43]. Bakhshipour et al. extracted wavelet features to successfully classify black tea [44]. Nevertheless, the above research is limited to shallow features and lacks wavelet semantic features. Wulandari et al. combined wavelet transform and a deep learning model to improve classification accuracy [45], which demonstrated the effectiveness of advanced wavelet semantic features. Therefore, the classification accuracy of this study using the combination of LL + LH + HL is 98.7%, which was 5.4% higher than that of the original hyperspectral image, and was higher than that of LH + HL + HH, LL + HH + HH, LL + LL + HH, and LL + LL + LL with 4.7%, 3.4%, 1.4%, and 2.0%, respectively. It was because LL + LH + HL included comprehensive wavelet information. On the one hand, the interference information was removed after wavelet transform, which could realize the extraction of multi-angle features. On the other hand, multi-angle wavelet features were used to extract high-level semantic features through deep learning models, which was more conducive to tea identification.

4.2. Compare Classification Results Based on Different Deep Learning Models

To evaluate the performance of the tea classification model proposed in this study, our method was compared with the typical classification models AlexNet and MobileNet v2 [46,47]. It could be seen from Table 6 that the classification accuracy based on our proposed method was 5.47% and 1.42% higher than that of AlexNet and MobileNet v2, respectively. In addition, SVM was used to improve the classification addition method. Compared with the classification methods of random forest (RF), K-nearest neighbor (KNN), and adaptive boosting (AdaBoost), the experimental results showed that the classification accuracy of our proposed method was improved by 0.71%, 2.03%, and 2.74% overall. In short, the experiment showed that the combination of SVM and lightweight model improved not only the classification accuracy of the model but also the generalization ability of the model. Therefore, the proposed model had a better classification effect on tea classification.

Table 6. Comparison of classification accuracy of different methods.

Method	Overall	Black Tea	Green Tea	Yellow Tea
Our method	0.987	1.000	0.980	0.980
MobileNet v2 + RF	0.980	0.980	0.980	0.980
MobileNet v2 + KNN	0.967	0.960	0.980	0.960
MobileNet v2 + AdaBoost	0.960	0.960	0.940	0.980
MobileNet v2	0.973	1.000	0.960	0.960
AlexNet	0.933	0.960	1.000	0.860

4.3. *Different Network Visualization Based on Grad-CAM*

To visually show the potential recognition capabilities of different models, a color visualization method was applied: gradient weighted class activation mapping (Grad-CAM) technology, which embedded the Grad-CAM layer into the convolutional neural network, thereby making the proposed method more easily observed and explained. Grad-CAM was used to generate activation heat maps for the classification of different tea samples, as shown in Figure 8. The first, third, and fifth rows of Figure 8a represented the original hyperspectral images of black tea, green tea, and yellow tea, and rows 2, 4, and 6 of Figure 8a represented the combination of wavelet components LL + LH + HL extracted from the hyperspectral images of black tea, green tea, and yellow tea. It could be seen from Figure 8 that the activated regions that are important for the classification result in different input images are different. Among them, the darker the color indicates that the pixel feature of the corresponding position in the original image had a greater impact on the classification result.

From the second, fourth, and sixth rows of Figure 8, it was found that the stronger regions in the activated heatmap of LL+LH+HL for tea identification were wider than those of the original hyperspectral image. In addition, it was found from Figure 8b,d,f that our proposed method for feature positioning in tea identification could stably find the relevant target area. Compared with the activated heatmaps of AlexNet and Inception v3, our method could locate key areas significantly.

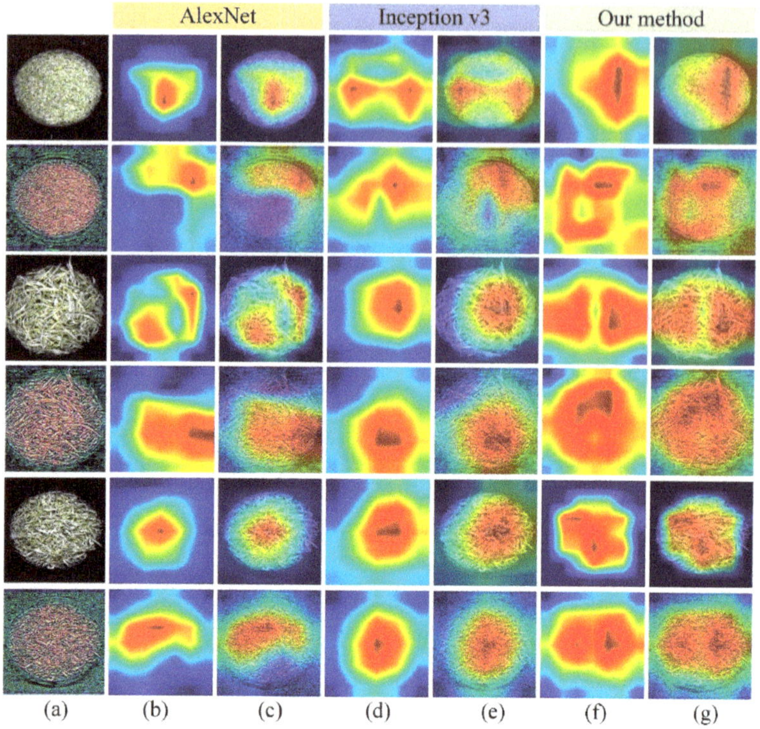

Figure 8. Samples visualization based on Grad-CAM. (**a**) represents the original hyperspectral image and wavelet component LL+LH+HL; (**b**,**d**,**f**) represent the activation heatmaps of tea sample recognition based on the AlexNet model, Inception v3 and our proposed method; (**c**,**e**,**g**) represent the activated heatmaps superimposed on the original image of tea sample recognition based on the AlexNet model, Inception v3 and our proposed method.

5. Conclusions

To improve the accuracy of traditional classification methods for tea based on hyperspectral images, the combined information of different components decomposed by redundant discrete wavelet transform was used to classify based on the lightweight CNN model. The experimental results showed that the combination of wavelet components LL+LH+HL based on the hyperspectral image of tea had the best classification effect, with an accuracy of 98.7%. In conclusion, the method of time-frequency signal decomposition and reconstruction based on hyperspectral images proposed in this study provided a new idea for tea identification, which will provide technical reference for identifying the grade and quality of tea.

Author Contributions: Methodology, B.Y. and B.L.; software, Q.C. and B.L.; data curation, Y.L.; writing, Q.C. and B.Y.; writing—review and editing, B.Y. and J.N. All authors have read and agreed to the published version of the manuscript.

Funding: This work was supported by the Major Science and Technology Projects in Anhui Province (202203a06020007), the Opening Project of Key Laboratory of Power Electronics and Motion Control of Anhui Higher Education Institutions (PEMC2001), the Open Fund of State Key Laboratory of Tea Plant Biology and Utilization (SKLTOF20200116).

Institutional Review Board Statement: Not applicable.

Informed Consent Statement: Not applicable.

Data Availability Statement: The raw data supporting the conclusions of this article will be made available by the authors, without undue reservation.

Conflicts of Interest: All the authors declare no conflict of interest.

References

1. Sereshti, H.; Samadi, S.; Jalali-Heravi, M. Determination of volatile components of green, black, oolong and white tea by optimized ultrasound-assisted extraction-dispersive liquid–liquid microextraction coupled with gas chromatography. *J. Chromatogr. A* **2013**, *1280*, 1–8. [CrossRef] [PubMed]
2. Yohei, S.; Masahito, S. Possible mechanisms of green tea and its constituents against cancer. *Molecules* **2018**, *23*, 2284.
3. Kujawska, M.; Ewertowska, M.; Adamska, T.; Jodynis-Liebert, J.; Ignatowicz, E.; Gramza-Michalowska, A. Protective effect of yellow tea extract on N-nitrosodiethylamine-induced liver carcinogenesis. *Pharm. Biol.* **2016**, *54*, 1891–1900. [CrossRef] [PubMed]
4. Kujawska, M.; Ewertowska, M.; Ignatowicz, E.; Adamska, T.; Szaefer, H.; Gramza-Michalowska, A.; Jodynis-Liebert, J. Evaluation of safety and antioxidant activity of yellow tea (*Camellia sinensis*) Extract for Application in Food. *J. Med. Food* **2016**, *19*, 330–336. [CrossRef]
5. Adhikary, B.; Yadav, S.K.; Roy, K.; Bandyopadhyay, S.K.; Chattopadhyay, S. Black tea and theaflavins assist healing of indomethacin-induced gastric ulceration in mice by antioxidative action. *Evid. Based Complement. Altern. Med.* **2011**, *8*, 546560. [CrossRef]
6. Qin, Z.H.; Pang, X.L.; Dong, C.; Cheng, H.; Hu, X.S.; Wu, J.H. Evaluation of Chinese tea by the electronic nose and gas chromatography–mass spectrometry: Correlation with sensory properties and classification according to grade level. *Food Res. Int.* **2013**, *53*, 864–874. [CrossRef]
7. Li, X.L.; Zhang, Y.Y.; He, Y. Study on Detection of Talcum Powder in Green Tea Based on Fourier Transform Infrared (FTIR) Transmission Spectroscopy. *Spectrosc. Spectr. Anal.* **2017**, *37*, 1081–1085.
8. Hong, Z.L.; Ze, J.L.; Hua, J.L.; Shu, J.S.; Feng, N.C.; Kai, H.W.; Da, Z.M. Robust Classification of Tea Based on Multi-Channel LED-Induced Fluorescence and a Convolutional Neural Network. *Sensors* **2019**, *19*, 4687.
9. Bhattacharya, N.A.; Tudu, B.B.; Jana, A.A.; Ghosh, D.A.; Bandhopadhyaya, R.B.; Bhuyan, M.C. Preemptive identification of optimum fermentation time for black tea using electronic nose. *Sens. Actuators B Chem.* **2008**, *131*, 110–116. [CrossRef]
10. Chen, Q.S.; Zhao, J.W.; Chen, Z.; Lin, H.; Zhao, D.A. Discrimination of green tea quality using the electronic nose technique and the human panel test, comparison of linear and nonlinear classification tools. *Sens. Actuators B Chem.* **2011**, *159*, 294–300. [CrossRef]
11. Yu, H.C.; Wang, Y.W.; Wang, J. Identification of Tea Storage Times by Linear Discrimination Analysis and Back-Propagation Neural Network Techniques Based on the Eigenvalues of Principal Components Analysis of E-Nose Sensor Signals. *Sensors* **2009**, *9*, 8073–8082. [CrossRef] [PubMed]
12. Sinija, V.R.; Mishra, H.N. Fuzzy Analysis of Sensory Data for Quality Evaluation and Ranking of Instant Green Tea Powder and Granules. *Food Bioprocess Technol.* **2011**, *4*, 408–416. [CrossRef]
13. Bhattacharyya, N.; Bandyopadhyay, R.; Bhuyan, M.; Tudu, B.; Ghosh, D.; Jana, A. Electronic Nose for Black Tea Classification and Correlation of Measurements with "Tea Taster" Marks. *IEEE Trans. Instrum. Meas.* **2008**, *57*, 1313–1321. [CrossRef]
14. Yu, H.C.; Wang, J.; Yao, C.; Zhang, H.M.; Yu, Y. Quality grade identification of green tea using E-nose by CA and ANN. *LWT—Food Sci. Technol.* **2008**, *41*, 1268–1273. [CrossRef]
15. Laddi, A.; Prakash, N.R.; Sharma, S.; Kumar, A. Discrimination analysis of Indian tea varieties based upon color under optimum illumination. *J. Food Meas. Charact.* **2013**, *7*, 60–65. [CrossRef]
16. Chen, Q.; Zhao, J.; Cai, J. Identification of tea varieties using computer vision. *Trans. ASABE* **2008**, *51*, 623–628. [CrossRef]
17. Kelman, T.; Ren, J.C.; Marshall, S. Effective classification of Chinese tea samples in hyperspectral imaging. *J. Artif. Intell. Res.* **2013**, *2*, 87. [CrossRef]
18. Zhao, J.W.; Chen, Q.S.; Cai, J.R.; Ouyang, Q. Automated tea quality classification by hyperspectral imaging. *Appl. Opt.* **2009**, *48*, 3557–3564. [CrossRef]
19. Tu, Y.X.; Bian, M.; Wan, Y.K.; Fei, T. Tea cultivar classification and biochemical parameter estimation from hyperspectral imagery obtained by UAV. *PeerJ* **2018**, *6*, e4858. [CrossRef]
20. Sohara, Y.; Ryu, C.; Suguri, M.; Park, S.B.; Kishino, S. Estimation of catechins concentration of green tea using hyperspectral remote sensing. *IFAC Proc. Vol.* **2010**, *43*, 172–177. [CrossRef]
21. Yang, B.H.; Gao, Y.; Li, H.M.; Ye, S.B.; He, H.X.; Xie, S.R. Rapid prediction of yellow tea free amino acids with hyperspectral images. *PLoS ONE* **2019**, *14*, e0210084. [CrossRef] [PubMed]
22. Ahmad, H.; Sun, J.; Nirere, A.; Shaheen, N.; Zhou, X.; Yao, K.S. Classification of tea varieties based on fluorescence hyperspectral image technology and ABC-SVM algorithm. *J. Food Process. Preserv.* **2021**, *45*, e15241. [CrossRef]
23. Ning, J.M.; Sun, J.J.; Li, S.H.; Sheng, M.G.; Zhang, Z.Z. Classification of five Chinese tea categories with different fermentation degrees using visible and near infrared hyperspectral imaging. *Int. J. Food Prop.* **2017**, *20*, 1515–1522. [CrossRef]
24. Sun, J.; Tang, K.; Wu, X.H.; Dai, C.X.; Chen, Y.; Shen, J.F. Nondestructive identification of green tea varieties based on hyperspectral imaging technology. *J. Food Process Eng.* **2018**, *41*, e12800. [CrossRef]

25. Yu, Y.J.; Wang, Q.Q.; Wang, B.Y.; Chen, J.; Sun, W.J. Identification of tieguanyin tea grades based on hyperspectral technology. *Food Sci.* **2014**, *35*, 159–163.
26. Ge, X.; Sun, J.; Lu, B.; Chen, Q.S.; Xun, W.; Jin, Y.T. Classification of oolong tea varieties based on hyperspectral imaging technology and BOSS-LightGBM model. *J. Food Process Eng.* **2019**, *42*, e13289. [CrossRef]
27. Matsuyama, E.; Tsai, D.Y.; Lee, Y.; Tsurumaki, M.; Takahashi, N.; Watanabe, H.; Chen, H.M. A modified undecimated discrete wavelet transform based approach to mammographic image denoising. *J. Digit. Imaging* **2013**, *26*, 748–758. [CrossRef]
28. Yang, B.H.; Qi, L.; Wang, M.X.; Hussain, S.; Wang, H.B.; Wang, B.; Ning, J.M. Cross-category tea polyphenols evaluation model based on feature fusion of electronic nose and hyperspectral imagery. *Sensors* **2020**, *20*, 50. [CrossRef]
29. Wu, X.; Yang, J.; Wang, S. Tea category identification based on optimal wavelet entropy and weighted k-nearest neighbors algorithm. *J. Multimed. Tools Appl.* **2018**, *77*, 3745–3759. [CrossRef]
30. Bakhshipour, A.; Zareiforoush, H.; Bagheri, I. Application of decision trees and fuzzy inference system for quality classification and modeling of black and green tea based on visual features. *J. Food Meas. Charact.* **2020**, *14*, 1402–1416. [CrossRef]
31. Borah, S.; Hines, E.; Bhuyan, M. Wavelet transform based image texture analysis for size estimation applied to the sorting of tea granules. *J. Food Eng.* **2007**, *79*, 629–639. [CrossRef]
32. Zhang, Y.D.; Muhammad, K.; Tang, C. Twelve-layer deep convolutional neural network with stochastic pooling for tea category classification on GPU platform. *J. Multimed. Tools Appl.* **2018**, *77*, 22821–22839. [CrossRef]
33. Hasan, M.M.; Chopin, J.P.; Laga, H.; Miklavcic, S.J. Detection and analysis of wheat spikes using convolutional neural networks. *J. Plant Methods* **2018**, *14*, 100. [CrossRef] [PubMed]
34. El-Latif, E.; Taha, A.; Zayed, H.H. A passive approach for detecting image splicing based on deep learning and wavelet transform. *J. Arab. J. Sci. Eng.* **2020**, *45*, 3379–3386. [CrossRef]
35. Qi, Z.; Jung, C.; Xie, B. Subband Adaptive Image Deblocking Using Wavelet Based Convolutional Neural Networks. *IEEE Access* **2021**, *9*, 62593–62601. [CrossRef]
36. Wang, Z.; Li, X.; Duan, H.; Zhang, X.; Wang, H. Multifocus image fusion using convolutional neural networks in the discrete wavelet transform domain. *Multimed. Tools Appl.* **2019**, *78*, 34483–34512. [CrossRef]
37. Masquelin, A.H.; Cheney, N.; Kinsey, C.M.; Bates, J.H.T. Wavelet decomposition facilitates training on small datasets for medical image classification by deep learning. *J. Histochem. Cell Biol.* **2021**, *155*, 309–317. [CrossRef]
38. Coffey, M.A.; Etter, D.M. Image coding with the wavelet transform. In Proceedings of the ISCAS'95-International Symposium on Circuits and Systems, Seattle, WA, USA, 30 April–3 May 1995; pp. 1110–1113.
39. Arrais, E.; de Medeiros Valentim, R.A.; Brandão, G.B. Real-time premature ventricular contractions detection based on Redundant Discrete Wavelet Transform. *Res. Biomed. Eng.* **2018**, *34*, 187–197. [CrossRef]
40. Borana, S.; Yadav, S. Comparison of model validation techniques for land cover dynamics in Jodhpur City. *Int. J. Emerg. Trends Technol. Comput. Sci.* **2017**, *6*, 215–219.
41. Zhao, J.; Jiang, H.; Di, J. Recording and reconstruction of a color holographic image by using digital lensless fourier transform holography. *Opt. Express* **2008**, *16*, 2514–2519. [CrossRef]
42. Lee, M.C.; Pun, C.M. Texture Classification Using Dominant Wavelet Packet Energy Features. In Proceedings of the 4th IEEE Southwest Symposium on Image Analysis and Interpretation, Austin, TX, USA, 2–4 April 2000; pp. 301–304.
43. Li, X.L.; Nie, P.C.; Qiu, Z.J.; He, Y. Using wavelet transform and multi-class least square support vector machine in multi-spectral imaging classification of Chinese famous tea. *Expert Syst. Appl.* **2011**, *38*, 11149–11159. [CrossRef]
44. Bakhshipour, A.; Sanaeifar, A.; Payman, S.H.; Guardia, M. Evaluation of data mining strategies for classification of black tea based on image-based features. *Food Anal. Methods* **2018**, *11*, 1041–1050. [CrossRef]
45. Wulandari, M.; Basari; Gunawan, D. Evaluation of wavelet transform preprocessing with deep learning aimed at palm vein recognition application. *AIP Conf. Proc.* **2019**, *2139*, 050005.
46. Krizhevsky, A.; Sutskever, I.; Hinton, G.E. ImageNet classification with deep convolutional neural networks. *Adv. Neural Inf. Process. Syst.* **2012**, *25*, 1097–1105. [CrossRef]
47. Kim, H.; Chung, Y. Improved Handwritten Hangeul Recognition using Deep Learning based on GoogLenet. *J. Korea Contents Assoc.* **2018**, *18*, 495–502.

MDPI
St. Alban-Anlage 66
4052 Basel
Switzerland
www.mdpi.com

Agriculture Editorial Office
E-mail: agriculture@mdpi.com
www.mdpi.com/journal/agriculture

www.ingramcontent.com/pod-product-compliance
Lightning Source LLC
Chambersburg PA
CBHW061510220926
51850CB00004B/217
* 9 7 8 3 0 3 6 5 8 8 4 8 3 *